OECD DOCUMENTS

Biotechnology for Water Use and Conservation

The Mexico '96 Workshop

The following tex

and do n

ORGANISATION ... NT

ORGANISATION FOR ECONOMIC CO-OPERATION AND DEVELOPMENT

Pursuant to Article 1 of the Convention signed in Paris on 14th December 1960, and which came into force on 30th September 1961, the Organisation for Economic Co-operation and Development (OECD) shall promote policies designed:

- to achieve the highest sustainable economic growth and employment and a rising standard of living in Member countries, while maintaining financial stability, and thus to contribute to the development of the world economy;
- to contribute to sound economic expansion in Member as well as non-member countries in the process of economic development; and
- to contribute to the expansion of world trade on a multilateral, non-discriminatory basis in accordance with international obligations.

The original Member countries of the OECD are Austria, Belgium, Canada, Denmark, France, Germany, Greece, Iceland, Ireland, Italy, Luxembourg, the Netherlands, Norway, Portugal, Spain, Sweden, Switzerland, Turkey, the United Kingdom and the United States. The following countries became Members subsequently through accession at the dates indicated hereafter: Japan (28th April 1964), Finland (28th January 1969), Australia (7th June 1971), New Zealand (29th May 1973), Mexico (18th May 1994), the Czech Republic (21st December 1995), Hungary (7th May 1996), Poland (22nd November 1996) and the Republic of Korea (12th December 1996). The Commission of the European Communities takes part in the work of the OECD (Article 13 of the OECD Convention).

FOREWORD

The OECD Workshop Mexico '96 on Biotechnology for Water Use and Conservation took place in Cocoyoc, Morelos, on 20-23 October 1996. It was hosted by Mexico's leading S&T policy organisation, Conacyt (National Council of Science and Technology), under its Director General, Mr. C. Bazdresch Parada.

Nearly 200 participants and speakers attended the workshop, approximately 120 from Mexico and the remainder from many other OECD Member countries. The discussions were conducted by the Scientific Chair, Mr. F. Bolivar Zapata, assisted by Deputy Chairs Mr. A. Noyola and Mr. D. Harper, as well as by the Co-chairs responsible for the two parallel sessions, Ms. R. Colwell, Mr. S. Miyachi, Mr. J. Ramos and Mr. W. Verstraete. The Mexican case study session was chaired by Mr. A. Jaime Paredes, and the workshop concluded with a speech by Ms. J. Carabias Lillo, Minister for the Environment, Natural Resources and Fisheries.

The workshop was prepared by Mr. E. Aceves Piña from Conacyt, assisted by Mexico's National Water Commission (CNA), and Mr. S. Wald and Mr. T. Hirakawa from the OECD Directorate for Science, Technology and Industry, assisted by a Mexico Steering Group set up by the Working Party on Biotechnology (WPB) to select speakers and approve the agenda (3rd and 4th Sessions on 8-9 November 1995 and 1-2 July 1996).

This was the third OECD workshop on bioremediation/bioprevention. The first, in Tokyo in 1994, reviewed the scientific aspects of bioremediation/bioprevention in all environmental media (OECD, 1995). The second, in Amsterdam in 1995, emphasized industrial aspects, but was limited to air and soil remediation (OECD, 1996). The Mexico workshop addressed water, considered by many as the most critical environmental issue, for ecological, as well as economic, political, social and strategic reasons. It is also perhaps the most complex to tackle scientifically.

For biotechnology, the Mexico workshop signals growing opportunities to help solve urgent and widespread water problems -- part of the impending "global water crisis" predicted by some experts. Thus, biotechnology/bioremediation is more than just another technical solution to specific local problems. It could become a strategic tool in addressing one of the key global issues of the future -- a tool policy-makers must increasingly incorporate in water policy.

For the OECD, this workshop, attended by Mexico's leading S&T and environmental policy-makers, marks the beginning of new initiatives in the fields of water technology and policy, with a view towards developing a co-ordinated approach to water problems.

These proceedings contain the Summary, Conclusions and Recommendations, drafted by the rapporteurs, Mr. M. Griffiths and Ms. D. Poole, the speakers' papers, summaries of the question-and-answer sessions, and the abstracts of the Mexican case studies. The Summary paper was revised by the speakers, chairs and co-chairs, and approved by the WPB and the Committee for Scientific and Technological Policy (CSTP). The latter agreed to derestriction of the proceedings, which are published on the responsibility of the Secretary-General of the OECD. They do not necessarily represent the official views of OECD Member countries.

Thanks are due to the Mexican government, and the many other OECD Member countries, as well as to the European Commission, whose generous financial contributions made the success of the Mexico workshop possible.

AVANT-PROPOS

La Réunion de travail de l'OCDE "Mexico 96" sur la biotechnologie appliquée à la préservation et à la qualité des ressources en eau s'est tenue à Cocoyoc (Morelos) du 20 au 23 octobre 1996. Elle a été accueillie par le Conacyt (Conseil national de la science et de la technologie), principal organisme chargé de la politique scientifique et technologique au Mexique, sous la conduite de son Directeur général, M. C. Bazdresch Parada.

Cette réunion a rassemblé près de 200 participants et orateurs, dont quelque 120 venaient du Mexique et le reste de nombreux autres pays Membres de l'OCDE. Les débats ont été menés par le Président scientifique, M. F. Bolivar Zapata, assisté par les Vice-présidents MM. A. Noyola et D. Harper, ainsi que par les coprésidents responsables des deux séances parallèles, Mme R. Colwell et MM. S. Miyachi, J. Ramos et W. Verstraete. M. A. Jaime Paredes a présidé la séance sur les études de cas mexicaines et la réunion s'est achevée par une allocution de Mme J. Carabias Lillo, ministre de l'Environnement, des Ressources naturelles et de la Pêche du Mexique.

La réunion a été préparée par M. E. Aceves Pina, du Conacyt, avec l'aide de la Commission nationale de l'eau du Mexique (CNA), et par MM. S. Wald et T. Hirakawa, de la Direction de la science, de la technologie et de l'industrie de l'OCDE, assistés par un Groupe directeur spécial mis sur pied par le Groupe de travail sur la biotechnologie (GTB) en vue de choisir les orateurs et d'approuver l'ordre du jour (3ème et 4ème réunions des 8-9 novembre 1995 et des 1-2 juillet 1996).

Il s'agit de la troisième réunion de travail de l'OCDE sur la biodépollution/bioprévention. La première réunion, tenue à Tokyo en 1994, a passé en revue les aspects scientifiques de la biodépollution/bioprévention dans tous les milieux de l'environnement (OCDE, 1995). La deuxième réunion, tenue à Amsterdam en 1995, a mis l'accent sur les aspects industriels, en se limitant à la dépollution de l'air et du sol (OCDE, 1996). La réunion de travail tenue au Mexique a porté sur l'eau, un milieu que beaucoup considèrent comme posant le problème d'environnement le plus grave, pour des raisons aussi bien écologiques qu'économiques, politiques, sociales et stratégiques. C'est peut-être aussi le milieu dont l'étude scientifique est la plus complexe.

Cette troisième réunion a fait ressortir que la biotechnologie est de mieux en mieux à même de contribuer à résoudre des problèmes urgents et fréquents liés à l'eau, qui préfigurent la "crise mondiale de l'eau" prévue par certains experts. La biotechnologie/biodépollution représente donc plus qu'une simple solution technique supplémentaire à des problèmes locaux précis. Elle pourrait devenir un instrument stratégique pour faire face à l'un des problèmes mondiaux majeurs de l'avenir -- un instrument que les décideurs devront de plus en plus souvent intégrer dans la politique de l'eau.

Pour l'OCDE, cette réunion de travail, à laquelle ont assisté les plus hauts responsables mexicains en matière de science, de technologie et d'environnement, marque le début de nouvelles initiatives dans les domaines de la technologie et de la politique de l'eau, en vue de définir une approche coordonnée des problèmes liés à l'eau.

Ce compte rendu contient les Résumé, Conclusions et Recommandations, rédigés par les rapporteurs, M. M. Griffiths et Mme D. Poole, ainsi que les exposés des orateurs, les résumés des séances de questions-réponses et les résumés des études de cas mexicaines. Le Résumé a été révisé par les orateurs, les présidents et les coprésidents, et approuvé par le GTB et par le Comité de la politique scientifique et technologique (CPST). Ce dernier est convenu de la mise en diffusion générale du compte rendu, qui est publié sous la responsabilité du Secrétaire général de l'OCDE. Les opinions exprimées ne représentent pas nécessairement la position officielle des pays Membres de l'OCDE.

Des remerciements doivent être adressés au gouvernement mexicain et aux nombreux autres pays de l'OCDE, ainsi qu'à la Commission européenne, dont les généreuses contributions financières ont permis la réussite de cette réunion.

TABLE OF CONTENTS

SUMMARY, CONCLUSIONS AND RECOMMENDATIONS 15

RÉSUME, CONCLUSIONS ET RECOMMANDATIONS 41

OPENING SPEECHES BY HOST COUNTRY

Mr. Francisco Bolívar Zapata, Director General of the Biotechnology Institute of the UNAM 71

Mr. Carlos Bazdresch Parada, Director General of CONACYT 71

GREETINGS BY OECD

Mr. Risaburo Nezu, Director, Directorate for Science, Technology and Industry, OECD 75

Mr. Bill Long, Director, Environment Directorate, OECD 79

GENERAL LECTURES

"US Environmental Protection Agency research on drinking water disinfection"
Mr. Robert Huggett, Assistant Administrator,
Environmental Protection Agency, United States 83

"The role of biotechnology for water use and conservation in Japan"
Mr. Kikuji Tateishi, Deputy Director-General, Basic Industries Bureau,
Ministry of International Trade and Industry, Japan 89

PARALLEL SESSIONS I AND II

I-A. BIOLOGICAL QUALITY OF WATER AND PUBLIC HEALTH

"Global climate change and cholera pandemics: the epidemiology of cholera"
Ms. R. Colwell, United States 99

Question-and-answer session

"New risks identified for viral and protozoan waterborne disease"
Ms. J. Rose, United States 109

Question-and-answer session

"Pathogenic viruses in the marine environment"
Mr. S. Myint, EC/United Kingdom 119

CLOSING SESSION

Mr. Adalberto Noyola, UNAM, Mexico 699

Mr. Bill Long, Director, Environment Directorate, OECD 701

Mr. Carlos Bazdresch Parada, Director General of CONACYT, Mexico 705

Ms. Julia Carabias Lillo, Minister for the Environment, Natural Resources and Fisheries, Mexico 707

LIST OF PARTICIPANTS 715

OECD WORKSHOP, COCOYOC MEXICO
SUMMARY, CONCLUSIONS AND RECOMMENDATIONS

Introduction

The OECD Workshop Mexico '96 on Biotechnology for Water Use and Conservation, hosted by CONACYT (National Council of Science and Technology of Mexico), was held in Cocoyoc, Morelos, on 20-23 October 1996. It was the third OECD Workshop devoted to issues of bioremediation and bioprevention.

The first of these, the OECD Workshop Tokyo '94 on Bioremediation, hosted by Japan, presented a wide-ranging review of both remediation and prevention technologies in all environmental media (see *Bioremediation: The Tokyo '94 Workshop*, OECD, 1995).

The second, the OECD Workshop Amsterdam '95 on Wider Application and Diffusion of Bioremediation Technologies, hosted by the Netherlands, focused on bioremediation of air and soil, from the perspective of wider industrial application (see *Wider Application and Diffusion of Bioremediation Technologies: The Amsterdam '95 Workshop*, OECD, 1996).

Following, and in connection with these two workshops, Mexico proposed to the OECD Working Party on Biotechnology (WPB) to host a workshop on biotechnology and water, covering the widest possible spectrum of issues. This would include water treatment, use and conservation, that is both quality and quantity aspects. Mexico also emphasized that the workshop would include, and would be addressed to, the key policy makers and institutions responsible for water in Mexico.

The WPB discussed and approved this proposal, and appointed a Steering Group to organise the workshop programme.

The Cocoyoc workshop was attended by approximately 180 participants from government, industry and universities of many OECD Member countries, more than 100 of whom were from Mexico, including, amongst others, the Minister of the Environment, Natural Resources and Fisheries.

In preparing the workshop, the scientific chairs and the steering group formulated three "**Key Questions**" for this workshop, which were communicated to all speakers and participants:

1. **What are the advantages of biotechnology (technological, economic, environmental) for water use and conservation?**

   - **immediately**
   - **in the short term**
   - **in the long term**

2. **What are the obstacles to the application of biotechnology for water use and conservation?**

   - **technological**
   - **financial/economic**
   - **training/manpower**
   - **government policy**
   - **public acceptance**

3. **What policies (including R&D support) should be recommended?**

The Summary of the workshop sessions, together with the Conclusions and Recommendations, present the answers which the workshop participants gave to these key questions.

Setting the scene

Water shortage and contamination are cause for increasing concern in many countries. While Mexico is amongst these countries, it is far from alone; many others, in both the developing and developed worlds, share these problems. Water contamination, and disagreement over the availability and use of water, are leading to tensions both within countries, and increasingly, between them. Water is more and more an international problem and is becoming a major economic issue as it becomes more expensive.

Of all environmental questions, those related to water are perhaps the most far-reaching in their long-term consequences, and the most difficult to tackle from a scientific point of view. It is for all these reasons -- the international, public health, economic, environmental and scientific dimensions of water -- that the OECD, together with Mexico, has devoted its attention to water and the role of biotechnology for water treatment, use and conservation.

Why the emphasis on biotechnology? Biotechnology has led to a revolution in health-care and pharmaceuticals, particularly in disease diagnostics and prevention. It is changing agricultural crops and food. It has also had many beneficial effects on the environment. Living organisms and systems have the capacity to reduce, degrade and even eliminate pollution in air, soil and particularly, water, where biological treatment is a standard technique the world over.

Mexico offered to host this workshop because it addressed some of Mexico's main national concerns. The OECD gathered some of the leading international experts in this field in an attempt to contribute to Mexico's own efforts to solve its water problems. In addition, Mexico's important contributions in this area were also discussed. Solving these water problems raises scientific and technological questions of great interest, not only to OECD Member countries, but widely beyond the OECD area.

Micro-organisms pose a serious threat to the safety of the world's drinking water; this is a growing peril with the potential to cause significant outbreaks of infectious disease even in industrialised nations. For the first time, through biotechnology, we have tools available to address such global public health concerns. Mexico, like many other countries, has immediate needs for reliable detection methods which only novel biotechnology can provide. The multinational nature of epidemic diseases means that no single country can address this global issue alone. The role of international organisations is therefore essential.

With regard to municipal water, the provision of high quality drinking water and efficient wastewater treatment must be priority targets for governments world-wide. While there is no single answer to technical questions, there are appropriate, low-cost and very reliable biotechnologies which can help achieve these targets in many cases. Often, this will require changes in financing mechanisms, and in people's daily habits. Therefore, individuals and communities must be involved at all stages of the planning and installation of new facilities. There is a great need for better public awareness and education with regard to the microbiological aspects of public health and sanitation. The wider use of measurement and control technologies is key to improvements in water processing. Modern biotechnology plays a major role in these. In developing countries, but also in the developed world, there are small communities which cannot afford advanced capital-intensive technologies. For them, biological technologies may well be the most appropriate.

To reduce the pollution of aquifers, a serious problem in some parts of Mexico as in other countries, bioremediation is often the only viable option. It is also an option of choice in many cases of surface water and coastal water pollution.

Summaries

Opening Session

The Opening Session began with welcoming speeches from the host country. Mr. Carlos Bazdresch Parada, Director of CONACYT, and the Scientific Chair, Dr. Francisco Bolivar Zapata, Director of the Biotechnology Institute of UNAM, emphasised the importance of this event for Mexico, and how much their country hoped to learn from the assembled international experts.

The greetings of the OECD were presented by Mr. Risaburo Nezu, Director of Science, Technology and Industry, Mr. Bill Long, Director of Environment, and Dr. David Harper in his capacity as chair of the OECD Working Party on Biotechnology.

Mr. Nezu placed the workshop into a broader context, that of the growing economic importance of biotechnology and of its impacts on trade, investment and employment. He underlined the need for regulatory reform, particularly in Japan and Europe, to facilitate the commercialisation of this technology. He expressed the wish to see the OECD's biotechnology work better known internationally, and to have it linked to OECD non-Member countries as well.

Mr. Long looked back to the 1970s when water issues were high on the international agenda. In the 1980s, environmental policy attention seemed to shift to problems of desertification, urbanisation and climate change. However, countries were now re-focusing on water problems. He mentioned the OECD expert workshop on "Sustainable Consumption of Water" planned in Sydney, Australia, on 10-12 February 1997, and expressed the hope that the outcome of the Mexico Workshop would make a contribution to the Australian meeting.

Keynote speeches were given by Dr. Robert Huggett, Assistant Administrator for Research and Development in the US Environmental Protection Agency, and Mr. Kikuji Tateishi, Deputy Director-General of the Basic Industries Bureau of MITI in Japan.

Dr. Huggett described clean water as one of our most valuable resources. However, despite the tremendous progress made in protecting surface and groundwater over the past several decades, substantial threats to water quality and quantity remain. These include not only waterborne pathogens, but also the by-products resulting from the use of chlorine and other disinfectants. Latin America, for example, has experienced severe cholera outbreaks in this decade, and the World Health Organisation estimates that two million children die annually in developing countries of diseases related to contaminated water. Population growth and development place ever-increasing demands on finite water resources, and per capita demand increases as standards of living rise.

Research is urgently needed on both quality and quantity aspects of water use; for example, the new strategic plan of the US EPA's Office of Research and Development identifies drinking water as one of its top six priorities. Current knowledge of pathogens is inadequate for assessing the risk to health. Both the infectivity of susceptible populations, and the virulence of different strains of pathogens, are uncertain.

Mr. Tateishi stressed that one of the major fields in which biotechnology could play an important role was in environmental applications, particularly in resource management and waste reduction. In Japan, high economic growth has caused serious water pollution problems in urban and industrial areas. Rapid population growth can likewise lead to volumes of wastewater that can overwhelm treatment facilities. However, legislative and financial measures, as well as practical ones, have been set in place to solve these problems.

While biotechnology shows considerable promise for water and land treatment, in Japan, there is as yet insufficient experience and very little data is available to demonstrate effectiveness; more knowledge is therefore required. To overcome this problem, MITI has launched programmes to test the effectiveness and safety of bioremediation technologies.

Session I-A. Biological quality of water and public health

A rapidly increasing number of countries can be considered as water-stressed. Many industrialised nations face growing problems associated with guaranteeing an adequate water supply. Waterborne pathogens represent a major and growing hazard. Pathogens of particular concern include enteric viruses and intestinal protozoa such as *Giardia* and *Cryptosporidium*. Contaminated water contributes to 1.6 billion cases of disease world-wide through the action of enteric viruses and protozoa. Both can result in severe and prolonged diarrhoea and additionally, some enteroviruses are now known to be associated with chronic effects such as diabetes and myocarditis. *Giardia* cysts and *Cryptosporidium* oocysts have been detected in source waters and drinking waters throughout South America and rotaviruses and coxsackie B virus have been isolated from treated drinking water in Mexico.

The origin and cyclical nature of cholera has intrigued scientists, and recently, the tools of molecular biology and immunology have permitted the demonstration of an association of *Vibrio cholerae* with zooplankton in marine and estuarine systems. With fluorescent monoclonal antibody and gene probes, coupled with PCR amplification, it has been possible to detect and monitor *Vibrio cholerae* in the environment and show that planktonic blooms are correlated with increased incidence of cholera. The use of similar molecular techniques is now permitting the semi-quantitative estimation of astroviruses and other enteric viruses.

The rapid detection of waterborne bacterial pathogens is crucial in order to prevent massive disease outbreaks. A key area of concern is the presence of these pathogens in water distribution systems. Several assays have been developed to detect low numbers of viable bacteria in drinking water, based on the use of 16S rRNA and on the *lux* gene. These methods are rapid, sensitive, specific, and can differentiate in some cases between viable and non-viable cells.

Current water quality assessments, based solely on bacterial load, are inadequate and there is a need for easier but reliable parasite indicators. Reports have suggested that some specific bacteriophages would be more reliable and economical indicators for evaluating viral health risks. Although a correlation may be found between levels of *Giardia* and *Cryptosporidium*, none has been found between other waterborne pathogens, whatever the type of water analysed. Therefore, neither specific phages nor bacteria can be indicators for *Giardia*, *Cryptosporidium* or the enteric viruses.

Even the best potable waters treated to international standards are not sterile and contain a diverse range of micro-organisms which form biofilms. These bacterial biofilms are resistant to the most current modes of decontamination and may harbour many different species of pathogens. Many of these pathogens have increased resistance to disinfection, and alternatives to chlorine, such as the more persistent monochloramine, have been advocated, but at levels which do not pose their own environmental problems.

Governmental efforts to chlorinate water, in Mexico as in many developing countries, are necessary. However, the cornerstone of better sanitation is better education in hygiene.

Issues remain to be addressed, such as the lack of a monitoring system, with data publicly available for epidemiological surveillance. Existing analysis and monitoring techniques for water are expensive, labour-intensive and time-consuming, and there is a need for on-line sensor equipment, capable of generating reliable information quickly and at a feasible cost. Only the regular monitoring of wastewaters, ambient waters, filtration and disinfection processes will provide the data needed to establish priorities for control options.

Session I-B. Bioremediation/biotreatment: aquifers

Water is an extremely scarce resource in semi-arid countries, and very rapid urban growth has produced increasing demands for potable water. While increasing recognition is being given to the value of wastewater as an important resource, and controlled disposal can provide significant additional resources of reasonable quality groundwater, improper disposal of untreated wastewater directly into aquifers can cause serious pollution problems. Performance standards and agreed techniques of risk assessment are still to be developed.

The effects of wastewater re-use on groundwater resources needs further study - adequate knowledge of the hydrogeology, infiltration processes and the movement and natural attenuation of pollutants is required for effective design and management of wastewater irrigation systems. The infiltration of wastewater may change the water chemistry throughout an aquifer and, in certain cases, the penetration of poor quality water to considerable depths can be inferred. In addition, a better picture of both groundwater resources and movement is required in order to allow evaluation of management options.

Groundwater quality has deteriorated in many places around the world as a consequence of fuels and chemicals spills in soils that occurred in the past. Techniques of *in situ* bioremediation, which show considerable promise for removing pollutants from aquifers, have been developed in several countries (including Japan, the United Kingdom and the United States) to clean up the volatile chlorinated compounds which contaminate groundwater and soil. The most important advantage of bioremediation is that organic contaminants can be transformed, and in some cases completely degraded. Compared to other remediation technologies, bioremediation is both safe and economical.

Widespread application of these techniques has faced difficulties however. The reason is that each site has its own characteristics, so each contamination problem may need to be regarded as unique. While generic know-how exists, the approach has to be specific to each site. There is also a need to demonstrate full-scale use rather than just at a laboratory level, and to involve physico-chemical techniques where appropriate.

The presence of natural micro-organisms at considerable depths raises the possibility of so-called intrinsic or natural bioremediation. This is an area that still requires considerable investigation.

Session I-C. Bioremediation/biotreatment: surface waters

The use of phytoremediation, and specifically wetlands or reedbeds, has been explored extensively for treating large volumes of water that contain relatively low concentrations of pollutants. Phytoremediation is a relatively low level technology and hence a low capital cost option and may be particularly suitable for developing countries.

A central question concerning the fate of pollutants in the environment is that of their biological degradation at low concentrations. For some pollutants, distinct "threshold" concentrations have been reported, below which these compounds are no longer degraded. However, it has also been shown that other xenobiotic compounds can be assimilated by micro-organisms even when they are present at concentrations of a few nanograms or even femtograms per litre, i.e. far below the "threshold" levels at which they can support growth. It may be presumed that these compounds are utilised simultaneously with uncharacterised dissolved organic materials that will either support growth or supply the energy required for maintenance.

Both inorganic and organic mercury compounds are important contaminants in many industrial locations world-wide, and native macrophytes, engineered to express the *merA9* gene which codes for an enzyme reducing mercury salts to elemental mercury, have been examined for their ability to remove mercury salts from polluted soil and water. A more serious problem is the organic mercury concentrated in the food chain, which poses a more immediate threat both to wildlife and humankind. Plants need to be engineered to transform this organic mercury.

Contaminated sediments may be a source of pollution for surface water. As yet, no proven technology exists for *in situ* bioremediation of sediments, and consequently, they need to be removed before being treated. Current clean-up techniques include combinations of physical, chemical, biological and thermal processes.

As the consequence of numerous oil spills, superficial waters are chronically polluted by hydrocarbons. On a world-wide basis, river pollution is the main source of sea pollution, far more than pollution caused by tanker disasters.

Session I-D. Bioremediation/biotreatment: marine and coastal waters

Anthropogenic pollution of marine ecosystems, including rivers, bays, estuaries, and other coastal areas, is a major global environmental issue that requires attention. A number of bioremediation options have been suggested, involving both marine plants and micro-organisms. Also, the use of macroalgae may provide a new and cost-effective technology for treating waterborne pollutants. A co-ordinated international effort is needed to demonstrate the feasibility of these concepts and to obtain operational data via *in situ* demonstration projects in polluted environments. Immediately following the Exxon Valdez accident, laboratory studies showed that indigenous hydrocarbon-degrading organisms were highly active, but nutrient limited. Seventy miles of oiled beaches were treated with oleophilic and slow release fertilisers, and this nutrient-enhanced bioremediation safely and effectively removed oil, both from the surface and subsurface of beaches.

There is a need for a more direct connection between biological parameters and the resulting chemical oxidation of hydrocarbons, to allow more effective application of nutrients and the prediction of the rate of chemical processes. Additionally, a better understanding of indigenous microbial consortia is needed. A major obstacle to more widespread use of bioremediation techniques is acceptance by regulatory agencies and preparation of contingency plans which permit implementation of bioremediation techniques in a timely fashion. One solution proposed for the problem of oil spills has been the transformation and preparation of contingency plans which permit implementation of bioremediation techniques in a timely fashion. One solution proposed for the problem of oil spills has been the transformation of an indigenous dominant microbial strain using a broad host range plasmid, which might contain a panel of catabolic operons, able to be quickly induced by the pollution.

Biotechnology is being used to monitor the ecotoxicology of heavy metals in the San Francisco Bay by developing biomarkers and assessing the reproductive effects at the molecular level. Selenium discharged into the San Francisco Bay from petrochemical sources is toxic to fish, water birds and small mammals. A promising application of environmental biotechnology is the removal of selenium using bacteria. The organisms are able to biotransform toxic forms of selenium into non-toxic selenium metal precipitates, which can be removed with the bacterial biomass. Another approach is the construction of engineered wetlands that employ specific plant species to bioabsorb the selenium.

Coastal marine waters are very sensitive to the discharge of the macro-nutrients, nitrogen and phosphorus. In particular, coral reef structures in tropical regions can experience significant degradation when exposed to these nutrients. Combination of biological nutrient removal technologies with oxygen-enriched bioreactors and membrane separation techniques for sludge recycling will both improve the effluent quality, and significantly reduce the risk.

Session II-A. Prevention of water pollution: municipal sources

In developed countries, very high proportions of municipal wastewater may be connected to the sewage system. Nevertheless, problems such as drought, over-exploitation, flooding, and overflows exist. A major threat to water quality arises from non-point source contamination. Agricultural activities, urban runoff, atmospheric deposition, land disposal, construction work and mining operations, all contribute to the non-point source pollution. New approaches are required, aimed at sustainability (as measured by reduced energy consumption, rational use of natural resources and limited production of contaminated sludge). Recycling wastewater as a second quality (grey or e-quality water), and the separation of rainwater from wastewater, are options.

A major problem for developing countries is the growth of mega-cities (more than 10 million inhabitants); 17 out of 22 will be in the developing world by 2000. An estimated 600 million people in urban areas of developing countries live in life- and health-threatening homes and neighbourhoods, primarily in peri-urban areas. Problems of clean water and sanitation are beyond the reach of most of these communities, which municipal authorities tend to ignore, as they find themselves overwhelmed by the sheer numbers of people.

Since the conventional approach to the disposal of human wastes -- centralised collection and treatment of sewage -- is very expensive, new approaches, located as close as possible to the user, are required, bearing in mind that changing individual and communal behaviour may be at least as important as new facilities.

Technical and non-technical solutions are required -- there is no single magic bullet. Citizen and community participation (and education) is essential. For example, in the last decade, a decentralised and modular system using decentralised ponds and wetlands combined with solar energy to treat and re-use wastewater has been developed in the city of Tijuana. The goal was to manage the limited water resources, improve the quality of life, and enhance the urban environment within the city, as well as to reduce the pollution of the Tijuana River.

Both high and low technology novel processes are available to improve standard facilities. The handling of sludge, the by-product of wastewater treatment plants, is a complex and expensive operation, and consequently, sludge biotechnologies are not commonly used in developing countries such as Mexico. However, a range of biological treatment possibilities exist, and recycling to land is the best option, if no metals or other toxic substances are present. Control of pathogens is crucial if this option is chosen.

Phytoremediation and the use of wetlands represent new, but not high technology, both for degrading organics and concentrating and removing (harvesting) metals. Both naturally occurring and constructed wetlands have offered significant treatment capacity as interceptors of contaminated water flows. Judicious selection and use of single plant species can offer significant advantages for the treatment of non-point source pollution. A major advantage of the wetland process is that it makes use of well understood agronomic practices.

Session II-B. Prevention of water pollution: industrial sources

In industry, water is often seen as just one of the production factors. Good process water is, however, becoming more scarce and more expensive. Both quality and quantity of industrial water must relate to purpose, in other words, be demand driven. The ultimate target should be sustainable use -- input and recycling leading ultimately to zero discharge. Measures can be adopted from the end-of-pipe to process substitution stages, and biotechnology, in combination with physico-chemical technologies, will play an important role in realising this goal.

The development of industry in towns and cities has given rise to toxic effluents, which, arriving at sewage treatment works, are inhibitory to the conventional biological processes. Monitoring technologies, which may permit the identification of pollutants and offending processes, will enable operators to police treatment works inflows. Novel on-line technologies for this purpose include the use of the bioluminescent response of naturally occurring and genetically engineered bacteria. Trade effluent cleaning should ultimately be the responsibility of the producer, and a system of levies based on monitored effluents may be the only way to achieve this.

Clean technologies will eventually lead to waste minimisation. However, if enzymes are to gain widespread use in industrial processes, they must be more stable than those currently commercially available. There exists a huge metabolic potential among extremophile organisms which can be exploited via screening and directed evolution.

For companies to meet more stringent waste disposal authorisations while maintaining competitiveness, flexible, efficient and more cost-effective environmental technologies will be required. Particular difficulties are associated with the clean-up of complex liquid effluent compositions. Hence, a multi-disciplinary, focused and integrated research programme to develop novel clean-up technologies is required, and industry must then be convinced to adopt these technologies.

Although many industries produce low levels of heavy metal wastes, the primary sources are often the relatively small number of industries involved in metal working/refinishing, circuit board and semi-conductor manufacturing, and photographic processing. Clean-up of metal-contaminated waste-streams may be achieved using a family of technologies that employ the active or passive accumulation of metals by micro-organisms, coupled with separation and recovery of those metals. A variety of different microbial processes are exploited, including bioprecipitation, redox transformations and biosorption. The full potential of micro-organisms to sequester or precipitate heavy metals, including radionuclides, can only be achieved through a more fundamental knowledge of microbial physiology. Processes need to be cost-effective, and the metals left in a recoverable form.

The Upflow Anaerobic Sludge Blanket (UASB) process is widely used to treat wastewater from food processing industries. Biogas from anaerobic wastewater treatment may contain hydrogen sulphide (H_2S), but this can be neatly removed in a bio-scrubber process, which utilises sulphur oxidising bacteria, effectively using the same activated sludge which polishes the final effluent.

Wider industrial use of bioprocesses requires a reduction in the levels of uncertainty associated with them. Improved predictability of design may be achieved by running a suite of treatability studies in which alternative treatment systems are evaluated through long-term pilot- or bench-scale studies. An alternative technique is to assess the fundamental kinetics of biodegradation and to use the resulting kinetic parameters in reactor engineering studies.

The lack of robust on-line sensors is another obstacle to the uptake of biotreatment technologies. If key parameters can be continuously monitored, the potential for improving treatment reliability, effluent quality and process economics is vast. Once data have been collected from plants operating under a range of inputs, fail-safe automatic control procedures can be developed to optimise plant performance.

Session II-C. Prevention of water pollution: agricultural sources

At least one third of the population in developing countries is still deprived of an adequate water supply. This shortage has become the driving force for the re-use of treated wastewater for numerous municipal applications. Both low and high technology systems exist to permit the re-use of water and nutrients, and the recovery of energy. Appropriate technology can yield high quality effluent at low capital and operating costs. Microfiltration, for example, can achieve effective removal of suspended solids, biological and chemical oxygen demand, and pathogens.

Wider coverage, particularly in rural areas, calls for a shift from high-cost technologies to low-cost alternatives. There are simple and reliable technologies for treatment and disinfection of surface water for rural communities and for individual use. For example, solar water disinfection has been successful using transparent plastic bags and bottles. The combination of roughing filters with slow sand filters represents an efficient and reliable treatment scheme for rural community water supplies. There is some debate as to whether these techniques are an alternative to chlorination or whether the latter would still be required.

The use of Rotating Biological Contactors (RBCs) followed by reed beds as an alternative technology for wastewater treatment was described. These represent another robust and "best environmental fit", low-cost option for small communities.

Pesticides applied to water can be a major environmental problem. There are a number of examples in which enzymes have been explored for end-of-pipe treatment and remediation. An enzyme has been used to degrade a harmful by-product without affecting the main product, so minimising waste generation, permitting longer product use and resulting in significant cost savings.

Farmers can save money and reduce waste by adopting closed cycle or integrated systems to reduce input from fossil energy and output of contaminants. Such systems link waste treatment to biogas production, composting and hydroponic ponds.

Excess nitrogen, leached into waterways following agricultural application, can cause severe eutrophication. Wetlands are very effective at nitrogen removal, but new agricultural policies are needed for appropriate use of fertiliser and manure. For sustainable agriculture, existing wetlands should be retained in the landscape, and in areas of intensive agriculture, drained wetlands should be re-established.

Constructed wetlands are not yet a fully developed technology. The choice of plants for any particular use is not yet optimised, and the mechanisms of pollutant degradation in the rhizosphere are largely unknown. While they cannot be expected to do everything, they certainly will produce an effluent suitable for unrestricted irrigation in developing countries.

Session II-D. Desertification and environmental modification

Plants adapted to arid and semi-arid areas exist, but most of the economic crops need a constant supply of water. We need to select relevant genes for expression in transgenic versions of crop plants.

There exist different defence mechanisms, including protecting proteins, sequestering proteins, ion export and exclusion and cell wall modification. Osmolytes are molecules that help cells retain water; trehalose is the most well-known. The so-called "resurrection" plants can survive very high levels of desiccation and are also the only plants to contain trehalose. Transgenic plants containing trehalose-producing enzymes have improved drought resistance.

Drought and salinity tolerance are very complex phenomena which depend on multiple genes rather than one unique gene. Maize embryos are very tolerant to drought, and attempts have been made to isolate the relevant genes. Although the mechanism of tolerance is largely unknown, transgenic plants have shown a significantly increased tolerance to salinity, and higher growth rates under stress, than control plants. Additionally, the transgenic plants showed improved recovery following the removal of the stress.

Toxic metal pollution of water and soils is a major environmental problem facing the modern world. The use of specifically selected and engineered metal-accumulating plants is an emerging technology.

Bioprevention -- the development of environmentally friendly products and processes -- is a key component of sustainability. An example of such a product is a bioabsorbent extracted from a mutant of *Alcaligenes latus*. This compound is a polysaccharide containing different sugar moieties; it has a very high water retention capacity even in the presence of salt. One major global ecological concern is the accelerated expansion of the world's deserts. Incorporation of this bioabsorbent into desert soils on an experimental basis has shown that it can protect growing plants at high stress levels.

Case studies

The fourth day of the workshop, devoted to four Mexican case studies, was opened by Dr. A. Jaime Paredes, Deputy Director-General of the National Water Commission (CNA). These case studies provided excellent examples of the problems and solutions discussed earlier and gave added weight to the conclusions and recommendations presented in the next section.

Dr. Paredes mentioned Mexico's difficulties to arrive at a correct evaluation, use and exploitation of its resources, particularly water. Water is very unevenly distributed in space and time. Long considered an inexhaustible resource, it must now be seen as limited because of the degradation of its quality. In Mexico, agriculture is the source of most (62 per cent) of the waste water and industry supplies 10 per cent, the remaining discharge coming from municipal activities. Deforestation and other unsustainable practices contribute to the reduction in water quality. Aquifers underlying agricultural areas are suffering from diffuse contamination, while over-exploitation of coastal aquifers is giving rise to an influx of salt water. Thus, the management of natural resources has to be improved, in line with the concept of sustainable development.

Of the wastewater discharged in the country, only approximately 15 per cent is treated, and of the 700 treatment plants, many are out of service. However, these plants also include approximately 37 anaerobic reactors, which provide some experience with biotechnology. Biotechnological solutions will gain increasing acceptance, in so far as they fulfil some key requirements: competitiveness with alternative methods, and operational reliability and safety.

Case 1. Contamination of the Lerma river

The Lerma River basin is one of the most developed urban/industrial areas in Mexico. Population and industrial growth has resulted in a deterioration of the ecosystem, reflected in reduced water availability and quality. An action plan was begun in 1989 to develop new sanitation. In 1994, in a reservoir which is part of the basin, there was an event resulting in high mortality among water birds. The characteristics and significance of this event were presented.

Technologies which are used for the analysis and collection of data in the basin were described. Strategies based on these data will eventually lead to the remediation of the ecosystem. These technologies include hydrodynamic modelling of techniques, such as forced circulation and aeration of the related coastal lagoons (see also Case 2). The use of biotechnology in areas such as aquifer remediation has been considered.

Case 2. Marine and coastal water contamination

The effects of pollution on both recreation, including tourism, and commercial fishing are of particular concern to the Mexican authorities. This was presented in the context of Mexico's strategy for improving water quality in coastal zones, where the main problems are due to point and diffuse pollution from industry, agriculture and municipal sources. Organic pollutants are mainly derived from local petroleum and the petrochemical industry.

The activities associated with a coastal petroleum refinery, and especially the effect on the marine environment of submarine discharge of wastewaters containing chemicals, have been studied. Pollutant levels have been related to animal and plant communities, and plankton and fish abundance and distribution analysed. The integration and systematic development of this data will allow future diagnosis of environmental conditions and provide a decision tool for the sustainable development of the area.

One feature of the Mexican coast is a series of interlinked lagoons. Nevertheless, these are shallow and there is little in the way of communication between them and with the ocean. The natural condition of these lagoons has been severely altered by human activity, including dredging and wastewater discharge. Based on data obtained from hydrodynamic studies and field measurements, a proposal has been made that, in order to reverse this situation, wastewater discharge should cease and forced circulation and aeration be adopted.

Case 3. Mexico Valley's wastewater treatment

A new scheme is being planned for the treatment and transportation of rain and wastewater for Mexico City and the whole of the Mexico Valley. Existing systems are sinking in the soft clay on which the city is built, with consequent loss in capacity, particularly in high rainfall (storm) conditions. The quality of the water is not exceptionally low; however, there are high concentrations of boron derived from the local rock, and parasites, faecal coliforms and helminths are found. Currently, 96 000 ha outside the Valley are irrigated with untreated water. There is seen to be considerable value in the nutrient content. The major problem is one of improving public health.

The scheme's planners perceive that it is too large for biotechnology to make an appropriate contribution, and consequently, chemical flocculation and sedimentation is the favoured route. The resultant sludge is to be conditioned with lime and used for land remediation. Effluent disinfection, after use of sand filters, will remove the pathogens. Several of these concepts were questioned by workshop participants. In particular, mention was made of existing biological treatments of equal or larger size, and it was pointed out that anaerobic treatment of the water would give rise to much less sludge of a considerably higher quality. There was some doubt that chemical sludge could be used as a soil remediation agent. Also, the use of faecal coliforms and helminths as marker organisms for the presence or absence of pathogens was not appropriate.

Case 4. Treatment of water in rural areas

Over 70 per cent of Mexican communities have less than 100 inhabitants. These communities lack even the most basic facilities, resulting in disposal of faeces to soil with the consequent spread of gastrointestinal infections. The construction of a centralised supply of drinking water and a sewage network in such rural areas would be both difficult and costly and for these situations, the National Water Commission has developed five technological packages for on-site treatment of wastewater and faecal solids. Even in extremely poor areas, the provision of essential supplies has provided sufficient incentive for the local population to participate and build their own facilities.

Education of the local population about what constitutes good sanitation and why they should want it, is the highest priority need. The technology is basic and available. Workshop delegates pointed out that there were other (bio)technologies at the same level of sophistication that could be used in addition to the ones described.

Summing-up of case studies

An independent summing up of the four cases was made by Dr. Ron Atlas from the University of Louisville in Kentucky, who noted that Mexico, like the rest of the world, has a diversity of water needs. The case studies discussed water re-use and technologies that could contribute to the sustainability of water supplies. In carrying out their assessments, the projects faced the difficulty of defining water quality standards in terms that were acceptable for Mexican applications. There was a lack of internal validation and a need for guidelines and technologies specific to Mexican requirements.

All of the case studies lacked significant application of biotechnology. Monitoring and treatment largely relied on physico-chemical methods; none of the modern biotechnological approaches for water quality monitoring, such as specific pathogen detection, had been examined, and none of the designs for major sewage treatment facilities incorporated secondary biological treatment. Although there are plans for sludge removal, what to do with the sludge remaining after physical treatments, and the resulting water quality, should be considered. Too often, cost and management issues seemed to block consideration of biotechnology. In addition, there seems to be a lack of communication between the scientific community and government authorities, and perhaps also a lack of communication with the public.

The four case studies spanned a wide range of water quality and usage issues in Mexico. One major project used a very innovative approach to an integrated study and involved monitoring water quality with limited use of biotechnology. A second project on coastal marine systems was largely descriptive and aimed at monitoring, rather than clean-up. The project on waste and water management in rural communities provided an interesting study in behaviour modification and novel uses of technology, where the main focus is on human health and public education rather than increased water need.

The Mexican Valley project gave rise to the greatest controversy, especially as to whether options were still open for the incorporation of biotechnology. There is a desire to retain high levels of nitrogen and phosphate in the water to eliminate the need for fertiliser applications, and this presents a novel challenge to biotechnology. The retention of nutrients may not be compatible with the elimination of pathogens. A holistic approach to waste management that incorporates biotechnology is likely to be the only cost-effective solution for long-term sustainability of environmental quality.

Closing plenary

Mr. Adalberto Noyola, Deputy Scientific Chair, opened the final plenary session by reviewing some of the highlights of the workshop, underlining the importance of biotechnology for wastewater treatment and public health. In the future, wastewater management has to be seen in the framework of sustainable development. The application of biotechnology has to take integrated and interdisciplinary approaches into account. New technologies for waste disposal should be efficient and low cost, and could thus find wide public acceptance.

In the following speech, Mr. Bill Long, Director for Environment of the OECD, reminded delegates of four recurring points from the workshop:

- the extent of the threat to public health from diseases spread by waterborne pathogens;
- the present and potential contribution of biotechnology, the benefits of which need to be expressed more widely to policy makers;
- the relationship between water recycling and re-use and the issues of water quality; and
- the growing link between climate change and waterborne disease.

Mr. Carlos Bazdresch Parada, Director General of CONACYT, expressed the benefits of holding the workshop in Mexico. Mexico is well aware of the importance of water quality and conservation. The co-ordination of policies for the rational use of water and for improving its quality will benefit from the recommendations made at the workshop. The international exchanges that took place between specialists from Mexico and many other countries on bioremediation and biotreatment of wastewater, surface and coastal waters and aquifers, will encourage the studies of the applicability and possible use of novel technologies in Mexico.

In bringing the workshop to a formal closure, Ms. Julia Carabias Lillo, Minister for the Environment, Natural Resources and Fisheries, highlighted the value to Mexico of membership in the OECD. Membership has provided the opportunity to organise this scientific meeting, the third after those in Tokyo and Amsterdam, which, using the combined efforts of academic specialists, water users, industrialists and government authorities, has permitted the analysis of current problems and a movement towards further co-operation on an international level. The scientific community has been brought closer to the decision making institutions and to participation in the development of strategies specifically for the use and conservation of water.

By developing some of Mexico's concerns regarding water, Minister Carabias explained how and why a workshop of this kind contributes to Mexico's national priorities. On the scale of water availability, Mexico is in an intermediate position, not as low as countries of the Middle East but not as high as the United States. However, the distribution within the country is very uneven with 80 per cent of the territory receiving only 20 per cent of the water. Localised flooding and surface run-off are also problems. Although Mexico is well equipped in terms of water infrastructure, irrigation and dams, etc., there is a problem of operational efficiency -- 50 per cent of agricultural water is lost through leakage. Twelve million Mexicans have no drinking water and 27 million have no sewage system; many of the largest basins are contaminated and aquifers suffer from over-exploitation.

Minister Carabias referred to a number of the Mexican programmes described during the workshop, including the clean-up of the Mexico Valley and also of the Lerma-Chapala basin where the greater part of industry is located. A proper planning structure is being created to regulate the use, re-use and cleaning of water. This requires agreement among federal, state and local authorities and among agricultural, industrial and domestic users. Like few other natural resources, water can generate powerful social and political conflict.

How can biotechnology help in these major challenges? It can provide valuable solutions in terms of quality, through the application of bioremediation and the consequent improvement of access. Additionally, it may solve problems at a technical level, before they enter the political, social, health or ecological spheres.

In concluding, Minister Carabias identified three important areas for study:

- The first is the regulatory framework. If regulations are clear and universal, then better ways of making common use of water can be found. Economic and social inequalities can be an important factor in the use of technology.

- Secondly, biotechnology should not become a new power, with some countries more dependent than others. Mexico must find a way of developing its own technologies and participating on the world scene.

- Third is the need to ensure that this activity in biotechnology is turned into concrete, specific projects which will benefit those who are most in need. Major diseases are basically due to the poor quality of the water, and this is a problem that can be solved if the necessary infrastructure and technology are provided.

Conclusions

The following conclusions, and the answers to the three questions, are based on the discussion sessions that took place after the speeches. The answers to the first two questions are presented below. The answers to the third question on policy recommendations are summarised in the "Recommendations" section.

General

There is a direct link between enhanced sustainability and the wider application of biotechnology in the environmental area which will also offer an opportunity for achieving better water quality. Biotechnology has the greatest capacity for success, whether applied to water treatment or public health problems, when applied in an interdisciplinary, holistic, manner.

The unique opportunities provided by biotechnology, combined with other chemical and physical methods as appropriate, are the only way forward for addressing quality of water and human health issues on a global scale.

The value of clean water has not been subjected to cost-benefit analysis, but should be. The OECD could be an appropriate body to carry out such studies. Networking and collaboration of governments, universities, industry, and the public is needed on a global basis.

Biotechnology offers significant opportunity for the elaboration of sustainable, cleaner, industrial processes through the development of improved biocatalysts. A few large-scale chemical synthetic processes based upon biocatalysis have been operating successfully for several years, resulting in waste reductions compared to the conventional chemical alternatives.

Water quality and public health

Improved health surveillance and the investigation of disease outbreaks are a priority. Biotechnology is providing increasingly sophisticated tools to survey, detect, identify, model, predict and assess the risk of both current and emerging waterborne disease-causing micro-organisms.

Further support of R&D for human health issues is essential for the global public good. For industry, direct economic opportunity lies in the provision of instruments, not in the information itself; hence support for gathering this information must rest with international organisations and governments. Within each country, there are different stages of development, so that R&D priorities may be different in different local environments. However, the priorities are basically similar for all nations.

The need for international case studies, along with support for the global collation, storage and dissemination of epidemiological information, is paramount. In this context, there is a need for the development of databases by international organisations such as the OECD.

Following the recognition that associated diseases such as cancer and myocarditis can be caused by waterborne micro-organisms, further studies are seen as essential.

The use of indicator organisms, which currently only approximate risk and are at best of limited predictive value, should now be abandoned, since these methods are being superseded by emerging technologies such as gene probes, DNA finger-printing and PCR. These sophisticated biotechnologies, which allow the identification and tracking of single pathogen cells (or viruses) via their nucleic acids, and are both faster and more specific than the methods they replace, are a major improvement. Molecular epidemiology integrated with microbial ecology is the way forward, offering the opportunity to monitor and survey waterborne diseases and their causal micro-organisms, and to assess and predict, on a global scale and with greater precision, associated risks.

Fate and transport studies on epidemic diseases, i.e. of pathogens in the aquatic environment, are emerging priorities. The multinational nature of epidemic diseases means that no single country can address this global issue alone. The role of the OECD in this context is essential and complementary to that of the WHO.

Water supply

It is of paramount importance that governments ensure the distribution of the highest quality drinking water to everyone. The next most immediate priority should be efficient processes for treatment and sanitation of wastewater. Only when this is achieved, might it be appropriate to consider installing a second distribution system for grey or e-quality water. The term e-quality refers to a more economically priced water composed of rainwater, reclaimed water, etc., to be used for a variety of commodity uses. Use of e-quality water has its dangers, especially in developing countries, and will require a high level of community education.

There is a major role for biotechnology in the clean-up of all types of water. For instance, slow filtration over sand or granular activated carbon (GAC) is a secure and economic unit process for drinking water. These technologies might also be applicable for e-quality water. Moreover, rapid on-line or off-line bio-monitoring procedures for water quality should be used to increase public assurance.

For third world countries, safe drinking water production can be greatly improved by roughing filtration followed by slow sand filtration. If further disinfection at the household level is required, the application of new techniques based on the use of solar energy (e.g. the UV content of sunlight) might be considered. In addition, it appears advisable that appropriate technologies to produce chlorine as a stand-by disinfectant be developed for third world countries.

Bioremediation

Biological treatment of wastewater is the most advanced of all bioremediation processes and has clear advantages over chemical processes even for the highest volumes of throughput.

The use of modern biotechnology no longer requires special justification -- its practical value has been amply demonstrated in bio-medicine. The promise in bioremediation is also being fulfilled. Biological processes are being introduced more widely as they are shown to be cost-effective. A proactive mode of action is now required, e.g. identifying high risk oil tanker routes, sites of potential viral disease/water quality degradation, etc., to prepare for future action. The transfer of technology from the laboratory to real, *in situ* field conditions is of great importance.

Anaerobic treatment of wastewater gives rise to less sludge than aerobic, and aerobic less than chemical processes. Both anaerobic and aerobic treatments yield sludges that are higher in quality than chemical processes.

Residual amounts of xenobiotics can seriously endanger the environment, and especially water conservation. Cradle-to-grave responsibility by the manufacturers of potentially polluting materials needs to be strongly encouraged. This might result, for example, in the provision of reagents together with a pesticide to neutralise any unused material. Chemicals banned by international regulatory bodies because of their detrimental eco-toxicology should be rigorously excluded from use in all countries, developed or developing.

There is a need for specific tailor-made process biocatalysts to reduce the level of pollutants entering the waste stream, such as the use of enzymes in animal feeds, which reduces the excretion of phosphate, and to remove specific toxic substances, such as heavy metals.

Wastewater treatment

In domestic sewage treatment, the emphasis should be first and foremost on primary sanitation, that is removal of pathogenic organisms, organics, and suspended solids. Subsequent to this, attention can be given to further treatment, more specifically to biological nutrient removal. However, in the latter case, the biological nutrient removal processes must be carefully judged against the merit and risk of recovering the nitrogen and phosphorus via crop growth, particularly in countries where treated sewage can be used for irrigation.

New urban development should examine the possibilities of newer wastewater management patterns that consider health protection, pollution prevention and resource recovery (water and nutrients) in the context of increasing demands for sustainability.

Wastewater treatment has to be affordable by the user. While this may involve government subsidy, people may need convincing at least to pay for operation and maintenance.

One successful approach to the treatment of wastewater especially in small communities is the combination of anaerobic or aerobic treatment and wetlands (reedbeds). The choice of the first stage technology depends on capital and operating cost considerations. The technology of wetlands requires further R&D in terms of plant communities, rhizosphere transformations occurring, degree of pathogen removal achieved and build-up of pollutants, e.g. phosphates.

The biotechnology of microbe-plant interaction in the removal of pollutants and the recovery of metals via ponds, wetlands and crops, has been field demonstrated in both developed and developing countries and is a low capital alternative, especially for rural areas. Further development should lead to improved efficiency.

Treatment of excreta

While waste treatment plants can be designed to cope with urban solid waste, localised treatment in peri-urban and rural areas, where collection facilities do not exist, poses a more difficult problem. For example, excreta in peri-urban areas represent a major hazard of faecal-oral contamination. To deal with them, diverse techniques must be developed. One of these, the dry latrine, drastically reduces the danger of contamination of ground- and surface water, but it might not always be accepted socially because the water flushed toilet with septic tank is considered the standard. Therefore, both R&D and education are necessary in order to make low-cost removal and disposal of excreta in peri-urban areas more efficient, and also to give it a higher public acceptance. Biotechnology can contribute in both peri-urban and rural circumstances by, for example, the development of sorptive matrices to add to the excreta and the use of composting technologies.

Sewage sludges - biosolids

Beneficial use of sewage bio-solids is becoming more difficult in industrialised countries. However, physico-chemical treatment of wastewaters produces more sludge than biotreatment and, moreover, chemical sludges are more difficult to dispose of. Hence, biotreatment of wastewater should not be disfavoured on the basis of its sludge production. However, two lines of action are needed with respect to bio-solids: the removal of heavy metals to prevent their entering the wastewater, and upgrading the biosolids, for example by enhanced composting.

Monitoring and control

Monitoring and control of environmental biotechnological processes is inadequate and too little practised. Facilities have been deliberately over-designed so as to dispense with the need for sophisticated controls. Less costly, smaller biotreatment facilities with higher throughput are increasingly required, and these could be designed if more intelligent use of monitoring and control devices were made. In recent years, a wide range of off-line and on-line sensors, using biotechnological and also physico-chemical systems, has become available, but because of cost or unreliability, few are in general use.

Novel technologies for environmental modification

Research on transgenic plants with increased tolerance to drought is proceeding at an increased pace through the application of advanced biotechnological techniques. In view of the water scarcity in many regions, the application of this research is urgently needed. New biosorbents, which can also contribute to improved plant growth under water stressed conditions, are now emerging from the laboratory.

Awareness

An informed public is fundamental to the advance of biotechnology and its practical application. Public information is no longer a national need but an international imperative for the sharing of research information, provision of educational information to the lay-person and of access to databases via the World Wide Web and other information resources.

Public policy makers need to be kept well informed of those areas of biotechnology which are not receiving high profile media attention but are nevertheless progressing successfully, e.g. bioremediation, classical biological wastewater treatment, and public health applications. Awareness of the value of environmental biotechnology in public (human) health needs to be developed on the part of governments.

Public awareness and acceptability are major determinants in the use of biotechnology, especially for applications concerning the use of GMO/novel organisms. Environmental biotechnology will advance most rapidly if a uniform regulatory framework for the application of genetically modified organisms (GMOs) in bioremediation and water treatment is in place world-wide.

Answers to the three questions

"What are the advantages of biotechnology (technological, economic, environmental) for water use and conservation?"

Immediate

The tools of biotechnology, as components of a multi-disciplinary approach, offer a powerful approach to addressing public health issues, biotreatment and bioremediation.

Biotechnology at last provides the tools to address global public health concerns associated with contaminated water, be it ground, surface, aquifer, or marine waters. Effective biotechnological monitoring and assessment techniques are now available to identify and track single cells or components of micro-organisms such as nucleic acids.

This monitoring and assessment for specific diseases will soon be extended to address a wider repertoire of disease-causing organisms, and these novel techniques will be particularly important for the surveillance, modelling and prediction of emerging new diseases.

For the present, biotechnology is predominantly used to treat contamination, be it of land, air or water, after it has occurred. Particularly in the area of water treatment, biological processes have been shown to be cost-effective and environmentally friendly. They are perceived to be natural, and thus more easily accepted. They may be used to degrade materials untreatable by physico-chemical means, and their flexibility in response to changing circumstances makes them uniquely valuable in the treatment of domestic and industrial mixed effluents. In addition to the removal of bulk pollutants, biological techniques are valuable for very specific contaminants that are costly to remove by other means.

Biological processes provide multiple options for single problems at different levels of technology. They may thus provide wider commercial opportunities for indigenous technologies and for local populations.

Biological treatment of water and waste involves little use of chemicals, and consequently does not affect the structure of sludges or sediments. These, unlike chemical sludges, can be reintegrated into the environment.

A further advantage of biotechnology lies in the treatment of non-point source (dispersed) pollution, for example polluted groundwater and aquifers. These environmentally friendly technologies are not only cost-effective in the majority of cases, but also provide clear opportunities for *in situ* application. Bioremediation technology is being employed now, and its use has been demonstrated by a number of high-profile case studies, e.g. Exxon-Valdez.

Short term

As the potential of biotechnology continues to expand over the next three to five years, further studies will give rise to an explosion of global epidemiological information. The collation, storage and dissemination of this information will comprise the basis of an international network.

Highly specific, on-line, rapid, sensitive, automated and portable process monitoring and control techniques are already coming into use, and in the short term these will improve further to allow remediation processes to become quicker, cheaper, and better.

The exquisite selectivity of biological systems for recognising, for example, specific metals, is being widely explored at laboratory scale and in a few years will be the basis for a number of generally accepted biotreatment processes.

There will be an ever growing demand for biological systems in bioremediation, sustainable development, and water quality maintenance.

Long term

In the long term, biotechnology will have a wider role to play in sustainability, profiting from rapid advances in molecular biology and deeper knowledge of directed evolution. Biological processes will be at the heart of new, cleaner industrial processes which will be inherently less polluting.

The ultimate advantage of biotechnology lies in its contribution to the goal of primary protection of the whole ecosystem -- an holistically based system for the prevention of contamination in the first place.

"What are the obstacles to the application of biotechnology for water use and conservation?"

Technological

Further studies are required to build on the emerging novel and sophisticated biotechnologies, to replace the current imprecise methods used for approximating risk. The technological obstacles differ for developed and developing countries. For example, disinfection, fate and transport studies in the water system are concerns for developed countries, whereas basic purification and distribution are of primary importance to developing countries.

Bioremediation technology is possible now in laboratory, pilot, demonstration and site-specific full-scale applications. However, further R&D is required to accelerate the development to widely applicable full-scale processes for use in all countries, both developed and developing.

Process optimisation and more precise process control both require better understanding of the fundamental biological mechanisms. Biotreatment can be carried out *in situ*, in landfill, by land-farming or in bioreactors, but current treatment is carried out on a case-by-case basis. Research to provide platforms of generic technology is necessary, which will be enhanced by further modelling and prediction studies. So far, there is a lack of credibility because of the insufficient number of good examples of successful remediation. Biological processes are seen by non-biologists as sludge producing -- essentially the production of one waste material from another. The problem of sludge production and minimisation (and disposal) needs to be resolved. This is not unique to biotechnology -- chemical treatment processes result in larger quantities of less manageable sludges.

Financial/economic

Biotechnology offers a less expensive alternative for remediation but can often be a long-term process compared to physico-chemical techniques. Studies to better understand the underlying mechanisms and thus accelerate the bioprocesses should alleviate this. Both the low cost and relative slowness of biological processes may be a disadvantage until the processes are more widely used -- they offer insufficient added-value and profit to small companies.

International co-operation is critical and will help to ease the burden on individual countries, but priorities must be set for R&D targets. A major obstacle is in the limited resources available for biotechnology R&D generally.

Since they are often small, biotechnology companies are at a disadvantage with respect to large chemical and conventional waste treatment companies.

A major enabler of biotechnology development for pharmaceuticals has been venture capital. However, the lower returns perceived to be associated with investments in environmental biotechnology can be expected to limit private capital as a significant resource for new companies in this area.

Training/manpower

As new and sophisticated techniques of biotechnology are applied in water quality and remediation, there is a demonstrable need for trained manpower in both the public and private sectors. Biological processes are perceived to be inherently more complex and therefore more difficult to comprehend and control. Additionally, where waste treatment processes are not given high priority or are seen as a cost, they are not allocated appropriate management skills.

There is a lack of support for fundamental microbiology education and also for cross-linking education -- biology for engineers and vice versa.

Government policy

A major obstacle is that the only measure of public impact used by governments is the outbreak of disease. However, the truly chronic and long-lasting effects of poor water quality and lack of safe food (i.e. the loss of working time and efficiency) are a far greater loss over any given period of time.

Fear of liability in the private sector has caused severe lack of information sharing by the bioremediation industry, with the result that the valuable experience accumulated to date in field trials is not available for improving bioremediation methods and applications. Equally, most treatment plants are designed and evaluated by engineers who operate only within their own skill base to avoid liability problems. Legislation needs to be framed in order to avoid these liability problems.

Public acceptance

Health risks are currently viewed as a health issue only. Effective policy needs to adopt a more holistic approach, combining public health with the consideration of the whole planetary ecosystem. Programmes for basic hygiene education to reduce the risk of contracting waterborne diseases are lacking in many areas of the world.

Public awareness and acceptance are major determinants in the use of biotechnology, especially for applications concerning the use of GMOs.

Biotechnology does not always give the immediate solutions which consumers, particularly industrial consumers, seem to prefer over better answers. Public misconceptions, for instance that biological processes are unreliable and need more space, and that the use of living organisms is essentially less predictable, must be resolved.

Recommendations: "What policies (including R&D support) should be recommended?"

While these Recommendations were formulated by the workshop participants in their personal capacities and are not binding on OECD Member countries, they nonetheless represent considerable intellectual and persuasive force.

General

- In order to move toward sustainability, governments should support, in the short term, the application of biotechnology for pollution prevention and resource recovery and, in the long term, for cleaner technologies.

- Since the wider use of biocatalysis will inevitably reduce waste generation (and thus lead to better water quality), government programmes should be encouraged to promote the use of this technology. There is a need for the development of better process engineering and also the education of process engineers and those responsible for choosing new processes, in the role that could be played by the new biotechnology.

- Biotechnology will penetrate further into the marketplace only when the cost and technical advantages are more widely understood. Governments have a role to play in undertaking the necessary cost-benefit analyses.

- The OECD is invited to organise a follow-up activity to address the problem of the lack of standardisation for techniques in water quality and the bioremediation of water, with the objective of developing a series of international guidelines.

- A strong recommendation is made for establishing a global database and information dissemination, on water quality and waterborne infectious organisms, and for the collection of strains of micro-organisms for reference studies in bioremediation and public health. Full use should be made of multi-disciplinary teams to address these areas.

- Industrial products, especially agrichemicals, are a major source of pollution. One approach is to encourage all major producers of potentially polluting materials to take full responsibility for their products from cradle to grave.

- Because of limited resources, governments must set priorities for support of R&D.

Water quality and public health

- Governments and public health authorities should leave behind the use of indicator organisms and replace them with more advanced technologies which permit the monitoring and prediction of waterborne disease and the global assessment of risk and exposure.

- International organisations should be encouraged to develop databases of epidemiological information and, by their use, to highlight and publicise the successful application of biotechnology to the improvement of public health; this in turn will require further effort on standardisation of methodology.

- Further studies on diseases associated with waterborne micro-organisms, such as cancer and myocarditis, are essential.

Water supply

- The universal provision of high quality drinking water and appropriate waste (water and solid) treatment facilities should be given first priority.

- Where drinking water is not available, appropriate techniques to purify and to disinfect surface water should be widely supported.

- Education at all levels is required concerning the role of biotechnology and new and alternative sanitation technologies.

- The OECD is recommended to carry out cost benefit studies on the value of clean water.

Bioremediation and waste treatment

- Treatment policies should focus on primary sanitation for everyone, followed by the optimisation of nutrient use and resource recovery.

- For small to moderate-sized communities, a combination of biological oxygen demand (BOD) and suspended solids removal, followed by wetland polishing, is recommended.

- For peri-urban areas in developing countries, attempts should be made to improve the social acceptability of alternative technologies, such as dry latrines and appropriate composting technologies.

- R&D on fundamental mechanisms of the interactions between plants, micro-organisms and pollutants should be encouraged.

- Demonstration projects which allow the evaluation of economics and long-term applicability of alternative processes, such as phytoremediation, should be encouraged.

- Governments are strongly encouraged to work towards a uniform regulatory framework concerning the use of GMOs in this sector.

- Further studies on risk assessment and the bioavailability of contaminants in water systems are necessary, both to advance our fundamental knowledge, and to feed into the legislative process.

- International organisations such as the OECD have a role in developing multinational demonstration sites and collecting and disseminating data on a uniform basis. This is a very high priority for bioremediation of water systems.

Monitoring and control

- Incentives should be provided to encourage the wide use of sensors and control mechanisms, for instance by applying levies to all discharges.

- Standards should be developed for low cost, robust sensors integrated within EMAs (Eco-Monitoring and Auditing Systems), and there should be comparative quality evaluation of equipment and processes.

Novel technologies for environment modification

- R&D in the areas of biosorbents and drought and salinity-resistant plants and micro-organisms must be accelerated.

Awareness

- Governments should give priority to provision, through all available avenues, of public information regarding the role and advantages of biotechnology. This recommendation for public awareness/education is of high priority.

- A strong programme for basic hygiene education is required, since this reduces the risk of contracting waterborne diseases -- the highest risk factor.

- Community education and participation in local projects to upgrade water supply and waste treatment is essential.

Governments should encourage the development of the skill base -- for example, plant operators and engineers need to acquire biotechnology skills, and biologists need a better understanding of engineering.

REUNION DE TRAVAIL DE L'OCDE TENUE A COCOYOC (MEXIQUE) RESUME, CONCLUSIONS ET RECOMMANDATIONS

Introduction

La Réunion de travail de l'OCDE "Mexico 96" sur la R-D en biotechnologie appliquée à la préservation et à la qualité des ressources en eau, accueillie par le CONACYT (Conseil National des Sciences et Technologies du Mexique), s'est déroulée à Cocoyoc (Morelos) du 20 au 23 octobre 1996. C'est la troisième Réunion de travail de l'OCDE consacrée à la biodépollution et à la bioprévention.

Au cours de la première d'entre elles, la Réunion de travail de l'OCDE "Tokyo 94" sur la biodépollution, accueillie par le Japon, un large éventail de technologies de biodépollution et de bioprévention, destinées à tous les milieux de l'environnement, a été passé en revue (voir *Bioremediation: The Tokyo '94 Workshop*, OCDE, 1995).

La deuxième, la Réunion de travail de l'OCDE "Amsterdam 95" sur l'application et la diffusion à plus grande échelle des technologies de biodépollution, accueillie par les Pays-Bas, portait sur la biodépollution de l'air et du sol dans la perspective du développement de son application industrielle (voir *Wider Application and Diffusion of Bioremediation Technologies: The Amsterdam '95 Workshop*, OCDE, 1996).

Comme suite à ces deux Réunions de travail, le Mexique a proposé au Groupe de travail de l'OCDE sur la biotechnologie (GTB) d'accueillir une réunion de travail sur la biotechnologie et l'eau, couvrant la gamme de sujets la plus large possible, notamment les aspects qualitatifs et quantitatifs du traitement, de l'exploitation et de la préservation des ressources en eau. Le Mexique a également fait ressortir que la réunion de travail se tiendrait en présence des principaux décideurs et institutions chargés de l'eau au Mexique et s'adresserait à eux.

Le GTB a examiné et approuvé cette proposition, et nommé un groupe de direction pour établir le programme de la réunion de travail.

La réunion de Cocoyoc a été suivie par quelque 180 participants représentant les gouvernements, les industries et les universités de nombreux pays Membres de l'OCDE, dont plus d'une centaine de Mexicains et notamment le Ministre de l'Environnement, des Ressources naturelles et de la Pêche.

En préparant la réunion de travail, les présidents scientifiques et le groupe de direction ont formulé trois "**questions fondamentales**" pour cette réunion, qui ont été communiquées à tous les orateurs et participants :

1. **Quels sont les avantages (techniques, économiques, écologiques) de la biotechnologie pour l'exploitation et la préservation des ressources en eau ?**

   - **dans l'immédiat**
   - **à court terme**
   - **à long terme**

2. **Quels sont les obstacles à l'application de la biotechnologie à l'exploitation et à la préservation des ressources en eau ?**

   – **techniques**
   – **financiers/économiques**
   – **liés à la formation et aux ressources humaines**
   – **liés à l'action des pouvoirs publics**
   – **liés à l'acceptation par le public**

3. **Quelles mesures (y compris le soutien à la R-D) faudrait-il recommander ?**

Le résumé des séances de la Réunion de travail, ainsi que les conclusions et les recommandations, fournissent les réponses que les participants à la réunion de travail ont apportées à ces questions fondamentales.

Exposé de la situation

La pénurie et la pollution des ressources en eau suscitent des inquiétudes croissantes dans de nombreux pays. Le Mexique compte parmi ces pays, mais il est loin d'être le seul ; beaucoup d'autres pays développés et en développement partagent cette préoccupation. La pollution de l'eau et les désaccords sur l'approvisionnement en eau et son utilisation induisent des tensions à l'intérieur des pays et, de plus en plus, entre eux. Le problème de l'eau se pose de plus en plus à l'échelle internationale et les ressources en eau sont en train de devenir un enjeu économique majeur, à mesure que leur prix augmente.

De toutes les questions ayant trait à l'environnement, la question de l'eau est probablement celle qui aura les répercussions les plus importantes à long terme et la plus difficile à traiter du point de vue scientifique. Pour toutes ces raisons, à savoir les aspects internationaux, économiques, écologiques et scientifiques du problème de l'eau, ainsi que ceux liés à la santé publique, l'OCDE et le Mexique ont consacré leur attention à l'eau et au rôle de la biotechnologie dans le traitement, l'exploitation et la préservation des ressources en eau.

Pourquoi mettre l'accent sur la biotechnologie? La biotechnologie a amené une révolution dans les soins de santé et les produits pharmaceutiques, en particulier dans le diagnostic et la prévention des maladies. Elle modifie les plantes cultivées et les aliments. Elle présente également quantité d'avantages pour l'environnement. Les organismes et les systèmes vivants sont capables de réduire, de dégrader et même d'éliminer la pollution de l'air, du sol et, en particulier, de l'eau, où le traitement biologique est une technique courante dans le monde entier.

Le Mexique a proposé d'accueillir cette réunion de travail parce qu'elle aborde quelques unes de ses principales préoccupations nationales. L'OCDE a réuni quelques-uns des plus grands experts internationaux dans ce domaine, afin de soutenir les efforts déployés par le Mexique pour résoudre ses problèmes d'eau. De plus, les contributions substantielles du Mexique en la matière ont aussi été examinées. La résolution de ces problèmes d'eau soulève des questions scientifiques et technologiques très intéressantes, pour les pays Membres de l'OCDE et bien au-delà de la zone OCDE.

Les micro-organismes font peser une lourde menace sur la salubrité des réserves mondiales en eau potable ; ce danger s'accroît et risque de déclencher d'importantes épidémies de maladies infectieuses, même dans les nations industrialisées. Pour la première fois, grâce à la biotechnologie, nous disposons d'outils nous permettant de faire face à ce problème mondial de santé publique. Le Mexique, à l'instar de nombreux autres pays, a besoin d'appliquer immédiatement des méthodes de détection fiables, qui ne se trouvent que parmi les innovations biotechnologiques. Le caractère international des maladies épidémiques implique qu'aucun pays ne peut résoudre ce problème planétaire à lui tout seul. Les organisations internationales jouent par conséquent un rôle essentiel à cet égard.

S'agissant de la distribution municipale de l'eau, les pouvoirs publics du monde entier doivent inscrire parmi leurs objectifs prioritaires, la fourniture d'une eau potable de haute qualité et le traitement efficace des eaux usées. Bien qu'il n'existe pas de réponse unique résolvant toutes les questions techniques, on dispose de biotechnologies appropriées, peu coûteuses et très fiables, pouvant contribuer à remplir ces objectifs dans bien des cas. Celles-ci entraîneront souvent des modifications dans les mécanismes financiers et dans les habitudes quotidiennes des gens. C'est pourquoi il faut faire participer les personnes et les collectivités à tous les stades de la planification et de l'installation des nouvelles infrastructures. Il est indispensable de renforcer l'éducation et la sensibilisation du public aux aspects microbiologiques de la santé publique et de l'assainissement. L'utilisation à plus grande échelle de techniques de mesure et de contrôle est capitale pour l'amélioration du traitement des eaux. La biotechnologie moderne joue un rôle majeur à cet égard. Dans les pays en développement, mais aussi dans le monde développé, certaines petites collectivités ne peuvent pas se permettre d'appliquer des techniques de pointe très coûteuses. Les technologies biologiques sont probablement celles qui leur conviennent le mieux.

La biodépollution est souvent la seule solution viable pour réduire la pollution des aquifères, qui représente un problème grave dans certaines parties du Mexique et dans d'autres pays. Cette solution est également applicable dans beaucoup de cas de pollution des eaux superficielles et côtières.

Résumés

Séance d'ouverture

La séance d'ouverture a débuté par les discours de bienvenue du pays hôte. M. Carlos Bazdresch Parada, Directeur du CONACYT, et le Président scientifique, le Dr Francisco Bolívar Zapata, Directeur de l'Institut de biotechnologie de l'UNAM, ont souligné l'importance de cet événement pour le Mexique, en précisant que leur pays comptait beaucoup sur cette réunion d'experts internationaux pour acquérir de nouvelles connaissances sur le sujet.

Les allocutions de bienvenue de l'OCDE ont été prononcées par M. Risaburo Nezu, Directeur, Direction de la science, de la technologie et de l'industrie, par M. Bill Long, Directeur, Direction de l'environnement, et par le Dr David Harper, en sa qualité de Président du Groupe de travail sur la biotechnologie.

M. Nezu a situé la réunion de travail dans une perspective élargie, celle de l'importance économique croissante de la biotechnologie et de ses impacts sur les échanges, l'investissement et l'emploi. Il a insisté sur la nécessité d'une réforme réglementaire, notamment au Japon et en Europe, pour faciliter la commercialisation de cette technologie. Il a souhaité que les travaux de l'OCDE relatifs à la biotechnologie soient mieux connus sur la scène internationale et qu'ils soient aussi associés à ceux de pays non Membres de l'OCDE.

M. Long a évoqué les années 70, au cours desquelles les questions liées à l'eau figuraient parmi les premiers points à l'ordre du jour des débats internationaux. Dans les années 80, l'attention des responsables de la politique d'environnement s'est, semble-t-il, tournée vers la désertification, l'urbanisation et les changements climatiques. Toutefois, le problème de l'eau revient en force dans les préoccupations des pays. Il a mentionné la réunion d'experts de l'OCDE sur "la consommation durable de l'eau", prévue à Sydney, en Australie, du 10 au 12 février 1997, et a souhaité que les conclusions des débats de la réunion de travail du Mexique puissent alimenter la réunion australienne.

Les exposés généraux ont été présentés par le Dr Robert Huggett, Administrateur adjoint pour la recherche et le développement à l'Agence pour la protection de l'environnement des Etats-Unis (US EPA), et par M. Kikuji Tateishi, Directeur général adjoint du Bureau des industries de base du MITI (ministère du Commerce international et de l'Industrie) du Japon.

Le Dr Huggett a placé l'eau pure parmi nos ressources les plus précieuses. Cependant, en dépit des progrès immenses qui ont été accomplis pour protéger les eaux superficielles et souterraines au cours des dernières décennies, de lourdes menaces pèsent encore sur la qualité et la quantité des ressources en eau. Elles comprennent non seulement les agents pathogènes aquatiques, mais aussi les produits dérivés du chlore et d'autres désinfectants. L'Amérique latine, par exemple, a connu plusieurs graves épidémies de choléra durant cette décennie et l'Organisation mondiale de la santé estime que deux millions d'enfants meurent chaque année dans les pays en développement de maladies liées à la pollution de l'eau. Les ressources épuisables en eau sont de plus en plus sollicitées du fait de la croissance de la population et du développement, et la demande par tête d'habitant augmente avec le niveau de vie.

Il est urgent de mener des recherches sur les aspects quantitatifs et qualitatifs de l'utilisation de l'eau. A titre d'exemple, le Bureau de recherche et de développement de l'US EPA a inscrit l'eau potable parmi les six grandes priorités de son nouveau plan stratégique. Les connaissances actuelles des agents pathogènes ne sont pas suffisantes pour évaluer les risques qu'ils font courir à la santé. Des incertitudes subsistent quant à l'infectivité des populations à risque et à la virulence des différentes souches d'agents pathogènes.

M. Tateishi a souligné que l'un des principaux domaines où la biotechnologie pouvait jouer un rôle important était l'environnement, notamment la gestion des ressources et la réduction des déchets. Au Japon, la forte croissance économique a entraîné de graves problèmes de pollution des eaux dans les zones urbaines et industrielles. Une croissance démographique rapide risque également de générer des volumes d'eaux usées outrepassant la capacité d'épuration des installations de traitement. Des mesures juridiques et financières, ainsi que des dispositions concrètes, ont cependant été prises afin de résoudre ces problèmes.

Bien que la biotechnologie offre des solutions très prometteuses pour le traitement de l'eau et du sol, le Japon manque encore d'expérience à ce sujet et possède très peu de données prouvant l'efficacité de ces techniques ; il doit par conséquent développer ses connaissances. En vue de faire face à ce problème, le MITI a lancé des programmes destinés à tester l'efficacité et la sécurité des technologies de biodépollution.

Séance I-A. Qualité biologique de l'eau et santé publique

On estime que le nombre de pays confrontés à des problèmes d'eau augmente rapidement. De nombreuses nations industrialisées éprouvent de plus en plus de difficultés à garantir la fourniture d'une quantité d'eau suffisante. Les agents pathogènes aquatiques représentent un danger majeur et croissant. Les entérovirus et les protozoaires intestinaux, tels que *Giardia* et *Cryptosporidium*, suscitent une inquiétude particulière. Les eaux polluées provoquent 1,6 milliard de cas de maladies à travers le monde à cause des entérovirus et des protozoaires. Les deux sont susceptibles d'engendrer des diarrhées graves et de longue durée ; par ailleurs, on sait que certains entérovirus sont corrélés à des effets chroniques comme le diabète et la myocardite. Des kystes de *Giardia* et des ookystes de *Cryptosporidium* ont été décelés dans des eaux de source et dans des eaux potables à travers l'Amérique du Sud ; des rotavirus et des virus coxsackies B ont été isolés dans de l'eau potable traitée au Mexique.

L'origine et le caractère cyclique du choléra ont intrigué les scientifiques et, récemment, les outils de la biologie moléculaire et de l'immunologie ont permis de mettre en évidence une association entre *Vibrio cholerae* et le zooplancton dans des systèmes marins et estuariens. Grâce à des anticorps monoclonaux fluorescents et à des sondes géniques, couplés à une amplification par PCR, il a été possible de détecter et de suivre *Vibrio cholerae* dans l'environnement et de démontrer l'existence d'une corrélation entre la prolifération de plancton et une incidence accrue du choléra. L'application de techniques moléculaires analogues permet à l'heure actuelle de procéder à des estimations semi-quantitatives d'astrovirus et d'autres entérovirus.

La détection rapide des agents pathogènes bactériens aquatiques est vitale pour empêcher l'apparition de grandes épidémies. La présence de ces agents pathogènes dans des réseaux de distribution d'eau est particulièrement préoccupante. Plusieurs essais ont été mis au point pour détecter un petit nombre de bactéries viables dans l'eau potable, moyennant l'utilisation de l'ARN recombiné 16S et du gène *lux*. Ces méthodes sont rapides, sensibles, spécifiques et peuvent parfois distinguer les cellules viables des cellules non viables.

Les évaluations actuelles de la qualité de l'eau, qui ne s'appuient que sur la charge bactérienne, ne sont pas suffisantes et on a besoin d'indicateurs de parasites d'un emploi plus aisé, mais fiables. Des rapports donnent à penser que certains bactériophages déterminés constitueraient des indicateurs plus fiables et moins coûteux pour évaluer les risques sanitaires d'origine virale. Bien qu'une corrélation puisse apparaître entre les concentrations de *Giardia* et de *Cryptosporidium*, aucune n'a été mise en évidence entre d'autres agents pathogènes aquatiques, quel que soit le type d'eau analysé. Aucun phage ni aucune bactérie déterminé ne peut donc servir d'indicateur pour *Giardia*, *Cryptosporidium* ou les entérovirus.

Même les eaux potables de la meilleure qualité, traitées selon les normes internationales, ne sont pas stériles et renferment divers spectres de micro-organismes qui forment des biofilms. Ces biofilms bactériens résistent aux méthodes de décontamination les plus courantes et peuvent abriter de nombreuses espèces différentes d'agents pathogènes. Bon nombre d'entre eux ont acquis une résistance accrue à la désinfection et des produits de remplacement du chlore, tels que la monochloramine, plus persistante, ont été préconisés à des teneurs qui ne présentent pas de danger pour l'environnement.

Les efforts déployés par les pouvoirs publics pour chlorer l'eau, au Mexique et dans nombre d'autres pays en développement, sont nécessaires, mais l'amélioration des conditions sanitaires se fonde avant tout sur le développement de l'éducation en matière d'hygiène.

Certaines questions n'ont pas encore été abordées, comme l'absence d'un système de surveillance épidémiologique dont les données seraient accessibles au public. Les techniques actuelles d'analyse et de surveillance de l'eau sont coûteuses et exigent beaucoup de temps et de travail ; on aurait besoin d'un dispositif de détection en temps réel, capable de donner rapidement des informations fiables à un coût acceptable. Seule la surveillance régulière des eaux usées, des eaux du milieu ambiant et des processus de filtration et de désinfection fournira les données nécessaires pour fixer des priorités entre les options de contrôle.

Séance I-B. Biodépollution/biotraitement : aquifères

L'eau est une ressource extrêmement limitée dans les régions semi-arides, et l'explosion urbaine a accru la demande d'eau potable. Les eaux usées sont considérées de plus en plus comme une ressource importante et tandis que leur évacuation contrôlée peut constituer un supplément non négligeable de ressources en eaux souterraines de qualité acceptable, le rejet direct d'eaux usées non traitées dans les aquifères peut entraîner de graves problèmes de pollution. On ne dispose pas encore de normes d'efficacité ni de techniques reconnues pour évaluer les risques.

Les effets de la réutilisation des eaux usées sur les ressources en eaux souterraines demandent des études supplémentaires : il est nécessaire d'avoir une connaissance suffisante de l'hydrogéologie et des processus d'infiltration, ainsi que de la circulation et de l'atténuation naturelle des polluants, pour concevoir et gérer efficacement des systèmes d'irrigation qui utilisent les eaux usées. L'infiltration des eaux usées peut modifier la qualité chimique de l'eau dans tout un aquifère et, dans certains cas, la pénétration d'une eau de mauvaise qualité à grande profondeur est à craindre. En outre, une meilleure visualisation des ressources en eaux souterraines et de leurs mouvements s'impose, afin de pouvoir évaluer les choix de gestion.

La qualité des eaux souterraines s'est dégradée en de nombreux endroits du globe, à la suite des rejets d'hydrocarbures et de produits chimiques dans les sols qui ont eu lieu dans le passé. Les techniques de biodépollution *in situ*, qui sont très prometteuses pour éliminer les polluants des aquifères, ont été mises au point dans plusieurs pays (y compris le Japon, le Royaume-Uni, et les Etats-Unis) pour éliminer les composés chlorés volatils qui contaminent les eaux souterraines et les sols. La biodépollution a pour principal avantage que les polluants organiques peuvent être transformés et, dans certains cas, complètement dégradés. Comparée à d'autres techniques de dépollution, la biodépollution est à la fois sans danger et économique.

L'application à grande échelle de ces techniques soulève toutefois des difficultés, chaque site possédant des caractéristiques propres, si bien qu'il risque d'être nécessaire de résoudre chaque problème de pollution au cas par cas. Un savoir-faire général existe, mais la méthode doit être adaptée à chaque site. Il y a lieu également de démontrer l'utilisation en vraie grandeur et pas seulement à l'échelle du laboratoire, et d'adjoindre des techniques physico-chimiques, selon les besoins.

La présence de micro-organismes naturels à grande profondeur permet de supposer l'existence d'une biodépollution dite "intrinsèque" ou naturelle. Ce domaine doit encore faire l'objet de vastes études.

Séance I-C. Biodépollution/biotraitement : eaux superficielles

Le recours à la phytodépollution, notamment dans des zones humides ou des roselières, a été exploré de manière approfondie pour traiter de grands volumes d'eau renfermant des concentrations relativement faibles de polluants. La phytodépollution est une technologie assez simple et donc peu coûteuse qui peut s'avérer particulièrement adaptée aux pays en développement.

Une question fondamentale en ce qui concerne le devenir des polluants dans l'environnement est leur dégradation biologique aux faibles concentrations. Pour certains polluants, on a fait état de seuils de concentration déterminés en dessous desquels ils ne sont plus dégradés. Toutefois, il a été démontré que d'autres substances xénobiotiques pouvaient être assimilées par les micro-organismes, même à des concentrations de quelques nanogrammes, voire de quelques femtogrammes, par litre, c'est-à-dire bien en dessous des seuils qui permettent à ces organismes de continuer à croître. On peut supposer que ces substances sont utilisées en même temps que des matières organiques dissoutes indéterminées qui permettent aux organismes de se multiplier ou du moins fournissent l'énergie nécessaire à leur survie.

Les composés organiques et inorganiques du mercure sont des polluants importants sur de nombreux sites industriels à travers le monde, et des macrophytes indigènes modifiés pour exprimer le gène *merA9* qui code pour une enzyme réduisant les sels de mercure en mercure élémentaire ont été testés pour leur capacité d'éliminer les sels de mercure des eaux et des sols pollués. Le problème du mercure organique concentré dans la chaîne alimentaire est plus grave : il menace de façon plus immédiate les espèces sauvages et l'homme. Il est nécessaire de modifier génétiquement les plantes, afin de les rendre capable de transformer ce mercure organique.

Les sédiments contaminés peuvent être une source de pollution pour les eaux superficielles. Jusqu'à présent, il n'existe aucune technologie éprouvée permettant de biodépolluer les sédiments sur place, on doit donc les prélever avant de les traiter. Les techniques de nettoyage actuelles combinent des procédés physiques, chimiques, biologiques et thermiques.

A cause des nombreux déversements de produits pétroliers, les eaux superficielles pâtissent d'une pollution chronique par les hydrocarbures. A l'échelle mondiale, la pollution des cours d'eau est la principale source de pollution marine, bien avant les catastrophes impliquant des pétroliers.

Séance I-D. Biodépollution/biotraitement : eaux côtières et marines

La pollution anthropique des écosystèmes marins, y compris les cours d'eau, les baies, les estuaires et d'autres zones côtières, est un problème d'environnement majeur à l'échelle de la planète, auquel il convient d'être attentif. Plusieurs options de biodépollution faisant appel à des plantes et des micro-organismes marins ont été proposées. L'utilisation d'algues macroscopiques peut également constituer une technologie nouvelle et efficace par rapport à son coût pour traiter les polluants en suspension dans l'eau. Un effort international coordonné s'impose pour démontrer la faisabilité de ces méthodes et obtenir des données utilisables, à l'aide de projets de démonstration *in situ* exécutés dans des environnements pollués. Juste après l'accident de l'Exxon Valdez, des études de laboratoire ont montré que des organismes indigènes capables de dégrader les hydrocarbures étaient très actifs, mais limités par le manque de nutriments. 112 kilomètres de plages couvertes de pétrole ont été traités avec des agents fertilisants oléophiles à libération lente et cette méthode de dépollution avec enrichissement en nutriments a permis d'éliminer efficacement et en toute sécurité le pétrole, tant à la surface que dans le sous-sol des plages.

Il faudrait établir une relation plus directe entre les paramètres biologiques et l'oxydation résultante des hydrocarbures, pour que l'on puisse appliquer les nutriments de manière plus efficace et prévoir la vitesse des réactions chimiques. Il est également nécessaire de mieux connaître les associations indigènes de micro-organismes. Un obstacle majeur à l'utilisation des techniques de biodépollution à plus grande échelle est l'acceptation par les organismes de réglementation et la préparation de plans d'urgence permettant d'appliquer en temps voulu les techniques de biodépollution. On a proposé une solution pour remédier aux marées noires, qui consiste à transformer une souche indigène dominante de micro-organismes à l'aide d'un plasmide capable de s'intégrer dans une large gamme d'hôtes et de contenir une série d'opérons cataboliques, rapidement activés par la pollution.

La biotechnologie mise en oeuvre pour surveiller l'écotoxicologie des métaux lourds dans la baie de San Francisco consiste à mettre au point des biomarqueurs et à évaluer les effets sur la reproduction au niveau moléculaire. Le sélénium déversé dans la baie de San Francisco par l'industrie pétrochimique est toxique pour les poissons, les oiseaux aquatiques et les petits mammifères. L'élimination du sélénium à l'aide de bactéries est une application prometteuse de la biotechnologie à la protection de l'environnement. Ces organismes sont capables de biotransformer des formes toxiques de sélénium en précipités non toxiques de sélénium métallique, qui peuvent être alors récupérés avec la biomasse bactérienne. Une autre méthode consiste à construire des zones humides artificielles, peuplées d'espèces végétales déterminées capables d'absorber biologiquement le sélénium.

Les eaux marines côtières sont très vulnérables aux rejets de deux macronutriments : l'azote et le phosphore. Les récifs coralliens des régions tropicales, notamment, peuvent subir des dégradations substantielles lorsqu'ils sont exposés à ces nutriments. La combinaison de technologies biologiques d'élimination des nutriments avec des bioréacteurs enrichis en oxygène et des techniques de séparation par membrane pour le recyclage des boues améliorera la qualité des effluents et réduira sensiblement les risques.

Séance II-A. Prévention de la pollution des eaux : sources municipales

Dans les pays développés, les réseaux d'égouts sont capables de collecter une très grande proportion des eaux municipales. Ils sont néanmoins touchés par des problèmes tels que la sécheresse, la surexploitation, les inondations et les débordements. La pollution par les sources diffuses menace fortement la qualité de l'eau. Les activités agricoles, le ruissellement urbain, les dépôts atmosphériques, la mise en décharge, les travaux de construction et les activités minières contribuent à la pollution par les sources diffuses. De nouvelles stratégies s'imposent, qui visent la durabilité (celle-ci se mesurant à la diminution de la consommation énergétique, l'utilisation rationnelle des ressources naturelles et la limitation de la production de boues contaminées). Il existe d'autres possibilités, comme le recyclage des eaux usées pour obtenir une eau de second choix (eau grise ou de qualité e) et la séparation de l'eau de pluie des eaux usées.

La croissance des mégapoles (plus de dix millions d'habitants) représente un problème important dans les pays en développement ; 17 mégapoles sur 22 se trouveront dans les pays en développement en l'an 2000. On estime que 600 millions de personnes habitant dans les zones urbaines des pays en développement vivent dans des conditions de logement ou dans un milieu environnant qui menace leur vie ou leur santé, principalement dans les zones péri-urbaines. La plupart de ces collectivités n'ont pas les moyens de faire face aux problèmes liés à la pureté de l'eau et à l'assainissement, que les autorités municipales tendent à ignorer, étant elles-mêmes débordées du fait du très grand nombre d'habitants.

Comme la méthode classique d'élimination des déchets anthropiques (collecte centralisée et traitement des eaux usées) est très onéreuse, il est nécessaire d'employer d'autres méthodes, dont la mise en oeuvre a lieu le plus près possible de l'usager, en tenant compte du fait que la modification des comportements individuels et municipaux peut être au moins aussi déterminante que les nouvelles infrastructures.

Il convient de faire appel à des solutions techniques et non techniques ; il n'existe pas de solution miracle universelle. La participation (et l'éducation) des citoyens et des collectivités est essentielle. Au cours de la dernière décennie, par exemple, la ville de Tijuana a élaboré un système modulaire utilisant des bassins et zones humides décentralisés combinés à l'énergie solaire pour traiter et réutiliser les eaux usées. Il s'agissait de gérer les ressources en eau limitées, d'élever la qualité de la vie et d'améliorer l'environnement urbain au sein de la ville, ainsi que de réduire la pollution du fleuve Tijuana.

Il existe des procédés innovateurs, relevant ou non de la haute technologie, qui permettent d'améliorer les infrastructures classiques. Le traitement des boues résiduaires des stations d'épuration des eaux usées est une opération complexe et coûteuse, c'est pourquoi les biotechnologies destinées à cette fin sont rarement utilisées dans les pays en développement comme le Mexique. Il existe cependant toute une série de traitements biologiques praticables et le recyclage dans le sol est la meilleure option, si les boues ne contiennent pas de métaux ou d'autres substances toxiques. L'élimination des agents pathogènes est indispensable si cette option est choisie.

La phytodépollution et l'utilisation des zones humides sont des technologies nouvelles, mais simples, qui permettent à la fois de dégrader les substances organiques et de concentrer et d'éliminer (collecter) les métaux. Les zones humides naturelles et artificielles ont montré leur remarquable capacité de traitement en interceptant les flux d'eau polluée. La sélection judicieuse d'espèces végétales précises et leur utilisation peuvent présenter des avantages certains dans le traitement de la pollution émanant des sources diffuses. Le point fort de l'exploitation des zones humides est le fait qu'elle repose sur des pratiques agronomiques bien comprises.

Séance II-B. Prévention de la pollution des eaux : sources industrielles

Dans l'industrie, l'eau n'est souvent considérée que comme l'un des facteurs de production. Une bonne eau de procédé devient cependant une ressource de plus en plus rare et chère. La qualité et la quantité de l'eau industrielle doivent être adaptées à son utilisation, autrement dit répondre à la demande. L'objectif ultime doit être l'utilisation durable (la charge de départ et le recyclage doivent conduire en fin de compte à un niveau de rejets nul). Les mesures peuvent s'appliquer en bout de chaîne ou tout au long du processus par la substitution d'étapes, et la biotechnologie, associée à des techniques physico-chimiques, jouera un rôle notable à cet égard.

Le développement de l'industrie dans les villes de toutes dimensions engendre des effluents toxiques qui, lorsqu'ils parviennent aux stations d'épuration des eaux usées, inhibent les processus biologiques courants. Les techniques de surveillance capables de détecter les polluants et les processus nuisibles permettront aux exploitants des stations d'épuration de maîtriser les flux entrant dans la station. Les nouvelles technologies en temps réel employées à cette fin exploitent la bioluminescence de bactéries naturelles et modifiées génétiquement. L'épuration des effluents industriels devrait, en fin de compte, être prise en charge par le producteur et un système de taxes établies sur la base de la surveillance des effluents est probablement le seul moyen d'y parvenir.

Des technologies non polluantes conduiront finalement à la réduction des déchets au minimum. Toutefois, pour que l'utilisation des enzymes puisse se généraliser dans les procédés industriels, elles devront être plus stables que celles que l'on trouve actuellement dans le commerce. Les organismes extrémophiles présentent un énorme potentiel métabolique, susceptible d'être exploité moyennant une sélection et une évolution dirigée.

Pour pouvoir répondre à des conditions plus strictes d'élimination des déchets tout en demeurant compétitives, les entreprises devront recourir à des technologies écologiquement rationnelles qui soient adaptables et plus efficaces par rapport à leur coût. L'épuration des effluents liquides de composition complexe soulève des difficultés particulières. C'est pourquoi il est nécessaire de mettre sur pied un programme de recherche pluridisciplinaire, bien ciblé et intégré destiné à la mise au point de technologies innovantes en matière d'épuration, d'une part, et de convaincre les industriels d'adopter ces technologies, d'autre part.

La plupart des industries produisent peu de résidus de métaux lourds ; les principaux rejets de métaux lourds proviennent souvent d'un groupe relativement restreint d'industries : travail et traitement de surface des métaux, fabrication de plaquettes de circuits imprimés et de semi-conducteurs et secteur de la photographie. L'épuration des flux de déchets pollués par les métaux peut être mené à bien au moyen d'un ensemble de techniques fondées sur l'accumulation active ou passive des métaux par les micro-organismes, couplée à la séparation et à la récupération de ces métaux. Toute une série de processus microbiologiques sont mis à profit, notamment la bioprécipitation, les transformations redox et la biosorption. La capacité des micro-organismes de séquestrer ou de précipiter les métaux lourds, y compris les radionucléides, ne pourra être pleinement exploitée que lorsqu'on aura une connaissance plus approfondie de leur physiologie. Les procédés doivent être efficaces par rapport à leur coût et les métaux laissés sous une forme récupérable.

Les réacteurs UASB (anaérobies à lit de boues à flux ascendant) sont couramment utilisés pour traiter les eaux résiduelles provenant des industries qui transforment les aliments. Le biogaz émanant du traitement anaérobie des eaux résiduelles risque de contenir du sulfure d'hydrogène (H_2S), mais celui-ci peut être aisément éliminé par un biolaveur, qui contient des bactéries oxydant le soufre et utilise en fait la même boue activée qui polit l'effluent final.

L'utilisation de processus biologiques à plus grande échelle dans l'industrie est subordonnée à la diminution du taux d'incertitude qui les caractérise. Il est possible d'améliorer la prévisibilité de la conception en effectuant une série d'études des possibilités de traitement afin d'évaluer divers systèmes de traitement par des essais à long terme à l'échelle pilote ou à celle du laboratoire. Une autre technique consiste à estimer la cinétique fondamentale de la biodégradation et à incorporer les paramètres cinétiques ainsi obtenus dans les études sur la conception des réacteurs.

Le manque de détecteurs en temps réel fiables représente un autre obstacle à l'adoption des technologies de biotraitement. En surveillant les principaux paramètres en continu, on augmente considérablement les possibilités d'améliorer la fiabilité du traitement, la qualité des effluents et la rentabilité du processus. Une fois qu'on aura recueilli des données auprès d'installations industrielles ayant recours à toute une série d'intrants, on sera en mesure de mettre au point des procédures de régulation automatique à sécurité intégrée, destinées à optimiser les performances de l'installation.

Séance II-C. Prévention de la pollution des eaux : sources agricoles

Un tiers au moins de la population des pays en développement ne dispose toujours pas d'un approvisionnement convenable en eau. Cette carence favorise le réemploi des eaux usées traitées pour de nombreuses utilisations municipales. Il existe des systèmes à haute ou basse technologie qui permettent de réutiliser l'eau et les nutriments et de récupérer de l'énergie. Des technologies appropriées peuvent produire des effluents de haute qualité pour un investissement et un coût d'exploitation faibles. La microfiltration, par exemple, élimine efficacement les solides en suspension, la demande biologique et chimique en oxygène et les agents pathogènes.

Afin d'assurer une couverture plus étendue, notamment dans les zones rurales, il est nécessaire d'abandonner les technologies coûteuses au profit de solutions meilleur marché. Il existe des techniques simples et fiables pour traiter et désinfecter les eaux superficielles à l'usage des collectivités rurales et des particuliers. La désinfection solaire de l'eau au moyen de sacs et de bouteilles en plastique transparent a par exemple prouvé son efficacité. La combinaison de filtres grossiers et de filtres à sable lents constitue un mode de traitement fiable et efficace pour alimenter en eau les collectivités rurales. La question de savoir si ces techniques peuvent remplacer complètement ou partiellement la chloration n'est pas encore tranchée.

Une autre technique de traitement des eaux usées a été décrite : l'utilisation de disques biologiques suivie par un passage dans une roselière. Elle représente une solution fiable, peu coûteuse et "optimale pour l'environnement" dans le cas de petites collectivités.

Les pesticides appliqués dans l'eau peuvent nuire sérieusement à l'environnement. On connaît plusieurs exemples d'enzymes étudiées en vue d'être exploitées dans les traitements et la dépollution pratiqués en bout de chaîne. Une enzyme a été mise à profit pour dégrader un dérivé toxique sans porter atteinte au produit principal, de façon à réduire au minimum la production de déchets, permettre une utilisation plus longue du produit et donner lieu à des économies appréciables.

Les agriculteurs peuvent réaliser des économies et réduire leurs déchets en adoptant des systèmes intégrés ou en cycle fermé, afin de réduire la consommation d'énergie fossile et la génération de polluants. Ces systèmes couplent le traitement des déchets à la production de biogaz, au compostage et à des bassins de culture hydroponique.

L'azote excédentaire qui est entraîné par lessivage dans les cours d'eau après son épandage dans les champs peut causer une eutrophisation grave. Les zones humides éliminent très efficacement l'azote, mais de nouvelles mesures agricoles s'imposent pour utiliser correctement les engrais et le fumier. Pour mettre en place une agriculture durable, il faudrait conserver les zones humides existantes et, dans les zones d'agriculture intensive, rétablir les zones humides naguère drainées.

L'aménagement de zones humides artificielles n'est pas encore tout à fait au point. Le choix d'une espèce végétale dans un but particulier doit encore être optimisé et les mécanismes qui commandent la dégradation de polluants dans la rhizosphère sont très mal connus. Bien que l'on ne puisse pas tout attendre des zones humides, elles sont sûrement capables de produire des effluents utilisables sans restriction pour l'irrigation dans les pays en développement.

Séance II-D. Désertification et modification de l'environnement

Il existe des plantes adaptées aux régions arides et semi-arides, mais la plupart des cultures commerciales demandent un arrosage constant. Il faut sélectionner les gènes que l'on souhaite voir s'exprimer dans les versions transgéniques des plantes cultivées.

Il existe différents mécanismes de défense, tels que les protéines protectrices, les protéines séquestrantes, l'exportation et l'exclusion d'ions, et la modification des parois cellulaires. Les osmolytes sont des molécules qui aident les cellules à retenir l'eau ; le tréhalose est la plus connue d'entre elles. Les plantes qui "ressuscitent" peuvent survivre à des taux de déshydratation très élevés et sont aussi les seules qui contiennent du tréhalose. Les plantes transgéniques qui contiennent des enzymes produisant du tréhalose résistent mieux à la sécheresse.

La tolérance à la sécheresse et au sel sont des phénomènes très complexes qui dépendent non pas d'un gène unique mais de plusieurs. Les embryons de maïs tolèrent très bien la sécheresse et on a tenté d'isoler les gènes à l'origine de cette capacité. Quoique le mécanisme de la tolérance soit fort mal connu, les plantes transgéniques ont accru sensiblement leur tolérance au sel et se développent plus rapidement dans des conditions de stress que les plantes témoins. De plus, les plantes transgéniques se remettent mieux après la suppression des conditions de stress.

La pollution des sols et des eaux par les métaux toxiques représente un problème d'environnement non négligeable dans le monde moderne. L'utilisation de plantes spécialement sélectionnées et modifiées pour accumuler les métaux se développe.

La bioprévention (la mise au point de produits et de procédés qui respectent l'environnement) est un facteur déterminant pour la durabilité. La substance bioabsorbante extraite d'un mutant d'*Alcaligenes latus* est l'un de ces produits. Cette substance est un polysaccharide formé de différents sucres ; elle possède une capacité de rétention d'eau très élevée, même en présence de sel. L'accélération de l'avancée des déserts représente l'un des grands problèmes écologiques de la planète. L'incorporation de ce bioabsorbant dans des sols désertiques à titre expérimental a démontré qu'il pouvait protéger les plantes en phase de développement dans des conditions de stress poussé.

Etudes de cas

Le quatrième jour de la réunion de travail a été consacré à quatre étude de cas mexicaines, et le Dr A. Jaime Paredes, Directeur général adjoint de la Commission nationale de l'eau (CNA), a ouvert la séance. Ces études de cas illustrent parfaitement les problèmes et les solutions abordés précédemment et donnent un poids supplémentaire aux conclusions et aux recommandations présentées dans la section suivante.

Le Dr Paredes a fait état des difficultés rencontrées par le Mexique pour évaluer, utiliser et exploiter correctement ses ressources, notamment en eau. L'eau est répartie de façon très inégale dans l'espace et le temps. Longtemps jugée inépuisable, cette ressource doit maintenant être considérée comme limitée, en raison de la dégradation de sa qualité. Au Mexique, l'agriculture est la source principale d'eaux résiduaires (62 pour cent), et l'industrie en est responsable de 10 pour cent, les déversements restants venant des activités municipales. La déforestation et d'autres pratiques qui épuisent les ressources contribuent à diminuer la qualité de l'eau. Les aquifères situés en dessous des zones agricoles sont atteints par une pollution diffuse, tandis que la surexploitation des aquifères côtiers provoque un apport d'eau salée. La gestion des ressources naturelles doit donc être améliorée, conformément au principe du développement durable.

Quelque 15 pour cent seulement des eaux usées déversées dans le pays sont traitées et parmi les 700 stations d'épuration, beaucoup sont hors service. Néanmoins, ces stations possèdent environ 37 réacteurs anaérobies, qui confèrent au pays une certaine expérience en matière de biotechnologie. Les procédés biotechnologiques seront de mieux en mieux acceptés s'ils obéissent à quelques conditions essentielles : la compétitivité par rapport aux autres méthodes, ainsi que la fiabilité et la sécurité.

Premier cas. Pollution du fleuve Lerma

Le bassin du Lerma est l'une des régions urbaines et industrielles les plus développées du Mexique. La croissance démographique et industrielle a entraîné une dégradation de l'écosystème, qui se reflète dans la diminution de la quantité d'eau disponible et de sa qualité. Un plan d'action a été lancé en 1989, en vue de développer de nouveaux moyens d'assainissement. En 1994, dans un réservoir faisant partie du bassin, il s'est produit un événement qui a provoqué une forte mortalité parmi les oiseaux aquatiques. Les caractéristiques et la portée de cet événement ont été présentées.

Les technologies employées pour l'analyse et la collecte de données dans le bassin ont été décrites. Les stratégies élaborées en fonction de ces données permettront finalement de dépolluer l'écosystème. Ces technologies comprennent la modélisation hydrodynamique de techniques telles que l'aération et la circulation forcées dans les lagons côtiers reliés à l'écosystème (voir aussi le deuxième cas). L'application de la biotechnologie à des domaines comme la dépollution des aquifères a été envisagée.

Deuxième cas. Pollution des eaux côtières et marines

Les effets de la pollution sur les activités récréatives, y compris le tourisme, et la pêche commerciale inquiètent particulièrement les autorités mexicaines. Cette question a été présentée dans le contexte de la stratégie adoptée par le Mexique pour améliorer la qualité de l'eau dans les zones côtières, qui sont affectées principalement par une pollution ponctuelle et diffuse issue de sources industrielles, agricoles et municipales. Les polluants organiques proviennent surtout des industries pétrolière et pétrochimique locales.

Les activités liées à une raffinerie de pétrole située sur la côte, et notamment l'impact sur l'environnement marin du rejet sous-marin d'eaux usées renfermant des produits chimiques, ont été étudiées. On a mesuré les concentrations de polluants dans les communautés animales et végétales et analysé l'abondance et la distribution des poissons et du plancton. L'intégration et le traitement systématique de ces données permettront à l'avenir de diagnostiquer l'état de l'environnement et constitueront une aide à la décision en matière de développement durable dans la région.

La côte mexicaine se caractérise par un chapelet de lagons. Cependant, ils sont peu profonds et ne communiquent guère entre eux et avec l'océan. L'état naturel de ces lagons a été fortement altéré par les activités humaines, notamment le dragage et le rejet d'eaux usées. Afin d'enrayer cette situation, on a proposé, compte tenu des études hydrodynamiques et des mesures effectuées sur le terrain, de mettre un terme aux rejets d'eaux usées et de pratiquer l'aération et la circulation forcées.

Troisième cas. Traitement des eaux usées dans le bassin de Mexico

Un nouveau programme est en préparation pour le traitement et l'acheminement des eaux pluviales et usées dans la ville et le bassin de Mexico. Les installations existantes s'enfoncent dans l'argile tendre sur laquelle la ville est construite, de sorte que leur capacité diminue, en particulier lorsqu'il pleut abondamment durant les orages. La qualité de l'eau n'est pas exceptionnellement mauvaise, mais on a relevé de fortes teneurs en bore issu de la roche locale, ainsi que la présence de parasites, de coliformes fécaux et d'helminthes. Actuellement, 96 000 hectares situés en dehors du bassin sont irrigués avec de l'eau non traitée. Cette eau semble particulièrement riche en nutriments. L'amélioration de la santé publique constitue le problème majeur.

Les concepteurs du programme estiment que le système est trop vaste pour que la biotechnologie puisse apporter une contribution appropriée et que la floculation chimique et la sédimentation sont par conséquent préférables. Les boues d'épuration devront être traitées à la chaux et servir à la remise en état des sols. La désinfection des effluents, après utilisation de filtres à sable, détruira les agents pathogènes. Plusieurs de ces idées ont été remises en question par les participants à la réunion. Ils ont notamment mentionné l'existence de traitements biologiques appliqués à une échelle identique ou supérieure, et fait remarquer que le traitement anaérobie de l'eau produirait des boues beaucoup moins abondantes et d'une qualité nettement supérieure. L'utilisation des boues d'épuration chimique comme agent de remise en état des sols a suscité quelques doutes. L'emploi des coliformes fécaux et des helminthes comme indicateurs de la présence ou de l'absence d'agents pathogènes a aussi été jugée inadéquate.

Quatrième cas. Traitement de l'eau dans les zones rurales

Plus de 70 pour cent des localités mexicaines ont moins de 100 habitants. Ces collectivités ne disposent même pas des installations les plus élémentaires, si bien que les excréments laissés à même le sol favorisent la propagation des infections gastro-intestinales. La construction d'un système centralisé d'adduction d'eau potable et d'un réseau d'égouts poserait trop de difficultés et coûterait trop cher à ces collectivités rurales. Aussi la Commission nationale de l'eau a-t-elle mis au point cinq formules techniques pour le traitement in situ des eaux usées et des matières fécales. Même dans les régions extrêmement pauvres, la fourniture d'installations élémentaires s'est avérée suffisante pour encourager la population locale à participer et à construire ses propres infrastructures.

Il y a lieu d'éduquer la population locale sur ce qui constitue un bon assainissement et de lui expliquer pourquoi elle devrait vouloir en disposer ; ce point constitue la priorité absolue. La technologie est simple et disponible. Les délégués présents à la réunion ont signalé l'existence d'autres (bio)technologies répondant au même degré de complexité qui pouvaient être utilisées en plus des techniques décrites.

Résumé des études de cas

Le Dr Ron Atlas, de l'Université de Louisville dans le Kentucky, a fait un résumé des quatre études de cas, dans lequel il remarque que le Mexique, comme le reste du monde, doit répondre à divers besoins en eau. Les études de cas ont examiné la réutilisation de l'eau et les technologies susceptibles de concourir à la durabilité des approvisionnements en eau. En pratiquant leurs évaluations, les responsables des projets ont éprouvé des difficultés à définir des normes de qualité de l'eau acceptables pour les applications mexicaines. Une procédure de validation nationale fait défaut au Mexique, de même que des lignes directrices et des technologies adaptées aux besoins du pays.

Les études de cas ne présentaient aucune application biotechnologique notable. La surveillance et le traitement s'appuyaient largement sur des méthodes physico-chimiques ; aucune méthode biotechnologique moderne pour la surveillance de la qualité de l'eau, comme la détection d'agents pathogènes déterminés, n'a été examinée et la conception des principales installations de traitement des eaux usées ne comportait jamais de traitement biologique secondaire. Il existe des plans pour l'élimination des boues, mais il faudrait aussi se pencher sur le sort à réserver aux boues issues d'un traitement physique et sur la qualité de l'eau ainsi obtenue. Trop souvent, des obstacles liés au coût et à la gestion ont, semble-t-il, exclu l'examen d'une solution biotechnologique. En outre, on relève apparemment un manque de communication entre la communauté scientifique et les pouvoirs publics et peut-être aussi avec le public.

Les quatre études de cas couvraient une large gamme de qualités et d'usages de l'eau au Mexique. Un grand projet s'appuyait sur une étude intégrée, suivant une approche très novatrice, et comportait un recours limité à la biotechnologie pour le suivi de la qualité de l'eau. Un deuxième projet, essentiellement descriptif, sur les systèmes marins côtiers portait plus sur la surveillance que sur la dépollution. Le projet concernant la gestion de l'eau et des déchets dans les collectivités rurales offrait une étude intéressante sur la modification des comportements et sur de nouvelles applications de la technologie, plus axée sur la santé humaine et l'éducation de la population que sur l'augmentation des besoins en eau.

Le projet relatif au bassin de Mexico a suscité les plus vives controverses, notamment sur le fait de savoir s'il est encore possible d'y inclure une application biotechnologique. On souhaite conserver des teneurs élevées en azote et en phosphates dans l'eau pour pouvoir se passer de l'épandage d'engrais, ce qui représente un nouveau défi pour la biotechnologie. La rétention des éléments nutritifs risque d'être incompatible avec l'élimination des agents pathogènes. Une stratégie globale de gestion des déchets incorporant la biotechnologie est, semble-t-il, la seule solution efficace par rapport à son coût pour maintenir à long terme la qualité de l'environnement.

Séance plénière de clôture

M. Adalberto Noyola, vice-président scientifique, a ouvert la séance plénière finale en passant en revue certains points marquants de la réunion et en soulignant l'importance de la biotechnologie dans le traitement des eaux usées et du point de vue de la santé publique. A l'avenir, la gestion des eaux usées devra être envisagée dans le cadre du développement durable. L'application de la biotechnologie doit s'inscrire dans une démarche intégrée et pluridisciplinaire. Les nouvelles technologies d'élimination des déchets devraient être efficaces et peu coûteuses, afin de remporter une large adhésion auprès du public.

Dans l'exposé suivant, M. Bill Long, Directeur, Direction de l'environnement de l'OCDE, a rappelé aux délégués quatre points récurrents de la réunion :

- l'ampleur de la menace qui pèse sur la santé publique en raison des maladies propagées par les agents pathogènes aquatiques ;
- la contribution actuelle et potentielle de la biotechnologie, dont les avantages doivent être plus largement portés à la connaissance des décideurs ;
- la relation entre le recyclage et la réutilisation de l'eau et les questions liées à la qualité de l'eau ; et
- la corrélation croissante entre les changements climatiques et les maladies d'origine hydrique.

M. Carlos Bazdresch Parada, Directeur général du CONACYT, a fait état des avantages qu'il y avait à tenir la réunion au Mexique. Le Mexique est très conscient de l'importance de la qualité et de la conservation des ressources en eau. La coordination des mesures visant l'utilisation rationnelle de l'eau et l'amélioration de sa qualité bénéficiera des recommandations formulées au cours de la réunion. Les échanges de vues qui ont eu lieu entre les spécialistes du Mexique et de bien d'autres pays à propos de la biodépollution et du biotraitement des eaux usées, des eaux superficielles et côtières et des aquifères stimuleront la réalisation d'études sur l'applicabilité et l'utilisation possible de nouvelles technologies au Mexique.

En clôturant officiellement la réunion, Mme Julia Carabias Lillo, Ministre de l'Environnement, des Ressources naturelles et de la Pêche, a souligné tout le bénéfice que le Mexique tirait de son adhésion à l'OCDE. Celle-ci lui a donné l'occasion d'organiser cette réunion scientifique, la troisième après celles de Tokyo et d'Amsterdam, qui a permis aux spécialistes, aux consommateurs d'eau, aux industriels et aux pouvoirs publics de conjuguer leurs efforts en vue d'analyser les problèmes actuels et de faire progresser la coopération internationale. Elle a conduit à un rapprochement entre les institutions chargées de prendre les décisions et la communauté scientifique, qui se trouve dans une situation plus propice pour participer à l'élaboration de stratégies visant expressément la préservation et la qualité des ressources en eau.

En exposant certaines préoccupations du Mexique concernant l'eau, la Ministre Carabias a expliqué comment et pourquoi ce type de réunion servait les priorités nationales du Mexique. Sur l'échelle de la disponibilité des ressources en eau, le Mexique se trouve dans une position intermédiaire, pas aussi basse que celle des pays du Moyen-Orient, mais pas aussi élevée que celle des Etats-Unis. Cependant la distribution de l'eau dans le pays est très inégale, 80 pour cent du territoire ne recevant que 20 pour cent de l'eau. Le pays est aussi confronté à des inondations localisées et au ruissellement superficiel. Il est convenablement équipé en infrastructures pour l'eau et l'irrigation, en barrages, etc., mais leur fonctionnement est peu efficace : 50 pour cent de l'eau destinée à l'agriculture se perd à cause des fuites. Douze millions de Mexicains n'ont pas accès à l'eau potable et 27 millions ne sont pas raccordés à un réseau d'égouts ; nombre des plus grands bassins sont pollués et les aquifères pâtissent d'une surexploitation.

La Ministre Carabias a rappelé plusieurs programmes mexicains décrits durant la réunion, notamment la dépollution du bassin de Mexico et du bassin du Lerma-Chapala, où sont implantées la majorité des industries. Une structure de planification appropriée destinée à régir l'utilisation, la réutilisation et l'épuration de l'eau est en cours de création. Elle nécessite une entente entre les autorités locales, des Etats et fédérales et entre les consommateurs agricoles, industriels et ménagers. L'eau, à l'instar de quelques autres ressources naturelles, est susceptible de déclencher d'âpres conflits sociaux et politiques.

En quoi la biotechnologie peut-elle aider à relever ces défis majeurs? Elle peut offrir des solutions de valeur sur le plan de la qualité, si elle est appliquée à la biodépollution et qu'elle concourt par conséquent à améliorer l'accès. En outre, elle est susceptible de résoudre des problèmes au niveau technique, avant qu'ils ne se posent au niveau politique, social, sanitaire ou écologique.

Pour conclure, la Ministre Carabias a cerné trois domaines d'étude importants :

– Le premier concerne le cadre réglementaire. Si les réglementations sont claires et universelles, on est alors mieux à même de trouver les moyens d'utiliser l'eau en commun. Les inégalités économiques et sociales peuvent être un facteur important dans l'utilisation de la technologie.

– Deuxièmement, la biotechnologie ne doit pas devenir une nouvelle forme de pouvoir qui rendrait certains pays plus dépendants que d'autres. Le Mexique doit trouver les moyens de développer ses propres technologies et d'occuper une place sur la scène mondiale.

- En troisième lieu, il faut veiller à ce que cette activité en biotechnologie se traduise par des projets concrets et spécifiques qui profiteront à ceux qui en ont le plus besoin. Des maladies très répandues sont essentiellement imputables à la mauvaise qualité de l'eau et ce problème peut être résolu en fournissant les infrastructures et les technologies nécessaires.

Conclusions

Les conclusions suivantes et les réponses aux trois questions s'appuient sur les débats qui ont eu lieu après les exposés. Les réponses aux deux premières questions sont présentées ci-dessous. Les réponses à la troisième question relative aux actions recommandées sont résumées dans les "Recommandations".

Généralités

Il existe un lien direct entre l'amélioration de la durabilité et l'application à plus grande échelle de la biotechnologie dans le domaine de l'environnement, qui permettra également d'élever la qualité de l'eau. La biotechnologie a le plus de chances d'être bénéfique, qu'elle s'applique au traitement de l'eau ou aux problèmes de santé publique, si sa mise en oeuvre s'inscrit dans une démarche pluridisciplinaire et globale.

Les possibilités particulières offertes par la biotechnologie, associée à d'autres méthodes chimiques et physiques le cas échéant, représentent le seul moyen de résoudre les problèmes de qualité de l'eau et de santé humaine à l'échelle mondiale.

La valeur de l'eau pure n'a pas été soumise à une analyse de coûts-avantages, mais devrait l'être. Les études pourraient être confiées à l'OCDE. La création de réseaux et la collaboration entre les pouvoirs publics, les universités, l'industrie et la population est indispensable à l'échelle mondiale.

La biotechnologie offre des possibilités remarquables pour l'élaboration de procédés industriels moins polluants et durables, grâce à la mise au point de biocatalyseurs plus performants. Quelques procédés chimiques synthétiques fondés sur la biocatalyse sont exploités à grande échelle depuis plusieurs années avec de bons résultats ; ils engendrent moins de déchets que leurs analogues chimiques classiques.

Qualité de l'eau et santé publique

L'amélioration de la surveillance sanitaire et l'étude de l'apparition des maladies sont prioritaires. La biotechnologie fournit des instruments de plus en plus perfectionnés pour étudier, détecter, identifier, modéliser, prévoir et évaluer les risques actuels et à venir posés par les micro-organismes pathogènes aquatiques.

Il est essentiel de soutenir davantage la R-D appliquée à la santé humaine pour le bien-être de la population mondiale. C'est la fourniture d'instruments qui offre des possibilités économiques directes à l'industrie et non l'information en elle-même. Par conséquent, l'aide à la collecte de cette information doit incomber aux organisations internationales et aux pouvoirs publics. Le stade de développement étant variable d'un pays et d'un domaine à un autre, les priorités en matière de R-D peuvent varier localement. Mais les priorités ne sont pas fondamentalement différentes d'un pays à l'autre.

Il est capital de réaliser des études de cas internationales et, parallèlement, de soutenir la collecte, le stockage et la diffusion des informations épidémiologiques à l'échelle mondiale. Cette nécessité appelle la création de bases de données au sein d'organismes internationaux tels que l'OCDE.

Le fait que des maladies associées, telles que le cancer et la myocardite, puissent être engendrées par des micro-organismes aquatiques étant reconnu, il est essentiel de procéder à des études approfondies.

L'utilisation d'organismes indicateurs, qui, à l'heure actuelle, ne donnent qu'une idée approximative du risque et qui, au mieux, autorisent des prévisions limitées, devrait être abandonnée, ces méthodes étant aujourd'hui supplantées par de nouvelles techniques comme les sondes géniques, la détermination de l'empreinte génétique et la PCR. Ces biotechnologies perfectionnées, qui permettent d'identifier et de suivre des organismes unicellulaires ou des virus pathogènes grâce à leurs acides nucléiques, sont plus rapides et plus spécifiques que les méthodes qu'elles remplacent et représentent un grand pas en avant. La marche à suivre consiste à intégrer l'épidémiologie moléculaire et l'écologie microbienne, en vue de pouvoir surveiller et étudier les maladies d'origine hydrique et leurs agents microbiens et d'évaluer et de prévoir avec plus de précision les risques correspondants à l'échelle mondiale.

De nouvelles priorités voient le jour dans le domaine des maladies épidémiques ; elles concernent l'étude du devenir et du transport des agents pathogènes dans le milieu aquatique. Le caractère international des maladies épidémiques implique qu'aucun pays ne peut s'atteler seul à ce problème. Le rôle de l'OCDE à cet égard est crucial et complète celui de l'OMS.

Approvisionnement en eau

Il est capital que les pouvoirs publics assurent la fourniture d'une eau potable de la plus haute qualité à tous les citoyens. La mise en oeuvre de procédés efficaces pour le traitement et l'assainissement des eaux usées vient immédiatement après dans l'ordre des priorités. C'est seulement lorsque ces deux conditions seront remplies que l'on pourra envisager d'installer un deuxième système de distribution pour l'eau grise ou de qualité e. L'appellation "qualité e" désigne une eau meilleur marché composée d'eau de pluie, d'eau de récupération, etc., destinée à divers usages pratiques. L'utilisation de l'eau de qualité e n'est pas sans danger, notamment dans les pays en développement, et devra être associée à une éducation poussée de la population.

La biotechnologie a un grand rôle à jouer dans l'épuration de tous les types d'eau. A titre d'exemple, la filtration lente sur du sable ou du charbon actif en granulés est un procédé sûr et économique pour obtenir de l'eau potable. Ces techniques pourraient aussi s'appliquer à l'eau de qualité e. Il faudrait en outre mettre en oeuvre des procédures rapides de biosurveillance de la qualité de l'eau, en temps réel ou non, afin de donner des assurances supplémentaires à la population.

Pour les pays du Tiers monde, il est possible d'améliorer considérablement la production d'eau potable salubre moyennant une filtration grossière suivie d'une filtration lente sur sable. Si une désinfection supplémentaire s'impose au niveau des foyers, l'application de nouvelles techniques fondées sur l'exploitation de l'énergie solaire (par exemple la fraction U.V. de la lumière solaire) est envisageable. Par ailleurs, il paraît souhaitable de développer des technologies appropriées qui permettront aux pays du Tiers monde de produire du chlore destiné à servir de désinfectant de secours.

Biodépollution

Le traitement biologique de l'eau usée représente le procédé de biodépollution le plus avancé et possède des avantages certains sur les procédés chimiques, même pour les débits les plus élevés.

Le recours à la biotechnologie moderne n'a plus besoin de justification particulière, ses avantages pratiques ayant été largement démontrés en biomédecine. Elle tient aussi ses promesses dans le domaine de la biodépollution. Les procédés biologiques sont appliqués à plus grande échelle à mesure qu'ils s'avèrent efficaces par rapport à leur coût. Il convient maintenant d'adopter des stratégies prévisionnelles, comme le repérage des itinéraires à haut risque des pétroliers, des sites où la qualité de l'eau est susceptible de se dégrader et d'être une source de maladies virales, etc., en vue de préparer les actions futures. Le transfert de technologies du laboratoire vers les réalités du terrain est indispensable.

Le traitement anaérobie des eaux usées produit moins de boues que le traitement aérobie et ce dernier en produit moins que les procédés chimiques. Les traitements anaérobies et aérobies donnent des boues de meilleure qualité que les traitements chimiques.

Les quantités résiduaires de substances xénobiotiques peuvent nuire gravement à l'environnement et en particulier à la préservation des ressources en eau. Il y a lieu d'encourager fortement les fabricants de produits potentiellement polluants à assumer la responsabilité de leurs produits sur la totalité de leur cycle de vie. Cette responsabilité pourrait se concrétiser, par exemple, par la fourniture de réactifs lors de la vente d'un pesticide afin que l'on puisse neutraliser le produit inutilisé. Les produits chimiques interdits par les organismes internationaux de réglementation en raison de leurs caractéristiques écotoxicologiques préoccupantes doivent être rigoureusement interdits à l'usage dans tous les pays, développés ou non.

Il convient d'avoir recours à des biocatalyseurs adaptés au procédé considéré pour réduire le volume de polluants entrant dans le flux de déchets (incorporation d'enzymes dans les aliments pour animaux afin de diminuer l'excrétion de phosphates, par exemple) et pour éliminer des substances toxiques déterminées, comme les métaux lourds.

Traitement des eaux usées

S'agissant du traitement des eaux usées ménagères, il faudrait d'abord et principalement assurer l'assainissement primaire, c'est-à-dire l'élimination des organismes pathogènes, des matières organiques et des solides en suspension. Ensuite, un traitement plus poussé est envisageable et plus précisément l'élimination biologique des nutriments. Toutefois, dans ce cas précis, il faudra comparer attentivement les avantages et les inconvénients respectifs de l'élimination biologique des nutriments et de la récupération de l'azote et du phosphore par l'intermédiaire de la croissance des cultures, notamment dans les pays où l'eau usée traitée peut servir à l'irrigation.

Les responsables des nouveaux aménagements urbains devraient examiner les possibilités offertes par les modes les plus récents de gestion des eaux usées, qui tiennent compte de la protection de la santé, de la prévention de la pollution et de la récupération des ressources (eau et nutriments), eu égard à l'exigence, de plus en plus souvent formulée, d'assurer un développement durable.

Le prix du traitement des eaux usées doit être abordable pour le consommateur. Les pouvoirs publics devront peut-être fournir une subvention, mais il faudrait aussi convaincre la population de prendre en charge au moins l'exploitation et l'entretien des installations.

Une méthode de traitement des eaux usées qui donne de bons résultats, en particulier au sein des petites collectivités, est la combinaison d'un traitement anaérobie ou aérobie avec un passage à travers une zone humide (roselière). Le choix de la technologie de la première étape dépend du montant de l'investissement et des frais de fonctionnement. S'agissant de la technique des zones humides, il est nécessaire que l'on pousse plus avant la R-D, notamment en ce qui concerne les communautés végétales, les transformations qui se déroulent dans la rhizosphère, le degré d'élimination des agents pathogènes et l'accumulation de polluants, par exemple les phosphates.

La biotechnologie fondée sur l'interaction entre les plantes et les micro-organismes, qui s'applique à l'élimination des polluants et à la récupération des métaux par l'intermédiaire de bassins, de zones humides et de cultures, a fait ses preuves sur le terrain dans les pays développés et en développement ; c'est une option à faible intensité de capital qui convient particulièrement aux zones rurales. Son développement devrait accroître son efficacité.

Traitement des excreta

Alors que les résidus urbains solides peuvent être pris en charge par les installations de traitement des déchets, le traitement localisé est plus difficile dans les zones péri-urbaines et rurales qui ne disposent pas d'infrastructure de collecte. Dans les zones péri-urbaines, les excreta, par exemple, comportent un risque élevé de contamination fécale-orale. Différentes techniques doivent être mises au point pour les traiter. L'une d'entre elles, la latrine sèche, diminue considérablement le danger de contamination des eaux souterraines et superficielles, mais elle n'est pas toujours bien acceptée par le public qui considère que la toilette à chasse d'eau reliée à une fosse septique est le système normal. C'est pourquoi il est nécessaire d'entreprendre des travaux de R-D et d'éduquer la population, afin d'accroître l'efficacité de techniques peu coûteuses d'enlèvement et d'élimination des excreta dans les zones péri-urbaines, et de les faire mieux accepter par les utilisateurs. La biotechnologie peut être appliquée dans les zones péri-urbaines et rurales, notamment par la mise au point de matrices sorbantes à ajouter aux excreta et le recours à des techniques de compostage.

Boues d'égout - biosolides

Il devient de plus en plus difficile de trouver une utilisation profitable des biosolides d'épuration dans les pays industrialisés. Cependant, le traitement physico-chimique des eaux usées engendre plus de boues que leur biotraitement et, de surcroît, les boues chimiques sont plus difficiles à éliminer. C'est pourquoi, il ne faut pas désavantager le biotraitement des eaux usées à cause des boues qu'il génère. Toutefois, deux lignes d'action doivent être suivies en ce qui concerne les biosolides : l'élimination des métaux lourds pour les empêcher d'aller contaminer les eaux usées et la revalorisation des biosolides, par exemple en améliorant le compostage.

Surveillance et régulation

La surveillance et la régulation des procédés biotechnologiques appliqués à l'environnement sont insuffisantes et trop peu pratiquées. On a volontairement doté les installations d'une capacité supérieure aux besoins pour ne pas devoir utiliser des systèmes de régulation complexes. Des installations de biotraitement moins chères, plus petites et capables d'absorber un plus grand débit sont de plus en plus demandées ; il est possible de les mettre au point à condition de faire un usage plus judicieux des dispositifs de surveillance et de régulation. Depuis quelques années, on dispose d'une large gamme de détecteurs, en temps réel ou non, basés sur des systèmes biotechnologiques et physico-chimiques, mais peu sont utilisés de façon générale en raison de leur coût ou de leur manque de fiabilité.

Innovations technologiques destinées à la modification de l'environnement

La recherche sur les plantes transgéniques dotées d'une résistance accrue à la sécheresse s'accélère grâce à l'application de techniques de pointe en biotechnologie. Cette recherche répond à un besoin urgent, étant donné la pénurie d'eau qui affecte de nombreuses régions. Les laboratoires mettent actuellement au point de nouveaux sorbants biologiques qui peuvent aussi favoriser la croissance des végétaux dans des conditions de stress hydrique.

Sensibilisation

L'information du public est indispensable au progrès de la biotechnologie et à son application pratique. Cette information n'est plus simplement une nécessité nationale, mais un impératif international pour la diffusion des résultats de la recherche, la fourniture de renseignements pratiques aux profanes et l'accès aux bases de données via le World Wide Web et d'autres sources d'information.

Les responsables de l'action publique doivent être tenus bien au courant des domaines de la biotechnologie qui ne sont pas sous les feux des médias, mais qui enregistrent néanmoins des progrès certains, par exemple la biodépollution, le traitement biologique classique des eaux usées et les applications à la santé publique. Les pouvoirs publics doivent faire connaître les bienfaits de la biotechnologie appliquée à l'environnement pour la santé publique.

La connaissance et l'acceptation par le public sont déterminantes pour le recours à la biotechnologie, notamment en ce qui concerne les applications qui utilisent des organismes génétiquement modifiés. La biotechnologie appliquée à l'environnement progressera plus rapidement si un cadre réglementaire uniforme pour l'utilisation des organismes génétiquement modifiés dans la biodépollution et le traitement des eaux est mis en place à l'échelle mondiale.

Réponses aux trois questions

"Quels sont les avantages (techniques, économiques, écologiques) de la biotechnologie pour l'exploitation et la préservation des ressources en eau?"

Dans l'immédiat

Les outils de la biotechnologie, intégrés à une démarche pluridisciplinaire, offrent de vastes possibilités dans le domaine de la santé publique, du biotraitement et de la biodépollution.

La biotechnologie fournit, enfin, des moyens de s'atteler aux problèmes mondiaux de santé publique liés à la pollution des eaux, qu'il s'agisse des eaux souterraines, des eaux superficielles, des nappes aquifères ou des eaux marines. Il existe maintenant des biotechnologies efficaces de surveillance et d'évaluation qui permettent d'identifier et de suivre des organismes unicellulaires ou des parties de micro-organismes comme les acides nucléiques.

Cette surveillance et cette évaluation de maladies déterminées seront bientôt étendues à une gamme plus large d'organismes pathogènes et ces nouvelles techniques seront particulièrement importantes pour surveiller, modéliser et prévoir l'apparition de nouvelles maladies.

Pour le moment, la biotechnologie s'applique surtout au traitement de la pollution terrestre, atmosphérique ou aquatique, une fois qu'elle a eu lieu. Dans le domaine du traitement des eaux, en particulier, on a pu montrer que les procédés biologiques étaient efficaces par rapport à leur coût et respectueux de l'environnement. Ils sont perçus comme naturels et donc plus facilement acceptés. Ils peuvent servir à dégrader des substances intraitables par des procédés physico-chimiques et leur adaptabilité à diverses conditions les rend irremplaçables dans le traitement des effluents ménagers et industriels mélangés. Les techniques biologiques peuvent non seulement éliminer le gros de la pollution, mais aussi des polluants très spécifiques dont l'élimination par d'autres moyens coûte très cher.

Les procédés biologiques offrent plusieurs réponses à un problème particulier, à différents niveaux de technologie. Aussi peuvent-ils fournir davantage de débouchés commerciaux pour les technologies nationales et les populations locales.

Le traitement biologique des eaux et des déchets utilise peu de produits chimiques et n'affecte donc pas la structure des boues ou des sédiments. Ceux-ci, contrairement aux boues chimiques, peuvent par conséquent être réincorporés à l'environnement.

La biotechnologie possède aussi un avantage dans le traitement de la pollution par les sources diffuses, notamment la pollution des eaux souterraines et des aquifères. Ces technologies sans danger pour l'environnement sont non seulement efficaces par rapport à leur coût dans la majorité des cas, mais offrent des possibilités réelles d'application *in situ*. La technologie de la biodépollution est employée à l'heure actuelle et elle a fait ses preuves dans un certain nombre de cas célèbres, comme la catastrophe de l'Exxon Valdez.

A court terme

Alors que les potentialités de la biotechnologie continueront à se développer au cours des trois à cinq années à venir, de nouvelles études donneront lieu à une explosion de données épidémiologiques mondiales. La collecte, le stockage et la diffusion de ces informations jetteront les bases d'un réseau international.

On commence déjà à utiliser des techniques très spécifiques, en temps réel, rapides, sensibles, automatiques et transportables pour la surveillance et la régulation de procédés ; elles se perfectionneront à court terme, si bien que les procédés de dépollution deviendront plus rapides, moins coûteux et meilleurs.

La sélectivité très poussée des systèmes biologiques, notamment pour la reconnaissance de métaux déterminés, est abondamment étudiée en laboratoire ; dans quelques années, elle sera à la base de plusieurs procédés de biotraitement largement acceptés.

La demande de systèmes biologiques appliqués à la biodépollution, au développement durable et au maintien de la qualité de l'eau ne cessera de s'accroître.

A long terme

A long terme, profitant des progrès rapides de la biologie moléculaire et d'une connaissance plus approfondie de l'évolution dirigée, la biotechnologie aura un rôle plus important à jouer dans la durabilité. Les procédés biologiques seront au coeur de nouveaux processus industriels intrinsèquement moins polluants.

L'ultime avantage de la biotechnologie réside dans sa contribution à l'objectif de protection primaire de l'ensemble de l'écosystème (c'est-à-dire un système global destiné avant tout à prévenir la pollution).

"Quels sont les obstacles à l'application de la biotechnologie à l'exploitation et à la préservation des ressources en eau?"

Techniques

Il est nécessaire d'étudier plus avant les biotechnologies les plus récentes, innovantes et perfectionnées, afin de remplacer les méthodes actuelles d'évaluation des risques qui sont imprécises. Les obstacles technologiques ne sont pas les mêmes dans les pays développés et dans les pays en développement. Les pays développés, par exemple, doivent étudier la désinfection, le devenir et le transport dans le réseau de distribution d'eau, tandis que les pays en développement se préoccupent avant tout de l'épuration de base et de la distribution.

La technologie de la biodépollution est à l'heure actuelle mise en oeuvre en laboratoire, dans des installations pilotes et de démonstration et dans des applications en grandeur réelle adaptées au site. Il est toutefois nécessaire de poursuivre la R-D, afin d'accélérer le développement en vraie grandeur de procédés largement applicables dans tous les pays, développés ou en développement.

L'optimisation des procédés et la régulation plus précise des processus demandent une meilleure connaissance des mécanismes biologiques fondamentaux. Le biotraitement peut être mené à bien sur place dans les décharges, par la biorégénération du sol, ou dans des bioréacteurs, mais les traitements actuels sont exécutés au cas par cas. Il est nécessaire d'effectuer des recherches pour établir des programmes de technologie générique, qui seront étayés par d'autres études de modélisation et de prévision. Jusqu'à présent, la crédibilité est insuffisante, parce que les bons exemples de dépollution réussie ne sont pas assez nombreux. Les non-biologistes considèrent que les procédés biologiques engendrent des boues, ce qui signifie en substance qu'ils transforment un déchet en un autre déchet. Le problème de la production de boues, de leur réduction au minimum et de leur élimination doit être résolu. Il n'est pas propre à la biotechnologie ; les procédés de traitement chimique produisent un plus grand volume de boues plus difficiles à traiter.

Financiers/économiques

La biotechnologie propose des méthodes de dépollution moins coûteuses, mais qui comportent souvent des procédés à long terme par rapport aux techniques physico-chimiques. Des études pour mieux comprendre les mécanismes sous-jacents et, par là, accélérer les processus biologiques devraient permettre de réduire cet inconvénient. Le coût faible et la lenteur relative des processus biologiques les désavantagent tant qu'ils ne sont pas appliqués à plus grande échelle : ils offrent une valeur ajoutée et un bénéfice trop faibles aux petites entreprises.

La coopération internationale est déterminante et permettra d'alléger la charge qui pèse sur chaque pays, mais il faut fixer des objectifs prioritaires pour la R-D. L'obstacle majeur à cet égard tient à la limitation des ressources qui sont allouées en général à la R-D en biotechnologie.

Généralement de petite taille, les entreprises biotechnologiques sont désavantagées par rapport aux grosses sociétés de traitement chimique classique des déchets.

Le capital-risque a été l'un des grands facteurs qui ont permis le développement de la biotechnologie appliquée aux produits pharmaceutiques. Toutefois, le fait que les investissements dans la biotechnologie appliquée à l'environnement sont considérés comme peu rentables devrait limiter l'apport de capitaux privés aux nouvelles entreprises qui se créent dans ce domaine.

Liés à la formation et aux ressources humaines

L'application de nouveaux procédés biotechnologiques perfectionnés à la qualité de l'eau et à la dépollution demande bien sûr un personnel qualifié dans les secteurs public et privé. Les processus biologiques, perçus comme intrinsèquement plus complexes, sont considérés comme plus difficiles à comprendre et à maîtriser. En outre, lorsque le traitement des déchets ne bénéficie pas d'une grande priorité ou est considéré comme un coût, les compétences requises ne sont pas affectées à sa gestion.

La formation en microbiologie fondamentale n'est pas suffisamment soutenue, de même que la formation réciproque (biologie pour les ingénieurs et vice versa).

Liés à l'action des pouvoirs publics

Le fait que les pouvoirs publics ne mesurent l'impact sur la population qu'en termes d'apparition de maladies constitue un obstacle non négligeable. En fait, les effets réellement chroniques et à long terme de la mauvaise qualité de l'eau et du manque d'aliments salubres (c'est-à-dire les pertes d'heures de travail et de rendement) représentent une perte infiniment plus grande sur n'importe quelle période de temps.

La peur de la responsabilité juridique dans le secteur privé a considérablement limité l'échange d'informations au sein de l'industrie de la biodépollution, si bien que l'on n'a pas accès aux résultats des expériences utiles accumulées jusqu'à présent sur le terrain, qui permettraient d'améliorer les méthodes et les applications de la biodépollution. De plus, la plupart des installations de traitement sont conçues et évaluées par des ingénieurs qui ne travaillent que dans le domaine qu'ils connaissent, afin d'éviter tout problème de responsabilité. La législation doit être conçue de façon à éviter ces problèmes de responsabilité.

Liés à l'acceptation par le public

Les risques ne sont envisagés à l'heure actuelle que sous l'angle de la santé. Une stratégie efficace doit reposer sur une approche plus globale qui intègre la santé publique dans l'ensemble de l'écosystème planétaire. Les programmes d'éducation sur l'hygiène de base, destinés à diminuer le risque de contracter des maladies d'origine hydrique, font défaut dans de nombreuses parties du monde.

La sensibilisation du public et l'acceptation par ce dernier sont déterminants pour l'utilisation de la biotechnologie, notamment en ce qui concerne les applications faisant appel à des organismes génétiquement modifiés.

La biotechnologie n'apporte pas toujours les solutions immédiates que les consommateurs, en particulier les consommateurs industriels, semblent préférer aux meilleures réponses. Les idées fausses qui circulent parmi le public, selon lesquelles, par exemple, les processus biologiques ne sont pas fiables et demandent plus d'espace, et les organismes vivants sont fondamentalement moins prévisibles, doivent être combattues.

Recommandations : "Quelles mesures (y compris le soutien à la R-D) faudrait-il recommander?"

Bien que ces recommandations aient été formulées par les participants à la réunion en leur nom personnel et qu'elles n'engagent pas les pays Membres de l'OCDE, elles n'en possèdent pas moins une valeur intellectuelle et un pouvoir de conviction considérables.

Généralités

- Afin de progresser vers la durabilité, les pouvoirs publics devraient favoriser l'application de la biotechnologie à la prévention de la pollution et à la récupération de ressources, à court terme, et à des technologies moins polluantes, à long terme.

- Comme l'utilisation de biocatalyseurs à plus grande échelle réduira incontestablement le volume de déchets (et améliorera par conséquent la qualité de l'eau), il faudrait encourager les responsables des programmes publics à promouvoir le recours à cette technologie. Il est nécessaire de perfectionner la conception des procédés et d'améliorer la formation des ingénieurs et de ceux qui sont chargés de choisir de nouveaux procédés au sujet du rôle que pourrait remplir la nouvelle biotechnologie.

– La biotechnologie ne fera une percée plus forte sur le marché que lorsque son coût et ses avantages techniques seront mieux compris. Les pouvoirs publics ont un rôle à jouer à cet égard en réalisant les indispensables analyses de coûts-avantages.

– L'OCDE est invitée à organiser une activité de suivi abordant le problème de l'absence de normalisation des techniques s'appliquant à la qualité de l'eau et à sa biodépollution, en vue d'élaborer une série de lignes directrices internationales.

– On a vivement recommandé la création d'une base de données mondiale et la diffusion d'informations sur la qualité de l'eau et les organismes aquatiques pathogènes, ainsi que la récolte de souches de micro-organismes destinées à des études de référence sur la biodépollution et la santé publique. Il faudrait faire appel à toutes les capacités d'équipes pluridisciplinaires pour traiter ces questions.

– Les produits industriels, en particulier les produits agro-chimiques, engendrent une pollution importante. Une stratégie pourrait consister à encourager tous les gros producteurs de substances potentiellement polluantes à assumer l'entière responsabilité de leurs produits sur la totalité du cycle de vie.

– Compte tenu de la limitation des ressources, les pouvoirs publics doivent établir des priorités concernant l'aide à la R-D.

Qualité de l'eau et santé publique

– Les gouvernements et les autorités responsables de la santé publique devraient renoncer à utiliser des organismes indicateurs et les remplacer par des techniques plus perfectionnées qui permettent de surveiller et de prévoir l'évolution des maladies d'origine hydrique et d'évaluer globalement les risques et l'exposition.

– Les organisations internationales devraient être encouragées à créer des bases de données épidémiologiques et, à travers leur utilisation, à mettre en lumière et à faire connaître les applications réussies de la biotechnologie à la santé publique ; cette tâche nécessite un développement de la normalisation des méthodes.

– Il est essentiel d'entreprendre d'autres études sur les maladies associées aux micro-organismes pathogènes aquatiques, telles que le cancer et la myocardite.

Adduction d'eau

– Il faudrait accorder une priorité absolue à la distribution universelle d'eau potable de haute qualité et à la mise en place d'installations appropriées de traitement des déchets (eau et solides).

– Il faudrait stimuler largement l'utilisation de techniques adéquates de purification et de désinfection des eaux superficielles dans les endroits non alimentés en eau potable.

- Une éducation doit être dispensée à tous les niveaux au sujet du rôle de la biotechnologie et des nouvelles techniques d'assainissement.

- On recommande à l'OCDE de réaliser des études de coûts-avantages sur la valeur de l'eau pure.

Biodépollution et traitement des déchets

- Les politiques de traitement doivent porter sur la mise en place d'un système d'assainissement primaire pour l'ensemble de la population et ensuite sur l'optimisation de l'utilisation des nutriments et de la récupération de ressources.

- Pour les petites et moyennes collectivités, on préconise de combiner l'élimination de la demande biologique en oxygène (DBO) et des solides en suspension, puis de pratiquer une épuration complémentaire à travers une zone humide.

- Dans les zones péri-urbaines des pays en développement, il faudrait tenter de faire mieux accepter les dispositifs de remplacement, comme les latrines sèches et les techniques de compostage appropriées.

- La R-D sur les mécanismes fondamentaux des interactions entre les plantes, les micro-organismes et les polluants devrait être encouragée.

- Il conviendrait de stimuler les projets de démonstration qui permettent d'évaluer les caractéristiques économiques et l'applicabilité à long terme des procédés de substitution, comme la phytodépollution.

- Les gouvernements sont vivement encouragés à établir un cadre réglementaire uniforme pour l'utilisation des organismes génétiquement modifiés dans ce secteur.

- Il est nécessaire d'effectuer des études supplémentaires sur l'évaluation des risques et la biodisponibilité des polluants dans les systèmes aquatiques, afin d'approfondir les connaissances fondamentales et d'alimenter le processus législatif.

- Les organisations internationales comme l'OCDE ont un rôle à jouer en mettant en place des sites de démonstration multinationaux et en organisant la collecte et la diffusion d'informations suivant une procédure uniforme. C'est une très grande priorité pour la biodépollution des réseaux de distribution d'eau.

Surveillance et régulation

- Il faudrait créer des incitations pour encourager l'utilisation à plus grande échelle des détecteurs et des mécanismes de régulation, par exemple en prélevant une taxe sur tous les déversements.

- Il faudrait établir des normes concernant des détecteurs peu coûteux et robustes intégrés dans les systèmes d'éco-surveillance et d'audit (EMAS - Eco-Monitoring and Auditing Systems) et on devrait procéder à l'évaluation comparative de la qualité du matériel et des procédés.

Innovations technologiques destinées à la modification de l'environnement

- Il convient d'accélérer la R-D s'appliquant aux biosorbants ainsi qu'aux plantes et micro-organismes résistant à la sécheresse et au sel.

Information

- Les pouvoirs publics devraient en priorité informer la population, par tous les canaux possibles, sur le rôle et les avantages de la biotechnologie. Cette recommandation qui concerne la sensibilisation et l'éducation du public est hautement prioritaire.

- Il est nécessaire d'appliquer un programme efficace d'éducation à l'hygiène de base, qui doit permettre de réduire le risque de contracter des maladies d'origine hydrique, car c'est là le facteur de risque le plus élevé.

- L'éducation de la population et sa participation à des projets locaux destinés à améliorer la distribution des eaux et le traitement des déchets sont essentielles.

Les pouvoirs publics devraient encourager l'acquisition d'une base de compétences ; à titre d'exemple, les exploitants des installations et les ingénieurs devraient recevoir une formation en biotechnologie et les biologistes devraient acquérir une meilleure maîtrise de l'ingénierie.

OPENING SPEECHES BY HOST COUNTRY

OPENING SPEECHES BY HOST COUNTRY

by

Dr. Francisco Bolívar Zapata
Director General of the Biotechnology Institute of the UNAM

Mr. Carlos Bazdresch Parada
Director General of CONACYT

The workshop was opened by Dr. Francisco Bolívar Zapata, who introduced the speakers of the opening session, conveyed the apologies of Dr. Martínez Palomo, who was prevented from attending at the last moment, and presented a number of key questions for discussion. He emphasized the importance of the workshop for Mexico and many other countries, as the applications of biotechnology to water problems looked to be increasingly promising.

Mr. Carlos Bazdresch Parada then underlined that improving the quality of water was one of Mexico's main national priorities, and that the workshop's conclusions and recommendations would have to be considered not only by the scientific community, but by the policy-makers as well. The dialogue between the two, made possible by the presence of the National Water Commission together with so many national and international experts, was a unique opportunity. He invited the workshop participants to identify important research topics and propose larger R&D programmes for possible international funding.

GREETINGS BY OECD

WATER AND BIOTECHNOLOGY: AN OECD PERSPECTIVE

by

Risaburo Nezu
Director, Directorate for Science, Technology and Industry, OECD

On behalf of the Secretary-General of the OECD, it is my pleasure to give this opening address to the OECD Workshop Mexico '96 on Biotechnology for Water Use and Conservation. The OECD's greetings and thanks go to all of you: to the Government of Mexico, and to Mrs. Carabias, Minister of the Environment, who will conclude the workshop. We thank also CONACYT, whose Director, Mr. Bazdresch, and his devoted staff have made this Conference possible; Mexico's National Water Commission, which has played a key role in the preparations; Professor Bolívar Zapata, who has agreed to be Scientific Chair; also the important delegations and outstanding scientific speakers who have arrived from many countries; and last but not least, the Chairman of OECD's Working Party on Biotechnology, Dr. David Harper. These are just a few out of many names I must mention.

Before getting to the substance, I want to place this workshop within a broader context and perspective. To my knowledge, it was about 20 years ago that the first industrial application of modern biotechnology was made possible in the United States for pharmaceutical products. Before that, biotechnology had remained within the research laboratory. Since then, this technology has grown rapidly in the pharmaceutical industry as a new solution to diseases. Some illnesses which had been considered incurable have been brought under our control, and there is hope for many more. However, this was not an easy road. The concept of this technology gave rise to emotional, cultural and ethical reactions from the general public and consumers. This technology was understood as an attempt to use living organisms to make or modify products, to improve or change plants and animals or to develop materials that mimic molecular structure or functions of a living organism. This was not incorrect. In a way, it was aiming to do just that. But in many instances, the story was exaggerated or distorted to an extent that it was feared that new plants and animals, including human beings were being created. It took many years of strenuous effort on the part of scientists and government officials, as well as industrial people, to dispel this groundless fear and to establish public acceptance of this technology.

The first challenge which OECD Member countries faced was to establish a scientific view about the safety of this technology. The OECD has grappled with this question by bringing together scientists and officials to examine various aspects of the technology and established guidelines in 1986, which, I assume, are familiar to all of you.

These guidelines were well accepted by the Member countries and functioned as a springboard to accelerate the commercial application of biotechnology, making the bioindustry one of the most active areas world-wide. Let me cite a few figures. According to a US government report, by the year 2000, the biotechnology industry is projected to have sales reaching $50 billion in the United States. This should be compared with $6 billion in 1992. The annual growth rate is about 40 per cent. For Japan, growth is estimated to be somewhat more modest, but still something like 30 per cent per annum, from $8 billion in 1995 to $30 billion in the year 2000. In Europe, according to a report prepared by a consultant, "the (European) market for goods is already extensive and is now worth an estimated 38 billion ECU. Approximately, 184 000 jobs already depend directly on the application of biotechnology." I find these European figures quite impressive*.

Looking to the future. Many OECD countries view biotechnology as one of the key emerging technologies and it is of strategic importance both to the competitiveness of their industries and to creating jobs. There is no doubt that, with technology advancing and the ageing of population progressing in OECD countries, the market for pharmaceutical products will be the fastest expanding segment of the economy. It is believed advances in health-related biotechnology could be of great value in containing health care costs through new diagnostic, prevention and treatment techniques.

There is a need to follow the response of the governments in the OECD to reap the immense benefit brought about by this technology. First, they are promoting R&D by mobilising the public research facilities and supporting the private firms. The average annual amount spent by the government in recent years is $4 billion in the United States, $1 billion in Japan, $160 million in Germany and $250 million in the United Kingdom. The OECD database on public support to industry identifies 269 R&D subsidy programs in the OECD Member countries, many of which include key technologies, such as micro/electronics, and of course, biotechnology. Six of them purport to advance solely biotechnology. I think it is not an overstatement to say that there is a global, head to head competition for supremacy and a leadership role in this technology.

Second, in light of the diverse and expanding potential of this technology, OECD countries are strengthening co-ordination among different agencies, such as Ministries/Departments responsible for industry, environment, agriculture, welfare and health, not to say science and technology. In the United States, for example, more than ten agencies are co-ordinating their approaches under the Biotechnology Research Subcommittee (BRS) of the Committee on Fundamental Science/National Science and Technology Council. I know similar efforts for co-ordination are being made among European countries, Japan and other OECD Members.

In the OECD, we have the ICGB, the Internal Co-ordination Group for Biotechnology, to ensure good communication and consistent approaches amongst the Directorates concerned: DSTI, Environment, Agriculture, Trade and the Development Centre, with one Deputy Secretary-General chairing the group.

Third, regulatory constraints, in particular, product approval procedures and patent protection mechanisms, are seen as key elements for the further promotion of commercialisation of this technology. This is notable particularly for Europe and Japan. "The available competitiveness indicators emphasise that, despite a strong research base and major core enterprises in the EU, the European location for research, production and investment is not highly favorable" (this is a quotation from Mr. Martin Bangemann, the Commissioner for Industry).

The Japanese government is also of the opinion that excessive regulations by the government are detrimental to business activities, are pushing firms to move abroad, and eventually hollowing out the

* The definitions of biotechnology reflected in these figures are not consistent.

Japanese industrial basis, particularly in newly emerging fields. Regulatory reform and government restructuring will be the top priority items for the new government to be formed after the election, which happens to be today. At present, the OECD is carrying out horizontal work on regulatory reform. An interim report will be produced by next spring, examining the impact of regulation on competitiveness, innovation, overall economic performance, trade, etc. The experience in biotechnology will be useful input to this work.

OECD work in respect of biotechnology comprises 1) human health-related biotechnology; 2) intellectual property rights (IPR); and 3) bioremediation. I will not go into details as I assume Dr. Harper, Chairman of the Working Party on Biotechnology, will discuss the OECD approaches after me.

Trade in biotechnologically produced or engineered goods is rapidly gaining importance on the international agenda. Biofoods, like tomato, potato, corn, soybean and rape-seed, are already produced for commercial sale. In spite of the concerns voiced at an initial stage, consumers in the United States, Canada, the United Kingdom and Japan, etc., are accepting such products as foodstuffs, and the market seems to be growing. It is a natural course of action for these companies to try to market their products world-wide. Trade issues arise due to differences in safety standards and procedures. This also implicates systems for the protection of intellectual property rights. We are working with the Trade Directorate to find an international solution to this problem.

This is where we stand today.

Now, let me make a few brief remarks about this workshop.

It is with remarkable success that Mexico has recently responded to a difficult economic challenge in its modern history. In the OECD Programme of Work, the Mexico Workshop is the third and last workshop devoted to bioremediation and bioprevention, that is, the beneficial contributions of biotechnology to the environment. It is the fruit of several years of OECD work. The first workshop in Tokyo in 1994 gave a general overview of bioremediation and prevention. The second, in Amsterdam in 1995, focused on how to remediate pollution of air and soil. Mexico is focusing on water. Mexico is a particularly appropriate venue for a workshop on science and technology to address water contamination and water shortage. All countries have water problems, but some have more difficult ones than others, and solutions to be discussed might be valuable for many others. The Mexico water workshop should open up new perspectives, in two directions.

First, the sheer magnitude of the problems which Mexico has to solve may call for new solutions. A distinctive feature of biotechnology, in various fields of application, has been to develop new or cheaper technological options able to replace or supplement more traditional methods. Will this be true for water use and conservation as well? In which cases, under which conditions, in which countries? What could biotechnology contribute to the remediation and prevention of water pollution, and to the saving and more efficient use of water? Through its multiple applications, biotechnology can be an intellectual prism through which to review all aspects of water: short-, medium- and long-term.

I expect this Workshop to raise questions at the frontiers of scientific and technological knowledge. Some of the answers you may elaborate could be relevant widely beyond Mexico's borders, and beyond the present OECD area. Let me assure you that both the CSTP and the Working Party will study your findings, conclusions and recommendations with the greatest interest.

This brings me to the second of the new perspectives to which I have alluded. For the OECD, Mexico is a gateway to Latin America. Many of Mexico's water problems are shared by these countries, and therefore, the solutions which Mexico may develop will be relevant to these countries as well, and are likely to be keenly studied by them. The OECD would be happy to "reach out" to Latin America through Mexico, and through this water workshop. We would be proud if we could make a contribution -- even if it were only a small one -- to the future of Latin America.

As water is becoming a major economic factor, it will be increasingly linked to economic performance and thus, could attract the interest of the OECD as an economic organisation. Water is primarily an environmental, agricultural and technological issue. This is why I am happy that my colleague, Mr. Bill Long, Director of the OECD Environment Directorate, has agreed to join us here and to address this workshop. The environmental and the technological approaches to water must go hand in hand; neither one can be effective without an intimate understanding of the other. Environmental policy sets framework conditions, through tariffs, standards and regulations, but these must be closely attuned to the moving frontier of science and the changing technological options and constraints.

Ladies and gentlemen, thank you very much. I wish you a very successful workshop.

WELCOMING REMARKS

by

Bill L. Long
Director, Environment Directorate, OECD

I want to join my colleague, Rizaburo Nezu, in welcoming you to this workshop. In the interest of time, I will not repeat his quite specific, and well-deserved, expressions of appreciation to our Mexican hosts and organisers of our meeting, and to those of you who are contributing as chairs of sessions, as speakers, and as participants.

For my perspective as OECD's Director of the Environment, this meeting comes at the right time -- and we are in the right placc.

Regarding the place, Cocoyoc has a special significance for those of us involved in international environmental affairs. In 1974, a major meeting was held here under UNCTAD sponsorship, to follow up on the UN Conference on the Human Environment convened two years earlier in Stockholm, Sweden. That meeting produced what came to be known as the "Cocoyoc Declaration on Patterns of Resource Use, Environment and Development Strategies". It succeeded in crystallising the concept of development without destruction, and raised international awareness of resource scarcities and the growing pressures natural resources were under.

So, holding our meeting here this week makes sense for historical reasons, and also because, two and one-half decades later, it is obvious that the challenge of protecting and conserving the earth's natural resources, and utilising them efficiently, is still with us.

I also observed that our meeting is coming at the right time. When the Cocoyoc conference was convened in 1974, water issues were high on the international agenda, with planning already underway for a United Nations Water Conference in 1978. Then, in the 1980s, attention seemed to shift, to problems of desertification, to the environmental problems of urban settlements, and then to the upper atmosphere ... first to ozone depletion and then to climate change.

While these latter problems clearly remain high on international agendas, it is striking to me how attention across OECD Member countries has been re-focusing over the past two to three years on water problems, both water quality and water quantity.

This meeting is one expression of a renaissance of OECD interest in the water field. Another is our plan to convene a high level experts workshop in Sydney, Australia next February (10-12) on the subject of "Sustainable Consumption of Water". We intend to examine a broad spectrum of strategies, policy instrument and technologies, to make water consumption more sustainable in domestic, agricultural and commercial and urban sector situations. I anticipate that the conclusions of this workshop will be a welcome contribution to our forthcoming meeting in Australia.

As Mr. Nezu mentioned in his remarks, my Directorate is also a participant in OECD's work on biotechnology, which involves a broad array of OECD components. The Environment Directorate has two interests: one is in helping our Member countries develop a harmonized regulatory framework for biotechnology -- a framework which provides adequate safety for human health and the environment while minimising barriers to biotechnology innovation and diffusion.

Second, the OECD Environment Directorate has a keen interest in the *application* of bioremediation and bioprevention, including the tools of modern biotechnology, to environmental management objectives. As we work with our Member countries on ways to achieve "clean" economic growth, and to help them cope with difficult problems of contaminated soils, air pollution and degraded water supplies, it is clear to us that biotechnology has a great, yet-unrealised potential for helping confront these challenges effectively and efficiently.

Thus, we continue to urge OECD Governments and the private sector to maintain strong support for the basic and applied science on which the essential technological breakthroughs depend. At the same time, we are mindful of the concurrent challenge for science, industry and government to work together to turn the scientific possibilities into successful, real-life applications.

In the conduct of our work programme, we attach particularly high priority to the sharing of information and ideas among our Members. We have learned that if one looks around, one will find that somewhere, someone has figured out how to do it better. Further, we are persuaded of the value of meetings like this for building new international networks of knowledgeable people who share common interests and goals.

For these reasons, my colleagues and I in the Environment Directorate attach great importance to this meeting. I am pleased to be able to be part of it.

GENERAL LECTURES

US ENVIRONMENTAL PROTECTION AGENCY RESEARCH ON DRINKING WATER DISINFECTION

by

Robert J. Huggett
Assistant Administrator for Research and Development
US Environmental Protection Agency, United States

Clean water is one of our most valuable resources, both as a source of drinking water and for fisheries and wildlife, agriculture, industry, and recreation. Despite the tremendous progress made in protecting surface and groundwater over the past several decades, substantial threats to water quality and quantity remain. This paper discusses the most serious threats to water quality in the United States and the research program that the Environmental Protection Agency (EPA) is undertaking to understand and reduce these risks. Biotechnology is an important element of the research program.

Progress in protecting water resources

The United States has made much progress in protecting water resources. Since 1970, the United States has invested billions of dollars in municipal sewage treatment facilities, nearly doubling the number of people who receive secondary treatment to 150 million. EPA and its state partners have required thousands of industrial facilities to obtain discharge permits and have established nation-wide discharge standards, dramatically reducing industrial pollution. Freshwater fish are recovering in many areas, the rate at which wetlands are being lost has been slowed, and many miles of formerly contaminated beaches are now safe for swimmers. Typhoid fever and amebiasis, the waterborne diseases of greatest concern in the United States early in this century, have largely been eliminated. EPA has established drinking water standards for more than 80 contaminants, such as microbes, metals, and organic chemicals. EPA and the states have also helped protect water quality by regulating high risk sources of contamination such as pesticides, underground storage tanks, underground injection wells, and landfills.

Threats to water resources in the United States and around the world

Nevertheless, important threats to water quality remain. In 1993, an outbreak of gastrointestinal disease caused by the protozoan *Cryptosporidium* occurred in Milwaukee, Wisconsin, resulting in an estimated 400 000 cases of acute illness and possibly about 100 deaths. This incident represented a clear warning that waterborne pathogens still pose a substantial threat to health in the United States. Other pathogens of concern include the protozoan *Giardia*; viruses such as hepatitis A virus and Norwalk virus; and bacteria such as *Mycobacterium*. At the same time, use of chlorine or alternative disinfectants in drinking water can produce a variety of by-products that may pose risks of cancer and other health effects.

Waterborne pathogens also pose substantial health risks in many other parts of the world. Latin America has experienced severe cholera outbreaks this decade caused by the waterborne organism *Vibrio cholerae*. The World Health Organization has estimated that two million children die each year in developing countries due to diarrheal disease related to contaminated water. Besides mortality, chronic childhood diarrheal disease can contribute to severe adverse health outcomes, such as malnutrition, immunodeficiency, and growth stunting.

Unfortunately, threats to water quality and quantity are likely to persist both in developed and developing countries. Not only are currently available funds and technology inadequate to fully address known threats, but population growth and development will place ever-increasing demands on finite water resources. Population growth increases the absolute amount of water needed, and per capita demands often increase as standards of living rise. Growth and development can also increase contamination burdens and alter the water cycle through deforestation, destruction of wetlands, and other land practices. In developing countries, dams and other complex water developments in some cases have increased breeding sites for vector-borne diseases such as schistosomiasis.

In the United States, the infrastructure of many water systems is in need of replacement; some estimates suggest that at current rates, a given pipe in water utility distribution systems will be replaced every 200 years. Deteriorating distribution components, by allowing greater infiltration of contamination from wastewater and other sources, can result in unsafe drinking water even if the water leaving the water treatment plant is safe. Future uses of new materials in water systems (such as plastic pipes) might pose other risks by changing habitats for new or modified microbes to flourish.

Finally, populations with reduced immune capacity, such as the elderly, AIDS patients, and organ transplant patients, are projected to increase in the coming decades in the United States. This will elevate the need for thoroughly disinfected drinking water supplies.

Research priorities

Clearly, research to understand and reduce these threats is urgently needed. The new strategic plan for EPA's Office of Research and Development (ORD) identifies drinking water disinfection as one of its top six research priorities for the next several years. EPA chose this topic because of the high risks and uncertainties associated with drinking water contaminants, the potentially high costs of further regulation of drinking water, and the potential for ORD to make a significant contribution to this topic. EPA hopes that this research will be useful not only in the United States, but will benefit other countries experiencing similar threats, and welcomes collaboration with researchers from other countries.

EPA's Draft Drinking Water Research Plan, currently under review by EPA's Science Advisory Board, contains four major research goals:

1) To identify the health effects caused by microbial pathogens and disinfection by-products (DBPs) in drinking water.

2) To determine the distribution of exposures to microbial pathogens and DBPs in drinking water.

3) To assess the risks caused by microbial pathogens and DBPs in drinking water.

4) To evaluate the effectiveness of options for reducing risks from microbial pathogens and DBPs.

The research plan is described in more detail in the following sections, with an emphasis on the risks of microbial contamination. The challenge is to reach an optimal balance that will control risks from both pathogens and DBPs to acceptable levels, while ensuring costs of water treatment are commensurate with public health benefits.

Health effects research

In the United States, infectious waterborne diseases have been attributed to a variety of bacteria, parasites, and viruses. *Giardia* has been responsible for about half the outbreaks where a causative agent was identified. *Cryptosporidium* outbreaks have been reported less frequently, but the number of cases associated with each outbreak has been much larger. A number of outbreaks of acute gastrointestinal illness have not been linked to specific pathogens. Viruses are thought to cause many of these outbreaks, but because of the inability to culture many viruses, it has been difficult to establish this definitively. In addition to reported outbreaks, many may be either unrecognised or unreported.

Current knowledge of waterborne pathogens is inadequate for assessing their health risks. While their disease symptoms are generally known, limited information is available on the doses and conditions that produce effects. The infectivity of susceptible subpopulations, and the variation in infectivity and virulence among different pathogen strains, also are uncertain. The infectious doses of protozoans and viruses, including the role of immunity in resisting infection, are high priority research topics for EPA.

The incidence of waterborne disease in the United States is highly uncertain. Further research is needed to determine rates of illness, and, where rates are elevated, determine whether illness is caused by inadequately treated water or by water quality that has deteriorated in the distribution system. EPA plans to develop immunological assays for exposure to *Cryptosporidium* and investigate water disease outbreaks in co-operation with the US Centers for Disease Control.

For disinfection by-products, EPA's research will focus on understanding the toxicity of individual chemical contaminants and mixtures of DBPs, and analysing the health effects in communities served by disinfected drinking water.

Exposure research

The sources of pathogenic microbes in drinking water associated with disease are usually related to fecal matter from warm-blooded animals, including humans. Humans are one of the main sources of pathogens, but animals are frequently implicated as in the outbreaks caused by *Giardia* and *Cryptosporidium.* Many bacteria can grow and persist in the distribution system, some of which can cause infections under certain conditions.

Current techniques lack the precision and specificity to measure low levels of pathogens. As a result, the levels of pathogens in source water and drinking water are not well known. Information is also needed on the survival and transport of pathogens in groundwater. EPA plans to address each of these areas of uncertainty.

For example, EPA's laboratory in Cincinnati, Ohio is working to develop a "gene probe" method to more easily detect *Cryptosporidium.* The gene probe is designed to bind to genetic sequences found only in the DNA of the *Cryptosporidium* organism. Attached to the gene probe is a fluorochrome that emits coloured light visible through a microscope, allowing organisms to be identified and counted. For the probe to function, however, it must penetrate the interior of the organism where the genetic material is located. Current research is attempting to devise a way to penetrate the protective outer wall of the life stage of *Cryptosporidium* known as the "oocyst." If successful, this research should provide a much more effective way of detecting *Cryptosporidium.* Similar research is being conducted for *Giardia.*

EPA is also attempting to improve methods to detect pathogenic viruses. Currently, many are very difficult to grow in tissue culture, which severely hampers efforts to determine whether viable viruses are present in water supplies. EPA research is aimed at identifying tissue cultures most hospitable for viral growth. If improved tissue cultures can be identified, virus detection should become more effective.

Exposure research for disinfection by-products will focus on developing and evaluating methods for measuring occurrence of DBPs in drinking water, and on investigating the levels of DBPs that people are actually exposed to via their drinking water supplies.

Risk assessment research

Assessing the actual risks of waterborne pathogens depends on adequate effects and exposure information, which is to be obtained as described for the research plans described in the preceding two sections. To make the best use of this data, a comprehensive risk assessment model is needed. Currently accepted risk assessment models were developed for assessing chemical risks, and do not fully accommodate issues important to the assessment of pathogenic risks. For example, a person exposed to a micro-organism, as opposed to a chemical contaminant, may develop an immunity to future infections. Additionally, someone infected from a drinking water source may spread the disease to others. EPA is sponsoring work to develop improved risk assessment methodologies for micro-organisms.

EPA research is also focusing on improved risk assessment for disinfection by-products. Research will proceed in a phased approach, beginning with improved risk assessments for single chemical contaminants, followed by evaluations of risks associated with mixtures of agents and development of methods to assess risk based on generalisations across classes of DBPs.

Risk management research

Better data on pathogen removal and inactivation efficiencies for various treatment processes are needed to improve risk estimates and select treatments with the most potential for risk reduction. Research efforts should simultaneously evaluate control of disinfection by-products. The greatest need is to address organisms that are most resistant to treatment. For example, because *Cryptosporidium* is much more resistant to disinfection than most other waterborne pathogens, treatments effective for inactivating *Cryptosporidium* are likely to also reduce other pathogens.

Because of the expense and difficulty of measuring *Cryptosporidium*, however, surrogate parameters (such as indicator organisms or particle size) for evaluating treatment effectiveness are needed. In groundwater, where protozoa are not expected to occur, viruses must be considered when defining adequacy of treatment.

Bacteria pose a special problem because of their ability to grow in the distribution system, sometimes forming "biofilms" on surfaces. Many bacterial species identified in distribution systems are opportunistic pathogens that can cause illness in certain subgroups. Research is needed to better understand factors that influence microbial growth in distribution systems and to develop strategies to control such growth.

To address these research needs, EPA plans to investigate optimising conventional treatment to remove pathogens and DBPs and to study new control technologies. One current area of research is the use of biological filters for treating drinking water. Though biofilters are widely used to treat wastewater in the United States, they are not commonly used for drinking water. Besides mechanically filtering water, biofilters employ living bacteria to consume and degrade organic compounds. Compounds of concern are those that react with chlorine to form disinfection by-products, and those formed by ozone that, if not controlled by biofilters, would serve as bacterial substrates in distribution systems. Working with the University of Cincinnati, EPA is testing the effectiveness of several different filter media, such as anthracite and granular activated carbon, under different operating conditions. Biofilters are an appropriate treatment process after ozone disinfection and before final disinfection with chlorine or chloramine.

Small system, low cost technologies

In the United States, more than 50 000 community water systems are considered small or very small (that is, they serve less than about 3 300 people). Such systems cannot afford advanced, capital-intensive technology and tend to have the most difficulty in complying with federal drinking water requirements. Drinking water systems in many other parts of the world face similar difficulties. EPA recognises these difficulties and is conducting research specifically to assist small systems in controlling both pathogens and disinfection by-products.

The most important requirements for small systems are low construction and operating costs, simple operation, adaptability to part-time operation, low maintenance, and reliable performance. EPA has developed a co-ordinated research program combining in-house and external field-scale research projects to develop alternative treatment technologies to address these requirements. "Package plants," which are assembled in a factory and transported to the site, may be a promising solution for many systems. EPA is evaluating a variety of filtration technologies that can substitute for conventional technologies in package plant systems, such as bag filters, cartridge filters, and ultrafiltration membranes.

EPA also has been working with several Latin American countries to improve health conditions by enhancing the management of drinking water and sanitary practices. Under an agreement with the US Department of Agriculture, and in consultation with Mexican officials, EPA is working with private companies to demonstrate drinking water treatment technology in Mexico. One system is being developed for demonstration at a public school near the city of Cordoba. Another demonstration project is planned for Ixhuacan de los Reyes, a community of about 2 600 people.

Conclusion

Existing scientific knowledge and technology are not sufficient to protect our essential water resources. Carefully designed research programs, including biotechnology research efforts, are needed to control existing problems cost-effectively and cope with emerging ones. Not only must researchers continue to share the results of their efforts, as this conference is designed to do, but they must strive to apply the knowledge gained to solve our pressing health and ecological problems.

References

In addition to EPA documents, important sources for this manuscript were the reports, *Safe Drinking Water: Future Trends and Challenges* by EPA's Science Advisory Board (March 1995), and *A Global Decline in Microbiological Safety of Water: A Call for Action* by the American Academy of Microbiology (1996).

THE ROLE OF BIOTECHNOLOGY FOR WATER USE AND CONSERVATION IN JAPAN

by

Kikuji Tateishi
Deputy Director-General, Basic Industries Bureau
Ministry of International Trade and Industry, Japan

Application of biotechnology to environmental problems

As you well know, biotechnology has already been applied in various fields including chemical, pharmaceutical, agricultural and food industries. It will continue to be applied in many other industries in the future.

Environmental application will perhaps be one of the major fields where biotechnology can play a significant and effective role. It can contribute not only to energy saving and resource conservation, but also to reduction in waste materials. I believe biotechnology will become a key technology to realise sustainable industries which can prosper in harmony with the global environment.

Based on these views, I would like to talk about how biotechnology can be applied to cope with environmental problems. Before that, I would like to comment on water pollution problems we faced in Japan in the past and what countermeasures we took, and finally about MITI's policies on the application of biotech to those environmental issues.

History of water pollution in Japan

In our country, water pollution has been recognised as a social problem over the past century. It started with river pollution that broke out in the "Ashio" copper mining area 100 years ago. Since the late 1950s, high economic growth caused serious water pollution in both urban and industrial areas. Rapid growth of economic activity and population produced a tremendous amount of wastewater from industries and households, resulting in water contamination problems. To make matters worse, public investment in sewage treatment facilities had fallen behind economic growth, so the quality of river water running near large cities and bays deteriorated.

Development of measures to prevent water pollution

The Basic Law for Pollution Control was legislated and enforced in 1967, and the Law for the Prevention of Water Pollution was enacted in 1970. In these laws, we set the effluent standards for discharged water. It depends upon each prefecture, but some prefectures adopted much stricter standards than we set.

In addition to this, we set a well-organised monitoring system which allows us to control the water quality on a regular basis. We also set penal regulations for non-compliance with the standards. In the case of violations, the government could issue a provisional order to stop discharge of the water.

Besides these legal steps, the government has been providing financing to small and medium enterprises at lower interest rates and lightening the tax burden on them.

Furthermore, we have been developing various types of activated sludge systems using current biotech. Fortunately, those efforts worked relatively well, and we were able to improve the water quality.

When we talk about recent improvement of the water quality, we have to take two things into consideration. One is how water quality has been improved in terms of human health concerns. The other is how water quality has been improved in terms of our living environment. In summary, water pollution in relation to our health has been greatly improved. However, our living environments are still less than satisfactory due to the low quality of water, particularly in less open ecosystems,

Decrease in water pollution in relation to protecting human health was extremely improved

Substances such as heavy metals found in industrial wastewater have harmful effects on human health. We have made much effort to get rid of them, and now the water quality has been greatly improved.

As you can see in Figure 1, the quality of wastewater not meeting the environmental standards has decreased quite a bit since 1971. Particularly since 1984, you can see a significant improvement in the quality. I would say we were able to attain our initial goal.

Figure 1. **Overview of water pollution in Japan**
Not meeting environmental standards: heavy metals and others

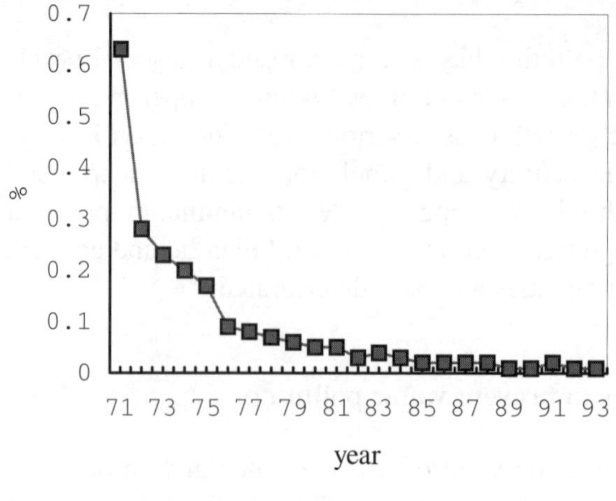

Source: MITI.

Decrease in water pollution in relation to our living environment was improved, but more improvement is needed

On the other hand, we still have water contamination problems in terms of our living environment, primarily due to organic substances discharged from households and industries. As a matter of fact, the water quality is getting better, but further improvement is still needed.

With regard to water pollution due to organic substances, the number of areas that satisfy environmental standards accounts for only 75 per cent based on BOD and COD measurements (Figure 2).

Figure 2. **Overview of water pollution in Japan**
Achievement of environmental standards: COD/BOD

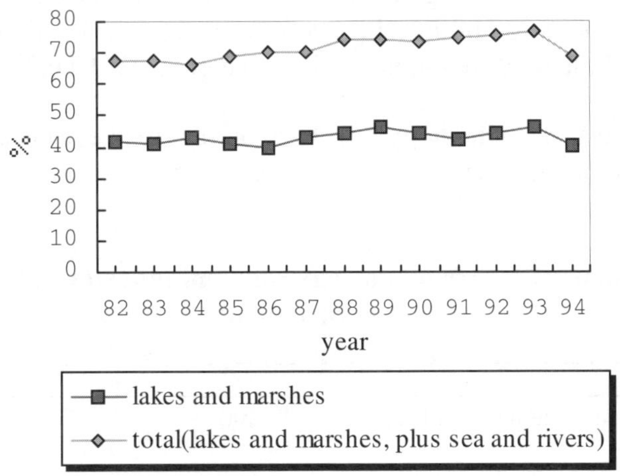

Source: MITI.

The situation is now better than it used to be, but the percentage is unfortunately still not high enough. This is because wastewater from households keeps the level of organic substances high in less open areas such as lakes, bays, and small-sized rivers running through urban areas. Our goal is to have beautiful rivers and beaches just like those in Cancun, Mexico.

Measures for recent problems

How can we solve the problem? There are two important points to be made here. The first point is how we can reduce the input of organic substances. Because of the concentration of population and industries in coastal areas, we keep discharging tremendous amounts of wastewater to rather closed waters. The water quality strictly depends upon the amount of organic substances. So the idea we came up with is to regulate the total volume of discharged water. We are now gradually introducing this idea to actual cases.

The second point is how we can control eutrophication. The major substances that cause serious eutrophication problems are phosphorus and nitrogen. Therefore, we established environmental standards in order to manage phosphorus and nitrogen in discharged waters. From a technological point of view, we are trying to apply not only physicochemical methods, but also biological techniques to reduce the levels of those chemicals.

Utilisation of biotechnology to solve environmental problems

As I have mentioned, biotech is one of the best ways to solve the water contamination problems in our country. As a matter of fact, it's been working fairly well. I expect biotech will be the technology which is the most promising in overcoming a large number of environmental issues.

As you know, the OECD report *Biotechnology for a Clean Environment*, or the so-called "Green Book," was issued in 1994. The report states that the market of environmental industries for the conservation and restoration of environments will expand from US$ 200 billion in 1990 to US$ 300 billion in 2000 (Table 1). MITI is also expecting that in Japan, the market of environment-related business including recycling of resources and energy saving technology will jump to approximately US$ 230 billion in 2000 from US$ 150 billion in 1994, although our estimation is based on a little bit different items from that of the OECD (Table 2).

There is no doubt that biotechnology will become one of the vital elements in environmental business. This is the reason why many private enterprises are becoming interested in the potential of biotechnology.

Table 1. **Forecasts of market trends for the environment industry in OECD Member countries** (US$ billion)

Item	1990	2000	Annual growth rate (%)
Water treatment, soil waste management, air pollution, services other	200	300	5.5

Source: OECD, 1994.

Table 2. **Current situation of eco-business and outlook in Japan** (US$ billion/¥100/$)

Item	1994	2000
Equipment for environment, environment recovery, environment enhancement, clean products, services, waste processing, recycling, clean energy	153	233

Source: MITI, 1994.

So far, biotechnology has been primarily used for manufacturing processes in pharmaceutical and food industries, because it has supplied us with better quality products at relatively lower cost. I believe this will also be true for the environmental business. The technology called "bioremediation" is also noteworthy as the next generation's biotech.

Considerations regarding application of biotechnology to environmental problems

The main issue to address is what we should take into consideration when applying biotechnology to environmental problems. The answer is effectiveness and safety of biotechnology.

With regard to effectiveness, the important point that should be made concerning the practical application such as bioremediation technology is that we have not yet experienced on-site bioremediation, so very little data is currently available to demonstrate the effectiveness of the technology. This means we have to gain lots of knowledge and experience to prove its effectiveness, as well as its safety.

Concerning safety, the OECD Member countries have shared the idea that we could not find any risks which are peculiar to recombinant DNA organisms. Needless to say, we have to think about the effect of recombinant micro-organisms on other living organisms and the environment when we apply them to environmental problems.

Those are also crucial factors not only to making the environment clean, but also to gaining public acceptance for dissemination of the technology.

MITI's activities

Speaking of the effectiveness of bioremediation, we are still not sure whether it works or not in actual contaminated sites, simply because we do not have sufficient experience in using this technology *in situ*. Japanese private sectors have been reluctant to adopt this technology in place of other conventional treatment processes for this reason. Lack of data also makes it very hard to make a comparative cost estimation. We also have to think about the fact that the limited data may also influence its public acceptability.

As is shown in Table 3, MITI launched a five-year program last year to test the effectiveness and safety of the technology at actual contaminated sites in Japan. Establishing a method to assess the effectiveness of the technology is one of the very important elements in this project. We think it very important to demonstrate that pollutants have been reduced as a result of microbial degradation, not due to physical movement. This cannot be achieved without conducting close, on-the-spot monitoring. It is also important to build up the lab-scale data to predict field performances.

Table 3. **Bioremediation project by MITI**

- Purpose: to prove the effectiveness and safety of bioremediation technology
- Site: Actual TCE contaminated site in Japan (in Chiba Pref.)
- Micro-organisms: methane utilising micro-organisms, etc.
- Period: FY 1995-1999
- Required total budget: US$ 12 million (US$ 1 = ¥ 100)

With regard to safety, in 1986, we instituted the "Guideline for Industrial Application of Recombinant DNA technology" based on the 1986 OECD Council recommendation, and it has been working very well in Japan. The MITI guideline for the deliberate release of microbes in the natural environment is now being developed based on the OECD Report on *Safety Considerations for Biotechnology* issued in 1992.

To institute the guideline, we are particularly focusing on the use of microbes for bioremediation under natural environments and trying to establish a general evaluation method. We plan to complete this research by the end of this fiscal year, as shown in Table 4.

Table 4. **Research by MITI for establishment of a general evaluation method of bioremediation**

- Focus: bioremediation in the natural environment
- Target: establishment of general safety evaluation method
- Period: FY 1993-1996

Furthermore, MITI is making continuous efforts to develop basic technologies to solve many kinds of environmental problems. MITI is supporting Research Institutes such as RITE (Research Institute of Innovative Technology for the Earth) to foster these environmental technologies (Table 5).

The first is carbon dioxide fixation technology using algae, etc. If this works, it might contribute to reducing global warming.

The second is technology for hydrogen production using the biological photosynthesis process. If this technology works, we would not need to use fossil fuels and discharge carbon dioxide to the environment anymore.

The third technology is recognised as one of the bioprevention technologies. We are also trying to make biodegradable plastics, and this is also based on the idea of bioprevention. This would allow us to lessen the impact on the environment and contribute to the reduction of wastes.

Table 5. **The development of basic technologies by MITI**

- CO_2 fixation and utilisation technology
    - CO_2 reduction utilising bacteria/algae
    - FY 1990 to 1999
- Environmentally-friendly technology for the production of hydrogen
    - Production of hydrogen using micro-organisms
    - FY 1991-1998
- Biodegradable plastics as bioprevention technology
    - Reduction of the total amount of wastes by micro-organisms
    - FY 1990 to 1997

Conclusion

Finally, I would like to emphasize that scientific and technological discussion and co-operation among OECD Member countries are valuable in order to promote bioremediation and other biotechnologies to solve environmental problems. In this sense, I believe that this workshop and the

OECD Working Party on Biotechnology's (WPB) activities on "Biotechnology for Clean Industrial Products and Processes" are very important. I strongly hope that this workshop will be a significant step for the furthering of the application of biotechnology.

REFERENCES

MITI (1994), *The Environmental Vision of Industries*, MITI, Tokyo.

OECD (1994), *Biotechnology for a clean environment*, OECD, Paris.

Parallel Session I

I-A. BIOLOGICAL QUALITY OF WATER AND PUBLIC HEALTH

GLOBAL CLIMATE CHANGE AND CHOLERA PANDEMICS -- THE EPIDEMIOLOGY OF CHOLERA

by

Rita R. Colwell
University of Maryland Biotechnology Institute
College Park, Maryland, United States

To develop an understanding of the relationship between water quality, climate, and public health, our recent findings concerning cholera are helpful in putting into perspective the role of water and climate in the transmission of infectious disease. The quantity of potable water available for human use is only a fraction of the total global water supply; that is, the amount of water not bound in glaciers or seawater is only a fraction of the total global water supply, and of the remainder, the amount available as a drinking water source is an even smaller fraction (Figure 1a, 1b). Thus, the assumption that there is an endless supply of water is not accurate. Furthermore, a clear association between availability of safe drinking water and child mortality has been established (Figure 2a, 2b). Rates of infection and disease are much higher, if there is not access to safe drinking water. With safe drinking water, the incidence of disease is lower.

Figure 1a. **The world supply of water**

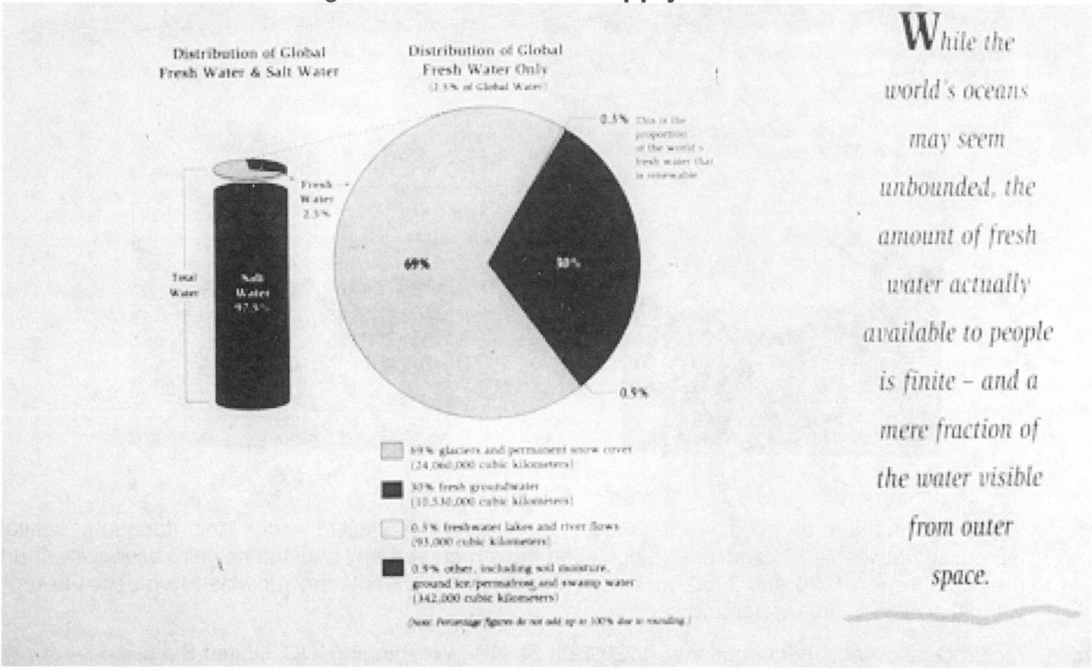

Source: Population Action International, 1120 19th St. NW, Washington, DC, United States.

Figure 1b. **World population and fresh water use, 1940-2000**

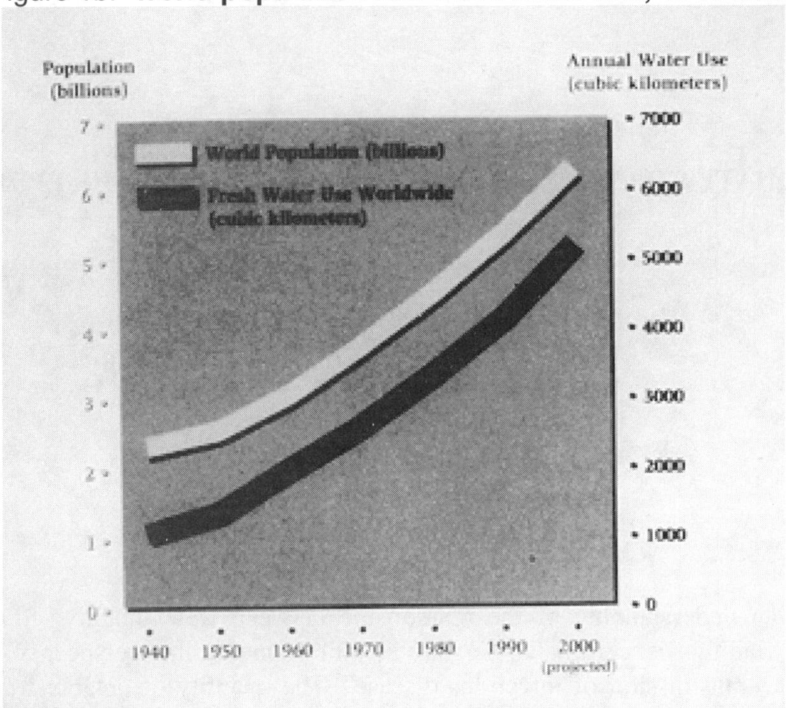

Note: Since 1940, the amount of fresh water used by humanity has roughly quadrupled as world population has doubled. Some water experts estimate the practical upper limit of usable renewable fresh water lies between 9 000 and 14 000 cubic kilometres yearly. That suggests a second quadrupling of world water use is unlikely.

Source: Peter H. Gleick, Pacific Institute for Studies in Development, Environment and Security.

Figure 2a. **Access to safe drinking water and sanitation in developing countries, 1980-2000**

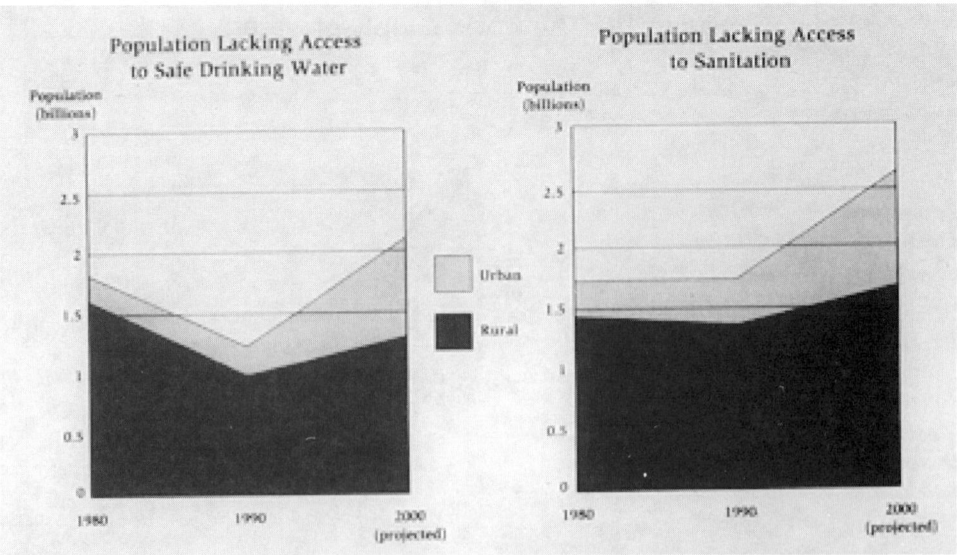

Note: While the number of people with access to safe and sufficient water and adequate sanitation increased between 1980 and 1990, population growth erased any substantial gain, especially in urban areas. Between 1990 and 2000, an additional 900 million people are projected to be born in regions without access to safe water and sanitation.

Source: Population Action International, 1120 19th St. NW, Washington, DC, United States.

Figure 2b. **Water quality and child survival**

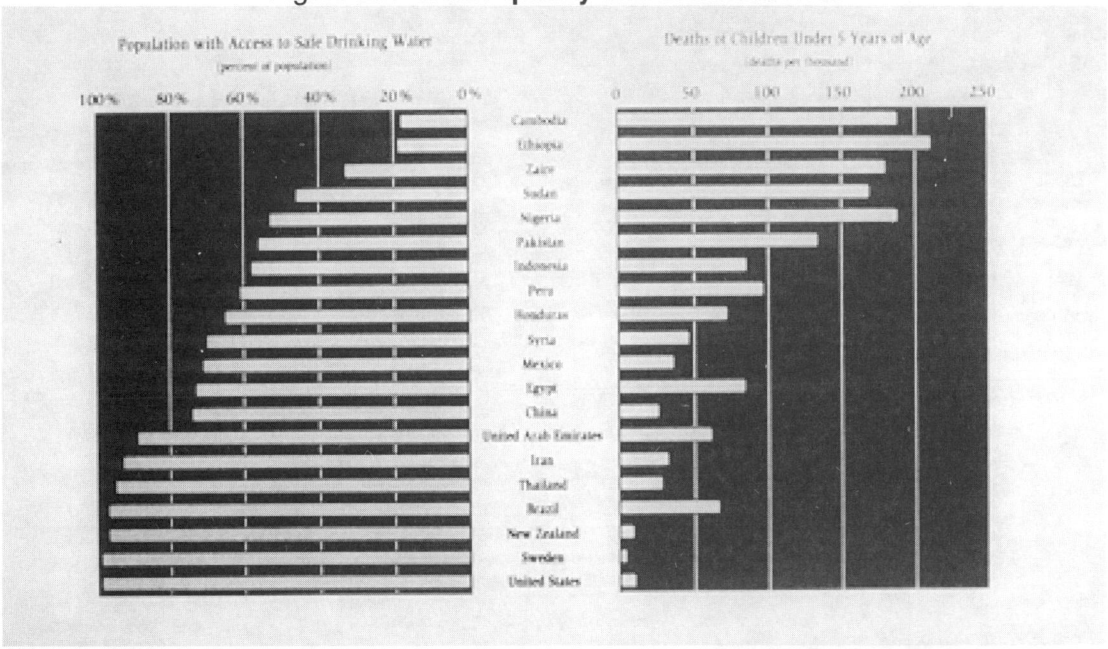

Source: Population Action International, 1120 19th St. NW, Washington, DC, United States.

Infectious disease is a major cause of death world-wide, even when compared to major diseases, such as cancer (Figure 3). Incidence and virulence of waterborne pathogens have been hypothesized to be greater than infectious agents transmitted by other routes (Figure 4). Association of virulence with water transmission has been suggested to be related to water transmission, i.e. the water route allows more rapid passage through a population, contributing to increased virulence (Ewald, 1994).

Figure 3. **Deaths by cause, world-wide, 1992**

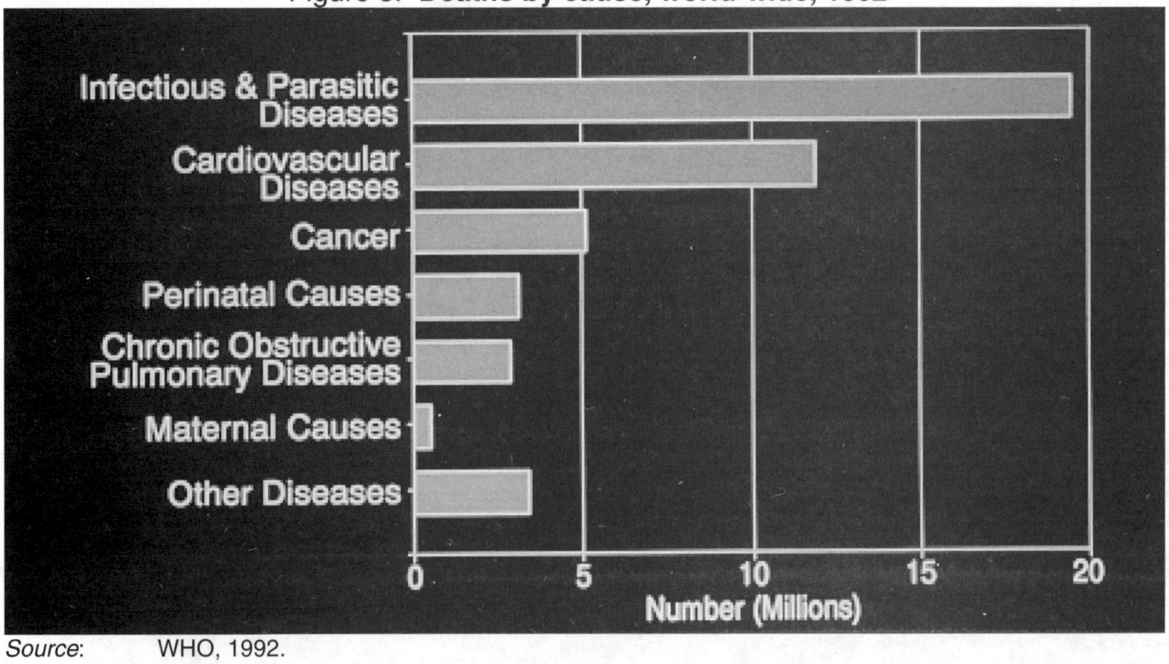

Source: WHO, 1992.

Figure 4. **Virulence of waterborne pathogens**

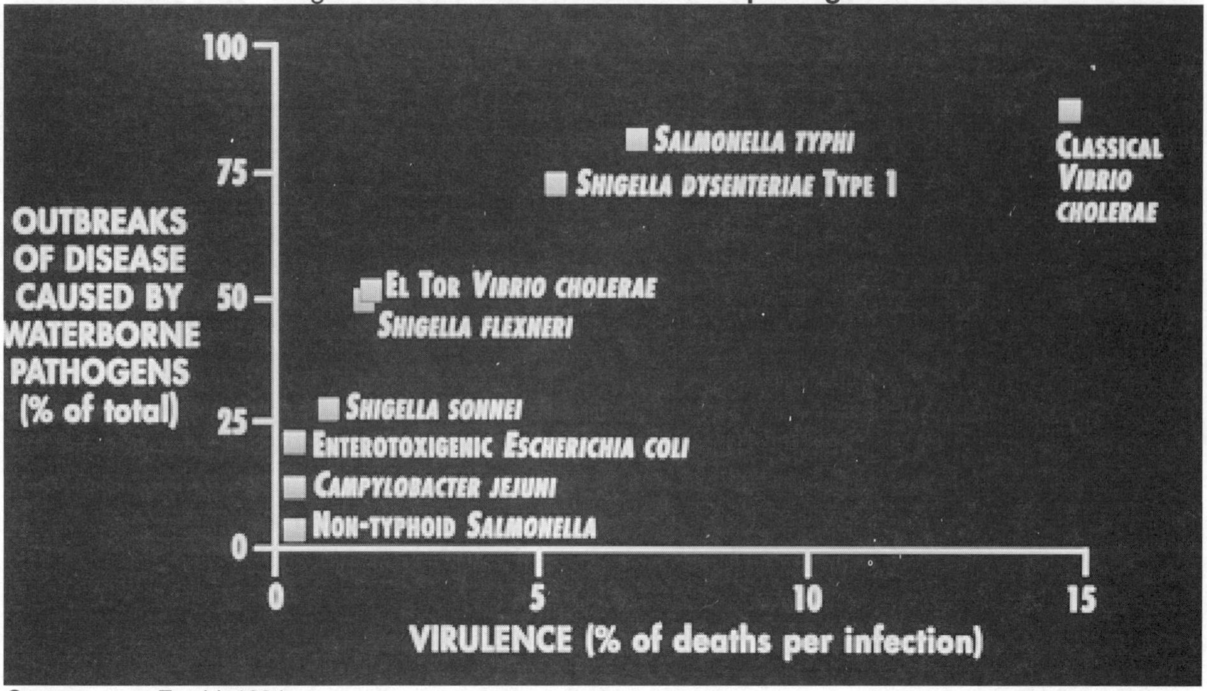

Source: Ewald, 1994.

Cholera is now in its seventh pandemic. As a disease, cholera has always been associated with coastal areas, with entry via harbours and brackish waters. The seventh pandemic was hypothesized to have migrated along the sea routes of Calcutta and Madras in India. However, our work has shown that the cholera bacterium is associated with plankton populations in seawater (Colwell and Huq, 1994; Colwell, 1996). Furthermore, recent outbreaks in South America, starting in 1991, have been illuminating in that the association with El Niño and climate has also been strongly indicated (Figure 5).

Figure 5. **Spread of epidemic cholera: Latin America, January 1991-March 1993**

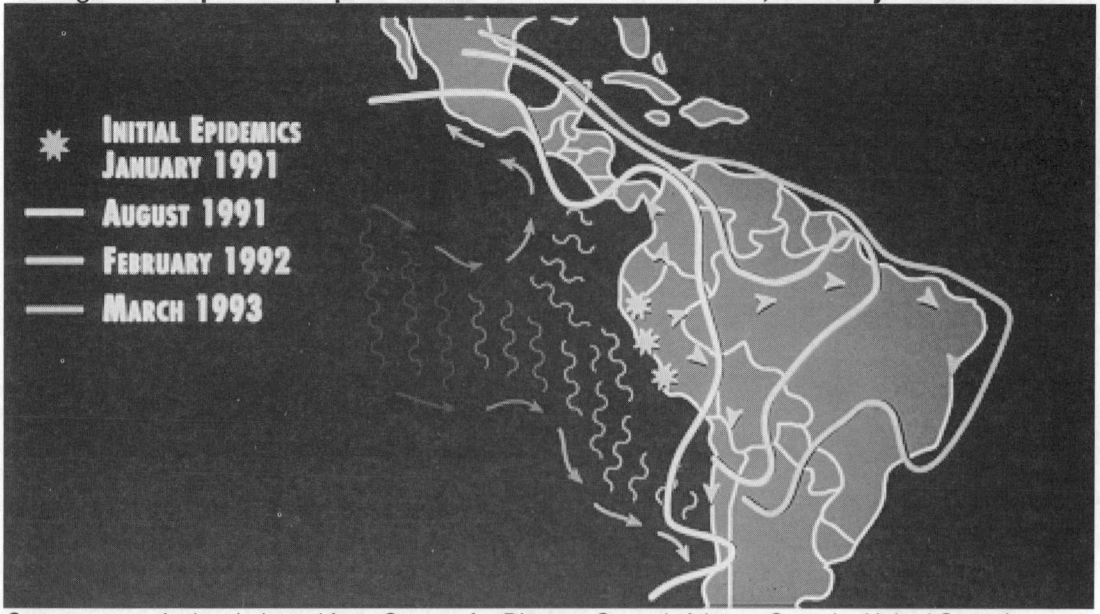

Source: Author (adapted from Centers for Disease Control, Atlanta, Georgia, United States).

In 1992, after the initial outbreak in Peru in 1991, countries affected by cholera included most of Latin America. Furthermore, areas of endemicity, such as India, Pakistan, Bangladesh, and Africa, also had cholera epidemics. However, the routes of cholera, that is, the association of cholera with coastal areas, showed patterns in the most recent epidemics that has not changed for almost 200 years (Colwell, 1996).

In the United States and Latin America during the last century, patterns of cholera epidemics show the disease emerging along coastal routes including the St. Lawrence River and Washington, DC, in the United States, where cholera outbreaks were common until 1900 (Figure 6).

Figure 6. **Historic routes of cholera epidemics**

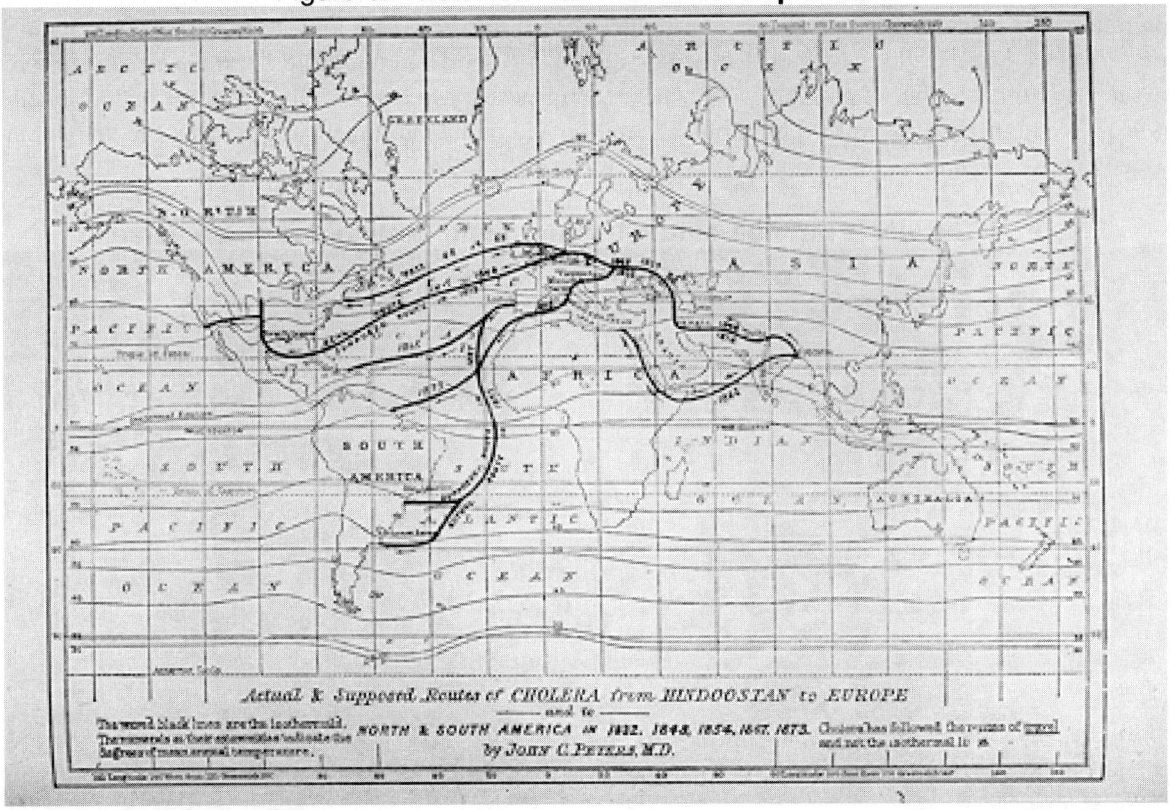

Source: Colwell, 1996.

Dr. Snow, a London physician, showed the association of deaths from cholera and a well in central London. In fact, the first careful epidemiological analysis showing the relationship of cholera with water was achieved by Snow in 1854. The suggestion, then, was provided as to the source, distribution, and relationship of cholera and the environment, namely contaminated water as the major source of epidemics.

A few years ago, a new epidemic strain, *Vibrio cholerae* 0139, entered the scene, originating in Madras, India. This strain soon appeared in coastal areas of Bangladesh, with penetration to the interior of that country along its rivers. The new strain of cholera, *Vibrio cholerae* 0139 serotype, supplanted *Vibrio cholerae* 01 serotype. *Vibrio cholerae* 0139 continued to cause disease in Calcutta and Madras in 1996, with fluctuation between the 01 and 0139 serotypes occurring most recently (B. Nair, personal communication).

The antigenic structure of cholera is important to understand, since the 01 gene-coding complex comprises a set of subunits. A monoclonal antibody to the "A" subunit of the 01 antigen was prepared in our laboratory (Tamplin *et al*., 1989). The "B" and the "C" subunits have been reported to occur in other marine vibrios, but the "A" subunit appears to be specific to the epidemic strain of *V. cholerae* 01, which provides a rapid diagnostic tool in the form of monoclonal antibody and gene probes specific to the "A" subunit (Colwell *et al*., 1990).

Another aspect of *V. cholerae* that is important to understand is that the bacterium is capable of entering a dormant or nonculturable stage (Figure 7) (Colwell *et al*., 1985). In this dormant state, the bacterium maintains viability, determined by a nalidixic acid and yeast extract incubation (Kogure *et al*., 1979, 1987). Survival strategies of *V. cholerae* include the starvation response, morphological alteration to a spore-like structure through reductive cell division, and long-term survival, including long-distance transport by seawater (Colwell and Huq, 1994). Nonculturable *V. cholerae* can survive in seawater for months, remaining viable during transport by ocean currents (Munro and Colwell, 1996). We have also determined by rabbit illeal loop and human volunteer studies that the viable but nonculturable *V. cholerae* remain infectious (Colwell *et al*., 1990).

Figure 7. **The viable but non-culturable (dormant) cycle of *Vibrio cholerae* 01**

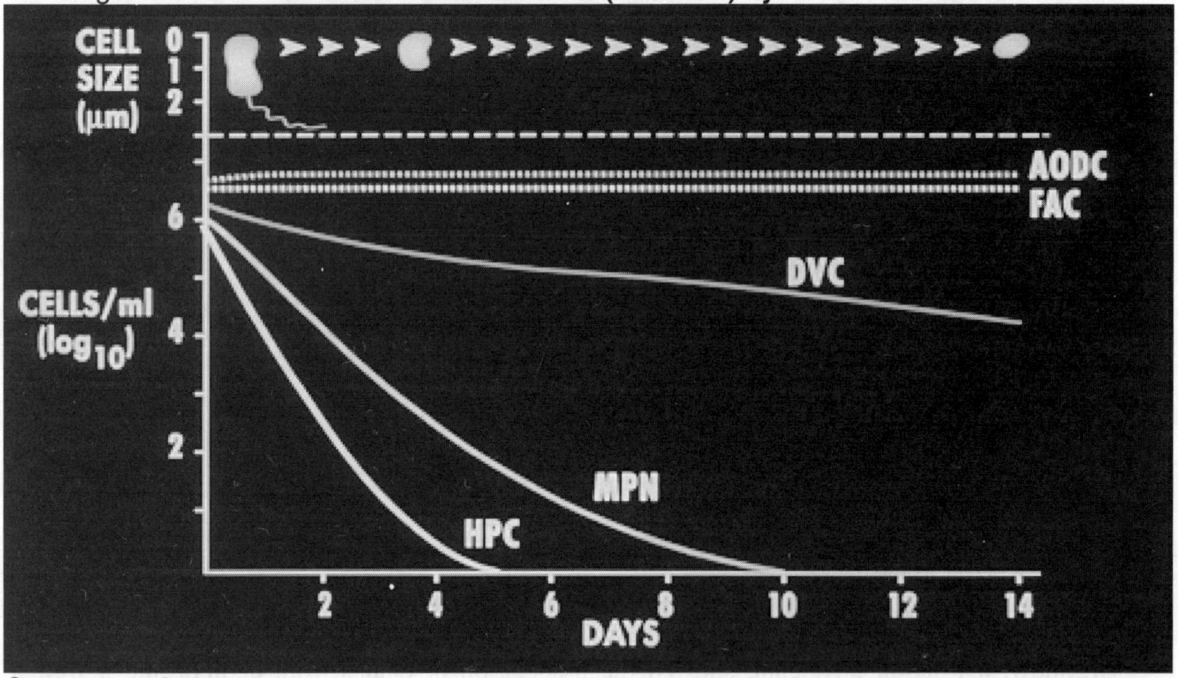

Source: Author.

Using the monoclonal antibody in a detection kit, the presence of the cholera bacterium in the environment can be detected (Huq *et al*., 1990). It was determined that crustaceans, especially zooplankton, i.e. copepods, are host organisms with which the vibrio is strongly associated (Huq *et al*., 1984). A three-year study in Bangladesh showed that the vibrios are associated with the egg and nauplii stages of the copepod. Interestingly, when cases of cholera occurred, only then was *V. cholerae* able to be isolated from plankton and water samples in relatively large numbers (Huq *et al*., 1990). Vibrios, including *V. cholerae*, are chitin digesting, and the copepod exoskeletal structure contains chitin, as is the case for crabs, shrimp, and other crustaceans.

In Bangladesh, the annual cycles of cholera are characteristic of the epidemics, epidemics in some years worse than in other years. A bimodal pattern of occurrence of cholera has been recorded in Bangladesh that appears to follow the distribution, i.e. "blooms" of plankton (Colwell *et al.*, 1981). Using the fluorescent antibody technique, we were able to detect a bimodal increase in *V. cholerae* even when the vibrios could not be cultured (Huq *et al.*, 1990). Thus, we can now hypothesize that the mode of transmission is via association with zooplankton, which are ingested when drinking unfiltered, untreated water. Interestingly, copepod distribution globally follows the "routes of entry" of cholera.

The association of cholera with El Niño, in Latin America, has proved very interesting. With El Niño, there are massive rainfalls and subsequent run-off into rivers and coastal waters, with increased turbidity occurring. Also, with El Niño, sea-surface temperature rises and changes in plankton populations occur, with selection of specific plankton species. In the Americas, the epidemic of cholera in Peru occurred in early 1991, and subsequent outbreaks occurred along the coastal areas (Figure 5). In Peru, the number of cases was catastrophic -- in the hundreds of thousands -- in February, March, and April of 1991. Major epidemics of cholera occurred in two successive years, 1991 and 1992, with cases continuing to occur in subsequent years, but at much lower numbers. Cases in Peru were more than ca. 200 000, Ecuador ca. 40 000, Columbia ca. 9 000, and Guatemala ca. 2 000. It is important to note that ca. 75 per cent of cholera cases are asymptomatic. Thus, with $\geq$ 200 000 cholera victims hospitalised, a potential epidemic of more than a million people infected may well have actually occurred in Peru in 1991.

The time course of cholera events in Latin America was fascinating -- first, in Peru, early 1991, then Ecuador, Columbia, Chile, Brazil, Mexico -- moving north, along with ocean current patterns. We hypothesize that with El Niño, warmer sea-surface temperatures are associated with cholera cases occurring along coastal regions. Plankton populations bloom under conditions of increased temperature and the correlation is compelling, based on remote sensing data (Colwell, 1996).

A rapid diagnostic kit, applying biotechnology methodology and employing a monoclonal antibody, allows diagnosis of the presence of *V. cholerae* 01 and/or 0139 within three to five minutes without culturing (Huq *et al.*, 1996; Hassan *et al.*, 1994*a*, 1994*b*).

In conclusion, we are now using remote sensing and retrospective data analysis employing satellite and clinical data archives to determine whether cholera outbreaks can be predicted and, eventually, global monitoring for cholera via satellite undertaken (Colwell, 1996). This example of cholera is useful in understanding the need for water treatment and modern methods for detection of human pathogens in drinking water supplies.

Questions, comments and answers

Q: Have you picked up any other pathogenicity determinants able to predict what kind of sub-strains might be involved?

A: We can't answer the sub-strain question yet, but other workers using 01 and toxin gene probes have been able to confirm the presence of 01 and toxin genes in environmental strains. When the numbers of organisms went up, as determined by gene probe, cases of disease began to occur and culturable *V. cholerae* were found. In addition, we have found 01 determinants in non-01 cultures, with a fluctuation between 01 and non-01.

Questions, comments and answers (cont'd.)

Q: How do you predict/model the entry of cholera into Mexico? Did the *Vibrio* that appeared in Louisiana in the United States play a role?

A: The organism can be found now when there is no epidemic -- even in the Amazon. Our hypothesis is that the organism is part of the natural flora of the environment -- estuarine, coastal and brackish water inhabitants. The warm currents such as El Niño trigger the explosive bloom of plankton species, and these allow the bacterial numbers to go up. We are attempting to establish if there is a specific plankton carrier. We have done studies which show that the ballast water of ships can carry *Vibrio*, but I don't believe that dumping ballast is the cause of the major epidemics that occurred nearly simultaneously in coastal cities thousands of kilometres apart. Also, we believe that the toxin may be an osmo-regulator which assists the host to withstand salinity change.

REFERENCES

COLWELL, R.R. (1996), "Global Climate and Infectious Disease: The Cholera Paradigm", *Sci.* 274, pp. 2 025-2 031.

COLWELL, R.R. and A. HUQ (1994), "Vibrios in the environment: viable but nonculturable *Vibrio cholerae*", in I.K. Wachsmuth, O. Olsvik, and P.A. Blake (eds.), *Vibrio cholerae and Cholera: Molecular to Global Perspectives*, pp. 117-133, Chapter 9, American Society for Microbiology, Washington, DC.

COLWELL, R., R. SEIDLER, J. KAPER, S. JOSEPH, S. GARGES, H. LOCKMAN, D. MANEVAL, H. BRADFORD, N. ROBERTS, E. REMMERS, I. HUQ, and A. HUQ (1981), "Occurrence of Vibrio cholerae O-Group 1 in Maryland and Louisiana estuaries", *Appl. Environ. Microbiol.* 41, pp. 555-558.

COLWELL, R.R., P.R. BRAYTON, D.J. GRIMES, D.R. ROSZAK, S.A. HUQ, and L.M. PALMER (1985), "Viable but nonculturable *V. cholerae* and related pathogens in the environment: implication for release of genetically-engineered microorganisms", *Bio/Technology* 3, pp. 817-820.

COLWELL, R.R., M.L. TAMPLIN, P.R. BRAYTON, A.L. GAUZENS, B.D. TALL, D. HARRINGTON, M.M. LEVINE, S. HALL, A. HUQ, and D.A. SACK (1990), "Environmental aspects of in *V. cholerae* in transmission of cholera", in R.B. Sack and Y. Zinnaka (eds.), *Advances in Research on Cholera and Related Diarrhoeas, 7th ed.*, pp. 327-343K, T.K. Scientific Publishers, Tokyo.

EWALD, P.W. (1994), "On Darwin, Snow, and Deadly Diseases", *Natural History* 103, pp. 42-45..

HASAN, J.A.K., M.A.R. CHOWDHURY, M. SHAHABUDDIN, A. HUQ, L. LOOMIS, and R.R. COLWELL (1994*a*), "Cholera toxin gene polymerase chain reaction for detection of non-culturable *Vibrio cholerae* 01", *World J. Microbiol. and Biotech* 10, pp. 568-571.

HASAN, J.A.K., D. BERNSTEIN, A. HUQ, L. LOOMIS, M.L. TAMPLIN, and R.R. COLWELL (1994*b*), "Cholera DFA: An Improved direct fluorescent monoclonal antibody staining kit for rapid detection and enumeration of *Vibrio cholerae* 01", *FEMS Microbiology Letters* 120, pp. 143-148.

HUQ, A., E. SMALL, P. WEST, and R.R. COLWELL (1984), "The role of planktonic copepods in the survival and multiplication of Vibrio cholerae in the environment", in R.R. Colwell (ed.), *Vibrios in the Environment*, pp. 521-534, John Wiley & Sons, New York.

HUQ, A., R.R. COLWELL, R. RAHMAN, A. ALI, M.A.R. CHOWDHURY, S. PARVEEN, D.A. SACK, and E. RUSSEK-COHEN (1990), "Detection of *V. cholerae* 01 in the aquatic environment by fluorescent monoclonal antibody and culture method", *Appl. Environ. Microbiol.* 56, pp. 2 370-2 373.

HUQ, A., J. HASAN, A. CHOWDHURY, L. LOOMIS and R.R. COLWELL (1996), "Direct Detection of *Vibrio cholerae* 01 Cells in Clinical and Environmental Samples Using Rapid Detection Kits", in N. Naraki, Y. Taya, and M. Mohri (eds.), *Proceedings of the 13th Meeting of the UJNR Panel on Diving Physiology, 23-25 October 1995, Miura, Kanagawa, Japan,* pp. 164-180, JAMSTEC (Japan Mariane Science & Technology Center), Yokosuka.

KOGURE, K., U. SIMIDU, and N. TAGA (1979), "A tentative direct microscopic method for counting living bacteria", *Can. J. Microbiol.* 25, pp. 415-420.

KOGURE, K., U. SIMIDU, N. TAGA, and R.R. COLWELL (1987), "Correlation of direct viable counts with heterotrophic activity for marine bacteria", *Appl. Environ. Microbiol.* 53, pp. 2 322-2 337.

MUNRO, P.M. and R.R. COLWELL (1996), "Fate of *Vibrio cholerae* 01 in seawater microcosms", *Water Research* 30, pp. 47-50.

TAMPLIN, M., M. AHMED, R. JALALI, and R. COLWELL (1989), "Variation in epitopes of the B subunit of El Tor and classical biotype *Vibrio cholerae* 01 cholera toxin", *J. Gen. Microbiol.* 135, pp. 1 195-1 200.

WORLD HEALTH ORGANIZATION (WHO) (1992), *Global Health Situations and Projections, Estimates 1992*, World Health Organization, Geneva.

NEW RISKS IDENTIFIED FOR VIRAL AND PROTOZOAN WATERBORNE DISEASE

by

Joan B. Rose
Department of Marine Science, University of South Florida, St. Petersburg, Florida

Introduction and background

Infectious diseases continue to affect populations throughout the world, but there has been a recent recognition of emerging threats to global environmental health beyond the ordinary. These threats are associated with the changing world, including several key areas:

i) There is a growing susceptible population including the elderly and immunocompromised (AIDS, transplant and cancer patients) (Gerba *et al.*, 1996).

ii) Water quantity and water quality issues are increasing; there is more reclaimed wastewater being used for irrigation and drinking purposes.

iii) Trade barriers have been dissolved, and food is coming from an international market.

iv) There are changes in technology for food processing and water treatment.

v) Antibiotic resistance in many bacterial pathogens has been increasing.

There appears to be a greater need for assessing and protecting environmental health programs which address water quality and food safety among others, and the greatest risk may be associated with the microbial contaminants, the enteric viruses and protozoa.

Health risks associated with viruses and groundwater

There are several hundred enteric viruses which are possibly important agents of waterborne disease. However, there is limited information regarding the incidence of virus infections in the United States populations, as well as throughout the world, and the role of contaminated water in acquiring these. Bennett *et al.* (1987) have reported 20 million cases of enteric viral infections and 2 010 deaths per year. Adenoviruses, which may be transmitted by the respiratory route as well, account for 10 million cases and 1 000 deaths per year, making this the most significant virus affecting US populations. Rotavirus cases were documented as the second most common virus infection and are particularly of concern for infants.

Diarrhea has been one of the risks associated with many of the enteric viruses such as Norwalk virus, but more serious chronic diseases have now been associated with viral infections, and these risks need to be better defined. Studies have now reported for example, that Coxsackie B virus is associated with myocarditis (Klingel *et al.*, 1992). This could be extremely significant, given that 41 per cent of all deaths in the elderly are associated with diseases of the heart. In recent studies, enteroviral RNA was detected in endomyocardial biopsies in 32 per cent of the patients with dilated cardiomyopathy and 33 per cent of patients with clinical myocarditis (Kiode *et al.*, 1992). In addition, there is emerging evidence that coxsackievirus B is also associated with insulin-dependent diabetes (IDD) and this infection may contribute to an increase of 0.0079 per cent of IDD (Wagenknecht *et al.*, 1991).

Contamination of groundwater with viruses is of great concern, due to the resistant nature of the viral structure and the colloidal size (20nm) which makes this group of micro-organisms easily transported through soil systems. Viruses also survive up to months in groundwaters and are more resistant to water disinfection than are the coliforms (Gerba and Rose, 1989; Yates and Yates, 1988). Studies in the United States have found viruses in 20 to 30 per cent of the groundwaters, and coliforms were not predictive of viral contamination (LeChevallier, 1996). Rotaviruses and Coxsackie B virus have been isolated from treated drinking water in Mexico and contamination was seen much more frequently during the rainy season at levels as high as 20 viral units/L (Deetz *et al.*, 1984; Rose *et al.*, 1985). Rotaviruses have also been detected on occasion in waters from well water, and enteroviruses from sewage impacted streams in Bolivia and Columbia (Toranzos and Gerba, 1988).

Newly identified risks associated with *Cryptosporidium* and *Giardia*

Groundwaters

The enteric protozoa have a world-wide distribution. *Entamoeba* and *Giardia* reportedly have similar global infection rates of about 10 per cent (Feachem *et al.*, 1983). *Entamoeba* infections are particularly prevalent in areas without sewerage treatment and with poor hygienic conditions, and infection rates may be as high as 72 per cent. Although chronic asymptomatic infections may occur, serious diseases such as liver abscesses are related to the duration of the *Entamoeba* infection. Of all the protozoa, *Entamoeba* has the greatest risk of mortality which ranges from 0.02 to 6 per cent, depending on the virulence of the isolate (Gitler and Mirelman, 1986).

Cryptosporidium was first diagnosed in humans in 1976. Since that time, it has been well recognised as a cause of diarrhea (Dubey *et al.*, 1990). Reported incidence of *Cryptosporidium* infections in the population ranged from 0.6 to 20 per cent, depending on the geographic locale, with greater prevalence in populations in Asia, Australia, Africa and South America. *Cryptosporidium* and *Giardia* are the most significant causes of waterborne disease in the United States today. Centers for Disease Control (CDC) have estimated that 60 per cent of all giardiasis cases are associated with contaminated water (Bennett *et al.*, 1987). The occurrence in surface waters has been reported in 4 to 100 per cent of the samples examined at levels between 0.1 to 10 000/100L, depending on impact from sewage works and animals (Lisle and Rose, 1995). Groundwater was thought to be a more protected source from the large oocysts (5um) and cysts (7umx15um); however, recent data have shown that between 9.5 and 22 per cent of the samples were positive (Table 1).

In addition, drinking water outbreaks associated with the enteric protozoa documented in the United States in 1993 and 1994 found that 40 per cent occurred in wells (Figure 1).

Cryptosoridium is of particular concern for three reasons. The oocyst is extremely resistant to disinfection and can not be killed with routine water disinfection procedures. The disease is not treatable and the risk of mortality ranges between 50 and 60 per cent in the immunocompromised populations (Rose, in press).

Table 1. Occurrence of *Cryptosporidium* and *Giardia* in groundwater in the United States

Protozoan	Detected (per cent of total)	Non-detected (per cent of total)	Cysts or oocysts/100L
Cryptosporidium	17 (23)	57 (77)	41
Giardia	7 (9.5)	67 (90.5)	16

Source: Hancock and Rose, 1996.

Figure 1. **Water source associated with protozoan waterborne outbreaks in the United States (1993-1994)**

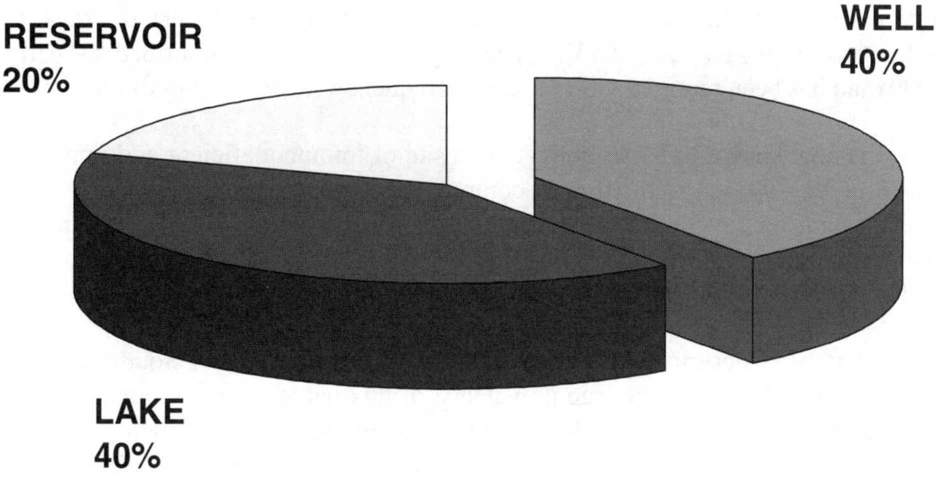

Source: MMWR, 1996*a*.

In South America

Over the last five years, surveys have been undertaken in various countries throughout South America to examine the possible contamination of water with *Cryptosporidium* oocysts (Table 2, Naranjo, 1994). Surface waters and treated waters have been found to be contaminated. It is difficult to show that there is a risk and that better filtration and source protection may be needed, unless this type of information is known. Of a total of 45 samples collected, 42 per cent were positive.

Table 2. **Average levels of *Cryptosporidium* in waters in South America**

Country	Sewage	Well water	Surface water	Drinking water
Argentina	ND	ND	5	0
Brazil	765	<1	ND	0
Chile	375	ND	ND	0
Costa Rica	ND	ND	ND	0
Mexico	ND	1	1	12
Panama	1 300	150	35	10

Notes: oocysts/100L
ND: Not Determined

Source: Naranjo, 1994.

The emerging protozoan *Cyclospora*

Cyclospora cayetanensis (previously termed cyanobacterium-like body) is a single cell coccidian protozoan that has been implicated as an etiologic agent of prolonged watery diarrhea, fatigue and anorexia in humans (Ortega *et al.*, 1993). The organism was first described as early as 1977 (Ashford, 1979) and has been reported with increased frequency since the mid-1980s.

Cyclospora is now known to be an obligate parasite of immunodeficient and immunocompetent humans (Ortega *et al.*, 1993). In an immunocompromised person, the parasite can cause profuse, watery diarrhea lasting for several months. The infection is much less severe in immunocompetent patients. Symptoms may range from no symptoms, to abdominal cramps, nausea, vomiting and fever lasting from three to 25 days (Goodgame, 1996).

Cyclospora has been described in patients from North, Central, and South America, Europe, Asia, and North Africa; however, the true prevalence of this parasite in any population is unknown (Soave and Johnson, 1995). In June 1994, several cases of diarrhea were detected among British soldiers and dependants stationed in a small military detachment in Pokhara, Nepal (Rabold *et al.*, 1994). The drinking water for the camp was a mixture of river and municipal water that was treated by chlorination. A candle filtration system was also used to remove particulates, bur was not guaranteed to filter *Cyclospora* sized particles (8-10*u*). *Cyclospora* was detected in 75 per cent of the diarrhea samples examined, and a water sample processed by membrane filtration taken from the camp also revealed the presence of *Cyclospora* oocysts.

In 1995, an outbreak occurred in eastern portion of Florida. The outbreak appeared to be linked to the consumption of strawberries. It was suspected that flooding that took place in Monterey County California that year could have been contaminated with sewage containing *Cyclospora*. The Monterey County agricultural commissioner estimated that up to 35 per cent of the strawberry crop was damaged that year by flooding (Artero, 1995; MMWR, 1996*b*).

In 1996, there were 17 outbreaks of *Cyclospora* in unrelated geographic areas in the United States. Cases occurred in Texas, Florida, New York, New Jersey, Massachusetts, Illinois, Pennsylvania, Ohio, South Carolina, Connecticut, Maryland, Vermont, New Hampshire, Virginia,

Washington, DC, Toronto, and Ontario, Canada. The suspected vehicle of transmission for these outbreaks was food, specifically Guatemalan raspberries. In retrospect, it is believed that the 1995 outbreak was also due to the consumption of contaminated raspberries and not strawberries.

Risks associated with produce and agriculture

The contamination of agricultural products that may be consumed without cooking may be an emerging risk which is currently not recognised. This may be associated with the use of untreated wastewater for the irrigation of crops in various countries which, with the international market, becomes a global environmental health risk. *Cryptosporidium* and other types of bacterial pathogens have been detected on fruits and vegetables (Figure 2, Monge and Chinchilla, 1996; Beuchat, 1996). Fresh vegetables from Costa Rica were found to be contaminated with oocysts in 1 to 8 per cent of the samples examined. There was no association with the coliform bacteria.

Figure 2. ***Cryptosporidium* contamination on fresh produce**

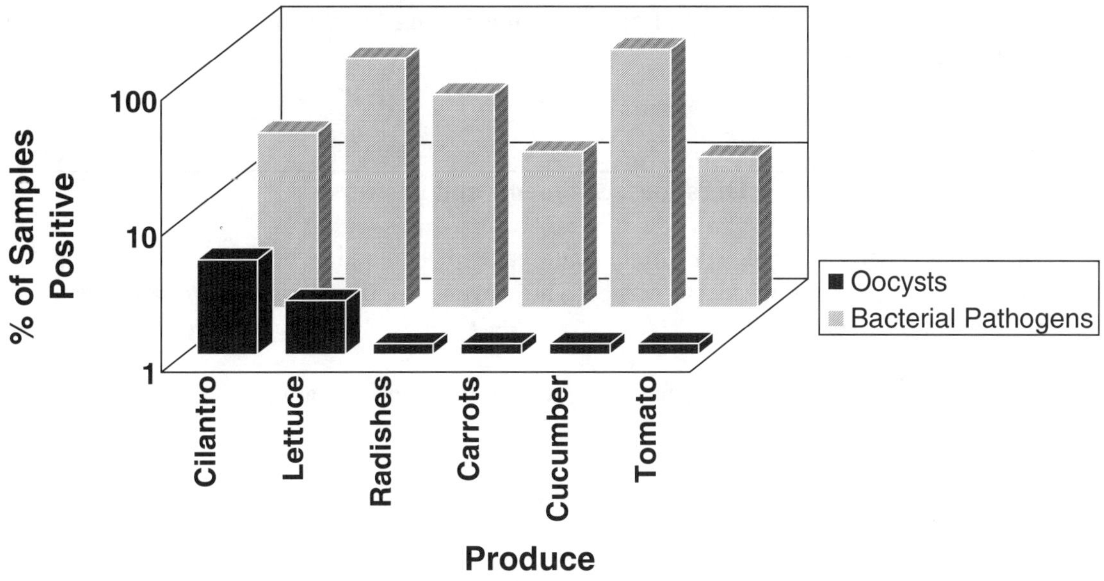

Staph, Listeria and Salmonella detected in 86% of the samples tested.

Source: Monge and Chinchilla, 1996; Beuchat, 1996.

Solutions and goals for better protection of environmental health

New methods are now available for determining what risks are associated with the contamination of water with enteric viruses and protozoa (Regli *et al.*, 1991; Rose *et al.*, 1996). This type of data will improve the decisions made in regard to wastewater management and reuse, protection and treatment of drinking water, and improved safety of the food supply. If these data are not available, then it will not be possible to know whether the risks are severe, or whether they can be easily controlled. Recommendations have been made by many groups most recently by a workgroup of experts organised by the American Academy of Microbiology (Ford and Colwell, 1996). It is clear that one of the key messages is that *reliance on the indicator system alone is no longer adequate for the protection of public health and determination of microbiological water quality.*

Molecular methods, immunological based methods, and better quantitative information provide the data needed to evaluate specific key risks associated with priority micro-organisms and allow for the development of scientific-based prioritisation based on public health benefits.

The types of data which are needed include:

1. Health surveillance in the community.

2. Environmental monitoring data and development of an occurrence database to determine exposure.

3. Evaluation of wastewater systems used for reclaimed water (coliforms and *Ascaris* are no longer appropriate recommendations).

4. Watershed and groundwater impact assessments and water treatment optimisation.

These data then can be used to improve i) the risk assessment methodology by providing better databases; ii) communication through educational programs for industry, community and government; and iii) the development of global approaches for interfacing policies at the local, state, national, and international levels.

Questions, comments and answers

C: There has been a *Cryptosporidium* outbreak in Holland where the water supply was contaminated with animal faecal waste. Thus, even countries with good quality water supply are susceptible.

A: There was some contamination from surface supply and they are also importing oocysts into Holland via the Rhine!

Q: You alluded to the use of so-called indicator organisms. Do you still see a place for these in risk assessment?

A: I do. They have been valuable and will be useful in the future. The key problem is the use of coliforms alone. There are many more indicators, such as *Clostridium*, coliphage, etc. They are valuable in treatment processes and should be used. They are much less valuable for monitoring ambient water and identifying the source of health risks.

REFERENCES

ARTERO, S. (1995), "Did a bug ruin your summer?", *Palm Beach Post Saturday*, 4 November.

ASHFORD, R.W. (1979) "Occurrence of an undescribed Coccidian in man in Papua New Guinea", *Ann. of Trop. Med. and Parasit.* 73, pp. 497-500.

BENNETT, J.V., S.D. HOMBERG, and M.F. ROGERS (1987), "Infectious and Parasitic Diseases", *American Journal of Preventive Medicine* 3, pp. 102-114.

BEUCHAT, L.R. (1996), "Pathogenic Microorganisms Associated with Fresh Produce", *J. Food Prot.* 59, pp. 204-216.

DEETZ, T.R., E.M. SMITH, S.M. GOYAL, C.P. GERBA, J.J. VOLLET III, L. TASI, H.L. DUPONT, and B.H. KESWICK (1984), "Occurrrence of Rota- and Enteroviruses in Drinking and Environmental Water in a Developing Nation", *Wat. Res.* 18, pp. 567-571.

DUBEY, J.P., C.A. SPEER, and R. FAYER (1990*), Cryptosporidiosis of Man and Animals,* CRC Press, Boca Raton.

FEACHEM, R.G., D.J. BRADLEY, and H. GARELICK (1983), "*Entamoeba histolytica* and Amebiasis", in *Sanitation and Disease: Health Aspects of Excreta and Wastewater Management,* pp. 337-347.

FORD, T.E. and R.R. COLWELL (1996), *A Global Decline in Microbiological Safety of Water: A Call for Action,* The American Academy of Microbiology, Washington, DC.

GERBA, C.P. and J.B. ROSE (1989), "Viruses in Source and Drinking Water", in *Advances in Drinking Water Microbiology Research,* G.A. McFeters (ed), Science Tech, Madison, Wisconsin.

GERBA, C.P, J.B. ROSE,and C.P. HAAS (1996), "Sensitive Populations: Who is at the Greatest Risk?", *Intern. J. Food Microb.* 30, pp. 113-123.

GITLER, C. and D. MIRELMAN (1986), "Factors Contributing to the Pathogenic Behavior of *Entamoeba histolytica*", *Annual Reviews in Microbiology* 40, pp. 237-261.

GOODGAME, R.W. (1996) "Understanding Intestinal Spore-Forming Protozoa: *Cryptosporidia, Microsporidia, Isospora, and Cyclospora*", *Ann. Intern. Med.* 124, pp. 429-441.

HANCOCK, C. and J.B. ROSE (1996), personal communication.

KIODE, H., Y. KITAURA, H. DEGUCHI, A. UKIMURA, K. KAWAMURA, and K. HIRAI (1992), "Genomic Detection of Enteroviruses in the Myocardium Studies on Animal Hearts with Coxsackievirus B3 Myocarditis and Endomyocardial Biopsies from Patients with Myocarditis and Dilated Cardiomyopathy", *Japanese Circulation Journal* 56, pp. 1 081-1 093.

KLINGEL, K, C. HOHENADL, A. CANU, M. ALBRECHT, M. SEEMANN, G. MALL and R. KANDOLF (1992), "Ongoing enterovirus-induced myocarditis is associated with persistent heart muscle infection: quantitative analysis of virus replication, tissue damage and inflammation", *Proceeding of the National Academy of Science* 89, pp. 314-318.

LECHEVALLIER, M.L. (1996) "What do Studies of Public Water System Groundwater Sources Tell Us?", Proceedings of the Groundwater Foundation's 12th Annual Fall Symposium Under the Microscope: Examining Microbes in Groundwater, 5-6 September, Boston, The Groundwater Foundation, Lincoln, Nebraska.

LISLE, J.T. and J.B. ROSE (1995), "*Cryptosporidium* Contamination of Water in the USA and UK: A Mini-Review", *J. Water SRT - Aua.* 44(3), pp. 103-117.

MORBIDITY MORTALITY WEEKLY REPORT (MMWR) (1996*a*), "Surveillance for Waterborne Disease Outbreaks - United States, 1993-1994", 45:SS-1, pp. 1-34.

MORBIDITY MORTALITY WEEKLY REPORT (MMWR) (1996*b*), "Outbreaks of *Cyclospora cayetanensis* Infection - United States, 1996", 45:25, pp. 549-551.

MONGE, R. and M. CHINCHILLA (1996), "Presence of *Cryptsporidium* Oocysts in Fresh Vegetables", *J. Food Prot.* 59, pp. 202-203.

NARANJO, J. (1994), "Monitoring for Protozoan Parasites in South America", Abstracts of 17th Biennial Conference and Exhibition of Intern. Assoc. Wat. Quality, 24-29 July, Budapest, Hungary, IAWQ, London, United Kingdom.

ORTEGA, Y.R., C.R. STERLING, R.H. GILMAN, C.A. VITALIANO, and F. DIAZ (1993), "*Cyclospora* Species - A new Protozoan Pathogen of Humans", *N. Engl. J. Med.* 328, pp. 1 308-1 312.

RABOLD, J.G., C.W. HOGE, D.R. SHLIM, C. KEFFORD, R. RAJAH, and P. ECHEVERRIA (1994), "*Cyclospora* Outbreak Associated With Chlorinated Drinking Water", *Lancet* 344, pp. 1 360-1 361.

REGLI, S., J.B. ROSE, C.N. HAAS, and C.P. GERBA (1991), "Modeling the Risk of *Giardia* and Viruses in Drinking Water", *J. Amer. Water Works Assoc.* 83, pp. 76-84.

ROSE, J.B., C.P. GERBA, S.N. SINGH, G.A. TORANZOS, and B. KESWICK (1985), "Isolating Viruses from Finished Water", *J. Amer. Water Works Assoc.* 78, pp. 56-61.

ROSE, J.B., J.L. LISLE, and C.N. HAAS (1996), "The Role of Pathogen Monitoring in Microbial Risk Assessment", in *Modeling Disease Transmission and its Prevention by Disinfection,* C. Hurst (ed.), Cambridge University Press, Cambridge, United Kingdom.

ROSE, J.B. (in press), "Environmental Ecology of *Cryptosporidium* and Public Health Implications", *Annual Reviews in Public Health.*

SOAVE, R. and W.D. JOHNSON (1995), "*Cyclospora:* conquest of an emerging pathogen" *Lancet* 345, pp. 667-668.

TORANZOS, G.A. and C.P. GERBA (1988), "Enteric Viruses and Coliphages in Latin America", *Toxicity Assessment* 3, pp. 491-510.

WAGENKNECHT, L.E., J.M. ROSEMAN, and W.H. HERMAN (1991), "Increased incidence of insulin-dependent diabetes mellitus following an epidemic of coxsackievirus B5", *American Journal of Epidemiology* 133, pp. 1 024-1 031.

YATES, M.V. and S.R. YATES (1988), "Modeling microbial fate in the subsurface environment", *Critical Reviews in Environmental Control* 17, pp. 307-343.

PATHOGENIC VIRUSES IN THE MARINE ENVIRONMENT

by

Steven Myint
Department of Microbiology & Immunology, University of Leicester, England

Standards of microbiological water quality are based on bacterial standards, particularly coliforms. There is increasing evidence, however, that these are not adequate markers for the presence of non-bacterial agents such as viruses and protozoa (Payment *et al.*, 1982).

The pursuit of viruses in the water environment is a relatively recent pastime: the first major paper not being published till 1947 by the eminent virologist, Dr. Joseph Melnick (Melnick, 1947). This early work did establish the importance of viruses and led, in 1974, to a meeting in Mexico City that established standards for enterovirus testing and the need for further research in the field. The next 20 years was beset with one major obstacle: the lack of good methods to detect viruses in waters.

Detection of viruses in water

The detection of viruses relies on three principle steps: concentration of the samples, removal of contaminants (including bacteria) and an actual detection step. Most commonly used methods rely on either filtration or flocculation/precipitation, and subsequent cell culture of virus present. Even in experimental studies, the yield of inoculated viruses might vary from 0-70 per cent (Sobsey and Jones, 1979). This results in a lack of both sensitivity and reproducibility of results. Consequently, attempts at quantification of viral loads are without meaning. It has been possible to identify factors, such as the pH of the water, that affect yield, but these factors appear to have complex effects on each other. Moreover, there are technical difficulties with cells, such as Buffalo-Green Monkey used for enterovirus culture, which limit reliability of the method.

Modern molecular methods would appear to offer the advantage of extreme sensitivity. Optimised PCR methods should theoretically be able to detect a single genome copy of a virus. In practice, however, the presence of inhibitors in natural waters, such as humic acids from the breakdown of plant material and the total DNA load, reduce this sensitivity. Indeed, these substances might totally inhibit the amplification induced by Taq polymerase. New concentration methods allied to removal of inhibitors have been utilised to overcome this difficulty. Ultrafiltration and vortex flow filtration offer more reproducible concentration of most viruses and the additional use of Chelex or Sephadex columns removes inhibitors (Cook and Myint, 1995). Although yield is not 100 per cent, it is greater than 70 per cent and is reproducible within a narrower spectrum than obtained by more standard filtration methods.

Does the detection of low numbers of genomes of virus mean anything? The presence of gene sequences does not equate to the presence of infectious virus. Indeed, theoretical consideration of the rate of loss of genome would suggest that the presence of small segments of genome would remain after considerable degradation of the original. An empirical calculation based on a genome of 7 000 bases and an amplification target of 100 base pairs would suggest that 70 breaks would have to be induced to abolish PCR-positivity. Experiments that have inoculated RNA or infectious virus into natural waters, however, would suggest that sufficient degradation does occur. PCR positivity is lost within two days (Tsai and Palmer, 1995).

Removal of viruses in sewage treatment processes

Most developed countries now utilise, at least, primary treatment of sewage before discharge into natural waters.

It has been shown by traditional cell-culture based methods that primary settling removes enteroviruses by an average 50 per cent (Rao *et al.*, 1977). Secondary treatment in trickling filters removes another 50 per cent of the remainder (Rao *et al.*, 1977). In our own laboratory, we have confirmed this low reduction of astroviruses using molecular methods (unpublished data). Thus, if we calculate that, conservatively, 1 per cent of the population is excreting a particular virus at any one time, and that the initial virus load in human faeces is of the order of 10^{10} viruses per gram, then we are left with 10^{7} viruses per litre of effluent. There is, obviously, considerable dilution when discharged into natural waters, but as most viruses survive as infectious agents for weeks in natural waters (Sobsey *et al.*, 1986), this would be expected to be counteracted by a cumulative build-up of viral load. Most of this ultimate load sediments, which protects it from natural inactivation of infectivity by sunlight, temperature, bacteria and physical pressures, and acts as a reservoir to seed overlying waters. Indeed, it is possible to show that there are high levels of virus in washings of sand found on beaches in the United Kingdom (unpublished data).

Presence of viruses in natural waters

The simple and rough calculation of expected viral loads in coastal discharges does not seem to be borne out by the use of traditional methods for detecting viruses in the environment. Despite high coliform counts, many water samples are deemed to be negative for enteroviruses on routine testing in the United Kingdom. Using a sensitive PCR-based assay coupled to vortex flow filtration and Chelex separation, we have been able to show that all water samples examined from 12 bathing resorts in the United Kingdom over a 12-month period in 1993-94 were positive for astroviruses. On the basis of limit-dilution PCR-positivity, which might be anticipated to underestimate virus load, the amount of virus varied from 10 to over 10 000 viruses per litre of water. The higher results were found, not surprisingly, closer to sewage outfalls, but in areas where there were bathers. Similar results could be found with rotaviruses (Myint *et al.*, in press). Other groups using PCR have also found significantly greater positivity rates for human pathogenic viruses in natural waters than when traditional methods were used (Puig *et al.*, 1994).

Is it worthwhile detecting viruses in water?

Most anecdotal reports and many of the outbreak reports of waterborne illness do not identify a pathogen (Herwaldt *et al.*, 1992). In general, bacterial and protozoal infections can be reliably determined using commonly adopted methodology. It could be inferred that viruses, or toxins, are responsible for most of the remainder. There are now many viruses found to be pathogenic for man. The most common disease manifestations associated with waterborne illness are gastrointestinal and respiratory. Of the viruses that cause these illnesses, it is more plausible that those that are associated with the latter illness readily find their way into bathing waters. These include rotaviruses, adenoviruses, astroviruses, Norwalk-like viruses and other caliciviruses. None of these can be detected using routine cell culture methods. Electron microscopy, which would detect them, is often not available or is implemented too late. When the best methods are used, all have been found in waters. Thus, the likelihood of under-reporting is great. The need for a method that detects these viruses, or a marker of them, is therefore important.

What do we need for the future?

Modern molecular methods have now given us sensitive tools to study the fate of viruses in the water cycle. We need to pursue this to develop better processes for removing viruses. In addition, we need to set up proper epidemiological studies, particularly prospective cohorts, to examine the role of viruses in causing illness.

We also need better means of detecting viruses in the environment. Molecular methods are expensive and may not be other than a research tool. It has been suggested that somatic coliphages may act as adequate markers of human viral pathogens. This needs to be better defined, as comparisons are usually made with enteroviruses which themselves may not be good markers of the hardier viruses which are found in faeces.

Perhaps the best approach is to introduce disinfection, so that the risks are minimised. Newer, cheaper disinfection methods are being developed, and it may be possible in the coming decade to introduce these in high risk areas such as bathing areas and waters where shellfish are harvested. Mexico 1974 was a landmark in water virology; Mexico 1996 might be the same.

REFERENCES

COOK, N. and S.H. MYINT (1995), "Modern methods for the detection of viruses in water and shellfish", *Rev. Med. Microbiol.* 6, pp. 207-216.

HERWALDT, B.L., G.F. CRAUN, S.L. STOKES, and D.D. JURANEK (1992), "Outbreak of waterborne disease in the United States: 1989-1990", *J. Am. Water Works Assoc.* 84, pp. 129-135.

MELNICK, J.L. (1947), "Poliomyelitis virus in urban sewage in epidemic and non-epidemic times", *Am. J. Hygiene* 45, pp. 240-253.

MYINT *et al.* (in press), "A survey of viruses in English bathing waters by RT-PCR", expected publication in 1997.

PAYMENT, P., M. LEMIEUX, and M. TRUDEL (1982), "Bacteriological and virological analyses of water from four fresh water beaches", *Water Res.* 16, pp. 939-943.

PUIG, M. *et al.* (1994), "Detection of adenoviruses and enteroviruses in polluted waters by nested PCR amplification", *Appl. Environ. Microbiol.* 60, pp. 2 963-2 970.

RAO, V. *et al.* (1977), "Virus removal in the activated sludge sewage treatment", *Prog. Water Technol.* 9, pp. 113-127.

SOBSEY, M. *et al.* (1986), "Survival and transport of hepatitis A virus in soils, groundwater and wastewater", *Wat. Sci. Technol.* 18, pp. 97-106.

SOBSEY, M.D. and B.L. JONES (1979), Concentration of poliovirus from tapwater using positively charged microporous filters", *Appl. Environ. Microbiol.* 37, pp. 588-595.

TSAI, Y-L. and C.J. PALMER (1995), "Analysis of viral RNA persistence in seawater by reverse-transcriptase-PCR", *Appl. Environ. Microbiol.* 61, pp. 363-366.

ON-LINE QUALITY MONITORING FOR RECLAIMED INDUSTRIAL WATER

by

I. Janssens and W. Verstraete
Center of Environmental Sanitation, University of Gent, Belgium

Introduction

Urban and industrial development highly depend on an abundant supply of water. Hydrologists designate water-stressed countries as those with annual supplies of 1 000-2 000 m^3 of unused water per person (Postel, 1992). In the coming decades, population growth will push many developing nations into conditions of water scarcity. By the turn of the millennium, two thirds of the countries of the world can be expected to fall in the category of low to very low water availability (Clarke, 1991).

A number of factors contribute to recent interest in research and implementation of water re-use projects. Many industrialised nations face growing problems associated with guaranteeing an adequate water supply. In addition, increasing costs of municipal and industrial wastewater disposal for water quality protection also catalyse interest in water re-use. Developing countries, particularly those in arid parts of the world, need reliable, low cost technology methods for acquiring new water supplies.

The contaminants in reclaimed wastewater that are of public health significance consist of biological and chemical agents. New biological agents of concern have been described for drinking waters such as aeromonads, for example (Kersters *et al.*, 1995). Similarly, there are chemical molecules which recently have been announced as potential environmental dangers, for example, oestrogenic pesticide residues (Arnold *et al.*, 1996). Where reclaimed wastewater is used for applications that have potential human exposure routes, the major acute health risks are associated with exposure to biological pathogens (Asano and Levine, 1996).

The purpose of this paper is to highlight the need for on-line quality monitoring of reclaimed wastewater. Besides quality criteria for reclaimed water and examples of water re-use applications, an overview of recently developed on-line sensor equipment is given.

Quality criteria for reclaimed wastewater

During the last quarter of the 20th century, the benefits of promoting wastewater re-use as a means of supplementing water resources have been recognised by most state legislatures in the United States, as well as by the European Union. In 1970, the California State Water Code stated that "it is the intention of the Legislature that the State undertake all possible steps to encourage development of water reclamation facilities so that reclaimed water be available to help meet the growing water requirements of the State" (California Water Code, 1989). In the same context, the European Communities Commission Directive (91/271/EEC) declared that "treated wastewater shall be re-used whenever appropriate (EEC, 1991). At present, the European directive nr 1836/93 concerning the Eco-Management and Audit-Scheme (EMAS) incites companies to implement environmental care systems in their industrial network. The latter also stresses the importance of considering and investigating new water sources, like re-use of process water (De Weerdt and Merckx, 1993).

The potential for water re-use depends mainly on the economic cost of natural water as compared to that of re-used water. Three applications offer opportunities of implementing wide re-use of water, i.e. crop irrigation, industrial cooling and processing and non-contact, non potable use (Verstraete, 1995).

In the use of reclaimed wastewater for irrigation, the degree of treatment required and the extent of monitoring necessary depend on the specific application. A higher degree of treatment is required for irrigation of crops that are consumed uncooked or for irrigation of locations that are likely to have frequent human contact. The main microbiological quality guidelines of the World Health Organization (1989) and the State of California's current Wastewater Reclamation Criteria (1978) are shown in Table 1. The California criteria rely on monitoring of the total coliform density for assessment of microbiological quality, whereas the WHO guidelines also require monitoring of intestinal nematodes.

Table 1. **Microbiological quality guidelines for irrigation by the World Health Organization (1989) and the State of California's current Wastewater Reclamation Criteria (1978)**

Category	Re-use conditions	Intestinal nematodes[1]	Fecal or total[2] coliforms
WHO	Irrigation of crops likely to be eaten uncooked, sport fields, ...	< 1 / L	< 1 000 / 100 mL
WHO	Landscape irrigation where there is public access, such as hotels	< 1 / L	< 200 / 100 mL
California	Spray and surface irrigation of food crops, high exposure landscape irrigation such as parks	No standard recommended	< 2.2 / 100 mL[2]

Notes:

1. Intestinal nematodes (*Ascaris* and *Trichuris* species and hookworms) are expressed as the arithmic mean number of eggs per litre during the irrigation period.
2. California Wastewater Reclamation Criteria is expressed as the median number of total coliforms per 100 ml.

Source: Asano and Levine, 1996.

Water (fresh or reclaimed) intended for human consumption or used for production, handling or preservation of food for human consumption must be, according to the European Directive 80/778/EEG (1980), of drinking water quality. This corresponds with the absence of total coli, fecal coli and fecal streptococci in 100 mL, absence of sulfite-reducing bacteria in 20 mL and a total count less than 10^2 per mL at 22°C or less than 10 per mL at 37°C. These are, considering the degree of contamination of the raw water, very stringent criteria. It should be reminded that when one tries to reduce the bacterial counts by dosing conventional disinfectants such as Cl_2, ClO_2, organic chloramines, etc., one produces a variety of side-products which are not compatible with the chemical drinking water standards. Hence, there is an urgent need for methodology capable of dramatically reducing bacterial counts, but not residing in the formation of undesired chemical residues. In this respect, developments in O_3, H_2O_2 and UV technology are necessary.

Despite the improvement and implementation of advanced water treatment technology, the question of safety of wastewater re-use is still difficult to define and delineation of acceptable health risks have been hotly debated. Asano *et al.* (1992) and Tanaka *et al.* (1993) studied the safety of wastewater re-use based on risk of human virus infection from exposure to reclaimed municipal wastewater.

The estimates of risk infection, expressed as annual risk, for different wastewater re-use situations, are shown in Table 2. The overall probability of infection due to ingestion of viruses is a combination of virus removal and inactivation by wastewater treatment, die-off in the environment and dose-response. Tanaka *et al.* (1993) used the US Environmental Protection Agency's Surface Water Treatment Rule (EPA SWTR) as a point of reference. A 10^{-4} infection risk criterion at least 95 per cent of the time was used to define acceptable risk. The risk of virus infection from exposure to reclaimed municipal wastewater was determined by applying risk assessment procedures to virus concentrations of 0.01 viral units (vu/L) and 1.11 vu/L, which are respectively estimates of the detection limit for enteric viruses and the maximum concentration found in tertiary effluents. Risk calculations based on these levels indicate that for golf course, food crop irrigation and groundwater recharge, the reliability of wastewater reclamation and re-use is such that more than 95 percent of the time, the criteria were met for all the effluents examined. However, for recreational impoundments, the reliability does not always appear satisfactory (Tanaka *et al.*, 1993).

Table 2. **Annual risk of contracting at least one infection from exposure to reclaimed wastewater at two different enteric virus concentrations**

Virus	Exposure scenarios			
	Landscape irrigation for golf courses	Spray irrigation for food crops	Unrestricted recreational impoundments	Groundwater recharge
Maximum enteric concentration of 1.11 vu / L[1] in chlorinated tertiary effluent				
Echovirus 12	1 E-03[2]	4 E-06	7 E-02[2]	6 E-08
Poliovirus 1	3 E-05	2 E-07	3 E-03[2]	5 E-09
Poliovirus 3	3 E-02[2]	1 E-04	8 E-01[2]	2 E-08
Maximum enteric concentration of 0.01 vu / L in chlorinated tertiary effluent[3]				
Echovirus 12	9 E-06	4 E-08	7 E-04	5 E-10
Poliovirus 1	3 E-07	1 E-09	2 E-05	5 E-11
Poliovirus 3	2 E-04	1 E-06	2 E-02[2]	2 E-10

Notes:

1. vu/L: viral units per litre
2. Case where risk is prominent
3. Limit of detection

Source: Asano *et al.*, 1992.

Fecal coli present in reclaimed water or cheese represent in principle the same human infection risk (Mara, 1995). When reusing reclaimed wastewater, vigilance is commonly required to be higher because most probably the latter *E. coli* are signalling the potential presence of a much larger diversity of unwanted propagules. Nevertheless, considering that in raw food, up to 100 000 fecal coli are allowed per 100 grams (ICMSF, 1974), and that in hard cheeses up to 10 000 000 fecal coli may be present in 100 grams (Publikatieblad, 1992), dramatising the infection risk is not necessary.

Existing methods for hygienic quality monitoring are archaic. The conventional plate count methods for bacteriological control are expensive (40 ECU/analysis), very labour-intensive and need 24 to 72 hours of incubation (Van den Broeke, 1996). Besides, there is a lack of cost-effective methods for detection and quantification of viruses (Asano and Levine, 1996).

The overall message is that on-line net-effect sensor techniques, capable of generating at feasible cost reliable information in terms of hygiene, need to be developed. Indeed, no health authority can take major risks with the bacteriological quality of reclaimed water, and yet at present, it is evident that if one reclaims water, one can hardly implement the classical expensive postfactum plate count measurements. Therefore, a possible approach can be the careful on-line monitoring of the wastewater treatment process. If the latter performs efficiently, there is a good reason to thrust the end-product, i.e. the reclaimed water.

Water recycling

The water re-used for agricultural and landscape irrigation includes agricultural, residential, commercial and municipal applications. Industrial re-use is a general category, encompassing water use for a diversity of industries including power plants, food processing, textile industry and other industries with high rates of water utilisation. In some cases, closed-loop recycle systems have been developed that treat water from a process stream and recycle the water back to the same process with some additional make-up water.

The degree of treatment required in wastewater reclamation facilities varies according to the specific re-use application and associated water quality requirements. The simplest treatments involve solid/liquid separation and disinfection, whereas more complex treatment systems involve combinations of physical, chemical and biological processes, employing multiple-barrier treatment approaches for contaminant removal. The most sophisticated treatment systems are associated with the production of potable water from reclaimed wastewater (Verstraete, 1995).

The potential use of reclaimed water as a source for potable water supplies has been investigated in several valuable studies. The first description in the literature of wastewater treated to produce drinking water concerned Windhoek in Namibia (Africa) (Isaacson *et al.*, 1984). Currently, both at Windhoek and Stander in South Africa, plants rated at 52 l/sec (4 500 m^3 per day) are producing effluents of reclaimed wastewater which are blended with existing freshwater resources to a maximum of 15 per cent of the total water supply. In 1980, the costs for reclaimed water were approximately 0.32 ECU/m^3, as compared with 0.21 ECU/m^3 for water produced from fresh water supplies. Cost was highly dependent on the chlorination demand associated with the ammonia concentration in the feed. The latter could be reduced by using biological N and P removal (Van Vuuren *et al.*, 1980). A 1984 study concludes that, within the limits of the epidemiological studies done, no adverse effects on health attributable to the consumption of reclaimed water could be demonstrated (Isaacson *et al.*, 1984).

In the United States during the 1980s, the City of Denver's Direct Potable Water Reuse Demonstration was initiated. An advanced treatment plant began operation in 1985 to investigate the feasibility of providing reliable drinking water quality using a multiple-barrier treatment concept based on advanced chemical and physical treatment operations and processes. Effluent from a secondary treatment plant was used as source of water. Advanced treatment included lime, clarification recarbonation, filtration and carbon adsorption followed by two parallel advanced treatment process trains: a) reverse osmosis, ozonation and chlorination, in parallel with b) ultrafiltration, ozonation and chlorination. The finished reclaimed water met existing drinking water quality regulations. Tests on animal health effects including chronic toxicity, carcinogenicity and reproductive toxicity, showed no adverse effect from either treatment stream (Lauer, 1993; Lauer and Rogers, 1994).

Sewage farming has a substantial record of the use of treated water. The wastewater -- raw or after treatment -- can be used to irrigate a variety of crops. Effluents of agro-industries can be recycled in this way. Effluents from the sugar-cane alcohol distilleries in Brazil are used to fertilize newly planted fields. Those from palm-oil factories in Malaysia are excellent liquid fertilizer for the standing trees after anaerobic digestion (Lim Kim Huan, 1988), and those from a factory that produces enzymes in Denmark are used to fertilize regular wheat crops after thermal sterilisation (personal communication). Some 500 000 hectares of cropland in some 15 countries, or 0.2 per cent of the world's irrigated area, are now being irrigated with municipal wastewater (Postel, 1992).

In Europe, reclaiming used water is of recent date. In certain food processing industries, the incoming raw materials are transported by means of water, and subsequently receive a thermal treatment that inactivates any possible microbial contamination. In the production of potato crisps (chips) for example, the potatoes are washed and steam-pealed and then sliced. Next, they are blanched to wash out the sugars and then fried. At present, a closed cycle treatment, in which the waste is treated as shown in Figure 1, is under development in a factory in Belgium. In the closed water cycle, the anaerobic-aerobic bio-treatment is completed by deep bed filtration and ozonation. Ozonation totally removes (below 0.1 µg/l) the anti-sprout herbicide chlorpropham which is applied to the potatoes in the storage chambers. This shows that it is possible to remove both macro- and micro-organics, and that the psychological barrier can be overcome if a highly trusted 'infallible' process is used.

The textile industry is also a major consumer of water. Recently, interest has turned towards water recycling in this sector as well. Dyestuffs constitute the most important hurdle. A schematic outline of a total recycling system tested and approved on pilot-scale in a jeans finishing plant in Belgium is given in Figure 2 (Liessens *et al.*, 1996; Verstraete, 1995). This application has shown the need to integrate waste treatment into downstream processing. Close co-operation between process engineers and environmental supervisors is needed.

On-line monitoring

With many countries facing severe water shortages, reusing water for irrigation and industrial purposes is gaining ground. Significant progress has been made in developing technical approaches to produce a qualitative and reliable water source from reclaimed wastewater. On the other hand, some key topics need further research: evaluation of microbiological, chemical and organic contaminants in reclaimed water, assessment of health risks associated with trace contaminants, and improved monitoring techniques to evaluate microbiological quality.

On-line monitoring of the overall quality of reclaimed wastewater is of main importance. There is an urgent need for on-line sensor equipment capable of generating reliable information at feasible costs.

Figure 1. **Water recycling in a potato crisp factory, Belgium**

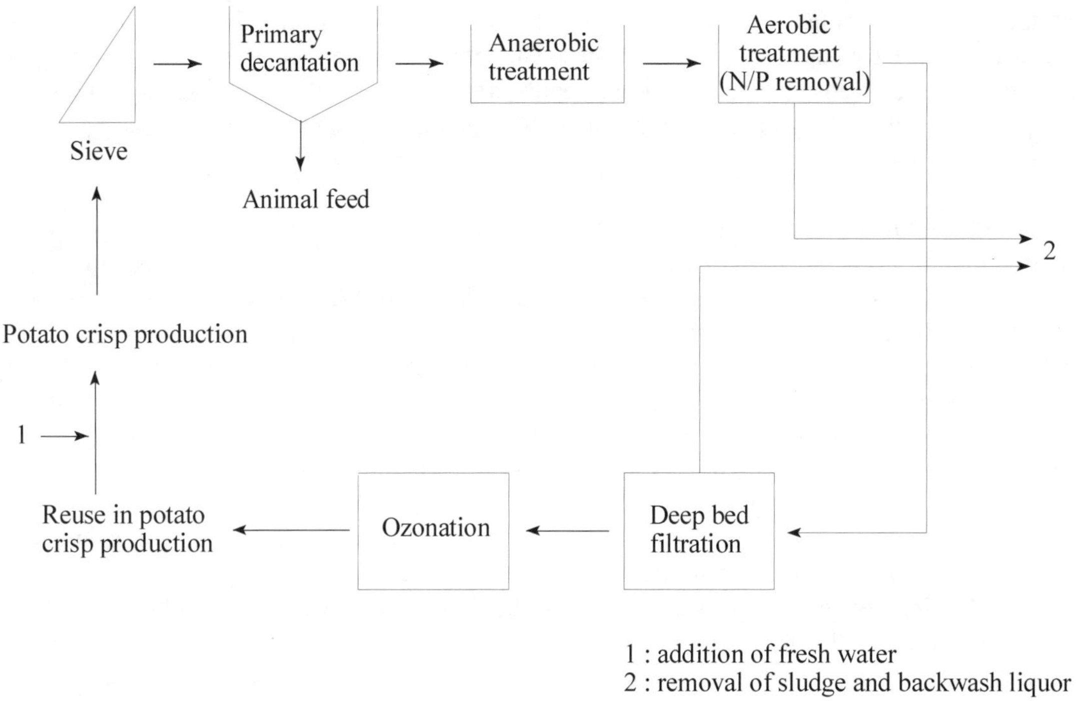

Source: Author.

Monitoring of performance of wastewater treatment

Sensors like dissolved oxygen (DO), pH and redox electrodes (ORP) have already proven their robustness and limited demand for maintenance. Gernaey *et al.* (1996) reviewed a number of biosensors capable of monitoring the performance of wastewater treatment systems based on simple measurements of pH and DO. Simple microbiological activity measurements can give information about the biodegradability of a substrate, toxicity of a wastestream or a chemical, etc.

The RODTOX is an apparatus which permits the monitoring of the biological available carbon in the water, as well as the nitrifiable nitrogen. Moreover, it can detect acute and chronic toxicity (Gernaey *et al.*, 1996; Kong *et al.*, 1996). It has now been some 10 years since it was established in practice, particularly in large-scale industries in Europe and Southeast Asia.

In the Biological Residual Ammonium Monitor (BRAM), the stoichiometric production of 2 H^+ per mole NH_4^+ is used to measure the residual NH_4^+-N concentration in mixed liquor samples (Massone *et al.*, 1995). For a measurement of the biosensor, an aliquot activated sludge from a nitrifying wastewater treatment plant is transferred to the reactor vessel of a titrator, and a batch experiment is performed. Base is added to maintain the pH within the "$pH_{setpoint} \pm \Delta$ pH interval". The amount of base needed to neutralise the protons produced in a mixed liquor due to the oxidation of NH_4^+, is recorded. It should be noted that NH_4^+-N has always been considered a powerful indicator of hygiene in drinking water technology. The fact that it now can be monitored on-line by a biosensor like the BRAM at a sensitivity of about 1 ppm is clearly of special interest, both with respect to the treatment of the wastewater, and to its subsequent re-use as process water.

Figure 2. **Water recycling for a jeans finishing plant, Belgium**

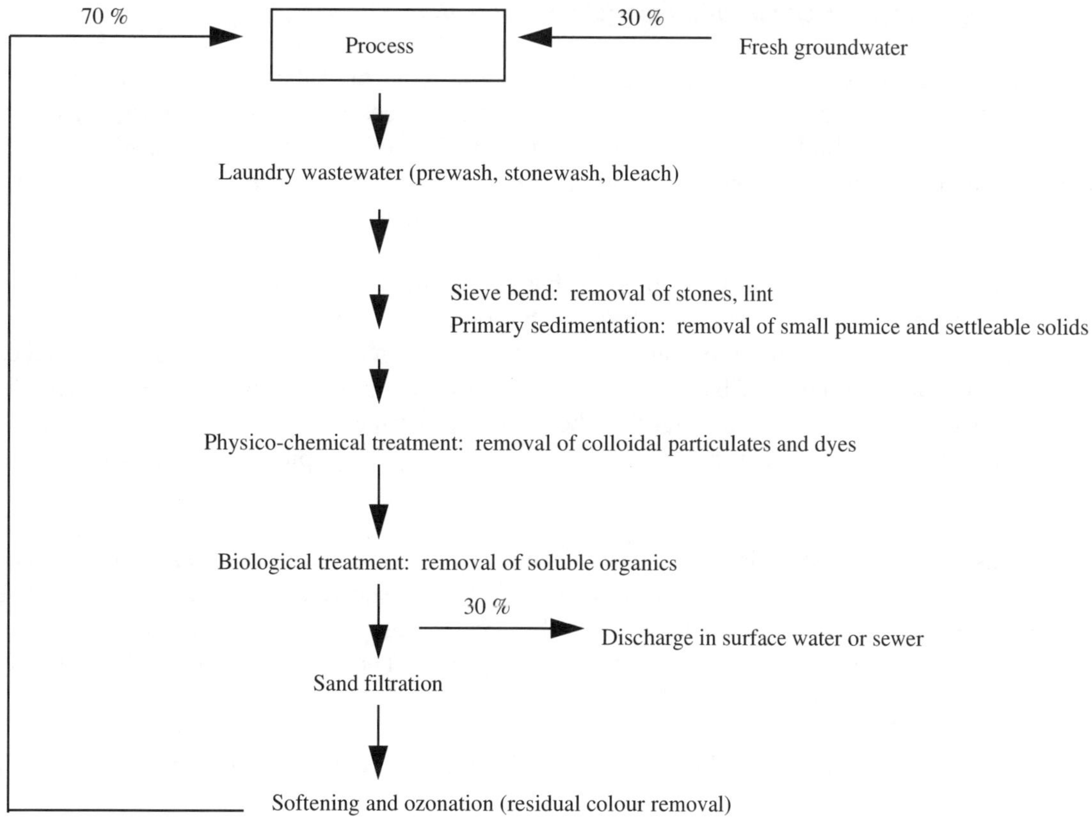

Source: Author.

The NITROX (NITRification tOXicity tester) detects toxicity towards nitrification on-line. It is based on the measurement of the activity of the nitrifiers as the difference between the total activity of oxygen uptake of a mixed culture, and the activity of the heterotrophs after inhibiting the nitrification by the addition of allylthiourea, a selective nitrification inhibitor (Stensel *et al.*, 1976, Kroiss *et al.*, 1992; Surmacz-Gorska *et al.*, 1995).

DECADOS (Denitrification Carbon source DOsage System; patent pending) (Vanderhasselt, 1995) is a biosensor for denitrification control in activated sludge plants. It is based on simple pH and ORP probes and, as for the BRAM, requires no filtration of the mixed liquor. It provides relevant information concerning the kinetics and stoichiometry of the denitrification process and, in some cases, concerning the concentration of nitrate. The concept is that of a 'titration' of nitrate with a readily biodegradable substrate (RBS) as a carbon source. The carbon source is added until all nitrate has been respired. This apparatus thus can guard against too high levels of nitrate in the reclaimed water. High NH_4^+ or NO_3^- levels signal improper functioning of the activated sludge communities, either by overloading or by intoxication.

The take home message is that NH_4^+, NO_3^-, DO and ORP are indirect but valuable parameters for wastewater treatment, as well as for hygienic monitoring of process water. As an example, positive ORP values can be considered to restrict the growth of *Clostridium botulinum*.

Monitoring of hygienic quality of industrial process water

Reusing reclaimed process water requires a rigorous control of the microbiological quality of the water. Implementing a hygienic quality control system necessitates the possibility of intervention when a critical microbiological level is reached. As already mentioned, the existing conventional plate counts need a long incubation period (24 to 72 hours). Hence, the development of alternative methods, yielding results in a faster way, is of main interest.

The ATP-bioluminescent-technique is based on the measurement of the energy-carrying molecule adenosine-tri-phosphate (ATP) in the biological cell. Since ATP is present in a relatively constant amount in the microbiological cell (1 fg ATP/CFU), measurement of this compound can give an indication of the amount of bacteria present in the water sample within one hour. The detection limit is 10^5 CFU/mL. The disadvantage of the ATP-measurement is the high analysis cost (30 ECU) and the interference with somatic ATP of non-microbial cells (e.g. pieces of food) (Van den Broeke, 1996).

Measurement of the hydrogen gas, produced by bacteria of the *Enterobacteriaceae*, is a rapid detection method for fecal coliforms. Wilkins *et al.* (1974) showed a linear correlation between the magnitude of the inoculum and the lag-phase before H_2 was produced. The latter varies from one hour for 10^6 CFU/mL, to seven hours for 10 CFU/mL. The disadvantage is the absence of an automated application (Van den Broeke, 1996).

An overview of rapid microbiological identification and detection techniques is given in Table 3 (Huis in 't Veld *et al.*, 1994).

Table 3. **Overview of rapid microbiological identification and detection techniques**

Measurement technique	Detection limit (CFU/mL)	Analysis time (h)
DEFT	10^4-10^5	≤ 1
ATP-bioluminescence	10^5	≤ 1
Impedance	10	≤ 24
Turbidity	10-100	≤ 24
Flow-cytometry	100	≤ 24
Infrared spectrophotometry	100	≤ 24

Note: DEFT: Direct Epiflorescence Filter Technique.

Source: Huis in 't Veld *et al.*, 1994.

Despite the significant progress in the development of a rapid microbiological detection system, an on-line quality monitor is not yet available.

Sampling techniques coupled to analysis methods which yield results within two hours need to be developed. A net-effect titrometric biosensor for hygienic quality monitoring of a process water was evaluated at the Laboratory for Microbial Ecology.

The titrometric biosensor is based on the measurement of the proton production due to bacterial growth in a process water sample. Growth and development of a microbial population result in pH changes of the environment. Both aerobic and anaerobic biodegradation give rise to acidification of the medium, as can be seen from equations (1) and (2):

Aerobic $C_6H_{12}O_6 + 6\,O_2 \Leftrightarrow 6\,HCO_3^- + 6\,H^+$ (1)

Anaerobic $C_6H_{12}O_6 \Leftrightarrow 3\,CH_3COO^- + 3\,H^+$; mixed volatile acids (2)

A pH controller keeps the process water at a constant pH. Base is added to compensate for the proton production due to substrate degradation. Figure 3 gives a scheme of the titrometric biosensor prototype constructed for hygienic quality monitoring.

Figure 3. **Scheme of the titrometric biosensor prototype for hygienic quality monitoring (A = 0.1 N HCl, B = 0.25 N NaOH)**

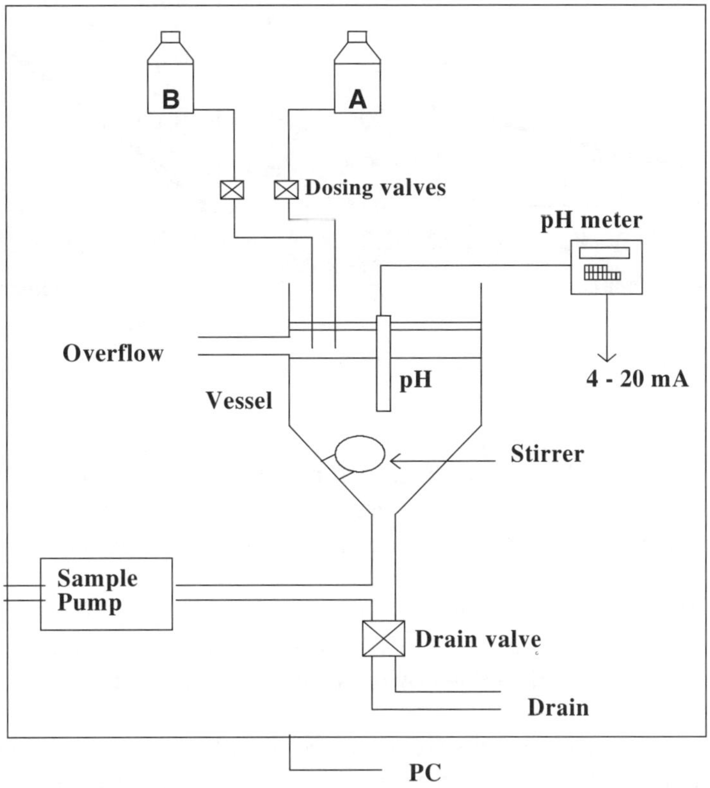

Source: Author.

The biosensor comprises a stirred measuring vessel (4 L) with a pH electrode. The two bottles (A and B) contain respectively acid (HCl, 0.1 N) and base (NaOH, 0.25 N). The acid and base solutions are added gravitairely to the measuring vessel by opening the appropriate valves. For each measurement of the biosensor, process water from the production unit is transferred to the reactor vessel of the titrator. During one hour, the rate of acidification of the process water is followed.

The performance of the titrometric biosensor as a hygienic quality monitor of process water has been tested on full scale in a factory of potato crisps production in Belgium, where at present a closed water cycle treatment is under development.

Figure 4 (A and B) shows the results of the measurements of two process waters with the titrometric biosensor.

Figure 4. **Measurements of two process waters with the titrometric biosensor**

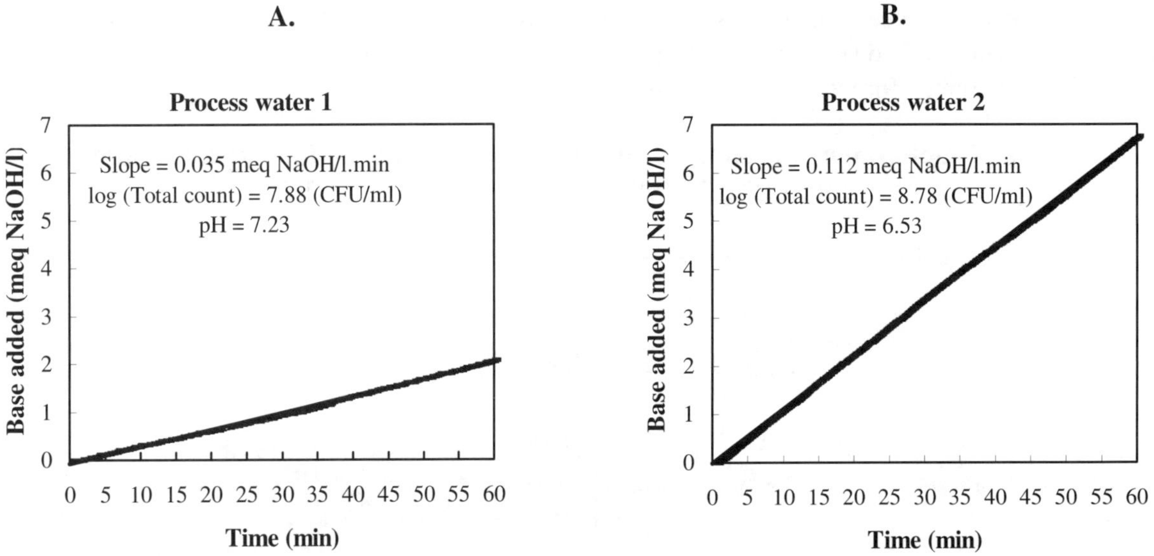

Source: Author.

During the measurement, there was no need for aeration, temperature control or addition of supplemental substrate. The carbohydrate substrate present in the process water satisfactorily allowed the growth and development of the microbial community present in the process water.

The influence of buffering components on the evolution of the pH during acidification of the process water is excluded by using a pH controller. This pH controller enables the monitoring of the acidification reaction at a constant pH, in other words, under the same buffering circumstances.

The titration data of the measurements are regressed linearly. The slope of the titration curves is proportional to the rate of acidification of the process waters. The slopes of process water 1 and 2 were respectively 0.035 and 0.112 meq NaOH/L.min. Obviously, process water 1 acidified slower than process water 2 (Figure 4). Taking into account the relation between the amount of substrate degraded and the amount of protons formed during biodegradation (equations 1 and 2), the difference in acidification rate could indicate that the bacterial population in process water 1 was smaller than in process water 2.

Figure 5 shows the relation between the slopes of the titration measurements of the titrometric biosensor and the logarithm of the respective total counts of three different process waters. The curve-fitted polynome has a correlation-determinant (R^2) of almost 70 per cent. The detection limit of the titrometric biosensor is 10^6 CFU/mL. Below this limit, no acidification is detected within one hour. Between the rate of acidification of the process water and the fecal indicator-organisms (fecal coliforms and fecal streptococci), no correlation has been observed ($R^2 < 10$ per cent). This indicates that the sensor reflects an overall level of microbial growth. Further research is required to make it reflect specific microbial populations such as the fecal bacteria.

Figure 5. **Relation between the slopes of the titration measurements of three different process waters with the titrometric biosensor and the logarithm of the respective total counts (CFU/mL)**

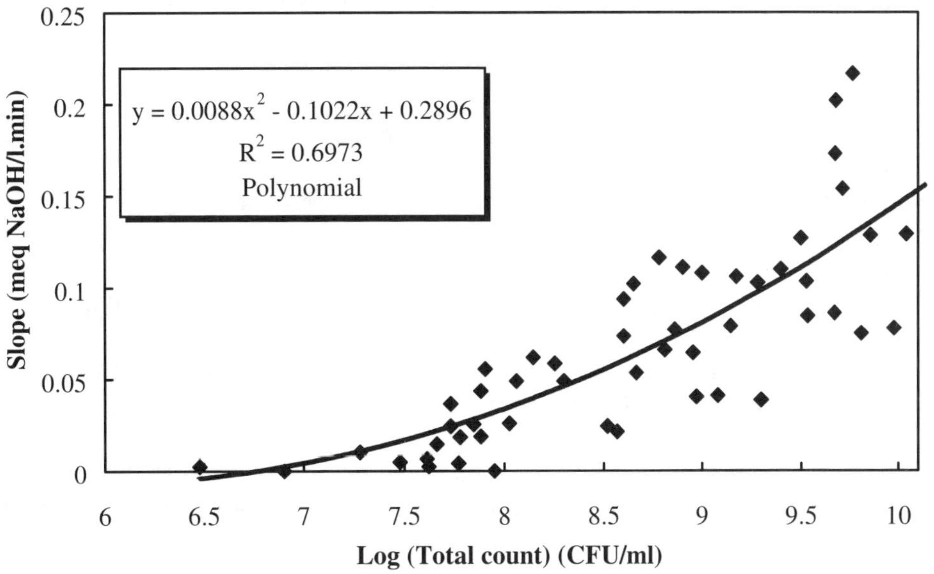

Source: Author.

Evaluation

With ever increasing demands for water and the reliability of its supply, it is time to recognise the significance of wastewater reclamation, recycling and re-use as vital and integral components in water resources management. Continued research and demonstration efforts have already resulted in additional progress in the development of water re-use applications. Actually, the focus is put on the assessment of health risks associated with biological pathogens.

When reclaimed wastewater is used, risk assessment is a must. Yet in water, quality standards are based on history and not on straightforward cause-effect tests. Furthermore, at present, water analysis and monitoring are costly and archaic.

For the near future, attention should be paid to disinfection methods leaving no residues in the reclaimed water, like UV-light, O_3 and H_2O_2. Measurements such as DO, NH_4^+, NO_3^-, ORP directly inform about the quality of the ongoing treatment process, and thus indirectly are most valuable with respect to overall hygiene.

There is an urgent need for new on-line net-effect hygienic monitoring devices such as, for example, the hygienic titrometric biosensor. It has been shown that simple microbiological activity measurements can give information about performance of wastewater treatment. Moreover, measurement of the proton production in a process water due to bacterial growth at constant pH, can be used to determine the magnitude of the bacterial density present in the process water, in other words, the hygienic quality of the process water. Compared to conventional plate countings, measurement of the hygienic quality of a process water with the titrometic biosensor is fast and inexpensive. The titrometric biosensor makes on-line measurement of the hygienic quality of process waters in the production unit possible.

Acknowledgement

Part of this work was supported by the IWT-project VLIM/O/9315.

Questions, comments and answers

Q: Do you have anything for trace metals, which are a real problem in industry?

A: I don't have a direct net effect sensor, but if a heavy metal affects the bioprocess, I see it immediately in a number of net effect sensors. There is an indirect internal quality assurance.

Q: Sometimes the problem is not the population, but specific micro-organisms which need not be present in large quantities to be dangerous.

A: We haven't solved this problem. We have worked with indicator organisms and probabilities. Net effect sensors don't work along these lines: their advantage is that they operate continuously and will trigger an immediate response.

REFERENCES

ARNOLD, S.F., D.M. KLOTZ, B.M. COLLINS, P.M. VONIER, L.J. GUILLETTE, and J.A. MC LACHLAN (1996), "Synergestic activation of estrogen receptor with combinations of environmental chemicals", *Science* 272, pp. 1 489-1 492.

ASANO, T., L.Y.C. LEONG, M.G. RIGBY, and R.H. SAKAJI (1992), "Evaluation of the California Wastewater Reclamation Criteria using enteric virus monitoring data", *Water Science and Technology* 26, pp. 1 523-1 524.

ASANO, T. and A.D. LEVINE (1996), "Wastewater reclamation, recycling and reuse: past, present and future", *Water Science Technology* 33, pp. 1-14.

CALIFORNIA WATER CODE (1989), "The Porter-Cologn Water Quality Control Act (1988 Amendments)", Chapter 7, Article 2, Section 13512, California Dept. of Water Resource, Sacramento, California.

CLARKE, R. (1991), *Water: the international crisis*, Earthscan Publications Ltd., London.

De WEERDT, H. and K. MERCKX (1993), "De EG-verordening inzake een 'Eco-management en audit-schema", *Leefmilieu* 5, pp. 175-181.

EUROPEAN COMMUNITIES COMMISSION DIRECTIVE (1991), "Council Directive regarding the treatment of urban wastewater (91/271/EEC)", *Official Journal of the European Communities*, No. L 135 of the 91.5.30, pp. 40-50.

GERNAEY, K., H. BOGAERT, P. VANROLLEGHEM, L. VAN VOOREN, and W. VERSTRAETE (1996), "Sensors for nitrogen removal monitoring in wastewater treatment", in *Proceedings EB '96: Environmental Biomonitoring: The biotechnology interface*, J.M. Lynch (ed.), 6-7 June 1996, Guilford, United Kingdom.

HUIS IN 'T VELD, J.H.J., F.K. STEKELENBURG, and J. DEBEVERE (1994), "HACCP en moderne microbiologische detectie en identificatietechnieken", *Voedingsmiddelentechnologie* 27, pp. 11-16.

INTERNATIONAL COMMISSION ON MICROBIOLOGICAL SPECIFICATION FOR FOODS (ICMSF) (1974), *Microorganisms in food. Sampling for microbiological analysis: principles and scientific applications*, University of Toronto Press, Toronto.

ISAACSON, M., A.R. SAYED, and W.H.J. HATTINGH (1984), "Studies on Health Aspects of Water Reclamation During 1974 to 1983 in Windhoek, South West Africa (Namibia)", *WRC Report* 38/1/87, Pretoria.

KERSTERS, I., L. VAN VOOREN, G. HUYS, P. JANSSEN, K. KERSTERS and W. VERSTRAETE (1995), "Influence of temperature and process technology on the occurrence of *Aeromonas* species and hygienic indicator organisms in drinking water production plants", *Microbial Ecology* 30, pp. 203-218.

KONG, Z., P.A. VANROLLEGHEM, P. WILLEMS, and W. VERSTRAETE (1996), "Simultaneous determination of inhibition kinetics of carbon oxidation and nitrification with a respirometer", *Water Research* 30, pp. 825-836.

KROISS, H., P. SCHWEIGHOFER, W. FREY, and N. MATSCHE (1992), "Nitrification inhibition - a source identification method for combined municipal and/or industrial wastewater treatment plants", *Water Science and Technology* 26(5/6), pp. 1 135-1 146.

LAUER, W.C. (1993), "Denver's Direct Potable Reuse Demonstration: Final Report", in T. Asano and A.D. Levine (1996), "Wastewater reclamation, recycling and reuse: past, present and future", *Water Science Technology* 33, pp. 1-14.

LAUER, W.C. and S.E. ROGERS (1994), "The demonstration of direct potable water reuse: Denver's Pioneer Project 1994", Water Reuse Symposium Proceedings, 27 February-2 March 1994, Dallas, Texas, pp. 779-798, American Water Works Association; in T. Asano and A.D. Levine (1996), "Wastewater reclamation, recycling and reuse: past, present and future", *Water Science Technology* 33, pp. 1-14.

LIESSENS J., M. BEECKMAN, I. JANSSENS and W. VERSTRAETE (1996), "Wastewater treatment: recent developments particularly in relation to reuse", in Proceedings Volume I of the International Conference on Environmental Pollution, 15-19 April 1996, Budapest, Hungary, pp. 24-35.

LIM KIM HUAN (1988), "Integrated systems for treating and utilizing plantation effluents: plam oil mill as a specific example", in E.R. Hall and P.N. Hobson (eds.), *Anaerobic Digestion*, pp. 303-314, Pergamon Press, Oxford.

MARA, D. (1995), "Faecal coliforms-everywhere (but not a cell to drink)", *Water Quality International* 3, pp. 29-30.

MASSONE, A., K. GERNAEY, A. ROZZI, P. WILLEMS, and W. VERSTRAETE (1995), "Ammonium concentration measurements", in Proceedings Part II of the Ninth Forum for Applied Biotechnology, 27-29 September 1995, Gent, Belgium, pp. 2 361-2 368.

POSTEL, S. (1992), *The last oasis: facing water scarcity*, W.W. Norton & Company, New York.

PUBLIKATIEBLAD van de Europese Gemeenschappen Nr. L168/32 (1992), Richtlijn EEG Nr. 46/92 van de raad van 16 juni 1992 inzake de hygiëneregels voor productie en verkoop van rauwe melk, warmte-behandelde melk en afgeleide producten.

STATE OF CALIFORNIA (1978), "Wastewater Reclamation Criteria, An Excerpt from the California Code of Regulations", Title 22, Division 4, Environmental Health, Dept. of Health Services, Sacramento, California.

STENSEL, H.D., C.S. MCDOWELL and E.D. RITTER (1976), "An automated biological nitrification toxicity test", *Journal of Water Pollution Control Federation* 48, pp. 2 343-2 350.

SURMACZ-GORSKA, J., K. GERNAEY, C. DEMUYNCK, P. VANROLLEGHEM, and W. VERSTRAETE (1995), "Nitrification process control in activated sludge using oxygen uptake rate measurements", *Environmental Technology* 16, pp. 569-577.

TANAKA, H., T. ASANO, E.D. SCHROEDER, and G. TCHOBANOGLOUS (1993), "Estimating the reliability of wastewater reclamation and reuse using enteric virus monitoring data", presented at the 66th Annual Conference & Exposion, 3-7 October 1993, Water Environment Federation; in T. Asano and A.D. Levine (1996), "Wastewater reclamation, recycling and reuse: past, present and future", *Water Science Technology* 33, pp. 1-14.

VAN DEN BROEKE, E. (1996), "Hygienische kwaliteitsbewaking in een gesloten watercircuit van een levensmiddelenbedrijf", Engineers thesis, Fac. Applied Agricultural and Biological Sciences, University of Gent, Belgium.

VANDERHASSELT, A. (1995), "Ontwikkeling van een on-line denitrifikatie sensor", Engineers thesis, Fac. Applied Agricultural and Biological Sciences, University of Gent, Belgium.

VAN VUUREN, L.R.J., A.J. CLAYTON, and D.C. VAN DER POST (1980), " Current status of water reclamation at Windhoek", *Journal of Water Pollution Control Federation* 52, pp. 661-672.

VERSTRAETE, W. (1995), "Role of biotechnology in water-cycle management", in *Bioremediation: The Tokyo '94 Workshop*, pp. 455-467, OECD, Paris.

WILKINS, J.R., G.E. STONER, and E.H. BOYKIN (1974), "Microbiological detection method based on sensing molecular hydrogen", *Applied Microbiology* 27, pp. 949-952.

WORLD HEALTH ORGANIZATION (1989), "Health Guidelines for the Use of Wastewater in Agriculture and Aquaculture", Report of a WHO Scientific Group, Geneva, Switzerland.

ALTERNATIVES TO MEMBRANE FILTRATION FOR THE ROUTINE MONITORING OF POTABLE WATER

by

C.R. Fricker
Thames Water Utilities, Reading, United Kingdom

Established methods for the detection and enumeration of coliforms and *Escherichia coli* from potable water provide for the presumptive isolation of these organisms when present with subsequent confirmation tests to exclude false positives. Membrane filtration has been the method of choice in most laboratories processing large numbers of samples. It is usual to filter two 100 ml volumes and incubate the membranes at 37 ºC and 44 ºC in order to give an early indication of the sanitary significance of any isolates encountered.

The detection of coliforms and *E. coli* is based on the ability of these organisms to ferment lactose. Fermentation is detected by the incorporation of a pH indicator in the medium (membrane lauryl sulphate broth), which turns yellow when acid is produced during fermentation. All colonies which appear yellow or colourless when growing on this medium are termed "presumptive" coliforms or *E. coli,* depending on the temperature at which they were recovered. Subsequent tests including production of acid from lactose at 37 ºC and 44 ºC, indole from tryptophan at 44 ºC, Gram reaction, and production of cytochrome oxidase are used to determine whether the organisms are coliforms or *E. coli*. The production of gas from lactose has long been thought not to be a satisfactory diagnostic feature for coliform bacteria, and this criterion was not included in the recently published sixth edition of Report 71 (Anon, 1994) in its definitions of either *E. coli* or coliforms. The production of indole from tryptophan at 44 ºC is not exclusive to *E. coli,* and some strains of *Klebsiella*, notably *K.oxytoca,* are known to produce a positive reaction in this test, whilst some strains of *E. coli* are indole-negative. Furthermore, not all strains of *E. coli* have the ability to grow at 44 ºC, most notably *E. coli* O157:H7, which can cause haemorrhagic colitis in humans. With respect to coliforms, the currently available tests are even more confusing. Whilst a significant proportion of coliforms are anaerogenic (fail to produce gas when fermenting lactose), the ability to produce gas is still used by some regulatory authorities in the definition of a "coliform". Approximately 10 per cent of coliforms isolated from potable source water within the Thames Water catchment area did not ferment lactose, probably due to the lack of the enzyme lactose permease (Fricker *et al.,* 1994). The inability to ferment lactose means that, under the definition outlined in the fifth edition of Report 71 (Anon, 1983), these organisms would not be classified as coliforms. It is, however, unreasonable to exclude organisms on the basis of a single physiological test which requires a suite of enzymes for full expression. There is little logic in assuming that a coliform such as a strain of *Enterobacter cloacae* which possesses the enzyme lactose permease is of any more health significance than a strain which does not possess the enzyme.

There has been a pressing need to reduce the time from sampling to result for many years. Recent work has concentrated on the provision of an early confirmed result through the incorporation of specific diagnostic enzyme tests into media. There have been calls to define the target organisms of *E. coli* and the coliforms on the basis of possession of a single characterising enzyme: ß-glucuronidase for *E. coli* and ß-galactosidase for coliforms (e.g. Fricker and Fricker, 1994). This need for improved definitions for coliforms was addressed in the "Guidance on Safeguarding the Quality of Potable Water Supplies" (Anon, 1989), which accompanied the Water Supply (Water Quality) Regulations 1989, where a key diagnostic feature was "normally possessing b-galactosidase". Whilst this approach does not entirely solve the problems with defining coliform bacteria as a group, it does allow for the development of simplified tests required to recover and identify the desired organisms. For *E. coli*, these tests have employed fluorogenic or chromogenic substrates such as methylumbelliferyl glucuronide (MUG) or 5-bromo-4-chloro-3-indolyl-ß-D-glucuronide (BCIG) for ß-glucuronidase, an enzyme which has been shown to be highly specific to this organism (Frampton and Restaino, 1993).

Sartory and Howard (1992*a*, 1992*b*) reported the use of an agar medium based on membrane lauryl sulphate broth (MLSB) incorporating BCIG. This medium, called membrane lactose glucuronide agar (MLGA) provides for the simultaneous enumeration of presumptive coliforms and *E. coli* using a single membrane filter. The medium was evaluated in several laboratories performing bacteriological tests on potable waters for Severn Trent Water and found to perform similarly to MLSB. Presumptive non-*E. coli* coliforms appear as yellow colonies due to the fermentation of lactose whilst *E. coli* are green through the combination of the yellow with the deposited blue insoluble 5,5'-bromo-4,4'-chloro-bisindigo. *E. coli* may occasionally produce blue colonies, this being consistent with the observation that some strains are lac negative. The fact that some *Shigella* and *Salmonella* have been demonstrated to be ß-glucuronidase-positive is not considered to be a problem as these organisms are of health significance in their own right. Sartory and Howard (1992*a*) identified some *Aeromonas hydrophila/caviae* from surface waters as the only other glucuronidase-positive bacteria present during their study. Of all the *E. coli* isolated, only 3.8 per cent were glucuronidase-negative. Confidence in the presumptive identification of *E. coli* was significantly improved with greater than 98 per cent of presumptive isolates confirming as *E. coli,* compared with approximately 70 per cent for MLSB incubated at 44 °C.

Used as described by Sartory and Howard (1992*a*, 1992*b*) the medium was noted to be less reliable for relatively polluted waters, where it was suspected that expression of glucuronidase may have been inhibited, due to overcrowding and the presence of increased concentrations of lactose catabolites.

Defined substrate technology (Edberg *et al.*, 1991) has been utilised for the detection of *E. coli* and coliforms in water in the United States and has become a widely adopted method, primarily as the commercially available product, Colilert. Initial studies with this product in the United Kingdom (Cowburn *et al.*, 1994) demonstrated that it gave similar recoveries to the UK standard method for detection of coliforms and *E. coli* in many types of water samples. After these initial studies were performed within the United Kingdom, a new product, Colilert 18, which gives results within 18 hours, has become available. Following further development, a simple MPN version (QuantiTray), which facilitates enumeration, has also become available.

Extensive studies comparing the use of QuantiTray, together with Colilert 18 and Colilert 24, have been undertaken (Fricker *et al.*, in press) and have demonstrated that within a particular geographical area, both formulations of Colilert, when used together with QuantiTray, give similar results to the use of duplicate membrane filters, although coliform detection was improved with

Colilert, probably due to the reasons given above. Examples of the data produced during trials of the product in the United Kingdom are shown in Tables 1-8.

Table 1. **Recovery of *E. coli* from water samples "spiked" to contain approximately one colony forming unit per 100 ml**

Colilert 24	Membrane filtration +	Membrane filtration -
+	106	35
-	28	181

Source: Fricker *et al.*, in press.

Table 2. **Recovery of coliforms and *E. coli* from treated drinking water using Colilert and membrane filtration**

	Colilert 24	Membrane filtration
Presumptive coliforms	296	363
Confirmed coliforms	296	257
Presumptive *E. coli*	70	102
Confirmed *E. coli*	70	71

Source: Fricker *et al.*, in press.

Table 3. **Contingency table showing the numbers of drinking water samples which gave identical and discrepant coliform results using membrane filtration and Colilert 24**

Membrane filtration	Colilert 24 +	Colilert 24 -
+	201	56
-	95	7 037

Source: Fricker *et al.*, in press.

Table 4. **Contingency table showing the numbers of drinking water samples which gave identical and discrepant *E. coli* results using membrane filtration and Colilert 24**

Membrane filtration	Colilert 24 +	Colilert 24 -
+	54	17
-	16	7 302

Source: Fricker *et al.*, in press.

Table 5. **Comparison of the number of samples found to contain coliforms using Colilert 24 and Colilert 18 from 1057 post filtration, pre-chlorination samples**

	Colilert 24	
Colilert 18	+	-
+	283	46
-	37	691

Source: Fricker *et al.*, in press.

Table 6. **Comparison of the number of samples found to contain *E. coli* using Colilert 24 and Colilert 18 from 1057 post filtration, pre-chlorination samples**

	Colilert 24	
Colilert 18	+	-
+	74	15
-	19	949

Source: Fricker *et al.*, in press.

Table 7. **Comparison of the number of partially disinfected sewage effluent samples found to contain coliforms using Colilert 18/QuantiTray™ and membrane filtration**

	Colilert 18/QuantiTray™	
Membrane filtration	+	-
+	598	56
-	49	1 760

Source: Fricker *et al.*, in press.

Table 8. **Comparison of the number of partially disinfected sewage effluent samples found to contain *E. coli* using Colilert 18/QuantiTray™ and membrane filtration**

	Colilert 18/QuantiTray™	
Membrane Filtration	+	-
+	286	34
-	37	2 106

Source: Fricker *et al.*, in press.

Other products have been developed which are based on similar technology. For example, LMX broth (Merck) is available for the detection of coliforms and *E. coli* using chromogenic substrates for the enzymes β-D-galactosidase and β-D-glucuronidase. However, in comparative studies with Colilert, the LMX medium did not perform as well. Whilst both methods gave similar recoveries of both coliforms and *E. coli*, many false positive reactions were seen for coliforms in particular (Fricker and Fricker, in press *a*). LMX gave positive reactions for the presence of β-D-galactosidase with 210 of 986 samples, but coliforms could be detected in only 177. By comparison, Colilert detected enzyme activity in 174 samples, and all were found to contain coliforms. For *E. coli*, LMX showed β-D-glucuronidase activity in 52 samples, but only 47 contained *E. coli*, whilst with Colilert all

49 samples with demonstrable β-D-glucuronidase activity were found to contain the target organism. In addition to the use of defined substrate technology for coliforms and *E. coli*, a product for the detection of enterococci (Enterolert, IDEXX) is available, which is based on detection of the enzyme β-D-glucosidase. Enterolert® detected enterococci in more samples than did membrane filtration, and had fewer false positive results. None of these differences were significant. Comparison of the counts of organisms obtained with QuantiTray™ with those obtained with membrane filtration showed that the correlation coefficient was 0.91 (Fricker and Fricker, in press *b*). Data from a comparative study of Enterolert and membrane filtration are shown in Table 9.

Table 9. **Strains of enterococci isolated using Enterolert® and membrane filtration**

Species	Enterolert Number of isolations	Membrane filtration Number of isolations
E. faecium	230 (54.2)	215 (55.8)
E. faecalis	113 (29.3)	102 (26.5)
E. casseliflavus	37 (8.7)	25 (6.5)
E. durans	19 (4.5)	16 (4.2)
E. gallinarum	6 (1.4)	3 (0.8)
Non enterococci	19 (4.5)	24 (6.2)

Note: Figures in parentheses are the percentage values.

Source: Fricker and Fricker, in press *b*.

Other methods based on culture which have been useful for detecting coliforms and *E. coli* include both direct and indirect impedance (Colquhoun *et al*., 1995; Timms *et al*., 1996). While these methods have been shown to be useful in certain situations, they are only semi-quantitative, and thus of limited value for routine monitoring.

At present, the use of defined substrate media appears to be the most suitable test for the detection of coliforms and *E. coli*, but while this test is simple to perform and cost-effective, results are only available after a period of 18 hours. While this may be adequate in many situations, there is often a need for a much more rapid result, and there have been many attempts to develop methods which give results within a few hours. The polymerase chain reaction has been applied to many areas of water microbiology with varying degrees of success. The ability of PCR to detect *E. coli* and coliforms in water has been investigated (Fricker and Fricker, 1994) and while suitable primers have been designed for *E. coli*, design of primers which were able to detect all coliforms was not possible. Furthermore, PCR has been shown to detect DNA from non-viable cells, either in ancient faecal material or in heat or chlorine treated material (Spigelman *et al*., 1995; Fricker and Fricker, in press *c*). Thus, it is not useful to directly examine water samples for bacteria using PCR, although the technique is useful after a period of culture which ensures that the DNA detected originates from viable cells.

An alternative to PCR for direct detection of bacterial cells is the use of 16S rRNA probes. These can be labelled with a variety of labels, including fluorescent tags such as FITC. We have designed 16S rRNA probes for *E. coli* and two probes for *Legionella*, one which is genus specific and the other specific for *L. pneumophila*. However, application of these probes for the direct detection of bacteria in water samples using currently available techniques relies on epifluorescence microscopy, and thus is time consuming and laborious. We are therefore developing an automated system (Chem*Scan*)for detection of labelled cells on membranes. The Chem*Scan* instrument has been designed for the detection and enumeration of fluorescently labelled micro-organisms which have been captured by membrane filtration. The filter is laser scanned, allowing labelled cells to be detected via a series of photo-multiplier tubes (detection channels). Finally the signals generated undergo a sequence of computer analyses which distinguish between labelled micro-organisms and autofluorescent debris. This permits an accurate enumeration to be made, down to one organism on a membrane. The Chem*Scan* records the location of each event on the membrane allowing a visual validation to be made using a epifluorescence microscope fitted with a motorised stage.

The laser spot is directed across a membrane by an array of oscillating mirrors, the arrangement being such that each scan line partially overlaps with the next. This results in all areas of the membrane being scanned at least twice, ensuring the detection of all labelled micro-organisms. The laser scanning of a membrane is complete in approximately 3-4 minutes.

During scanning, signals from fluorescent events are detected by a series of detection channels, each recognising fluorescence from different wavelength ranges. These signals are then analysed to distinguish between labelled cells and autofluorescent particles. Numerous discrimination parameters are used, the most significant being fluorescence ratio and specific intensity. Fluorescence ratio utilises the characteristic emission spectrum of the labelling fluorochrome(s). Labelled cells display a defined ratio between specific wavelengths whereas autofluorescent debris generally have a different spectrum and thus ratio. Also, labelled micro-organisms tend to be smaller than autofluorescent particles, even though they may have a similar brightness of fluorescence. This results in many particles having a lower specific intensity (light per unit area) than cells. Labelled cells have been well characterised, which allow further discrimination on the basis of size and signal amplitudes. The use of these discrimination parameters removes autofluorescent particles from the results, which are then displayed in the form of detailed scan maps. Because the location of each event is recorded, the membrane may be transferred from the instrument to an epifluorescence microscope fitted with a motorised stage. This stage can be driven to each event by the Chem*Scan* for a visual confirmation of the findings. Scanning, analysis and visual validation can be completed in approximately 5-6 minutes.

At present, we have used the Chem*Scan* for the detection of cells labelled with fluorescent viability markers, and the instrument is able to discriminate between bacteria and autofluorescent bacteria. The detection of 16S rRNA labelled bacteria relies on an adequate amount of target material being present. Thus, 16S rRNA probes are more useful for determining the viability of organisms, since it is much more labile than DNA. Nonetheless, in order to be certain that cells detected with our 16S probes are indeed viable, we have designed assays which incorporate both a nucleic acid probe and a viability probe (CTC). Initial results with this combined approach are promising and allow detection of viable *E. coli* within four hours.

Questions, comments and answers

Q: You said that PCR was not of great use for treated water because of its inability to distinguish between viable and non-viable cells. Are there any developments that will allow this?

A: There are technologies that can be put on the front end, but even for treated drinking water, they are not as sensitive as they might be, especially where there is iron or humic acids.

Questions, comments and answers cont'd.

C: Preliminary results show that when you chlorinate bacteria you don't really kill them, so that PCR is not necessarily detecting dead organisms. Also, although organics can interfere, a student of mine has been developing techniques that will achieve clean-up by extracting nucleic acids from sediments, for example.

A: It is true that chlorination doesn't kill everything. However, when *E. coli* is shown to be dead by a wide range of markers, it can still be detected by PCR. We have used a range of clean-up techniques especially for *Legionella*. However, we are looking for rapid detection in a utilities setting. From the point of view of simplicity, membrane separation and direct detection are better.

Q: You said that your system distinguishes between living and dead cells. I presume this is because β-galactosidase activity is lost. Is this the case?

A: You can distinguish between viable and non-viable cells on the basis either of CTC (electron transport) or esterase activity. After treating drinking water with chlorine (1 ppm for 30 minutes), most bacteria are killed. The vast majority cannot be cultured, but a few remain viable. If we take these and keep them, they maintain their transient state.

Q: When dealing with drinking water, is there any good news regarding *Cryptosporidium*, such as an easy detection system, especially for developing countries?

A: Right now the answer is "no", although there are potential indicators. There is some correlation between physical removal of bacterial spores and *Cryptosporidium* removal. At present, we have to look for the organisms.

REFERENCES

ANON (1983), "The Bacteriological Examination of Drinking Water Supplies 1982. Methods for the Examination of Water and Associated Materials", *Reports on Public Health and Medical Subjects* 71, HMSO, London.

ANON (1989), "Guidance on Safeguarding the Quality of Public Water Supplies", Department of Environment/Welsh Office, HMSO, London.

ANON (1994), "The Microbiology of Water 1994. Part 1 -- Drinking Water. Methods for the Examination of Waters and Associated Materials", *Reports on Public Health and Medical Subjects* 71, HMSO, London.

COLQUHOUN, K.O., S. TIMMS, and C.R. FRICKER (1995), "Detection of *Escherichia coli* in potable water using direct impedance technology", *Journal of Applied Bacteriology* 79, pp. 635-639.

COWBURN, J.K., T. GOODALL, E.J. FRICKER, K.S. WALTER, and C.R. FRICKER (1994), "A preliminary study of the use of Colilert for water quality monitoring", *Letters in Applied Microbiology* 19, pp. 50-52.

EDBERG, S.C., M.J. ALLEN, and D.B, SMITH (1991), "Defined substrate technology method for rapid and specific enumeration of total coliforms and *Escherichia coli* from water. Collaborative study", *Journal of the Association of Official Analytical Chemists* 74, pp. 526-529.

FRAMPTON, E.W. and L. RESTAINO (1993), "A Review. Methods for *Escherichia coli* identification in food, water and clinical samples based on beta-glucuronidase detection", *Journal of Applied Bacteriology* 74, pp. 223-233.

FRICKER, E.J. and C.R. FRICKER (1994), "Application of the polymerase chain reaction to the identification of *Escherichia coli* and coliforms in water", *Letters in Applied Microbiology* 19, pp. 44-46.

FRICKER, E.J. and C.R. FRICKER (in press *a*), "Use of two presence/absence systems for the detection of *E.coli* and coliforms from water", *Water Research.*

FRICKER, E.J. and C.R. FRICKER (in press *b*), "Use of defined substrate technology and a novel procedure for estimating the numbers of enterococci in water", *Journal of Microbiological Methods.*

FRICKER, E.J. and C.R. FRICKER (in press *c*), "Use of the polymerase chain reaction to detect *Escherichia coli* in food and water", in D. Kay and C.R. Fricker (eds.), *Coliforms and E.coli: Problem or Solution*, Royal Society of Chemistry, London.

FRICKER, C.R., J. COWBURN, T. GOODALL, K.S. WALTER, and E.J. FRICKER (1994), "Use of the Colilert system in a large U.K. water utility", Proceedings of the Water Quality Technology Conference, 1993, Florida.

FRICKER, E.J., K.S. ILLINGWORTH, and C.R. FRICKER (in press), "Use of two formulations of Colilert and QuantiTray for assessment of the bacteriological quality of water", *Water Research.*

SARTORY, D.P. and L. HOWARD (1992*a*), "A medium detecting ß-glucuronidase for the simultaneous membrane filtration enumeration of *Escherichia coli* and coliforms from drinking water", *Letters in Applied Microbiology* 15, pp. 273-276.

SARTORY, D.P and L. HOWARD (1992*b),* "A simple membrane enumeration medium for coliforms and *E.coli* utilising a chromogenic glucuronide substrate", Proceedings of the American Waterworks Association Water Technology Conference, November 1992, Toronto, Canada.

SPIGELMAN, M., C.R. FRICKER, and E.J. FRICKER (1995), "Extracting DNA from Lindow Man's gut contents. Modern technology looking for answers in ancient tissues", in R.C. Turner and R.G. Scaife (eds.), *Bog Bodies. New discoveries and new perspectives*, British Museum Press.

TIMMS, S., K.O. COLQUHOUN, and C.R. FRICKER (1996), "Detection of *Escherichia coli* in potable water using indirect impedance technology", *Journal of Microbiological Methods* 26, pp. 125-132.

LUMINESCENCE-BASED ASSAYS TO DETECT AND CHARACTERISE CHEMICAL AND BIOLOGICAL WATER CONTAMINANTS

by

Michael S. DuBow
Department of Microbiology & Immunology, McGill University, Montréal, Québec, Canada

Introduction

The continued monitoring for biological and chemical contaminants in water is of crucial importance for water quality. Waterborne diseases are still a major cause of concern throughout most of the world. Moreover, even in highly developed countries, the efficacy of water purification and treatment plants must be continually monitored to ensure contaminant elimination via cost-effective treatments, and water distribution networks assayed for bacterial contaminants from biofilms and leakage (Robinson *et al.,* 1995).

Chemical pollutants require a number of strategies for their measurement. These can often involve complex analytical procedures that may take days to perform by highly trained personnel. Moreover, major concerns in the monitoring of chemicals are understanding their structure/activity relationships, their bioavailability, and the determination of "safe" levels.

Recent excitement in biotechnology has come from the use of enzymes which emit photons of light as byproducts of their reactions (Stewart and Williams, 1992). These enzymes can allow the determination of the presence and physiological status of living organisms. Moreover, when coupled to the expression of individual genes, they can be used as "reporter assays" to transduce the signal of gene expression into easily-assayed light emission. Examples of these enzymes include the luciferase enzymes from bacteria *(lux)* and eukaryotes *(luc)*. More recently, green fluorescent protein (GFP) has been used (Kremer *et al.,* 1995), as has the development of fluorescent substrates for well-studied enzymes such as β-galactosidase (Eguchi *et al.,* 1994). The luciferase enzymes from bacteria are heterodimeric mixed function oxidases which utilise a short chain aldehyde and reduced NAD with the concomitant emission of a photon of light in the range of 490 nm. In eukaryotic luciferases, luciferin is used as a co-factor along with ATP. Because of this co-factor use, these enzymes can also act as important "reporters" of reducing power or energy levels within a cell.

One of the first uses of luciferase as a sentinel of water quality involved the development of the Microtox assay system (Bulich, 1984). In this procedure, a constitutively luminescent bacterium *(Photobacterium phosphoreum)* is exposed to a water sample, and its luminescence measured versus a control (untreated) sample. Any decrease in light output from the luminescent bacteria would be suggestive of cytotoxicity and therefore worthy of further investigation. Numerous studies have correlated the presence of toxic chemicals in water with a decrease in light output in the Microtox assay. This simple assay requires a single instrument, can be performed in minutes, can be automated for multiple assays on microtiter plates, or even connected to a computer for "on-line" measurements (Blaise *et al.,* 1994). Advances in genetic engineering have allowed the extrapolation of this simple assay to more precise measurements, as described in the following two sections.

Luciferase reporter phages

The detection of living micro-organisms in post-treatment water is a complex problem compounded by the need for extremely accurate and rapid results so that proper treatment adjustments can be made. Moreover, due to the presence of biofilms in potable water distribution systems, rapid assays for viable micro-organisms are crucial to ensure their continued absence in tap water. A variety of methods have been developed to measure viable organisms in water, including measurements of ATP (using luciferase/luciferin), respirometry, and traditional culture methods, to name but a few. Recent excitement has come from the use of luciferase-encoding genes incorporated into the genomes of bacteriophages specific for a given micro-organism. In this case, the gene for luciferase *(lux* or *luc)* is cloned within the genome of a double-stranded DNA bacteriophage under the control of a phage or host-specific promoter. These "luciferase reporter phages" are then used to infect cultures of bacteria, where the double-stranded DNA genome of the phage will enter the cell and luciferase will be expressed. Upon the addition of the appropriate co-factors, viable cells will begin to emit photons of light. The amount of light emitted can then be correlated to the number of viable cells. Thus, a water sample can have luciferase reporter phages added to it and then, a short time (minutes) after infection, the sample is placed within a luminometer. After correction for background luminescence of the sample, the quantity of viable organisms (if any) can then be estimated from the amount of light emission that is measured.

This technology was first used with great success to measure the antibiotic sensitivity of *Mycobacterium tuberculosis* (DuBow, 1993). Under ordinary conditions, it can take several weeks to determine the antibiotic sensitivity of this organism due to its extremely slow growth. The use of luciferase reporter phages has allowed the time necessary for measurements of cell viability (before and after antibiotic exposure) to be reduced to several hours. These luciferase reporter phages have also been used to measure the presence of viable *Salmonella* or *Listeria* in food prior to its sale (Stewart and Williams, 1992). Efforts are under way to create and optimise luciferase reporter phages for the detection of both common and disease causing bacteria in water. Many bacteriophages are very stable and can even be lyophilised. Moreover, the availability of portable, battery-operated luminometers ensures that this assay can be performed in a variety of settings. One of the major obstacles that remains to be overcome is the sensitivity of light measurement with the low numbers of bacteria that one would expect to find in drinking quality water. However, the use of strong promoters upstream of the luciferase gene, plus highly infectious bacteriophages, may ultimately allow this assay to be commercialised for use, including the generation of commercial kits for use in homes by ordinary consumers. The identification of broad, host-range, double-stranded DNA bacteriophages will ultimately allow the measurement of a wide variety of bacterial species using a single phage preparation. Lastly, the capacity of phages to infect biofilms can be exploited to study the efficacy of different treatments or prevention strategies to control biofilm growth and development (Doolittle *et al.,* 1995).

Detection of chemical contaminants

All organisms have evolved the capacity to measure changes in their environment and, in many cases, strategies to deal with these changes. These strategies frequently involve the induction (or repression) of specific genes in order to reorient cell physiology to cope with these environmental changes. The very first genetic system whose regulation was elucidated (the *lac* operon of *Escherichia coli)* is induced to allow *E. coli* to detect and utilise a molecule (in this case, the sugar lactose) in its environment. Many bacteria have evolved these genetically-programmed responses to a wide variety of molecules in their environment. The proteins whose expression is altered in these responses may play a role in the detoxification of toxic environmental compounds, or a role in augmenting or repressing particular enzymatic reactions to either overcome the stress or prevent further damage. The capacity to quantitatively measure these responses can be utilised to determine the presence and amounts of these chemical contaminants in water. One of the earliest examples of these assay systems exploited the knowledge of the "SOS response" of *Escherichia coli* to mutagens (Quillardet *et al.,* 1982). This response occurs when *E. coli* is exposed to a mutagen, and results in a suite of genes being released from repression (by the LexA repressor) to cause the expression of enzymes which repair damage to DNA (Kenyon and Walker, 1980). One of these genes (*sulA)* was fused to the *lacZ* gene (minus its own transcriptional regulatory sequences) such that the expression of the *lacZ* gene and its gene product, β-galactosidase, was now under the control of the bacterium's "SOS response". If this strain of *E. coli* is exposed to a mutagen, then β-galactosidase expression is induced, and its presence easily assayed using colorimetric substrates. The assay, called the SOS Chromostest, is rapid (less than one hour) and inexpensive to perform.

The gene fusion techniques used in the creation of this "biosensor" (or "bioreporter") clone can be extended and improved to exploit the genetically-programmed responses of bacteria to toxic environmental compounds for the creation of a plethora of clones to measure a variety of toxic chemicals. Over the past decade, a number of "gene fusion" biosensors have been developed which have been used not only to discover these responses, but also to measure their cognate chemical inducers in aqueous conditions (Table 1).

Table 1. **Selected gene fusion clones for environmental stressors**

Environmental contaminant (or deficiency)	Reference
Alkalinity	Bingham *et al.,* 1990
	Slonczewski *et al.,* 1987
Arsenic	Diorio *et al.,* 1995
Dimethylsulfoxide	Briscoe *et al.,* 1996
Mercury	Selifonova *et al.,* 1993
Mutagens	Quillardet *et al.,* 1982
Napthalene/salicylate	Heitzer *et al.,* 1994
Nickel	Guzzo and DuBow, 1994*b*
Oxygen	Kogama *et al.,* 1988
Phosphate deficiency	Metcalfe *et al.,* 1990
Tributyl tin	Briscoe *et al.,* 1996

In order to systematically search for these genetically-programmed responses in *E. coli*, our laboratory has created a collection of over 3 000 gene fusion clones, each of which contains a single insertion of the luciferase gene (minus its own transcriptionary regulatory sequences) from the luminescent bacterium *Vibrio harveyi* (Guzzo and DuBow, 1991). Luciferase was chosen as the "reporter gene" for these studies as its measurement is several orders of magnitude more sensitive (Meighen, 1991) than that of β-galactosidase (the *lacZ* gene product). These clones were exposed, via replica plating, to a variety of different chemical compounds, and several clones were isolated whose luminescence is induced in the presence of a particular (class of) toxic chemical. These clones can now be used to determine the bioavailable (as opposed to total) quantities of these chemicals in water, as well as assays to determine the structure/activity relationships of the chemicals under a variety of conditions. Moreover, due to the gene "tagging" used in this research, the gene(s) whose expression is augmented by the particular chemical can be isolated (using *lux* as a probe) and the compound's role and effects on cell physiology discerned.

With this technology, our laboratory has identified several previously unknown genes which are induced in the presence of toxic metals and metalloids (Guzzo *et al.*, 1991; Guzzo and DuBow, 1994*a*), tributyl tin and dimethyl sulfoxide (Briscoe *et al.*, 1996). For example, one of our clones has allowed us to identify genes whose expression is induced in the presence of arsenic, as arsenite or arsenate (Diorio *et al.*, 1995; Cai and DuBow, 1996). These toxic arsenic oxyanions were found to induce the expression of an operon consisting of a repressor (*ars*R), a membrane-located arsenite ion specific export pump (*ars*B), and an arsenate reductase (*ars*C). These genes were found to be present in a wide variety of gram-negative bacteria, including *Pseudomonas aeruginosa,* and were also found to be homologous to previously studied plasmid-borne (*ars*) operons isolated from both gram positive and negative bacteria resistant to high levels of arsenic. It is conceivable that multiple copies of the *ars* operon (as would be found with genes located on multiopy plasmids) could allow higher levels of arsenic oxyanion resistance due to increased quantities of the expressed genes. To test this hypothesis, the chromosomal *ars* operon was cloned into a high copy number plasmid, and elevated levels of arsenate or arsenite resistance were observed in *E. coli.* Moreover, mutants in the chromosomal *ars* operon were found to be 10-100 fold more sensitive to arsenic oxyanions than wild type *E. coli.* These results suggest that the chromosomal *ars* operons may be the progenitor of the plasmid-borne *ars* operons, and also that other resistance determinants may ultimately be found to have their origin in chromosomal genes. Moreover, these results also point to the fact that single gene differences can confer enormous variation in sensitivity or resistance to environmental contaminants.

The induction of *ars* gene expression was found to occur within five minutes after cellular exposure to toxic arsenic oxyanions. Non-toxic forms of arsenic (e.g. cacodylic acid) were found not to induce the operon, even at extremely high levels (greater than 1 microgram per millilitre) of arsenic (as cacodylic acid). Maximal enzymes levels were reached within 60 minutes after cellular exposure and, coupled to the use of portable luminometers, suggest that this assay can be rapidly performed in the field. The construction of on-line monitoring systems for luciferase expression have also been reported (Heitzer *et al.*, 1994). Lastly, this clone is a very sensitive biosensor for the determination of arsenic in the environment, as induction of reporter gene expression (and thus luminescence) was found to occur at levels of arsenic (5 parts per billion; micrograms per litre) only slightly above the environmental background. It is always important in these assays to have a control biosensor, containing luciferase constitutively expressed, in order to measure cytotoxic effects (reduced luminescence) or effects on the levels of the required cellular co-factors for the reaction (Guzzo *et al.,* 1992). Finally, because amplification of the *ars* genes conferred resistance to high levels of arsenic, these genes may prove useful in engineering bacteria for bioremediation in the presence of high levels of toxic arsenic oxyanions.

Extrapolations

An increasing variety of these gene-fusion biosensor clones are being prepared in *E. coli* and other bacterial species. They will allow for the identification of a number of toxic environmental compounds or conditions (e.g. low or high nitrogen, low or high phosphorus). These clones can be grown very inexpensively in large quantities. This approach, though is still in its infancy, nevertheless holds great promise for the future. The use of a wide variety of bacterial clones, each detecting different (classes of) chemical compounds, will be increasingly used in the "battery of microbiotests" approach that is gaining increasing favour in toxicological assessment (Blaise, 1991). Moreover, genetic engineering can allow changes in the specificity and selectivity of these clones to be prepared such that they can detect a wider (or narrower) variety of chemical and biological contaminants under many different conditions.

It should also not be forgotten that this biotechnological approach allows the identification of genes, and their regulatory circuits, which respond to these toxic environmental pollutants. This information can ultimately be used to search for analogous/homologous responses in higher organisms. The identification of these pollutant-inducible genes will allow the capacity to directly measure organism (including human) exposure using assays for their expression. In addition, information concerning these responses, and the presence of antagonistic or synergistic chemicals, will allow the creation of better modelling and treatment protocols. Ultimately, it may prove possible to test individuals and/or species to determine their risk to these chemicals and/or conditions *a priori* and to institute precise corrective measures due to the knowledge obtained on the effects of these chemical compounds on cell and organismal physiology.

Acknowledgements

I would like to thank all the members of my laboratory for their wonderful suggestions, discussions, and terrific science. Work in my laboratory has been supported by a grant (C96035R2) from the Centre for the Alternatives to Animal Testing (US).

Questions, comments and answers

Q: Arsenite is a thio-group reagent. How much free arsenite is there, and can it act as a regulator?

A: We don't know how much there is. It has no observable effect on viability. The operon kicks in at concentrations three to four orders of magnitude lower than the LD_{50}.

Q: The cell responds to the presence of the metal and finds a way around it?

A: In the case of arsenic, this is a survival and detoxification mechanism. In the case of aluminium, we don't know what it does. We can't achieve a concentration high enough to kill the organism. *E. coli* has at least two genes inducible by tributyl tin but we don't know what the effect is. In some cases, we know that the protein is membrane-bound but have no idea to what family it belongs.

REFERENCES

BINGHAM, R.J., K.S. HALL, and J.L. SLONCZEWSKI (1990), "Alkaline induction of a novel gene locus, *alx* in *Escherichia coli*", *J. Bacteriol* 172, pp. 2 184-2 186.

BLAISE, C. (1991), "Microbiotests in aquatic ecotoxicology: characteristics, utility and prospects", *Environ. Toxicol. and Water Qual.* 6, pp. 145-155.

BLAISE, C., R. FORGANI, R. LEGAULT, J. GUZZO, and M.S. DUBOW (1994), "A bacterial toxicity assay performed with microplates, microluminometry, and Microtox reagent", *BioTechniques* 16, pp. 932-937.

BRISCOE, S.F., C. DIORIO, and M.S. DUBOW (1996), "Luminescent biosensors for the detection of tributyltin and dimethyl sulfoxide and the elucidation of their mechanisms of toxicity", in M. Moo-Young, W.A. Anderson, and A.M. Chakrabarty (eds.), *Environmental Biotechnology: Principles and Applications*, pp. 645-655, Kluwer Academic Publishers, Dordrecht.

BULICH, A.A. (1984), "Microtox-a bacterial toxicity test with several environmental applications", in D. Liu and B.J. Dutka (eds.), *Toxicity Screening Procedures Using Bacterial Systems*, pp. 55-64, Marcel Dekker, New York.

CAI, J. and M.S. DUBOW (1996), "Expression of the *Escherichia coli* chromosomal *ars* operon", *Can. J. Micro.* 42, pp. 662-671.

DIORO, C., J. CAI, J. MARMOR, R. SHINDER, and M.S. DUBOW (1995), "An *Escherichia coli* chromosomal *ars* operon homologue is functional in arsenic detoxification and conserved in gram-negative bacteria", *J. Bacteriol.* 177, pp. 2 050-2 056.

DOOLITTLE, M.M., J.J. COONEY, and O.E. CALDWELL (1995), "Lytic infection of *Escherichia coli* biofilms by bacteriophage T4", *Can. J. Micro.* 41, pp. 12-18.

DUBOW, M.S. (1993), "Antituberculosis drug screening", *The Lancet* 342, pp. 448-449.

EGUCHI, H., S. HAYAHSI, J. WATANABE, O. GOTOK, and K. KAWAJINI (1994), "Molecular cloning of the human AH receptor gene promoter", *Biochem. Biophys. Res. Comm.* 203, pp. 615-622.

GUZZO, A., C. DIORIO, and M.S. DUBOW (1991), "Transcription of the *Escherichia coli fli*C gene is regulated by metal ions, *Appl. Env. Micro.* 57, pp. 2 255-2 259.

GUZZO, A. and S. DUBOW (1991), "Construction of stable single-copy luciferase gene fusions in *Escherichia coli*", *Arch. Micro.* 156, pp. 444-448.

GUZZO, A. and M.S. DUBOW (1994*a*), "Identification and characterization of genetically programmed responses to toxic metal exposure in *Escherichia coli*", *FEMS Micro. Rev.* 14, pp. 369-374.

GUZZO, A. and M.S. DUBOW (1994*b*), "A *lux*A,B transcriptional fusion to the *cel*F gene of *Escherichia coli* displays increased luminescence in the presence of nickel", *Mol. Gen. Genet.* 242, pp. 455-460.

GUZZO, J., A. GUZZO, and M.S. DUBOW (1992), "Characterization of the effects of aluminum on luciferase biosensors for the detection of ecotoxicity", *Tox. Lett.* 64/65, pp. 687-693.

HEITZER, A., K. MALACHOWSKY, J.E. THONNARD, P.R. BIENKOWSKI, D.C. WHITE, and G.S. SAYLER (1994), "Optical biosensor for environmental on-line monitoring of naphthalene and salicylate bioavailability with an immobilized bioluminescent catabolic reporter bacterium", *Appl. Env. Micro.* 60, pp. 1 487-1 494.

KENYON, C.J. and G.C. WALKER (1980), "DNA-damaging agents stimulate gene expression at specific loci in *Escherichia coli*", *Proc. Natl. Acad. Sci. USA* 77, pp. 2 819-2 823.

KOGAMA, T., S.B. FARR, K.M. JOYCE, and D.O. NATVIG (1988), "Isolation of gene fusions *(soi::lacZ)* inducible by oxidative stress in *Escherichia coli*", *Proc. Natl. Acad. Sci. USA* 85, pp. 4 799-4 803.

KREMER, L., A. BAULARD, J. ESTERGRIER, O. POULAIN-GODEFROY, and C. LOCHT (1995), "Green fluorescent protein as a new expression marker in *Mycobacteria*", *Mol. Micro.* 17, pp. 913-922.

MEIGHEN, E.A. (1991), "Molecular biology of bacterial luminescence", *Micro. Rev.* 55, pp. 123-142.

METCALFE, W.M., P.M. STEED, and B.L. WANNER (1990), "Identification of phosphate starvation-inducible genes in *Esherichia coli* K-12 by DNA sequence analysis of *psi::lacZ(*Mu d1) transcriptional fusions", *J. Bacteriol.* 172, pp. 3 191-3 200.

QUILLARDET, P., O. HUISMAN, R. D'ARI, and M. HOFNUNG (1982), "SOS Chromotest, a direct assay of induction of an SOS function in *Escherichia coli* K-12 to measure genotoxicity", *Proc. Natl. Acad. Sci. USA* 79, pp. 5 971-5 975.

ROBINSON, P.J., J.T. WALKER, C.W. KEEVIL, and J. COLE (1995), "Reporter genes and fluorescent probes for studying the colonisation of biofilms in a drinking water supply line by enteric bacteria", *FEMS Micro. Lett.* 129, pp. 183-188.

SELIFONOVA, O., R. BURLAGE, and T. BARKRAY (1993), "Bioluminescent sensors for detection of bioavailable Hg(ll) in the environment", *Appl. Env. Micro.* 59, pp. 3 083-3 090.

SLONCZEWSKI, J.L., T.N. GONZALEZ, F.M. BARTHOLOMEW, and N.J. HOLT (1987), "Mu d-directed *lacZ* fusions regulated by low pH in *Escherichia coli*", *J. Bacteriol.* 169, pp. 3 001-3 006.

STEWART, G.S. and A.B. WILLIAMS (1992), "*Lux* genes and the application of bacterial luminescence", *J. Gen. Microbiol.* 138, pp. 1 289-1 300.

WATERBORNE PARASITES AND VIRUSES STILL WAITING FOR RELIABLE INDICATORS

by

Ph. Vilaginès, B. Sarrette, A. Pezzana, M. Le Guyader, C. Cun and R. Vilaginès
Centre de Recherche et de Contrôle des Eaux de Paris, Paris, France

Introduction

The microbial quality of drinking water sources is of growing health concern: waterborne outbreaks continue to be reported (Craun, 1988; Levine *et al.,* 1990), some of unknown ethiology, many of which are of suspected viral origin (Payment *et al.*, 1991; Sobsey *et al.*, 1993) and more recently, several due to the intestinal protozoa, *Giardia* and *Cryptosporidium* (Lisle and Rose, 1995). In many instances, such as the Milwaukee cryptosporidiosis that sickened over 400 000 people in 1993 (MacKenzie *et al.*, 1994), there are questions concerning the effectiveness of drinking water treatment plants in preventing passage of these micro-organisms in the public water supply and the reliability of bacteriological indicators used world-wide to guarantee a microbiologically safe water to consumers. *Cryptosporidium* is of particular concern to the water industry because of its widespread prevalence in surface water supplies, its extreme resistance to chlorine, and its potentially fatal consequences for immunocompromised individuals.

The ideal index of pathogenic micro-organism pollution is its own absence or presence. Considering their very low concentration in waters, virus and parasite detection remains at present the prerogative of sophisticated laboratories and skilled staff. Besides, methods are time consuming for routine monitoring, and, at least for parasites, need improvement in terms of recovery sensitivity.

In light of these difficulties, there is a cyclic tendency among scientists to forward bacteriophages as viruses indicators (Guelin, 1952; Kott *et al.*, 1974; Figueroa *et al.*, 1978; Wentsel *et al.*, 1982; Stetler, 1984; Havelaar, 1987), taking into account that their size and resistance in the environment resemble those of enteric viruses (Yates *et al.*, 1985; Borrego *et al.*, 1990; Havelaar *et al.*, 1995; Jofre *et al.*, 1995; Sobsey *et al.*, 1995*b*). The ecology and detection of three of them -- somatic coliphages, F-specific RNA phages and *Bacteroïdes fragilis* phages -- have been well developed (IAWPRC Study Group, 1991). Besides, their enumeration is easy and does not require specific devices. However, comparative epidemiological studies with the F-specific RNA phages do not lead to a general agreement on this choice.

Currently, water standards in France take in account enteroviruses assays to assess drinking water as well as recreative water quality. So, before substituting enterovirus for phages, authorities are concerned with their real significance in regard to public health.

To address this question, we investigated the occurrence and relationship of enteroviruses, F-specific RNA phages, parasites and usual bacterial indicators in different types of water. Results of a 15-month survey in the Paris area are presented here.

Material and methods

Environmental samples

Three different water types were examined. Wastewater samples, both raw and treated, were taken from a plant located on the Seine upstream from Paris, designed to treat 300 000 m^3 sewage per day based on biological process. Surface water was collected in the receiving river (Seine) at the intake, located about 3 km downstream from the wastewater discharge, for potabilization purposes. Water was collected every fortnight on the same day for each trial in sterilised plastic containers (25 litres), and samples were examined within six hours for the presence of seven micro-organisms.

Microbial analysis

Each water sample was thoroughly mixed and sterily distributed in sub-samples for each micro-organism determination. All tests were run on the same day, using the following methods:

- Faecal coliforms: membrane filtration of a 100 ml sample (eight dilutions by trial) according to French standard protocol: AFNOR - NF T 90-414 (1985).

- Faecal *enterococci*: membrane filtration of a 100 ml sample (eight dilutions by trial) according to French standard protocol: AFNOR - XP T 90-416 (1996).

- *Salmonella* were detected from a one litre sample by a presence absence test consisting of pre-enrichment in buffered peptone, followed by enrichments in Rappaport-Vasliadis and cystein selenite media, and isolation on BGA and Hektoën agar media (ISO†6340, 1995).

- F-specific RNA phages were numerated in 1 ml aliquots (eight dilutions by trial) according to ISO 10705-1 by the double agar technique with the host strain WG49 (1995).

- Enteroviruses were recovered from water samples (five to 20 litres for raw water and 20†litres for treated and surface waters) concentrated by adsorption-elution through glass wool (Vilaginès *et al.*, 1993*a*) and organic flocculation (Katzenelson *et al.*, 1976). The totality of the concentrates were inoculated to BGM cells (Barron *et al.*, 1970) and virus numbered by plaque assay. Each plaque was cloned as previously described (Vilaginès *et al.*, 1989), and the virus characterised after coloration with hematoxylin-eosin (Maurin, 1965).

- *Giardia* and *Cryptosporidium* were recovered in the pellet obtained by centrifugation of a one litre sample, and were purified by percoll sucrose gradient flotation. Cysts and oocysts were numbered (eight dilutions by trial) after staining with fluorescent labelled monoclonal antibody.

Except for *Salmonella*, determined by a presence/absence test and viruses numbered in the whole concentrate, each micro-organism concentration was evaluated in each experience by the most probable number method, taking into account for each assay all numeric data available, whatever the number of usable dilutions (Maul, 1991).

Statistical analysis

Two statistical tests were used:

- Principal Components Analysis (PCA), a descriptive method showing the magnitude and direction of each variable, thus indicating their potential association. The major step in a PCA is the extraction of the eigenvectors from the variance-covariance matrix to obtain uncorrelated new variables called Principal Components, which span the maximum variance in the data set. The Principal Components are linear combinations of all the original variables.

- Simple regression procedure with statistics for the fitted linear model, the estimate of the slope and the intercept, including the standard error, t-statistic (Student test on the slope), F-statistic (Fisher test on the model by an ANOVA method) and the p-value for each estimate. A leverage analysis and a distribution fitting study of the model residuals were realised.

Results

Results (raw data) obtained in all samples analysed from March 1995 to July 1996 are presented in Table 1a (raw wastewater), Table 1b (treated wastewater) and Table 1c (surface water). *Giardia* and *Cryptosporidium* concentrations in the first 13 trials are not given as they were determined by a different technique which proved to yield very poor recovery leading to unreliable data. These would have interfered with statistical analysis.

Table 1a. **Micro-organisms concentration in raw wastewater**

Assay N°	FC (100 ml)	FE (100 ml)	PHAGES (1 ml)	VIRUS (20 L)	GIARDIA (1 L)	CRYPTO (1 L)	SA (1 L)
1	1.82E+07	1.12E+06	6.13E+03	37	NC	NC	0
2	2.45E+07	5.63E+06	3.04E+02	5	NC	NC	0
3	6.71E+07	2.65E+06	2.69E+04	1	NC	NC	0
4	5.40E+06	1.03E+06	3.75E+03	T	NC	NC	0
5	1.81E+07	7.53E+06	1.94E+05	T	NC	NC	0
6	2.21E+07	1.46E+06	5.42E+03	37	NC	NC	0
7	1.10E+07	1.90E+06	1.40E+04	0	NC	NC	0
8	1.44E+07	1.35E+06	6.94E+03	8	NC	NC	0
9	1.50E+07	1.83E+06	3.87E+04	T	NC	NC	(+)
10	2.33E+07	2.22E+06	7.20E+04	T	NC	NC	0
11	2.80E+07	1.64E+06	1.02E+05	13	NC	NC	0
12	1.57E+07	1.09E+06	2.22E+04	190	NC	NC	0
13	1.90E+07	1.77E+06	2.34E+04	8	NC	NC	(+)
14	1.94E+07	1.48E+06	5.87E+03	8	1.49E+04	1.45E+03	0
15	3.75E+07	1.29E+06	9.84E+03	20	4.19E+04	6.45E+02	(+)
16	2.05E+07	2.39E+06	5.76E+03	56	1.52E+05	1.45E+04	(+)
17	2.23E+07	2.15E+06	6.03E+03	630	1.75E+05	0.00E+00	(+)
18	2.23E+08	1.32E+06	8.69E+03	47	8.31E+05	3.07E+04	(+)
19	4.26E+07	1.36E+06	9.17E+03	T	8.09E+05	1.05E+05	(+)
20	8.82E+06	1.76E+06	9.97E+03	16	1.46E+06	6.62E+04	0
21	9.90E+06	1.95E+06	8.90E+03	200	5.39E+05	2.94E+04	0
22	1.24E+07	9.17E+05	3.02E+04	20	2.37E+05	0.00E+00	0
23	1.62E+07	2.57E+06	4.11E+03	4	1.55E+05	1.32E+03	0
24	6.90E+06	9.74E+05	8.81E+03	12	9.15E+04	0.00E+00	(+)
25	3.48E+06	1.22E+06	4.25E+03	16	2.54E+05	2.60E+04	0
26	1.57E+07	4.70E+04	3.47E+03	4	2.13E+03	0.00E+00	0
27	8.49E+06	1.29E+06	9.22E+03	24	1.43E+05	1.12E+04	0
28	2.22E+07	1.67E+06	6.03E+03	8	1.23E+04	0.00E+00	0
29	1.95E+07	1.76E+06	ND	12	5.15E+04	0.00E+00	0
30	2.19E+07	1.84E+06	1.35E+04	644	2.56E+04	1.40E+03	0
31	4.05E+07	1.32E+06	1.13E+04	136	5.69E+04	7.50E+02	0
32	1.94E+07	2.24E+06	1.29E+04	112	3.95E+04	2.11E+02	0
33	3.74E+07	1.41E+06	1.99E+04	4 800	2.52E+04	1.82E+02	(+)
34	3.01E+07	1.10E+06	2.62E+03	4	4.59E+04	9.25E+03	0
35	3.68E+07	1.24E+06	4.00E+04	52	1.37E+04	1.65E+03	0
MEAN	2.16E+07	1.84E+06	2.20E+04	237.5	2.35E+05	1.36E+04	

Notes: FC: faecal coliforms. FE: faecal *enterococci*. PHAGES: F-specific RNA phages. VIRUS: enteroviruses. GIARDIA: *Giardia* cysts. CRYPTO: *Cryptosporidium* oocysts. SA: *Salmonella*. T: Toxic. C: Contamination. ND: not done. NC: data not considered.

Source: Author.

Table 1b. **Concentration of micro-organisms detected in treated wastewater**

Assay N°	FC (100 ml)	FE (100 ml)	PHAGES (1 ml)	VIRUS (20 L)	GIARDIA (1 L)	CRYPTO (1 L)	SA (1 L)
1	9.31E+04	1.31E+04	63.4	0	NC	NC	0
2	3.78E+04	1.10E+05	57.1	T	NC	NC	0
3	2.82E+05	1.38E+04	49.2	1	NC	NC	0
4	2.99E+04	1.45E+04	12.7	4	NC	NC	0
5	8.17E+04	3.26E+04	39.7	T	NC	NC	0
6	2.02E+04	7.97E+03	20.6	7	NC	NC	0
7	1.23E+04	7.97E+03	42.8	1	NC	NC	0
8	1.67E+04	5.79E+03	30.2	0	NC	NC	0
9	1.44E+05	1.52E+04	30.3	0	NC	NC	(+)
10	8.17E+04	1.01E+04	4.8	2	NC	NC	0
11	1.23E+04	2.67E+03	20.6	7	NC	NC	0
12	2.99E+04	4.68E+03	3.2	2	NC	NC	0
13	3.51E+04	8.28E+03	14.3	0	NC	NC	(+)
14	5.18E+04	9.82E+03	25.4	0	1.00E+02	0.00E+00	0
15	2.55E+04	1.05E+04	19.1	0	3.16E+02	4.51E+01	(+)
16	1.58E+04	1.01E+04	25.4	4	6.43E+02	0.00E+00	(+)
17	4.74E+04	2.90E+04	47.2	C	2.35E+03	0.00E+00	(+)
18	5.54E+04	4.91E+03	58.3	27	3.23E+03	3.87E+02	(+)
19	1.05E+04	8.04E+03	56.7	0	1.62E+04	1.19E+04	(+)
20	9.67E+03	1.30E+04	100.8	0	2.96E+04	2.28E+03	0
21	1.67E+04	1.01E+04	226.8	0	3.70E+03	0.00E+00	0
22	3.16E+04	8.89E+03	190.6	0	2.59E+04	0.00E+00	0
23	2.02E+04	9.49E+03	107.1	0	2.80E+04	4.43E+03	0
24	1.37E+05	2.00E+04	69.3	0	0.00E+00	0.00E+00	(+)
25	7.83E+05	2.10E+04	88.2	T	5.78E+03	0.00E+00	0
26	4.39E+04	1.57E+04	69.3	7	3.47E+02	0.00E+00	0
27	6.74E+05	1.18E+04	25.2	15	2.30E+03	0.00E+00	0
28	4.92E+04	4.12E+03	85.0	3	1.46E+02	0.00E+00	0
29	2.02E+04	5.00E+03	40.9	1	0.00E+00	0.00E+00	0
30	3.87E+04	3.13E+03	20.5	5	0.00E+00	0.00E+00	0
31	2.55E+04	6.50E+03	20.6	0	6.10E+02	0.00E+00	0
32	3.25E+04	9.31E+03	70.9	4	1.86E+03	0.00E+00	0
33	4.83E+04	4.58E+03	12.6	12	2.42E+03	7.06E+01	(+)
34	1.73E+05	1.38E+04	12.6	5	6.93E+03	0.00E+00	0
35	1.73E+05	8.72E+04	568.2	12	8.41E+03	0.00E+00	0
MEAN	9.60E+04	1.61E+04	66.5	3.8	6.31E+03	8.68E+02	

Notes: FC: faecal coliforms. FE: faecal *enterococci*. PHAGES: F-specific RNA phages. VIRUS: enteroviruses. GIARDIA: *Giardia* cysts. CRYPTO: *Cryptosporidium* oocysts. SA: *Salmonella*. T: Toxic. C: Contamination. ND: not done. NC: data not considered.

Source: Author.

Table 1c. **Concentration of micro-organisms detected in surface water**

Assay N°	FC (100 ml)	FE (100 ml)	PHAGES (1 ml)	VIRUS (20 L)	GIARDIA (1 L)	CRYPTO (1 L)	SA (1 L)
1	8.02E+04	9.76E+03	4.4	0	NC	NC	0
2	1.82E+03	8.17E+02	0.7	T	NC	NC	0
3	1.46E+04	3.03E+02	1.8	0	NC	NC	0
4	6.38E+03	8.10E+03	8.0	0	NC	NC	0
5	1.18E+04	3.81E+03	4.8	0	NC	NC	0
6	8.20E+03	1.90E+03	1.2	0	NC	NC	0
7	7.29E+03	3.10E+03	2.3	0	NC	NC	0
8	1.28E+04	3.87E+02	4.4	0	NC	NC	(+)
9	8.20E+02	1.40E+02	5.2	0	NC	NC	(+)
10	8.20E+03	6.07E+01	0.3	1	NC	NC	(+)
11	2.92E+04	6.82E+02	1.1	0	NC	NC	0
12	1.55E+04	4.76E+02	1.5	0	NC	NC	0
13	6.38E+03	1.03E+02	0.7	0	NC	NC	0
14	1.28E+04	5.18E+02	3.1	0	0	0	0
15	2.00E+04	1.00E+03	6.8	0	4.0	0.5	(+)
16	1.73E+04	1.23E+03	3.5	1	8.0	2.5	(+)
17	3.64E+03	1.19E+03	6.7	0	0.7	0	0
18	1.09E+04	1.02E+03	12.0	0	12.4	12.4	0
19	9.11E+03	1.00E+03	17.4	0	135.5	379.5	0
20	2.73E+03	7.19E+02	4.1	0	67.6	105.2	(+)
21	1.82E+03	9.24E+02	6.6	2	15.6	37.3	0
22	9.11E+03	2.46E+03	13.3	0	0	0	0
23	3.64E+03	9.15E+02	6.6	1	0	0	0
24	7.29E+03	1.04E+03	4.6	10	0	0	(+)
25	3.64E+03	2.80E+02	2.5	2	0	0	0
26	9.11E+03	2.89E+02	10.4	1	0	0	0
27	9.00E+02	2.99E+02	5.4	0	0	0	0
28	1.15E+04	4.79E+03	2.1	0	344.4	0	0
29	3.15E+03	2.61E+02	1.7	0	0	0	0
30	2.88E+03	2.57E+02	0.8	0	19.3	0	0
31	1.01E+04	8.40E+01	0.8	4	0	0	0
32	5.68E+03	9.34E+01	1.7	0	0	0	(+)
33	4.50E+02	1.03E+02	3.3	1	0	0	0
34	1.37E+04	2.85E+02	0	0	77.4	0	0
35	1.05E+04	1.45E+02	0	4	0	31.0	0
MEAN	1.07E+04	1.39E+03	4.3	0.8	31.1	25.8	

Notes: FC: faecal coliforms. FE: faecal *enterococci*. PHAGES: F-specific RNA phages. VIRUS: enteroviruses. GIARDIA: *Giardia* cysts. CRYPTO: *Cryptosporidium* oocysts. SA: *Salmonella*. T: Toxic. C: Contamination. ND: not done. NC: data not considered.

Source: Author.

The average concentrations calculated in one litre in the three different types of waters are summarised in Table 2.

Table 2. **Mean concentrations of micro-organisms per litre in raw and treated wastewaters and surface water**

Type of water	Faecal coliforms (per litre)	Faecal *enterococci* (per litre)	F-specific RNA phages (per litre)	Enteroviruses (per litre)	*Giardia* cysts (per litre)	*Cryptosporidium* oocysts (per litre)
Raw	2.16 x 10^8	1.84 x 10^7	2.2 x 10^7	11.87	2.35 x 10^5	1.36 x 10^4
Treated	9.60 x 10^5	1.61 x 10^5	6.65 x 10^4	0.19	6.31 x 10^3	8.68 x 10^2
Surface	1.07 x 10^5	1.39 x 10^4	4.28 x 10^3	0.04	31.1	25.8

Source: Author.

They were of the same order of magnitude as those reported in most other studies, except for enteroviruses, which were found at a rather low concentration, particularly in surface water by comparison with earlier data (Vilaginès *et al.*, 1993*b*). In any water type, F-specific phages always outnumbered viruses and parasites.

As expected, concentrations of all germs decreased after sewage treatment, with over 2 logs reduction for faecal coliforms, faecal *enterococci* and F-specific RNA phages, and about 1.5 log for the three other germs (Table 3).

Table 3. **Micro-organisms elimination by the wastewater treatment plant**

Reduction	Faecal coliforms	Faecal *enterococci*	F-specific RNA phages	Enteroviruses	*Giardia* cysts	*Cryptosporidium* oocysts
Percentage	99.6	99.1	99.7	98.4	97.3	93.6
log 10	2.35	2.06	2.52	1.80	1.57	1.20

Source: Author.

All water samples were positive for bacterial indicators and F-RNA phages (except in surface water: 94 per cent), whereas the occurrence of the other micro-organisms was highly variable, depending on the water quality. Alone, the percentages of positivity of *Salmonella* were almost constant (24 per cent) in the three types of water (Table 4).

Table 4. **Percentage of positive samples**

Micro-organisms	Assays (number)	Raw water (per cent)	Assays (number)	Treated wastewater (per cent)	Assays (number)	Surface water (per cent)
Faecal coliforms	35	100	35	100	35	100
Faecal *enterococci*	35	100	35	100	35	100
F-specific RNA phages	34	100	35	100	35	94.3
Enteroviruses	30	96.7	31	58.1	34	29.4
Giardia cysts	22	100	22	86.4	22	45.5
Cryptosporidium oocysts	22	72.3	22	27.3	22	31.8
Salmonella	35	25.7	35	25.7	35	22.9

Source: Author.

Results obtained from statistical tests by Principal Component Analysis (PCA) are shown in Table 5. Only correlation coefficients greater than 0.4 have been displayed, this level classically considered as the minimum to be statistically significant.

Table 5. **Principal Component Analysis: correlation coefficients > 0.4**

Micro-organisms	Raw wastewater			Treated wastewater			Surface water		
	PC1	PC2	PC3	PC1	PC2	PC3	PC1	PC2	PC3
Faecal coliforms	0.72				0.48		-0.76	0.53	
Faecal *enterococci*			0.61	-0.90			-0.87		
F-specific RNA phages			-0.75	-0.87				-0.68	
Enteroviruses	0.47	-0.65			0.60	-0.50	0.45		0.52
Giardia cysts	-0.83	-0.45			-0.80		-0.44		
Cryptosporidium oocysts	-0.87				-0.62			0.78	
Salmonella		-0.68				-0.78			0.72

Note: PC: Principal Component.

Source: Author.

In raw water, the variables faecal coliforms, *Giardia* and *Cryptosporidium* are well correlated to the first Principal Component, positively for the first one, negatively for the two others, whereas enterovirus and *Salmonella* are negatively correlated to the second Principal Component, and faecal *enterococci* and phages to the third one, the former positively, the latter negatively. So the best representation of all variables will be the CP1/CP3 plan.

In treated wastewater, variables faecal *enterococci* and phages are well correlated to the first Principal Component, variables faecal coliforms and viruses positively to the second Principal Component, and negatively for *Giardia* and *Cryptosporidium.* The variable *Salmonella* is only significantly correlated to the third Principal Component. So, in this case, the best representation of all variables will be the CP1/CP2 plan.

In surface water, all variables are well correlated to the first or second Principal Components, except *Salmonella* and *Giardia*, which are more closely correlated to respectively the third and fourth Principal Component.

These results are well illustrated by the following biplots (Figures 1, 2 and 3) respectively for raw wastewater, treated wastewater and surface water.

Figure 1. **Raw wastewater**

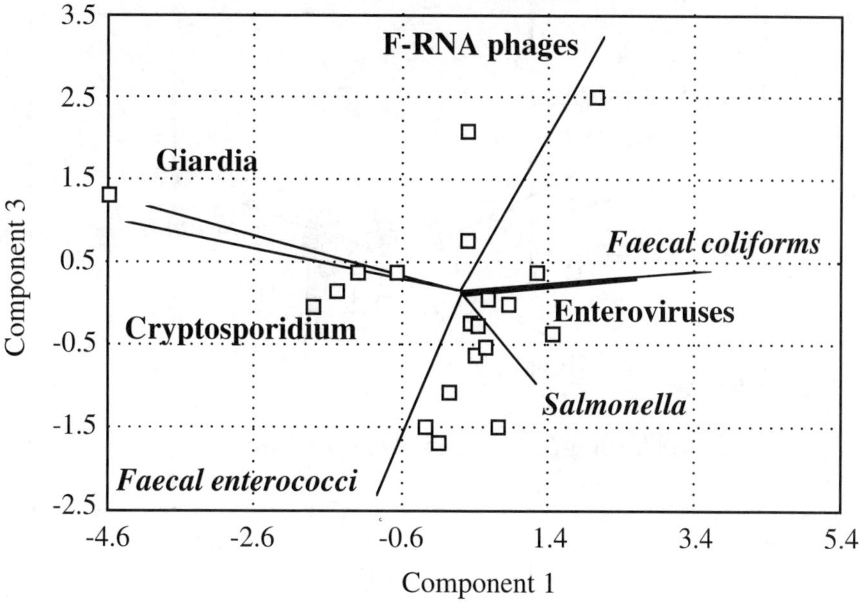

Source: Author.

Figure 2. **Treated wastewater**

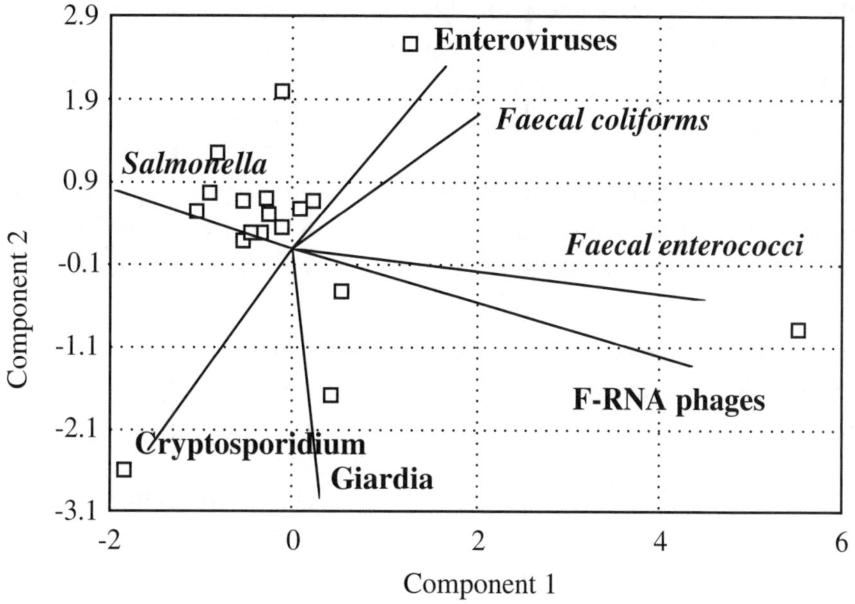

Source: Author.

Figure 3. **Surface water**

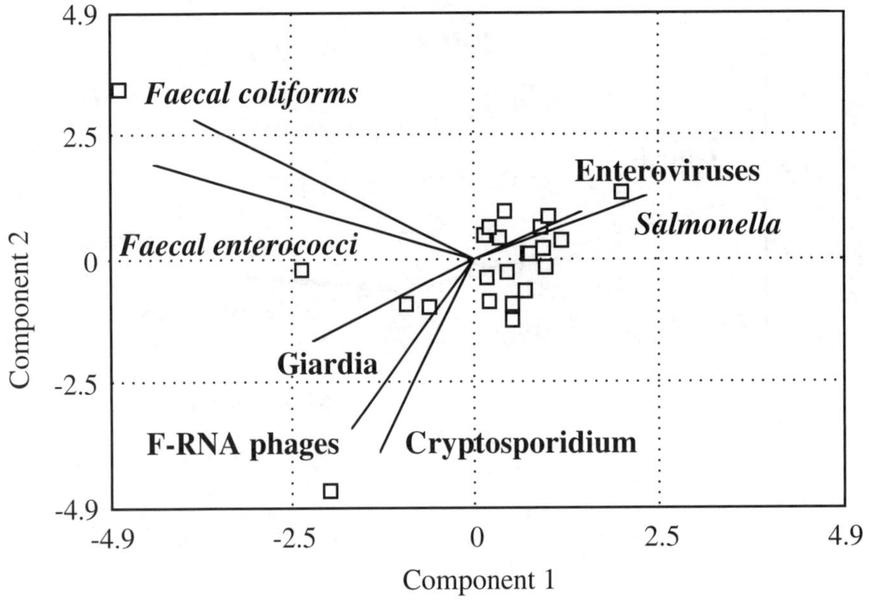

Source: Author.

They show that the behaviour of each micro-organism compared to the others is highly variable depending on the water quality. Solely the variables *Giardia* and *Cryptosporidium* seem to fluctuate similarly in raw and treated wastewaters. On the other hand, phages and viruses behave systematically differently. Consequently, this analysis, although descriptive, precludes any covariation between these germs, as well as with all other variables pairs.

Though unnecessary in regard to these observations, the subsequent regression analysis resulted in the only potentially significant correlation coefficients (i.e. > 0.4) shown in Table 6, with the p-values corresponding to the t-test on the slope and to the Fisher test on the model.

The analysis of the leverage points and the study of the distribution fitting of the model residuals showed that a statistically significant correlation existed only between *Giardia* and *Cryptosporidium*, and this only in raw water.

In all the other cases, the model was not any more appropriate either after analysis of the leverage points, such was the case for the faecal *enterococci*/phages pair in treated water, or the *Giardia/Cryptosporidium* and the faecal coliform/faecal *enteroccoci* pairs in surface water, or because the correlation value resulted from an artifice due to numerous zero values of the variables: *Giardia/Cryptosporidium* pair in treated water, and *Cryptosporidium*/phages pair in surface water.

Table 6. **Regression analysis**

Water type	Micro-organisms pairs	Correlation coefficient	Slope	Test t / Test F p value	Leverages and residue analysis
Raw wastewater	[GIA-CRY]	$r^2 = 0.821$	a = 11.49	0.000	* Leverages: trials n°19 & n° 21 but fitted the model when eliminated.
Treated wastewater	[FE-PHA]	$r^2 = 0.516$	a = 113.1	0.015	* Leverages: trials n°35 & n° 2
	[CRY-GIA]	$r^2 = 0.495$	a = 1.772	0.019	** Artifice: high number of "0"
Surface water	[CRY-PHA]	$r^2 = 0.568$	a = 10.21	0.006	* Leverage: trial n°19 ** Artifice: high number of "0"
	[GIA-FE]	$r^2 = 0.772$	a = 0.058	0.000	* Leverage: trial n°28
	[FC-FE]	$r^2 = 0.612$	a = 3.806	0.0001	* Leverage: trials n°1, 4 & 28

Notes: GIA: *Giardia* cysts.
CRY: *Cryptosporidium* oocysts.
FE: faecal *enterococci.*
FC: faecal coliforms.
PHA: F-specific RNA phages.

Source: Author.

Discussion-conclusion

The principal purpose of this study was to evaluate the appropriateness of phages in predicting the presence of human pathogens such as viruses, and eventually of the emerging protozoan parasites *Giardia* and *Cryptosporidium.* The occurrence of these micro-organisms was investigated parallel to the one of classical faecal indicators. This experimental protocol allowed the study of variable arrangements in regard to the percentage of positivity of the micro-organisms, thus providing information in the different schemes found in the environment. Among the candidates as alternate indicators of virus presence, F-specific RNA phages were tested together because the method for their numeration was the only one standardized, and because of their high resistance to disinfection processes. Besides, most studies on the subject were undertaken with this phage.

Field data obtained in this study indicated, in agreement with other reports, that faecal coliforms and *enterococci* failed in the role of virus (Wyer *et al.*, 1995) or parasite (Smith and Rose, 1990) indicators. In addition, they were found to behave differently from one another, and therefore the presence of one did not necessarily indicate the presence of the other (Shadix *et al.*, 1995). This observation enforces the actual tendency to limit the number of indicators to be tested, or at least to establish a choice, taking account of their density in function of the sampling site. However, they still remain useful in assessing faecal pollution.

Concerning F-specific RNA phages, they were found, surprisingly, to be less resistant to wastewater treatment process than other organisms. Besides, they largely outnumbered enterovirus concentration in the three water types studied, but they were never predictive of their presence as observed in other studies (Palmateer *et al.*, 1991; Abbaszadegan *et al.*, 1995; Schwab *et al.*, 1995; Sobsey *et al.*, 1995*a*). Since this conclusion is in opposition to several published reports, questions concerning that divergence arise. It is likely that this controversy results principally from the difference in the statistical analysis applied to data. Factors that influence the statistical interpretation are mainly the choice of the risk levels α (Student and Fisher tests), and the confirming tests to validate the model. Regression analysis is the most widely used statistical test to evaluate relationship between micro-organisms. Yet, in majority, reports fell in providing indications concerning the different tests, if used, for model validation. This lack of information, as well as the medley between the predictive limits of the model and its confidence interval may lead to erroneous conclusions. Classically, a regression analysis should imply a systematic validation of parameters in the model by the Student test (α=5 per cent) and of the model quality by the Fisher test (α=1 per cent), as well as a leverage analysis and a distribution fitting study of the model residuals.

Applied in these conditions to our data, the statistical analysis demonstrated there was no correlation between phages and enteroviruses, nor between phages and *Giardia* and *Cryptosporidium*. These finding first observed by Principal Component Analysis, were logically confirmed by the regression analysis. Therefore, F-specific RNA cannot be adopted as viruses nor parasites indicators.

This research was supported by the French Ministry of Health.

Questions, comments and answers

Q: Did you take any account of recovery efficiency for *Cryptosporidium* and *Giardia*?

A: We know that, with the method we use, it varies widely. For other organisms, we use standard techniques and for viruses our efficiency is about 50 per cent.

C: It seems very difficult to glean useful information in order to look for indicators if we don’t know what the recoveries are, since they are very variable. Until we have a more reliable method of detection, it may be premature to say that these are not indicators for these organisms.

REFERENCES

ABBASZADEGAN, M., P. STEWART, M. LECHEVALLIER, M. YATES, and C. GERBA (1995), "Occurrence of enteroviruses in groundwater and correlation with water quality parameters", WQTC Proceedings, American Water Works Association, pp. 2 099-2 114.

AFNOR NF T 90-414 (1985), "Essais des eaux. Recherche et dénombrement des coliformes et des coliformes thermotolérants. Méthode générale par filtration sur membrane".

AFNOR XP T 90-416 (1996), "Essais des eaux. Recherche et dénombrement des entérocoques. Méthode générale par filtration sur membrane".

BORREGO, J.J., R. CORNAX, M.A. MORINIGO, E. MARTINEZ-MANZANARES, and P. ROMERO (1990), "Coliphages as an indicator of faecal pollution in water: their survival and productive infectivity in natural aquatic environment", *Wat. Res.* 24, pp. 111-116.

BARRON, A.L., C. OLSHEVESKY, and M.N. COHEN (1970), "Characteristics of the BGM line of cells from African Green Monkey Kidney", *Arch. ges. Virusforsch.* 32, pp. 389-392.

CRAUN, G.F. (1988), "Surface water supplies and health", *Am. Water Works Assoc. J.* 80, pp. 40-52.

FIGUEROA, R., A. SOPULVEDA, M.A. SOTO, and J. TEHA (1978), "Informational analysis of MS2 and ΦX 174 virus genomes", *J. Theor. Biol.* 74, pp. 203-207.

GUELIN, A. (1952), "Application des bactériophages à l'étude des eaux polluées. I La survie des enterobacteriacées dans les eaux. II Bactériophages des eaux à grandes et petites plages", *Ann. Inst. Pasteur Paris* 82, pp. 78-89.

HAVELAAR, A.H. (1987), "Bacteriophages as Model Organisms in Water Treatment", *Microbiol. Sci.* 4, pp. 362-364.

HAVELAAR, A.H., M. VAN OLPHEN, and J.F. SCHIJVEN (1995), "Removal and inactivation of viruses by drinking water treatment processes under full scale conditions", *Wat. Sci. Tech.* 31, pp. 55-62.

IAWPRC STUDY GROUP ON HEALTH RELATED WATER MICROBIOLOGY (1991), "Bacteriophages as model viruses in water quality control", *Wat. Res.* 25, pp. 529-545.

ISO 6340 (1995), "Qualité de l'eau. Recherche de *Salmonella*".

ISO 10705-1 (1995), "Water quality. Detection and enumeration of F-Specific RNA bacteriophages".

JOFRE, J., E. OLLÉ, F. LUCENA, and F. RIBAS (1995), "Bacteriophage removal in water treatment plants", *Wat. Sci. Tech.* 31, pp. 69-73.

KATZENELSON, E., B. FATTAL, and T. HOSTOVESKY (1976), "Organic flocculation: an efficient second step concentration method for the detection of viruses in tap water", *Appl. Environ. Microbiol.* 32, pp. 638-639.

KOTT, Y., N. ROZE, S. SPERBER, and N. BETZEN (1974), "Bacteriophages as viral pollution indicators", *Wat. Res.* 8, pp. 165-171.

LEVINE, W.T., W.T. STEPHENSON, and G.F. CRAUN (1990), "Waterborne disease outbreaks", *CDC Morbidity and Mortality Weekly Report* 39, (SS-1) 1.

LISLE, J.T. and J.B. ROSE (1995), "*Cryptosporidium* contamination of water in the USA and UK: a mini review", *J. Water SRT-Aqua* 44, pp. 103-117.

MACKENZIE, W.R., N.J. HOXIE, M.E. PROCTOR, M.S. GRADUS, K.A. BLAIR, D.E. PETERSON, J.J. KAZMIERCZAK, D.G. ADDISS, K.R. FOX, J.B. ROSE, and J.P. DAVIS (1994), "A massive outbreak in Milwaukee of *Cryptosporidium* infection transmitted through the public water suppy", *New England J. Med.* 331, pp. 161-167.

MAUL, A. (1991), in Lavoisier (ed.), "Virologie des milieux hydriques", coordinateur L. Schwartzbrod, pp. 149-152.

MAURIN, J. (1965), "Application des cultures cellulaires à l'isolement et l'identification des virus. Isolement et étude des virus dans l'œuf embryonné et en cultures cellulaires", Cateigne G., Maurin J., Edition de la Tourelle, St.Mandé, pp. 157-171.

PALMATEER, G.A., B.J. DUTKA, E.M. JANZEN, S.M. MEISSNER, and M.G. SAKELLARIS (1991), "Coliphage and bacteriophage as indicators of recreational water quality", *Wat. Res.* 25, pp. 355-357.

PAYMENT, P., L. RICHARDSON, J. SIEMIATYCKI, R. DEWAR, EDWARDES, and E. FRANCO (1991), "A randomized trial to evaluate the risk of gastrointestinal disease due to consumption of drinking water meeting current microbiological standards", *Am. J. Public Health.* 81, pp. 703-708.

SCHWAB, K.J., R. DE LEON, and M.D. SOBSEY (1995), "Comparison of processing methods for RT-PCR detection of enteric viruses in water", American Water Works Association 1995 WQTC Proceedings, pp. 2 037-2 057.

SHADIX, L.C., B.S. NEWPORT, S.R. CROUT, and R.J. LIEBERMAN (1995), "Occurence of microbial indicators in various groundwater sources", American Water Works Association 1995 WQTC Proceedings, pp. 239-248.

SOBSEY, M.D., A.P. DUFOUR, C.P. GERBA, M.W. LECHEVALLIER, and P. PAYMENT (1993), "Using a conceptual framework for assessing risks to health from microbes in drinking water", *Jour. Am. Water Works Assoc.* 85, pp. 44-88.

SOBSEY, M.D., A. AMANTI, and T.R. HANDZEL (1995*a*), "Detection and occurence of coliphage indicator viruses in water", American Water Works Association 1995 WQTC Proceedings, pp. 2 087-2 097.

SOBSEY, M.D., D.A. BATTIGELLI, R. ARMON, and M. TRUDEL (1995*b*), "Male specific coliphages as indicators of viral contamination of drinking water", Amer. Water Works Assoc. Research Foundation, Denver, Colorado.

SMITH, H.V. and J.B. ROSE (1990), "Waterborne Cryptosporidiosis", *Parasitology Today* 6, pp. 8-12.

STETLER, R. (1984), "Coliphages as indicators of enteroviruses", *Appl. Environ. Microbiol.* 48, pp. 668-670.

VILAGINÈS, Ph., B. SARRETTE, and R. VILAGINÈS (1989), "Viral multicloning procedure and replicates technique", *Wat. Sci. Tech.* 21, pp. 85-92.

VILAGINÈS, Ph., B. SARRETTE, G HUSSON, and R. VILAGINÈS (1993*a*), "Glass wool for virus concentration from water at ambient pH levels", *Wat. Sci. Tech.* 27, pp. 299-306.

VILAGINÈS, Ph., B. SARRETTE, and R. VILAGINÈS (1993*b*), "Improvement in water detection by using glass wool in the concentration", Contamination of the Environment by Viruses and Methods of Control, *Band 112-Wien*, pp. 11-13.

WENTZEL, R.S., P.E. O'NEILL, and J.F. KITCHENS (1982), "Evaluation of coliphage detection as a rapid indicator of water quality", *Appl. Environ. Microbiol.* 43, pp. 430-434.

WYER, M.D., J.M. FLEISHER, J. COUGH, D. KAY, and H. MERRETT (1995), "An investigation into parametric relationships between enterovirus and faecal indicator organisms in the coastal waters of England and Wales", *Wat. Res.* 29, pp. 1 863-1 869.

YATES, M.V., C.P. GERBA, and L.M. KELLY (1985), "Virus Persistence in Groundwater", *Appl. Environ. Microbiol.* 49, pp. 778-781.

BACTERIAL AND PROTOZOAL PATHOGEN SURVIVAL IN BIOFILMS AND POSSIBILITIES FOR CONTROL WITHOUT NITRITE FORMATION

by

C. William Keevil
Centre for Applied Microbiology & Research, Porton Down, Salisbury, Wilts., United Kingdom

Brief historical review

The evolution of man from a hunter-gatherer to primarily farmer and urban dweller has had a profound influence on the environment. This is because the sharp rise in the world population has necessitated intensive agricultural practises to provide sufficient food, coupled with the need to dispose of increasing quantities of waste and wastewater from homes and industries. Consequently, the normal water cycle has been perturbed by the introduction of toxic organic compounds and heavy metals from disposal of waste to land and leaching into water courses. Eutrophication of rivers and lakes has occurred due to ingress of nitrogen and phosphorus resulting from excessive use of detergents, with the associated production of toxic by-products from cyanobacterial blooms, such as microcystin hepatotoxins and anatoxin neurotoxins (Keevil, 1991). Moreover, intensive animal rearing and poor sewage disposal has increased the transmission of microbial pathogens in water supplying industrial and domestic premises (Table 1).

Table 1. **Man and the water cycle**

Location	Cause	Pollution
Land	Landfill leakage Sewage sludge disposal	Heavy metals, organics
River/borehole	Cows Chickens	*Cryptosporidium, Giardia, E. coli* 0157 *Campylobacter, Salmonella*
Lakes	Detergent eutrophication (N, P)	Blue-green algal blooms and toxins
Distribution supplies	Ingress; poor treatment	Coliforms, *Aeromonas, Ps. aeruginosa, Acinetobacter*
Buildings/cooling towers	Contaminated supplies Corrosion	*Legionella, Mycobacteria avium*, Amoebae Metal salts
Sea	Poor wastewater treatment	Viruses, Coliforms, *Vibrio cholerae*

Source: Keevil, in press.

An important facet of the life cycle of many micro-organisms is the benefit gained from aggregating at solid/liquid, air/liquid or liquid/liquid interfaces (Ellwood *et al.*, 1982). Attachment and growth on environmental and clinical surfaces, as well as various man-made materials, results in biofilm formation (Table 2; Costerton *et al.*, 1987). Biofilm is a heterogeneous system, comprising clinical or environmental components such as salivary proteins, fats or metals and salts from associated corrosion processes. A range of micro-organisms can be present with a high species diversity, particularly in the mesophilic environment (Rogers *et al.*, 1994*a*, 1994*b*). Copious exopolymeric substance (EPS) formation is frequently encountered, together with extracellular enzymes which may be involved in EPS formation, such as glucosyl- and fructosyl-transferases catalysing glucan and fructan formation from sucrose in dental plaque (Keevil *et al.*, 1984). Fermentation by-products include acid formation to decrease pH, ammonia to increase pH and hydrogen sulphide gas which is toxic. All of these products can drive corrosion processes, such as sulphide accelerated steel corrosion (Hamilton, 1985), and can markedly affect drinking water quality (Keevil *et al.*, 1989).

Table 2. **Biofilm composition**

Heterogeneous system	Aqueous phase containing:
Environmental components	Peat derived humic acids Metals/salts Oils/fats Salivary proteins Epithelial mucin, fibrinogen, etc.
Micro-organisms	Bacteria, yeast, fungi Protozoa, plankton
Microbial products	Extracellular enzymes EPS Acids, NH_3, H_2S

Source: Keevil, in press.

Potable waters are treated to European Union and World Health Organisation standards to prevent the spread of pathogens and indicator bacteria; however, this water is not intended to be sterile and, consequently, contains a diverse range of micro-organisms (Geldreich *et al.*, 1972). Biofilms exist in all water distribution and plumbing systems at temperatures below 60 °C. By contrast to most clinical biofilms, environmental biofilms are comprised of complex consortia of micro-organisms and can contain aerobic and anaerobic bacteria, amoebae, protozoa, nematodes and fungi (Table 3).

Biofilm formation comprises a dynamic flux of processes involving reversible and irreversible attachment, colonisation and maturation, active detachment or passive sloughing, and predator grazing. A pseudo-steady state is reached for biofilms in potable water where the numbers of recoverable viable bacteria vary between 10^5 to 10^7 per cm^2, dependent on the nutrient availability, shear force and physicochemistry of the substratum. By contrast, the numbers of viable bacteria recovered from the planktonic phase are typically only 10 to 10^3 per ml. Thus, considering the many kilometres of distribution main and plumbing pipe supplies from the treatment works to the tap, the biofilm represents a much more significant environmental reservoir of micro-organisms than the water phase. The concerted metabolic activities therein may also affect water quality by providing a locus for processes, such as nitrite formation by ammonia oxidising autotrophs in waters containing excess ammonia (Mackerness and Keevil, 1991).

Table 3. **Microbiology of water and biofilms**

Heterotrophic spp.	Other species
Acinetobacter	*Nitrosomonas, Nitrobacter*
Aeromonas	Iron oxidising bacteria
Alcaligenes	Sulphate reducing bacteria
Flavobacterium	
Methylobacterium	
Pseudomonas	Yeasts, *Actinomycetes*
Legionella	*Chladysporium, Aspergillus*
Enterobacter	
Klebsiella, Proteus, etc.	
Micrococcus	*Acanthamoeba, Hartmanella*
Staphylococcus	*Paramecium, Tetrahymena*
Bacillus	*Lachrymaria, Vorticella*
Corynebacterium	

Source: Keevil, in press.

Attempts to control dissemination and regrowth of pathogens and indicator bacteria in potable water through the use of oxidative disinfectants such as chlorine or ozone have led to subsequent biofilm problems. This is because these disinfectants oxidise refractory carbon to assimilable organic carbon (AOC) which fuels biofilm growth. Van der Kooij (1992) has suggested that AOC concentrations of >10 $\mu g\ L^{-1}$ enable growth and biofilm formation in potable water systems. The implication of these findings is that it is essential to maintain low AOC concentrations for water in distribution to prevent regrowth and the multiplication of indicator bacteria and potential pathogens such as *Legionella*, *Aeromonas* and *Pseudomonas* spp. With ozonation treatment of potable water becoming widespread, oxidising refractory dissolved organic carbon to AOC, it is essential to install something like granular activated carbon and/or sand filtration to reduce the AOC post-ozonation.

Eukaryotes should not normally be present in potable water but ingress into the main supply by breakthrough at the works entry via poor integrity of the mains supply, cross connections or backsiphonage at outlets allows their colonisation, particularly where there is low residual chlorine disinfection and/or high chlorine demand. These circumstances also permit ingress of indicator bacteria and potential pathogens which, despite chlorine and nutrient shock, may survive in the biofilm and exhibit regrowth (McFeters *et al.*, 1984; Mackerness *et al.*, 1991, 1993). The adherent species therefore provide a haven for waterborne pathogens against extremes of water chemistry, temperature and disinfection (LeChavellier *et al.*, 1988; Keevil *et al.*, 1989, 1990).

Field studies may be cumbersome, longwinded, expensive and not necessarily reproducible due to changes in weather patterns. Developments in biotechnology have permitted the design of laboratory fermentation systems and microcosms, together with probes to track species *in situ*, to reproducibly model specific environments and defined physico-chemical parameters. Some of these developments will now be reviewed.

Current activities

Molecular probes

An important advance in biotechnology has been the development of molecular probes which can be used to track pathogens in the environment. This powerful approach permits an assessment of water quality and the suitability of current treatment strategies. Probes which fluoresce when hydrolysed with specific enzymes have been used to track coliforms, producing β-galactosidase activity, and *E. coli*, producing β-glucuronidase activity (Robinson *et al.*, 1995). Genetically engineered constructs of *E. coli* containing the *nirB* promoter linked to β-galactosidase have also been used to track the micro-organism. In addition, the constructs act as reporter genes which reveal at the µm level the microenvironments, in this case low oxygen concentration, where the organism is residing. This provides more precise measurements than microelectrodes which tend to exceed 10 µm diameter, are not robust and are sensitive to electrical noise.

Monoclonal antibodies linked to fluorochromes such as FITC or Texas Red provide specific, high sensitivity probes for use with fluorescence plate readers, fluorescence microscopy and scanning laser confocal microscopy (SCLM). Antibody probes have been used to track *Legionella pneumophila, Campylobacter jejuni, Pseudomonas aeruginosa, Cryptosporidium parvum* and *Microcystis aeruginosa* in water samples and biofilms with a high species diversity (Keevil *et al.*, 1995).

Nucleic acid chemistry is now facilitating the development of fluorescently labelled RNA oligomers complementary to specific sequences in ribosomal 16S and 23S RNA of micro-organisms (Amann *et al.*, 1995). Using this approach, probes have been developed to successfully track *Campylobacter, Legionella* and *Mycobacteria* spp.

Environmental models

The complexity of biofilm ecosystems has made reproducible study difficult, and in previous work, we have developed a continuous culture ecological system in the laboratory to model biofilms in the environment (Figure 1; where *F* denotes the flow of media, such as potable water, into each chemostat vessel; *G* denotes flow of sparging gases, such as air or N_2/CO_2, for aerobic or anaerobic growth, respectively; *C* denotes flow of spent culture through a weir to another vessel or to waste; *CHX* denotes continuous or pulse addition of disinfectants such as chlorhexidine or chlorine; and *AT* denotes insertion of metal, plastic or epoxy tiles for cell attachment; Keevil, 1989). A primary seed vessel ensures the reproducible maintenance of complex microbial consortia and supplies subsequent growth vessels in an open flow system for biofilm experimentation in defined environments. Tiles of known physico-chemistry can be inserted and removed aseptically after hours or months, for microbiological and microscopy image analysis. By contrast to field studies, the model is cheap, reproducible and easily manipulated. It can be used to model the effects of different growth rates, temperatures, environmental chemistry and disinfectant concentration. The shear rate imposed on surfaces by water velocity can be determined by manipulation of the stirrer speed: typical water velocities of 0.2-3.0 m sec^{-1} have been investigated. Importantly, the model can be assembled in Class III containment cabinets for the study of more infectious pathogens such *Legionella pneumophila* and *Cryptosporidium parvum.*

Figure 1. **Two-stage continuous culture apparatus for biofilm formation**

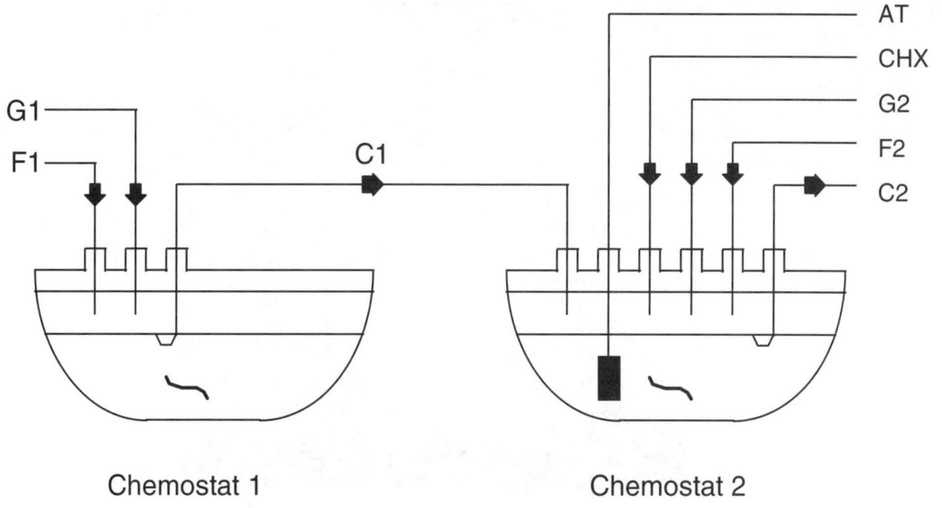

Source: Keevil *et al.*, 1987; Keevil, 1989.

The biofilm model has facilitated important breakthroughs in our knowledge of biofilm structure, function and ecology. The majority of so-called biofilms are actually extremely heterogeneous in structure with many water channels penetrating to the substratum, which permit ingress of nutrients and, presumably, antibiotics or disinfectants (Rogers *et al.*, 1991; Lawrence *et al.*, 1991). The 'biofilm' consists of a patchy basal layer, 5-10 μm thick, covered with stacks or fronds of microcolonies rising 100-200 μm above the substratum surface (Figure 2; Keevil and Walker, 1992). Variations in cell morphology and colour within the microcolonies of the stacks suggest close physiological associations of consorting species (Keevil *et al.*, 1995). There are inevitably many water channels but the apparent spaces between the stacks is dependent on the physicochemistry of the substratum and its conditioning pellicle layer, the availability of nutrients and the activity of grazing predators. Indeed, we have demonstrated that motile bacteria, protozoa and nematodes can be seen passing through the channels. The convective flow of microscopic fluorescent beads through the channels has also been described (Lewandowski *et al.*, 1993; DeBeer *et al.*, 1994). In nutrient rich nutrient environments the biofilm may thicken and consolidate, with the water channels narrowing to resemble a sponge. The structure can also appear confluent due to the production of copious EPS gel within the channels. Any observation technique which requires even partial dehydration will result in shrinkage phenomena generating artefacts. The structures described are still largely microscopic in scale. In the presence of sunlight, however, photosynthetic species can proliferate, producing secondary metabolites, EPS and intracellular storage polymers to subsequently enrich the other members of the adherent consortium. This photosynthetic stimulation of biofilm metabolism can result in the proliferation of green macro-blooms, obvious to the naked eye (Keevil, 1995).

The surface of the so-called bio*film* may mop up reactive biocides such as chlorine, acting sacrificially to protect the cells below. However, diffusion may not be rate limiting due to accessibility through the water channels which penetrate to the substratum. This penetrability might explain how less reactive monochloramine appears to be a better residual disinfectant than chlorine to control biofilm micro-organisms (see below; LeChavellier *et al.*, 1988; Keevil *et al.*, 1990).

Figure 2. **Open architecture structure of biofilm with fronds and water channels**

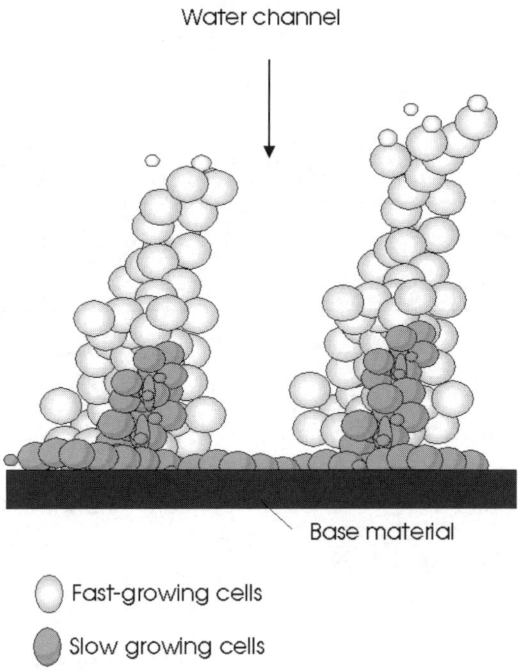

Source: Rogers *et al.*, 1991; Keevil and Walker, 1992; Walker *et al.*, 1995*a*.

To date, the model has been adapted to show that biofilms consisting of aerobic heterotrophs and autotrophs provide an environmental reservoir or safe haven for the microaerophiles, *L. pneumophila* and *Campylobacter jejuni,* coliforms such as *Escherichia coli* and *Klebsiella oxytoca*, and *Aeromonas hydrophila*, *Pseudomonas aeruginosa*, and *C. parvum* (Rogers and Keevil, 1992, 1995; Mackerness *et al.*, 1991, 1993; Packer *et al.*, in press; Robinson *et al.*, 1995). Individual cells of these micro-organisms could be observed *in situ* in the biofilm community using gold or fluorochrome-labelled specific antibodies or utilising β-galactosidase or glucuronidase activities of the coliforms to hydrolyse fluorochrome-substituted substrates. Species development and succession in biofilm consortia can be followed temporally on different substrata (Tables 4 and 5; Rogers *et al.*, 1994*a*, 1994*b*). Interestingly, indicators of pollution such as *Ps. aeruginosa* and *Klebsiella* spp. were naturally present in very low numbers in the inoculum but were enriched to form a significant proportion of the biofilm consortia growing on several of the different substrata.

Many of the potential pathogens investigated have shown increased resistance to chlorine-based disinfectants in biofilms (Keevil *et al.*, 1990) but are more susceptible to bromine, copper or silver (Thurman and Gerba, 1989; Rogers *et al.*, 1995; Walker *et al.*, 1994*a*, 1994*b*). In particular, *L. pneumophila* colonises a variety of biofilms on distribution supply bitumen-painted iron (C.W. Mackerness, unpublished data), and plumbing materials in cold or warm potable water, but is sensitive to copper and galvanised iron surfaces (Tables 4-6; Rogers *et al.*, 1994*a*, 1994*b*). The pathogen is able to grow in amoebae, other micro-organisms and lung macrophages, which has led many to speculate that it is an obligate intracellular parasite. However, we have shown that this is not so: it can proliferate in the microaerophilic niches of the biofilm, in the absence of environmental hosts, utilising its considerable metabolic versatility (Mauchline and Keevil, 1991) to grow on the consortium's secondary metabolites (Rogers and Keevil, 1992; Rogers *et al.*, 1994*a*, 1994*b*; Surman *et al.*, in press).

Table 4. **Colonisation of plumbing materials after 24 hours at 30 °C**

	Steel	SS	Latex	E-P	PP	PE	PVCu	PVCc
L. pneumophila	4	0.1	15	9	1.5	0.5	2.9	1.5
P. aeruginosa					1.9	260		
P. acidovorans							37	880
P. diminuta				1 000	22			
P. fluorescens					26	380	1	
P. maltophila	80	10	8 000	100		20	3	
P. mendocina	60	25	4 000	3	3		4.7	210
P. stutzeri	90	1						
P. testosteroni	410	12	9 000		1	80	2	150
P. vesicularis						6.2	59	
P. xylosoxidans		36 000					310	
S. paucimobilis	100	10		1 000		0.2	1.2	300
Actinomyctes sp.	10	6	34 000		1		19	200
Aeromonas sp.			8 000					
Alcaligenes sp.	70			2 000	4	0.2		
Flavobacterium sp.	40	7	800	1 000	3	0.3	0.2	290
Methylobacterium sp.	110	4			3		5	
Klebsiella sp.	130							
Acinetobacter sp.	300	26	40 000	10 000	17	440	0.6	

Note: SS denotes stainless steel substratum, E-P ethylene-propylene copolymer, PP polypropylene, PE polyethylene, PVCu unpolymerised polyvinyl chloride, PVCc chlorinated polyvinyl chloride.

Source: Rogers *et al.*, 1994*b*.

Table 5. **Colonisation of plumbing materials after 21 days at 30 °C**

	Steel	SS	Latex	E-P	PP	PE	PVCu	PVCc
L. pneumophila	17	13	150	500	37	13	11	7.9
P. aeruginosa	30							
P. acidovorans						40		11
P. diminuta						2		
P. fluorescens				1 000				3
P. maltophila	10	11	3 000			10	10	
P. mendocina			13			40	0.01	
P. stutzeri	140	70	2 000					0.1
P. testosteroni		180				20		8
P. vesicularis	250							3
P. xylosoxidans						7	40	
S. paucimobilis	30	36	5 000	1 600	790	170	140	362
Actinomyctes sp.	130	2	7 000	8 000		9	0.01	2.8
Aeromonas sp.		6 000						
Alcaligenes sp.	10	10			320	80		30
Flavobacterium sp.	41		15 000	2 400		0.2	90	50
Methylobacterium sp.	20	150			140	30	60	60
Klebsiella sp.			31 000					
Acinetobacter sp.	70	39	22 000	3 100	400	180	40	60

Note: SS denotes stainless steel substratum, E-P ethylene-propylene copolymer, PP polypropylene, PE polyethylene, PVCu unpolymerised polyvinyl chloride, PVCc chlorinated polyvinyl chloride.

Source: Rogers *et al.*, 1994*b*.

One of the technological advances in urban development has been the design of the cooling tower to keep homes and large institutional buildings at cool temperatures in hot climates and to cool process waters in industrial plants. A particular problem of wet evaporative cooling towers is that the sump temperatures are ideal for microbial growth (typically 30-35 °C), and this can lead to biofilm formation and colonisation by *L. pneumophila* in the trough and gutter design of the towers (Colborne and Dennis, 1985). In fact, some of the most serious outbreaks of Legionnaires' disease have been caused by inhalation of contaminated aerosol released from poorly maintained cooling towers. We have been able to extend the biofilm model studies to the study of *L. pneumophila* survival in cooling towers by designing an exact replica of a typical evaporative cooling tower in Class III containment. This unique assembly functions with a normal heat load and is designed to incorporate tiles at various positions in the sump, trough and gutter, and upper packing to study biofilm formation. It is currently being used to study *Legionella* colonisation and control, operational practices such as continuous and discontinuous heat load, and disinfection and cleaning.

Table 6. **Colonisation of plumbing materials by aquatic flora and *Legionella pneumophila***

Temp	Material	Microflora	L. pneumophila
20 °C	Copper	2.2×10^5	0
	Polybutylene	5.7×10^5	665
	PVCc	1.8×10^5	2 130
40 °C	Copper	8.1×10^4	2 000
	Polybutylene	1.2×10^6	112 000
	PVCc	3.7×10^5	68 000
50 °C	Copper	2.3×10^4	0
	Polybutylene	3.2×10^6	890
	PVCc	1.2×10^5	60
60 °C	Copper	4.5×10^2	0
	Polybutylene	4.3×10^4	0
	PVCc	5.2×10^3	0

Source: Rogers *et al.*, 1994*a*.

C. parvum is the protozoal Coccidian parasite of man responsible for many waterborne outbreaks of debilitating Cryptosporidiosis; indeed, 400 000 people were affected in a single outbreak in Milwaukee, United States in 1993 (Mackenzie *et al.*, 1994). The pathogen's oocyst appears resistant to chlorine based disinfectants but more susceptible to ozone (Presdee *et al.*, 1995). Work at CAMR has shown that the 5 µm round oocysts can survive in biofilms of the chemostat model for many weeks in an infectious state (Rogers and Keevil, 1995; Keevil *et al.*, 1995). Higher shear rates are required for their removal and transient sloughing from the biofilm might explain the many sporadic cases of unknown origin. This persistence in the biofilm safe haven therefore presents an interesting challenge for the disinfectant industry and the safeguarding of the public health.

The presence in potable water of coliform bacteria in general, and *E. coli* in particular, is a public health concern because it is assumed that these organisms are there as a result of faecal contamination of human or animal origin. Their presence is believed to be due to (i) a loss of residual disinfectant (e.g. chlorine); (ii) back-siphonage, cross connections, line breaks, and/or repair of the distribution main; (iii) survival and recovery of injured organisms; or (iv) failure of the treatment plant (Mackerness *et al.*, 1993). The presence of coliforms in drinking water, when there are no known breaches in treatment barriers, and the presence of these organisms in the absence of any evidence of faecal contamination, continues to be a major problem for the water industry, and has emerged as a critical regulatory issue. The chronic presence of coliform bacteria in drinking water supplies is characterised by the absence of coliforms from water that leaves the treatment plant; the routine presence of coliforms in distribution supply samples at various points; the persistence of coliforms in the system despite the maintenance of a disinfectant residual that is assumed to be effective; and the persistence of the problem over a long time. *E. coli* has attracted the greatest interest as an indicator of the microbiological quality of water, but other bacteria have recently been considered as waterborne pathogens. For example, aeromonads are ubiquitous in freshwater environments and *A. hydrophila* is being increasingly recognised as an aetiological agent of gastro-intestinal disease associated with chlorinated water supplies (Janda, 1991).

The possibility that coliforms and aeromonads can become part of the biofilm community, and show regrowth or aftergrowth, was investigated using the continuous culture model (Mackerness *et al.*, 1993). Bitumen-painted mild steel is a material used for the construction of many potable water distribution supplies in the United Kingdom. Heterotrophic biofilms were developed on tiles of this material in potable water at 25 °C and challenged with *E. coli* and *A. hydrophila* isolated from distribution supplies. These organisms established in the planktonic population, rapidly incorporated into the biofilms and survived many weeks at approximately one per cent of the population (Table 7). Monochloramine has been advocated as a superior disinfectant for potable water systems rather than chlorine, due to greater longevity in long distribution supplies with a high chlorine demand and believed ability to penetrate biofilms. When monochloramine was added to the model system at 0.3 mg l^{-1} (a concentration regularly used by water undertakings) there was no decrease in the biofilm viable counts although *E. coli* was eliminated from the aqueous planktonic phase. Rather, the numbers of *E. coli* and *A. hydrophila* recovered from the biofilms rose slightly. Higher concentrations were required for their eradication. Thus, *E. coli* and *A. hydrophila* can become part of the autochthonous heterotrophic biofilm and resist chemical disinfection. Therefore, the occurrence of coliforms in a distribution system which has no faecal contamination might be due to the detachment of these organisms from the biofilm, from where they can be transported around the system and colonise new sites. The only course available for their control by the supplier is to use a residual disinfectant which persists and is non-toxic to man. The biofilm model system is ideal for deciding which disinfectant is the most appropriate under the conditions of use. For example, we have some recent evidence that *A. hydrophila* is more susceptible to eradication with a low dose of monochloramine at five or 15 °C.

Using the model system, a strain of *E. coli* was introduced containing a *lacZ* reporter linked to the *nirB* promoter (Robinson *et al.*, 1995). This promoter is susceptible to low oxygen concentrations, and it was possible to see β-galactosidase expression identifying a mosaic of microenvironmental niches of low oxygen concentration in the biofilm where the *E. coli* colonised. Current studies are making constructs with green fluorescent protein which does not need exogenous substrates to track its expression.

Table 7. **Effect of monochloramine on colonisation of a heterotrophic biofilm on bitumen-painted mild steel by *Aeromonas hydrophila* and *Escherichia coli***

Time	Heterotrophs	*A. hydrophila*	*E. coli*
1	6.1 (4.9)	4.8 (3.3)	5.6 (2.8)
4	5.8 (5.9)	3.3 (3.0)	5.0 (5.1)
7	4.9 (5.3)	3.7 (4.3)	3.4 (4.6)
14	5.8 (5.6)	3.8 (4.2)	3.3 (4.2)
21	5.6 (5.3)	3.8 (4.5)	3.1 (4.3)

Note: The biofilm viable counts are expressed as $\log_{10}$ cfu cm^{-2}. Cultures were grown aerobically at 25 °C at a dilution rate of 0.2 h^{-1}. Values in parentheses indicate viable counts after addition of monochloramine to the culture medium at a concentration of 0.3 mg l^{-1}.

Source: Mackerness *et al.*, 1993.

There has previously been concern in the United Kingdom over the increasing concentrations of nitrate and nitrite in drinking water in some areas. This is due in part to worries of nitrite causing blue baby syndrome (methaemoglobinemia) and also its strong links with bowel cancer and leukaemia when used in food. EC Directive 80/778/EEC has set the maximum permissible concentration of nitrate at 50 mg l^{-1} and nitrite at 0.1 mg l^{-1} in water at point of use. Some water utilities have experienced problems of non-compliance with the EC Directive for the maximum permissible concentration of nitrite in potable water, which are higher than the concentrations in the source (raw) water. The formation of nitrite in the drinking water supply can be from two potential mechanisms, either chemical or microbiological, involving the oxidation of ammonia or the reduction of nitrate:

$$NH_3 \underset{\text{reduction}}{\overset{\text{oxidation}}{\rightleftharpoons}} NO_2^- \underset{\text{reduction}}{\overset{\text{oxidation}}{\rightleftharpoons}} NO_3^-$$

There is no evidence to suggest that in practise nitrite in water is the result of a purely chemical process. Moreover, there are too few bacteria in the water to account for the nitrite production. A strong possibility was therefore that nitrite appearance in water was due to the action of bacteria in surface films. To investigate this, a continuous culture model of an affected distribution supply was set up, using a biofilm recovered from the pipeline and the water it supplied (Mackerness and Keevil, 1991). A continuous addition of exogenous nitrate to the model had no effect on the nitrite or ammonia concentrations detected. By contrast, addition of a low concentration of ammonia resulted in its complete utilisation, but with no change in the nitrate or nitrite concentrations. After several days, however, whilst free ammonia could still not be detected, there now was a stoichiometric conversion to nitrite. Analysis of biofilm samples from the model and affected distribution supplies identified the presence of autotrophic nitrifiers, such as *Nitrosomonas* spp. These were in similar numbers to the heterotrophs in the biofilms, approximately 10^6 cfu cm^{-2}. Nitrifying bacteria have been shown to be resistant to disinfectants. An early report by Larson (1939) showed that an incomplete nitrification, i.e. the formation of nitrite, in a distribution system containing 1 mg l^{-1} monochloramine caused the loss of residual disinfectant, and an increase in heterotrophic plate counts. We surmised therefore that many of the nitrite non-compliance problems in the United Kingdom might be due to excessive use of ammonia added to chlorinated water to give a monochloramine residual for maintaining the microbiological quality of the water. Gratifyingly, nitrite was not formed in the distribution supply when the ammonia addition was tightly controlled at the treatment works, and heterotrophic numbers were kept in check. Thus, small scale laboratory biofilm experiments were shown to have a major impact on the quality of drinking water supplied to many homes in the United Kingdom.

Such studies show that biofilms provide a locus of concerted catalytic activities, producing not only nitrite, but also carbon dioxide and methane which can cause anoxia and explosions. Furthermore, products such as hydrogen sulphide (from sulphate reducing bacteria) and nitric oxide (from denitrifiers) cause localised corrosion and can be released as toxic agents. The footprint caused by the biofilm structural and electrochemical heterogeneity has also been linked to cuprosolvency and copper pitting corrosion, releasing high concentrations of copper into the water to affect its quality (Walker *et al.*, 1994*b*).

Sequelae

It is essential to provide the necessary treatment procedures to ensure the supply of large quantities of wholesome water to the population. However, the studies described here beg the question of how biofilms, their metabolic activity, and colonisation by indicator bacteria and pathogens can be controlled to ensure the safe transmission of potable water. The relative resistance of biofilms and associated pathogens to disinfection has required the use of higher concentrations of chlorine and monochloramine, and not always with success. Moreover, some countries now worry about the use of chlorine in drinking water causing formation of trihalomethanes and also chemical corrosion. Many now advocate the use of ozone, not only to breakdown pesticides in source waters, but also to kill resistant pathogens such as *C. parvum*. Nevertheless, some countries are concerned about formation of bromate ion from bromide and its possible deleterious affect on health. More recently, chlorine dioxide has been advocated for use in disinfecting water for potable use, but here too, there are concerns that it cause by-products and also corrosion which may affect water quality (Walker *et al.*, 1995*b*).

Another strategy, at least for control of ingressing pathogens, would be to ensure effective water treatment barriers are in place. It is easy to say that protecting the watershed will eliminate waterborne transmission of pathogens, but is becoming more difficult to achieve as farming practise intensifies, and microbiological and chemical pollution infiltrates rivers and aquifers. Coagulation and filtration processes have been demonstrated to be very efficient at removing contaminants, especially using slow sand beds. However, this is slow and occupies expensive space in urban environments. Sudden pulses of contamination, associated with heavy rainfall, can also overload the flocculation/filtration process: this has happened on several occasions with outbreaks of Cryptosporidiosis and Giardiasis. Chlorine remains the prime treatment barrier and is cheap to use, but, as mentioned above, some have questioned its use in developed countries on health grounds. Ozone is now being used routinely for specialist applications such as pesticide removal, but it does not provide a residual disinfectant activity, is very expensive and has caused health concerns. Interest is now increasing in the use of UV irradiation of water at short and long wavelengths (265, 365 nm). Recent design configuration of the lamp and flow path have greatly increased the efficiency of energy transfer to microbial cells such as *C. parvum* oocysts. It remains to be seen whether biofilm adapted pathogens which have sloughed off singly or as clumps will be resistant to this approach. The final treatment barrier, and probably the most effective, is to boil water for potable use. This is obviously tedious and expensive in developed urban societies, but is still practised when water utilities issue a boil notice when waterborne outbreaks of Cryptosporidiosis are suspected. Boiling water can be very beneficial in isolated areas of developing countries, but it should not be forgotten that it requires large quantities of wood to boil modest volumes of water. Countries such as Bangladesh do not have much wood, and this is not an option.

Future considerations

In many parts of the world, water has been a scarce resource for centuries. This has caused many wars, and tensions over securing wholesome supplies of water for expanding populations are bound to continue. The situation is also being exacerbated by global warming introducing semi-arid climates to areas which previously have enjoyed respectable levels of rainfall to supply their rivers and aquifers. Increased temperatures lead to increased demand for water, and also make survival and possible growth of some microbial pathogens in the environment more likely. Thus, biotechnology has an important role to play in conserving supplies of wholesome water, protecting rivers and aquifers from infiltration by toxic chemicals and pathogens, and improving treatment processes to render wastewater safe for reuse.

Inevitably, treatment procedures will depend on the standards set by each country. Developed countries would probably expect higher standards than the Third World and certainly have more disposable income to bear the increased treatment costs. A major problem, however, is that outbreaks of waterborne disease have occurred, such as the Milwaukee *Cryptosporidium* infection, where there was no evidence of coliform failures indicative of faecal contamination (Mackenzie *et al.*, 1994). One possibility might be that the indicator bacteria were killed during the treatment process, but the protozoa and other pathogens were not. Thus, alternative indicators must be considered when proposing standards for the microbiological quality of potable water.

Standards and risk assessment are also difficult to determine when considering the potential for a pathogen's virulence to be modulated by its environment and, hence, determine the minimum infective dose. There are several examples, such as with *L. pneumophila*, where virulence is attenuated at low temperatures, and therefore higher concentrations can be tolerated by healthy people when inhaling contaminated aerosols (Mauchline *et al.*, 1994). Similarly, is *C. parvum* less virulent when its oocysts have been stressed in potable water supplies or biofilms for many days? Do the stressed oocysts become more susceptible to routine disinfectants such as chlorine? Cyanobacterial (blue-green algal) blooms can produce hepatotoxins and neurotoxins which are relatively easy to assess in terms of acute exposure, but much more difficult when considering carcinogenic risk due to long-term, low level chronic exposure (Ueno *et al.*, 1996).

Despite the difficulties of setting appropriate standards, how can biotechnology contribute further to the problem of providing wholesome water, fit for bathing and drinking? New disinfectant strategies must be considered, taking into account the relative resistance of adherent cells on substrata. With current worries about chlorine, ozone and chlorine dioxide, it is important to identify new molecules which do not share the problems of the former, and which are effective not only against planktonic cells, but also biofilm populations. Even 2 000 years ago, the Romans advocated dropping silver coins into drinking water to keep it sweet. Today, silver is still being utilised, together with copper, in ionisation systems to treat domestic water. Perhaps we should also consider use of probiotics: can we identify safe bacterial species which can be seeded into distribution supplies to colonise potable water biofilms and inhibit potential pathogens by producing antibacterial agents, e.g. bactericins or colicins; similarly can we use phage, lytic to specific pathogens; or are there protozoal species which can be seeded to graze the biofilms and remove pathogens safely?

Is it also possible to prevent initial cell attachment to surfaces and consolidation as structured communities? Recent research has pointed to physiological adaptation mechanisms for cells at surfaces. One such mechanism provides an electro-chemical description of how cells become chemi-osmotically polarized and sense a surface (Ellwood *et al.*, 1982; Keevil *et al.*, 1995) and how this might be modulated by application of mild electric currents (Costerton *et al.*, 1994). Another involves expression of global regulators such as σ factors which interact with RNA polymerase-mediated transcription of essential genes when adjacent to a surface, e.g. alginate expression in *Ps. aeruginosa* (Deretic *et al.*, 1994). Another involves quorum sensing molecules, such as autoinducing homoserine lactones for consolidating cell structures (Williams and Stewart, 1994). These mechanisms all offer potential molecular biology strategies to control cell attachment and proliferation of the biofilm consortium.

Much has been said about the deleterious effects of biofilms. Nevertheless, they have also played a key role for many years in wastewater treatment processes to reduce C, N, P and heavy metal loading before disposal or reuse. Adherent biofilms on particles play an important role in the work of slow sand filter beds. One advantage of biotechnology would be to design new bed matrices with appropriate physico-chemical characteristics and flow properties to enhance biofilm formation, structure and physiological activity: these could be used to trap and break down not only potential chemical hazards in source waters but also microbiological hazards such as protozoal cysts and viruses, which afflict many parts of the developing world.

Questions, comments and answers

Q: Is there any difference between biofilms on steel and those on steel coated with plastics such as polypropylene and so on?

A: There are marked differences. Copper and zinc are the best materials to suppress biofilm formation. Stainless steel and mild steel are quite good for biofilm formation and *Legionella* actively likes a rusting steel surface. Some of the plastics are very good substrates for biofilm formation.

Q: We have seen that if you first establish a biofilm with organism A then organism B cannot invade, and vice versa. Is your experience similar?

A: Many of the laboratory experiments are flawed, because they examine one or two species biofilms which do not occur in the real world. Most biofilms are complex multi-species. Also, most of these experiments take place in an artificial laboratory medium which has no relevance to the natural environment.

REFERENCES

AMANN, R.I., W. LUDWIG, and K.-H. SCHLEIFER (1995), "Phylogenetic identification and *in situ* detection of individual microbial cells without cultivation", *Microbiol. Rev.* 59, pp. 143-169.

COLBOURNE, J.S. and P.J. DENNIS (1985), "Distribution and persistence of *Legionella* in water systems", *Microbiol. Sci.* 2, pp. 40-43.

COSTERTON, J.W., K.-J. CHENG, G.G. GEESEY, T.I. LADD, J.C. NICKEL, M. DASGUPTA, and T.J. MARRIE (1987), "Bacterial biofilms in nature and disease", *Ann. Rev. Microbiol.* 41, pp. 435-464.

COSTERTON, J.W., B. ELLIS, K. LAM, F. JOHNSON, and A.E. KHOURY (1994), "Mechanism of electrical enhancement of efficacy of antibiotics in killing biofilm bacteria", *Antimicrob. Agents Chemother.* 38, pp. 2 803-2 809.

DeBEER, D., P. STOODLEY, and Z. LEWANDOWSKI (1994), "Liquid flow in heterogeneous biofilms", *Biotechnol. Bioeng.* 44, pp. 636-641.

DERETIC, V., M.J. SCHURR, J.C. BOUCHER, and D.W. MARTIN (1994), "Conversion of *Pseudomonas aeruginosa* to mucoidy in cystic fibrosis: environmental stress and regulation of bacterial virulence by alternative sigma factors", *J. Bacteriol* 176, pp. 2 773-2 780.

ELLWOOD, D.C., C.W. KEEVIL, P.D. MARSH, C.M. BROWN, and J.N. WARDEL (1982), "Surface-associated growth", *Phil. Trans. R. Soc. London B* 297, pp. 517-532.

GELDREICH, E.E., H.D. NASH, D.J. REASONER, and R.H. TAYLOR (1972), "The necessity of controlling bacterial populations in potable water: community water supply", *J. Am. Water Works Assoc.* 64, pp. 96-102.

HAMILTON, W.A. (1985), "Sulphate-reducing bacteria and anaerobic corrosion", *Ann. Rev. Microbiol.* 39, pp. 195-217.

JANDA, J.M. (1991), "Recent advances in the study of the taxonomy, pathogenicity and infectious syndromes associated with the genus *Aeromonas*", *Clin. Microbiol. Rev.* 4, pp. 397-410.

KEEVIL, C.W. (1989), "Chemostat models of human and aquatic corrosive biofilms", in T. Harrori *et al.* (eds.), *Recent Advances in Microbial Ecology*, pp. 151-156, Japan Scientific Societies Press, Tokyo.

KEEVIL, C.W. (1991), "Toxicology and detection of cyanobacterial (blue-green algal) toxins", *PHLS Microbiology Digest* 8, pp. 91-95.

KEEVIL, C.W. (1995), "The value of *in situ* biofilm investigations", in J.W.T. Wimpenny, P. Handley, P. Gilbert, and H.M. Lappin-Scott (eds.), *The Life and Death of Biofilms*, pp. 17-20, Bioline, Cardiff.

KEEVIL, C.W. (in press), "Pathogens and metaboliets associated with biofilms", in C.W. Keevil *et al.* (eds.), *Biofilms in Aquatic Systems*, Royal Society of Chemistry, Cambridge.

KEEVIL, C.W. and J.T. WALKER (1992), "Normarski DIC microscopy and image analysis of biofilms", *Binary: Computing in Microbiology* 4, pp. 93-95.

KEEVIL, C.W., A.A. WEST, N. BOURNE, and P.D. MARSH (1984), "Inhibition of extracellular glucosyl- and fructosyltransferase synthesis and secretion in *Streptococcus sanguis* by sodium ions", *J. Gen. Microbiol.* 130, pp. 77-82.

KEEVIL, C.W., D.J. BRADSHAW, A.B. DOWSETT, and T.W. FEARY (1987), "Microbial film formation: dental plaque deposition on acrylic tiles using continuous culture techniques", *J. Appl. Bacteriol.* 62, pp. 129-138.

KEEVIL, C.W., A.A. WEST, J.T. WALKER, J.V. LEE, J.P.L. DENNIS, and J.S. COLBOURNE (1989), "Biofilms: detection, implications and solutions", in D. Wheeler, M.L. Richardson and J. Bridges (eds.), *Watershed 89: The Future of Water Quality in Europe* 2, pp. 367-374, Pergamon Press, Oxford.

KEEVIL, C.W., C.W. MACKERNESS, and J.S. COLBOURNE (1990), "Biocide treatment of biofilms", *Int Biodeter* 26, pp. 167-179.

KEEVIL, C.W., J. ROGERS, and J.T. WALKER (1995), "Potable water biofilms", *Microbiology Europe* 3, pp. 10-14.

LARSON, T.E. (1939), "Bacterial corrosion and red water", *J. Am. Water Works Assoc.* 31, pp. 1 186-1 196.

LAWRENCE, J.R., D.R. KORBER, B.D. HOYLE, J.W. COSTERTON, and D.E. CALDWELL (1991), "Optical sectioning of microbial biofilms", *J. Bacteriol.* 173, pp. 6 558-6 567.

LECHAVELLIER, M.W., C.D. CAWTHORNE, and R.G. LEE (1988), "Inactivation of biofilm bacteria", *App. Environ. Microbiol.* 54, pp. 2 492-2 499.

LEWANDOWSKI, Z., S.A. ALTOBELLI, and E. FUKUSHIMA (1993), "NMR and microelectrode studies of hydrodynamics and kinetics in biofilms", *Biotech. Progress* 9, pp. 40-45.

McFETERS, G.A., M.W. LECHEVALLIER, and M. DOMEK (1984), "Injury and improved recovery of coliform bacteria in drinking water", *EPA-600/-84-166*, US Environmental Protection Agency, Cincinnati, Ohio.

MACKENZIE, W.R., N.J. HOXIE, M.E. PROCTOR, M.S. GRADUS, K.A. BLAIR, D.E. PETERSON, J.J. KAZMIERCZAK, D.G. ADDISS, K.R. FOX, J.B. ROSE, and J.P. DAVIS (1994), "A massive waterborne outbreak of *Cryptosporidium* infection transmitted through the public water supply", *N. Engl. J. Med.* 331, pp. 161-167.

MACKERNESS, C.W. and C.W. KEEVIL (1991), "Origin and significance of nitrite in water", in M.J. Hill (ed.), *Nitrate and Nitrite in Food and Water*, pp. 77-92, Ellis Horwood, London.

MACKERNESS, C.W., J.S. COLBOURNE, and C.W. KEEVIL (1991), "Growth of *Aeromonas hydrophila* and *Escherichia coli* in a distribution system biofilm model", in R. Morris, L.M. Alexander, P. Wyn-Jones, and J. Sellwood (eds.), *Health Related Water Microbiology*, pp. 131-138, IAWPRC, London.

MACKERNESS, C.W., J.S. COLBOURNE, P.J. DENNIS, T. RACHWAL, and C.W. KEEVIL (1993), "Formation and control of coliform biofilms in drinking water distribution systems", *Soc. Appl. Bacteriol. Technical Series* 30, pp. 217-226.

MAUCHLINE, W.S. and C.W. KEEVIL (1991), "Development of the BIOLOG substrate utilisation system for identification of *Legionellae pneumophila*", *Appl. Environ. Microbiol.* 57, pp. 3 345-3 349.

MAUCHLINE, W.S., B.W. JAMES, R.B. FITZGEORGE, P.J. DENNIS, and C.W. KEEVIL (1994), "Growth temperature reversibly modulates the virulence of *Legionella pneumophila*", *Infect. Immun.* 62, pp. 2 995-2 997.

PACKER, P.J., D.M. HOLT, J.S. COLBOURNE, and C.W. KEEVIL (in press), "Does *Klebsiella oxytoca* grow in the biofilm of water distribution systems? The effect of different source waters on coliform growth in a chemostat model," in D. Kay *et al.* (eds.), *Coliforms and E. coli: Problem or Solution*, Royal Society of Chemistry, Cambridge.

PRESSDEE, J.R., T. HALL, and E. CARRINGTON (1995), "Practicalities of disinfection for control of *Cryptosporidium* and *Giardia*", in W.B. Betts, D. Casemore, C. Fricker, H. Smith, and J. Watkins (eds.), *Protozoan Parasites and Water*, pp. 206-208, Royal Society of Chemistry, Cambridge.

ROBINSON, P.J., J.T. WALKER, C.W. KEEVIL, and J.A. COLE (1995), "Reporter genes and fluorescent probes for studying the colonisation of biofilms in a drinking water supply line by enteric bacteria", *FEMS Microbiol. Letters* 129, pp. 183-188.

ROGERS, J. and C.W. KEEVIL (1992), "Immunogold and fluorescein immunolabelling of *Legionella pneumophila* within an aquatic biofilm visualised by using episcopic differential interference contrast microscopy", *Appl. Environ. Microbiol.* 58, pp. 2 326-2 330.

ROGERS, J. and C.W. KEEVIL (1995), "Survival of *Cryptosporidium parvum* in aquatic biofilms", W.B. Betts, D. Casemore, C. Fricker, H. Smith, and J. Watkins (eds.), in *Protozoan Parasites and Water*, pp. 209-213, Royal Society of Chemistry, Cambridge.

ROGERS, J., J.V. LEE, P.J. DENNIS, and C.W. KEEVIL (1991), "Continuous culture biofilm model for the survival and growth of Legionella pneumophila and associated protozoa in potable water systems", in R. Morris, L.M. Alexander, P. Wyn-Jones, and J. Sellwood (eds.), *Health Related Water Microbiology*, pp. 192-200, IAWPRC, London.

ROGERS, J., A.B. DOWSETT, P.J. DENNIS, J.V. LEE, and C.W. KEEVIL (1994*a*), "Influence of temperature and plumbing material selection on biofilm formation and growth of *Legionella* *pneumophila* in potable water systems containing complex microbial flora", *Appl. Environ. Microbiol.* 60, pp. 1 585-1 592.

ROGERS, J., A.B. DOWSETT, P.J. DENNIS, J.V. LEE, and C.W. KEEVIL (1994*b*), "Influence of plumbing materials on biofilm formation and growth of *Legionella pneumophila* in potable water systems containing complex microbial flora", *Appl. Environ. Microbiol.* 60, pp. 1 842-1 851.

ROGERS, J., A.B. DOWSETT, and C.W. KEEVIL (1995), "A paint incorporating silver to control mixed biofilms containing *Legionella pneumophila*", *J. Indust. Microbiol.* 15, pp. 377-383.

SURMAN, S.B., L.H.G. MORTON, and C.W. KEEVIL (in press), "Growth of *Legionella pneumophila* in aquatic biofilms is not dependent on intracellular multiplication", in C.W. Keevil *et al.* (eds.), *Biofilms in Aquatic Systems*, Royal Society of Chemistry, Cambridge.

THURMAN, R.B. and C.P. GERBA (1989), "The molecular mechanisms of copper and silver ion disinfection of bacteria and viruses", *CRC Crit. Rev. Environ. Control* 18, pp. 295-315.

UENO, Y., S. NAGATA, A. HASEGAWA, M.F. WATANABE, H.D. PARK, G.C. CHEN, G. CHEN, and S.Z. YU (1996), "Detection of microcystins, a blue-green hepatotoxin, in drinking water sampled in Haimen and Fusui, endemic areas of primary liver cancer in China by highly sensitive immunoassay", *Carcinogenesis* 17, pp. 1 317-1 321.

VAN DER KOOIJ, D. (1992), "Assimilable organic carbon as an indicator of bacterial regrowth", *J. Am. Water Works Assoc.* 84, pp. 57-65.

WALKER, J.T., J. ROGERS, and C.W. KEEVIL (1994*a*), "An investigation of the efficacy of a bromine containing biocide on an aquatic consortium of planktonic and biofilm microorganisms including *Legionella pneumophila*", *Biofouling* 8, pp. 47-54.

WALKER, J.T., D. WAGNER, W.R. FISCHER, and C.W. KEEVIL (1994*b*), "Rapid detection of biofilm on corroded copper pipes", *Biofouling* 8, pp. 55-63.

WALKER, J.T., C.W. MACKERNESS, J. ROGERS, and C.W. KEEVIL (1995*a*), "Heterogeneous mosaic biofilm - a haven for waterborne pathogens" in H.M. Lappin-Scott and J.W. Costerton (eds.), *Microbial Films*, pp. 196-204, Cambridge University Press.

WALKER, J.T., C.W. MACKERNESS, D. MALLON, T. MAKIN, T. WILLIETS, and C.W. KEEVIL (1995*b*), "Control of *Legionella pneumophila* in a hospital water system by chlorine dioxide", *Journal of Industrial Microbiology* 15, pp. 384-390.

WILLIAMS, P. and G.S.A.B. STEWART (1994), "Cell density dependent control of gene expression in bacteria - implications for biofilm development and control", in W. Nicholls, J.W.T. Wimpenny, D.J. Stickler, and H.M. Lappin-Scott (eds.), *Bacterial Biofilms and their Control in Medicine and Industry*, pp. 9-12, Bioline, Cardiff.

WATER CONTAMINATION AND POPULATION'S HEALTH; SUPPORTING OUR EFFORTS

by

Carlos Santos-Burgoa and Rocío Alatorre Eden Wynter
National Institute of Public Health, Cuernavaca, Morelos, México

Summary

This paper deals with the health effects from water pollution. We start with a general review of the human interaction with water, and its implications in the health transition. Then we approach bacteriological pollution and how it has been controlled, as reflected by the drop in gastro-enteritis mortality. We then review our past five years research in water related problems. We first address the existence of basic hygiene malpractice by the population. Then we confront the toxics in water, showing the difficulty in its documentation. We show data of a risk assessment in Coatzacoalcos showing a 1.15 rate ratio (RR) in total cancer mortality, with an excess of 2/1 000 in childhood cancers. We also show an adjusted odds ratio of two in the Tula District for low birth weigh associated to wastewater usage for agricultural purposes. Then we show the use of health indicators for environmental surveillance, showing excess leukemia and neural tube defects in relation to water polluted areas. We conclude that our understanding of the health effects from toxics in the water in Mexico is understudied, and poorly documented; that we need a good monitoring and surveillance system for it; and that we should perform health risk assessment with local characterisation in order to develop standards for human exposure.

Introduction

As air, water is the basic element for population's survival. Human's alteration of the natural hydrological dynamics in order to satisfy their needs means that the water has to be treated, stored and conducted to the residents, used for various purposes and disposed of as wastewaters (Anton, 1993). Therefore, water can become a vector for hazardous agents that impact human's health through skin and digestive tube absorption, ocular and respiratory mocosa contact, as well as through inhalation.

Emphasis in Mexico has been given first to the provision of water, and in a much later stage to its quality, mainly considering bacteriological safety and physical constitution. However, little attention has been paid to its physicochemical quality. It is not our intention to give you a general assessment of the health consequences of water quality and sufficiency.

In this paper, we first provide a quick overview of the epidemiological conditions in Mexico. Then we will show some efforts of our group and closely related researchers in understanding the risks associated to water pollution in Mexico. We end by identifying our current challenges suggesting an agenda for research and for action.

Epidemiological transition

Our country has gone through several transitions that have impacted the population. These include social and economic polarisation, the changing role of the state, urbanisation, the energy use, cultural and nutritional changes, and a profound environmental and labour change (Santos-Burgoa *et al.*, 1995*a*). Concordant with this, we have observed a fast and downward trend in mortality from gastro-enteritis and infectious diseases with a concurrent increase of accidents, cardiovascular diseases and cancer mortality.

This is not a uniform process; this data shows the distribution of the country's 32 states, as their mortality distribution, clearly showing the southern poor states as Chiapas and Oaxaca in one extreme, while the urbanised and industrialised Federal District and the Northern states are at the other extreme. This is a pattern of rapid protracted epidemiological transition, and we believe that a most important component that explains such a change is the environmental changes associated to it. Water pollution is to have a major role in the so-called infectious diseases epidemiological profile, being a major mean for transmission of infectious material; it has and will have a major role in the more "modern" non-transmissible diseases.

As we mentioned that water provision has been the governmental priority, the map in Figure 1 deploys the distribution of home drinking water. Many studies have clearly stated the importance of having water inside the house in order to provide proper hygiene. One can observe that Mexico is still an heterogeneously provided country, having much to do with urban infrastructure and socio-economic conditions. The most provided areas are those with an epidemiological pattern with less infectious diseases (INEGI, 1992) in the north, while the southern states are those with the lesser provision of drinking water in homes.

Cholera: a case study

We do not want to say that waterborne infectious diseases are not a problem. It is true that after we managed to cover the lag in water chlorination of our water sources in the early 1990s because of the cholera epidemic, the drop in diarrhea mortality was dramatic.

However, we tried to understand how areas well provided with water, and with an acceptable level of chlorination, were still hit with severe diarrheic diseases. Using the opportunity provided by the cholera epidemic, we have studied the cultural and conceptual issues related to the use of water and its disposal in semi-urban and semi-rural conditions.

A case-control study was carried out in the State on Puebla, Mexico. The results confirm the persistence of hygiene deficiencies and adverse socio-economic conditions as the key factors in the transmission of cholera. The evaluation of the independent effects of these factors, using a logistic regression model, shows that the risk increases up to three times with overcrowding, uncovered water deposits, not washing hands before eating, eating outside the home in open places, not treating the drinking water, living together with sick people, and the absence of a sanitation system to deposit excreta. These results show that the problem of cholera transmission could be deterred by basic hygiene promotion beyond chlorination of water supply sources (Alatorre *et al.*, in press) (Table 1).

Figure 1. **Potable water distribution**

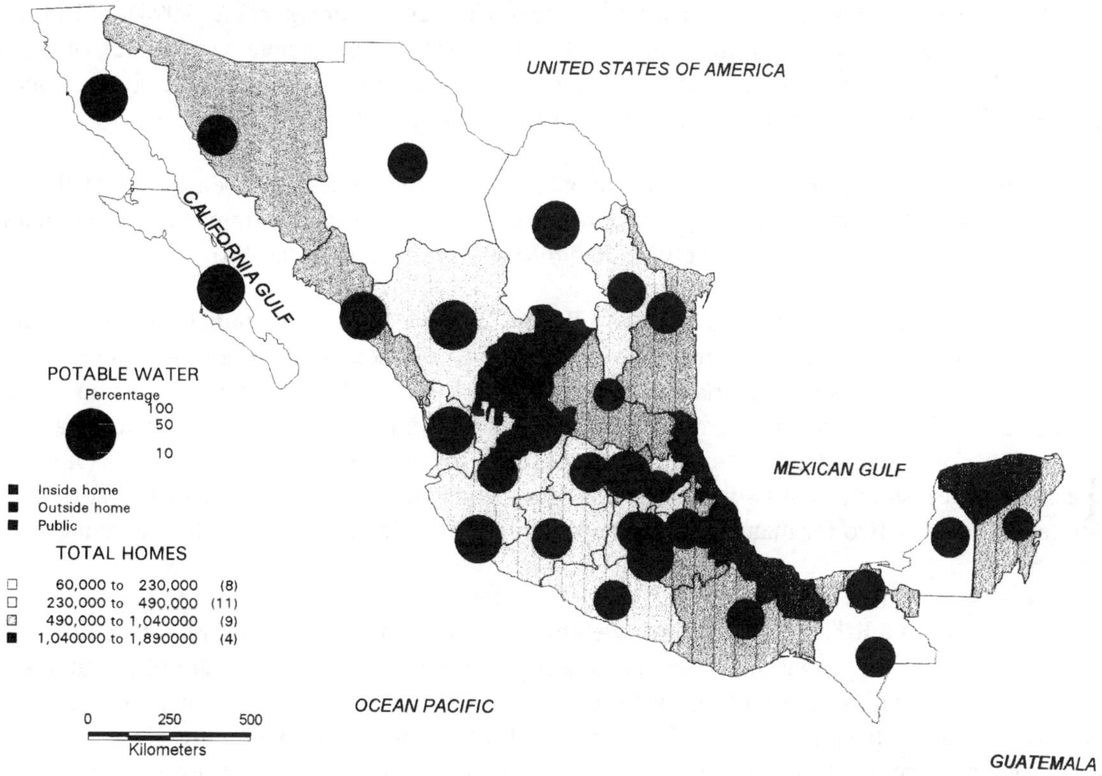

Source: HIES, 1996.

Table 1. **Multivariate odds rations for cholera risk estimated by logistic regression**

Variable	Rate ratio (RR)	95% Confidence interval (CI)
Very overcrowded	1.60	0.98-2.63
Uncovered water depository	1.52	1.07-2.14
Lack of handwashing before eating	2.82	1.83-4.36
Eat outside in open places	1.50	1.07-2.10
Lack of treating drinking water	1.59	1.14-2.23
Living with people sick with cholera	1.55	1.09-2.19
Lack of basic sanitation	1.39	0.80-1.62

Source: Author.

Approaching the health effects from water toxics

We believe that most of the water born bacterial diseases are well understood both abroad and in Mexico. Therefore, we have been looking into issues related to surface water, water tables, their potential exposure to toxic agents and their subsequent potential health effects.

We performed with Dr. Cebrián and Dr. Albores from the Cinvestav an exploratory comparative risk assessment of the hydrologic basins where information was available. With this exercise we were able to have a first hand experience of how the nationally gathered data is currently available for decision making. This provided a first overview of the level of risk shown by these areas, and I will not focus on this. But our biggest learning was that standardized, routine, and quality information is difficult to obtain, and still harder to relate to health effects (Albores *et al.*, 1994). Cebrián has studied arsenic health effects from natural high concentrations in ancient watertables of northern Mexico. We have done collaborative research using epidemiological methods to validate and interpret biomarkers of metabolism of arsenic (del Razo *et al.*, 1996).

Trying to have a closer look at health impacts from water toxics, we identified that the use of wastewater for agricultural purposes, a world-wide practice, has subtle impacts beyond parasitic diseases. The latter has been extensively studied by Siebe and Cifuentes (1995).

But we wanted to study the toxics. With Cortes (1993), we studied the trace metals' concentration in water, blood and urine. We surveyed the presence of lead, cadmium and manganese, as well as the metabolism biomarker of xincprotoporphyrin. This study was done in volunteer agricultural workers of the Tula Irrigation District. This semi-desertic area uses wastewater from Mexico City, only treated by sedimentation ponds, to irrigate extensive crop lands. Our logistic regression models showed risks adjusted for exposure, consumption and duration of irrigation of up to 20 times for lead, of two for manganese, and up to three for cadmium, with levels above international standards.

It has been established that low birth weight is a good indicator of saturation of detoxification processes, and we believe that this irrigation area has an impact of a large mixture of pollutants. In a case-control study with Carreon (1993) we searched for mothers with new-born babies as detected in the local general hospitals (Table 2). We adjusted our multiple regression models for socio-economic, age, medical care and other related factors. In spite of all that, we identified a 20 per cent increased risk of low birth weight for mothers living in the wastewater irrigated area; this represented a double risk for term low birth weight babies in this population.

Table 2. **Adjusted estimators of relative risk for the association between living in the irrigation district (ID) 063 area and the different effect outcomes**

Dependent variable	Cases	Controls	Rate ratio (RR)	95% Confidence interval (CI)
Total cases				
Live in ID 063	32	56	1.2	0.727-2.152
Outside the area	60	123		
Preterm cases of normal weight				
Live in ID 063	3	56	0.4	0.121-1.668
Outside the area	15	123		
Low birth weight cases borne to term				
Live in ID 063	15	56	2.2	1.002-4.893
Outside the area	17	123		
Low birth weight cases born prematurely				
Live in ID 063	8	56	1.1	0.427-2.705
Outside the area	22	123		

Source: Author.

Using data from Botello and collaborators (1995) and our own (Sanchez Rios, 1996), we are developing a Risk Assessment (RA) of the Coatzacoalcos area. Coatzacoalcos has a concentration of policyclic aromatic hydrocarbons similar to that of the East River in New York in 1975. In these areas, they collected seafood with concentration of PAHs up to 0.88 ppm; sediments get above 70 ppm of hydrocarbons. We have observed a 15 per cent excess risk for the 1992-1993 time period for all cancers; we estimate in our RA an excess of 4/1 000 in childhood cancers just related to benzo(a)pyrene exposure either through seafood, water ingestion or soil ingestion. In fact, Botello over the last 20 years has measured toxics in several hydrologic systems in Mexico. In a recent compilation of data at the Coatzacoalcos area he has identified concentrations in water of mercury 60 times the allowable levels (30 ± 10), of lead twice the allowable (12 ± 3), and in sediments a concentration of chromium 10 times the allowable (71.8 ± 27.9) (Botello *et al.*, 1995). Therefore, this is a problem area not only to carcinogenic but to neurologic, renal and reproductive effects.

Health Indicator for Environmental Surveillance (HIES)

Our work has taken us to the development of a study on health indicators for environmental surveillance*. We started this project with an observation in Mexico City about the distribution of sentinel diseases, using the following indicators:

- low birth weight;
- congenital malformations of neural tube;
- leukaemia <5 years;
- hepatitis;
- chronic bronchitis;
- asthma;
- mesothelioma.

We gathered the death certificates of Mexico City, and analysed their computerised database. After adjusting for age and sex, we showed a 50 per cent increased risk of ten year mortality in childhood lymphomas (Figure 2) and leukemias in three of the 16 delegations of the federal district. Furthermore, they showed a statistically significant increased trend. We also showed a statistically significant increased trend for the ten year period of neural tube defects in one delegation, sharing the risk with that of leukemias and lymphomas. The concordance is ecologically associated with one oil refinery, and two very old municipal and industrial waste sites (Santos-Burgoa, 1996). This raised the hypothesis yet to test of a chemically polluted water table.

Having had this experience, we thought that we should systematise this process. There is a world-wide effort to generate a health effects monitoring system that is able to measure the environmental impact in health terms. We thought that knowledge should not be spotty, and if in other areas, epidemiological surveillance has helped to detect severe health problems, we should apply this to environmental health. Furthermore, we thought that we should look for indicators that are reasonably related to environmental toxic and infectious exposures, that could be rather easily gathered, and useful for timely identification of risky areas and the orientation of focused research. We gathered an expert group to prioritise tracer and sentinel environmental health events. From a list of more than 50 indicators, we selected seven in order to develop methods for its analysis. We did a test first in the state of Morelos, and then in a highly urbanised and in a very rural area.

---

* An indicator is any preventable disease, disability or premature death that could be used as a signal or indicator of an environmental agent level, to take some action to prevent such disease.

We have developed geographic and cluster analytic methods to check on these relations (Alatorre and Santos-Burgoa, 1996; Santos-Burgoa *et al.*, 1996). We have shown that in rural, semi-urban and urban areas, using secondary data available from health and municipal sources, clustering of events can be identified, including chronic conditions such as neural tube defects and leukemias, as well as viral diseases such as hepatitis, that tell us about the persistence of fecal pollution that can not be taken by chlorination.

Figure 2. **Lymphoma mortality trends in Mexico City (1980-1990)**

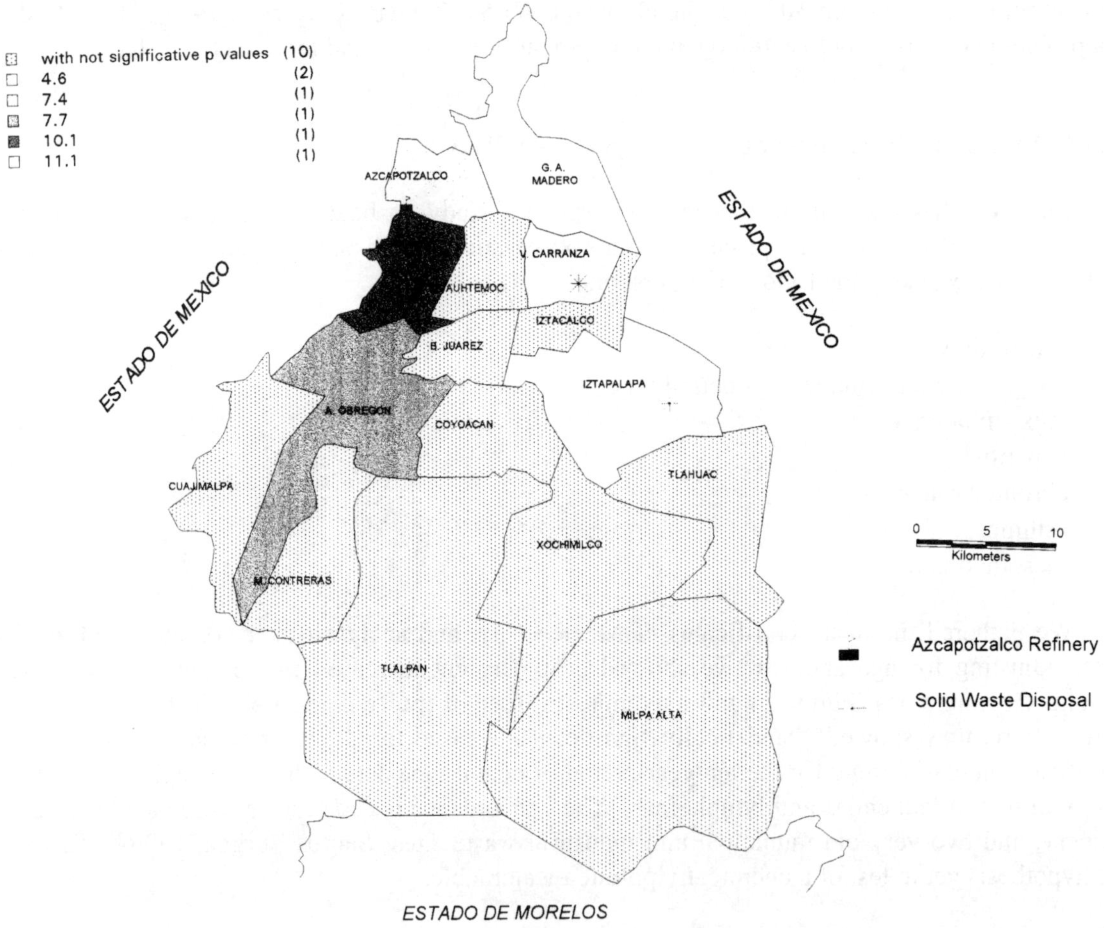

Source: HIES, 1996.

In the map in Figure 3, we show the graphical association of fecal coliform data in 1993-1994 in Cuernavaca with cases of severe viral hepatitis. The map shown in Figure 4 demonstrates the agglomeration of neural tube defects in new-born children for the same place and time period. We have analysed the clustering of the mentioned health indicators, but these two have been the most relevant ones in demonstrating a statistically significant distribution, pinpointing sources of pollution in semi-urban and rural areas. We are currently analysing these data for Cuernavaca, but in Morelos we have been able to associate it to the industrial corridor and the wastewater irrigation districts.

Figure 3. **Fecal coliforms in wells and hepatitis morbidity cases (1993-1994) in the metropolitan area, Cuernavaca City, Morelos, Mexico**

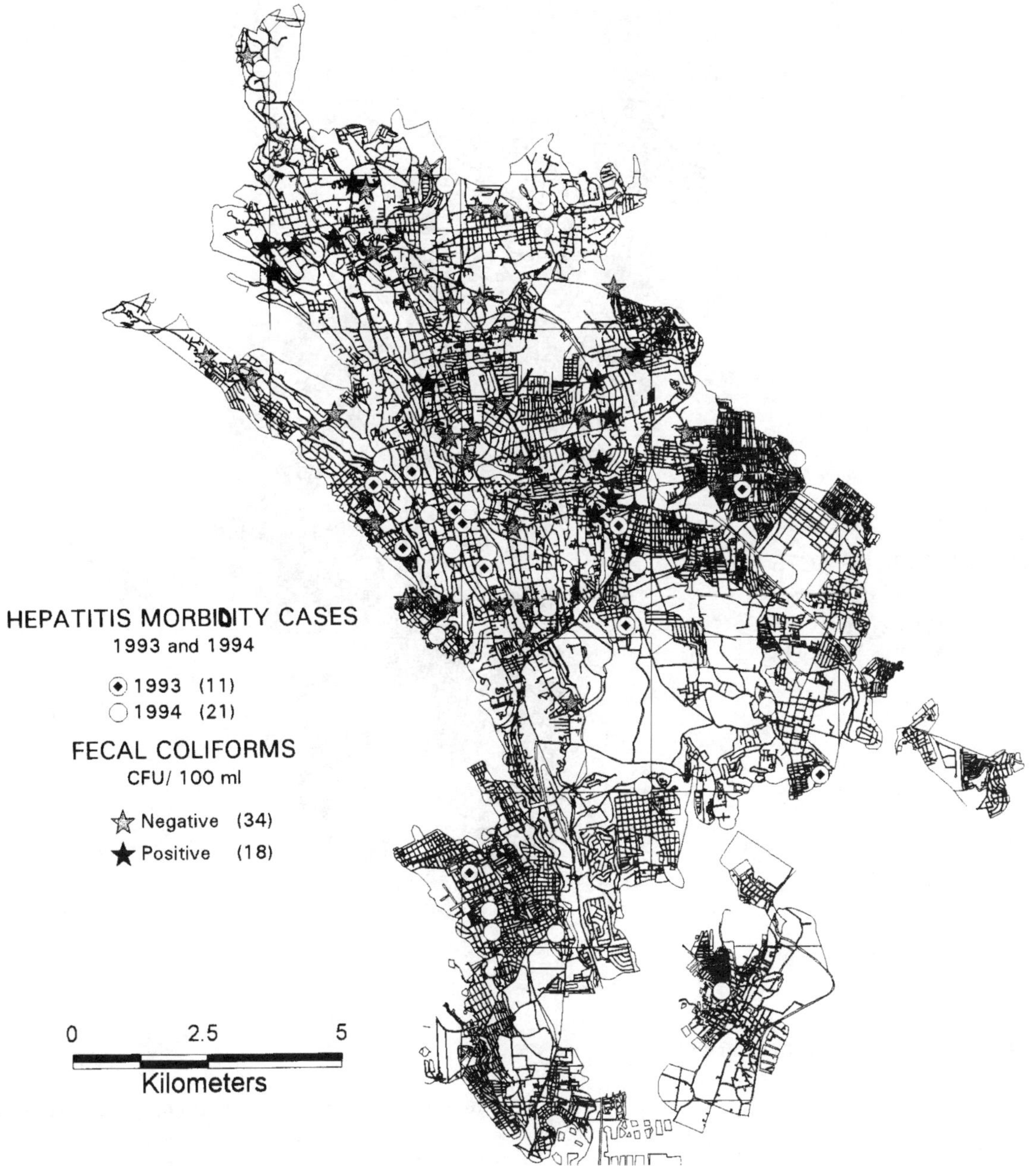

Source: HIES, 1996.

Figure 4. **Neural tube defect cases (1993-1994) in the metropolitan area, Cuernavaca City, Morelos, Mexico**

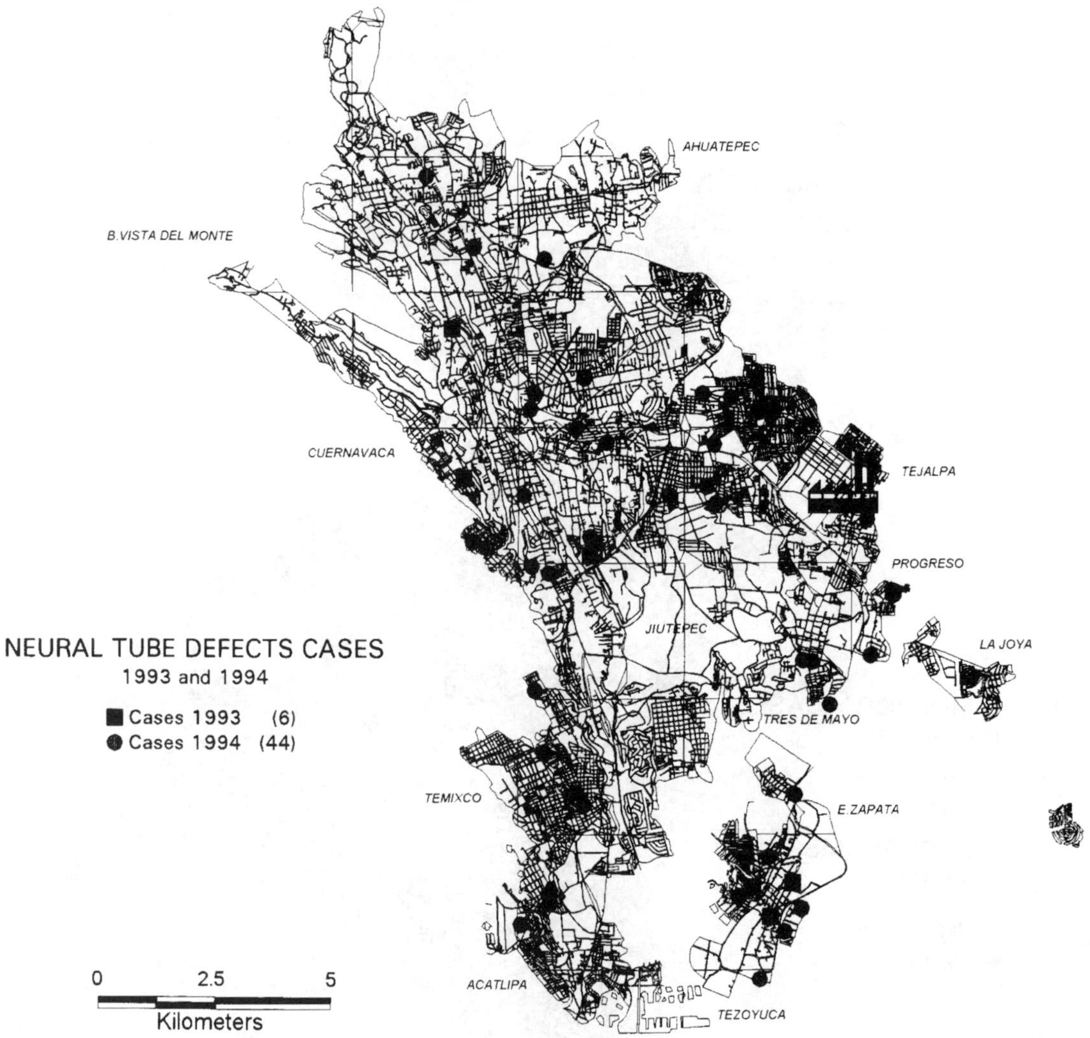

Source: HIES, 1996.

Conclusions and recommendations

Water supply was the main target in Mexican water policy for over 50 years, and we have succeeded.

Water bacteriologic safety has been its target over the last five years, and our recent successful story is that we have been able to increase residual chlorine dramatically.

Water safety and renewal should be its target over the next 25 years, and I am sure that we could be as successful if we only manage to really focus, and open this priority in our national and social agenda.

But, how safe is the water that we are in contact with, through skin or drinking, or inhalation? We do not know for sure, but there are reasons to worry about it.

- Does it have hydrocarbons? Yes, in areas where we have measures of it.

- Does it have trace metals? Yes, in the same manner.

- Do we still have waterborne infectious diseases? For sure, either bacterial, although decreasing, as well as viral, with no specific trend.

- Do we have chronic diseases that could be related to water pollution? Yes, and with some suspicion of its association. But we have to be sure.

The studies mentioned above were developed with a total budget of US$ 350 000 over a period of five years. Obviously we have advanced, but this has been an area understudied, and underfunded. We lack a great deal of knowledge, on exposure assessment, on biomarkers of exposure, of early damage or of disease.

Issues remain, and I want to mention four of them:

- I would like to make a priority that we lack a monitoring system oriented towards human exposure, within the urban areas, as well as with the surface waters throughout the country, with data publicly available, managed by organisms technically accountable for the data, and useful for epidemiological surveillance and research (Reichard, 1991). Biomarkers could be most useful for this system, especially if monitoring data could be related to human health indicators of pollution effects. I do insist in the public availability of this data, that so far, in spite of being paid by public taxes, are not publicly available. History has shown that data is required for decision-making; that public availability increases its quality and reliability; and that public input on the data is useful. Take what has been gained from public availability of air pollution data; the best research and contribution in health has been made by non-government organisations.

- Challenges for research are mainly related to the presence of toxic agents in water sources, and its means to detect their effects, especially sensitive and predictive biomarkers.

- We have to develop health Risk Assessment appropriate to our population needs. If we are to develop norms we have to do that in relation to the effects of toxics in relation to reproductive events, a major challenge in a society like ours.

However an integral plan has to be made that includes population's health and its direct and social costs (Avila-Burgos *et al.*, 1996) as its major criteria for a rigorous policy of technology innovation for pollution prevention (Santos-Burgoa *et al.*, 1995*b*), cleaning, and priority-setting, based on health risk assessment with local characterisation to base human exposure standards.

Questions, comments and answers

Q: In your correlation, these contaminants in water have effects other than just infectious disease. How do you plan to deal with chemically contaminated water supplies?

A: What I showed were hypothetical associations. We have not been able to pinpoint past exposure, so it is hard for us to make a hard and fast statement. We need to have a good grasp of past exposure of these people in order to support an intervention strategy.

REFERENCES

ALATORRE, R. and C. SANTOS-BURGOA (1996), "Selection of Health Indicators for an Environmental Surveillance System: report of an international panel of experts", *Epidemiology* 7:4, p. S54.

ALATORRE, E.W.R., C. SANTOS-BURGOA, L.H. SANIN, and V. BORJA (in press), "Persistence of Hygiene Factors in the Risk of Cholera Transmission in Mexico", *International Journal of Epidemiology*.

ALBORES, A., M.E. CEBRIÁN, T. REYES, M.C. OLIVO, E. PORRAS, E. VERA, M. ROJAS, C. SANTOS-BURGOA, and S. ESTRADA (1994), "Un estudio mexicano sobre valoracion de riesgos: limitaciones y perspectivas", Conference of Health and Toxic Chemical Waste related to the USA-Mexico Border, 27-29 August, Tucson, Arizona.

ANTON, D. (1993), *Thirsty Cities. Urban Environments and Water Supply in Latin America*, International Development Research Center, Ottawa, Canada.

AVILA-BURGOS, L., C. GUTIÉRREZ-ZÚÑOGA, P. HERNÁNDEZ-PEÑA, C. SANTOS-BURGOA, and L. SILVA-AYCAGUER (1996), "El costo social de la bronquitis crónica en la ciudad de México: una experiencia piloto", *Salud Pública de México* 37:4, pp. 354-362.

BOTELLO, V.B., S. VILLANUEVA, G. PONCE, L. RUEDA, I. WONG, and G. BARRERA (1995), "La contaminación en las zonas costeras de México", in I. Restrepo (ed.), *Agua, Salud y Derechos Humanos*, pp. 53-122, Comisión Nacional de Derechos Humanos, México D.F.

CARREON, V.T., (1993), "Riesgo Perinatal por Exposición a Contaminantes Ambientales en la Zona de Riego con Agua Residual, Hidalgo", final thesis submitted to the Escuela de Salud Pública de México.

CORTES-MUÑOZ, J. (1993), "Metales Pesados en agricultores expuestos a aguas residuales crudas en el distrito de riego de Tula", final thesis submitted to the Escuela de Salud Pública de México.

DEL RAZO, L.M., G. GARCÍA-VARGAS, M. HERNÁNDEZ, A. ALBORES, M. CEBRIAN, and C. SANTOS-BURGOA (1996), "Relation of cummulative individual arsenic exposure, the urinary excretion profile or arsenic species and the presence of arsenic diseases", International Seminar on Arsenic Exposure, 8-10 October, Santiago de Chile.

HEALTH INDICATORS FOR ENVIRONMENTAL SURVEILLANCE (HIES) (1996), Final Report of the Research Project, submitted to the National Institute of Ecology.

INEGI (1992), *Censo Nacional 1990*, Instituto Nacional de Estadística, Geografía e Información, México D.F.

REICHARD, E.G. (1991), "Health risks from ground water contamination: current issues and research needs", Paper presented at the Santa Fe Workshop of Research Priorities for Environmental Health, [DATES PLEASE], New Mexico.

SANCHEZ RIOS, G. (1996), "Evaluacion de Riesgo a la salud por exposición a benzo(a)pireno en la población de Coatzacoalcos, Veraqcruz", thesis submitted to the Escuela de Salud Pública de México.

SANTOS-BURGOA, C. (1992), "Salud y Seguridad Social", in J. Narro and J. Moctezuma (eds.), *La seguridad social y el estado moderno*, pp. 299-333, IMSS-ISSSTE, Fondo de Cultura Económica, México.

SANTOS-BURGOA, C., R. ALATORRE, and R. RASCÓN-PACHECO (1995*a*), "Riesgos a la salud asociados a contaminantes químicos presentes en el agua del acuifero de la Ciudad de México", in *El Agua y la Ciudad de México*, Academia de la Investigación Científica, A.C. México.

SANTOS-BURGOA, C., N. ASHFORD, and P. HERNÁNDEZ-PEÑA (1995*b*), "Regulaciones ambientales y laborales como estimulo a la competitividad", *Comercio Exterior* 45:8, pp. 615-622.

SANTOS-BURGOA, C. *et al.* (1996), "Pilot testing of health indicators for environmental surveillance in Mexico", *Epidemiology* 7:4, p. S54.

SIEBE, C. and E. CIFUENTES (1995), "Environmental impact of wastewaters irrigation in central Mexico: an overview", *International Journal of Environmental Health Research* 5, pp. 161-173.

Parallel Session I

I-B. BIOREMEDIATION/BIOTREATMENT: AQUIFERS

THE INFLUENCE OF WASTEWATER ON THE GROUNDWATER QUALITY OF THE MEZQUITAL VALLEY, STATE OF HIDALGO, MEXICO

by

Rubén Chávez Guillén
National Water Commission, Mexico

Introduction

In many basins, the freshwater availability is low, due either to the environment itself, or to the increasing demand in the area, limiting existing or future sustainable development. On the other hand, volumes of wastewater have been increasing at almost the same rate as water use, because only a fraction of this represents consumptive use. For example, around 60 per cent of the water in the pipelines of the municipalities is discharged to the sewage system, and from 30 to 50 per cent of irrigation water is returned to the rivers or the ground as return-flow. Thus, this study addresses a complex problem encompassing the different aspects as the supply, protection and pollution of water.

The use of wastewater for non-drinking purposes should be taken into account in those areas of scarce water resources. This practice could increase the availability of the water resources. However, the use of wastewater requires effective management to prevent serious pollution problems. One of the most delicate problems Mexico faces in the water resources sector is that in the Mexico City Metropolitan Area (MCMA), the current water demand of 60 m^3/sec for domestic, urban and industrial purposes, produces a river of wastewater which is discharged into the Mezquital Valley (MV).

In this paper, some preliminary results of the joint project of the Comision Nacional del Agua (National Water Commission, Mexico) and the British Geological Survey are presented. The main objective of this study is to determine the impact of wastewater re-use on the groundwater quality in the Mezquital Valley.

Hydrogeological setting

The Mezquital Valley is located in the south-western part of Hidalgo State, about 50 km north of the MCMA (Figure 1). It belongs to the Tula River Basin, where the Tula Irrigation District is located. It is considered the largest and oldest region in the world whose irrigation waters largely come from wastewater -- it receives about 70 per cent of the sewage effluent (about 40 m^3/sec) from Mexico City (Figure 2).

Figure 1. **Location of the study area within Mexico**

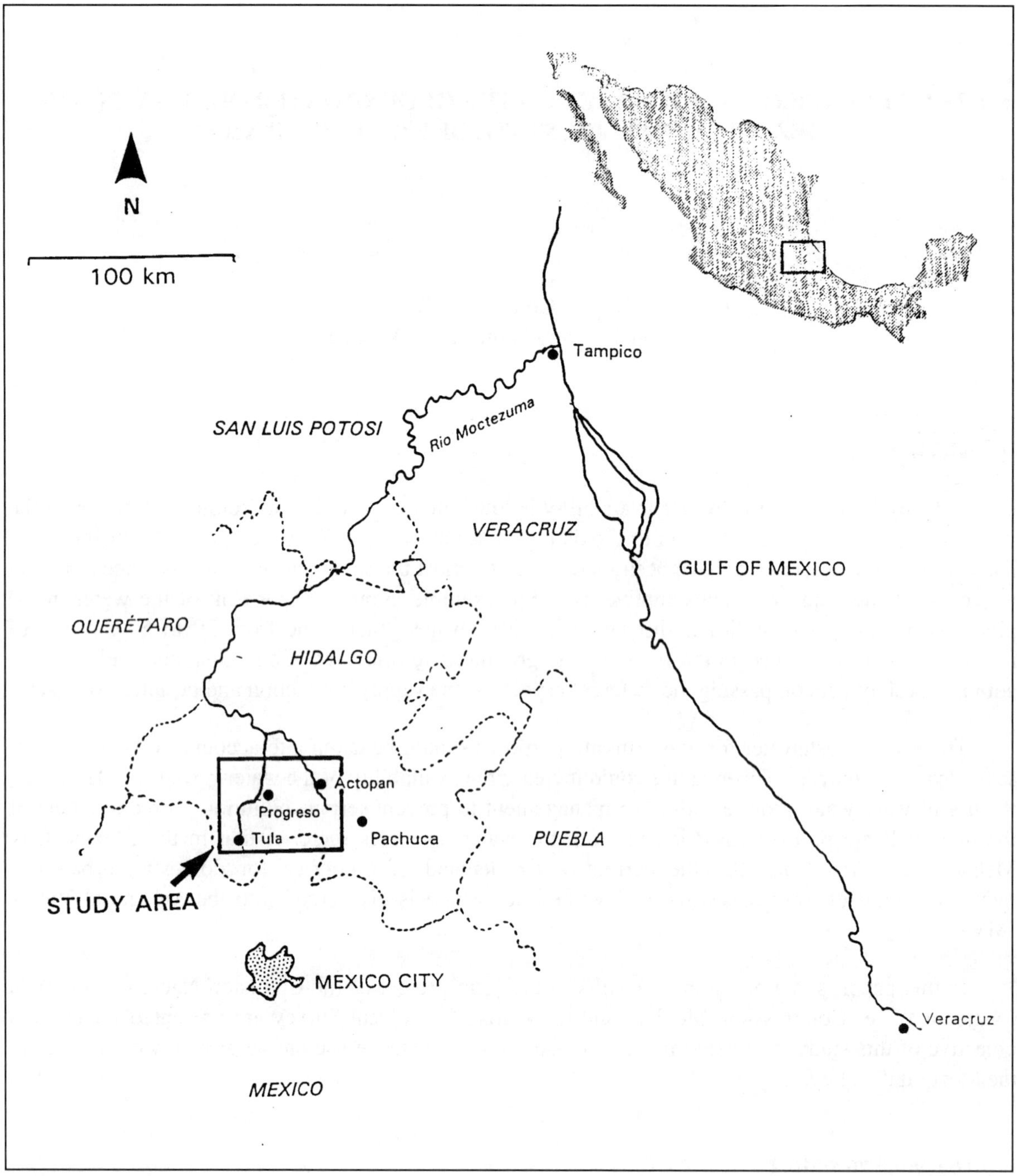

Source: Author.

Figure 2. **Sketch map of Mezquital Valley showing the area irrigated with wastewater**

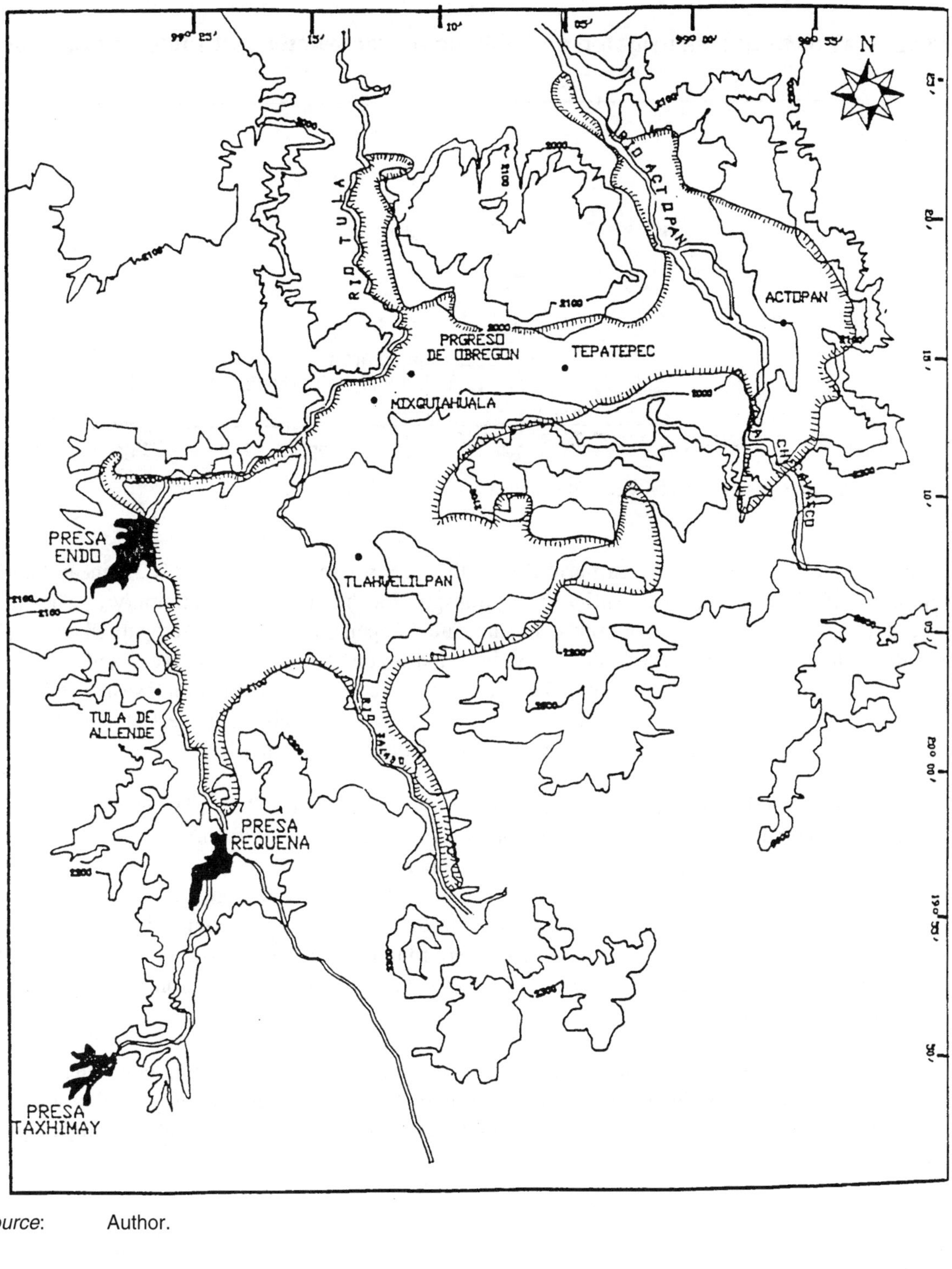

Source: Author.

The huge and steadily increasing quantity of wastewater effluent allowed agriculture to flourish in a semiarid zone with scarce natural water resources. Agriculture is the main economic activity in the Valley, with a cropping area of 45 000 ha. Other economic activities are industry, and raising of livestock and poultry (Figure 3). The population of the area has been estimated at about 500 000 inhabitants.

Figure 3. **Growth of area irrigated and volume of wastewater distributed since 1968**

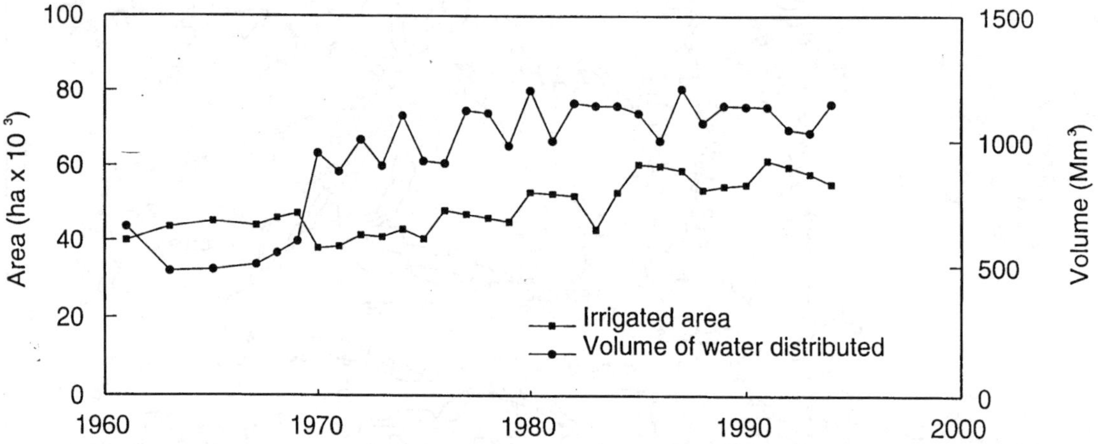

Source: British Geological Survey and National Water Commission, 1995.

The climate of the region is semi-arid-dry, with a mean temperature of 18 °C, a mean annual precipitation of 450 mm, and a potential evapo-transpiration higher than 2 100 mm/yr. The main surface drainage system of the area comprises the Tula (Tepeji), Salado and Actopan Rivers. The Tepeji River drains the western portion of the Valley, and is controlled by the Requena and Taxhimay Dams, which hold freshwater. The Endho Dam, in contrast, receives freshwater and wastewater, from the Salado River and through a deep canal from the MCMA, respectively. The western portion of the basin is drained by the Actopan River, a tributary of the Tula River (Figure 3).

The Tula Basin is part of the Trans-Mexican Volcanic Belt. The MV is elongated and surrounded by mountains and hills. The outcropping rocks range in age from Lower Cretaceous to Quaternary. The former correspond to sedimentary rocks such as limestones, dolomites and shales, which outcrop on some mountains. The sedimentary rocks in this area are generally covered by volcanic rocks of Tertiary age (lava flows, and pyroclastic rocks varying in composition from basalts to andesites). The alluvial and lacustrine sediments are found in the lower portions of the basin. The alluvial deposits range from gravels, sands, and silts to clays. Lacustrine limestones and lava flows are also found within the alluvial deposits.

Basin fill overlays the volcanic and sedimentary rocks of the basement, configured as an echelon-type structure with depths up to 300 m (Figure 4). The tectonic activity of this area has created the complex geologic structure, developing faults and fractures of hydrogeological importance.

The aquifer

A heterogeneous and complex system constitutes the regional aquifer. The permeable units are represented by the fractured lava flows and the alluvial deposits. In the higher portion, the aquifer is unconfined, changing to confined conditions at the centre of the valley. Groundwater also exists in the karst limestone. When less permeable units are present, a multilayer system is observed with changes in the hydraulic head (up to 30 m).

Figure 4. **Natural and actual hydrogeological conditions**

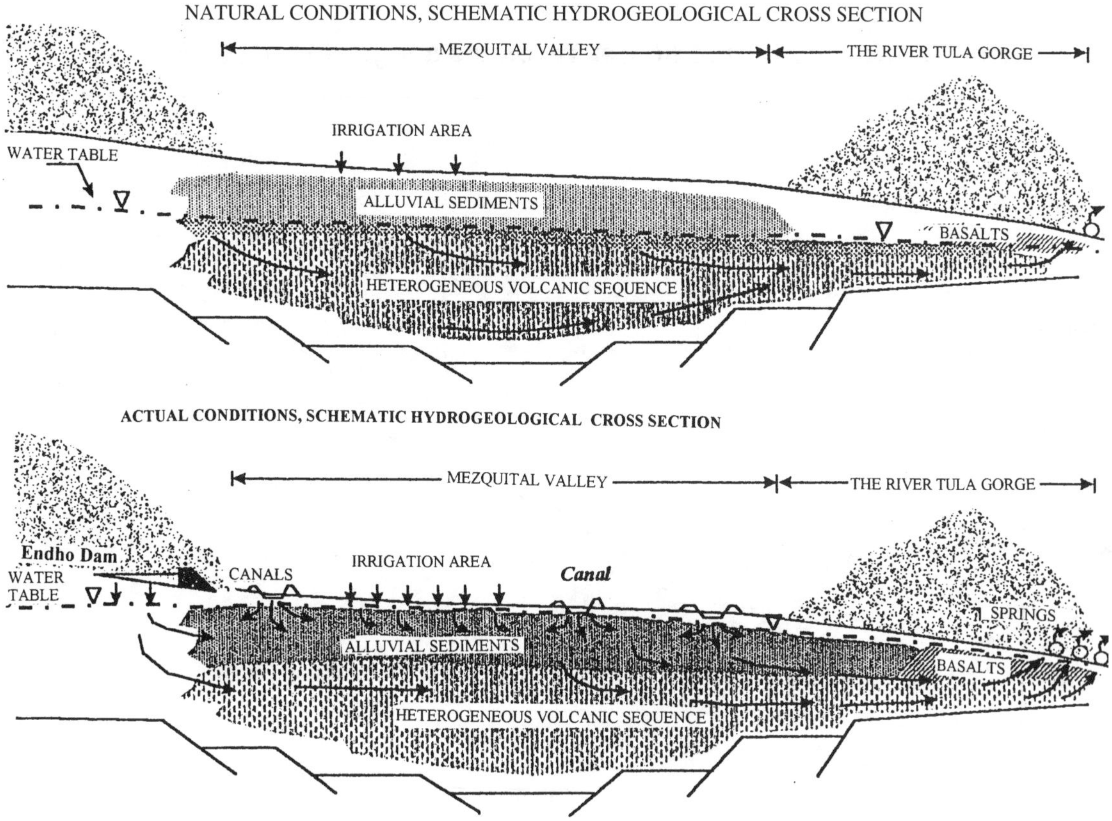

Source: Author.

Under natural conditions, at the beginning of the century, there were two independent groundwater flow subsystems: the Tula River Valley and the Actopan Valley. The groundwater recharge was limited at that time. The base flow of the rivers indicated a deep groundwater level, probably more than 50 m (Figure 4).

During the last 80 years, the use of wastewater for irrigation modified the regional hydrogeology, increasing the size of the original aquifer as a result of infiltration of excess irrigation water from the fields. The distribution system of the area has a total length of 575 km of major and lateral wastewater canals of which only 40 per cent are lined. The gauging station readings suggest a leakage of the canals ranging from 15 to 25 per cent of the total wastewater. Another factor of additional recharge is the excess irrigation of the fields. The two components contribute with 800 Mm^3/yr to the aquifer, which is more than 10 times the estimate of the natural recharge (Figure 4).

Subsequently, the water table rose so close to the surface that it reached the new canals at the lower portions of the Valley, creating springs and flooding the surface. This recharge may have been discharged into the Tula River, whose base flow rose from 50 to 400 Mm^3/yr in the period 1945-1995 (Figure 5). A high yield spring (Cerro Colorado) appeared 50 years ago in the central portion of the

Valley. Since that time, the spring was used as water supply of the communities (600 l/s). In the San Salvador area at the north-eastern portion of the valley, the water table reached the surface, covering about 1 300 ha. The waterlogging has produced an increase in the salinity of water and soil, along with the presence of aquatic vegetation. Vertical ascendant discharge of the aquifer has been identified through the artesian wells, some of them discharging up to 100 l/s.

Figure 5. **Tula River base flow at Tecolotes gauge station (1945-1974)**

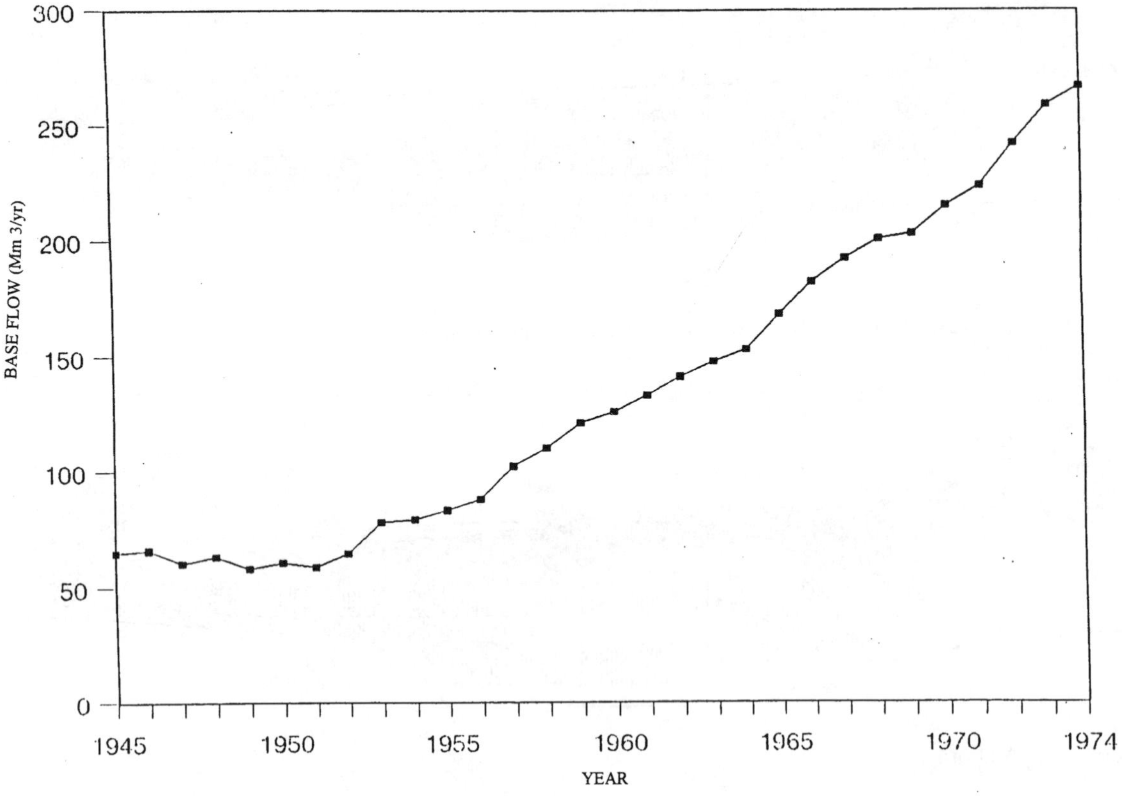

Source: British Geological Survey and National Water Commission, 1995.

The aquifer is the main source of water supply for industrial, urban, domestic purposes. The volume of abstracted groundwater is about 163 Mm^3/yr. Approximately 25 per cent of the volume is used by agriculture, 33 per cent by industry, 17 per cent by domestic purposes, 12 per cent by other sectors, and the remaining 13 per cent is not used. The main crops are corn, alfalfa, beans, chilies, squash, oat, barley, and vegetables.

Wastewater irrigation

At the end of the last century, the practice of using wastewater for irrigation purposes was introduced in the MV as a result of increasing effluent discharge from Mexico City. Between 1925 and 1934, agricultural infrastructure was developed in the MV. Wastewater was diverted to the irrigation district through three main water works: the Central Canal, the Salto River Aqueduct, and the Tequixquiac Tunnels. Industrial wastewater was also discharged into the area through the Endho Dam.

The wastewater of the MCMA contains high concentrations of contaminants. Geochemical analyses have shown a high salt content, BOD, coliforms, organic compounds, detergents, fats, and oils. On the other hand, the heavy metal concentrations are moderate to low. Along with these contaminants, the wastewater also carries the industrial, agricultural and municipal wastes produced in the Valley itself. Thus, the water quality of surface and groundwater has deteriorated. In spite of its low water quality, the amount of nutrients carried by the irrigation water has contributed to development of the crops with a high yield.

In the Endho Reservoir, the main receptor of the wastewater from the MCMA, salinity has increased, while its nitrogen species have decreased. This last phenomenon has been attributed to the processes of ammonia volatilisation, denitrification of the water body, and uptake of nitrogen by aquatic plants, which had spread over the reservoir's water surface, producing a reduction of nutrient availability for the crops. Furthermore, a low concentration of heavy metals compared to the wastewater of MCMA has been observed due to heavy sedimentation at the bottom of the reservoirs.

Impact of wastewater irrigation

Soil impact

Soils in the area consist of a mixture of clay and loamy textures. They are slightly alkaline with medium to high cation exchange capacity. The impact of wastewater irrigation on these soils has been identified by an increase in its sodium content. Measurements of soil metal contents showed an increase in heavy metal contents with time. However, the soil heavy metal contents are within the international standards.

Aquifer pollution

Under natural conditions, before intensive irrigation with wastewater had taken place, the groundwater quality in the Valley was good, with a salinity lower than 700 ppm. At that time, the aquifer was less vulnerable to contamination because water levels were deep enough (tens of meters). The thickness of the unsaturated zone allowed the attenuation of pollutants.

The groundwater quality has changed over time. In 1962, the groundwater pollutants were only detected in some shallow wells. In 1985, qualities of surface and ground waters were similar. Since that time, the high concentrations of boron, nitrogen compounds, and total dissolved solids (TDS) (1 100 to 1 700 ppm), as well as mercury, lead and lithium, suggest an increasing wastewater influence into the aquifer.

Hydrogeochemical and isotopic studies have identified three different types of water: groundwater, wastewater, and thermal water (Figure 6). The groundwater of the southern and western portions of the valley are not influenced by the return flow. This groundwater has been classified as bicarbonate type (calcic to mixed) with a low salinity (<500 ppm).

On the other hand, the central portion of the valley has groundwaters with high concentrations in sodium, chloride and nitrate (mixed-sodium type). The TDS range from 700 to 2 000 ppm. The concentrations are similar to those of the wastewaters, springs, and canals, which confirms the influence (infiltration) of the wastewater use in agriculture. The infiltration of wastewater into the aquifer has been homogeneous, from the top to the bottom (400 m), probably reaching the regional flow system (Figure 6).

Figure 6. **Piper diagram of groundwater from the central valley**

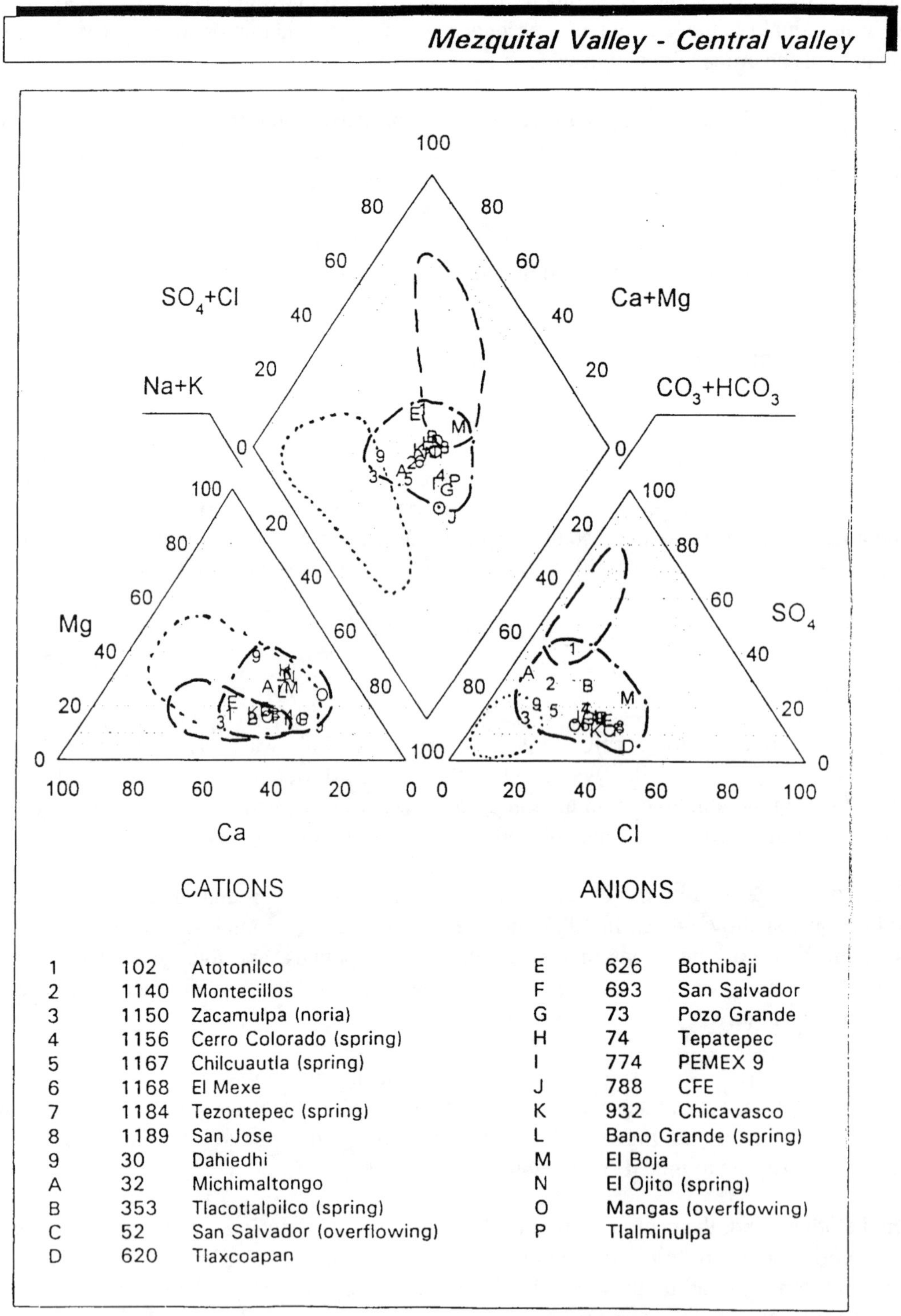

Source: British Geological Survey and National Water Commission, 1995.

The southern portion of the valley has high temperatures, with high concentrations in boron, lithium and cesium, which has been associated to deep flow systems and geothermal components (Figures 7 and 8). There is no evidence of infiltration of wastewater in this area. The shallow water is influenced by the uprising geothermal water. In the same area, a well with high concentrations of aluminum, manganese, copper, zinc, and fluorine was identified.

Figure 7. **Speciation of groundwaters using sulphate and chloride concentrations**

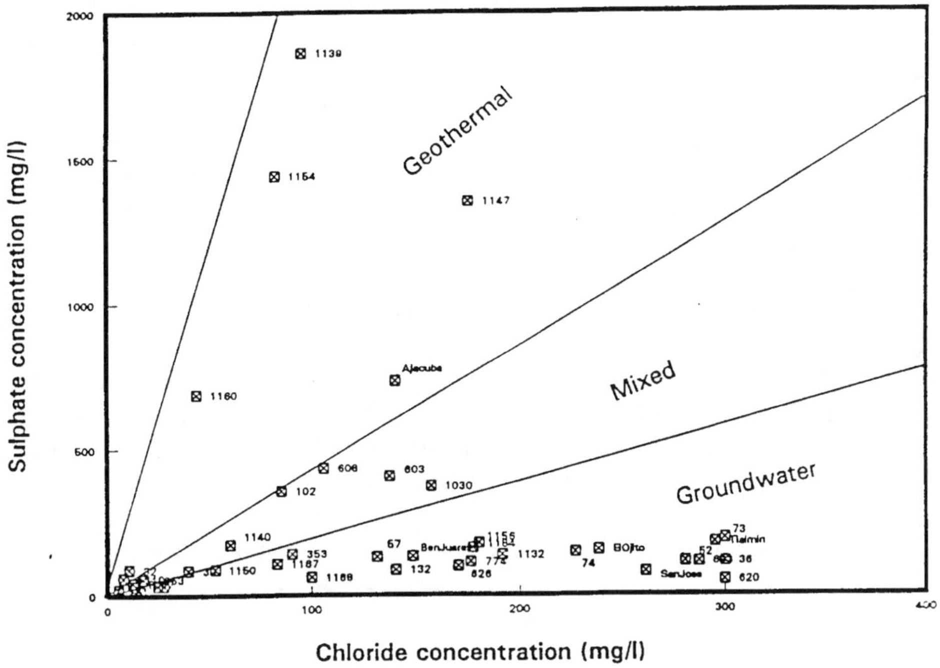

Source: British Geological Survey and National Water Commission, 1995.

Figure 8. **Lithium and cesium as geothermal indicators**

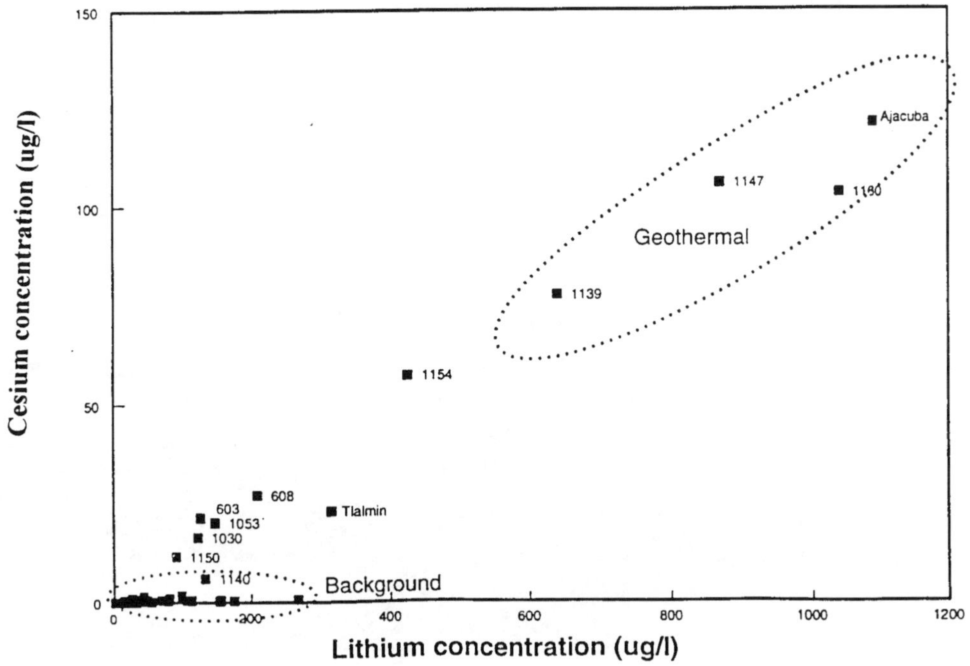

Source: British Geological Survey and National Water Commission, 1995.

Impact on human health

One of the main goals of this study was to determine the risk of wastewater irrigation practice on human health. The highest risk on human health is the consumption of contaminated water. However, other sources should be considered, such as contaminated food and direct contact with wastewater or soil.

The lack of accurate health statistics make it difficult to determine the influence of wastewater on human health. Not withstanding, available data indicate a high and increasing risk to gastrointestinal diseases related to wastewater contamination.

Most of the water coming from the shallow aquifer is usually not suitable for drinking purposes. The groundwater quality is one of the main concerns in the MV because it represents the water supply. Generally, the groundwater is treated before the municipalities use it. On the other hand, the small communities used untreated water, tapped from shallow wells.

Generally, the aquifers have a natural barrier against biological contamination. The micro-organisms are retained and destroyed in the vadose zone (Figure 9). However, in the MV, this natural barrier is not present due to different factors: poor water quality (Figure 10) of the main recharge source, a high excess of irrigation, and a shallow water table. All these factors have produced an aquifer with a high vulnerability to contamination. Besides, the fractured rocks in the area provide paths to biological contaminants into the aquifer.

Figure 9. **The ground: a natural treatment plant**

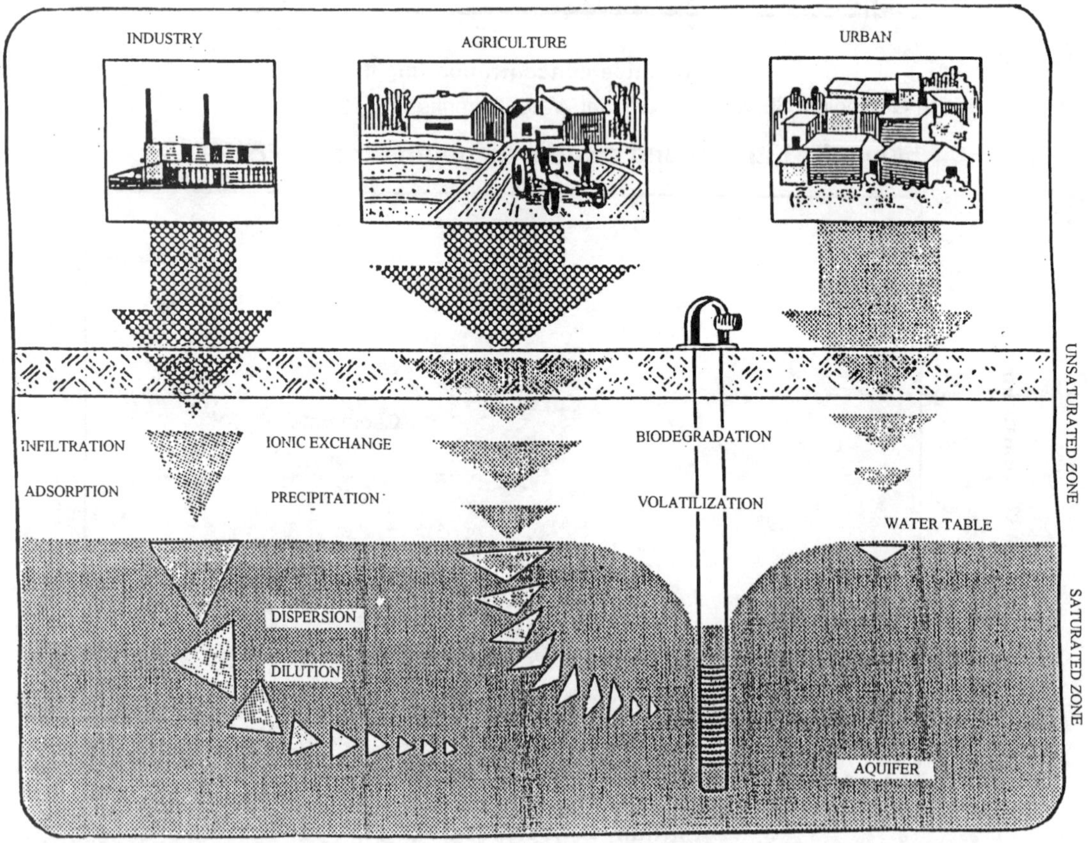

Source: Foster and Hirata, 1991.

Figure 10. **Water quality along the main canals and in related drainage water**

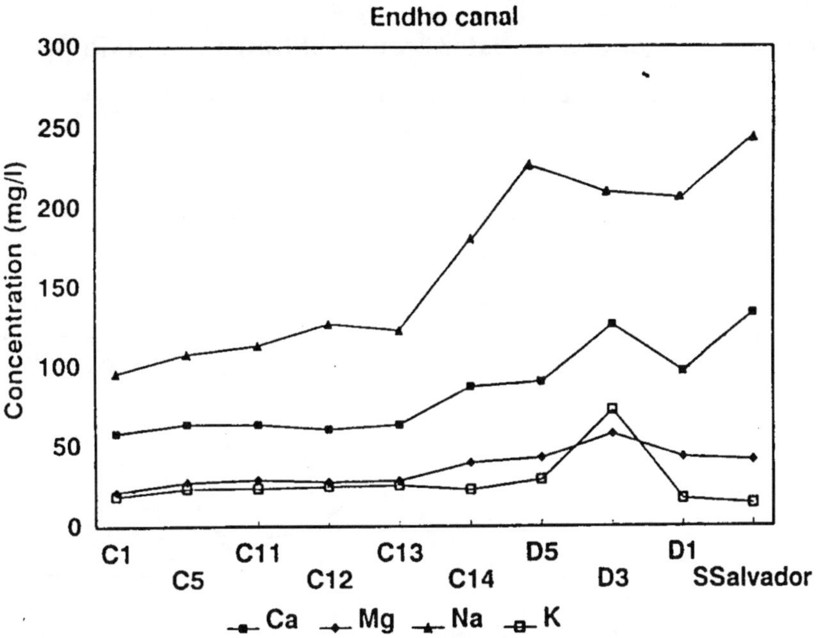

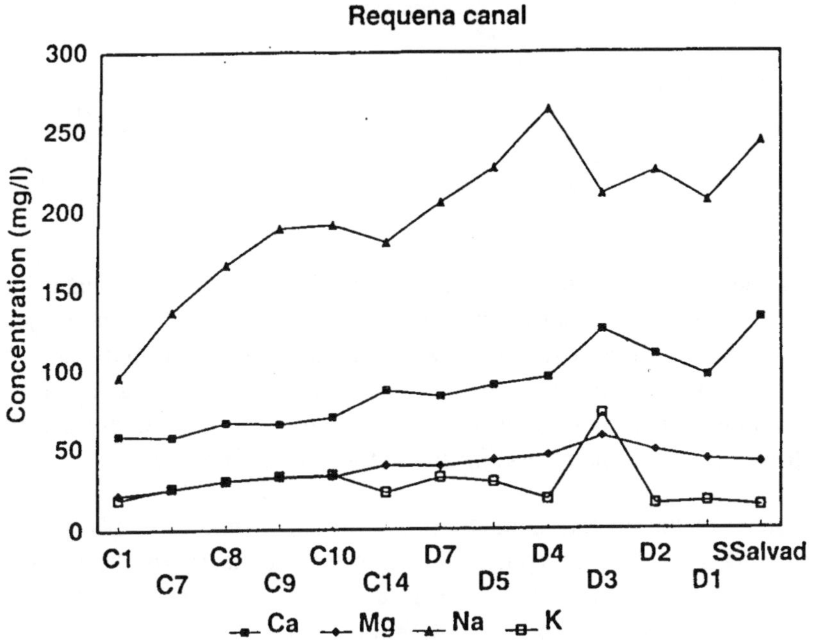

Source: British Geological Survey and National Water Commission, 1995.

Changes in the water quality are illustrated by the evolution in the Cerro Colorado Spring. When this spring first appeared, it was reported to be of good quality and was developed for public supply. A gradual deterioration then occurred; present water quality is similar to wastewater and all groundwater in the area (Figure 11). This presumably represents the displacement of the original meteoric water by increased infiltration of wastewater.

Figure 11. **Water quality deterioration of Cerro Colorado Spring**

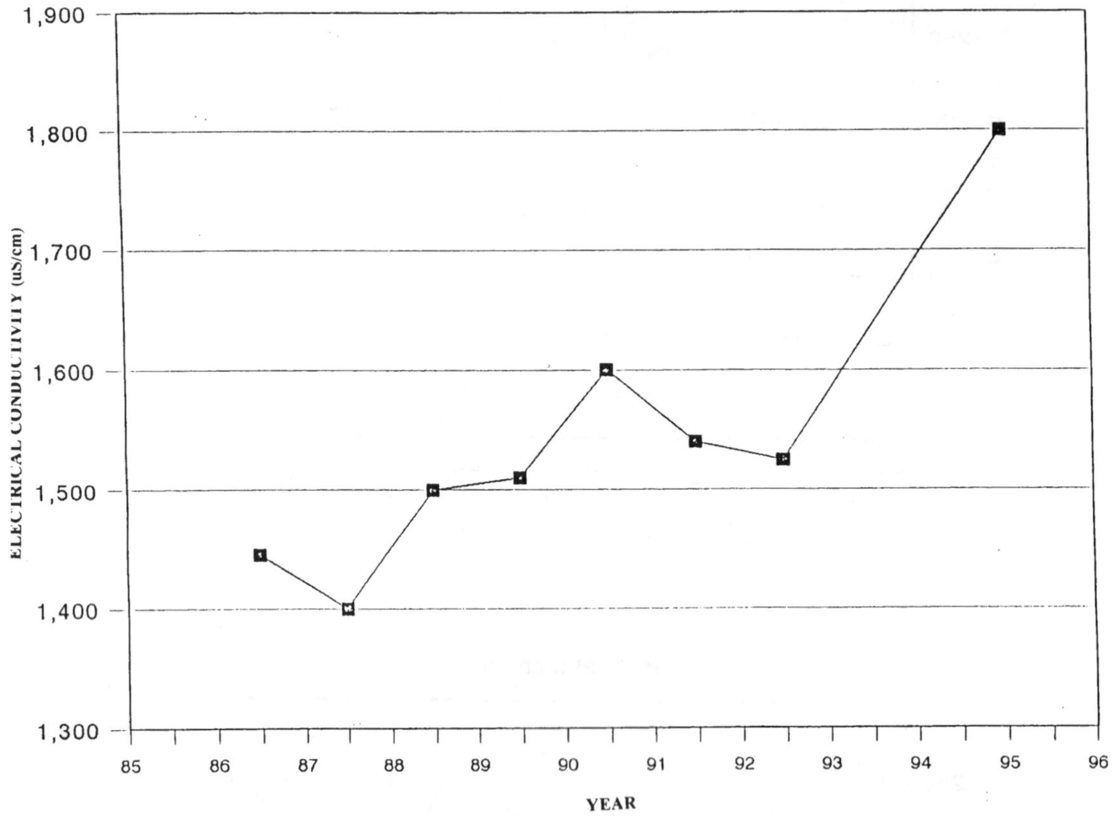

Source: British Geological Survey and National Water Commission, 1995.

The risk assessment of wastewater on human health is not easy to evaluate due to the lack of information; however, the high concentration of TDS, organic compounds, and nitrates (among other contaminants) should have a direct correlation with human health. Groundwaters in the MV have elevated concentrations of heavy metals compared to unpolluted groundwaters. However, none are present at concentrations higher than the drinking water standards. It is likely that heavy metals are accumulating in the sediments of the reservoirs and in the agricultural soils.

Even though some of the contaminants that may produce some type of disease on humans are attenuated or destroyed in the vadose zone, one of the main concerns is the deterioration of water quality in the whole basin for both surface and groundwaters (Figure 12). The water quality may slow the agricultural development of the area in the short term.

Figure 12. **Groundwater contamination due to agricultural activities**

Source: Author.

The future goals of this project will be oriented to define, design and implement a protection programme. Also, there is a proposal to install a treatment plant to reduce the pollutants of the MCMA effluents, enhancing the quality of the water used for irrigation at the MV. The following measures are being implemented nation-wide under the CNA Clean Water Programme to safeguard public health:

- systematic chlorination of all potable water supplies;
- control of the quality of bottled water and the ice manufacturing;
- restriction and control of the use of wastewater for irrigation, particularly for crops that are eaten raw;
- the promotion of treatment plant construction and efficient operation;
- sanitary protection for potable supply sources;
- discharge control from hospitals, laboratories and other centres of high risk.

Conclusions

The wastewater of the MCMA has been an important hydrological factor in the development of agriculture and industry in the Mezquital Valley. Without the availability of wastewater for irrigation, no agricultural development would have been possible in the MV.

Irrigation with wastewater has radically changed the hydrological regime of the basin by increasing the groundwater availability. The recharge of the regional aquifer has been increased to more than 15 times the natural recharge rate.

The impact on groundwater quality has been adverse, in spite of the contaminant attenuation effects in the reservoirs, and within the vadose zone. The induced recharge has essentially the same water quality as the wastewater, which has led to a deterioration of the groundwater quality in the irrigation areas, due to a rapid increase in salinity.

The practice of wastewater irrigation should be carefully controlled, delineating the areas under irrigation, irrigation methods and permissible crops, to avoid health problems. From a hydrogeological point of view, it is very important to define the mechanisms that govern and control the fate and transport of contaminants in the MV aquifer.

In the developed basins, the volume of wastewater increased proportionally to the volume of used water, in spite of more efficient practices. Therefore, the discharge produces problems of protection and risk of contamination of water supply.

The use of wastewater for non-drinking purposes has to be taken into account in those areas of scarce water resources. This practice could increase the availability of the water resources, reducing the competition for the resource. However, the use of wastewater requires treatment and effective management to prevent serious pollution problems, particularly the contamination of water supply facilities.

Questions, comments and answers

Q: Are there any planned aquifer recharge projects in Mexico, and if so with what quality of water?

A: In one sense, we have an incidental recharge from irrigation. A very high percentage of waste water applied to ground goes to the aquifers. One objective is to locate limestone aquifers in order to reserve fresh water for drinking purposes.

Q: What are you doing with regard to carcinogenic organics? Are you looking at these as a possible risk to the population?

A: Of course we are worried about organics and also about nitrates and heavy metal. We have fewer problems with organics in deep aquifers. There the problem is high salinity. We are only at the characterisation stage for waters from different aquifers. We know there is stratification of pollutants.

REFERENCES

BRITISH GEOLOGICAL SURVEY and NATIONAL WATER COMMISSION (Mexico) (1995), *Impact of Wastewater Reuse on Groundwater in the Mezquital Valley, Hidalgo State, México.*

FOSTER, S. and R. HIRATA (1991), *Determinación del Riesgo de Contaminación de Aguas Subterráneas*, 2nd revised edition, Centro Panamericano de Ingeniería Sanitaria y Ciencias del Ambiente (CEPIS), Lima, Péru.

MICROBIAL ROLE IN REGULATION OF PAH BIOAVAILABILITY

by

Elizabeth A. Guthrie and Frederic K. Pfaender
Department of Environmental Sciences and Engineering, University of North Carolina
Chapel Hill, North Carolina

Introduction

A major issue in assessing the quality of water for recreational, industrial or drinking water use is the presence of organic contaminants that can have adverse human and ecological health effects. These contaminants range from materials of anthropogenic origin like industrial solvents, pesticides, and PCBs to compounds that have a variety of sources, both natural and man-made, like polynuclear aromatic hydrocarbons (PAH). The presence of these materials in water can occur directly through intentional or fugitive discharges, or indirectly by movement through groundwater or surface run-off. In either case, these contaminants will interact with the biological community of the soil or sediments, which can be impacted by or alter the contaminants. A common feature of many organic pollutants is their highly hydrophobic nature. The non-polar nature of these compounds and their binding to soil or sediment (Harvey, 1991) reduces their mobility and can lead to bioaccumulation of these compounds in exposed biota. Because the observed distribution of hydrophobic organic contaminants is broad, and many are toxic or carcinogenic, at fairly low concentrations, they may constitute a significant risk to human and/or ecological health. A significant question for assessing the risk of these tightly sorbed contaminants is availability of the chemical when associated with surfaces or particles. We have conducted research over the last several years to examine the interaction of several PAHs with the organic carbon matrix of soil and the soil microbial community, including communities of active biodegraders (Guthrie and Pfaender, in press *a*, in press *b*, in press *c*). We are using PAHs as models of how many other classes of non-polar organic contaminants might behave.

PAH(s) are multi-ring aromatic compounds that are by-products of the incomplete combustion of organic matter, and thus can originate from both natural and anthropogenic sources (Blumer, 1976; Suess, 1976). While PAH(s) are virtually everywhere in the environment, high concentrations are usually associated with disposal of combusted materials or petroleum residues (wood preserving, coal-tar disposal, oil recycling). They exist in the environment as mixtures of compounds with different numbers of rings, from the simple two ring materials to those with five to seven condensed rings. Figure 1 illustrates the complex interactions of PAH in the environment and how the fate of the parent and products may be impacted by microbes. Numerous studies have addressed the environmental fate of PAH, mostly dealing with the lower molecular weight compounds. It is clear that micro-organisms can utilise low molecular weight PAH(s) (Bauer and Capone, 1988; Cerniglia, 1992; Heitkamp and Cerniglia, 1987; Mueller *et al.*, 1989). Higher molecular weight PAH(s) are more slowly utilised, and few appear to result in the formation of mineral products (Carmichael and Pfaender, 1996; Heitkamp and Cerniglia, 1987; Manilal and Alexander, 1991; Sutherland *et al.*, 1995). Most metabolism of high molecular weight PAH(s) probably occurs by cometabolic mechanisms that result in the formation of stable, intermediate products, (Keck *et al.*, 1989). These products have different fates than the parent materials and are known to have different environmental effects. In some cases, intermediate products may be more toxic that the parent compounds (Lambert *et al.*, 1995). The conditions under which different products form and how long they persist is largely unknown. The common measurement of mineralization as an experimental end-point in metabolism studies reveals little about these other potential fates.

The hydrophobic nature of the high molecular weight PAH(s) (HMWPAH) and their early degradation products contributes to strong interactions between these materials with surfaces and organic phases, such as soil organic matter (SOM) (Weissenfels *et al.*, 1992). Adsorption of PAH to organic matter and their low water solubility has been shown to influence microbial growth on and metabolism of PAH(s) (Carmichael and Pfaender, 1996; Volkering *et al.*, 1992; Weissenfels *et al.*, 1992; Wodzinski and Coyle, 1974). Limited growth and partial degradation can lead to stable metabolic products, that may be further metabolised or bound to the SOM. The chemical, biological and environmental factors that control this interaction are largely unstudied. Several workers have shown that in soils with long period of exposure to PAH and other non-polar organics, the fraction either recovered or available for degradation declines over time (Carmichael *et al.*, 1997; Hatzinger and Alexander, 1995). These observations are most commonly explained as being due to "irreversible adsorption" to the soil matrix. Adsorption equilibria of PAH and other non-polar organics is clearly complex and may involve several different mechanisms over time (Pignatello and Xing, 1996), but there currently is no satisfactory theoretical explanation for "irreversible adsorption". The situation becomes even more complex if we introduce potentially stable degradation products and their interactions with the SOM. To understand and perhaps minimise the risk from non-polar organic contaminants requires examination of how PAH intermediates interact with SOM in studies that simulate real world conditions. Experimentally this requires studies with "aged" contaminants. Freshly added soil contaminants have not had time to partition to the same extent as "aged" materials not to select and establish a contaminant degrading microbial community and so do not behave at all like aged contamination (Carmichael *et al.*, 1997).

We can envision at least three different types of PAH:SOM interactions. The best studied are adsorption and/or partitioning of the PAH or their degradation product into or onto the SOM or mineral matrix of soil (Guthrie and Pfaender, in press *a*; Kan *et al.*, 1994; Karickhoff *et al.*, 1979; Weissenfels *et al.*, 1992). This is an abiotic process based on the most stable thermodynamic state, which for most of these materials is association with the more non-polar SOM. Whether adsorption equilibrium is ever achieved in nature is questionable. Other processes may be constantly changing the concentrations of contaminants that are subject to the equilibrium forces. Examples of these processes include microbial degradation, heterogeneous sorptive sites, and biotic/abiotic changes to the soil matrix itself. However, it is certainly a major driving force for the partitioning. With the vast majority of most HMWPAH(s) sorbed onto the soil, desorption is often cited as the driving force in regulating bioavailability (Carmichael *et al.*, 1997; Hatzinger and Alexander, 1995; Weissenfels *et al.*, 1992). The important point is that adsorption/desorption is probably not the only interaction and may be influenced by other abiotic and biological processes.

A second interaction would be based on physical trapping of some PAH or degradation products within the SOM. There is evidence to suggest that soil humics and potentially other organics change molecular size and shape based on changes in pH, ionic strength, etc. Any PAH adsorbed on surfaces that become covered by other surfaces, or buried within the carbon matrix, would no longer be in contact with either water or the solvents used to extract the SOM and therefore would appear unavailable. Certainly rainfall and soil recharge would produce regular alterations in ionic strength, and in many soils, the degradation of annual leaf fall generates organic acids that could change soil pH. The extent and frequency of such processes in nature is likely highly variable both spatially and temporally. The subsequent availability of these entrained materials has not been examined, but at least conceptually, when the configuration of the SOM changes the molecules may return to their previous shape, making the PAH again amenable to the sorption/desorption processes

The third interaction involves the formation of covalent bonds between the PAH, or more likely the degradation products, and components of the SOM. It has been shown that soil enzymes can lead to the formation of polymeric materials and humification of pesticides (Dec and Bollag, 1993; Bollag *et al.*, 1992). The incorporation of aromatic lignin degradation products, with functional group structures not unlike known PAH degradation products (Cerniglia, 1992), into the humic/fulvic acids of soil, has been shown to occur (see Pfaender, 1988 for review), although the mechanisms are not clear. Covalent binding could result in the molecular identity of the parent compound or product disappearing as it becomes part of the polymeric soil organic matrix. Once part of the SOM matrix, the compound becomes subject to the fate processes of the soil carbon, slow biologically-driven degradation (or burial), and would probably not be bioavailable again.

An important question is what impact these different fate processes and PAH:SOM interactions have on the subsequent toxicity or carcinogenicity of the PAH. It has been observed that some of the microbial degradation products of HMWPAH are actually significantly more acutely toxic than the parent PAH (Carmichael and Pfaender, in press). The acute and chronic toxicity of hydrophobic organic contaminants to fish and invertebrates has been found to relate to the available concentration, or the amount dissolved in the soil water (or pore water of sediments), not the total amount present in soil or sediment, as determined by chemical extraction and analysis (DiToro *et al.*, 1991). Microbial degradation of PAH and other non-polar compounds is related to the rates of PAH desorption, and therefore the water phase concentrations as well (Carmichael *et al.*, 1997; Hatzinger and Alexander, 1995; Weissenfels *et al.*, 1992). The obvious conclusion is that the bioavailable concentration is the same for both animals and microbes, or stated differently, that dissolved concentrations are at least a major part of the bioavailable fraction. From a bioremediation perspective, increasing the dissolved concentration is desirable, since more rapid rates of biodegradation are possible. From a risk perspective, increasing dissolved concentrations generates greater risk since the toxin concentration increases. If interactions between the PAH and SOM determine what fraction of the contaminant is available to desorb and enter the dissolved state, then the nature of the PAH:SOM interaction will dominate the evaluation of PAH risk and perhaps bioremediation.

This paper summarises research conducted over the last several years to address how the environmental fate of HMWPAHs are impacted by microbial degradation and interaction with SOM. These studies have examined how microbes impact the nature of the interactions of PAH and SOM (Guthrie and Pfaender, in press *a*) using a mass balance approach combined with a SOM fractionation scheme. This approach allows us to determine how a radiolabel initially associated with the parent PAH distributes itself among soil components in the presence and absence of a variety of natural and manipulated soil microbial communities. While not directly related to the subject of this paper, we have also determined how the toxicities of PAH(s) relate to the microbiological fate and interaction with the SOM (Guthrie and Pfaender, in press *b*), and how the PAH:SOM:MICROBE interaction might be manipulated by addition of plants (Guthrie and Pfaender, in press *c*).

Figure 1. **Conceptual model of the complex interactions of HMWPAH, SOM, and micro-organisms in soil**

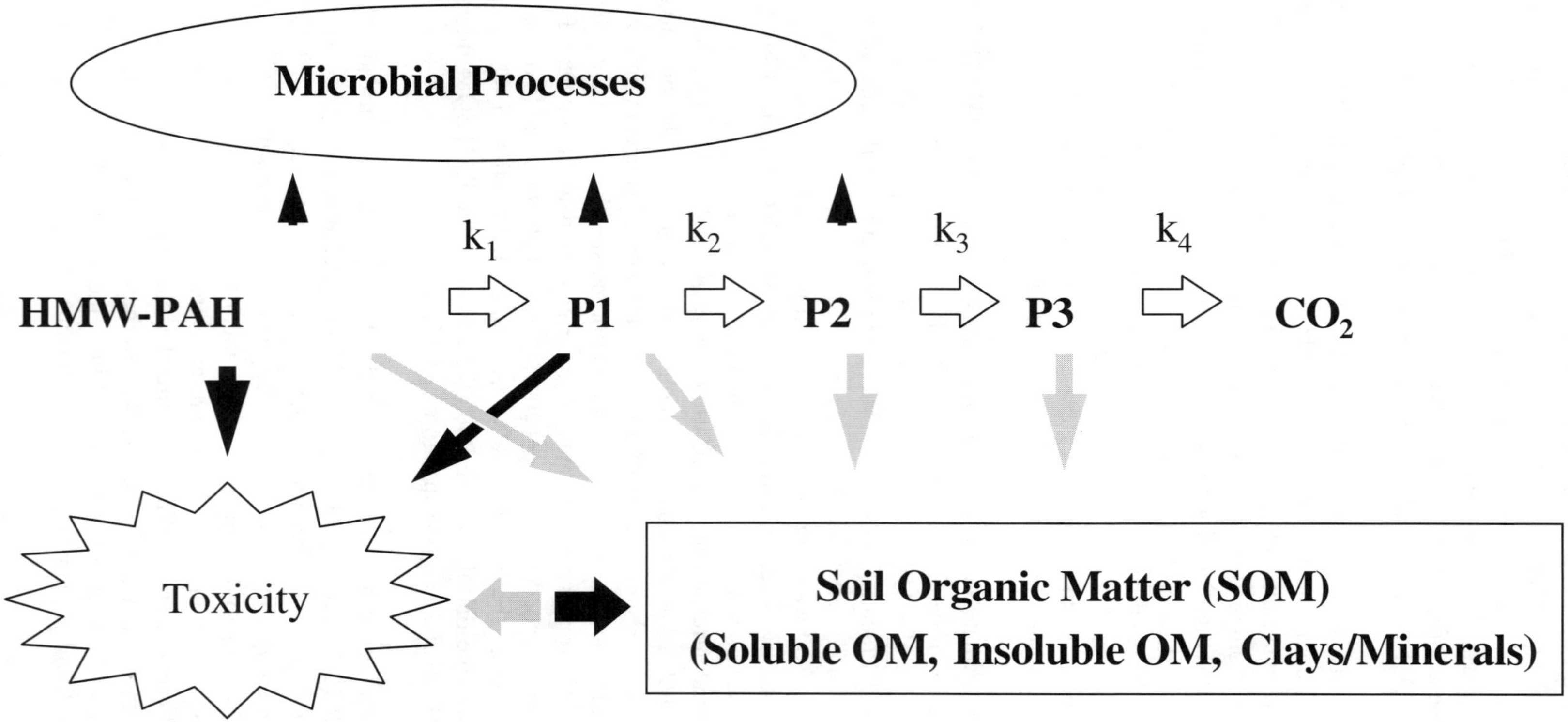

Source: Author.

Experimental methods

Soils

Surface soils with no known history of HMWPAH exposure were synthetically contaminated with pyrene and [4,5,9,10 ^{14}C]-pyrene and incubated with several separate and distinct microbial communities: (1) the natural, non-adapted microbial community; (2) a microbial community adapted to pyrene mineralization; (3) and a metabolically-inhibited microbial community. Surface soils were collected to a depth of 10 cm from a hardwood forest (North Carolina State University Schenck Forest, Raleigh, NC). Microbial communities were manipulated by adding amendments, including a PAH degradation adapted contaminated soil (MN soil), plants and compost. Soils containing a known pyrene-mineralizing community were collected from the Reilly Tar and Chemical Company Superfund Site (MN) (St. Louis Park, MN). These soils have been contaminated with wood treating wastes containing mixed PAHs (total PAH = 1 571 mg/kg) for over 50 years. Previous work has demonstrated that these soils contain an active pyrene mineralizing community (Carmichael and Pfaender, 1996). Abiotic, control soils were amended with 0.5 per cent (v/v) of sodium azide (NaN_3). Treatment with NaN_3 inhibits bacterial metabolism via electron transport disruption and provided the best means for metabolic inactivity without disrupting the soil organic matrix. All soils were sieved (2 mm) and stored at 4 °C in the dark prior to use. Soils were analysed for pH, cation exchange capacity, and mineral content by the North Carolina Department of Agriculture, Raleigh, NC. Total per cent organic carbon (per cent TOC) analyses of soils were conducted by Huffman Laboratories (Golden, CO) using a modified method of SW-846 Method 9060. Characteristics of all soils are detailed in Guthrie and Pfaender (in press *a*).

Chemicals

Pyrene and [4,5,9,10 ^{14}C]-pyrene (greater than 98 per cent purity, specific activity = 1.2 x 1 012 Bq/mol) were purchased from Sigma Chemical Company (St. Louis, MO). Pyrene and ^{14}C-pyrene stock solutions were maintained in 95 per cent ethanol/water solutions and were diluted in 95 per cent methanol (500 uL) prior to addition to soils. Sodium azide (NaN_3) was purchased from Sigma Chemical Company (St. Louis, MO). Sodium hydroxide (NaOH) and sodium pyrophosphate ($Na_4P_20_7$) were purchased from Aldrich Chemical Company (Milwaukee, WI). Hydrofluoric acid (HF), hydrochloric acid (HCl), and potassium hydroxide (KOH) were purchased from Fisher Scientific (Pittsburgh, PA). Deionized, carbon filtered, sterile filtered water was used for all solutions and soil rinses. Solvents [methylene chloride (DCM), methanol (MeOH), and acetonitrile (ACN)] were of the highest quality available (Malinckrodt Specialty Chemical Company (Paris, KY). Scintisafe Scintillation Cocktail (Fisher Scientific, Pittsburgh, PA) was use for liquid scintillation spectroscopy (LSC) analyses and to trap organic volatiles in aerated chambers.

Soil preparation prior to incubation

Soil samples were first sieved (2 mm) then mixed well in a clean plastic bag. Soils were gravimetrically divided into 100 g aliquots and mixed with soil amendments, if necessary, prior to addition of ^{12}C-pyrene and ^{14}C-pyrene. Three replicates were prepared for each soil series. Pyrene [100 (ug/g)] and [4,5,9,10 ^{14}C]-pyrene [0.1 (ug/g)] were added to soils via a methanol carrier (500 uL). Sterile water (15 ml) was immediately added to soils, and the soils were mixed well to distribute both methanol and water. Containers were then left open for four hours to evaporate the methanol. After evaporation, soils were mixed again, and a 10 g sample was removed to determine the amount and distribution of ^{14}C-activity at the beginning of incubations (t = 4 hours).

Aerated systems for soil incubations

All soils were incubated in glass jars with Teflon septa lined polypropylene caps and 0.015 mm Teflon tubing for connections to chambers and volatile trapping solutions (Figure 1). Chambers were aerated with a Whisper 100 aquarium pump (Oakland, NJ) at a rate of approximately 1 mL/min. Air was filtered through activated charcoal, glass wool, and 1N KOH to limit microbial and volatile organic contamination, then delivered separately to each chamber by a manifold. Soil moisture was maintained at 30 per cent field capacity by adding sterile water to each individual chamber via a water inlet and a syringe. Soils were incubated for 120 or 135 days. Over the course of incubations, trapping solutions were continuously monitored for $^{14}CO_2$ and ^{14}C volatile products with 1N KOH and Scintisafe Cocktail solutions, respectively.

Analytical procedures

Fractionation of soil organic matter

Soil subsamples (5-10 g) were periodically removed from aerated jars with clean, steel spatulas and gravimetrically weighed into 40 mL EPA vials. Soil aliquots were fractionated to determine the distribution of ^{14}C-activity in water soluble extracts, an organic solvent extract (DCM), humic and fulvic acids fractions, and non-extractable material in residual soil (humin) according to a modified version of the soil organic matter fractionation proposed by Schnitzer and Schuppli (1989) (Figure 2). Soils were first extracted with acidified water (pH 2) for 24 hours, washed with neutral water (pH 7), then air dried prior to soxhlet extraction with DCM. Humic/fulvic acids were removed from soils by sequentially extracting soils with 10 mL of 0.1 M $Na_2P_4O_7$, 0.1 M NaOH, and an H_2O rinse. Base solutions were decanted from soils prior to addition of the next base solution. Base solutions were mixed together (total 30 mL) before removing aliquots for LSC analysis or extracting with DCM. Aliquots of the remaining soils (residual soils) were either combusted to determine remaining ^{14}C-activity in residual soils or extracted with DCM/MeOH for HPLC analyses. For HPLC analyses, soils, humic/fulvic acids, and residual soils were extracted with DCM/MeOH (4:1:1 v/v/w). DCM extracts were decanted and aliquots were syringe filtered for LSC and HPLC analyses. Some residual soils were also acid digested with a 0.1 per cent HCl/HF (Schnitzer and Schuppli, 1989) solution for seven days. Acid digested soils were extracted again with DCM/MeOH (4:1:1 v/v/w) to determine if any pyrene or ^{14}C-activity could be recovered from the humin fraction of soils. All fractions were analysed for ^{14}C-activity by mixing aliquots of samples (1 mL) with 9 mL of Scintisafe Scintillation Cocktail (Fisher Scientific, Pittsburgh, PA) then analysed by liquid scintillation spectroscopy (LSC) on a Packard 1900TR Scintillation counter (Downers Grove, IL). The clarity of extracts did not require quench correction for samples quantified by LSC. Soils were not directly analysed by LSC analysis but were first combusted with a Harvey Biological Oxidizer (R.J. Harvey Instrument Corporation, Hillsdale, NJ) to collect $^{14}CO_2$, that was then quantified by LSC analysis. Efficiency recoveries of $^{14}CO_2$ from biological oxidation were greater than 96 per cent.

Figure 2. **Schematic of soil fractionation procedure**

SOM Fractionation Protocol

$^{14}CO_2$ efflux	→	PAH Mineralization
^{14}C - volatile products	→	Volatilized PAH or Organic Products
^{14}C - water extracts	→	Water soluble PAH Products
^{14}C - MeOH extracts	→	PAH or Polar Organic Products
^{14}C - Methylene Chloride extracts (Soxhlet Extraction - 24 hrs)	→	PAH or Non-polar Organic Products
^{14}C - Humic/Fulvic Acids (base extraction with NaOH)	→	PAH or Product Association with Extractable Organic Matter
^{14}C - Residual Soil Acid Digestion with (0.1% HF/HCl - DCM)	→	PAH or Product Association with Non-Extractable Organic Matter

Source: Adapted from Schnitzer and Schuppli, 1989.

HPLC chromatography

Soils, humic/fulvic acids, residual soils, and HF/HCl-digested soil fractions were extracted with DCM/MeOH (4:1:1 v/v/w) to quantify residual pyrene and to estimate the presence of pyrene products in these fractions. After shaking fractions in DCM/MeOH for 24 hours, extracts were analysed by reversed phase HPLC chromatography. DCM extracts were solvent exchanged to ACN/MeOH (2.5:1 v/v) prior to HPLC analysis. Extracts of soil fractions were analysed by HPLC chromatography with a Waters 600E System controller and 717 autosampler. Separation of products was achieved using a Supelco LC-PAH column (250 mm x 4.6 mm, i.d., Bellefonte, PA) and elution gradient from 65 per cent acetonitrile to 35 per cent 0.1 per cent trifluoracetic acid to 100 per cent acetonitrile at 18 minutes, with a flow of 1 mL/min. Both flourescence [Waters 470 Scanning Flourescence; (260;374 nm)] and UV absorbance (Spectroflow 757 Absorbance Detector [Kratos Analytical; (254 nm)] were used for detection. HPLC chromatography data was analysed with Millennium 2010 software (Millipore Corporation, Milford, MA).

Bioavailability of residual ^{14}C-activity in soils and SOM fractions after 270 days

Static microcosms

The bioavailability of the ^{14}C-activity in soils or SOM fractions was estimated by quantifying the amount of $^{14}CO_2$ evolved following addition of a community of known pyrene-mineralizing micro-organisms (MN soils) to soils or SOM fractions. For these experiments, mineralization by this known community of degraders indicated that the PAH or product was bioavailable. Soils and soil fractions were incubated in static microcosms for four weeks. Only soils and SOM fractions from the Day 270 SOM fractionation procedures were used in these experiments. Prepared solutions were incubated in 40 mL of EPA glass vials fitted with Teflon buckets and 1N KOH soaked chromatography paper to trap evolved $^{14}CO_2$. Mineralization ($^{14}CO_2$) of the residual ^{14}C-activity was the only endpoint measured. Control vials (MN standards) were prepared by adding 1 g of the MN soils to 10 mL of sterile, deionized water that was amended with a known amount of ^{14}C-pyrene in DCM (100 uL). These controls verified pyrene mineralization activity by the MN soils.

Soil and SOM fraction reparation for static microcosms

Aliquots of soil organic matter fractions were analysed for ^{14}C-activity bioavailability by adding 1 g of MN soil directly to the soils, SOM fractions, or SOM extracts. Prior to addition of soils or extracts, aliquots were removed and analysed by LSC to determine initial ^{14}C activity. DCM extracts were reduced to a volume of 100 mL, then mixed with 10 mL of sterilised, deionized water and 1 g of MN soils. A 10 mL sample of humic and fulvic acid fractions was mixed directly with 1 g of MN soils and adjusted to pH 8 (0.1M HCl), the natural pH of MN soils. Subsamples of residual soils (2 g) were mixed with 1 g MN soils in 10 mL of sterile, deionized water. After four weeks, all vials were acidified with 0.5 mL of 20 per cent v/v H_3PO_4 solution and shaken for 24 hours to liberate $^{14}CO_2$. The 1N KOH saturated chromatography paper was removed and analysed by liquid scintillation spectroscopy to determine the amount of $^{14}CO_2$ evolved.

Results and discussion

Soil fractionation studies

An important consideration in examining the interactions of PAH and soil constituents is the ability of the experimental methodology to account for the distribution of all added material. In our experiments, we used radiolabeled pyrene so that mass balances were possible. Additionally, we followed the interactions over extended periods of time to allow for important processes to approach equilibrium, or for the contaminants to “age”. Figure 3 presents the mineralization of pyrene in SF soil with a variety of amendments we have examined. The SF soil and its metabolically inhibited analog show no significant mineralization over the day incubation period. Examination of this data alone would lead to the conclusion that this soil did not metabolise pyrene significantly. It is also clear that addition of soil (5 per cent) from a superfund site in Minnesota (MN soil), containing a known PAH degrader community, results in immediate and significant mineralization of the pyrene. Adding domestic compost or having plants (sweet clover) grow in the soil during the incubation leads to more rapid mineralization after a brief period of adaptation (Guthrie and Pfaender, in press *b*).

A quite different perspective emerges if we examine Table 1 and the distribution of labelled pyrene into different fractions of the soil after a 270 day incubation period. In the metabolically inhibited systems (SF + Azide), the label is primarily distributed between solvent extractable material (DCM extract) and compound(s) associated with the residual soil. In live systems, even ones that show no mineralization (SF), there are statistically significant increases in the amount present as water soluble products, the amount associated with humic and fulvic acids, and the amount remaining in the residual soil after it had been extracted with water, solvent and base. There is a concomitant decrease in material extractable by DCM, relative to the metabolically inhibited control. When a community of active degraders, that has been exposed to high concentrations of PAH for over 50 years, is added to SF soil (SF + MN), yet another pattern emerges with much more in mineralization products and associated with humic and fulvic acids, and less remaining in the residual soil. An important conclusion from this data is that fairly extensive alteration of the parent PAH is possible, even in the absence of significant mineralization. It is assumed that the DCM fraction contains largely unmetabolised and easily extracted parent pyrene and perhaps non-polar degradation products. Water soluble label is only explained reasonably as products of extensive degradation. The nature of the material associated with humic and fulvic acids and residual soil is not clear from this data, and could be either degradation products or parent pyrene. This redistribution of the label originally present as pyrene is clearly biologically catalysed. As metabolic products are formed (CO_2, water soluble products) the amount of label in the DCM extract declines. The differences between the distribution with an unexposed community and one with active degraders suggests different kinds of interactions with the soil matrix, as will be discussed later.

In all cases, a major compartment for label is the residual soil. This is the soil that has been through all the extractions and contains what is generally called humin as the major organic fraction remaining (Schnitzer and Schuppli, 1989). The amount of label in this fraction is determined by combusting the soil, and so does not reveal how it is associated with the soil. This fraction could contain sorbed parent materials or degradation products, organically bound parent or products, or pyrene that has been incorporated into cellular constituents and these cellular products degraded, or a combination of all the above. It is quite likely that the different treatments would contain a different mix of these possible materials.

Figure 3. **Mineralization of pyrene by soil microbial communities**

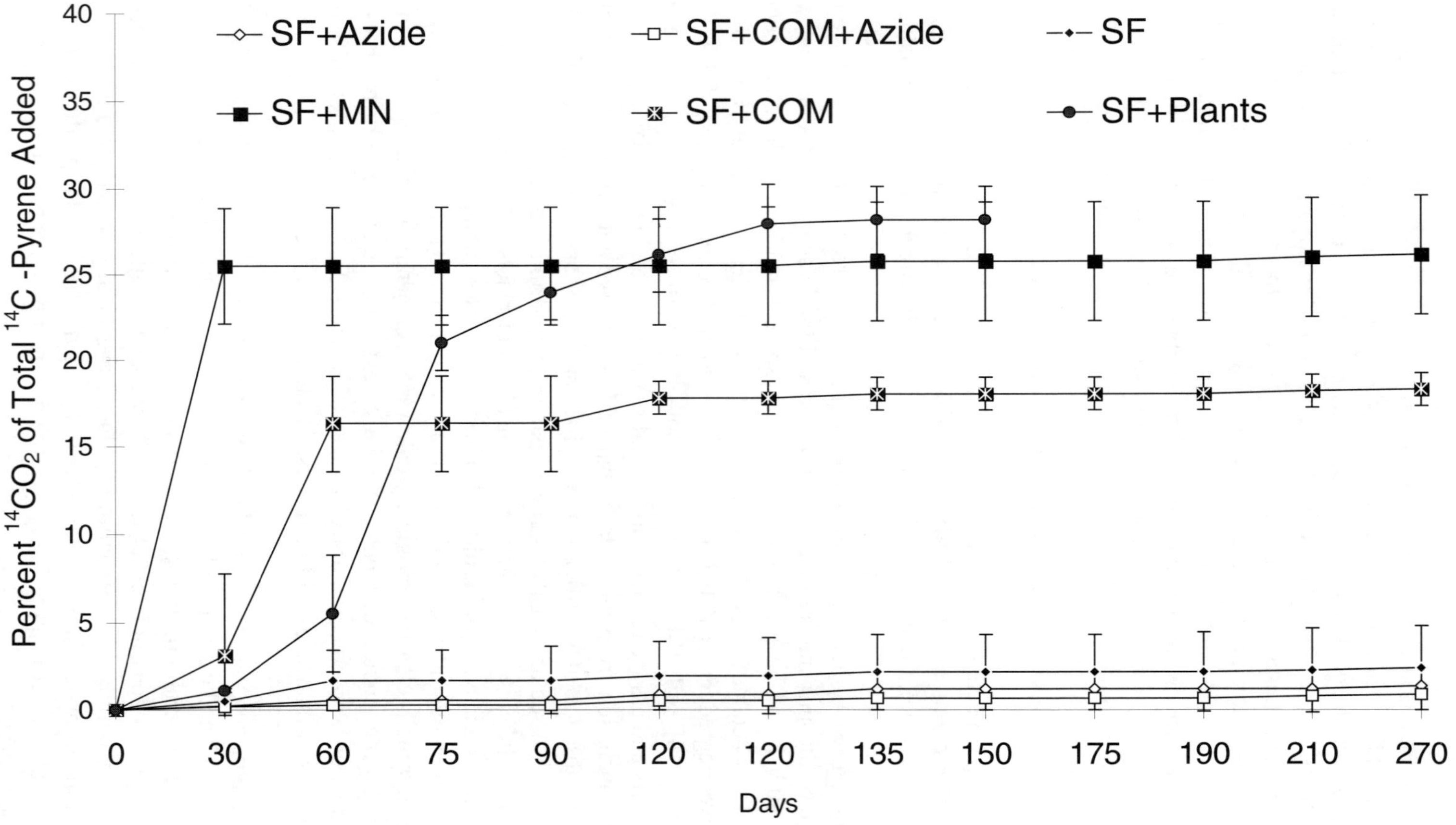

Source: Author.

Table 1. **Mass balance of labelled pyrene distribution into soil fractions**

SOM fractions	SF + NaN_3	SF	SF + MN
$^{14}CO_2$	0.6 ± 0.3	1.2 ± 0.2	27 ± 2.9
^{14}C-H_2O	1.3 ± 0.1	4.9 ± 0.1	2.8 ± 0.5
^{14}C-DCM	46 ± 2.8	30 ± 5.7	31 ± 13
^{14}C-HA/FA	7 ± 0.5	15 ± 1.1	24 ± 9.5
^{14}C-Residual soil	40 ± 5.5	61 ± 2.9	29 ± 18
^{14}C-HF/HCl	6.2 ± 0.3	22 ± 1.1	NA
^{14}C-DCM/HF-HCl	1.2 ± 0.1	4 ± 0.3	NA
^{14}C-Remaining residue	33 ± 0.1	35 ± 0.3	NA
^{14}C-Total recovery	95 ± 3.8	108 ± 8.3	109 ± 8.4

Source: Author.

We attempted to further fractionate the residual material by digesting it with a HF/HCl mixture that should dissolve the mineral components of the soil, leaving the humin behind. Table 1 shows that only 6.2 per cent of the label originally added is removed in the HF/HCl extraction and 1.2 per cent extract of the HF/HCl extract in metabolically inhibited systems. When live soils are extracted, 22 per cent is in the acid extract and 4 per cent in the solvent extract of the acid digestion. This suggests that somewhat more of the material is associated with mineral portions of soil in the live systems. The acid digestions and extractions of both systems indicate that a large portion of the material in the residual soil is not released, even by harsh acids. Acid digestion should liberate non-ionic compounds adsorbed to interior surfaces (residual parent pyrene and products) by inactivating active sites and hydration (Pignatello and Xing, 1996). The nature of the material bound to the remaining soil organics is quite important, since it represents approximately one third of the pyrene present. It is unlikely to be loosely sorbed materials, since the sorptive components (organics, humics, etc.) have largely been removed, as has the physical mineral matrix of the soil. It seems likely that this material represents pyrene degradation products that have been incorporated into soil humin carbon, at least in the live systems.

Bioavailability studies

We attempted to address the bioavailability of the label bound to residual soil in a series of experiments in which known PAH mineralizers from the Minnesota superfund site soil (MN soil) were added to soil fractions after extended incubations. The production of $^{14}CO_2$ was monitored and used as an indication of the bioavailability of the label remaining in the different fractions. We recognise this is a empirical definition of bioavailability. As a test of the method, fresh, radiolabeled pyrene was also added to the same fractions. The data in Table 2 shows that the labelled materials remaining in the residual soil fraction of SF soil, SF soil plus domestic compost, and SF soil amended at the beginning of the incubation with known degraders, was not available to be mineralized after the 135 day incubation period. When fresh pyrene was added, there was extensive mineralization of the new pyrene. This indicates that the added community was capable of detecting available pyrene, and if the residual soil material been available, it would have been mineralized. The obvious conclusion is that the material bound to the residual soil from all the incubations was not bioavailable.

Table 2. **Bioavailability of activity initially associated with pyrene after 135 days incubation**

Soil type	$^{14}CO_2$ evolved residual soil	$^{14}CO_2$ evolved after addition of ^{14}C-pyrene
Abiotic soil	3 ± 1.0	1 ± 0.2
Soil (SF)	1 ± 0.5	17 ± 8.2
Compost	2 ± 1.0	47 ± 4.5
5%MN	1 ±0.4	31 ± 6.9

Source: Author.

These experiments support the fractionation studies finding that a large fraction of the materials bound to residual soils cannot be removed in recognisable forms. It is also consistent with the findings of others, that ageing leads to materials that can no longer be utilised by micro-organisms (Hatzinger and Alexander, 1995; Weissenfels *et al.*, 1992). In our studies, a possible explanation for these observations is that a portion of the material has become part of the soil matrix. This data still does not allow differentiation between what may be organically bound degradation products or cellular constituents, and what may be very tightly sorbed into the humin matrix. This question was partially addressed by HPLC analysis of the materials in the different fractions to determine how much of what remained was pyrene or other recognisable degradation products.

HPLC analysis of soil fractions

Table 3 presents the relative amounts of pyrene detected in the fractions including, water phase products, materials extractable from soil by DCM, and that extracted from the HF/HCl digestion of residual soil by DCM. We are presenting the values as relative amounts for clarity of presentation. In the metabolically inhibited controls most of the pyrene was recovered in the DCM extract of the soil, which is also where the majority of the label was found. The other major reservoir of pyrene in the abiotic soil was the DCM extract of the HF/HCl digestion of residual soil. This suggests that in the abiotic systems the label found in most fractions is parent pyrene. In the live soil, the DCM extract also contained the largest amount of pyrene, although 61 per cent of the label was associated with the residual soil fraction. In the soil amended with known degraders (MN), that pattern is similar, in that most of the pyrene is in the DCM extract, and relatively little is in the residual soil. In the two microbially active systems, the label distribution does not necessarily match with the distribution of the pyrene, which again suggests that the material in the residual soil, and perhaps other fractions, is not parent material, but microbial degradation products.

Table 3. **Relative concentrations of pyrene in soil fractions as determined by HPLC**

HPLC-fluorescence	SF + NaN_3	SF	SF +MN
^{14}C-H_2O	4x	4x[1]	4x
^{14}C-DCM	200x[1]	50x	45x
^{14}C-HF/HCl-DCM	200x	12x[1]	4x

1. Indicates greatest percentage ^{14}C-activity compared to other soils.

Source: Author.

Attempts are underway to identify the specific compounds in the extracts of residual soil. It is known that the peaks found do not correspond to parent pyrene, but their identification is still in process. The analytical studies involve the DCM extract of the residual soil after the HF/HCl digestion, which is only a small portion of the material associated with the residual soil, perhaps one third in the SF samples. The nature of the rest of the material is currently unknown, but under investigation.

Conclusions

There are many possible mechanisms for the interaction of PAH and their degradation products with soil. It is also likely that different processes and mechanisms may dominate in soils of different properties (see Pignatello and Xing, 1996, for recent review). The results of our experiments with pyrene and forest soil suggest that microbial metabolism of the PAH over time may profoundly impact the fate of the parent materials. In microcosms that are biologically inhibited, after 270 days of contact, most of the label remains as pyrene and is present either in a form readily extractable from the soil with organic solvents or within the organic matrix of the soil. The nature of the PAH association with the matrix must involve other mechanisms in addition to adsorption, since destroying the matrix with solvent, base, and extensive acid digestion does not release the bound material. Data not shown suggests that this association with the soil matrix is time dependent with significantly increased amounts between 60 and 120 days of incubation (Guthrie and Pfaender, in press *a*). It is possible that this resistant material is pyrene that has migrated into inaccessible pore spaces, or that abiotic reactions or those involving soil enzymes may have altered the pyrene such that it becomes covalently bound to the soil. Similar processes have been shown to occur with pesticides and phenolics (Dec and Bollag, 1993; Bollag *et al.*, 1992).

The incubations in which there were metabolically active microbes show some significant differences from the abiotic system, even in soils where no mineralization was detected. It is clear that mineralization is not necessarily a part of PAH metabolism, at least in soil not adapted to PAH (Carmichael and Pfaender, 1996; Guthrie and Pfaender, in press *a*). There are increased amounts of label in water soluble products, humic and fulvic acids, and bound to the residual soil when microbes are present. There is also a time component to these associations, with most dramatic changes between 60 and 120 days. In this case, much more of the label associated with the soil matrix is recoverable after acid digestion, but not as pyrene. Again, a large fraction appears to be bound to the soil carbon since it is not released by destroying the soil matrix and is not biologically available.

When a degrader community capable of extensive mineralization is added to the forest soil a third pattern emerges. As seen in Figure 3, the mineralization occurs quickly, usually in the first 30 days of incubation. After that point, the amount that remains solvent extractable is approximately the same as the non-adapted system (approximately 30 per cent), while the amount associated with humic and fulvic acids is increased, and the amount remaining in residual soil decreased. Again, the material associated with the residual soil appears not to be pyrene. More active metabolism could result in the formation of partially metabolised products that are subsequently humified, accounting for the larger amount of material in this fraction. It does seem clear that something happens in the first 30 days of active mineralization that impacts the subsequent fate of pyrene.

Bioavailability is undoubtedly controlled by a number of complex processes that have physical, chemical and biological components. Our work has shown that the microbial role in this process is critical and perhaps dominant. Microbial degradation processes involve much more than simple formation of intermediates that are subject to further metabolism. These intermediates can react with soil carbon leading to a very different fate than mineralization. The study of the nature of these intermediates and the mechanisms of their interaction with the soil carbon provide the potential for generation of new remediation strategies in which the PAH can be made unavailable.

REFERENCES

BAUER, J.E. and D.G. CAPONE (1988), "Effects of co-occurring aromatic hydrocarbons and degradation of individual polycyclic aromatic hydrocarbons in marine sediment slurries", *Appl. Environ. Microbiol.* 54, pp. 1 649-1 655.

BLUMER, M. (1976), "Polycyclic aromatic hydrocarbons in nature", *Sci. Am.* 3, pp. 35-45.

BOLLAG, J.M., C. MYERS, and R.D. MINARD (1992), "Biological and chemical interactions of pesticides with soil organic matter", *Sci. Total Environ.* 123, pp. 205-217.

CARMICHAEL, L.M., R.F. CHRISTMAN, and F.K. PFAENDER (1997), "Desorption and mineralization kinetics of phenanthrene and chrysene in contaminated soils", *Environ. Sci. Technol.* 31:1, pp. 126-132.

CARMICHAEL, L.M. and F.K. PFAENDER (1996), "Polynuclear aromatic hydrocarbon metabolism in soils: Relationships to soil characteristics and preexposure", *Environ. Toxicol. Chem.,* in press.

CARMICHAEL, L.M. and F.K. PFAENDER (in press), "The effect of inorganic and organic supplements on the microbial degradation of phenanthrene and pyrene in soils", *Biodegradation,* in press.

CERNIGLIA, C.E. (1992), "Biodegradation of polycyclic aromatic hydrocarbons", *Biodegradation* 3, pp. 351-368.

DEC, J. and JM. BOLLAG (1993), "Dehalogenation of chlorinated phenols during oxidative coupling", *Environ. Sci. Technol.* 28, pp. 484-490.

DITORO, D.M., C.S. ZARBA, D.J. HANSEN, W.J. BERRY, R.C. SWARTZ, C.E. COWAN, S.P. PAVLOU, H.E. ALLEN, N.A. THOMAS, and P.R. PAQUIN (1991), "Technical basis for establishing sediment quality criteria for nonionic organic chemicals using equilibrium partitioning", *Environ. Toxicol. Chem.* 10, pp. 1 541-1 583.

GUTHRIE, E.A. and F.K. PFAENDER (in press *a*), "Microbially-Mediated Association of ^{14}C-Pyrene with Soil Organic Matter", *Environ. Sci. Technol.,* manuscript submitted.

GUTHRIE, E.A. and F.K. PFAENDER (in press *b*), "Comparative Bioavailability of Biologically and Abiotically Aged ^{14}C-Pyrene in Soils", *Environ. Sci. Technol.,* manuscript submitted.

GUTHRIE, E.A. and F.K. PFAENDER (in press *c*), "Comparative Fate of ^{14}C-Pyrene and ^{14}C-Chrysene in Vegetated and Non-vegetated Soils", *Environ. Toxicol. Chem.*, manuscript submitted.

HARVEY, R.G. (1991), *Polycyclic Aromatic Hydrocarbons, Chemistry and Carcinogenicity*, Cambridge University Press, New York.

HATZINGER, P.B. and M. ALEXANDER (1995), "Effects of aging of chemicals in soil on their biodegradability and extractability", *Environ. Sci. Technol.* 29, pp. 537-545.

HEITKAMP, M.A. and C.E. CERNIGLIA (1987), "Effects of chemical structure and exposure on the microbial degradation of polycyclic aromatic hydrocarbons in freshwater and estuarine ecosystems", *Environ. Toxicol. Chem.* 6, pp. 535-546.

KAN, A.T., G. FU, and M. TOMSON (1994), "Adsorption/desorption hysteresis in organic pollutant and soil/sediment", *Environ. Sci. Technol.* 28, pp. 859-867.

KARICKHOFF, S.W., D.S. BROWN, and T.A. SCOTT (1979), "Sorption of hydrophobic pollutants on natural sediments", *Water Res.* 13, pp. 241-248.

KECK, J., C. SIMS, M. COOVER, K. PARK, and B. SYMONS (1989), "Evidence for cooxidation of polynuclear aromatic hydrocarbons in soil", *Water Res.* 12, pp. 1 467-1 476.

LAMBERT, M., S. KREMER, and H. ANKE (1995), "Antimicrobial, phytotoxic, nematocidal, cytotoxic, and mutagenic activities of 1-hydroxypyrene, the initial metabolite in pyrene metabolism by the basidiomycete *Crinipellis stipitaria*", *Bull. Environ. Contam. Toxicol.* 55, pp. 251-257.

MANILAL, V.B. and M. ALEXANDER (1991), "Factors affecting the microbial degradation of phenanthrene in soil", *Appl. Microbiol. Biotechnol.* 35, pp. 401-405.

MUELLER, J.G., P.J. CHAPMAN, and P.H. PRITCHARD (1989), "Creosote-contaminated sites, their potential for bioremediation", *Environ. Sci. Technol.* 23, pp. 1 197-1 201.

PFAENDER, F.K. (1988), "Generation in controlled model ecosystems", in F.H. Frimmel and R.F. Christman (eds), *Humic Substances and Their Role in the Environment*, pp. 93-104, Wiley-Interscience, Chichester.

PIGNATELLO, J.J., and B. XING (1996), "Mechanisms of slow sorption of organic chemicals to natural particles", *Environ. Sci. Technol.* 30, pp. 1-11.

SCHNITZER, M. and P. SCHUPPLI (1989), "The extraction of organic matter from selected soils and particle size fractions with 0.5M NaOH and 0.1M $Na_4P_2O_7$ solutions", *Can J. Soil Sci.* 69, pp. 253-262.

SUESS, M.J. (1976), "The environmental load and cycle of polycyclic aromatic hydrocarbons", *Sci. Total Environ.* 6, pp. 239-250.

SUTHERLAND, J.B., F. FAFII, A.A. KHAN, and C.E. CERNIGLIA (1995), "Mechanisms of polycyclic aromatic hydrocarbon degradation", in L.Y. Young and C.E. Cerniglia (eds), *Microbial Transformation and Degradation of Toxic Organic Chemicals,* pp. 269-306, Wiley-Liss, New York.

VOLKERING, F., A.M. BREURE, A. STERENBERG, and J.G. VAN ANDEL (1992), "Microbial degradation of polycyclic aromatic hydrocarbons: effects of substrate availability on bacterial growth kinetics", *Appl. Microbiol. Biotechnol.* 36, pp. 548-552.

WEISSENFELS, W.D., H.J. KLEWER, and J. LANGHOFF (1992), "Adsorption of polycyclic aromatic hydrocarbons (PAHs) by soil particles: influence on biodegradability and biotoxicity", *Appl. Microbiol. Technol.* 36, pp. 689-696.

WODZINSKI, R.S. and J.E. COYLE (1974), "Physical state of phenanthrene utilization by bacteria", Appl. Microbiol. 27, pp. 1 081-1 084.

BIOREMEDIATION OF POLYCHLORINATED COMPOUNDS

by

Osami Yagi, Kazuhiro Iwasaki and Akiko Hashimoto
National Institute for Environmental Studies
Japan Environment Agency, Japan

Introduction

Volatile aliphatic chlorinated compounds such as 1,1,1-trichloroethane (TCA), trichloroethylene (TCE) and tetrachloroethylene (PCE) have been detected in groundwater throughout Japan. These compounds are suspected carcinogens and hepatotoxins. Before 1988, water from 2.7 per cent of wells exceeded the 0.03 mg/l drinking water quality standard for TCE, and that from 0.2 per cent exceeded the 0.3 mg/l standard for TCA. In 1993, 15 wells were reported to be polluted by TCE, and 24 wells by PCE. Now, more than 1 000 wells are known to be polluted by volatile chlorinated organic compounds.

Various soil and groundwater clean-up technologies are now being developed, studied and evaluated in Japan. For chlorinated compounds, soil vapour extraction, pumping up and treatment, and digging up and drying are very common approaches. Physical and chemical methods for remediation are expensive. Therefore, less expensive but more complete pollutant destruction technology is required.

Bioremediation could be one of the most promising new technologies for cleaning up the groundwater contamination, because of its low cost and the complete destruction of the pollutants it facilitates. For these reasons, we isolated TCA and TCE degrading bacteria and determined their fundamental characteristics for bioremediation.

Isolation of TCA and TCE degrading bacteria

Several studies have reported on aerobic TCA degrading bacteria (Oldenhuis *et al.*, 1989; Malachowsky *et al.*, 1994; Lefever *et al.*, 1994). We tried to isolate TCA degrading bacteria under aerobic conditions from various soils using an enrichment culture method. TCA, small amounts of various soils and various carbon sources, such as CH_4, C_2H_6, C_3H_8, and C_2H_4, were added to 15 ml of a mineral salt medium in 69 ml serum bottles with rubber caps and aluminum seals, and cultured aerobically at 30 °C. The fourth enrichment cultures in bottles in which TCA was degraded were plated on a mineral salt agar medium and incubated and purified under aerobic conditions with ethane as a carbon source.

Two strains of TCA degrading bacteria, strains TA5 and TA27, were isolated. The cells were rod-shaped, non-motile, non-spore forming, gram positive, catalase positive, oxydase negative and menaquinone 9(H_2). Strains TA5 and TA27 can utilise ethane, ethanol, and various carbon compounds as their energy source. Table 1 shows the assimilation of various carbon sources by strains TA5 and TA27. From these characteristics, it seemed that these strains are *Mycobacterium*. These strains can not grow on TCA as their sole carbon source, and they degrade TCA cometabolically.

Table 1. **Characteristics of TCA degrading bacteria**

Characteristics	TA5	TA27
Morphology	rods	rods
Motility	-	-
Gram stain	+	+
Catalase	+	+
Oxidase	-	-
O-F test	-	0
Spore	-	-
Menaquinone	MK-9(H_2)	MK-9(H_2)
Carbon assimilation		
methane	-	-
ethane	+	+
propane	-	+
ethanol	+	+
acetic acid	-	-
acetone	-	-
toluene	-	-
1-butanol	+	+
glucose	+	+
citrate	+	-
mannitol	+	+
xylose	+	-
L-arabinose	+	-

Source: Author.

Degradation characteristics

We studied the characteristics of TCA and TCE degradation by strains TA5 and TA27. Since TCA volatizes easily, we used a closed culture system. Specifically, TCE or TCA, strains TA5 or TA27, and ethane were added to 69 ml serum bottles containing 15 ml of culture medium. The bottles were sealed with rubber caps and aluminum rings, and incubated under aerobic conditions at 30 °C. TCA and TCE concentrations were determined by gas chromatography. The TCA concentration in groundwater is usually less than 1 mg/l. We determined the concentrations at which TCA was degradable.

Relative degradation of TCA by strains TA5 and TA27 was affected by TCA concentration (Figure 1). The rate of degradation by strain TA5 was high at low concentrations; at 0.1 mg/l, 60 per cent of TCA was degraded within seven days. Moreover, at an initial concentration of 50 mg/l,

40 per cent of TCE was degraded within 14 days. In the case of strain TA27, more than 95 per cent of initial TCA was degraded within seven days, and at an initial concentration of 150 mg/l, 10 per cent of TCA was degraded within 14 days. Strain TA27 has a higher degradation activity than TA5.

Figure 1. **Effect of TCA concentration on TCA degradation**

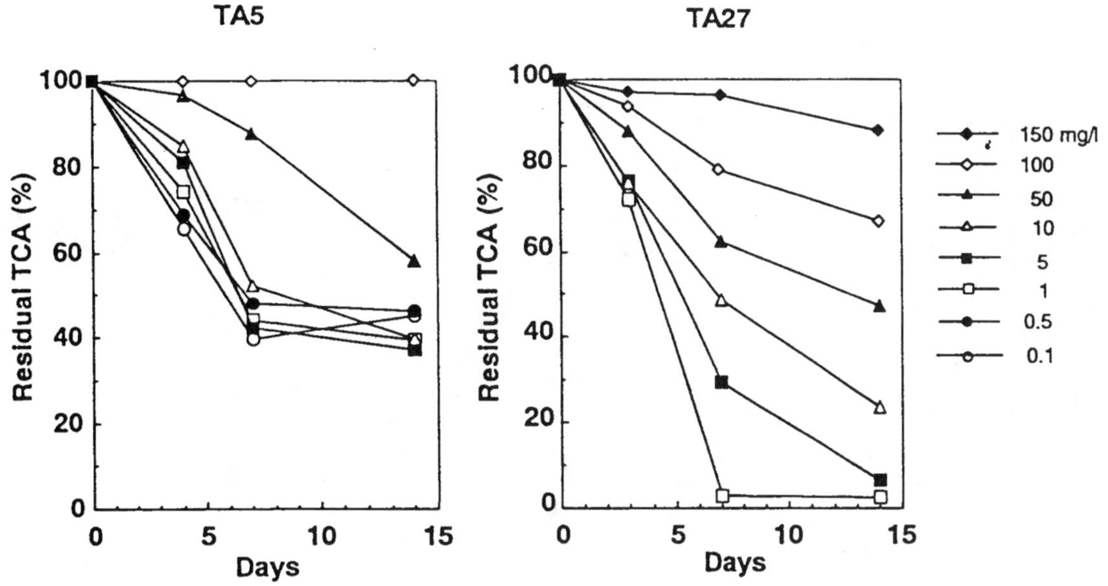

Source: Author.

Strains TA5 and TA27 also degraded TCE. Relative degradation of TCE by strains TA5 and TA27 was affected by TCE concentration (Figure 2). The relative rate of degradation by strain TA5 was more than 90 per cent, and 60 per cent at initial concentrations of 1 mg/l and 10 mg/l TCE, respectively, within seven days. For strain TA27, more than 90 per cent and 70 per cent of initial concentrations of TCE at 1 mg/l and 30 mg/l respectively were degraded within seven days. Twenty per cent of an initial 50 mg/l of TCE was degraded within seven days. Strain TA27 has higher degradation ability than strain TA5. Various TCE degrading micro-organisms have been found among groups such as methane (Oldenhuis *et al.*, 1989; Uchiyama *et al.*, 1989; Little *et al,*. 1988), propane (Malachowsky *et al.*, 1994; Wackett *et al.*, 1989; Vanderberg *et al.*, 1995), toluene (Wackett and Gibson, 1988), phenol (Nelson *et al.*, 1987) and ammonium (Vanelli *et al.*, 1990) oxidising bacteria. However, the maximum degradable TCE concentration was less than 30 mg/l. Strain TA27 was significantly more capable of TCE degradation than was strain TA5.

Figure 3 shows the effect of ethane concentration on degradation of TCA at an initial concentration of 1.2 mg/l. TCA degradation by strain TA5 was not affected by ethane concentration and stopped within three days. Under a headspace gas with five per cent ethane, strain TA27 degraded about 50 per cent of initial TCA within three days. Headspace ethane concentrations greater than five per cent inhibited TCA degradation. Ethane competitively inhibited TCA degradation.

Figure 4 shows the time course of TCA, oxygen and ethane concentrations, cell growth and pH for an initial concentration of 1.2 mg/l. In the case of strain TA5, TCA, oxygen and ethane concentrations decreased with cell growth. Cell growth reached a maximum at three days, and subsequently TCA degradation continued. For strain TA27, cells reached a maximum density at two days, by which time 20 per cent of the initial TCA had been degraded. After seven days, about 80 per cent of the initial TCA had been degraded, and pH had decreased significantly. Strain TA27 had a higher degradation ability than strain TA5.

Figure 2. **Effect of TCE concentration on TCE degradation**

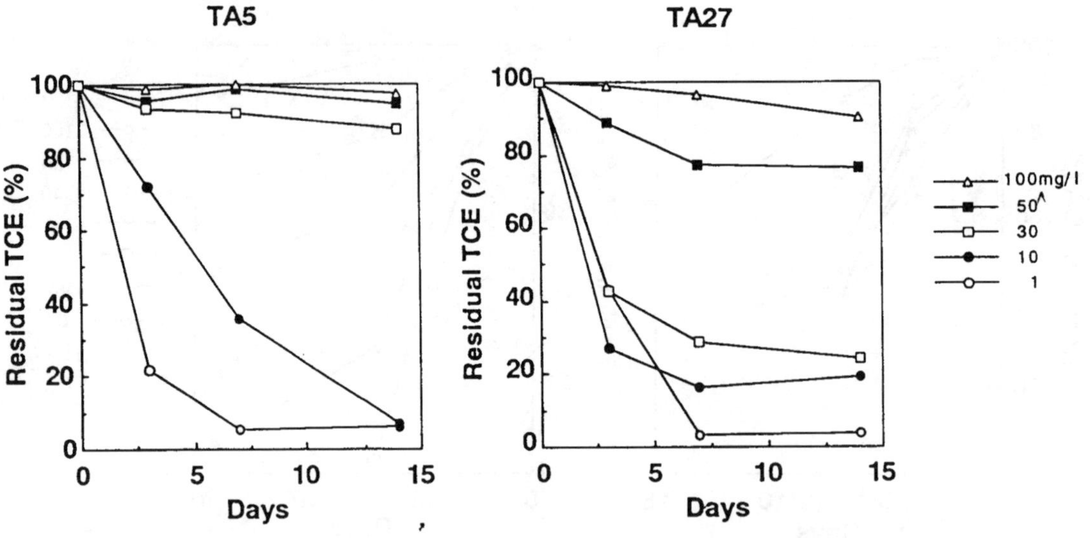

Source: Author.

Figure 3. **Effect of ethane concentration on TCA degradation by strains TA5 and TA27**

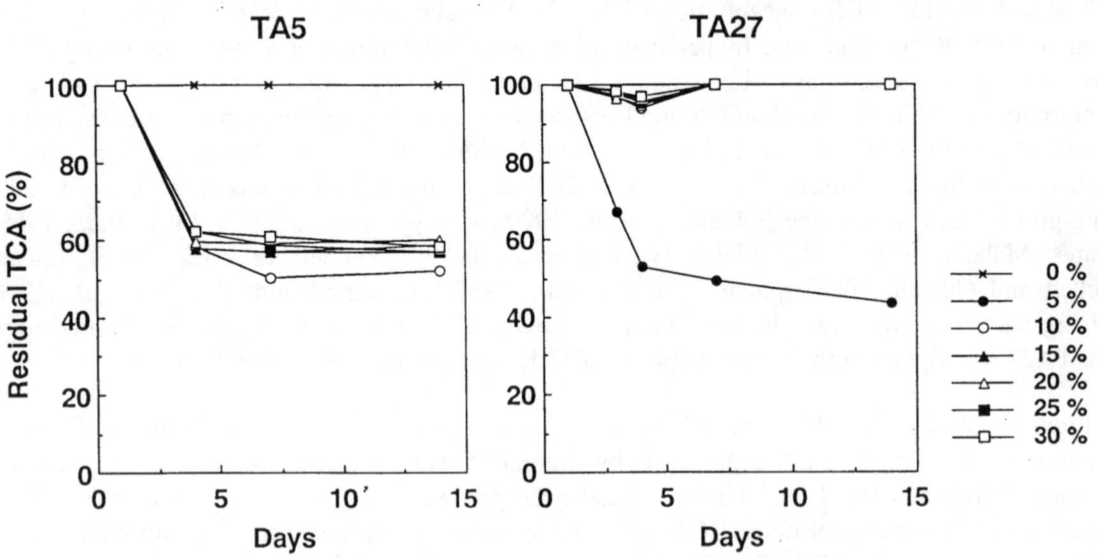

Source: Author.

Figure 4. **Time courses of TCA degradation by strains TA5 and TA27**

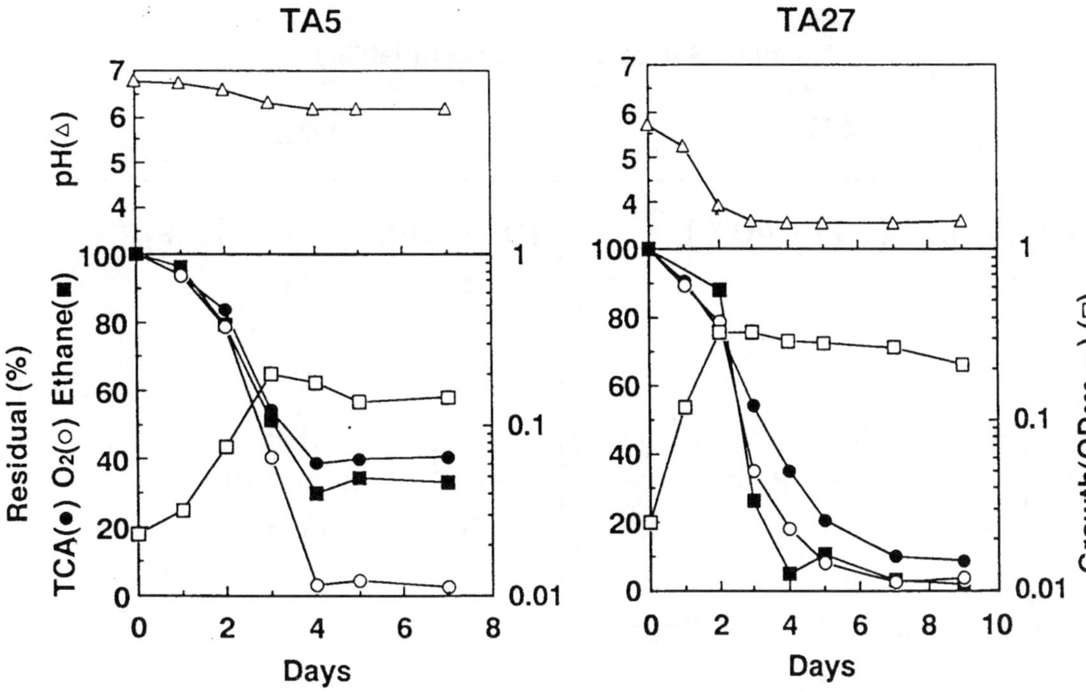

Source: Author.

Biodegradability of chlorinated compounds

The biodegradability of chlorinated methane, ethane and ethylene derivatives by strains TA5 and TA27 was determined, in seven-day cultures with ethane as the primary carbon source. Biodegradability was classified by the value of 40 per cent degradation of test substances (Figure 5). In the case of methane derivatives, more than 40 per cent of initial dichloromethane and chloroform concentrations of more than 1 mg/l was degraded by strain TA5 or TA27, while carbon tetradchloride was not degraded by either strain.

More than 40 per cent of the ethane derivatives, 1,1- and 1,2-dichloroethane, TCA, 1,1,2-trichloroethane, 1,1,1,2- and 1,1,2,2,-tetrachloroethane at initial concentrations of more than 1 mg/l were degraded by strain TA5 or TA27. More than 40 per cent of the ethylene derivatives, cis-dichloroethylene and TCE, at initial concentrations of more than 1 mg/l, were degraded by both strains. Carbon tetrachloride, freon 113 and tetrachloroethylene were not degraded by either strain.

Degradation pathways

The metabolic pathway of TCA was determined. The culture broth of strains TA5 and TA27 at seven days was centrifuged and extracted with ether. The extracts were analysed by the GC-MS method. One of the resulting peaks observed by gas chromatography for which the mass spectrum was analysed was identical with that of 1,1,1-trichloroethanol. In the case of strain TA27, TCA was completely degraded, and half of the initial TCA was converted to 1,1,1-trichloroethanol. 1,1,1-trichloroethanol was further metabolised to other compounds. We have isolated and characterised *Mycobacterium* strains TA5 and TA27 which can degrade TCA and TCE. These strains may be very useful for cleaning up TCA- and TCE-contaminated soils.

Figure 5. **Summary of biodegradability**

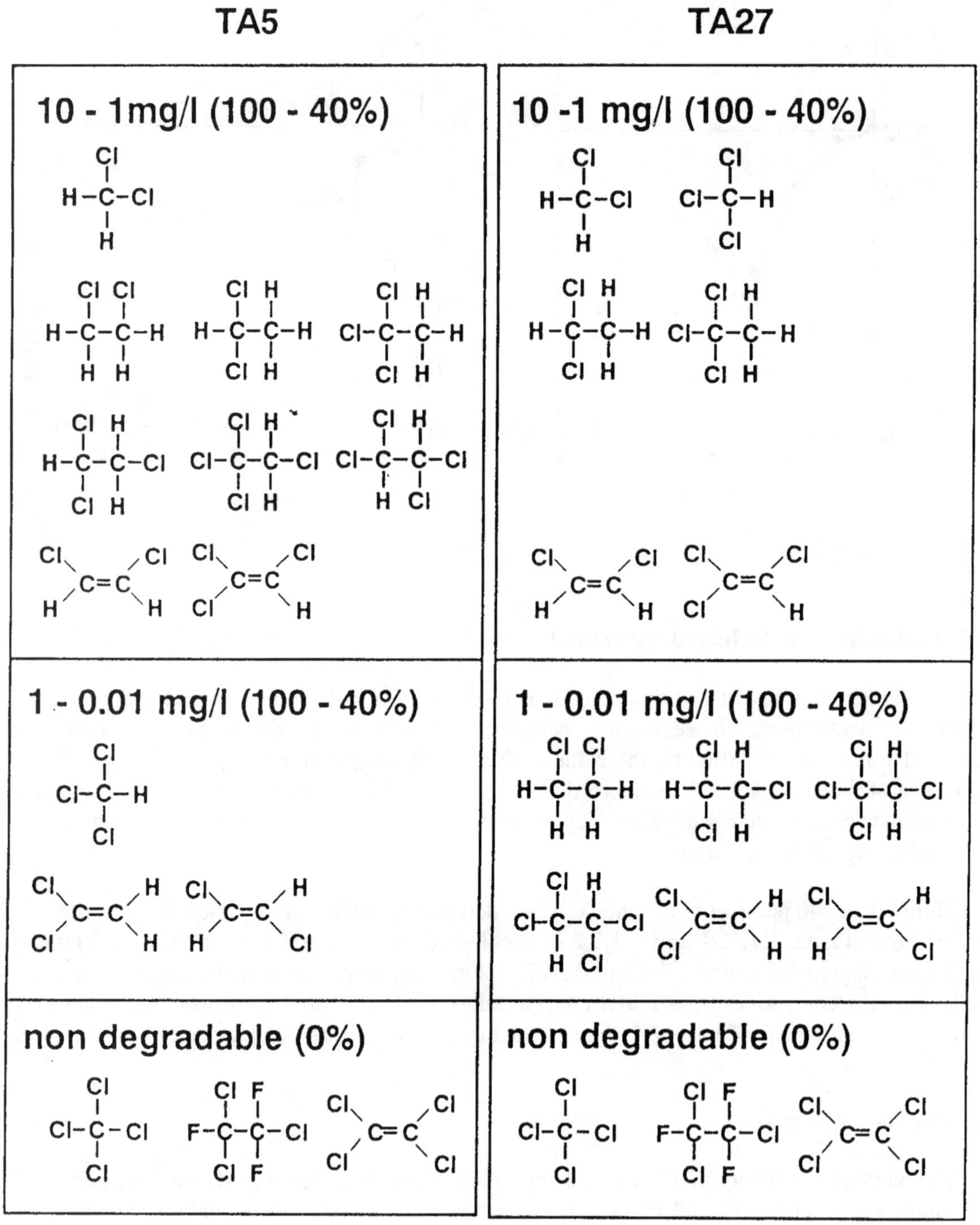

Source: Author.

REFERENCES

LEFEVER, M.R. and L.P. WACKETT (1994), "Oxidation of low molecular weight chloroalkanes by cytochrome P450 CAM", *Biochem. Biophys. Res. Commun.* 201, pp. 373-378.

LITTLE, C.D., A.V. PALUMBO, S.E. HERBES, M.E. LIDSTROM, and R.L. GILMER (1988), "Trichloroethylene biodegradation by a methane-oxidizing bacterium", *Appl. Environ. Microbiol.* 54, pp. 951-956.

MALACHOWSKY, K.T., T.J. PHELPS, A.B. TEBOLI, D.E. MINNIKIN, and D.C. WHITE (1994), "Aerobic mineralization of trichloroethylene, vinyl chloride and aromatic compounds by Rhodococcus species", *Appl. Environ. Microbiol.* 60, pp. 542-548.

NELSON, J.S., S.O. MONTGOMERY, W.R. MAHAFFEY, and P.H. PRITCHARD (1987), "Biodegradation of trichloroethylene and involvement of an aromatic biodegradative pathway", *Appl. Environ. Microbiol.* 53, pp. 949-954.

OLDENHUIS, R., R.L.J.M. VINK, D.B. JANSSEN, and B. WITHOLT (1989), "Degradation of chlorinated aliphatic hydrocarbons by Methylosinus trichosporium OB3b expressing soluble methane monooxygenase", *Appl. Environ. Microbiol.* 55, pp. 2 819-2 826.

UCHIYAMA, H., T. NAKAJIMA, O. YAGI, and T. TABUCHI (1989), "Aerobic degradation of trichloroethylene by a new type II methane-utilizing bacterium, strain M", *Agric. Biol. Chem.* 53, pp. 2 903-2 907.

VANDERBERG, L.A., B.L. BURBACK, and J.J. PERRY (1995), "Biodegradation of trichloroethylene by Mycobacterium vaccae", *Can. J. Microbiol.* 41, pp. 298-301.

VANNELLI, T.M., M. LOGAN, D.M. ARCIERO, and A.B. HOOPER (1990), "Degradation of halogenated aliphatic compounds by the ammonia-oxidizing bacterium Nitrosomonas europaea", *Appl. Environ. Microbiol.* 56, pp. 1 169-1 171.

WACKETT, L.P. and D.T. GIBSON (1988), "Biodegradation of trichloroethylene by toluene dioxygenase in whole-cell studies with Pseudomonas putida F1", *Appl. Environ. Microbiol.* 54, pp. 1 703-1 708.

WACKETT, L.P., G.A. BRUSSEAU, S.R. HOUSEHOLDER, and R.S. HANSON (1989), "Survey of microbial oxygenases:trichloroethylene degradation by propane-oxidizing bacteria", *Appl. Environ. Microbiol.* 55, pp. 2 960-2 964.

SEVERAL FACTORS AFFECTING THE SPECIFIC DENITRIFICATION RATE OF HYDROGEN OXIDISING BACTERIA

by

Y. Miyake*, H. Myoga* and Y. Magara**
* ORGANO Corporation, ** The Institute of Public Health
Tokyo, Japan

Introduction

During the last decade, the presence of nitrate in groundwater has become of increasing concern (Strebel *et al.*, 1989), and in some cases, nitrate concentration above the maximum acceptable limit (10 mgN/l in Japan) was reported (Kawashima *et al.*, 1993). It is said that massive use of fertilizers causes the nitrate build-up in groundwater (Dries *et al.*, 1988). The effects of nitrates in drinking water on human health are in dispute; infant methemoglobinemia (the blue baby syndrome) and stomach cancer are said to be related to high nitrate concentrations in water (Bouchard *et al.*, 1992). Several treatment methods have been proposed to remove the nitrates from drinking water; some of these treatments are based upon physico-chemical processes, such as ion exchange (Lauch and Guter, 1986), reverse osmosis and electrodialysis (Kopp *et al.*, 1987), which only displace nitrates from the contaminated water supply to a concentrated brine solution. Biological denitrification permanently removes nitrates by reducing them to nitrogen gas.

The biological process examined in this study is based on autotrophic denitrification. Inorganic bicarbonate ions and carbonic acid dissolved in the water are used as the carbon source, and hydrogen gas acts as the electron donor (Kurt *et al.*, 1987; Gros *et al.*, 1988). Under conditions in which the concentration of dissolved oxygen is low or zero, nitrate acts as the electron acceptor. The chemical reactions of the nitrate reduction of hydrogen oxidising bacteria can be shown stoichiometrically as follows:

$$2NO_3^- + 2H_2 = 2NO_{2-} + 2H_2O \quad (1)$$
$$2NO_2^- + 3H_2 = N_2 + 2H_2O + 2OH^- \quad (2)$$
$$2NO_3^- + 5H_2 = N_2 + 4H_2O + 2OH^- \quad (3)$$

Stoichiometrically, 0.35 mg/l of H_2 is required for complete denitrification of 1.0 $mgNO_3^-N/l$. Some kinds of bacteria such as *Pseudomonas pseudoflava*, *Alcaigenes eutrophus* and *Paracoccus denitrificans* seem to have the ability of autotrophic denitrification, using hydrogen gas as the hydrogen donor (Schmidt *et al.*, 1989; Selenka and Dressler, 1990). But the denitrifying characteristics of each micro-organism and their distribution in nature are not well known.

In this paper, we report some of our experimental results concerning several factors affecting the specific denitrification rate of hydrogen oxidising bacteria.

Materials and methods

Distribution of hydrogen oxidising denitrifiers

A batch experiment was carried out to examine the distribution of hydrogen oxidising denitrifiers. The experiment was performed in a serum vial bottle with a working volume of 124 ml. Initially, a small amount of sludge obtained from various sources was mixed with synthetic water shown in Table 1. Fifty ml of the mixture was introduced to the vial and oxygen gas remaining in the vial was purged with N_2 gas. Then N_2 gas was replaced by 40 ml of hydrogen gas and capped. The vials were placed in a reciprocal shaker (Figure 1) at 25 °C for three days. Sludge used in this experiment was obtained from various sources such as soil in flower beds, the sludge of a laboratory-scale activated sludge plant for BOD removal and the sludge of a laboratory-scale nitrification-denitrification plant treating inorganic synthetic wastewater using methanol as the hydrogen donor.

Table 1. **Composition of synthetic water used in the experiments of the distribution of denitrifiers**

Component	Concentration
$NaNO_3$	3030 mg/l
Phosphate buffer	0.01 M (pH 7)
diluted with tap water	

Source: Myoga *et al*., 1994.

Figure 1. **Batch experimental apparatus**

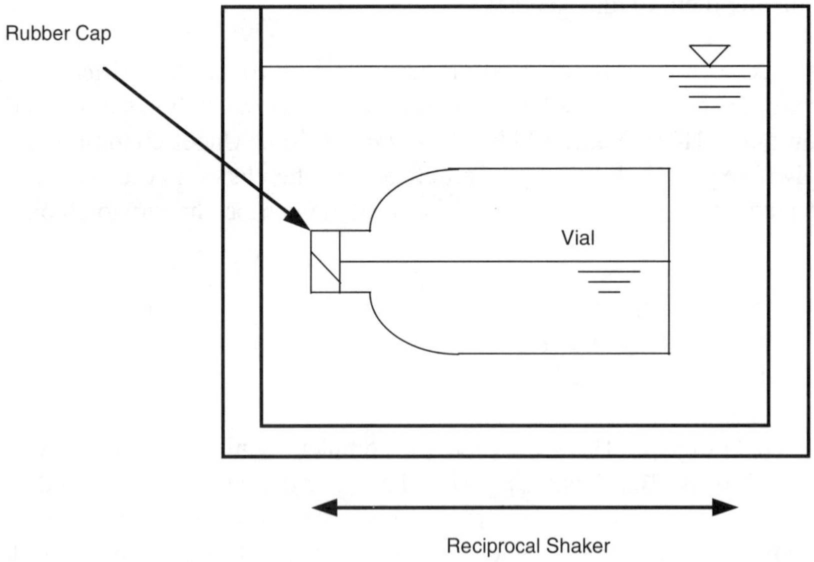

Source: Myoga *et al*., 1994.

As shown in equation (3), when two moles of nitrate were reduced to one mole of nitrogen gas as the result of autotrophic denitrification, five moles of hydrogen gas was consumed. So we can estimate the denitrifying activity by checking the drop of gas pressure in the closed vial. After detection of the drop of gas pressure in the vial, the concentration of nitrate, nitrite, ammonium and pH of samples were analysed.

Growth of bacteria

The hydrogen oxidising bacteria used in this experiment were *Pseudomonas pseudoflava* (ATCC 33668), *Alcaligenes eutrophus* (ATCC 17697), *Paracoccus denitrificans* (IFO 13301) and a mixed culture of autotrophic denitrifying organisms obtained from the results of the distribution study shown above. Each culture of *Pseudomonas*, *Alcaligenes* and *Paracoccus* was routinely grown in 500 ml of serum vial bottle. In the bottle, a nutrient medium containing 2.3 g/l of KH_2P0_4, 2.9 g/l of $NaHPO_4\ 2H_20$, 1.0 g/l of NH_4Cl, 0.5 g/l of $MgSO_47H_20$, 0.5 g/l of $NaHCO_3$, 0.01 g/l of $CaCl_2\ 2H_2O$, 0.5 g/l of $Fe(NH_4)$citrate and 0.2 g/l of KNO_3 was prepared and the gas mixture, made from commercial cylinder gases consisting of nine parts (by volume) of hydrogen, one part of CO_2 and ten parts of N_2, was filled in the gas phase of the bottle. The temperature for growth was maintained at 30 °C by means of a thermostated bath.

Characteristics of hydrogen oxidising denitrifiers

A batch experiment was performed to study the several characteristics of hydrogen oxidising denitrifiers using the apparatus shown in Figure 1. The effects of pH, temperature and partial pressure of hydrogen gas on the specific denitrification rate of autotrophic denitrifiers were examined. A small amount of the denitrifiers was mixed with a synthetic water containing 121 mg/l of $NaNO_3$, that was controlled at the given pH value by a phosphate buffer. A certain volume of the mixture was introduced into the vial and oxygen gas was purged with N_2 gas. Then N_2 gas was replaced by a certain volume of hydrogen gas to control the partial pressure of H_2 gas, and the vials were placed in a reciprocal shaker controlled at the given temperature for a certain period. Specific denitrification was determined by analysing the water quality and the biomass concentration at the start and at the end of the batch experiment.

Identification of hydrogen oxidising denitrifier

Isolation of bacteria from mixed cultured samples was conducted by distributing samples onto an agar plate containing 15 g/l of agar, 10 g/l of meat extract, 10 g/l of peptone and 5 g/l of sodium chloride. The identification of isolated hydrogen oxidising denitrifiers was based on criteria given by Bergy's Manual using the metabolisms of several kinds of organic substances.

Analytical methods

Nitrate, nitrite and ammonium were analysed photometrically according to the Japanese standard method (JIS K 0102). Biomass concentration was estimated using suspended solids concentration.

Results and discussion

Distribution of hydrogen oxidising denitrifier

From every source examined in this study, the existence of hydrogen oxidising denitrifiers was recognised. In three days of incubation, a decrease of the gas pressure in the closed vials was detected, and the decrease of nitrate concentration in the liquid phase was also recognised. This result shows that hydrogen oxidising denitrifier is widely distributed. It means that we can obtain hydrogen oxidising denitrifiers not only from type culture, but from various sources, such as soil, activated sludge and groundwater.

Characteristics of hydrogen oxidising denitrifier

It is recognised that *Pseudomonas sp., Alcaligenes sp.* and *Paracoccus sp.* all have the ability of autotrophic denitrification using hydrogen gas as the electron donor. The effects of several factors on the denitrification rate of each bacteria are summarised in Table 2.

Table 2. **Characteristics of hydrogen oxidising denitrifiers**

	unit	*Pseudomonas pseudoflava*	*Alcaligenes eutrophus*	*Paracoccus denitrificans*	Mixed culture of denitrifiers
Optimum pH	(-)	7.2-7.6	7.4-7.8	7.0-7.4	6-8
Optimum temperature (°C)		45	40	45	35
Arrnenius-constant	(kJ/mole)	63.2	44.1	56.5	50.9
Michaelis-constant for H_2 gas	(per cent)	3	3	3	3
Optimum pH for NO_2 reduction	(-)	4.0-7.6	6.9-7.3	6.8-7.4	7.1

Source: Author.

Effect of pH

The experiments were carried out at 30 °C, and the volume of hydrogen gas in the vial was six times that of the stoichiometrically required volume. The effect of pH on the specific denitrification rate of *Pseudomonas pseudoflava* is shown in Figure 2. This figure indicates that a pH value around 7.4 is optimum for the autotrophic denitrification of *Pseudomonas sp.* The optimum pH value of the mixed culture of denitrifiers was approximately 6.8, relatively lower than others, as shown in Table 2. The calculated specific denitrification rate at pH 6.8 was 0.36 mgN/mgSS/d.

Effect of temperature

The effect of temperature was examined at optimum pH value, and the volume of hydrogen gas in the vial was controlled at six times the required volume. Figure 3 shows the effect of temperature on the specific denitrification rate of *Pseudomonas sp.*

Figure 2. **Effect of pH on the specific denitrification activity of *Pseudomonas pseudoflava***

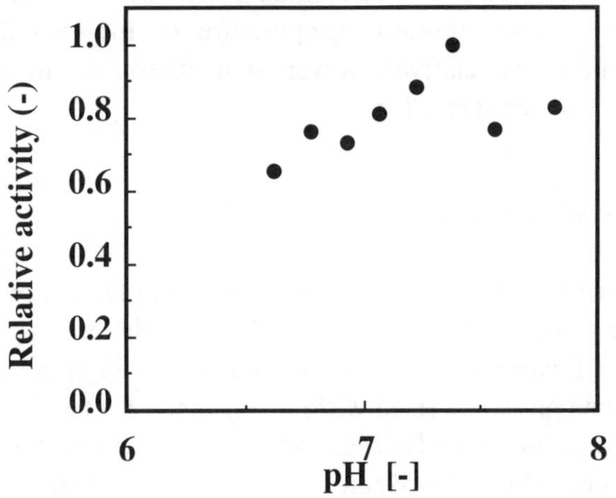

Source: Myoga *et al.*, 1994.

Figure 3. **Effect of temperature on the specific denitrification rate of *Pseudomonas pseudoflava***

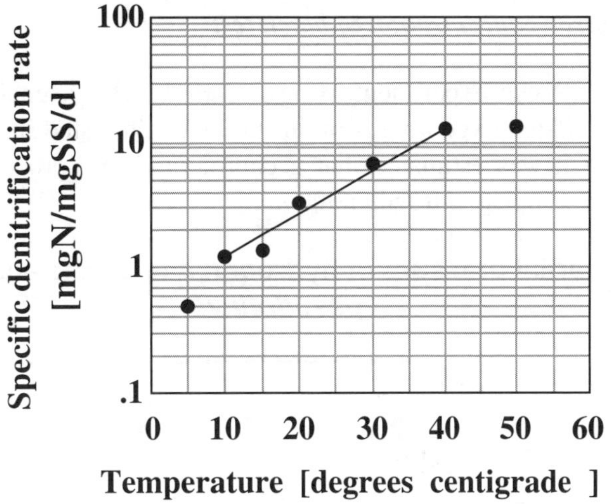

Source: Myoga *et al.*, 1994.

As shown in Figure 3, the optimum temperature was around 45 °C, and the specific denitrification rate at 10 °C was above 1.0 mgN/mgSS/d. This result indicates that this micro-organism has a relatively high specific denitrification rate at lower temperatures. The relationship between the specific denitrification rate and temperature is shown as Equation 4.

$$Kt = Kt'.\theta^{(t-t')} \quad (4)$$

where Kt: specific denitrification rate at t °C [mgN/mgSS/d]

Kt': specific denitrification rate at t' °C [mgN/mgSS/d]

θ: temperature dependency coefficient

t: temperature [°C]

Evaluation of the θ values of *Pseudomonas sp., Alcaligenes sp., Paracoccus sp.* and the mixed culture of autotrophic denitrifiers yielded 1.080 ($10<t<45$), 1.062 ($20<t<45$), 1.080 ($15<t<40$) and 1.072 ($15<t<30$) respectively, and the calculated Arrhenius-constant was 63.2, 44.1, 56.5 and 50.9 kJ/mole respectively. The optimum temperature of the mixed culture of denitrifiers was approximately 35 °C, which was relatively lower than others, as shown in Table 2, and at 40 °C a sharp decrease in activity was recognised.

Effect of partial pressure of hydrogen gas

The experiments were carried out at 30 °C and optimum pH, and partial pressure of hydrogen gas was varied within the range of 3 to 87 kPa. Figure 4 shows the effect of partial pressure of hydrogen gas on the relative denitrification rate of *Pseudomonas sp.* Figure 4 and Table 2 indicate that the partial pressure of hydrogen gas was varied within the range of 3 to 87 kPa. Figure 4 shows the effect denitrification rate was constant above 10 kPa, and that the effect of partial pressure of H_2 gas was not significant. The calculated Michaelis-Menten half-velocity constant of each micro-organism was found to be less than 3 per cent of saturation. Taking into consideration that the cost of hydrogen gas makes up the major part of operating costs, the high affinity between hydrogen gas and denitrifiers shown in the experiments is desirable in terms of economy.

Profile of nitrate and nitrite concentration in batch test

A typical result of a batch experiment using *Pseudomonas sp.* under autotrophic conditions is represented in Figure 5. Batch experiments on all the bacteria tested in this study always exhibited nitrite accumulation, but it was estimated that a continuous flow process having a relatively long retention time resulted in complete nitrogen removal.

Figure 4. **Effect of partial pressure of hydrogen gas on the activity of *Pseudomonas pseudoflava***

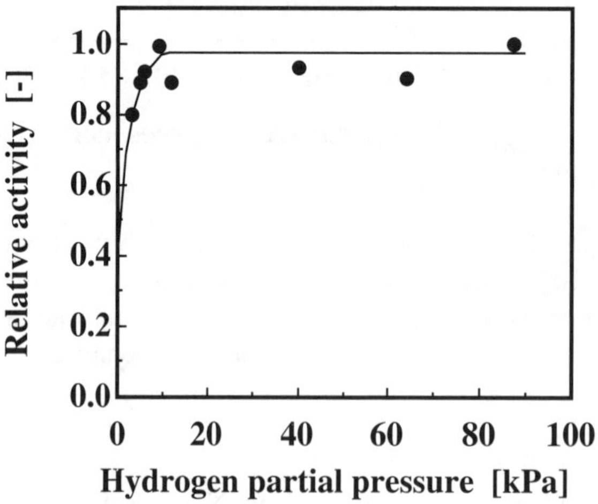

Source: Myoga *et al.*, 1994.

Figure 5. **Profile of nitrate and nitrite concentration in batch experiment of *Pseudomona pseudoflava***

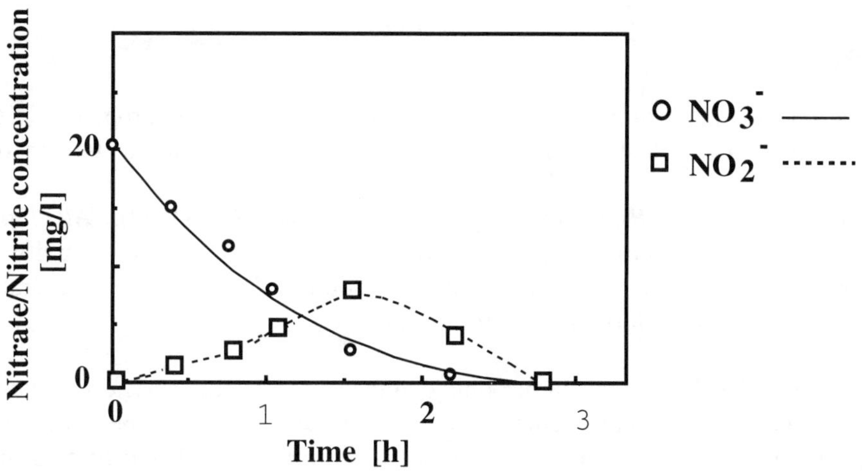

Source: Myoga *et al.*, 1994.

Reduction of nitrite

The effect of pH on nitrite reduction was examined. The experiments were carried out at 30 °C, and the volume of hydrogen gas in the vial was controlled at six times the required volume. Figure 6 shows the effect of pH on the nitrite removal rate of *Pseudomonas sp.*, and the optimum pH value was found to be approximately 6.8-7.3. The results of the experiments on the other bacteria are shown in Table 2.

Figure 6. **Effect of pH on the nitrite removal rate of *Pseudomonas Pseudoflava***

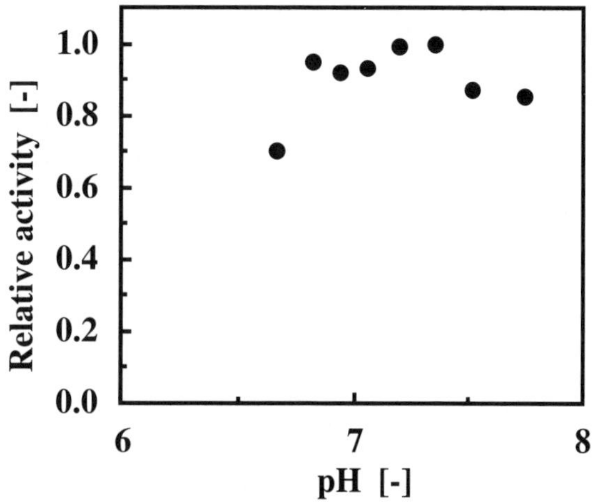

Source: Myoga *et al.*, 1994.

Identification of denitrifiers in the mixed culture

According to the results of morphological and biochemical tests of the mixed culture of denitrifiers, the following bacteria were identified as autotrophic denitrifying bacteria: *Pseudomonas pseudoflava* and *Alcaligenes eutrophus*. These results coincided with those of Selenka's report (Selenka *et al.*, 1990).

Conclusions

From various sources examined in this study, the existence of hydrogen oxidising denitrifiers was recognised, and this result shows that hydrogen oxidising bacteria are widely distributed.

It is recognised that *Pseudomonas pseudoflava* (ATCC 33668), *Alcaligenes eutrophus* (ATCC 17697) and *Paracoccus denitrificans* (IFO 13301) all have the ability of autotrophic denitrification using hydrogen gas as the hydrogen donor. The effects of several factors on the denitrification rate of each bacteria were as follows:

The optimum pH value for the autotrophic denitrification of these bacteria was within the range of 6.8 to 7.8, and the optimum temperature was around 40 °C. The Arrhenius-constants of *Pseudomona sp*., *Alcaligenes sp*., and *paracoccus sp*., were 63.2, 44.1 and 56.5 kJ/mol respectively. The specific denitrification rate of each micro-organism was constant above 10 kPa of partial pressure of hydrogen gas, and the Michaelis-Menten half-velocity constant for hydrogen gas was found to be less than three per cent of saturation.

Questions, comments and answers

Q: Among the three species, which showed the highest denitrification capacity?

A: *Alcaligenes eutrophus* has the highest activity.

REFERENCES

BOUCHARD, D.C., M.K. WILLIAMS, and R.Y. SURAMPALLI (1992), "Nitrate contamination of groundwater: Sources and potential health effects", *Journal AWWA* 9, pp. 85-90.

DRIES, D., J. LIESSENS., W. VERSTRAETE, P. STEVENS, P. DE VOS, and O.J. DE LEY (1988), "Nitrate removal from drinking water by means of hydrogenotrophic denitrifiers in a polyuretane carrier reactor", *Wat. Supply* 6, pp. 181-192.

GROS, H., G. SCHNOOR, and P. RUTTEN (1988), "Biological denitrification process with hydrogen-oxidizing bacteria for drinking water treatment", *Wat. Supply* 6, pp. 193-198.

KAWASHIMA, H., A. TSUMURA, N. KIHOU, S. YAMASAKI, and K. FUJII (1993), "Mechanism of nitrate concentration change in the ground water under the paddy field", *Journal JSWE* 16:2, pp. 108-113.

KOPP, R.R.W., G. OPBERQEN, and R. HELLEKES (1987), "Nitrate reduction of well water by reverse osmosis and electrodialysis", *Desalination* 65, pp. 241-258.

KURT, M., J. DUNN, and J.R. BOURNE (1987), "Biological denitrification of drinking water using autotrophic organisms with H_2 in a fluidized-bed biofilm reactor", *Biotechnol. & Bioeng*. 29:3, pp. 493-501.

LAUCH, R.P. and G.A. GUTER (1986), "Ion exchange for the removal of nitrate from well water". *Journal AWWA* 5, pp. 83-88.

MYOGA, H., F. KAKUDA, M. YANG, Y. MIYAKE, and Y. MAGARA (1994), "Several Factors Affecting the Specific Denitrification Rate of Hydrogen Oxidizing Denitrifiers", *Journal of Japan Society on Water Environment* 19, pp. 669-675.

SCHMIDT, I., J. BEYERSDORF, U. DIRKS, P. REMMERS, and K.D. VORLOP (1989), "Development of a process for denitrification of drinking water", *Dechema Biotechnology Conference* 3, pp. 833-836.

SELENKA, F. and R. DRESSLER (1990), "Microbiological and chemical investigations on a biological, autotrophic denitrification plant using hydrogen as an energy source", *J.Water SRT-Aqua* 39:2, pp. 107-116.

STREBEL, O., W.H.M. DUYNISVELD, and J. BOTTCHER (1989), "Nitrate pollution of ground water in western Europe", *Agriculture, Ecosystems and Environments* 26, pp. 189-214.

[illegible]

[illegible]

[illegible]

[illegible]

[illegible]

[illegible]

[illegible]

[illegible]

[illegible]

[illegible]

[illegible]

[illegible]

BIOREMEDIATION: CONSTRAINTS AND CHALLENGES

by

Susana Saval
Instituto de Ingeniería, Autonomous University of Mexico (UNAM), México, D.F., Mexico

Introduction

In the course of the two last decades, it has been demonstrated that quality of groundwater resources in industrialised countries and in those under development has been progressively deteriorating. This deterioration has been closely related to industrial growth and to unsuitable practices for final disposal of all kinds of wastes. The uncontrolled spills of fuels and lubricants in shops, service stations and industrial plants, together with the disposal of industrial wastewaters in open sewers, have also played an important role in groundwater pollution.

On a world-wide basis, a large variety of technologies has been developed for remediation of contaminated soils and aquifers. In terms of their performance, they can be classified as physicochemical, thermal and biological procedures. Among biological technologies, bioremediation has evolved in the last few years as a very attractive alternative, because contaminants are actually transformed and some of them become completely degraded. When compared with other technologies, bioremediation is safe and economical, and after treatment, soils are likely to recover their capacity to sustain the growth of plants, and groundwater becomes suitable for exploitation.

In bioremediation technologies, advantage is taken of the important role played by micro-organisms in biogeochemical cycles and of their wide biochemical versatility to become adapted to extreme conditions within their own micro-environment. An example of extreme conditions is evident in the spills of hydrocarbons, which are chemicals alien to micro-environment and in high concentrations. One way to evaluate the degree of adaptation of micro-organisms is the survival rate and the inherent ability to benefit from such hydrocarbons as a source of carbon and energy just like in their natural habitat. However, when micro-organisms are isolated from their habitat for the purpose of exporting them to a new environment with different characteristics, it is not easy for them to become adapted. This fact has been demonstrated in Mexico when trials have been made with limited success to apply foreign technologies to remediation of soils and aquifers.

This paper includes an analysis of the challenges, opportunities and constraints posed by bioremediation technologies of soils and aquifers, and an overview is presented of the needs for research and support as applied to the specific conditions of Mexico.

Historical review of remediation in Mexico

Progress made in analytical chemistry throughout the world at the beginning of the 80s contributed to a large extent to detection and identification of chemical compounds as groundwater pollutants, mostly fuels, industrial solvents and heavy metals that are regarded as hazardous materials because of their high degree of toxicity. These findings promoted the development and innovation of equipment to drill wells for prospecting surveys, as well as detection and quick, accurate quantification of groundwater contaminants. Thanks to these developments, the number of contaminated sites, and therefore the remediation needs, have been determined in industrialised countries. For instance, Kavanaugh (1995) mentions that after more than a decade of work as of 1993, more than 300 000 contaminated sites had been characterised in the United States.

It was not until 1988-1989 that in Mexico, the presence of hydrocarbons floating on water table became evident, and as a result, the prospecting of underground fuel tanks of gas stations was started. Detection of aromatic compounds in the subsoil evidenced that a large percentage of inspected tanks were in bad shape. After the explosion that occurred in Guadalajara City in 1992, caused by gasoline leaks that seeped into the primary sewer system, brigades were formed for the continuous monitoring of gas stations in Mexico City through the Civil Protection and the Hydraulic Construction and Operation Division (DGCOH: Dirección General de Construcción y Operación Hidráulica) of the city government (DDF: Departamento del Distrito Federal); the prospecting of subsoil in industrial areas was also started by means of environmentally-oriented auditings.

Environmental auditings constituted the instrument created in 1992 by federal authorities (Federal Environmental Protection Agency, PROFEPA: Procuraduría Federal de Protección al Ambiente) for the purpose of verifying if all industries have the systems or devices necessary to comply with an environmental regulation, as well as the capability to prevent and to respond when confronted with an emergency. In view of the fact that environmental auditings are relatively recent in Mexico, it is not yet possible to give an assurance that industries are already complying with the corresponding legislation; the most important benefit has been the start of a awareness programme and the major involvement of industry to reduce environmental deterioration (Bojórquez-Tapia and García, 1995). At present, all industries shall comply periodically with environmental auditings which can be regarded as an initial stage of the implementation of an environmental management plan, or as a subsequent verification of such a plan (Carmona, 1995).

A large number of foreign companies started to come to Mexico since the end of 1994, with the purpose of offering bioremediation technologies for soil decontamination. Some of them sought for strategic alliances with local environmental consulting firms, but they encountered several obstacles.

In 1995, remediation activities were started in the subsoil of gas stations that included in most cases pumping out of hydrocarbons floating in the water table or excavation of contaminated geological material (Lesser, 1995). Nowadays, several gas stations in Mexico City are being remodelled, and a search has been carried out for the best cleaning alternatives that offer immediate positive results at a low cost.

Recently, the National Institute of Ecology (INE: Instituto Nacional de Ecología, 1996) started the instrumentation of an administrative process to be followed by environmental service companies interested in participating in remediation projects for contaminated sites in Mexico. The purpose of that procedure is to identify those companies devoted to remediation activities, but with a true scientific and technological background.

Technological and environmental advantages of bioremediation

Bioremediation can be regarded as a clean technology, in which contaminants can actually be transformed, and some of them completely mineralised. Furthermore, since it is recommended to use micro-organisms that have soils as their natural habitat, they die when the nutrient source becomes exhausted.

Bioremediation is a versatile technology because it can be adapted to the particular needs of each site. Therefore, biostimulation can be applied if only addition of nutrients is sought; bioaugmentation can be applied when there exists the need to increase the proportion of microbial flora that degrades contaminants. Otherwise, bioventing can be applied when it becomes necessary to supply the oxygen from the air. In addition, bioremediation can be performed outside the site when pollution is superficial, although necessarily *in situ* once contaminants have reached the aquifer.

In the case of topsoil, and when treatment is not performed at the site, biocells or biopiles can be used on top of impervious surfaces that allow the collection of leachates in such a way that clean space is spared from contamination. In addition, once bioremediation is completed, the soil can be used to grow vegetation plants suitable to reincorporate it to its better known biological function.

When groundwater is dealt with, the most popular technology is the so-called pumping-treatment-recharging that comprises extraction of groundwater for further treatment at surface, followed by reinjection into the aquifer. Although no official data are available, Kavanaugh (1995) assumes the existence of more than 3 000 systems of this type operating in the United States. In spite of being the most frequently applied technology, there are certain constraints that determine the success of its implementation, such as for instance, the fact that contaminants may be strongly adsorbed into geological strata or otherwise, when they are found in low-permeability areas that induce constraints in mass transfer. In other circumstances, it becomes more difficult to reach the clean-up levels required, because low concentrations of contaminants that micro-organisms use as substrate are not sufficient to support their microbial activity and they start to die. Furthermore, the treatment by itself becomes quite costly because of high power consumption demanded by pumping operations. In those cases, other types of treatment can be incorporated to facilitate operations and to achieve the recovery of groundwater quality.

Economical advantages of bioremediation

An important feature of bioremediation is its low cost with respect to other treatment technologies. As indicated by Alper (1993), the cost of bioremediation is at least six times lower than incineration and three times cheaper than confinement. This low cost is due to a lower consumption of energy, lower needs of nutrients and operation under environmental conditions. However, it should be mentioned that such cost comparisons cannot be generalised, and they only apply to each particular case.

Technological obstacles of bioremediation

Bioremediation cannot be applied when:

- radioactive compounds are present;
- organic contaminant compounds are highly halogenated;

- heavy metals are present in concentrations that inhibit microbial activity;
- there are extreme micro-environmental conditions.

Since each micro-organism has its own characteristics, a very particular tolerance to each of the situations can be expected. For example, inhibition phenomena can be observed when the concentrations of the organic contaminants are very high.

Bioremediation is not recommended when the geological material is purely cohesive because low permeability of the clay constrains mass transfer in the system. This also holds true when contaminants have reached the water table and an *in situ* treatment is pursued. This problem can be overcome in the case of topsoil if sand is added, or if some agro-industrial residues are incorporated to increase permeability and expedite mass transfer.

Even though micro-organisms have been able to become adapted to conditions of the site, biodegradation of contaminants will take place over a relatively long period when compared to other remediation techniques.

Another aspect that should be mentioned is the chemical complexity of contaminants at the different sites, and the fact that in some cases, contaminants already evidence a high degree of weathering with a resulting difficulty in treatment procedures.

To be able to participate in bioremediation projects, it will be necessary to become fully familiar with the problem, to have access to the contaminated site and to carry out a comprehensive investigation of the history of the site and the causes that led to the problem of contamination. It will be then possible to set the bases for the full characterisation of the site and to outline a research strategy that leads to the solution of a real problem, through application of an integral technological development suitable to each particular case.

When a direct reference is made to biotechnology, researchers should keep in mind that to implement bioremediation solutions it is not enough to work with ideal cases in the laboratory, but it is also very important to maintain the conditions that prevail at the site, so that the study becomes representative of the scope intended in field.

In bioremediation and in environmental biotechnology in general, it is necessary to become aware of the following aspects encountered in practice:

- Micro-organisms responsible for degrading the contaminants are microbial consortia with specific characteristics that depend on the micro-environmental conditions available at the site where they are grown;

- It is preferable to stimulate the native microbial population instead of adding exogenous micro-organisms that are likely to die because of competition established in the natural medium;

- Micro-organisms are generally adhered to soil particles and receive from them a contribution of micro-elements that enable their survival and even their metabolic activity; therefore, they should be cultivated in conditions as similar as possible to those prevailing in their natural habitat;

- If it is decided to isolate micro-organisms and to hinder their massive growth, they will not necessarily become adapted to the new micro-environment in which they are placed;

- Sources of nitrogen and of phosphates should preferably be added in limited concentrations, so as to allow only a degrading activity, that is not necessarily related to the increase of microbial population;

- Special care should be taken in bioremediation during the selection of chemical compounds to be added as nutrients or co-substrates;

- Micro-environmental conditions of the site should be adopted, even though they are not optimal;

- The bioreactor should be "built" at the treatment site; if a biopile or a biocell is involved, it will be partly confined, but in the case of an aquifer bioreactor, it should be left with no walls, cover and bottom;

- The existence of a contaminated site is a risk in itself, and the application of bioremediation involves an additional risk. Therefore, the idea of adding genetically engineered micro-organisms as part of bioremediation should remain on hold until a clear picture of transport mechanisms and of fate points that govern a natural medium is obtained, provided the additional risks inherent to their use are fully understood.

A large number of foreign companies have been coming to Mexico for the purpose of selling their technologies applicable to bioremediation of soils and aquifers contaminated with oil-based hydrocarbons; however, the results obtained did not fulfil the expectations. Among the reasons for this failure, mention can be made of low reduction efficiencies of contaminants, an increased environmental deterioration due to the addition of chemicals, and high operating costs (Saval, 1995). A special feature that should be considered for successful application of any of the technologies is the need to carry out studies on their suitability and innovation; this could even lead to new developments. In the case of bioremediation technologies, it should be considered that characteristics of each soil are not always the same, and the assumption of micro-organisms becoming readily adapted to any particular habitat cannot be regarded as a rule of thumb. Also, Mexican soils have very peculiar physical, chemical and biological characteristics, to such a degree that they are quite different from the soils encountered in other countries.

Financial obstacles

Generally speaking, it can be asserted that possibilities of obtaining economic support to finance research activities in Mexico are quite limited. They are usually sponsored by the National Council of Science and Technology (CONACyT: Consejo Nacional de Ciencia y Tecnología), by branches of different institutions, and in some cases by international organisations, but in practice the funds are insufficient to cover the needs. It is extremely difficult to secure funds applicable to investment expenses, which are basically oriented to the acquisition of equipment from foreign suppliers most of the time, and therefore very costly. However, this financial assistance benefits a large number of research projects that are not necessarily implemented in practice.

In the case of full-scale bioremediation projects, their cost and duration depend on the degree of clean-up sought for, and in turn their scope is different in each case. It is obvious that the main parties involved in these projects are the contaminating industry, the consultant of environmental services, and the authorities responsible for the environment. Universities are newcomers in this field of activity, and their participation could lead to development of technologies with a truly scientific support and could also serve as a source of income to finance the needs of each individual project. Under this scheme, two aspects should be taken into account: the objectives and goals set for the project shall be achieved within the times agreed upon, and the confidentiality of the results shall be honoured. In this way, the ideal university-industry relation could become a reality.

Training

In spite of the importance of environmental problems and the need to find solutions suitable to our living standards, there are no undergraduate curricula favourable for training of practitioners with a reasonable degree of expertise, and with their efforts oriented towards control of the environmental pollution; even fewer opportunities are available in what concerns soils and aquifers. The areas where universities should strengthen the training of practitioners with a very high integral environmental approach are biotechnology, geohydrology, soil sciences, engineering, physicochemistry, and chemistry, among others.

Better options exist at graduate level for the training of expert personnel because students have already approved the undergraduate courses and higher learning becomes an easier task. Nevertheless, there are very few professors with practical experience that could deliver graduate courses. The scheme that has proved successful in Mexico nowadays is the enrolment in intensive continuing education courses, in which universities have played a leading role.

Another way to orient human resources under development toward environmental-related activities is during preparation of the thesis work. It is worth mentioning that from an analysis of the statistics of papers submitted to national conferences of biotechnology and bioengineering since 1983, an important growth of the environmental field has been experienced in the last few years, and this fact reflects the interest of dealing with these topics (Saval, 1996). A large proportion of papers are presented by students as partial fulfilment of their thesis, and in many cases, it is evident that the investigations go far beyond a practical application.

Government

During the last administration, Mexican government developed a policy framework for environment and sustainable development, covering different areas. Priority was given to biotechnology as a relevant area for sustainable development, particularly for prevention and control of soil pollution, the appropriate handling of hazardous substances, and preserving the quality of water and its optimal use. The present government administration is trying to give continuity to the approach of the last one and, according to the National Programme of the Environment, progress will be made in several areas (Solleiro and Castañón, 1996). One of the most important aspects is the approval of amendments to the General Law for Ecological Equilibrium and Environmental Protection enacted in 1988. The purpose of these amendments is basically to set up the legislative framework for preservation of ecological equilibrium, and to perform clean-up activities on contaminated sites.

Environmental authorities have regarded bioremediation technologies as a sound alternative for clean-up operations of contaminated sites, although they are fully aware that not all technologies are readily applicable and that not all companies which render environmental services have the experience to implement them. This is why they are currently opening new possibilities for the universities, through their research staff, to become involved in full-scale projects.

Another aspect directly related to Mexican government concerns the semi-state organisms responsible to a large extent for contaminated soils and aquifers. As a part of actions to persuade the oil industry of the imminent need to join forces for clean-up of contaminated sites, the National Autonomous University of Mexico (UNAM), through the Institute of Juridical Research, published in 1995 a book in Spanish that incorporates the experience and reflections of its authors under the title *PEMEX: Environment and Energy. Challenges Posed by the Future.* A similar effort was also made by the electrical industry and the results obtained so far have proved favourable.

The first attempts to integrate the entities involved in remediation projects are already under way, although further commitments from the National Water Commission (CNA: Comisión Nacional del Agua), the federal organism responsible for water control and management in Mexico, are necessary in what concerns groundwater. As part of these actions, a Forum in Mexico on Ground Water Remediation was organised in December 1995; evidence was presented that in Mexico, attention has been focused on contamination of water in urban/industrial zones, although a great deal of effort has been placed on rural areas where biological contamination of groundwater is of common occurrence due to the lack of basic health facilities and the deficient supply of potable water, conditions likely to generate public health problems of an epidemic nature (Chávez-Guillén, 1995). On the other hand, it is a known fact, although not officially accepted, that several water wells located in areas where the urban growth has encroached upon the industrial zones, have been plugged.

Public acceptance

Mexico has publicly acknowledged the application of environmental biotechnologies to the solution of real problems. In fact, two excellent examples of technological development have been recorded: namely, the anaerobic treatment of wastewater and the gas treatment in biofilters (Revah and Noyola, 1996). Both have been developed as a result of an important research work carried out by Mexican biotechnologists, and they evidence a not-so-common potential of relating universities with the industrial sector. These findings have been transferred to industry and have been successfully applied to full-scale projects, not only in Mexico, but in foreign countries as well, with the advantage that they have opened new opportunities in the course of their increasing acceptance by customers. It is now possible to implement new biotechnological developments in other fields of application, such as bioremediation of soils and aquifers.

On the other hand, as part of the organisational scheme that Mexican environmental authorities are planning for remediation of contaminated sites, the participation of third parties in the decision-making process becomes particularly important; this is why attention has been paid to the involvement of universities. It will be thus possible to achieve a better understanding of the points of view of various disciplines, together with an increased credibility of the actions to be undertaken.

Policies

Concerning entities likely to support research activities, serious reflection is appropriate before implementing new policies for assessment of projects suitable for financing. There are many evident environmental problems in Mexico that could be used as experimental prototypes, even though they do not constitute typical research models on a world-wide basis. Solving local problems is the duty of the educational institutions, to convey their commitment to the human resources that are being developed.

To enhance the relationships among universities and industries and promote development of environmental biotechnologies suitable for real problems, it could be appropriate to go back to the scheme of the 80s, when biotechnology was ranked as a priority area in Mexico, and the trend for higher-learning institutions to become centres for technological development was a reality. The current policies, related to areas in which the technological development is of utmost importance, misuse the potential of all the researchers involved in biotechnology.

Concluding remarks

- The need to confront Mexico's difficult environmental problems based on the principles of sustainability has become part of a framework in which environmental biotechnology can make an important contribution to clean-up of soils and aquifers in the short and medium term.

- Development of bioremediation technologies is by necessity an inter- and multidisciplinary activity in which biotechnology plays a leading role, but without the collaboration of experts in geohydrology, chemistry, physicochemistry, engineering and soil sciences, it will not be possible to warrant the success of an environmental project.

- A growing demand exists for bioremediation technologies applied to soils and aquifers, and therefore, a good opportunity for universities to become involved in projects of immediate implementation can be anticipated.

- It is urgent to establish policies of research and development in biotechnology, with the purpose of funnelling higher economic resources into real research projects in which the main objective becomes feasible solutions to environmental problems through technological developments.

- Even though the analysis of the potential and constraints of bioremediation has been carried out as applicable to the conditions of Mexico, the concepts discussed herein can be shared with other countries where development models of environmental preservation requirements are similar to those found in Mexico.

Special mention should be made that Institute of Engineering of UNAM has been carrying out studies related to contamination of soils and aquifers. In addition, research activities have been oriented towards clean-up of sites contaminated with oil hydrocarbons through the use of simple and economical bioremediation technologies suitable to each particular case. Some of these projects can be summarised as follows:

1. Biostimulation and composting of topsoils contaminated with gasoline;

2. Geohydrologic, physicochemical and microbiological characterisation of sites (soils and groundwater) contaminated with mixtures of diesel-fuel and gasoline;

3. Characterisation and biotreatability studies in mesocosms (biostimulation and bioaugmentation) of soils contaminated with diesel-oil;

4. Evaluation of soil and aquifer remediation technologies that are commercialised in Mexico;
5. Development of standardized procedures for characterisation and bioremediation of soils and aquifers;
6. Isolation and characterisation of bacterial consortia degraders of gasoline, diesel-fuel and diesel-oil;
7. Development of techniques for detection of bacteria capable for degrading hydrocarbons;
8. Biodegradation of volatile monoaromatic hydrocarbons (BTEX);
9. Jointly with the Center for Ecology of UNAM, the characterisation of canal sites where industrial wastewaters from the metropolitan area of the Valley of Mexico are discharged.

It should be mentioned as well that the first five projects are being financed by semi-state and private corporations, and they have been implemented in the field.

In addition, courses on Contamination of Aquifers and on Bioremediation of Soils and Aquifers with a multidisciplinary approach have been offered to graduate students and practitioners who work for government agencies, as well as to semi-state and private institutions. During such courses, it has been possible to confirm the concern of all these sectors to find solutions to Mexico's environmental problems, and the acceptance of the use of bioremediation technologies at all levels.

Questions, comments and answers

Q: You mentioned that the application is mainly for gasoline remediation. Do you have any information on other toxic chemicals?

A: No, our only experience is with hydrocarbons. We are just getting started.

Q: In Japan, we have regulations for the use of GMOs. Do you have any guidelines?

A: We can apply any kind of remediation depending on site characteristics. Many companies come with "magic potions", and we don't know what is in them. Bioaugmentation is another option.

C: You said that bioremediation cannot be applied to radioactive contamination, but in fact there are a number of organisms that concentrate uranium and strontium and offer very good possibilities for bioremediation. Also, plants can be used to concentrate these in roots and leaves and then be harvested.

Q: You showed a picture of an area where the grass was being taken up. What was the contaminant, and when do you decide that you need to apply bioremediation?

A: There the soil was contaminated with petroleum waste. The example shown was a demonstration plot where the company demonstrates their technology.

REFERENCES

ALPER, J. (1993), "Biotreatment firms rush to marketplace", *Bio/Technology* 11, pp. 973-975.

BOJÓRQUEZ-TAPIA, L.A. and O. GARCÍA (1995), "Aspectos metodológicos de la auditoría ambiental" (Methodological Aspects Related to Environmental Auditing), in *PEMEX: Ambiente y Energía. Los Retos del Futuro* (PEMEX: Environment and Energy. Challenges Posed by the Future), pp. 59-72, UNAM-Petróleos Mexicanos, Mexico.

CARMONA, L.M.C. (1995), "Aspectos jurídicos de la auditoría ambiental en México" (Legal Aspects of Environmental Auditings in Mexico), in *PEMEX: Ambiente y Energía. Los Retos del Futuro* (PEMEX: Environment and Energy. Challenges Posed by the Future), pp. 73-109, UNAM-Petróleos Mexicanos, Mexico.

CHÁVEZ-GUILLÉN, R. (1995), "Strategies and Programs for Ground Water Quality Preservation", *Proceedings of Forum in Mexico on Ground Water Remediation*, 5-6 December, Mexico City, Mexico.

INSTITUTO NACIONAL DE ECOLOGÍA (1996), *Protocolo de los requisitos que deben cumplir los promoventes de servicios para la remediación de sitios contaminados con residuos peligrosos* (Formal registration of the requirements to be fulfilled by the promoters of services related to the remediation of sites contaminated with hazardous residues), Dirección General de Residuos, Materiales y Actividades Riesgosas, Mexico.

KAVANAUGH, M.C. (1995), "Remediation of Contaminated Ground Water: A Technical and Public Policy Dilemma", *Proceedings of Forum in Mexico on Ground Water Remediation*, 5-6 December, Mexico City, Mexico.

LESSER, J.M. (1995), "Background and Current State of Aquifer Contamination by Hydrocarbon Spills in Mexico", *Proceedings of Forum in Mexico on Ground Water Remediation*, 5-6 December, Mexico City, Mexico.

REVAH, S. and A. NOYOLA (1996), "El mercado de la biotecnología ambiental en México y las oportunidades de vinculación universidad-industria" (The Potential Market of Environmental Biotechnology in Mexico and the Opportunities to Develop University-Industry Relationships), in E. Galindo (ed.), *Fronteras de Biotecnología y Bioingeniería,* pp. 121-133, Sociedad Mexicana de Biotecnología y Bioingeniería, A.C., Mexico.

SAVAL, S. (1995), "Remediación y restauración" (Remediation and Restoration), in *PEMEX: Ambiente y Energía. Los Retos del Futuro* (PEMEX: Environment and Energy. Challenges Posed by the Future), pp. 151-189, UNAM-Petróleos Mexicanos, Mexico.

SAVAL, S. (1996), "La Sociedad Mexicana de Biotecnología y Bioingeniería: un resumen de su historia" (Mexican Biotechnology and Bioengineering Society: a summary of its history), in

BioTecnología: Libro del año 1996. (BioTechnology: Book of the Year 1996), pp. 1-10, Sociedad Mexicana de Biotecnología y Bioingeniería, A.C., Mexico.

SOLLEIRO, J.L. and R. CASTAÑÓN (1996), "Environmental Biotechnologies in Mexico: Potential and Constraints for Development and Diffusion", in *Biotechnology for Sustainable Development*, University of Tennessee, Knoxville, United States.

Parallel Session I

I-C. BIOREMEDIATION/BIOTREATMENT: SURFACE WATERS

BIODEGRADATION OF POLLUTANTS AT EXTREMELY LOW CONCENTRATIONS AND IN THE PRESENCE OF EASILY DEGRADABLE ORGANIC CARBON OF NATURAL ORIGIN

by

Thomas Egli
Swiss Federal Institute for Environmental Science and Technology (EAWAG)
Dübendorf, Switzerland

Introduction

In the laboratory, physiological and kinetic studies on the degradation of pollutants by heterotrophic micro-organisms are usually carried out with pure cultures, which are supplied in batch cultures, with high concentrations of a particular chemical that serves as the sole source of carbon and energy. This contrasts strongly with the conditions microbes experience in ecosystems where growth occurs in a dilute, usually carbon/energy-limited environment and where -- in addition to a pollutant -- a multiplicity of other alternative, easily degradable carbon compounds of natural origin are present[1]. There is now much evidence that under carbon-limited environmental conditions micro-organisms do not specialise on one particular carbon substrate, but that they simultaneously utilise as many of the different available carbon compounds as possible, a behaviour usually referred to as "mixed substrate growth" (Egli, 1995). However, although it has been assumed that the presence of natural carbon substrates affects the biodegradation of pollutants in the environment, there is very little known with respect to the consequences of this interaction (Alexander, 1994). Only recently, general physiological and kinetic patterns have begun to emerge from laboratory studies with carbon-limited chemostat cultures fed with mixtures of pollutants and additional carbon substrates. Here, an overview of this new information is presented.

Simultaneous utilisation of alternative carbon substrates and pollutants

The fact that pollutants and alternative substrates are consumed simultaneously under carbon-limited growth conditions has been demonstrated convincingly for several pure culture systems, such as for *Chelatobacter heintzii* growing with glucose plus nitrilotriacetic acid (NTA), for *Comamonas testosteroni* with acetate plus *p*-toluenesulfonate (*p*-TS), or for *Methylobacterium* DM4 cultivated with acetate plus dichloromethane (DCM) (Bally *et al.*, 1994; Tien, 1997). But many other, less well documented examples can be found in literature, demonstrating that this is the rule

---

1 Systems that are heavily polluted with carbonaceous compounds, and where the availability of either terminal electron acceptors or other nutrients, such as nitrogen or phosphorus, become limiting for the degradation, will not be considered here.

and not an exception (compiled in Egli, 1995). From these mixed substrate studies, two aspects, namely those concerning the expression of pollutant-degrading enzymes and the kinetics of growth, are particularly interesting because they seem to be widely applicable.

Patterns of expression for pollutant-degrading enzyme systems

Because many of the enzyme systems involved in the degradation of pollutants are inducible, the question arises as to how the expression of pollutant-catabolising enzymes is regulated under environmental growth conditions. Data obtained in our laboratory from carbon-limited chemostat cultures supplied with mixtures of pollutants and alternative carbon substrates, indicate that essentially two different regulation patterns can be distinguished for pollutant-degrading enzyme systems:

1. Pollutant-catabolising enzyme systems that are essentially inducible, but which, nevertheless, are expressed at a low background level (derepression) also during growth in the absence of the pollutant (e.g. during cultivation with carbon sources such as glucose, acetate, or even synthetic sewage).

2. Pollutant-degrading enzyme systems that are not expressed during growth in the absence of the pollutant, but are dependent on a certain (threshold) concentration of this compound for expression.

Derepressed systems

Examples for the first pattern have been reported for the NTA-, *p*-TS-, and DCM-utilising bacterial strains mentioned above (Bally *et al.*, 1994; Tien, 1997). For all these micro-organisms, a low background level of pollutant-degrading enzymes was detected when they were cultivated in carbon-limited chemostat culture at a low growth rate with an easily degradable carbon substrate. Whereas in the case of *C. heintzii* this background level was hardly detectable (Bally *et al.*, 1994), expression levels of *p*-TS and DCM-degrading enzyme systems in *C. testosteroni* and *Methylobacterium* DM4 were in the range of 1 to 3 per cent of those found in fully induced cells (Tien, 1997).

The consequences of this background expression of pollutant-degrading enzyme systems was that cells cultivated on media that did not contain the pollutant were able to degrade the pollutant when it was added to the growth medium in small amounts. This is demonstrated in Table 1 for the case of *C. heintzii* growing in continuous culture with mixtures of glucose and NTA. Even if NTA contributed only a small amount of the cell's total carbon consumed, it was degraded together with glucose. As long as NTA contributed less than approximately 1 per cent of the total carbon handled by the cell no significant additional induction over the derepression level of NTA-monooxygenase was detected. When NTA contributed between 1-20 per cent of the simultaneously utilised carbon, NTA-monooxygenase expression increased with increasing proportions of NTA in the feed. It was in this range where expression of NTA-monooxygenase was most strongly stimulated. Similar enzyme expression and pollutant degradation patterns were observed for *C. testosteroni* and *Methylobacterium* strain DM4 during growth in carbon-limited chemostat culture with mixtures of acetate, plus the *p*-TS or acetate plus DCM, respectively (Tien, 1997). In the case of *Methylobacterium*, this pattern was even confirmed for growth with mixtures of DCM plus synthetic sewage.

Table 1. **Regulation of nitrilotriacetate (NTA) degradation in *Chelatobacter heintzii***

NTA/glucose conc. in feed (mg C/L)	NTA/glucose conc. in culture (mg C/L)	Expression of NTA-mono-oxygenase
0 / 727	0.017 / <0.01	<1%
0.262 / 727	0.012 / <0.01	<1%
2.62 / 724	0.009 / <0.01	<1%
26.2 / 693	0.012 / <0.01	~3%
131 / 596	0.027 / <0.01	~10%
727 / 0	0.050 / <0.01	100%

Note: Cells were cultivated in carbon-limited continuous culture ($D = 0.06\ h^{-1}$) with mixtures of glucose and NTA. Expression of the NTA-mono-oxygenase protein was quantified immunologically.

Source: Adapted from Bally *et al.*, 1994.

Threshold systems

The second regulation pattern was documented recently in our laboratory for the utilisation of the aromatic compound 3ppa by a culture of *E. coli* (Kovárová *et al.*, in press, 1997*a-b*). In contrast to the examples discussed above, cells cultivated in a glucose-limited chemostat exhibited no detectable 3ppa-degrading activity. This is illustrated in Figure 1 for a culture growing in a chemostat with glucose as the only substrate (100 mg L^{-1} in the feed) and where, at time zero, 3ppa was added to the feed medium. When the concentration of 3ppa added to the feed was 0.3 and 3 mg L^{-1}, respectively, no degradation of 3ppa was observed, resulting in a perfect wash-in curve of 3ppa in the culture medium. This indicates that the cells were unable to induce the 3ppa catabolic pathway. However, when the feed concentration of 3ppa was raised to 5 mg L^{-1} or more, degradation of 3ppa started (Kovárová *et al.*, 1997*a*). The 3ppa threshold concentration required for induction of enzymes involved in the degradation of this compound was approximately 3 mg L^{-1}. It should be stressed that once induced, the cells were able to degrade 3ppa below the threshold concentration required for induction, and that the degradation continued although the actual concentration of 3ppa in the reactor was always below 1 mg L^{-1} (Kovárová *et al.*, 1997*b*).

Mixed substrate growth kinetics

First evidence for the fact that this simultaneous utilisation of carbon compounds has kinetic consequences was obtained in the mid-70s (reviewed in Egli, 1995). This information has recently been extended and firmly established (Lendenmann *et al.*, 1996; Kovárová *et al.*, 1997*b*). Lendenmann *et al.* (1996) demonstrated for *Escherichia coli* growing in carbon-limited chemostat culture with mixtures of up to six different sugars that, compared to growth with a single sugar at a particular growth rate, the steady-state concentrations were consistently lower during simultaneous utilisation of mixtures of sugars. The steady-state concentrations of particular sugars depended approximately linearly on their contributions to the total carbon consumption rate. Although the experimental data presently available are still limited (see Egli, 1995), one can already predict that the reduction of steady-state concentrations of individual compounds during mixed substrate growth is most probably a general phenomenon that will also be applicable to growth with mixtures of easily degradable carbon compounds plus pollutants. Results obtained recently for the bacterium *E. coli* cultivated in carbon-limited continuous culture with mixtures of glucose plus 3-ppa, indicate that this kinetic behaviour does not only hold for mixtures of structurally similar compounds such as sugars, but can be extended to mixtures of carbon sources that are degraded via distinctly different catabolic pathways (Kovárová *et al.*, 1997*b*).

Figure 1. Degradation of 3-phenylpropionic acid (3ppa) by a culture of *Escherichia coli* growing in a chemostat with glucose as the primary substrate

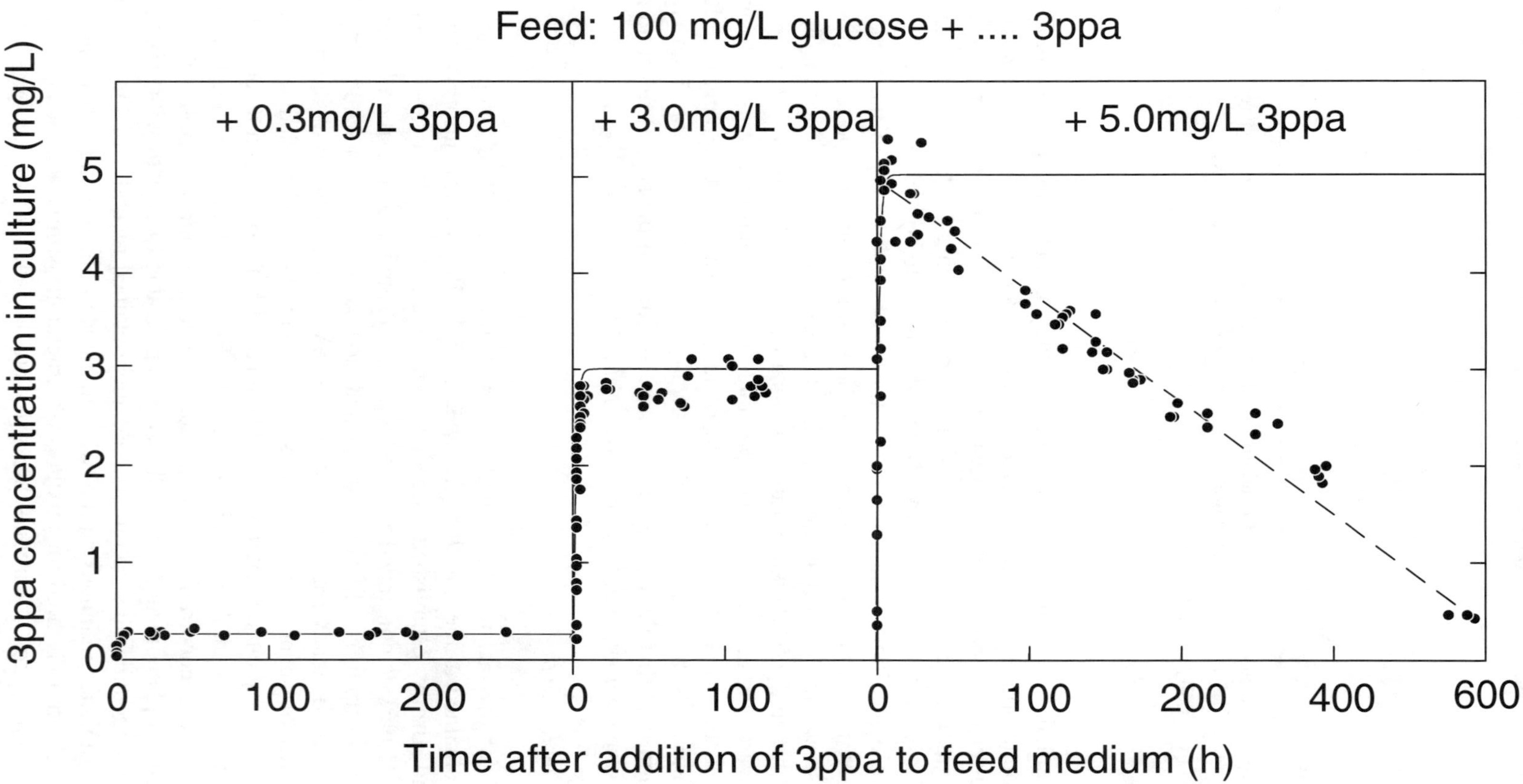

Note: Initially, the culture was grown with glucose as the only limiting carbon substrate at a dilution rate of 0.6 h^{-1}. At time zero, 3ppa at the concentration indicated was added to the inflowing medium and the resulting concentration of 3ppa in the culture vessel was monitored. The solid lines indicate the theoretically expected concentration of 3ppa in the bioreactor assuming that 3ppa was not utilised.

Source: Adapted from Kovárová *et al.*, 1997*a*.

Conclusions and some speculations

The fact that under environmental conditions, pollutants are always present together with carbon compounds of natural origin has consequences for the residual pollutant concentrations that will be observed (and may be achieved) during biodegradation processes; the experimental data presented here indicates that especially the regulation pattern for pollutant-degrading enzymes plays a crucial role. For example, no threshold concentrations for utilisation should, in principle, be observed for pollutants which are degraded via enzyme pathways that are always expressed to a certain degree (even if only at a low, hardly detectable background level). Even if present at concentrations far below the threshold concentrations that are able to support cellular growth as single compounds, such pollutants should still be utilised. This is a direct result of the fact that carbon compounds of natural origin will continue to supply energy for growth and metabolism of the microbial cell. The experimental data presented here give a reasonable explanation for the fact that many compounds are utilised in the nanogram or even femtogram per litre concentration range, i.e. far below the threshold concentrations typically found for the utilisation of individual compounds (see Alexander, 1994). Contrasting to this is the observation that the expression of some pollutant-degrading biochemical pathways are dependent on a distinct (threshold) concentration of this compound before their induction is triggered and degradation can occur. Fortunately, in our example, degradation of 3ppa was possible below the threshold concentration for induction. Whether this is a common property of such systems remains to be established. However, this behaviour may explain why some compounds -- although strains able to degrade them can easily be enriched from the environment -- remain untouched in ecosystems.

It can be envisaged that the two phenomena described in this contribution, i.e. the reduction of steady-state concentration during mixed substrate growth and the knowledge on regulatory patterns of pollutant-degrading pathways, do not only supply basic information, but can be used to achieve reduced concentrations of harmful pollutants, e.g. in the production of drinking water or the purification of industrial wastewater.

Acknowledgements

The author thanks K. Kovárová and A. Tien for allowing me to use their data. The work has been financially supported by EAWAG, the Swiss National Science Foundation (Projects Nos. 31-30004.90 and 90/2, SPP Environment project No. 5001.35285), the Research Commission of ETH Zürich (Project No. 0330.089.86), and by Lever (Switzerland) / Unilever Merseyside (England).

Questions, comments and answers

C: It is difficult to know what is meant by concentration when referring to biopolymers such as polysaccharides. Also, some of the pollutants mentioned are lipophilic and penetrate the cell membrane very easily.

A: We are aware of these problems. We are just trying to see what happens in an ecosystem and to deduce principles of utilisation with simple systems. Similar experiments done with synthetic wastewater and pollutants show the same regulatory patterns. This shows that the patterns we have observed are similar in more complex systems. We can only apply this knowledge when we know the principles.

REFERENCES

ALEXANDER, M. (1994), *Biodegradation and Bioremediation*, Academic Press, San Diego.

BALLY, M., E. WILBERG, M. KÜHNI, and T. EGLI (1994), "Growth and enzyme synthesis in the nitrilotriacetic acid (NTA)-degrading *Chelatobacter heintzii* ATCC 29600", *Microbiology* 140, pp. 1 927-1 936.

EGLI, T. (1995), "The ecological and physiological significance of microbial growth with mixtures of substrates", *Adv. Microb. Ecol.* 14, pp. 305-386.

LENDENMANN, U., M. SNOZZI, and T. EGLI (1996), "Kinetics of the simultaneous utilization of sugar mixtures by *Escherichia coli* in continuous culture", *Appl. Enviro. Microbiol.* 62, pp. 1 493-1 499.

KOVÁROVÁ, K., A. KÄCH, V. CHALOUPKA, and T. EGLI (in press), "Cultivation of *Escherichia coli* with mixtures of 3-phenylpropionic acid and glucose: Dynamics of growth and substrate consumption", *Biodegradation* (accepted).

KOVÁROVÁ, K., V. CHALOUPKA, and T. EGLI (1997*a*), "Threshold substrate concentrations required for induction of the catabolic pathway of 3-phenylpropionic acid in *Escherichia coli*", *Appl. Enviro. Microbiol.* (submitted).

KOVÁROVÁ, K., A. KÄCH, A.J.B. ZEHNDER and T. EGLI (1997*b*), "Cultivation of *Escherichia coli* with mixtures of 3-phenylpropionic acid and glucose: Steady-state growth kinetics", *Appl. Enviro. Microbiol.* (submitted).

TIEN, A. (1997), "The Physiology of a Defined Four-membered Mixed Bacterial Culture During Continuous Cultivation with Mixtures of Three Pollutants in Synthetic Sewage", PhD thesis No. 11905, Swiss Federal Institute of Technology, Zürich, Switzerland.

SEDIMENT REMEDIATIONS: TECHNOLOGY DEVELOPMENT AND PILOT PROJECTS

by

Marijke M.A. Ferdinandy - van Vlerken, Willem A. Bruggeman and Gerard N.M. Stokman
Ministry of Transport, Public Works and Water Management
Institute for Inland Water Management and Waste Water Treatment (RIZA)
Lelystad, the Netherlands

Introduction

Polluted sediments represent a risk for contamination of surface waters and groundwaters. Diffusion of pollutants, or (bio)chemical processes, may lead to transport from the sediment into the water. In order to minimise this risk, polluted sediments may be isolated *in situ*, or dredged followed by disposal in landfills or by remediation.

In order to stimulate the development of dredging and remediation techniques, the Ministry of Transport, Public Works and Water Management, started a national R&D program in 1989. The program is called 'Development Program for Treatment Processes of Contaminated Sediments', with the Dutch abbreviation POSW.

This paper will give an introduction on the problem of contaminated sediments in the Netherlands. Results from POSW (treatment technologies and pilot projects) will be summarised. Most attention will be paid to the development of biological treatment technologies. Finally, the route to large-scale treatment of contaminated sediments will be depicted.

Contaminated sediments in the Netherlands

The name 'the Netherlands' refers to the low position of the country. The Netherlands are located at the downstream parts of several European rivers (Rhine, Scheldt, Meuse), in a flat area at the North Sea. A large part of the country lies below sea level, so that many polders and protections against flooding were constructed.

In the rivers, water velocity is low, so that sedimentation of suspended matter takes place. Especially in lakes and harbours, this results in relatively large quantities of sediment. In order to keep the main waterways and ports at sufficient depth for traffic and transport, the sediment must be removed by means of dredging. This sediment used to be highly appreciated and used in building dikes and polders.

In the beginning of the 80s, the first problems of pollution in the sediment were recognised in the region of the Rotterdam Harbour. Since dredging was necessary for nautical reasons, a solution for the contaminated dredged material (CDM) had to be found. A start was made with the construction of landfills. Also, the thought arose that maybe pollution could occur in other sediments, where dredging was not necessary. Therefore, a national monitoring program was started. It turned out that many other sediments in rivers, but mostly lakes, were polluted. A distinction was made between sediment that had to be removed for nautical reasons ('maintenance material') and for environmental reasons ('sanitation material').

Several 'quality classes' of (polluted) sediments are distinguished. In the beginning of the 80s, these classes were based on 'areas of reference' with different degrees of pollution. In 1988, a start was made with ecotoxicological motivation of different classes. Known aquatic toxicity effects of various compounds (NOEC-levels: NO Effect Concentrations) were the starting point. The NOEC-levels were translated to sediment concentrations, based on partitioning coefficients (water/solids equilibrium). The Ministry of Transport, Public Works and Water Management and the Ministry of Health, Spacial Planning and Environment, are working on an integral environmental policy (Ministry of Transport, Public Works and Water Management, 1991). This policy will be used to formulate legislation in the coming years.

In Table 1, a summary is given of the quality frontiers between five classes, allowed destinations and the estimation of quantities to be dredged in the Netherlands. Most sediments contain cocktails of contaminants (heavy metals and various organic pollutants). The quality frontiers were formulated for many individual components (see Table 2). The quality class of sediments is determined by the component in the highest class.

Dredging activities because of environmental reasons are considered for sediments of class III and IV. For class IV sanitation is urgent, while for sediments of class III investigations will point out the urgency. For maintenance material of class III quality, disposal in landfills or remediation is necessary.

Table 1. **Present policy on contaminated sediments in the Netherlands**

Class/frontier	Destination and quality frontier	Quantities Main	Quantities San
0	free		
	Target value; negligible risks (concentrations<NOEC)		
I	free	345	0
	Target 2000 value; maximal tolerable risks (5% of species possibly affected)		
II	until 2000; dispersal in water allowed	110	0
	Testing value; until 2000; "area of reference" Lake IJsselmeer		
III	disposal in landfill or remediation	31	40
	Sanitation value; intervention level (50% of species possibly affected)		
IV	disposal in landfill or remediation	8	47

Notes: Quantities: estimation of quantities (in million cubic meters) to be dredged in the Netherlands between 1991 and 2010. Main: Maintenance material; San: Sanitation material.

Source: Ministry of Transport, Public Works and Water Management, 1991.

The total amount of CDM (both nautical and sanitation), to be dredged between 1991 and 2010 in the Netherlands is estimated at 581 million cubic meters (Ministry of Transport, Public Works and Water Management, 1989). Legislation is in full preparation at the moment and is expected to be ready in a few years. Much discussion is taking place about the destination of class II maintenance material. At present, it is still possible to disperse this material in the water system. Possibly the frontier between class II and III sediments will disappear, so that class II sediment would have to be treated or disposed in landfills.

Besides destinations in the water system, remediation might also lead to products to be re-used as construction materials. Legislation on construction materials, as described in 'Decision Building Materials' (DBM), will be official in 1997 (CUR, RIZA and RIVM, 1995). The DBM distinguishes almost the same outer frontiers (Target value and Sanitation value) as the policy on sediments. However, between those limits, the DBM allows application as construction material if leaching of heavy metals is below certain standards.

Table 2 gives a summary of the target value and the upper frontier of the DBM for a selection of components.

Table 2. **Summary of target values and sanitation values in the Netherlands (in mg/kg dry matter, correction for 10% organic matter and 25% lutum)**

Compound		Target value (mg/kg dm)	Sanitation value DBM
PAHs	sum of 10	1	40
Mineral oil		50	500
Cl-benzenes	hexa or penta	0.0025	
	tetra, tri or di	0.01	
	sum		5
PCBs	each of 6	0.001 - 0.004	
	sum		0.5
Metals	Cd	0.8	12
	Hg	0.3	10
	Cu	35	190
	Ni	35	210
	Pb	85	530
	Zn	140	720
	Cr	100	380
	As	29	55

Notes: sum of PAH: naphthalene, phenanthrene, anthracene, fluoranteen, chryseen, benzo(a) anthracene, benzo(a) pyrene, benzo(k) fluorantheen, benzo(ghi)peryleen, indeno(1,2,3 cd) peryleen.

Source: Ministry of Transport, Public Works and Water Management, 1991; CUR, RIZA and RIVM, 1995.

Besides dredging and sanitation, the Dutch policy is also directed at reduction of sources of pollution. At the moment, depending on the type of the source, up to 95 per cent of the sources present in 1985 have been tackled. The remaining five per cent is more difficult to cope with, since it mainly concerns diffuse sources. Sanitation of already existing contaminated sediments still must be carried out. The policy aims at treatment of 20 per cent of the expected dredged material in the year 2000.

In order to stimulate the development of dredging and remediation techniques, the Ministry of Transport, Public Works and Water Management, started a national R&D program in 1989. The program is called 'Development Program for Treatment Processes of Contaminated Sediments', with the Dutch abbreviation POSW.

The aim of the POSW is to come forward with proven techniques, applicable not only from a technological, but also an economic and environmental point of view. These techniques should be applicable for the bulk of the Dutch sediments, leading to product qualities which meet legal standards for re-use or application as building material. In this way, the risks of contaminated sediments and the required space in landfills are reduced.

The POSW is co-ordinated by the Institute for Inland Water Management and Waste Water Treatment (RIZA), while projects are carried out by many different universities, consulting agencies, technology vendors, consulting engineers and contractors.

The activities within POSW are divided into 1) technology development; 2) assessment of costs and environmental effects; and 3) demonstration of treatment chains in pilot projects.

Starting with a large number of possible technologies, POSW has now selected a smaller number of promising techniques (POSW, 1995; Stokman, 1995). At the moment most attention is paid to:

- development of methods for assessment of environmental effects;
- improved system for prediction of costs;
- decision-system for optimal treatment chains for different sediments;
- feasibility study on large scale treatment of sediments.

Technology development

Site investigation and environmentally friendly dredging

Contrary to maintenance dredgings, environmental dredging activities must be performed in a very accurate way. When too little of the contaminated material is removed, the site will still be polluted. When too much material is removed, the costs for transport, treatment and/or disposal will be unnecessarily increased.

Also, dispersion of dredged material must be minimised, in order to prevent contamination of the surface water. Conventional dredging equipment has been adjusted to these prerequisites, resulting in several different types of environmental friendly dredging facilities.

In order to achieve an optimal dredging result, the exact location of the contaminated material is analysed during the Site Investigation. The location can be identified by means of seismic technologies, which visualise differences in grain size *in situ*. Since the distribution of contaminants often is related to specific fractions of the sediment, seismic surveys give a good impression of the location. Additional chemical analysis of core-samples is still required. During the dredging operation, the exact location can be monitored on board the ship.

Treatment of the material can be based on the principles of physical (classification), biological, chemical or thermal processes. Often, combinations of different processes are applied, resulting in treatment chains.

The efficiency of different techniques is highly dependent on the characteristics (both grain size and pollutants) of sediment. These parameters are subject of study during the Site Investigation. In order to chose optimal treatment chains, laboratory or pilot scale research with several techniques is included in the Site Investigation.

POSW has concluded that it is essential and worthwhile to put quite a lot of effort into the site characterisation. It optimises the amount of material to be dredged and the input will be re-earned in the further treatment process.

Classification: the first step of treatment

The applicability of techniques, as well as the resulting product quality, is dependent on both texture and contaminants of the sediment to be treated. Also, contaminants are not often distributed homogeneously over the material. Mostly the fine fraction (smaller than 63 um) contains the largest part of the organic matter, to which the contaminants are strongly attached. The course fraction, consisting of sand, often is clean or can be cleaned easily.

Therefore, classification of sediments into fine and course fractions, often is an effective start of the treatment chain. The produced fine and course fraction can be remediated with different technologies, depending on their characteristics.

Classification by means of hydrocyclones is applicable at the moment. Development is directed towards improved accuracy, especially at lower cut-points. In order to predict the results of hydrocyclonage, the so-called 'fingerprint method' has been developed. This method consists of a series of hydrocyclones, and generates detailed information on the distribution of solids and pollutants. Supported with chemical analysis, the optimum separation diameter (cut-point) and hydrocyclone configuration can be selected.

Hydrocyclonage can result in several sediment fractions:

- a sandy fraction, which can be polished and re-used;
- a fine fraction, which can be remediated;
- a residue fraction, which will be disposed.

Treatment of (fractions of) sediments

Sandy fractions mostly are clean or can be relatively simply polished by means of washing or flotation to remove remaining pollutants. Flotation is applied on a large scale in the mining industry. Specific chemicals (solubilisers) are added to the material to dissolve pollutants. Due to active aeration, the flotation foam can hence be removed at the top of the reactor.

For treatment of the fine fractions, several biological, chemical and thermal technologies are investigated and optimised.

Chemical processes were studied for removal of both heavy metals and organic compounds. No single technique was suitable for all types of contaminants. The chemical treatment of flotation (for sand polishing) was also tested for treatment of fine fractions. Bench- and pilot-scale tests resulted in a concentration of PAH and oil in 10 to 30 per cent of the feed. Selective removal of heavy metals was poor.

Heavy metal removal with acid extraction methods was effective, but the matrix was destroyed and no re-usable products could be obtained. EDTA-extraction of heavy metals was very promising on lab scale but chemical (regeneration difficulties) and technical complications disturbed the upscaling.

Wet oxidation for removal of organic pollutants was studied on laboratory and semi-practical scale. Wet oxidation is based on increased solvability of oxygen at high temperature and pressure, through which organic pollutants are degraded and destroyed. The advantage of wet oxidation lies in its high removal efficiency and the possibility of combined treatment with sewage sludge. Development is now focusing on process optimisation and market possibilities of the products.

Solvent extraction and thermal desorption are chemical treatments (for oil and PAH removal) which have been studied on laboratory scale. Evaluation of the results is taking place.

Bioremediation is directed at degradation of organic pollutants (mineral oil, PAH, organo-chloric compounds) by means of micro-organisms. Four different technological concepts have been developed: treatment *in situ*, treatment in landfill, treatment in landfarms, and treatment in reactors. More details on the biological techniques are given below.

Thermal immobilisation offers a solution for sediments contaminated with a cocktail of pollutants (inorganic and organic). Thermal immobilisation can be performed by sintering or by smelting.

Sintering is carried out at temperatures of about 1 200 °C, after dewatering, drying and pellet-pressing. Organic pollutants are completely incinerated, whereas the non-volatile metals are immobilised in the sintered product. The volatile metals are intercepted in the flue gas treatment facility. The artificial gravel product can be applied in concrete or asphalt.

Smelting is carried out at temperatures over 1 400 °C, producing a liquid melt. The melt cooling rate determines whether a glassy or crystalline product is formed. One form is artificial basalt that can replace a primary raw material.

Air emissions can be controlled by oven circumstances. In an oxidative surrounding a large part of the metals will remain in the formed products. Still, a thorough flue gas treatment is required.

For both sintering and smelting products, leachability tests proved that the products meet the environmental standards.

Evaluation and demonstration

Treatment technologies that are currently being considered for large scale treatment include landfarming and ripening, sand separation (followed by washing or froth flotation) and immobilisation. A number of treatment technologies that look promising but are still being developed include flotation of the fine fraction, wet oxidation, solvent extraction, and thermal technologies like desorption and reduction. Some of these technologies will be demonstrated in pilot remediation projects. An indication of the costs of applicable technologies is presented in Table 3.

Table 3. **Applicable technologies and costs**

Technology	Level of pollution	Sediment type	Costs (Gld/t dm)	Product
Operational				
Ripening	lightly	clay	35-70	clay
Landfarming	moderately in oil/PAH	sandy	50-100	dry soil
Classification/polishing	heavily	all	25-100	sand
Immobilisation	heavily	residuals	350-700	gravel/basalt
Promising				
Thermal treatment	heavily in organics	all	115-150	dry soil
Biological treatment	heavily in oil/PAH	all	80-150	dry soil
Flotation	heavily in organics	fine fraction	65-75	fine sand
Wet oxidation	heavily in oil/PAH	fine fraction	160-250	clay
Solvent extraction	heavily in organics	all	?	sludge

Note: t dm: ton dry matter.

Source: Stokman, 1995.

Costs

The costs, which have to be paid for by the government, are an important factor in realisation of remediation, since they have to be comparable to the costs for disposal in landfill. Therefore, during the optimisation of technologies much attention was given to reduction of costs. A realistic estimation of costs requires testing on a large scale. Dredging and classification are applied on full-scale, but most treatment technologies up have only been tested on smaller scale. The pilot remediations will help in estimating the costs for these techniques. In general, the costs will decrease when application scale increases.

A cost system for supporting detailed estimations and preparation of contractor's specifications is being prepared by POSW. A manual for cost estimations of sediment remediation has been written, aiming for systematically collecting and arranging financial data on sediment remediation. Secondly, a cost system for sediment remediation is being developed for policy and planning aspects. Indicators will be the basis of this system in support of scenarios for treatment of sediment.

At the moment, POSW thinks that cost-effective treatment can only be accomplished by using a few large-scale treatment facilities (instead of in a large amount of small treatment plants). Large-scale implementation requires considerable investments, in addition to several basic choices in policy. Feasibility of this approach is currently being investigated in a project called "Feasibility Study for Large Scale Treatment of Contaminated Dredged Material". This study should come forward with a description of technological, financial, environmental and organisational aspects of various scenarios for large scale treatment (Roeters and Bruggeman, 1996).

Assessment of environmental effects

Instruments for assessment of environmental effects of remediation chains and remediation technologies are being developed using a two-track approach, one track being a five-criteria tool (sediment quality, product quality, quantity, expenditure, inconvenience), with quantifiable parameters. Individual criteria can be assessed rather well; total assessment has not yet been possible. A Life Cycle Analysis (LCA) is expected to enable total assessment of environmental effects.

The first track will be further developed for assessment of individual technologies. Assessment of toxicity and leaching of sediments, sludge and products will be built into this track. The LCA track will be further developed for use in general studies on the problems of contaminated sediments.

Demonstrations in pilot projects

Pilot remediations are carried out for testing and demonstrating combinations of environmentally sound dredging and treatment technologies in an integral clean-up chain. The knowledge and experience gained from these demonstrations should provide a sound base for future sediment remediation projects by facilitating rapid and effective composition of clean-up chains.

The first pilot remediation was carried out in 1994 in Elburg Harbour, a location heavily contaminated with PAH. Selection of this location was due not only to the danger of contamination spreading to an adjacent lake, but also because the location was considered representative for a large number of small Dutch harbours.

The techniques used were classification, followed by polishing of the sandy fraction, and ripening of the fine fraction. The results of the remediation were a 93 per cent reduction of PAH contamination and 84 per cent of the dredged material re-used after treatment.

This demonstration clearly indicated that an intensive survey of the location combined with a thorough characterisation of sediment are prerequisites for a successful execution of the remediation. In addition, it was learned that critical factors for a remediation project include, amongst others, complex administrative and regulatory issues; a thorough preparation of organisation, phasing and decision making; formulating operational project goals; selection of contractors, with special attention to quality in dredging and treatment.

Elburg Harbour remediation has shown that environmentally sound dredging within ten centimetres, and separation of sand using hydrocyclones, are operational.

The second pilot remediation was carried out in 1995/1996 in a small area along the Nieuwe Merwede, a river south of Rotterdam. This location was selected since it is representative of Dutch river sediments: a high amount of fine particles, a 'cocktail' of heavy metals and organic contamination. Project objectives were on the one hand dredging and ecologically rebuilding the river bank zone, and on the other hand, thermal immobilisation of contaminated sediment to make at least 300 tons of reusable products. Additionally, the project will provide information on remediation methodologies for large water systems.

The third pilot remediation took place in Autumn 1996 in the Petroleum Harbour in Amsterdam, a location considered representative for highly industrialised harbours. The location is heavily contaminated with oil and PAH. Therefore, biological technologies are suitable.

Development of bioremediation techniques

Principles and possibilities

Biological remediation techniques make use of micro-organisms to degrade contaminants. The micro-organisms (mostly bacteria and fungi) are able to convert organic contaminations into compounds they can use for their growth and/or energy supply. The contamination is converted into harmless compounds, sometimes resulting in total mineralisation into H_2O and CO_2.

An advantage of bioremediation techniques is that generally, little energy and chemicals are required. Also biological treatment does not interfere in a severe way with the natural structure of the sediment, so that re-use as soil or sediment would be possible after treatment.

Contaminants that can be converted by micro-organisms are:

- Organic compounds (PAHs and mineral oil)
  These compounds are known to be relatively easily biodegradable by bacteria. Good results with bacterial degradation on laboratory and pilot scale have been obtained. The applicability of biodegradation by (exo-enzymes of) fungi is being studied at laboratory scale.

- Organic chlorinated compounds (HCB and PCBs)
  Chlorinated compounds have long been thought to be persistent. In the last years, laboratory and *in situ* field research have proven that biodegradation of these contaminants is possible.

Heavy metals and other inorganic contaminants are not biodegradable and will have to be removed by extraction or by immobilisation. Most of the sediments in the Netherlands contain several types of contaminants. Bioremediation is applicable for sediments in which the pollution is mostly caused by contamination of PAHs, mineral oil, HCB and PCBs. In the Netherlands, the quantity of these types of sediments represent about 27 (class III and IV) and 96 (class I and II) million cubic meters of sediment that has to be removed by the year 2010.

Technological concepts

When the contaminants are in principle biodegradable, environmental conditions still have to be met. Micro-organisms are living creatures and have requirements with regard to their surroundings, like presence of nutrients (N, P, K), the absence of toxic compounds and pH. Temperature affects outdoor treatments, since micro-organisms are more active (in the summer) at temperatures between 20 and 35 °C.

The most important environmental factor, however, is the oxygen concentration. Oxygen is necessary for the degradation of mineral oil and PAHs and for the mineralisation of chlorinated compounds. Since the sediment has an anaerobic origin, the introduction of oxygen without increasing the costs too much, has been a main point of interest in developing techniques.

Of course, it is also necessary for the right types of micro-organisms to be present. It is possible to inoculate the sediments with bacterial or fungi cultures. However, often the introduction of the appropriate environmental conditions is sufficient to induce growth and degradation from naturally occurring bacteria from the *in situ* population.

In the past years, these prerequisites for bacterial activity have led to the development of biological remediation techniques, within four different technological concepts (Figure 1).

Figure 1. **Technological concepts of bioremediation of sediments**

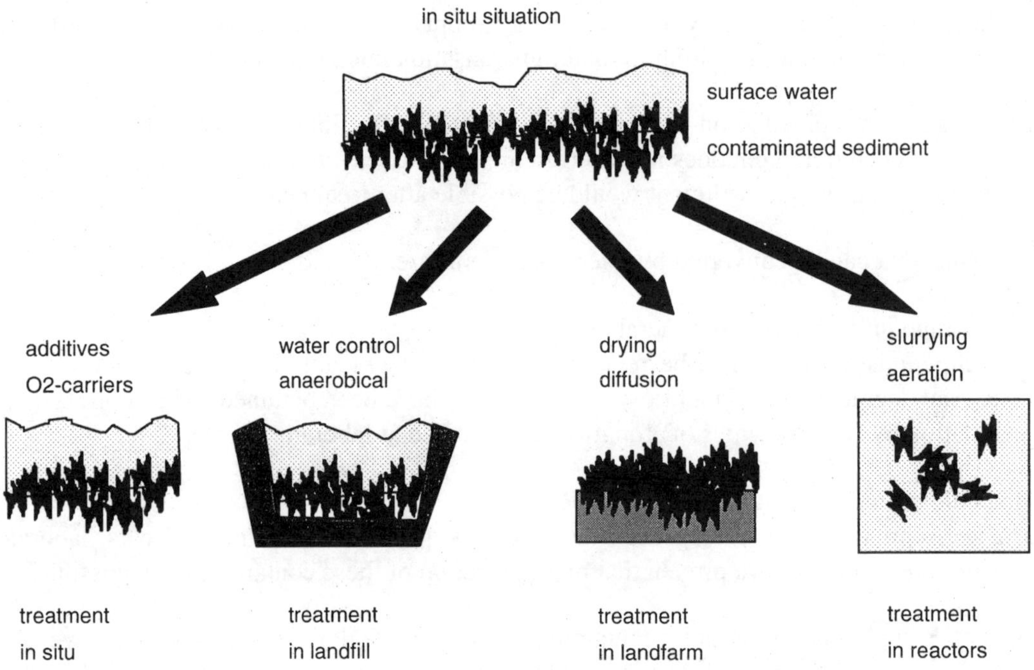

Source: Ferdinandy van Vlerken, 1996.

Research goals

In the period between 1989 and 1995, many studies have been carried out on biodegradation in sediments (Bioclear 1994, 1995; Bird Engineering, 1995; DHV, 1995; De Vries en van de Wiel, 1996; Ferdinandy van Vlerken, 1996; Kleijntjens *et al.*, 1995; Staring Centrum, 1995; TNO, 1993, 1995; Wieggers and Bezemer, 1995). These studies have shown that bioremediation generally takes place in two phases: a period of quick degradation is followed by slow degradation, ending in a 'platform' representing the rest-concentration (Figure 2).

Bioremediation always results in a two-phased degradation: high degradation rates in the first phase, followed by low rates in the second phase resulting in rest-concentrations. Both degradation rates and levels of rest-concentrations vary with the type of sediment being treated.

This phenomenon can be explained by (variations in) sorption of pollutants to the matrix of the sediment (Figure 3). The extent and mechanism of the sorption is varying, depending on the interaction between sediment and pollutant.

Figure 2. **Characteristics of bioremediation in sediments**

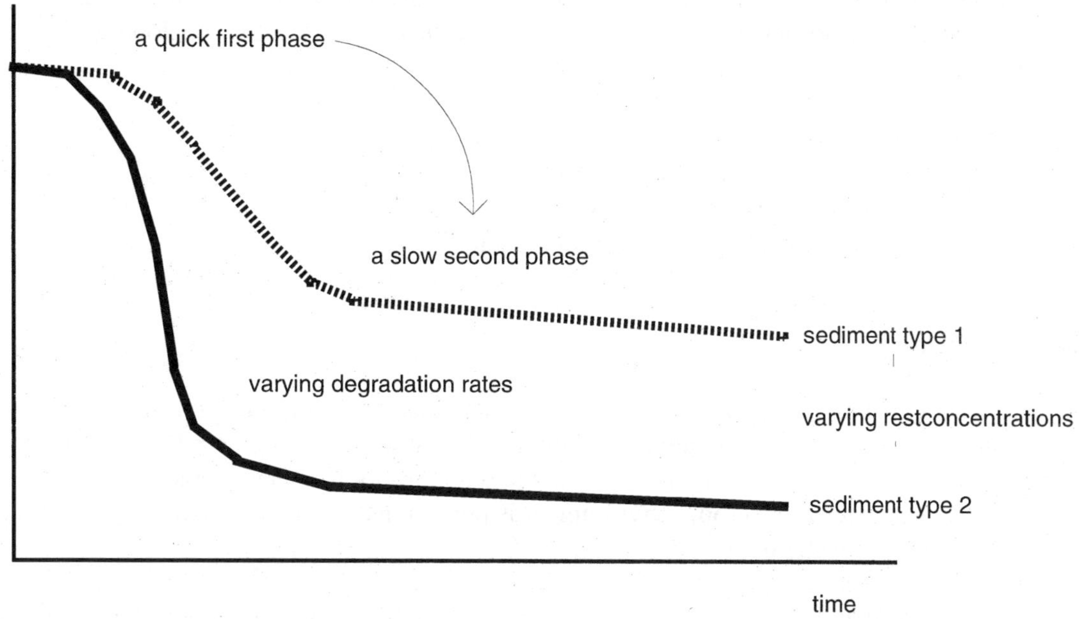

Source: Ferdinandy van Vlerken, 1996.

Figure 3. **Hypothesis: sorption from pollutant to sediment**

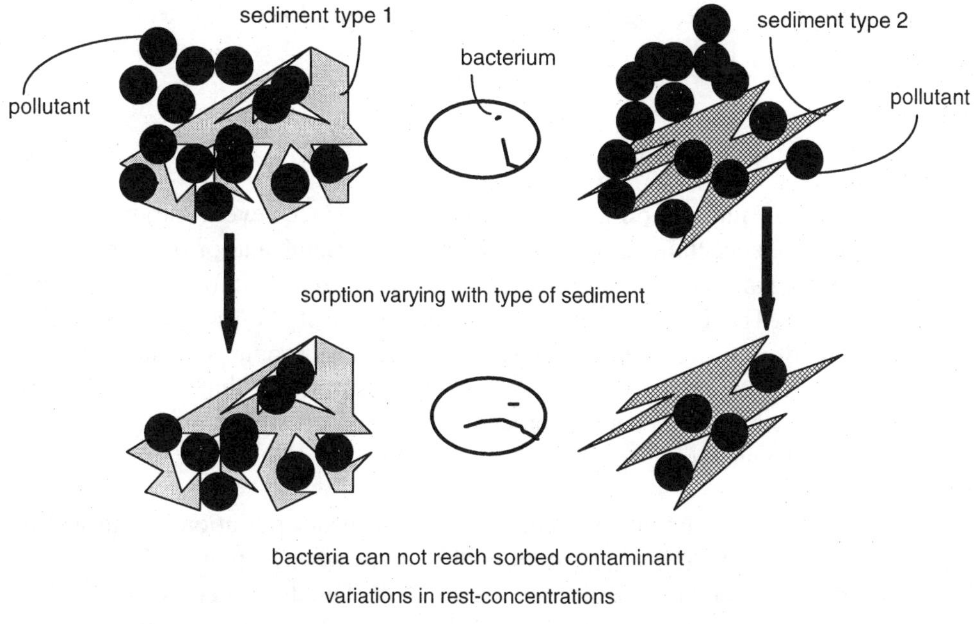

Source: Ferdinandy van Vlerken, 1996.

In the first phase of bioremediation, the lightly sorbed pollutants are degraded. In the second phase, the strongly sorbed pollutants are degraded. The rest-concentrations will finally be converted at a rate equal to the diffusion rate out of the sediment.

The degradation rates directly affect the costs of the remediation. The rest-concentrations determine whether or not legal standards for re-use are met. Technology development nowadays is directed towards:

- improvement of oxygen supply systems (reducing costs);
- decreasing residence times (reducing costs);
- decreasing rest-concentrations (meeting legal standards);
- measurement of actual risks from rest-concentrations (leaching and toxicity).

Treatment in landfill

Treatment in landfills could be suitable for degradation of organo chlorinated contaminants (HCB and PCBs). These compounds require a two-phase remediation: an anaerobic phase for dechlorination, followed by an aerobic phase for mineralisation. Due to the large water content (50-60 per cent), a natural anaerobic environment is present and can be easily maintained in landfills, by keeping a layer of water on it.

HCB dechlorination into tri- and di-chlorobenzenes occurs at anaerobic conditions. *In situ*, half-life times between seven and 20 years do occur, while under optimal conditions in the laboratory, between 10 and 15 days are required. In 1994, a project started in which extensive stimulation (by adding substrates and bacteria) is tested in sediment in a land-fill. Half-life times between one and three years are expected. In April 1995, laboratory and land-fill results showed that the HCB (20 mg/kg dm) was totally sorbed to the sediment. This means that dechlorination rates will be equal or lower than the desorption rate. Now desorption rates and effects on lower organisms are being tested, while the frequency of monitoring in the land-fill has been drastically decreased (Bioclear, 1995).

Treatment in situ

The *in situ* degradation of mineral oil and PAHs was tested, since several products had entered the Dutch market. These products claimed to contain bacteria, nutrients and or oxygen-carriers. The dried product could be spread out over the surface water or mixed into the sediment, where biodegradation would start. The POSW studied the composition of the products as well as their effects on laboratory and true-to-life scale (Bioclear, 1994). The products contained hardly any degrading bacteria and no oxygen carrier. Therefore, it was not surprising that no effects were found on PAHs or oil degradation, which is an aerobic process. The only possible way for *in situ* treatment is to introduce oxygen into the anaerobic sediment.

Since application in field experiments, includes the risks of water pollution by nutrient addition or by (if degradation of organic matter is successful) diffusion of heavy metals, RIZA advised the water boards to gather these methods under the Wvo Act (Dutch Water Act). Before field experiments, laboratory tests would have to indicate the harmless effectiveness of the method.

In 1995, a system that includes mechanical air addition was tested in the lab and in the field. Degradation of PAH was poor, most probably the result of insufficient oxygen supply.

Treatment in landfarms and bioreactors

Biological degradation of PAHs and mineral oil requires the presence of oxygen. Therefore, much attention has been paid to introduce oxygen into the anaerobic sediment, without increase of energy demand and costs. In principle, two concepts are possible:

1. drying the sediments (in landfarms);
2. slurrying the sediments (in bio-reactors).

The principle of landfarming is that the first demand for aerobic degradation, i.e. oxygen, is met by drying the sediment which is spread over land. To prevent emission to the groundwater, a drainage system and impermeable foil are legally required. Only when sufficient oxygen is present, further stimulation (addition of micro-organisms, nutrients, etc.) can occur. This means that research aims mainly at finding ways to obtain and maintain a loose structure in the sediment. Within the POSW, three different variations on landfarming are currently being investigated:

1. Intensive landfarming
   The outdoor version in which about 40 cm layer of dried sediment is farmed for about one to two years, depending on the sediment type.

2. Greenhouse farming
   In this artificial environment, process conditions can be improved through heating, aeration or artificial rain. This type of landfarming is increasingly used for sandy soil remediation, where residence times from 6-12 weeks are usual.

3. Intrinsic landfarming
   This outdoor version is meant to decrease rest-concentrations in sediments that are pre-treated with other techniques, like intensive landfarming. The idea is to give a lot of time (around 10 years) in which little processing is necessary.

When using reactors, the necessary oxygen is supplied by slurrying the sediment under active aeration. Reactors have the advantage that process conditions can be controlled. Three types of reactors are being investigated:

1. Bioreactor
   This reactor, consisting of a rotating vessel, with oxygen supply, was tested only on laboratory scale.

2. Aeration basin
   Designed for the fine fraction (<63 mm) which can be kept in suspension (10 per cent dm) in an energy-extensive way by means of moving aeration elements, resulting in low costs. Suitable for sediments in which the course fraction is (easily) clean(ed).

3. Slurry Decontamination Process (SD-process)
   Originally designed for soil remediation, the SDP consists of different compartments for separation, biodegradation and dewatering, for treatment of all soil and sediments in a high density

slurry (20-45 per cent dm). The idea of this technique is that partitioning would result in an increased bio-availability of contaminants.

Table 4 summarises the POSW-projects with different *ex situ* techniques for oil and PAH removal, which were tested with different sediments.

Table 4. **POSW projects for biological, *ex situ* treatment for oil and PAH removal**

Technique	Goal	Sediment	Scale
Intensive landfarming	general applicability	GEU, ZIE	lab - 500 m^3
	improving dewatering ripening	PETROL, WEM	250 m^3
Intrinsic landfarming	removing rest-concentrations after pre-treatment	GEU, ZIE	500 m^3
Greenhouse farming	increasing reaction rates	PETROL, WEM	12-40 m^3
Bioreactor	general applicability	GEU, ZIE	lab
Aeration basin	oxygen supply system	GEU, ZIE	lab, 8 m^3
	oxygen supply system and % dry matter	PETROL	4 m^3
SD-process	oxygen supply system, partitioning	PETROL	3 m^3

Note: GEU: Geul Harbour; ZIE: Zierikzee Harbour; PETROL: Petrol Harbour; WEM: Wemeldinge Harbour.

Source: Ferdinandy van Vlerken, 1996.

In April 1994, the choice was made to continue further optimisation with the severely contaminated sediment from the Petrol Harbour: 500 ppm PAHs and 10 000 ppm mineral oil. Using the same sediment for optimisation of the different techniques enabled the POSW to compare and assess the techniques. The following techniques were tested with Petrol Harbour sediment: intensive landfarming (with and without plantation), greenhouse farming (with and without pre-dewatering), aeration basin (fine fraction), and the SD-process.

The degradation rates increased with the intensiveness of the technique, from intensive landfarm to greenhouse farm, to aeration basin, to the SD-process. The SD-process resulted within 14 days in a decrease of the PAH content from 450 to 30 ppm and of the oil content from 10 000 to 3 000 ppm.

Surprisingly, in all techniques the PAH-degradation started before oil-degradation, and rest-concentrations of oil below 3 000 ppm were not reached. Probably the low bio-availability of this oil caused this effect.

Several interesting effects related to technology development were discovered.

For the outdoor landfarm, a new strategy for ploughing was developed, starting off with light ploughing of the top layer, working towards rigorous ploughing of the entire layer. Heavy rainfall, before ripening of the sediment, resulted in water on top of the sediment. By applying the sediment in slopes, the water was transported to a ditch, and dewatering could proceed.

Practical problems occurred in the landfarm where plantation was tested: the reed was eaten by rabbits in the first summer. While looking for new types of plants (metal absorbing types), natural vegetation was developed within a year.

The greenhouse farm without pre-dewatering was not cost-effective, since dewatering in Petrol Harbour sediment was not limited by temperature. After natural dewatering in the outdoor landfarm, greenhouse farming resulted in a relatively quick degradation.

In both reactors, the mixing and handling of the sediment resulted in severe foam production. This problem was successfully dealt with by decreasing the air supply (aeration basin) and by addition of anti-foam (SDP process). For the aeration basin, the decrease in aeration resulted in sub-optimal degradation, since the oxygen still was the limiting factor.

The SD-process indicated the possibilities for the other techniques, when given sufficient time. Based on the achieved degradation rates, an extrapolation of the required time and space was made in 1995. Based on the achieved degradation rates, an extrapolation was made for all techniques, to reach a PAH level of 30 ppm in Petrol Harbour sediment (Ferdinandy van Vlerken, 1995). The results are summarised in Table 5.

Table 5. **Comparison of applicability and costs of biological techniques, estimation of required volume and time for degradation of 3 000 tons dry matter Petrol Harbour sediment**

Technique	Suitable for	Costs (dfl/ton dm)	Volume (m^3)	Time (weeks)
Landfarming	CF or CS	40-80	5 000	100-150
Greenhouse farming	CF or CS	50-100	5 000	50-60
Aeration basin	FF	80-130	30 000	25-35
SD-process	all	110-150	700	12-14

Notes: FF = Fine Fraction; CF = Course Fraction; CS = course sediment.
Residence times and costs are still indicative and will vary with type of sediment and required product quality.

Source: Ferdinandy van Vlerken, 1995.

In Autumn 1996, after 110 weeks, the outdoor landfarm had reached a PAH-level of around 50 ppm, indicating that the extrapolation was correct. The conclusion is that a choice between those techniques for remediation of Petrol Harbour sediment, is dependent on the available time, space and money.

Within the third pilot project of POSW, bioremediation of Petrol Harbour sediment is demonstrated. The sediment is classified by hydrocylones, after which the course fraction (>20 um) is polished with flotation, resulting in re-usable building material. The fine fraction (<20 um) is treated with bioremediation in reactors. The product of bioremediation will either be fit for re-use as construction material, or will have to be disposed in a landfill. In the case of disposal, the bio-treatment will have resulted in a large decrease in disposal costs: from around Gld 300 per ton dry matter (untreated material is chemical waste) to around Gld 20 per ton dry matter. The pilot project was concluded by March 1997.

Risks of rest-concentrations

Sediment legislation is based on general ecotoxicological data, translated in to sediment concentrations (see also section above on contaminated sediments in the Netherlands). The re-use in building materials allows higher concentrations, if heavy metal leaching is below certain standards.

The generally accepted hypothesis regarding rest-concentrations after bioremediation is based on sorption. Since bioremediation is limited to bio-available components, the remaining pollutants should always -- independent of the concentrations -- be strongly sorbed to the sediment. This would mean that rest-concentrations do not diffuse out of the sediment matrix, and would be harmless. This thought was investigated in leaching tests (diffusion in to water phase) and bio-assays (ecotoxicological effects), before and after treatment in three bioreactors.

The leaching tests were performed under rigorous conditions, after which the concentrations of heavy metals, PAH and mineral oil were measured in the water phase. For the ecotoxicity, the effects on shrimps, oyster and the luminescent bacterium *Vibrio* were tested. A summary of the results is depicted in Figure 4.

Figure 4. **Concentrations, leaching and toxicity: remaining percentages after biological treatment**

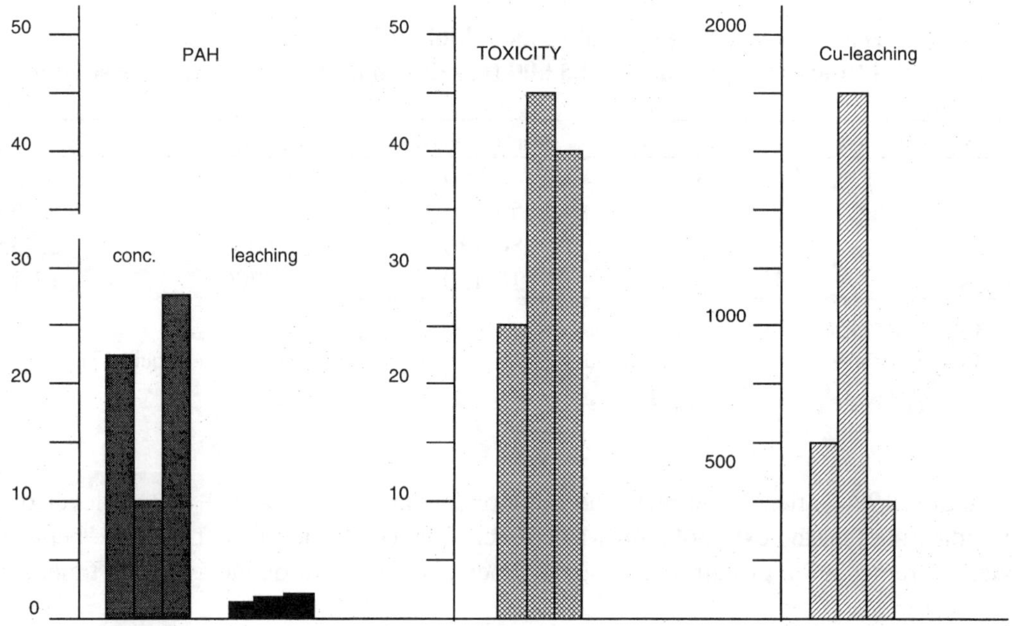

Source: Ferdinandy van Vlerken, 1996.

These results are in coherence with the hypothesis. The concentration of PAH decreases by 60-85 per cent, while the leaching of PAH decreases by 99.9 per cent. Not depending on the rest-concentrations, the absolute PAH-leaching after bio-treatment is comparable and extremely low. The results for mineral oil (not shown in Figure 3) were different. The concentrations decreased by 40-90 per cent, while leaching decreased by around 85 per cent.

On the other hand, leaching of metals did increase extremely (up to a factor 17) during the bioremediation. This might be the result of oxidation processes, caused by the aerobic environment required for biodegradation.

Finally, the toxicity did decrease, but to a smaller extent than PAH and oil leaching. Probably this toxicity is caused by the mobility of the heavy metals. This possibility is currently being tested in leaching-test and bio-assays, in anaerobic and aerobic samples.

Conclusions

The policy target to remediate 20 per cent of the contaminated sediments by the year 2000 probably will be reached.

An extensive site investigation is the first step in sanitation. The costs and efforts will be re-earned in effectiveness of both dredging and remediation.

Classification of sediments in fine and course fractions is operational and often very efficient.

Sandy fractions can be polished in a relatively simple way (scrubbing, washing, flotation).

Fine fractions can be treated by means of chemical or biological treatment.

Sediments contaminated with cocktails of pollutants, can be dealt with by thermal immobilisation.

Instruments for assessment of costs and environmental effects will be developed for use in the water management.

Results from POSW will be used for policy development and for advice on future sediment remediation projects.

Bioremediation offers an environmentally friendly and relatively cheap solution for contaminated sediments. Intensive landfarming is ready for full scale application. Treatment in reactors is being demonstrated at 5 000 cubic meter scale.

Research on biotechnology should be directed towards:

- decreasing rest-concentrations (surfactants, fungi);
- combination of metal removal by plants (phytoremediation) and landfarming;
- assessment of the risks of rest-concentrations.

Questions, comments and answers

Q: How well do your slurry basins with the pontoons that move back and forth work?

A: They were only tested at 8 cu. m scale, and they worked well. We could only choose one technology for scale-up and that was not the one chosen.

Q: When the boom moves back and forth in the lagoon for aeration, what was the period, that is how fast did it move?

A: That was some years ago and we were just concerned with how much oxygen needed to be delivered. We are now looking at impeller aeration.

Questions, comments and answers cont'd.

Q: Have you ever considered amending landfarming, for example, by adding compostable materials such as food waste to stimulate breakdown?

A: We tested this but it didn't make much difference. Now we measure to see if some nutrients are needed and for the rest, it is a question of delivering enough oxygen. We have a strategy of increasingly intensive ploughing. We tried planting reeds, but rabbits ate them! There is now a spontaneous plantation with deep roots.

Q: Is the regulatory endpoint based on toxicity, bulk chemical concentration, or bioavailability? How is the level determined?

A: It is based on bioassays and risk tests and converted to general chemical composition. This may not be subtle enough.

REFERENCES

BIOCLEAR MILIEUBIOTECHNOLOGIE, POSW (1995) "Interim rapport, Dechlorering van hexachloorbenzeen in depot", POSW internal report.

BIOCLEAR MILIEUBIOTECHNOLOGIE, GRONTMIJ, POSW (1994) "Onderzoek naar produkten voor *in situ* biologische reiniging van waterbodems", POSW-fase II, deel 1.

BIRD ENGINEERING, POSW (1995) "Biologische reiniging Petroleumhavenspecie met Slurry Decontamination Process", POSW internal report.

CUR, RIZA, RIVM (1995), "Onderweg naar het bouwstoffenbesluit", CUR-publikatie 95-1.

DE VRIES EN VAN DE WIEL (1996) "Kasfarming met Petroleumhavenspecie", POSW internal report.

DHV, POSW (1995) "Landfarming van baggerspecie: laboratorium- en praktijkonderzoek, eindrapport", POSW-fase II, deel 2.

FERDINANDY VAN VLERKEN, M. (1995), "Development of biological remediation techniques in the Netherlands", in *Proceedings of 'Sediment Remediation '95'*, 8-10 May, Windsor, Canada, the Wastewater Technology Centre.

FERDINANDY VAN VLERKEN, M. (1996), "Biological remediation of contaminated sediments: development and operationalization in the Netherlands", in De Schutter and Vanbrabant (eds.) *CATS III Congress 1996, Proceedings,* pp. 121-131, 18-20 March, Oostende, Belgium.

KLEIJNTJENS, R., I. OOSTENBRINK, G. MIJNBEEK, and K. LUIJBEN (1995), "Biotechnological sediment clean up using the slurry decontamination process", Satelite Seminar Remediation of Contaminated Sediments, Fifth International FZK/TNO Conference on Contaminated Soil, 1 December, Maastricht, the Netherlands.

MINISTRY OF TRANSPORT, PUBLIC WORKS AND WATER MANAGEMENT (1991), "National policy document on water management, Water in the Netherlands: a time for action".

POSW (1995), "Interim report, Development Programme for Treatment Processes for Polluted Sediments", POSW stage II (1992-1996).

ROETERS, P. and W. BRUGGEMAN (1996), "Feasibility of large scale treatment of contaminated sediments in the Netherlands", in De Schutter and Vanbrabant (eds.), *CATS III Congress 1996, Proceedings*, pp. 15-24, 18-20 March, Oostende, Belgium.

STARING CENTRUM, POSW (1995), "Intensieve landfarming: ontwatering en rijping", internal POSW report.

STOKMAN, G. (1995), "The Netherlands development programme for treatment for polluted sediments (POSW)", Satelite seminar Remediation of contaminated sediments, Fifth International FZK/TNO conference on contaminated soil, pp. 43-50, 1 December, Maastricht, the Netherlands.

TNO, POSW (1995) "Biologische reiniging: toepassing en opschaling van beluchtingsbassin", POSW internal report.

TNO, POSW (1993) "Biologische reiniging van de fijne fractie van vervuilde baggerspecie in een beluchtingsbassin", POSW-fase I, deel 11.

WIEGGERS, H. and H. BEZEMER (1995) "Intensive landfarming of contaminated sediments", Satelite seminar Remediation of contaminated sediments, Fifth International FZK/TNO conference on contaminated soil, pp. 135-136, 1 December, Maastricht, the Netherlands.

GENETIC ENGINEERING APPLIED TO HYDROCARBONOCLAST BACTERIA

by

Claude Danglot, François Artiguenave, Michel Delecu and Roland Vilaginès
Mairie de Paris, Centre de Recherche et de Contrôle des Eaux, Paris, France

Aquatic pollution by hydrocarbons

At various degrees, in every country, most underground or superficial waters are chronically polluted by hydrocarbons. On a world-wide basis, it is noteworthy to recall that river pollution is the main source of sea pollution. With approximately 1.9 million metric tons/year, river pollution represents 30 per cent of all marine pollution, far above that caused by wrecks or clandestine cleansings of supertankers (1.0 million metric tons/year) (Floodgate, 1984).

Because of the gathering of polluting industries along the banks, and because of the river flows, oil slicks move towards the sea, being progressively diluted, and a quantitative and qualitative pollution gradient is established between the source and the river estuary. This hydrocarbon gradient imposes a selective pressure on indigenous bacteria as important as their location by the estuary. Figure 1 shows the correlation between the metabolic ability of the local bacterial flora and its geographical location along the Seine river (Danglot, 1991).

Figure 1. **Sampling alkanoclast bacteria along the Seine River (panel A). Metabolic properites of sampled populations according to their geographical location in the river (panel B)**

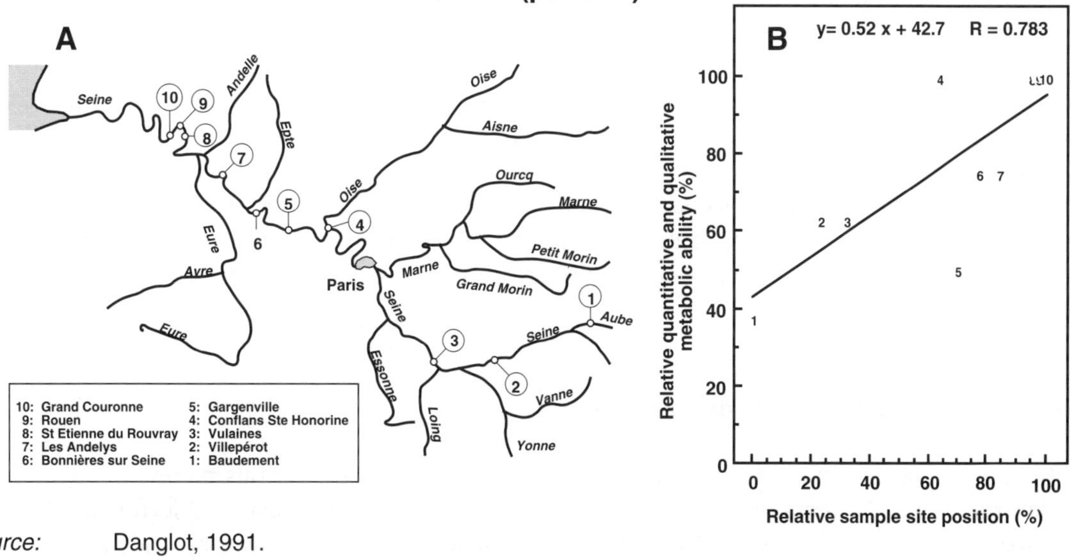

Source: Danglot, 1991.

The genetic study of the bacterial strains isolated from these sampling sites demonstrated that the different phenotypes identified corresponded actually to different genotypes, the bacteria isolated near the Seine estuary exhibiting the most extended genetic abilities. Because hydrocarbon pollution has increased considerably in the past 150 years, it is envisionable that this genetic heterogeneity in bacteria from aquatic environment is a relatively recent phenomenon, driven by the increasing level and diversity of the river hydrocarbon pollution.

Beside their noxious environmental effects, hydrocarbons threaten public health because some aromatic compounds are potent carcinogens. They also hamper waterworks production because their complete removal during casual pollutions consumes large amounts of granulated activated carbon (GAC), and because in the last stage of water treatment, trace amounts of hydrocarbons can react with chlorine to produce a very bad taste.

Biodegradation of hydrocarbons in rivers

Fortunately, hydrocarbons are biodegraded in rivers, but their biodegradation is a relatively slow process for several general reasons. First of all, biodegradation of hydrocarbons is mainly due to bacterial consortia, and the complete mineralisation is as slow as the slowest bacteria in the degradative chain. Also, the hydrocarbonoclast bacteria present increasing catabolic properties from the source to the river estuary, and this gradient mimics the one of pollution, both in quality and quantity. Furthermore, the enzymatic catabolic systems are often slowly inducible and are not fully active in less than 24 to 36 hours when the hydrocarbon slick is gone. Finally, in rivers, the growth of hydrocarbonoclast bacteria is far from optimal because of limited nitrogen, magnesium and oxygen supplies. Other local factors impede the biodegradation rate. As mentioned above, the hydrocarbonoclast bacteria are not evenly distributed in rivers; they are mainly located around the chronically polluted sites, and in low polluted places, their concentration can be very low.

Identity of polluting hydrocarbons

A crude oil can contain as many as a million different compounds, and refined products such as kerosene can contain from 5 000 to 10 000 components. Classically, after chloroform extraction, the compounds are separated by liquid-solid chromatography in saturated and aromatic fractions, resins and asphaltenes. These fractions are then analysed by gas-liquid chromatography. A heavy fuel, like fuel n °2 contains approximately 37 per cent saturated hydrocarbons, 35 per cent aromatic hydrocarbons, 10 per cent resins and 18 per cent asphaltenes. The saturated fraction includes n-alkanes, and branched alkanes like pristane and cycloalkanes, whereas the aromatic fraction includes monocyclic aromatic hydrocarbons (benzene, toluene, xylene, etc.) and polycyclic aromatic hydrocarbons (naphthalene, anthracene, phenanthrene, etc.). Resins are heterocyclic compounds like pyridines, quinoleines, sulfoxides or amides. Asphaltenes are high molecular weight compounds, with a complex structure, and can contain sulphur, nitrogen, oxygen, nickel or vanadium atoms in addition to a polyaromatic backbone.

Biological systems degrading hydrocarbons

Numerous bacteria able to metabolise specific hydrocarbons by various pathways have already been described, but the complete description of the genetic structures responsible for the described phenotypes is most often lacking. However, some genetic systems are now well known, and among them the best studied for aromatic compounds metabolism are plasmid-encoded. These systems include the TOL plasmid family conferring to their host the ability to metabolise toluene and xylene, but also methyl, dimethyl and trimethylbenzene (Williams and Murray, 1974; Spooner *et al.*, 1986; Burlage *et al.*, 1989; Harayama *et al.*, 1989; Harayama and Rekik, 1990; Delecu *et al.*, 1993), the NAH plasmid family involved in mineralisation of naphthalene, anthracene, biphenyls and phenanthrene; the DMP family responsible for phenol and dimethylphenol catabolism (Shingler *et al.*, 1992); the TCB plasmid family implied in the degradation of chlorosubstituted aromatics (Gosal and You, 1989; Van der Meer *et al.*, 1991). The degradation of alkanes has been less studied, and the only system to have been sequenced is the *alk* operon located on the pOCX plasmid of *Pseudomonas oleovorans* and responsible for the degradation of short linear alkanes (C_6 to C_{12}) (Kok *et al.*, 1989*a*; Kok *et al.*, 1989*b*; Van Beilen *et al.*, 1992). But this system does not seem to be very representative because most systems implied in alkane metabolism were shown to be chromosome-encoded, as the *lal* system implied in large alkane catabolism (C_{10} to C_{24}) and found mainly in *Acinetobacter sp.* (Danglot, 1991). Finally, genetic systems encoding for the degradation of cycloalkanes remains to be described.

Strategies to improve hydrocarbons biodegradation

Until now, only two main strategies have been used to improve biodegradation in the environment: the seeding and the feeding strategies.

The first one to be tested was the seeding strategy. It consists of seeding a polluted site with a laboratory strain selected for its ability to degrade specific pollutants. In most cases, this strategy gave few or no results, and the seeded strain was completely superseded by indigenous strains.

The feeding strategy is a more recent approach. It consists of supplying the indigenous hydrocarbonoclast bacteria with nutrients able to restore a well-balanced growth. In Alaska, Inipol EAP22 and Customblen were used with some success to treat gravel shorelines polluted by the Exxon-Valdez oil spill (Prince, 1994). Inipol EAP22 is a stabilized micro-emulsion of urea and organic phosphate in oleic acid. Customblen consists primarily of ammonium nitrate, calcium phosphate, encapsulated in a polymerised vegetable oil covering. As oleic acid and oil are soluble in hydrocarbons, a high concentration of nutrients is therefore maintained near the compounds to be degraded.

Requirements to further improve hydrocarbons biodegradation

In light of our present knowledge, it is possible to assess what is required to further improve *in situ* hydrocarbon biodegradation.

First of all, the concerned bacterial strains should be perfectly adapted to the polluted environment. This requirement was not met in seeding strategies but was fulfilled in feeding experiments.

Secondly, the bacteria should be resistant to the physical contact with hydrocarbons and should be able to internalise them efficiently. This requirement was probably met partially in seeding and feeding experiments.

Thirdly, the bacteria should be genetically well adapted to the pollution. This third requirement was fulfilled in seeding experiments, but probably not totally in feeding experiments.

We therefore propose to use a "seed-and-feed strategy" with the following steps:

- A water sample is withdrawn from the polluted site;
- The dominant bacterial strain is isolated;
- This dominant strain is electrotransformed with a broad-host-range synthetic plasmid harbouring all the genetic information needed;
- The transformed strain is isolated and grown at large scale in a selective medium;
- It is then seeded at the polluted site along with a metabolic enhancer.

Construction of pOIL, a polyvalent plasmid

The construction of such a synthetic plasmid must fulfil many requirements, in order to give it the best theoretical chances to be efficient.

- It should contain a broad host range replication origin, to be introduced and maintained efficiently in various environmental species;
- It should be modular and allow the easy insertion of several "catabolic cassettes";
- It should contain a cassette for aromatic compounds catabolism;
- It should contain a cassette for small alkanes (C_6 to C_{12}) catabolism;
- It should contain a cassette for large alkanes (C_{12} to C_{24}) catabolism;
- It should contain a cassette for production of a biosurfactant.

To find all the necessary genetic elements for the construction of such a polyvalent plasmid (we call it pOIL), we have analysed hydrocarbonoclast bacteria in Seine river, and we have isolated more than 150 different strains able to metabolise many different hydrocarbons.

Among all the species isolated, we have selected Ma2, a *Pseudomonas putida* strain harbouring a 185 kbp TOL plasmid and able to metabolise toluene and xylene at very high concentration, without any cytotoxic effect (upper operon *xylUWCMABN*, meta operon *xylXYZLTEGFJQHIH*, and two regulatory genes *xylR* and *xylS*). We have also selected L6.5, a *Pseudomonas fluorescens* strain, harbouring a 105 kbp TOL broad-host-range plasmid, because of the small size of its origin of replication (*oriV*). For small alkanes (C_6 to C_{12}), we have chosen the well described *Pseudomonas oleovorans* IL4 strain harbouring the 410 kbp pOCX plasmid (*alkST* genes and *alkBFGHJKL* operon), and for large alkane (C_{12} to C_{24}), we used an *Acinetobacter sp.* harbouring a chromosomal *lalABC...* operon and producing large amounts of biosurfactant (*bsfABC...* operon).

At the present time, several cassettes are ready (oriV cassette, *alkST/alkBFGHJKL* operon cassette) some are under construction (*xylUWCMABN* operon cassette, *xylXYZLTEGFJQHIH* operon cassette, *xylRS* cassette) and some are under preparation, the genes being under study and cloning (*lalABC...* cassette, *bsfABC...* cassette). Figure 2 summarises the construction steps used to construct the synthetic pOIL plasmid.

In parallel with the construction of pOIL, assays have been conducted to optimise the introduction of large plasmids into environmental bacterial strain (Artiguenave *et al.*, 1996) and the results have shown that it was possible to electrotransform indigenous bacteria with plasmids up to 120 kbp without difficulty. However this is a lower limit for pOIL, because the introduction of other its size to approximately 150 kbp which is the upper limit we found for electrotransforming indigenous bacterial strains.

Figure 2. **Construction of the pOIL plasmid from individual cassettes**

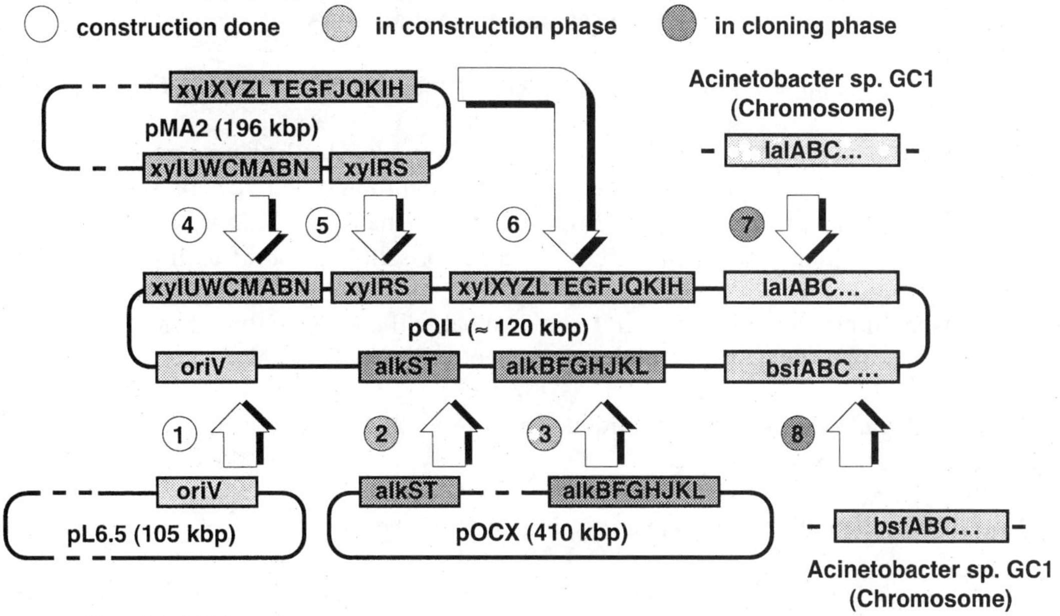

Source: Danglot, 1996.

The GEM release problem

Before constructing the pOIL plasmid in the environment it was necessary to take care of potential problems linked to the release of genetically modified DNA in environment.

- All the genetic systems to be assembled in pOIL are already present in the environment.

- All the genetic systems used were isolated from non-pathogens.

- The DNA fragment used were not modified, except some restriction sites that were removed or introduced.

- Every non-coding fragment was eliminated during construction to avoid unexpected side effects and also to reduce the pOIL size.

- Every transposon and IS in the donor strains was removed

- pOIL is not conjugative and does not seem to contain any *oriT*-like sequence.

- The possibility to introduce a suicide system in pOIL was considered, but a satisfactory solution remains to be found.

Of course, after construction and before testing its efficiency in waterworks, the use of pOIL in the environment will be submitted to the French Commission de Génie Génétique for an official clearance.

Questions, comments and answers

Q: What is the legal process when introducing GEMs into the environment in France?

A: A complete detailed report must be sent to the Biomolecular Engineering Commission. For laboratory work, every genetic engineering experiment has to be declared to the Commission on Genetic Engineering, and is classified according to the level of hazard concerned. We are in the lowest level (non-pathogenic bacteria) and authorisation is not difficult to obtain.

Q: You haven't done any field tests yet, I think. What will guarantee that the new super-strain will survive in the environment when it has to compete with the parental strains?

A: There is no guarantee, but if you introduce a plasmid, there is no reason for it to disappear from a site to which it is well-adapted.

Q: You have been modifying it so that it is, in effect, another strain.

A: That is not very important. We have experience with the genetic burden of this plasmid.

REFERENCES

ARTIGUENAVE, F., R. VILAGINES, and C. DANGLOT (1996), "High-efficiency transposon directed mutagenesis by electroporation of a *Pseudomonas fluorescens* strain", *FEMS Microbiol. Lett.*, (submitted for publication).

BURLAGE, R.S., S.W. HOOPER, and G.S. SAYLER (1989), "The TOL (pWW0) catabolic plasmid", *Appl. Environ. Microbiol.* 55, pp. 1 323-1 328.

DANGLOT, C. (1991), "Dégradation des alcanes par les bactéries de l'environnement", Mairie de Paris, Centre de Recherches et de Contrôle des Eaux de Paris, Etude pour l'Agence de Bassin Seine-Normandie, Convention n° 9099034.

DANGLOT, C. (1996), "Le génie génétique appliqué aux bactéries hydrocarbonoclastes", *Houille Blanche* 5, pp. 39-48.

DELECU, M., R. VILAGINES, and C. DANGLOT (1993), "Rapid mapping of TOL megaplamids", *Methods Mol. Cell. Biol.* 4, pp. 68-80.

FLOODGATE, G.D. (1984), "The fate of petroleum in marine ecosystems", in R.M. Atlas (ed.), *Petroleum microbiology*, pp. 355-397, MacMillan Publishing Company, New York, United States.

GHOSAL, D. and I.-S. YOU (1989), "Operon structure and nucleotide homology of the chlorocatechol oxidation genes of plasmids pJP4 and pAC27", *Gene* 83, pp. 225-232.

HARAYAMA, S. and M. REKIK (1990), "The meta-cleavage operon of TOL degradative plasmid pWW0 comprises 13 genes", *Mol. Gen. Genet.* 221, pp. 113-120.

HARAYAMA, S., M. REKIK, M. WUBBOLTS, K. ROSE, R.A. LEPPIK, and K.N. TIMMIS (1989), "Characterization of five genes in the upper-pathway operon of TOL plasmid pWW0 from *Pseudomonas putida* and identification of the gene products", *J. Bacteriol.* 171, pp. 5 048-5 055.

KOK, M., R. OLDENHUIS, MP.G. VAN DER LINDEN, P. RAATJES, J. KINGMA, P.H. VAN LELYVELD, and B. WITHOLT (1989*a*), "The *Pseudomonas oleovorans* alkane hydroxylase gene: Sequence and expression", *J. Biol. Chem.* 264, pp. 5 435-5 441.

KOK, M., R. OLDENHUIS, M.P.G. VAN DER LINDEN, C.H.C. MEULENBERG, J. KINGMA, and B. WITHOLT (1989*b*), "The *Pseudomonas oleovorans* alkBAC operon encodes two structurally related rubredoxins and an aldehyde dehydrogenase", *J. Biol. Chem.* 264, pp. 5 442-5 451.

PRINCE, R.C. (1994), "Monitoring the efficacy of shoreline bioremediation after the Exxon-Valdez oil spill", in *Bioremediation: The Tokyo '94 Workshop*. pp. 215-221, OECD Documents, OECD, Paris.

SHINGLER, V., U. MARKLUND, and J. POWLOWSKI (1992), "Nucleotide sequence and functional analysis of the phenol/ 3,4-dimethylphenol catabolic operon of *Pseudomonas sp.* CF600", *J. Bacteriol.* 174, pp. 711-724.

SPOONER, R.A., K. LINDSAY, and F.C.H. FRANKLIN (1986), "Genetic, functional and sequence analysis of the xylR and xylS regulatory genes of the TOL plasmid pWW0", *J. Gen. Microbiol.* 132, pp. 1 347-1 358.

VAN BEILEN, J.B., G. EGGINK, H. ENEQUIST, R. BOS, and B. WITHOLT (1992), "DNA sequence determination and functional characterization of the OCT-plasmid-encoded alkJKL genes of *Pseudomonas oleovorans*", *Mol. Microbiol.* 6, pp. 3 121-3 136.

VAN DER MEER, J.R., R.I. EGGEN, A.J. ZEHNDER, and W.M. DE VOS (1991), "Sequence analysis of the *Pseudomonas sp.* strain P51 tcb gene cluster, which encodes metabolism of chlorinated catechols: Evidence for specialization of catechol 1,2-deoxygenase for chlorinated substrates", *J. Bacteriol.* 173, pp. 2 425-2 434.

WILLIAMS, P.A. and K. MURRAY (1974), "Metabolism of benzoate and the methylbenzoates by *Pseudomonas putida* (*arvilla*) mt-2: evidence for existence of a TOL plasmid", *J. Bacteriol.* 120, pp. 416-423.

PHYTOREMEDIATION OF HEAVY METAL POLLUTION: IONIC AND METHYL MERCURY

by

Richard B. Meagher* and Clayton L. Rugh**
*Department of Genetics, **Daniel B. Warnell School of Forest Resources, University of Georgia
Athens, Georgia, United States

Introduction and background

Heavy metal contamination has reached toxic levels in the air, land, and water of many parts of the world (Nriagu, 1988), and clean-up has become an urgent problem. Over the last three decades (Nriagu and Pacyna, 1988), the contaminated air, water, and sludge produced by industries and population centres (Adriano, 1986; Alloway, 1990) has steadily increased. High on the list of heavy metal pollutants from man-made sources are lead, mercury, cadmium, copper, and arsenate. Heavy metal ion pollution problems are particularly difficult to solve. Whereas toxic hydrocarbons and chlorinated aromatic hydrocarbon compounds, for example, can be degraded into harmless levels of CO_2 and Cl_2 gases, metals are immutable (Cunningham *et al.*, 1995) by any technology short of nuclear fission. Toxic metals persist in the water, soil, sediment, and, in a few cases, transiently in the atmosphere even when they are biologically converted to less toxic forms. Current electrolytic, chemical-leaching, vitrification, and *in situ* immobilisation technologies for cleaning heavy metal contaminated sites are all extremely expensive and questionably effective, particularly given the size of some sites (U.S. Army Toxic and Hazardous Materials Agency, 1987). The common practice of burying heavy metal contaminated land just defers the problem and may result in further leaching into groundwater.

Phytoremediation, the use of plants to extract, sequester, and/or detoxify heavy metals and other pollutants, may offer a cost-effective and ecologically sound alternative (Raskin, 1996). Phytoremediation appears to be a relatively less expensive, less invasive, and potentially more effective means of addressing existing heavy metal contamination than those currently practised. Phytoremediation also should be more efficient for the emergency treatment of newly contaminated sites and for the treatment of the metal ion contaminated effluent from small manufacturing plants. The long-term goal of our research is to manipulate single-gene traits in plants to enhance their phytoremediation ability. We have modelled our molecular genetic approach to finding solutions for ecosystem-wide problems on the pharmaceutical industry. This industry searches for treatments of complex human diseases with pleiotropic symptoms by investigating key genes and proteins. This model suggests that successful solutions can be rapidly obtained by focusing phytoremediation research on genes directing the degradation or processing of one environmental pollutant at a time. Target host plants include wild grasses, shrubs, and trees that control the energy in most ecosystems and can have the greatest impact. This article and our research focus on the use of phytoremediation for the detoxification and removal of heavy metals, particularly mercury (Rugh *et al.*, 1996).

The man-made sources of mercury pollution

Natural and man-made mercury contamination occurs throughout the United States, but it is a particular problem in the Southeast (Alberts *et al.*, 1990; Becker *et al.*, 1993; Leigh, 1994). Mercury is a common pollutant at government production sites, where it is used in energy- and defence-related activities (e.g. as a coolant in reactors, as ballast in submarines) (Adriano, 1986). At one government site in the United States, a single crack in a storage tank dumped over a million pounds of mercury in one night. Industrial and agricultural sources account for thousands of square miles of land, rivers, lakes, and estuaries contaminated with millions of pounds of mercury in the United States alone. The largest industrially contaminated sites have resulted from its use in the bleaching industry (chlorine and NaOH production, paper, textiles). A single broken electrode at a $NaOH/Cl_2$ plant in the United States nearly 30 years ago resulted in 100 000 pounds of mercury spreading over 50 square miles of wetlands (Holmes, 1977). Mercury's use as an agricultural pesticide directed against bacteria and fungi is also a major source of mercury pollution. The production of mercury-based fungicides was responsible for the notorious Minamata Bay incident off Kyushu Island, Japan. Other man-made sources of mercury pollution include mining operations, urban waste disposal, the use of mercury as a catalyst, as a pigment in paints, and as a seed and bulb dressing.

Under today's stricter guidelines for industry, relatively few enormous accidents occur with mercury in the United States or most other developed countries, and most of the Superfund sites are 20 to 30 years old. However, the available literature indicates that none of the large mercury-contaminated sites in the United States or anywhere in the world have been reclaimed.

Sources and effects of methyl mercury

Although much of the ionic mercury at polluted sites is bound in stable thio-organics in the soil, organic forms of mercury, particularly methyl mercury, pose a major threat to wildlife and humans. The prevailing theory, supported by mounting evidence, is that most methyl mercury is produced catalytically (Figure 1, equation 1) by the sulfate-reducing bacteria living at the aerobic/anaerobic interface, particularly in aquatic or marine environments (Compeau and Bartha, 1985; Gilmour *et al.*, 1992). Methyl mercury is the primary mercury compound that is concentrated as mercury is biomagnified up the food chain from sulfate bacteria in sediments to bacteria living on detritus at the sediment-water interface to bottom feeders like crabs and then on to fish, birds, and mammals (Gardner *et al.*, 1978). Methyl mercury is extremely toxic (Clarkson, 1994) and has had a tragic impact on humans and other animals (D'Itri and D'Itri, 1978). The first to show symptoms in Minamata Bay incident were birds and cats, followed by humans. The mercury found in all three species can be traced directly to the consumption of methyl mercury-contaminated fish from the Bay. The characterised neurological diseases and proposed immunological diseases produced by methyl mercury in animals and humans are so far untreatable. Although there has been extensive monitoring of methyl mercury at some sites, little basic biological research has been directed at its remediation.

Existing mechanisms of heavy metal resistance and sequestration in plants and other eukaryotes

Some parts of the globe, such as the western United States (Cannon, 1960) and Africa (Brooks and Malaisse, 1985), are naturally contaminated with high levels of a variety of toxic metals including arsenic, cadmium, copper, cobalt, lead, mercury, selenium, and/or zinc. These regions are often characterised by scrubby, heavy metal-tolerant flora (Wild, 1978; Brooks and Malaisse, 1985). Many of these hyperaccumulate large amounts of heavy metals. These plants have been found in a wide variety of habitats, and they accumulate malate and/or citrate chelates of particular metals to levels equal to 1-8 per cent of the dry weight of their tissues (Baker, 1989; Baker *et al.*, 1992; Baker *et al.*, 1994*a*; Baker *et al.*, 1994*b*). These hyperaccumulators do offer exciting potential solutions to the remediation of metal-ion contaminated sites. However, many of these plants have bizarre metal ion requirements, most grow poorly even in less contaminated habitats than those where they are found, and few are, as yet, of any agronomic utility as crop, forest, or conservation species. Furthermore, the genetic or biochemical complexities of these systems are essentially unknown (Ernst *et al.*, 1990; MacNair, 1993). Many more years of research may be required before plant breeding or genetic engineering can adapt hyperaccumulating systems to specific contaminated sites and more productive plant species.

Phytoextraction and rhizofiltration are two new technologies that enlist the natural nutrient up-take systems of crop and forest species to remove heavy metals from soil and water. The simple addition of chelating agents, which solubilize metals away from humic acids in the soil, has resulted in significant gains in the phytoextraction of lead into grasses (Cunningham *et al.*, 1995; Cunningham *et al.*, 1996). Radionucleids can be extracted directly from heavily contaminated reactor water into root systems of sunflowers growing on floating mats (Raskin *et al.*, 1994; Raskin, 1996). Rhizofiltration systems make use of the tremendous surface area and exchange capacity of plant roots (see below) and are a blend of previous engineering technologies and phytoremediation. The full potential of these technologies is not yet clear, but they will undoubtedly contribute to our ability to clean the environment.

Only a few eukaryotic systems of metal ion resistance and/or sequestration have been characterised at the molecular level in enough detail to be genetically manipulated through only a few genes. Among these, the best characterised mechanism of heavy metal ion resistance in plants is based on the phytochelatins (PCs), a group of cysteine-rich peptides that are the product of a complex biosynthetic pathway (Scheller *et al.*, 1987). Heavy metals are sequestered in this system when they are bound by the thiol (-SH) groups of these cysteine residues in an insoluble complex that gets dumped into plant vacuoles. Plants deficient in PC synthesis are sensitive to heavy metals like cadmium (Howden *et al.*, 1995). The metallothioneins (MT) are genetically encoded, cysteine-rich polypeptides usually of about 60 amino acids that sequester metals by a mechanism similar to PCs, and they are often considered a class of PCs. Transformed plant tissues expressing animal MT have a higher binding capacity for cadmium and increased tolerance to transient exposure to cadmium ions over control plants, but are not truly resistant to cadmium ions (Lefebvre *et al.*, 1987; Misra and Gedamu, 1989). A gene family encoding MT homologues has been identified in *Arabidopsis* and other plant species. Some of them can confer heavy metal resistance (Cu^{++}) when expressed in yeast (Zhou and Goldsbrough, 1994), and there are increased levels of these MTs in plant cells selected for Cd^{++} tolerance (Chen and Goldsbrough, 1994). These systems will undoubtedly be of use to phytoremediation strategies.

Using prokaryotic (bacterial) genes in plants

Our research has successfully focused on the introduction of the bacterial mercury-processing genes into the model plant, *Arabidopsis*. If there are existing plants systems for dealing with heavy metals, why use bacterial genes? Multicellular eukaryotes typically sequester bound heavy metals in organelles (e.g. vacuoles) and in different organs (e.g. roots, vascular tissues, leaf hairs). Eukaryotic cells' electron transport systems for energy conversion (e.g. chloroplasts, mitochondria) are protected by a layer of cytoplasm. In contrast, bacteria (prokaryotes) live with their outer membrane and its associated electron transport systems in direct contact with the environment. A toxic metal that poisons any of these electron transport systems kills the bacterial cell. Thus, to survive, bacteria have evolved more assertive solutions to the presence of many toxins, including heavy metals. Bacteria are capable of reducing toxic heavy metal cations (e.g., Hg^{++}, Ag^{+}, Fe^{+++}, Cu^{++}) and oxyanions (e.g. AsO_4^{-3}, $Cr_2O_7^{-2}$, TeO_3^{-2}) to less toxic or more easily disposed of metal species (Moore and Kaplan, 1992; Moore and Kaplan, 1994; Ji and Silver, 1995). Providing this capability to plants could vastly increase their effectiveness as remediation tools.

Only a few bacterial genes with activities directed toward toxic heavy metals have been characterised at the molecular level required for their genetic manipulation. The best characterised bacterial system, that for mercury resistance, is found in gram-negative bacteria and encoded by a set of genes called the *mer* operon (Summers, 1986). Bacteria with the *mer* operon live competitively in naturally contaminated and polluted sites containing ionic mercury (Barkay *et al.*, 1992; Nakamura and Silver, 1994). All sets of *mer* genes encode a mercury responsive regulatory protein (merR), a protein to transport mercury through the cell membrane and into the cell (merT), a protein to sequester mercury in the periplasmic space (merP), and a mercuric ion reductase to reduce ionic mercury to less toxic metallic mercury (merA) (see Figure 1, reaction 4). Together these genes provide what is called "narrow spectrum resistance" to ionic mercury. A subset of the mercury resistant bacteria that have been characterised contains an organo-mercurial lyase gene (*merB*) (see Figure 1, reaction 2). All these genes together provide "broad spectrum resistance" to a variety of organomercurial compounds.

Figure 1. **Reactions involving ionic and methyl mercury**

spontaneous and uncharacterised enzymes
1. ionic mercury ---> methyl mercury (sulfate reducing bacteria in soil and sediment)

MerB
2. methyl mercury ---> mercury ion (broad spectrum mercury resistant bacteria)

spontaneous
3. mercury ion + sulfur compounds ---> thiol bound ionic mercury
(in soil or any organism)

MerA
4. thiol bound ionic mercury + NADPH (reduced) ---> metallic mercury + $NADP^+$ (oxidised)
(all mercury resistant bacteria)

Source: Summers, 1986.

Methyl mercury is formed spontaneously in sulfate-reducing bacteria living in the aerobic/anaerobic transition zone (equation 1). Methyl mercury is detoxified by two bacterially encoded enzymes. In a coupled reaction, organomercury lyase (MerB) catalyses the protonolysis of the carbon-mercury bond in methyl mercury (equation 2) to release free mercury ions (Hg^{++}). Ionic mercury (Hg^{++}) reacts rapidly with the sulfur groups on any available thiol-organic compound, forming very stable thiol bound ionic mercury (equation 3). Mercuric ion reductase (MerA, equation 4) removes mercury from these stable thiol-organic compounds by electrochemically reducing it to a much less toxic form, metallic mercury, Hg°. This reduction reaction is coupled to the oxidation of one molecule of the redox currency of all cells, NADPH.

Advantages of using phytoremediation to clean the environment

In our research, we have extended the metal ion reduction portion of the bacterial system for mercury detoxification to transgenic plants. The following section outlines the advantages of using higher plants as phytoremediation agents and addresses seven of the most commonly expressed concerns about their use (Table 1).

1) The release of genetically engineered higher plants into the environment can be carefully controlled (Stomp *et al.*, 1993).

2) A number of plant species in diverse plant families already hyperaccumulate heavy metals, suggesting great flexibility in adapting to metals in their environment.

3) Plants can break down many organic pollutants into harmless constituents like CO_2, NH_4^+, and Cl_2. This is important because many sites are contaminated with organic as well as heavy metal pollutants.

4) Plant root systems cover extensive areas; typical estimates are as high as 100×10^6 miles of roots per acre, resulting in a vast amount of surface area.

5) Plants have the genetic capacity (using hundreds if not thousands of genes) to extract at least 16 nutrients from the soil and groundwater. This capacity can be chemically and genetically manipulated to extract environmental pollutants.

6) Plants are photosynthetic, trapping most of the new energy that enters an ecosystem, and they govern as much as 80 per cent of the energy at any time in most ecosystems. Most other organisms can tap only a small part of this energy somewhere up the food chain. Thus, plants have the greatest amount of energy to devote to detoxification and/or sequestration of pollutants.

   - Through photosystem I (a system not found in photosynthetic bacteria), plants use light energy to generate vast amounts of reducing power that can be used to efficiently reduce metal ions (e.g. redox currency--NADPH) (Rugh *et al.*, 1996).

   - Plants photosynthetically fix CO_2 and reduce it to make their own carbon/energy source. This reduced carbon energy is transported to roots to support their growth and some is converted back to redox currency, NADPH, which can be used to reduce toxic metal ions.

– Many plants produce large amounts of biomass that can be harvested annually to remove metal ions from the soil, and their root systems enrich the soil.

7) It is estimated that phytoremediation will cost between *two and four orders of magnitude less* than existing remediation technologies. The high cost of current remediation efforts is preventing most existing sites from being cleaned up.

8) Plants are aesthetically pleasing.

Table 1. **Seven myths about using genetically engineered plants for phytoremediation**

	Myths	Reality
Myth #1	Mutant engineered plants will spread uncontrollably, destroying the environment.	Genetically engineered plants are already in use to lower pesticide and herbicide use and will soon produce edible vaccines against cholera and other serious childhood diseases (Mason and Arntzen, 1995). There are no longer significant concerns in most US government regulatory agencies about the release of most genetically engineered plants. Our results indicate that some plants engineered to grow on mercury cannot, in fact, grow without it. Thus, they are self-limiting in their ability to spread in an ecosystem.
Myth #2	Genetic engineering will lower genetic variability.	It is true that standard agricultural plant breeding techniques have had genetic uniformity and yield as their goal for hundreds of years. However, this is now viewed as problematic and it is not related to genetic engineering. We can engineer numerous genetically diverse individuals within a species. Most forest and tree species, for example, are already propagated by methods that insure maximum genetic variability because they are not viable when inbred.
Myth #3	Engineered plants will strip the soil and ecosystem of vital nutrients.	On the contrary, most plants enrich the soil and increase both its fertility and tolerance (binding capacity) for contaminants. Engineered plants are no different.
Myth #4	Engineered plants are full of unnatural chemicals and hazardous to humans.	Engineered plants do contain one or two new proteins, but they are neither toxic nor unnatural. In fact, 99.9 per cent of the proteins an engineered plant expresses remain unchanged.
Myth #5	Plants alone will not achieve the level of decontamination required by the EPA.	Plants offer a low-cost, low-impact solution that should be the first line of attack at a polluted site, even when it is not known if they can achieve complete remediation. Plants are experts at mining the soil and current research is very encouraging. Relatively harmless soil amendments or pH adjustment can drastically alter the bioavailability of metal ions. More research on the mechanisms by which plants make metals available will lead to even greater improvements.
Myth #6	The problem has not really been solved because the plants now are contaminated with metals.	True metals are immutable (they can not be broken down), but once the metals are efficiently concentrated up from the soil and water, the plants can be harvested and the metals safely processed out.
Myth #7	Plants engineered to harvest heavy metals will pass them on to consumers and into the food chain.	Phytoremediation strategies would not utilize food-source plants. Insects generally will not feed on plants with high levels of toxic metals and thus may pose no danger to the food chain. Clearly, food-source animals, such as cattle or pigs, should be kept away from these plants.

Source: Author.

Recent work demonstrating the utility of phytoremediating mercury pollution

The majority of the *merA* data, which are briefly summarised in this section, are described more fully in Rugh *et al.* (1996).

Expressing the bacterial mercuric ion reductase gene in plants

The natural bacterial *merA* gene contains DNA sequences that are particularly unfavourable to efficient expression in plants. The first constructs we made for plant expression were completely inactive in numerous transgenic petunia plants (Thompson, 1990). We constructed a highly mutagenized *merA* sequence, *merA9,* in which nine per cent of the DNA sequence is optimised for plant expression (Stack, 1991). We put this synthetic *merA9* sequence under control of plant regulatory elements (Rugh *et al.*, 1996) in a model plant, *Arabidopsis thaliana.*

Transgenic Arabidopsis seeds expressing *merA9* germinated and these seedlings grew, flowered, and set seed on media containing 5-20 ppm available ionic mercury ($HgCl_2$), levels toxic to most plant species (10 ppm, Figure 2). Root growth of the transgenic plants was normal and vigorous on high levels of mercury (Rugh *et al.*, 1996). The parent plant line and transgenic controls died rapidly when grown with these concentrations of mercury. The mercury resistance trait follows Mendellian inheritance and has been stable for several generations. When seedlings were submerged in liquid with sublethal concentrations of mercury and monitored on a mercury vapour analyser, transgenic *merA9* seedlings evolved considerable amounts of metallic mercury vapour (Hg^o) relative to control plants (Figures 3A and B). The Hg^o evolved at a rate of 10 ppm per min by wet weight of plant tissue. The rate of mercury volatilisation and the level of resistance were directly proportional to the steady-state messenger RNA level (Figure 3C), confirming that resistance was due to expression of the *merA9* gene (see Rugh *et al.*, 1996). Messenger RNA is the information molecule between gene and protein that encodes the merA enzyme. Surprisingly, the three independent plant lines expressing the highest levels of *merA9* mRNA and evolving Hg^o most rapidly (Figure 3C, arrows) actually required ionic mercury in the media for optimal growth and did poorly on media with no mercury. This suggests that plants expressing this transgene will have a natural biological containment within mercury-contaminated sites.

Engineering plants that degrade methyl mercury

In our current research we are engineering the *merB* gene encoding methyl mercury lyase (Figure 1, reaction 2) for plant gene expression. We hope to demonstrate that transgenic plants expressing both *merA* and *merB* are resistant to methyl mercury and other methyl mercury derivatives. These plants should process methyl mercury to ionic mercury (Hg^{++}) and the salt to Hg^o (Figure 1, reactions 2-4). By adjusting the levels of these and other genes, we plan to adjust and balance detoxification, volatilisation, and sequestration of mercury pollutants.

Figure 2. **Transgenic Arabidopsis containing the *merA9* construct were resistant to mercury ions**

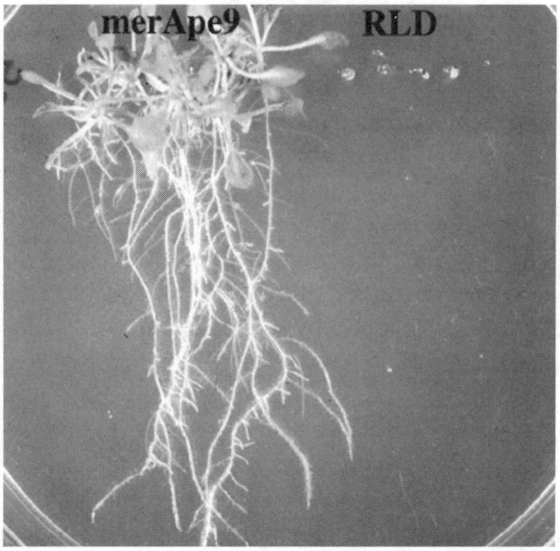

Source: Rugh *et al.*, 1996.

Seeds from Arabidopsis transgenic plant line 1 expressing the *merA9* gene (left hand side of plate) germinated and grew on plant growth medium containing 50 uM (10 ppm) available mercury ion (Hg^{++}). Transgenic controls expressing unrelated genes (not shown) and nontransgenic RLD control seeds (right hand sector of plate) seldom germinated, and if so, subsequently died on the medium with 50 uM Hg^{++}. The plate was incubated vertically at 22 °C for three weeks.

Figure 3. **Transgenic plants with the *merA9* construct volatilised Hg^{O} and the activity is proportional to *merA* mRNA levels**

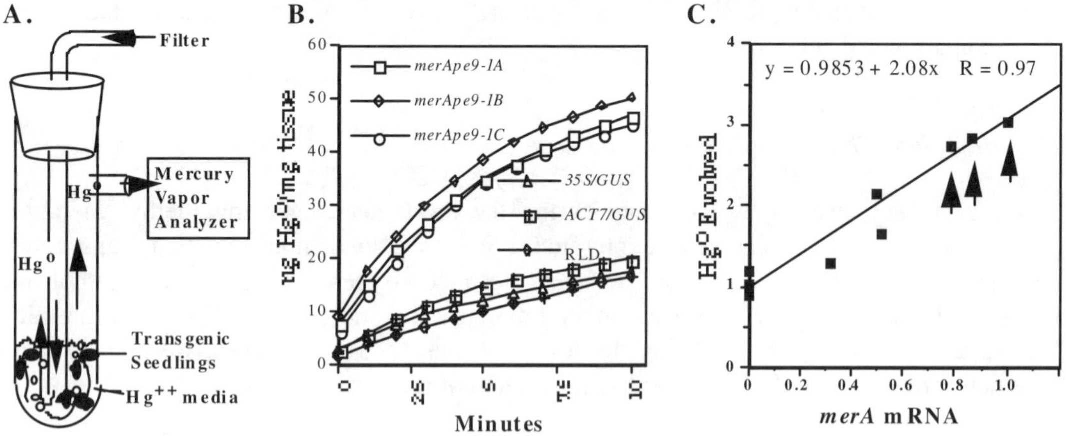

Source: Rugh *et al.*, 1996.

In (A), seedlings were placed in a small sidearm test tube in assay media. Hg^{++} was added to start the reaction, air was bubbled through the tube, and the level of volatilized Hg^{o} was measured in a mercury vapour analyser. In (B), transgenic seedlings expressing the *merA9* gene (plant lines 1A, 1B, 1C) catalysed significant reduction of ionic (Hg^{++}) to Hg^{o} relative to the background of chemical reduction seen in control RLD plants, or transgenic plants expression other genes *(35S/GUS* and *ACT7/GUS)*. For each of the three *merA9* expressing transgenic samples shown above, ~10 mg of seedlings evolved ~500 ng of Hg^{o} during the 10 min assay period. In (C), the levels of *merA9* RNA quantified on a phosphoimager and normalized to 18S rRNA levels are plotted against the relative levels of Hg^{o} vapour evolved by the individual transgenic lines. The highest level of mRNA in line 6E was set to one to obtain the relative levels shown in the figure. The three lines indicated with arrows require mercury for optimum growth.

Developing this technology for cleaning soil and water on a large scale

The next step in our research will be to move from the model system to large native plants that would be useful in environmental clean up. We will be placing the *merA* and *merB* genes into trees, shrubs, and grasses and testing these same variables. Initial work demonstrates that shrub and tree species expressing *merA* are mercury resistant. We hope to move these genes into aquatic and marine grasses in the next year and demonstrate efficient processing of mercury in these environments.

A number of basic questions remain unanswered from our initial research in a model plant, for example: How efficient is the system at removing mercury from soil and groundwater? Can we block the flow of methyl mercury from sediment into the environment and food chain, and how efficiently?

Vaporisation vs. sequestration of mercury in engineered plants

The dangers associated with methyl mercury are so high that transgenic plants expressing *merB* that even moderately reduce the biomagnification of methyl mercury would be considered advantageous. However, plants actively processing methyl and ionic mercury may eventually release some of it into the air as non-toxic Hg^{o}, which can itself be converted to toxic forms, when levels are high enough. Transgenic model plants expressing *merA* and suspended in liquid media do, in fact, release large amounts of Hg^{o} when suspended in a solution of Hg^{++} in the laboratory (see Recent work demonstrating the utility of phytoremediating mercury pollution above), on the order of 10 ppm by wet weight of plant tissue per min. An important issue is how this test tube technology will play out in large plants growing on contaminated soil and water. After ionic mercury has been reduced to Hg^{o} in plant roots, the Hg^{o} should be transported up through the vascular system and transpired through the leaves (i.e. this is what plants do with most inert or waste gases). Although we expect Hg evolution in nature to be orders of magnitude less efficient than in a test tube, it still should be possible to greatly accelerate the release of mercury from these sites. Therefore, an important question is, What rate of Hg^{o} release into the environment will be acceptable? Current state-of-the art technology (i.e. steam cleaning soil) also releases mercury as vapour into the environment. The safety of this and the planned release of mercury vapour from plants needs to be carefully considered.

At present, natural bacterial reduction is the presumed major source of Hg^{O} at contaminated sites (Barkay *et al.*, 1992). The levels of atmospheric Hg^{o}(0.004 ug/m^{3}) (Lindberg *et al.*, 1995) at one heavily contaminated site in Oak Ridge, TN are about *four orders of magnitude below federal regulated levels* (50 ug/m^{3}), whereas the mercury levels in soil, water, and wildlife at this and most other sites exceed acceptable levels. The Oak Ridge site, which contains 80 000 kg of Hg in a 250 ha area, has a maximum efflux rate of 10 kg/y for the whole site. At this rate, even if all the forms of mercury became available to endogenous reduction mechanisms, it will take 8 000 years for the site to lose most of its mercury. If *merA* expressing transgenic plants increased the rate of Hg release

200-fold, the level of Hg^{o} in the air would still be 50-fold below most government guidelines and the site would take only 40 years to clean up. For small sites and small waste streams from manufacturing operations, volatilisation would undoubtedly be found acceptable and well within current standards.

When large plants are grown on mercury contaminated soil, the soluble mercury vapour produced will spend some time in an extensive vascular system. Although Hg^{o} is soluble and non-toxic, it is not biochemically inert. Endogenous peroxidases and catalases in all organisms can convert ingested or absorbed metallic mercury back to toxic ionic mercury at some low rate (Ogata and Aikoh, 1984). Plants are known to have high levels of peroxidases and catalases in many tissues. It is likely that the enzymatic oxidation of metallic to ionic mercury makes very high levels of metallic mercury toxic to plants (Zimmerman and Crocker, 1934; Hitchcock and Zimmerman, 1957; Waldron and Terry, 1975; Goren and Siegel, 1976). We propose that as large plants transport soluble Hg^{o} gas through their vascular system, much of the Hg^{o} will be reoxidized to ionic mercury (Hg^{++}). The ionic mercury is very reactive and thus should be sequestered as thio-organics in above-ground plant tissues, which can be harvested annually from a site. This would obviate our concerns about volatilisation of Hg^{o}. Current research examines the balance between movement, sequestration, and volatilisation of mercury in *merA* transgenic plants.

A call to action

Less expensive, more effective technologies are urgently needed to clean up heavy-metal polluted sites. Government agencies and industry have repeatedly postponed the clean-up of larger pollution sites because the size of the sites and the cost involved simply render current technologies unworkable (Cunningham *et al.*, 1995). Compared to accepted practices such as literally digging up tons of polluted earth and dumping it elsewhere, phytoremediation appears to offer a much less invasive and much more effective means of isolating and sequestering heavy metals. Plants are very well-suited for remediation of both soil and water pollution problems, and research is showing that most concerns about the release of transgenic plants are invalid. Over the last decade, the pharmaceutical industry has demonstrated the power of modern molecular genetics and technology to solve human health problems. We can -- and should -- apply this power to cleaning the environment. Researchers in molecular genetics can focus their efforts on one class of pollutants and probably one gene at a time. Such directed research will have a tremendous impact on efforts to clean up the environment and design ecologically sound phytoremediation systems for processing industrial wastes.

What can public policy makers do to help?

A combination of government regulations, funding policies, and lack of public awareness about modern science may slow the application of this technology. Below are a few specific examples of problems and action that would help:

1) Jump-start remediation by planting plants. Make phytoremediation the first implementation, even before full assessment is completed since this often takes years. Even engineered plants will be relatively inexpensive, can only improve the environment, and are physically attractive.

2) Set realistic levels of "acceptable" contamination. Some EPA standards for the cleanup of effluent water are more stringent than those for local tap water. Unable to meet the standards

required for EPA approval of their efforts, companies are reluctant to initiate clean-up procedures.

3) A "clean soils" act is needed in the United States. Regulations that only protect air and water encourage the dumping of pollutants on soil. There are few government mandated clean soil standards, and yet polluted soils are one of the major sources of polluted water. At industrial sites, removed from wetlands and population centres, set levels for soil pollutants based on their bio-availability in the soil they contaminate. Factor in the natural attenuation process. Some contaminants are so tightly bound that they are not a danger to the environment, but they cannot be extracted by any but the most costly methods. No company will invest in the remediation of soil pollutants unless the expense will result in a site that meets regulatory standards.

4) Fund basic research to advance phytoremediation (e.g. university grants, competitive federal grants, research centres). The US pharmaceutical industry has a long history of NIH-subsidized basic research and the result has been a strong foundation of high-quality scientific literature. This is not the case for research on the detoxification and sequestration of heavy metal or many organic pollutants, nor is it the case for phytoremediation technologies. Most of the successful genetic engineering projects in the pharmaceutical industry are backed by 20-50 years of basic research performed in academic institutions. Government needs to be encouraged to fund the basic research behind solutions to ecological problems. Currently, funding tends to be aimed at the application of existing engineering technologies or ecosystem biology, but little is focused on molecular mechanisms and technology that might lead to solutions to ecosystem problems.

5) Fund and reward new business ventures to encourage development of the phytoremediation market [e.g. Small Business Innovative Research Grants (SBIRs), positive public support]. Just a small number of moderately funded applied research efforts have already produced significant breakthroughs in the phytoremediation of metal and organic pollutants from soil and water. For the moment, large gains are to be had with only moderate investment, particularly if you contrast phytoremediation with the cost of most engineered solutions.

6) Speed up the approval process for new technologies.

7) Provide a forum for testing new technologies (e.g. make demonstration and test sites available at brownfields).

8) Focus public attention on the potential of genetically engineered plants in medicine (i.e. oral vaccines) and in environmental remediation efforts (Mason and Arntzen, 1995; Rugh *et al.*, 1996).

9) Encourage engineers and agricultural engineers to work together with molecular biologists and ecologists to handle polluted properties and water systems, particularly those associated with the large Superfund sites. Currently, engineers have the attention of regulators and governments, but it is quite possible that molecular geneticists working with plants will have many of the most efficacious solutions.

What can scientists do to speed progress?

1) Keep public policy makers informed of progress; help them understand the value of good basic science to the future of phytoremediation technology.

2) Encourage communication among scientists from different disciplines: remediation microbiologists, plant scientists, molecular geneticists, and ecologists. Small annual meetings organised across disciplines and focused on planning out research and designing solutions to particular problems such as lead or mercury pollution or the clean up of radionucleids.

3) Encourage basic scientists to work side by side with specialists who develop and implement technologies in the field, such as environmental and groundwater engineers, soil chemists, and agricultural engineers. They need to work together to handle both point-source and Superfund sites. We could start at universities by having joint education and training programs for undergraduate and graduate students and postdoctoral fellows.

4) Develop transgenic plant technology for dominant and indigenous higher plants that are important to the environment, such as trees, shrubs, and native grasses.

5) Intensify basic research on the bacterial genes and their encoded proteins that degrade or detoxify environmental pollutants.

6) Pursue basic research on plant genes and proteins responsible for mobilising elemental nutrients in soil and water, bringing them in, and transporting them throughout the plant.

7) Again: Keep public policy makers informed of progress and educated about the technology!

Summary

Plants have a number of advantages in the remediation of heavy metal contamination of soil and water, not the least of which are the control of most of the energy in an ecosystem, excess reducing power generated by solar energy, an extensive root system already designed to extract minerals from the soil, the production of a large biomass of tissues that can be harvested annually to remove metal contaminants, and low cost. These same features and the natural beauty of plants will help ensure strong public support for phytoremediation solutions. Our initial efforts to use a metal ion reduction strategy to remove mercury from contaminated growth media and water have been very successful. In the laboratory, Arabidopsis seedlings expressing a reconstructed bacterial mercuric ion reductase gene detoxify 10 ppm of mercury per minute by weight of plant tissue. Our present research is focused on a scale-up of this technology and blocking methyl mercury from entering the environment. Many other bacterial genes exist that can be used to detoxify and sequester heavy metals. Molecular genetics has the ability to focus research on a manageable number of variables. This approach can probably solve many of our most difficult pollution problems if action is taken soon enough. Combined with present engineering and phytoextraction technologies, molecular genetics has few limits on its potential contributions to cleaning the environment.

Acknowledgements

We want to thank Anne Summers, Debra Devedjian, George Boyajian, and Gay Gragson for their input and helpful suggestions for the manuscript. This work has been supported by recent grants from the U.S. National Science Foundation and the Department of Energy and past grants from the University of Georgia Research Foundation's Biotechnology Development Program.

Questions, comments and answers

Q: How do you anticipate processing the plants after stripping the soil of mercury?

A: By fermenting the mercury back out or using a heat process and trapping it in a cold trap. We've really done very little at this stage.

Q: You showed the conditions only for mercury. What about other metals?

A: We've just started on a project to detoxify copper, cadmium, lead and arsenic. We are hoping to precipitate the metals in the cell.

Q: Would these other metals interfere with the mercury?

A: We have no evidence of that unless the levels are toxic.

Q: What about chromate? Would it be preferentially reduced?

A: This has no activity towards chromate. A couple of chromate reductases are reported in the literature, but there is insufficient information to encourage us to work on the genes.

Q: What about plant species, soil pH, etc.? Are these important? Are the sites important?

A: We are seeing very little effect from media or soil. In soil, there is very little mercury available, so we do not see a huge difference in toxicity between the controls. We would like to have some grasses such as *Spartina,* but we will need new promoter genes because this is a monocot.

Q: Have you examined where the mercury is? Is it mainly in the stems or the root zone?

A: There is good literature on this, but these are first experiments. This is what we want the hydroponics for. Most of the literature suggests that plants trap mercury in the roots. We think that when our plants reduce it to the zero form it will be a soluble gas and move up the transpiration stream. The longer the transpiration stream is, the more chance there is that the mercury will be re-oxidised and fixed in tissue. My prediction is that most of the mercury will remain in the plant.

Q: What about US legislation on genetically modified plants in the field?

A: We picked these trees because they are self-sterile. We are trying to focus on plants that do not spread pollen. I do not think there will be much resistance, as public opinion is far more positive than it was a few years ago.

C: There have been more than 1 000 releases -- experimental plantings and release of micro-organisms -- and there are engineered products on the market.

REFERENCES

ADRIANO, D.C. (1986), "Trace elements in the terrestrial environment", in *Trace elements in the terrestrial environment*, Springer Verlag, New York.

ALBERTS, J.J., M.T. PRICE, and M. KANIA (1990), "Metal concentrations in tissue of Spartina alterniflora (Loisel.) and sediments of Georgia salt marshes", *Estuarine, Coastal & Shelf Sci.* 30, pp. 47-58.

ALLOWAY, B.J. (1990), "Heavy metals in soils", in *Heavy metals in soils*, John Wiley & Sons, New York.

BAKER, A.J.M. (1989), "Terrestrial higher plants which hyperaccumulate metallic elements-A review of their distribution, ecology and phytochemistry", *Biorecovery* 1, pp. 81-126.

BAKER, A.J.M., S.P. MCGRATH, C.M.D. SIDOLI, and R.D. REEVES (1994*a*), "The possibility of *in situ* heavy metal decontamination of polluted soils using crops of metal accumulating plants", *Resources, Conservation and Recycling* 11, pp. 41-49.

BAKER, A.J.M., J. PROCTOR, M.M.J. VAN BALGOOY, and R.D. REEVES (1992), "Hyperaccumulation of nickel by the ultramafic flora of Palawan, Republic of the Philippines", in J. Proctor, R.D. Reeves, and A.J.M. Baker (eds.), *The Ecology of Ultramafic (Serpentine) Soils*, pp. 291-304, Intercept Ltd., Andover, Hampshire, United Kingdom.

BAKER, A.J.M., R.D. REEVES, and A.S.H. HAJAR (1994*b*), "Heavy metal accumulation and tolerance in British populations of the metallophyt *Thlaspi caerulescens* J. & C. Presl. (*Brassicaceae*)", *New Phytol.* 127, pp. 61-68.

BARKAY, T., R. TURNER, E. SAOUTER, and J. HORN (1992), "Mercury biotransformations and their potential for remediation of mercury contamination", *Biodegradation* 3, pp. 147-159.

BECKER, D.S., G.N. BIGHAM, and M.H. MURPHY (1993), "Distribution of mercury in a lake food web", *Soc. Environ. Toxic. Chem. 14 Annual Meeting*, pp. 1-6.

BROOKS, R.R. and F. MALAISSE (1985), "The Heavy metal-tolerant flora of south central Africa", in *The Heavy metal-tolerant flora of south central Africa*, A.A. Balkema Press, Boston.

CANNON, H.L. (1960), "The development of botanical methods of prospecting for uranium on the Colorado Plateau", *U.S. Geol. Surv. Bull.* 1085A, pp. 1-50.

CHEN, J. and P.B. GOLDSBROUGH (1994), "Increased activity of g-glutamylcysteine synthetase in tomato cells selected for cadmium tolerance", *Plant Physiol.* 106, pp. 233-239.

CLARKSON, T.W. (1994), "The toxicology of mercury and its compounds", in C.J. Watras and J.W. Huckabee (eds.), *Mercury Pollution Integration and Syntehesis*, pp. 631-642, Lewis Publishers, Ann Arbor, Michigan.

COMPEAU, G.C. and R. BARTHA (1985), "Sulfate-reducing bacteria: Principal methylators of mercury in anoxic estuarine sediment", *Appl. Environ. Micro.* 50, pp. 498-502.

CUNNINGHAM, S.D., T.A. ANDERSON, P. SCHWAB, and F.C. HSU (1996), "Phytoremediation of soils contaminated with organic pollutants", *Advances in Agronomy* 56, pp. 55-114.

CUNNINGHAM, S.D., W.R. BERTI, and J.W. HUANG (1995), "Phytoremediation of contaminated soils", *Trends in Biotech.* 13, pp. 393-397.

D'ITRI, P.A. and F.M. D'ITRI (1978), "Mercury contamination: a human tragedy", *Environ. Management* 2, pp. 3-16.

ERNST, W.H.O., H. SCHAT, and J.A.C. VERKLEIJ (1990), "Evolutionary biology of metal resistance in Silene vulgaris", *Evol. Trends in Plants* 4, pp. 45-51.

GARDNER, W.S., D.R. KENDALL, R.R. ODOM, H.L. WINDOM, and J.A. STEPHENS (1978), "The distribution of methyl mercury in a contaminated salt marsh ecosystem", *Environ. Pollut.* 15, pp. 243-251.

GILMOUR, C.C., E.A. HENRY, and R. MITCHELL (1992), "Sulfate stimulation of mercury methylation in freshwater sediments", *Environ. Sci. Tech.* 26, pp. 2 281-2 287.

GOREN, R. and S.M. SIEGEL (1976), "Mercury-induced ethylene formatin and abscission in *Citrus* and *Coleus* explants", *Plant Physiol.* 57, pp. 628-631.

HITCHCOCK, A.E. and P.W. ZIMMERMAN (1957), "Toxic effects of vapors of mercury and of compounds of mercury on plants", *Ann. NY Acad. Sci.* 65, pp. 474-497.

HOLMES, C. (1977), "Effects of dredged channels on trace-metal migration in an estuary", *Journal Research US Geol. Survey* 5, pp. 243-251.

HOWDEN, R., P.B. GOLDSBROUGH, C.R. ANDERSEN, and C.S. COBBETT (1995), "Cadmium-sensitive, *cadl* mutants of *Arabidopsis thaliana* are phytochelatin deficient", *Plant Physiol.* 107, pp. 1 059-1 066.

JI, G. and S. SILVER (1995), "Bacterial resistance mechanisms for heavy metals of environmental concern", *J. of Industrial Microbiology* 14, pp. 61-75.

LEFEBVRE, D.D., B.L. MIKI, and J.F. LALIBERTE (1987), "Mammalian metallothiionein functions in plants", *Biotechnology* 5, pp. 1 053-1 056.

LEIGH, D.S. (1994), "Mercury contamination and floodplain sedimentation from fromer gold mines in north Georgia", *Water Res. Bull.* 30, pp. 739-748.

LINDBERG, S.E., K.-H. KIM, T.P. MEYERS, J.G. OWENS (1995), "A micrometeorological gradient approach for quantifying air/surface exchange of mercury vapor: Tests over contaminated soils", *Envir. Sci. Technol.* 29, pp. 126-135.

MACNAIR, M.R. (1993), "The genetics of metal tolerance in vascular plants", *New Phytol.* 124, pp. 541-559.

MASON, H.S. and C.J. ARNTZEN (1995), "Transgenic plans as vaccine production systems", *Trends Biotechnol.* 13, pp. 388-392.

MISRA, S. and L. GEDAMU (1989), "Heavy metal tolerant transgenic *Brassica napus* L. and *Nicotiana tabacum* L. plants", *Theor. Appl. Genet.* 78, pp. 161-168.

MOORE, M.D. and S. KAPLAN (1992), "Identification of intrinsic high-level resistance to rare earth oxides and oxyanions in members of the class *Proteiobacteria*: characterization of tellurite, selenite, and rhodium sesquioxide reduction in *Rhodobacter sphaeroides*", *J. Bacteriol.* 74, pp. 1 505-1 514.

MOORE, M.D. and S. KAPLAN (1994), "Members of the family *Rhodospirillaceae* reduce heavy-metal oxyanions to maintain redox pois during photosynthetic growth", *ASM News* 60, pp. 17-23.

NAKAMURA, K. and S. SILVER (1994), "Molecular analysis of mercury-resistant *Bacillus* isolates from sediment of Minamata Bay, Japan", *Appl. Environ. Micro.* 60, pp. 4 596-4 599.

NRIAGU, J.O. (1988), "A silent epidemic of environmental metal poisoning?", *Environ. Pollut.* 50, pp. 139-161.

NRIAGU, J.O. and J.M. PACYNA (1988), "Quantitative assessment of worldwide contamination of air, water and soils by trace metals", *Nature* 333, pp. 134-139.

OGATA, M. and H. AIKOH (1984), "Mechanism of metallic mercury oxidation in vitro by catalase and peroxidase", *Biochem. Pharm.* 33, pp. 490-493.

RASKIN, I. (1996), "Plant genetic engineering may help with environmental cleanup", *Proc. Natl. Acad. Sci. USA* 93, pp. 3 164-3 166.

RASKIN, I., P. NANDA KUMAR, S. DUSHENKOV, and D.E. SALT (1994), "Bioconcentration of heavy metals by plants", *Current Opin. Biotech.* 5, pp. 285-290.

RUGH, C., D. WILDE, M. WALLACE, N. STACK, A. SUMMERS, and R. MEAGHER (1996), "Mercuric ion reduction and resistance in transgenic *Arabidopsis thaliana* plants expressing a modified bacterial *merA* gene", *Proc. Natl. Acad. Sci. USA* 93, pp. 3 182-3 187.

SCHELLER, H.V., B. HUANG, E. HATCH, and P.B. GOLDSBROUGH (1987), "Phytochelatin synthesis and glutathione levels in response to heavy metals in tomato cells", *Plant Physiol.* 85, pp. 1 031-1 035.

STACK, N. (1991), "The Reconstruction of the Bacterial Gene *merA*", Biology Honors Thesis, University of Georgia, Athens, Georgia.

STOMP, A.-M., K.-H. HAN, S. WILBERT, and M.P. GORDON (1993), "Genetic improvement of tree species for remediation of hazardous wastes", *In Vitro Cell. Dev. Biol.* 29P, pp. 227-232.

SUMMERS, A.O. (1986), "Organization, expression, and evolution of genes for mercury resistance", *Ann. Rev. Microbiol.* 40, pp. 607-634.

THOMPSON, D. (1990), "Transcriptional and Post-Transcriptional Regulation of the Genes Encoding the Small Subunit of Ribulose-1,5-Bisphosphate Carboxylase", PhD Thesis, University of Georgia, Department of Genetics, Athens, Georgia.

U.S. ARMY TOXIC AND HAZARDOUS MATERIALS AGENCY (1987), "Heavy metal contaminated soil treatment. Interim Technical Report. AMXTH-TE-CR-86101", in *Heavy metal contaminated soil treatment. Interim Technical Report. AMXTH-TE-CR-86101*, Roy F. Weston Inc., West Chester, Pennsylvania.

WALDRON, L.J. and N. TERRY (1975), "Effect of mercury vapor on sugar beets", *J. Environ. Qual.* 4, pp. 58-60.

WILD, H. (1978), "The vegetation of heavy metal and other toxic soils", in M.J.A. Werger (ed.), *Biogeography and Ecology of Southern Africa*, pp. 1 301-1 332, Junk, The Hague.

ZHOU, J. and P.B. GOLDSBROUGH (1994), "Functional homologs of animal and fungal metallothionein genes from Arabidopsis", *Plant Cell* 6, pp. 875-884.

ZIMMERMAN, P.W. and W. CROCKER (1934), "Plant injury caused by vapors of mercury and compounds of mercury", *Boyce Tompson Inst. Contrib.* 6, pp. 167-187.

BIOREMEDIATION OF MARINE ECOSYSTEMS AND THE CULTIVATION AND REFORESTATION OF MACROALGAE: DEMONSTRATION PROJECT DEVELOPMENT

by

William S. Busch
National Oceanic and Atmospheric Administration (NOAA), Maryland, United States

Anthropogenic pollution of marine ecosystems including rivers, bays, estuaries, and other coastal areas is a major global environmental issue requiring immediate attention. Ongoing efforts by researchers at universities and scientific organisations, both domestic and international, have indicated that use of marine biomass may provide a new and cost-effective technology for sequestering and/or preventing dissemination of waterborne pollutants. These contaminants may include heavy metals, radio nuclides, high nutrient concentrations from sewage and fertilizer runoff, bacterial and viral pathogens, and some toxins. Groups in Scandinavia, Europe, Russia, and Japan are also exploring this new technology.

A co-ordinated international focus is needed to implement *in situ* demonstration projects in polluted environments to show the feasibility of this concept and to obtain operational "pilot-farm" data. On 29 April - 1 May 1996 a workshop entitled "Design Workshop for the Tijuana River and Punta Banderas Outfall Macro-Algae Bioremediation Demonstration Project" was held at the Center for Scientific Research and Higher Education of Ensenada (CICESE) in Ensenada, Mexico. More than 30 researchers, engineers, and managers from both the United States and Mexican governments, industry, and academia participated. The original charge set before this workshop was:

> "To obtain all the information, data, requirements, and other items needed to prepare a detailed funding proposal and program development plan (PDP) that would address the experimental design, work break-down structure, milestones, schedules and funding requirements needed in describing this project and its implementation. Emphasis was to be placed on other related ongoing and planned projects having outputs and collaborative activities that would complement this project."

While this workshop was to focus on placing a demonstration coastal macroalgal farm directly off the coast of Tijuana at the Tijuana River outfall, the results of the workshop would be applicable, generically, to other potential demonstration sites being considered world-wide. These areas include Amursky Bay and Gulf of Finland off the coasts of Valdivostok and St. Petersburg, Russia respectively; Chesapeake and Florida Bays, United States; Sao Paulo, Brazil; and Marseille, France.

From an administrative perspective, it was emphasized by both the United States and Mexican sides that it would be important to have the Inter-American Institute (IAI) play a major role in this program. Representing more than 15 Latin American Countries, IAI could be instrumental in disseminating the results of these efforts to the other member countries and, where appropriate, initiating new and complementary activities. In addition, the involvement of the Organizacion Latino Americana de Energia (OLADE), headquartered in Ecuador, was also viewed as being important. A specific action that should be initiated through both IAI and the US-Mexico Foundation for Science is to have a Mexican researcher detailed to NOAA's Office of Global Programs in Washington, DC, for an extended period of time (one to two years) to work on implementing this program.

Based upon the initial discussions and presentations, the consensus of the participants was that the originally selected site of the Tijuana River outfall was, at the present, not necessarily appropriate for a number of reasons. It was highlighted that the pollution problem will be mitigated by an upstream treatment plant that is presently under construction and planned to be operational in two years. Since the Tijuana River estuary is extremely fragile, it would be a monumental task -- if not impossible -- to wade through and satisfy all of the legal, environmental, and social requirements and concerns from private organisations and at all governmental levels -- federal to local, and international. The proposed demonstration project could also impact ongoing and future projects already planned for both the outfall and the sanctuary. In addition, the Mexican participants indicated that from their perspective and needs, other sites such as those near Ensenada and Punta Banderas would be far more beneficial and of critical interest to them. They emphasized that their economic, health, and sustainable development needs were best served at these other sites.

Because of these concerns, it was decided to change the focus of the workshop from the development of a Tijuana border-area marine bioremediation demonstration project plan to exploring two different but complementary programs. One would address marine bioremediation, both fresh and salt water, while the other would deal with macroalgae cultivation and reforestation. Based on these new areas, the participants were divided into two discussion groups that were to develop outlines of activities that each area would entail. Summaries of the respective groups discussions are provided below. These will be used as bases for developing proposals for the World Bank and the Inter American Development Bank to secure funding for initiating respective demonstration projects. Note that while two independent proposals will be submitted, the activities proposed will be highly complementary and require close interactions. An initial listing of generic tasks that may be included in a project framework is also given.

A key point resulting from this workshop was that both US and Mexican researchers and organisations will work together collaboratively. This interaction will range from close one-to-one relations between researchers, to international governmental agencies involvement. The expansion of the program's scope to include multiple sites and to focus on both bioremediation and cultivation/reforestation provides numerous opportunities and needs for research, technology development, and economic and sustainable applications.

Bioremediation of marine effluents by production of algal products: a biotechnology approach

Problem

Marine shrimp farming operations and seafood processors produce effluents with high organic and inorganic nutrient loads. These effluents adversely affect nearby water quality, and are in part responsible for the decline of shrimp farming in several areas of the world. Efforts to encourage conventional wastewater treatment technologies for treatment of marine effluents have usually failed, since operators see no economic benefit in such investment, and environmental enforcement is weak in many developing countries. Environmental clean-up technologies that can produce revenue streams have a better chance of being incorporated into aquaculture technologies, as opposed to conventional wastewater treatment processes.

Rationale

Marine algae can provide excellent wastewater treatment of marine waters while also creating revenue streams. Marine algae provides various sources of products valued at $2.5 billion per year world-wide with continued annual growth anticipated to be on the order of 3-6 per cent. A large number of different algal products allows an operator a choice in product streams that can be developed, thus ensuring each region will retain a comparative advantage in economic returns. Marine algae has already been shown to be effective in treating a number of different waste streams. Algae can be used to reduce nutrient loadings in effluents by substantial amounts.

Program needs

Most of the pilot scale trials that use marine algae for wastewater treatment have been fragmented, with some studies focusing on inorganic nutrient removal (nitrates, phosphates), organic nutrient removal (Biological Oxygen Demand, Dissolved Oxygen), or production of commercial products. Such trials have usually been conducted in cooler temperature waters, rather than the warmer waters of developing regions. Hence, a detailed program is needed to develop a demonstration project that will produce an integrated design allowing shrimp farmers and other producers of marine effluents to bioremediate wastewater effluents in warmer waters and that can be used as a "turn-key" approach. Furthermore, marketing infrastructure of algal products needs to be clearly explained so that producers will know where to sell their products.

Approach

A demonstration project will be located at a commercial shrimp farm in Mexico, and will be co-ordinated through CICESE (Centro de Investigacion Cientifica y de Educacion Superior de Ensenada), IAI (InterAmerican Institute) and NOAA (National Oceanic & Atmospheric Administration). CICESE will provide in-country scientific expertise and co-ordination with UABC (Universidad Autonoma de Baja California), and CIB (Centro de Investigaciones Biologicas del Noroeste). IAI will provide co-ordination with consultants and scientists from the countries of Ecuador, Brazil, Guatemala, Panama, Honduras, Costa Rica, Columbia, Belize, Chile and Peru. Many of these countries have shrimp farming operations, and all are investing in marine aquaculture or seafood processing. All have concerns about marine effluents. NOAA will co-ordinate US scientific expertise from private mariculture companies such as Coastal Plantations Inc., and Neushul Mariculture Inc., as well as the University of North Carolina at Wilmington.

Goal

To develop integrated design choices that will work over a variety of aquaculture effluent volumes, and that will lead to significant improvement in marine wastewater treatment while allowing for the production of product streams.

Components of design choices

In situ wastewater treatment -- In some instances, operators can use macroalgae to reduce nutrient and pollution loading right in the shrimp ponds themselves. Such a system has to be compatible with shrimp farming operations. Several systems designed for cultivation of macroalgae in bays and estuaries may be adapted quite easily to operations within shrimp ponds.

Ex situ wastewater treatment -- Some aquaculture and food processing operations collect all the effluent at a single point before discharge. These types of operations can take advantage of treating this effluent after it has passed completely through the facility. Several approaches include well-documented cultivation of algae in raceways, a system capable of handling large volumes of effluents. Other approaches include effluent treatment through cyanobacterial mats or photobiological reactors, systems that are very compatible with effluents from hatcheries and seafood processing operations. These smaller volumes of marine effluents are often ignored, yet their large number can create a significant source of "diffuse" pollution points.

Processing residue and by-product disposition -- The disposal and/or use of outputs from marine biomass processes is a critical problem. Exploring alternative uses and developing new technologies to make them economically valuable are required. Of special note is the identification of new by-products (chemical, biological, or physical) that heretofore have not been considered.

Performance requirements

In order for marine algae wastewater treatment systems to be adopted by marine industries in developing countries, the demonstration project needs to show that this approach can effectively treat waste problems while generating salable products, and be useful from a social and economic perspective. The demonstration project needs to evaluate the following:

Wastewater treatment improvements -- Biological Oxygen Demand, Dissolved Oxygen, Suspended Solids, nitrate, ammonium, phosphate and trace metals (nutrients and metals measured in sediments and waters), sludge accumulation, total organic matter, agrochemicals and organic pollutants, pathogens such as bacteria and viruses.

Crop production characteristics -- Shrimp growth and yield per hectare, marine algal growth and yield per hectare, shrimp health as measured by biomarkers such as heat shock proteins, cytochrome P450, glutathione-S-transferase, sequestration of nutrients and metals in shrimp and marine algae.

Engineering, economic and social issues -- Engineering designs, process economics, labour requirements, social and regional economic benefits from improved wastewater treatment, product streams from algae, marketing infrastructure for algal products.

Other regional applications

Eastern Africa (such as Tanzania and Mozambique), India, Southeastern Asia (such as Thailand, Vietnam, Indonesia, Malaysia and the Philippines).

Cultivation and reforestation of macroalgae

The goal of this project is to provide economically feasible methods of cultivating and developing new and existing kelp beds for use as food sources for high value species and for marine pollution remediation.

Economic benefits

By determining and establishing proven methods of developing kelp beds, specific economic benefits can be realised. The primary benefit will be the improvement of Mexican abalone and sea urchin farms by providing a readily available and inexpensive supply of nutrients. The ability to develop sustainable kelp beds near existing high value species farms will be of tremendous benefit to the farmers and to the economy of Mexico. Increased employment of Mexican abalone and sea urchin farmers will result as these crops become more widespread, due to improved farming methods and more reliable sources of nutrients from kelp.

It is well known that kelp are effective in absorbing waterborne pollutants, for providing physical barriers to pollutant dispersion, and in diverting coastal supplied pollution to deep water. By deploying kelp farms near pollution sources and in contaminated marine ecosystems, marine environments will be enhanced. This in turn will stimulate the recreation and tourism sector of the Mexican economy as well as provide increased employment. In addition, pollution abatement would improve public health.

Rationale

The candidate seaweeds under consideration for this project are macrocystis, egregia, gracilaria, and porphyra. The initial activities will consist of establishing a joint venture between the United States and Mexico to develop algal culture growout facilities in Mexico by transferring technology from the United States to Mexico. Refurbishing existing facilities and establishing new facilities in Mexico will be considered.

In order to maximise the overall efficiency of this project, an economic parametric model and system analysis will be first developed to include all factors that contribute to the successful cultivation of the various kelp species. By adjusting the various factors in the model such as water quality, ambient light, water nutrient levels, planting and anchoring techniques, the plant growth can be maximised at the lowest cost to provide the greatest return on investment.

Initially an extensive baseline database will be obtained, looking at all initial parameters, including atmospheric, oceanographic, chemical, geographical, and anthropogenic. An environmental monitoring program will also be initiated to determine the relationship between the kelp and its surroundings throughout the program. Chemical compositions of the seaweed and the ambient water will be measured at predetermined sample intervals to determine the exchange between the two. The condition and health of the crops will be observed for plant stress indicators. The role of bacteria and micro-organisms in kelp development will also be closely monitored, since this relationship is poorly understood. Additional elements to be researched will include bed dynamics and associated stresses caused by ocean currents and how these factors influence kelp development.

Additional analyses of the role and effectiveness of the seaweeds in sequestering certain pollutants will be conducted. The specific processes and relationships between plant health conditions, pollutant content, and micro-organisms will be investigated to determine the optimum seaweed for various types of pollutants.

Potential areas of research, development and technology transfer

While seaweed growout facilities are relatively well advanced in the United States, the level of sophistication of these facilities in Mexico is insufficient to undertake a large-scale project such as that being proposed. To remedy this, existing technology will be transferred from US cultivation facilities to upgrade existing Mexican facilities and to develop new research laboratories in Mexico. Both equipment and intellectual talent will be transferred to accomplish these goals.

Major research and development advancements in the reforestation of seaweeds are necessary in the area of field planting methods. Research will be conducted in determining the most effective ways to transplant seaweeds from one site to another so that the survival rate of the plants is high. Once transplanted, how the plants are initially and remain attached to the bed become an issue. Currently, there are several options for anchoring plants to the beds. These will be compared for varying bed conditions and substrates to determine which provide stability and best growth potential.

Optimum site locations will be determined to correlate plant growth with environmental factors, such as ambient light, and with physical oceanographic parameters, such as bed energy. Various species of seaweed will be transplanted into different environments to establish baseline parameters and factors that may limit kelp growth. This research area will also enable researchers to decide which methods are best to restore and enhance existing seaweed beds.

Generic project framework

Tasks

- Establishment of program framework and staff that will co-ordinate all project tasks and activities including: begin permitting process; prepare environmental impact statements; establish public relations activity; identify and take steps to meet requirements of applicable legislation and regulatory standards; develop close interactions with international, local, and state partners as required; address potential legal, liability, and ownership issues; and explore alternative approaches.

- Synthesis of available information by conducting an extensive literature search, interacting with persons, organisations, and governments known to have experience, facilities, and/or future interests; and identify ongoing activities.

    1) Identify associated and complementary studies that should be conducted in parallel to the farm development;

    2) Explore needs of developing countries that such a farm may satisfy.

- Detailed site survey of demonstration site, collecting *in situ* data (i.e. oceanographic, geological, biological, hydrological, atmospheric, and benthic as well as pollutants and their related sources).

- Selection and evaluation of candidate plant species that may include investigating the prospective plants' characteristics, culturing requirements, nutrient needs, pollutant sequestering capabilities, transplantation requirements, and maintenance needs.

- Identify, describe, and quantify the various environmental impacts both "on and by" the marine biomass farm, such as:

  1) *On* the farm: grazers, wave climate, sediments (storms, deposition, and sediment movement), light intensity, currents, fouling, water quality, nutrients, temperature, and anthropogenic (i.e. chlorination, and fertilization, and substration).

  2) *From* the farm: littoral cell (sediments), orientation to currents, water influx to estuary, food webs (concentration of toxic elements), biodiversity, algae drift, societal (surfing, diving, navigation, accidents, storm damage, ...), construction, and harvesting/processing, disposal, and alternative uses.

- Design and scale model testing of an algae system to evaluate performance specifications and identify both capabilities and limitations in a controlled test and evaluation environment. Specific activities would include initial design of system; computer simulation of structure and plant hydrodynamics; construction and testing of scale model of system and plants in tow tank/test pool simulating adverse wave, water current, and surface wind conditions; and finalisation of performance specifications.

- Design and construction of system including production of performance specifications, contract plans, and production drawings and documents; manufacturing of hardware; and integration of system.

- Site preparation: collect background environmental data to establish baselines.

- Installation of farm system, testing and evaluation, and implementing operations support facilities.

- Long-term monitoring and evaluation of system performance, operational processes and procedures, efficiencies, capabilities, and limitations.

- Using the farm as an *in situ* marine biome and laboratory, conduct research and development activities on areas such as:

  1) appropriate methods for processing plants;

  2) pollutant extraction from plants and disposition;

  3) plant physiology in sequestering pollutants;

4) optimum growing conditions and methods of harvesting; and

5) identifying new by-products.

Criteria for measurement of success

- Positive efficiencies of pollutant removal, growth potential, and as a source of food;
- Effective design and implementation of moored grid system in subject marine environment and survivability in adverse weather and sea state conditions;
- Positive maintainability and growth rate of algae species in this system; and
- Successful processes for operating the system (i.e. viability, harvestability, and processing of plants, disposition of removed pollutants).

Partnerships and roles of participants

Throughout the project, numerous groups and persons will be involved from government, industry, and academia, through mechanisms such as formal grants and contracts, "in-kind" relationships, interagency agreements, ad hoc detailing and personnel exchanges, consulting, volunteers and interested observers, and other complementary/appropriate ongoing research programs.

Questions, comments and answers

Q: Have you figured out which species you would work with in the project?

A: We are looking at a number of alternatives. In each project, we want to use only indigenous plants. In the Baltic Sea , for example, they are focusing on sea grasses.

Q: Are there any problems in cultivating the plants?

A: This is one area where there is strong potential for collaboration, especially between the US and Mexican science communities. The scientists know how to cultivate but don't have the facilities to do so on a large scale.

Parallel Session I

I-D. BIOREMEDIATION/BIOTREATMENT: MARINE AND COASTAL WATERS

BIOREMEDIATION OF CRUDE OIL AS EVALUATED IN MESO-SCALE BEACH-SIMULATING TANKS

by

Shigeaki Harayama, Hideaki Maki, Tetsuya Sasaki, Hiroaki Nagashima, Keiji Sugiura, Etsuro Sasaki and Masami Ishihara
Marine Biotechnology Institute, Iwate, Japan

Introduction

Oil pollution in the sea attracts public attention because of environmental and economic concerns. Major oil pollution in the sea is mostly provoked by tanker accidents. Spilled oil is subject to a variety of physico-chemical weathering processes, and often affects land on the shoreline. The clean-up of oil-contaminated beaches is a difficult task. Physical removal of the oil by a variety of methods only results in partial cleaning. Bioremediation was used to clean up oil-contaminated beaches in the Gulf of Alaska after the oil spill from Exxon Valdez in 1989 (Atlas, 1984; Prince, 1993; Bragg *et al.*, 1994). Since then, bioremediation is considered to be a preferred method for beach cleaning. However, biological mechanisms underlying this new technology are still poorly understood. In order to develop bioremediation technologies for the clean-up of spilled oil, we investigated the biodegradation of crude oil in meso-scale beach simulating tanks.

Analysis of crude oil

In this study, we used Arabian light crude oil heated at 230 °C. By this treatment, volatile and readily biodegradable components of the crude oil were removed. This petroleum, like other crude oil samples, is a mixture of thousands of compounds which are generally classified into four chemical groups: saturates, aromatics, resins and asphaltenes (Leahy and Colwell, 1990). The compositional analysis of the heat-treated Arabian light crude oil is shown in Table 1.

Thin layer chromatography (TLC) in combination with a frame ionisation detection (FID) was used for the quantification of the saturated and aromatic compounds (Goto *et al.*, 1994), which account for 88 per cent of the total crude oil. Analysis of the crude oil by gas chromatography (GC) was conducted in a fused silica capillary column as has been described previously (Sugiura *et al.*, 1997). GC in combination with mass spectrometry (MS) was used for the separation and identification of specific compounds in crude oil (Wang *et al.*, 1994). Hopanes are constituents of petroleum, and resistant to biodegradation. These compounds therefore are used for internal standards to quantify petroleum (Prince *et al.*, 1993). In this study, amounts of hopanes were determined by GC/MS.

Table 1. **Characteristics of the heat-treated Arabian light crude oil**

Gravity, API		33.4
Viscosity (cp) at 20 °C		64
Aromaticity		0.08
n-C_{17}/Pristane		3.64
	Saturates	50
Content (%)	Aromatics	38
	Others	12

Note: API: American Petroleum Institute gravity; aromaticity: the ratio of aromatic carbons to total carbons; C_{17}/pristane: ratio of *n*-heptadecane to pristane in the saturated fraction.

Source: Sugiura *et al.*, 1997.

Biodegradation of crude oil in batch cultures

An oil-degrading consortium called SM8 has been isolated from a sediment in Shizugawa Bay, Japan (Venkateswaran *et al.*, 1991). Strain 4L was isolated from the SM8 consortium. A taxonomic study demonstrated that this gram-negative bacterium belongs to a new genus. The degradation of crude oil by the SM8 consortium and by strain 4L was examined in batch cultures. Both SM8 and strain 4L degraded 60 per cent of the saturated fraction in 30 days. As for the aromatics fraction, SM8 degraded 35 per cent, while strains 4L degraded 20 per cent in 30 days.

Crude oil biodegradation in beach-simulating tanks

The beach-simulating tank consists of a double-walled plastic tank, a reservoir and a level controlling device. Sea water in the reservoir was aerated by bubbling, temperature-controlled at 20 °C, and introduced into the tank at a flow rate of 60 l hr^{-1}. The internal volume of the tank was 1.5 m^3, and it was filled with 1 m^3 of gravel of 2 to 8 mm in diameter. The level of sea water was adjusted by the level controlling device which moves between 20 and 80 cm high from the bottom two cycles per day. Excess sea water overflowed from the level controlling device.

1.5 kg of crude oil was poured into each tank, and either fertilizers and/or the cells of the SM8 consortium were added to the tank. In the first experiment, 3 l of the SM8 cultures (approximately 10^8 cells ml^{-1}) and two types of chemical fertilizers, 300 g of a slow releasing solid granular nitrogen fertilizer (Super 1B, Mitsubishi Kasei, Tokyo, Japan) and 60 g of a solid granular phosphorus fertilizer (Linstar 30, Mitsubishi Kasei), were added in one (amended) tank, while these agents were not added in the other (control) tank. Aliquots (500 g) of gravel in the tanks were sampled at appropriate intervals, and crude oil attached to the gravel was extracted by chloroform and analysed by GC-FID. In the samples from the amended tank, the amount of GC-resolved alkanes, major constituents of crude oil, decreased continuously with time, and reached an undetectable level in 90 days. On the contrary, no biodegradation of the alkanes was observed in the control tank in 90 days (data not shown). The results of this experiment showed that the microbial activity to degrade crude oil was insignificant in the natural sea water, and the addition of fertilizers and the SM8 cells was effective in increasing the speed of oil degradation.

Effect of fertilizers and seeding of oil-degrading micro-organisms on crude oil degradation

We then examined the effect of the addition of the fertilizers alone, and that of the addition of the fertilizers and the SM8 cells on the oil degradation. In the experiment shown in Figure 1, the fertilizers were added to the first tank, while the fertilizers and the SM8 cells were added to the second tank. The crude oil was extracted from gravel as described above, and the amounts of the saturated and aromatic fractions as well as that of that of hopanes, were determined. The amount of the saturated or aromatic fraction in a sample was divided by the amount of hopanes in the same sample, and these normalised values were compared. The biodegradation patterns in both tanks were very similar: more than 80 per cent of the saturated fraction was degraded, while the degradation of the aromatic fraction was not significant. The seeding of the SM8 cells in this experiment did not increase the extent of the biodegradation.

Figure 1. **Effect of chemical fertilizers and seeding on the biodegradation of crude oil**

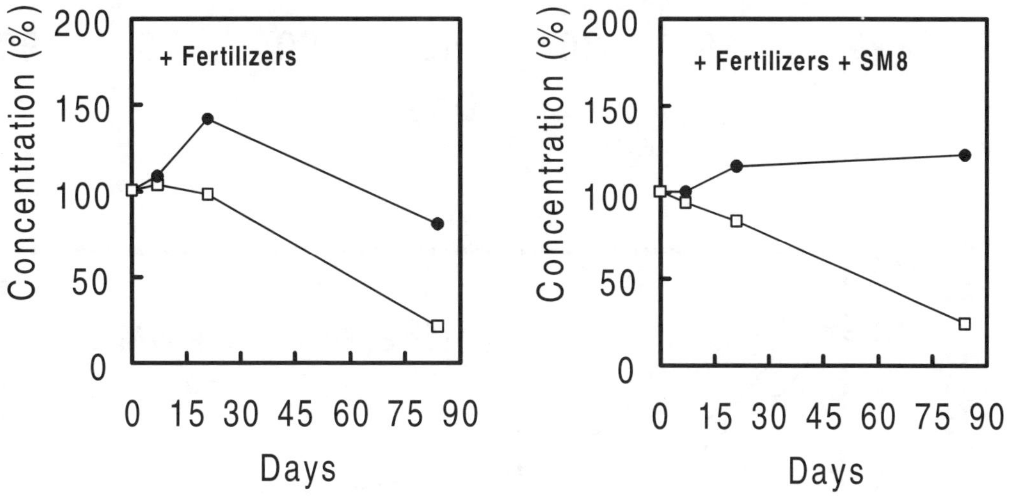

Note: The initial concentrations of the saturated and aromatic compounds in crude oil were taken as 100. Closed circle: concentration of aromatic compounds; open square: concentration of saturated compounds.

Source: Maki, H., M. Ishihara, and S. Harayama, unpublished.

Biodegradation of specific compounds in crude oil was analysed by means of GC/MS. The target compounds were (C_{10}-C_{34}) *n*-alkanes, (C_0-C_4) naphthalene, (C_0-C_4) dibenzothiophene, (C_0-C_7) phenanthrene, (C_0-C_7) anthracene and (C_0-C_2) fluorene. The concentrations of these compounds normalised by the concentration of hopanes are shown in Figure 2. In this figure, it is shown that different compounds were degraded with different speeds. The susceptibility to biodegradation of these compounds was in the order of (C_0-C_4) naphthalene > (C_{10}-C_{34}) *n*-alkanes > (C_0-C_2) fluorene > (C_0-C_4) dibenzothiophene > (C_0-C_7) phenanthrene + anthracene. The speeds of the biodegradation of these compounds were not different between two treatments, namely the addition of the fertilizers alone and that of the fertilizers and the SM8 cells.

Figure 2. **Effect of chemical fertilizers and seeding on the biodegradation of various components in crude oil**

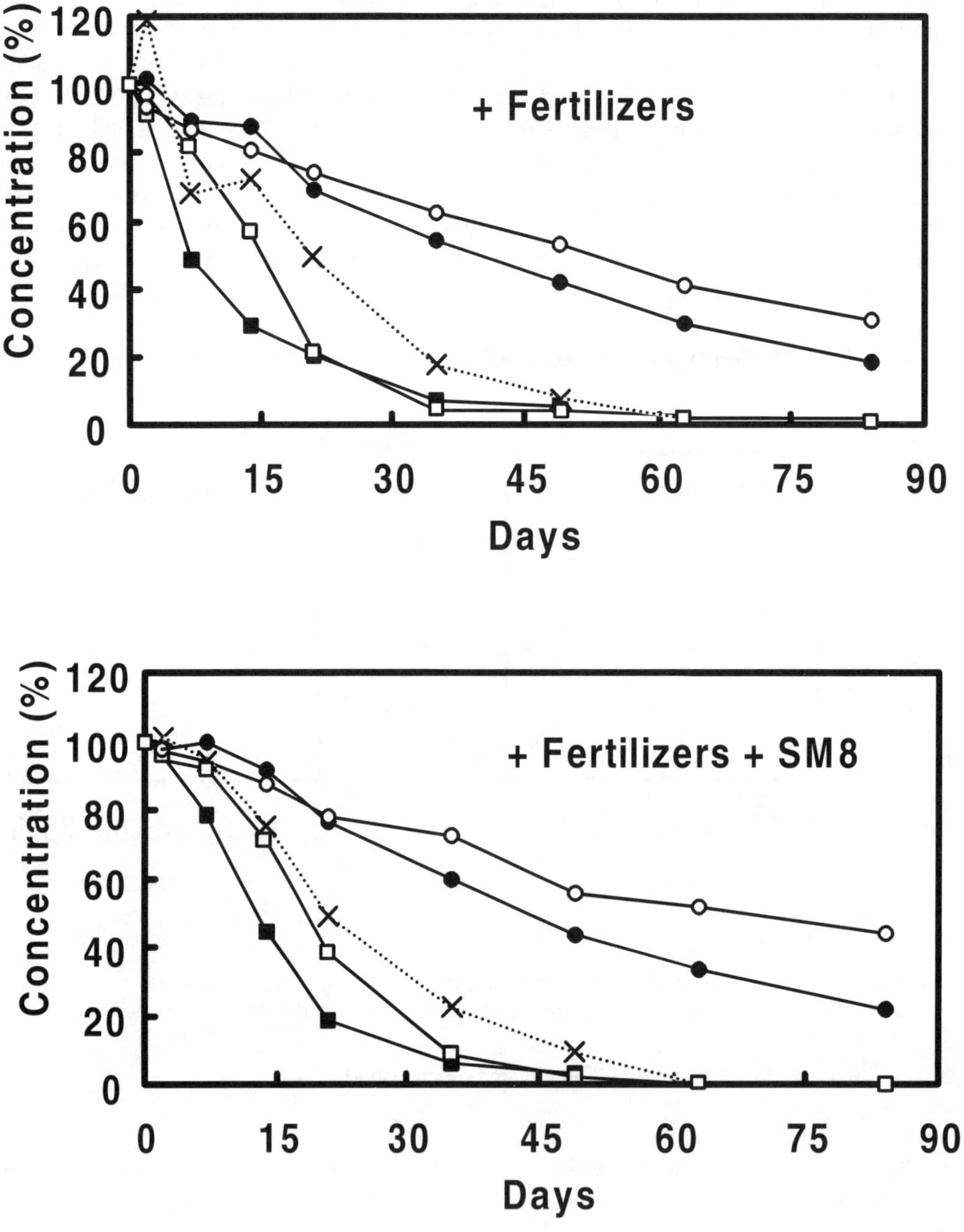

Note: Compounds examined were (C_0-C_4) alkylnaphthalenes (closed square), (C_{10}-C_{34}) *n*-alkanes (open square), (C_0-C_2) fluorene (cross with dotted line), (C_0-C_4) dibenzothiophene (closed circle), and (C_0-C_7) phenanthrene and anthracene (open circle).

Source: Maki, H., M. Ishihara, and S. Harayama, unpublished.

Fate of an oil-degrading bacterium introduced in beach-simulating tank

As described above, the addition of the SM8 cells did not enhance the crude oil degradation. One of the possible reasons for the failure to stimulate the biodegradation was that micro-organisms in the SM8 consortium could not compete with indigenous populations. To test this possibility, the survival of strain 4L, one of the constituents of the SM8 consortium, was examined. To enumerate strain 4L in sea water, we used the sequence of the *gyrB* gene that encodes the sub-unit B protein of DNA gyrase to distinguish this strain from other bacteria. In the first method to enumerate strain 4L, bacteria in sea water from the beach-simulating tank were spread on Marine broth plates, and the colony hybridisation against strain 4L-specific *gyrB* probe was carried out. From this experiment, it was shown that strain 4L was not colonised efficiently in the beach-simulating tank: strain 4L was detected in sea water of the amended tank one day after the seeding but not later. In the second method to enumerate strain 4L, bacteria in the sea water from the beach-simulating tanks were collected, and DNA isolated from the bacteria was used as a template for the PCR amplification using strain 4L-specific primers. The results of the PCR amplification demonstrated that strain 4L was first detected several days after the seeding, but its number sharply decreased, and 30 days after the seeding, it could not be detected even using the PCR amplification.

Effect of the addition of sludge on the crude oil biodegradation

We were interested in using sludge from a sewage treatment plant as nutrients. In this experiment, liquid- or dehydrated sludge in an amount equivalent to 40 ppm nitrogen was applied to tank 4 and tank 3, respectively. As a control, the chemical fertilizers described above were added to tank 2. Furthermore, neither fertilizer nor sludge was added to tank 1. As shown in Figure 3, no oil degradation was observed in tank 1, where no nutrients were added. In tank 2, to which the chemical fertilizers were applied, the saturated fraction but not the aromatic fraction was degraded as has been observed in Figure 1. The addition of liquid sludge or dehydrated sludge was as effective as the addition of the solid fertilizers: the degradation of the saturated fraction, but not the aromatic fraction, was observed.

Discussion

It is difficult to predict, from the results of laboratory experiments, the effect of various treatments on the biodegradation of crude oil in the natural environment. As field experiments are impossible to carry out, we constructed the meso-scale beach-simulating tanks to collect data with which methods to improve bioremediation technology would be perceived.

When nutrients were supplemented into sea water, most of saturated components in crude oil were degraded. On the contrary, the degradation of the aromatic fraction was not significant, although the degradation of alkylnaphthalene and other polyaromatic hydrocarbons was demonstrated in the GC/MS analysis. In batch culture experiments, the SM8 consortium and its constituent, strain 4L, degraded 15 to 35 per cent of the aromatic fraction. We do not know yet the reason why aromatics in crude oil were not significantly degraded in the beach-simulating tanks.

The molecular monitoring of microbial ecosystems has recently received increasing attention in order to understand, for example, the fate of micro-organisms introduced in natural environments, and the effect of the inocula on indigenous microbial populations. DNA probes in combination with PCR technology have been used to detect specific bacteria in various habitats with or without cell cultivation (Amann *et al.*, 1995). However, the application of molecular techniques to study the microbial ecology is still limited due to the difficulty in developing probes that are highly specific to each type of bacteria. The *gyrB*-based probing system developed in this study may be useful because of its specificity and generality.

Figure 3. **Effect of various nutrients on the biodegradation of crude oil**

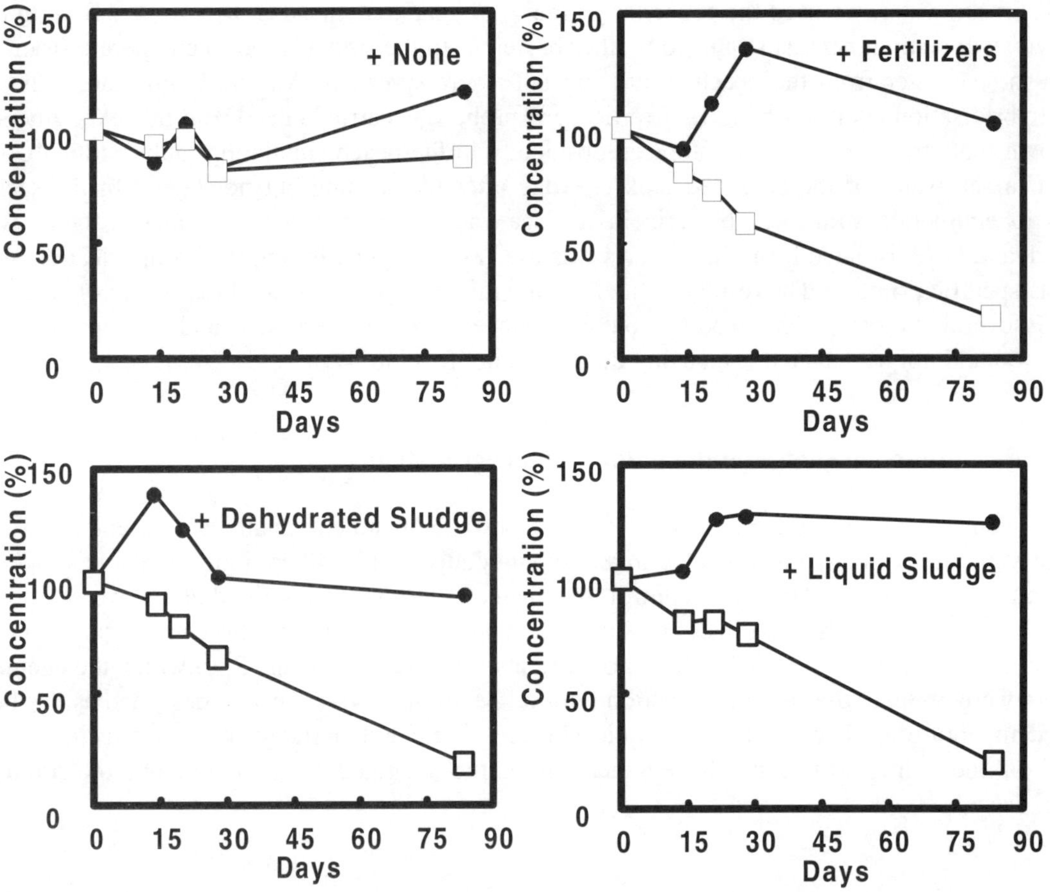

Note: Closed circle: concentration of aromatic compounds; open square: concentration of saturated compounds.

Source: Maki, H., M. Ishihara, and S. Harayama, unpublished.

Seeding of the SM8 consortium was not effective in stimulating the biodegradation of crude oil. One of the reasons for the failure may be due to the ineffective colonisation of these bacteria. We showed experimentally that this was the case at least for the strain 4L.

Several reports have indicated that conditions of nutrient supplementation strongly influence the rate of hydrocarbon contaminants. We therefore tested two types of nutrients, chemical fertilizers and sludge from a municipal sewage treatment plant. The pattern or the crude oil degradation was not affected by the type of nutrients.

In any case, the effectiveness of the bioremediation of crude oil was demonstrated in the beach-simulating tanks. One direction of our future studies will be the development of methods to colonise aromatic-degrading micro-organisms in the simulation system.

Acknowledgement

This work was supported by a grant from New Energy and Industrial Technology Development Organization.

Questions, comments and answers

Q: When you talk about degradation, do you mean total degradation or the disappearance of the original compound?

A: It is not necessarily mineralisation. The peak disappears on mass spectrograph. However, in parallel, we carried out *in vitro* experiments in flasks, and we estimated that there was more than 60 per cent mineralisation. We don't know the results in the tank.

REFERENCES

AMANN, R.I., W. LUDWIG, and K.-H. SCHLEIFER (1995), "Phylogenetic identification and *in situ* detection of individual microbial cells without cultivation", *Microbiol. Rev.* 59, pp. 143-169.

ATLAS, R.M. (1984), *Petroleum Microbiology*, Macmillan, New York.

BRAGG, J.R., R.C. PRINCE, E.J. HARNER, and R.M. ATLAS (1994), "Effectiveness of bioremediation for the Exxon Valdez oil spill", *Nature (London)* 368, pp. 413-418.

GOTO, M., M. KATO, M. ASAUMI, K. SHIRAI, and K. VENKATESWARAN (1994), "TLC/FID methods for evaluation of the crude-oil degrading capability of Marine Microorganisms", *J. Marine Biotech.* 2, pp. 45-50.

LEAHY, J.G. and R.R. COLWELL (1990), "Microbial degradation of hydrocarbons in the environment", *Microbiol. Rev.* 54, pp. 305-315.

PRINCE, R.C. (1993), "Petroleum spill bioremediation in marine environments", *Crit. Rev. Microbiol.* 19, pp. 217-242.

PRINCE, R.C., D.L. ELMENDORF, J.R. LUTE, C.S. HSU, C.E. HALTH, J.D. SENLUS, G.J. DECHERT, G.S. DOUGLAS, and E.L. BUTLER (1993), "17α(H),21β(H)-Hopane as a conserved internal marker for estimating the biodegradation of crude oil", *Environ. Sci. Technol.* 28, pp. 142-145.

SUGIURA, K., M. ISHIHARA, and S. HARAYAMA (1997), "Comparison of the biodegradability of four crude oil samples by a single bacterial species and by a microbial population", *Environ. Sci. Technol.* 31, pp. 45-51.

VENKATESWARAN, K., T. IWABUCHI, Y. MATSUI, H. TOKI, E. HAMADA, and H. TANAKA (1991), "Distribution and biodegradation potential of oil-degrading bacteria in North Eastern Japanese coastal waters", *FEMS Microbiol. Lett.* 86, pp. 113-122.

WANG, Z., M. FINGAS, and K. LI (1994), "Fractionation of a light crude oil and identification and quantitation of aliphatic, aromatic, and biomarker compounds by GC-FID and GC-MS, Part 1", *J. Chromatogr. Sci.* 32, pp. 361-366.

NUTRIENT ENHANCED BIOREMEDIATION: FROM LABORATORY TO ALASKAN BEACHES, TO REFINERIES

by

Russell R. Chianelli
University of Texas at El Paso, Texas, United States

Historical

Bioremediation is the process in which micro-organisms are used to biodegrade oil, and nutrient enhanced bioremediation employs application of nitrogen and phosphorous or other nutrients that may be limiting the degradation process. Biodegradation is a natural process in which bacteria consume petroleum and break it down to biomass and carbon dioxide. The bacteria that consume the petroleum are "hydrocarbon-oxidisers" that destroy petroleum molecules by adding oxygen to them. These bacteria have been discovered in all marine environments which have been studied (Zobell, 1973). The newly formed oxygenated molecules are further consumed until only the final products, biomass and carbon dioxide, remain. The process of converting the oil to biomass and carbon dioxide is called mineralization.

Biodegradation, together with physical processes such as evaporation, is a major process by which petroleum is removed from the environment following an oil spill. The role of these processes for a large spill such as that of the Amoco Cadiz was described by Gundlach *et al.* (1983). This study showed that biodegradation was acting as rapidly as evaporation, even during the first days following the spill. It is thought that the high rate of biodegradation along the coast of Brittany was due to runoff of agricultural fertilizers from farms near the coast. The addition of nitrogen and phosphorous from this agricultural runoff enhanced the biodegradation on the beaches presumably by removing the nitrogen and phosphorous limitation resulting from low naturally occurring levels of these elements in sea water (Atlas, 1990).

Although nutrient-enhanced bioremediation had been used extensively for land farming and groundwater applications, its use in marine spills had been limited prior to the Exxon Valdez oil spill. However, since the large Amoco Cadiz oil spill in 1978, efforts have been made to develop enhanced bioremediation techniques. In 1985, an oleophilic, or "oil-loving", fertilizer INIPOL EAP 22 (Tellier *et al.*, 1984) was applied to a small spill of diesel oil in Ny-Alesund, Spitsbergen, Norway. The results of this application were inconclusive due to the high rate of natural cleaning (Sveum and Ladousse, 1989). Other small-scale field trials (Sendstad *et al.*, 1982; Sendstad *et al.*, 1984; Halmo, 1985; Lee and Levy, 1989; Lee and Levy, 1991; Lee *et al.*, 1991) prior to the Exxon Valdez spill successfully demonstrated bioremediation potential, showing that addition of fertilizers could stimulate microbial growth and oil degradation. But it is difficult to quantify degradation for oil samples from the open marine environment, and most studies were unable to show statistically significant effects of fertilizer addition.

Exxon Valdez bioremediation project

The frequent occurrence of oil spills underline the need for effective and environmentally safe clean-up treatments. The Exxon Valdez spill of approximately 11 million gallons of Alaskan North Slope crude oil in Prince William Sound, Alaska in March 1989 resulted in contamination with oil of approximately 1 300 miles of rocky intertidal shorelines within Prince William Sound and the Gulf of Alaska. Bioremediation was used extensively to accelerate the natural degradation of beached oil. Nitrogen containing fertilizers were applied to stimulate growth of indigenous hydrocarbon-degrading micro-organisms within the intertidal zones. Shorelines were treated with INIPOL EAP 22 (7.4 per cent N, 0.7 per cent P), an oleophilic liquid fertilizer, and Customblen (28 per cent N, 3.5 per cent P), a slow-release granulated fertilizer. Approximately 50 000 kg of nitrogen and 5 000 kg of phosphorus were applied to shorelines over the summers of 1989-1992, making this the largest bioremediation project to date (Chianelli, 1994).

In 1989, field tests on Exxon Valdez beached oil by the US Environmental Protection Agency (USEPA)/Exxon provided evidence that bioremediation cleaned beaches (Pritchard and Costa, 1991). Treated areas were visibly cleaner than adjacent untreated areas within a few weeks of fertilizer application (Pritchard *et al.*, 1991; Stone, 1992). However, gravimetric analyses on heterogeneous sediments were statistically unconvincing. Laboratory testing confirmed that Inipol EAP 22 was not an effective agent for physically removing oil from sediments, and that oil removal occurred through biological activity (Chianelli *et al.*, 1991; Prince *et al.*, 1993).

Field analysis of bioremediation

In 1990, a statistically designed field test was conducted that included extensive chemical analyses of the oil. Results demonstrated that measured changes over time in the oil composition, relative to a stable, high-molecular weight hydrocarbon present in the oil, allowed quantification of the rate and extent of the oil biodegradation with high levels of statistical confidence. The authors found that hopane, a multi-ring saturated hydrocarbon, was not biodegraded and was a useful conserved standard. Monitoring hydrocarbon losses relative to hopane provided benchmark confirmation of oil biodegradation, and showed that adding fertilizers could accelerate the rate of oil removal by a factor of five or more. It was also found that the rate of oil biodegradation is a function of the nitrogen concentration maintained in the pore water of the intertidal sediments. These results suggested that bioremediation application can be further improved by monitoring nutrient level in real time and maintaining the nutrient level at an optimum and safe level (Bragg *et al.*, 1994). On-site monitoring of nutrients in sediment pore waters would provide practical, real-time guidance on amounts and frequency of fertilizer applications.

A concurrent monitoring program conducted in Alaska jointly by the USEPA, Alaskan Department of Environmental Conservation (ADEC) and Exxon, incorporated comprehensive environmental monitoring, including tests for toxicity and potential stimulation of photosynthetic plankton (Prince *et al.*, 1990). No adverse effects were found, supporting the safety of bioremediation techniques and that more frequent fertilizer application would have been equally safe and even more effective. The ultimate rate of nutriated bioremediation will be limited by the amount of nutrients which may be added without adversely affecting the indigenous biota.

Bioremediation has inherent limitations. Bioremediation is not likely to be effective on extensively degraded oil. Once the polar content of the oil residue reaches 60-70 per cent of the total mass, nutrient availability will no longer be the limiting factor. Further, adequate oxygen must be available; bioremediation within anaerobic fine sediments such as mud flats or marshes would probably be effective only at shallow depths, but oil penetration in such sediments is also restricted. Nevertheless, based on the statistically significant field tests along with a substantial body of laboratory testing, bioremediation is an important treatment for oil spills on rocky intertidal shorelines of the type found in Alaska.

Recent activities

A recent study has further confirmed and extended the effectiveness of nutrient enhanced bioremediation on sandy beaches (Venosa *et al.*, 1996). The study concluded that significant intrinsic biodegradation of petroleum hydrocarbons can take place naturally if sufficient nutrients already exist in the impacted area. The study also demonstrated that statistically significant rate enhancement occurs in the presence of an already high intrinsic rate by supplementing natural nutrient levels with inorganic mineral nutrients. It was also shown that bioaugmentation will likely not significantly contribute to clean-up of an oil spill. The study developed first-order biodegradation rate constants for the resolvable normal and branched alkanes and two- and three-ring aromatic hydrocarbons present in light crude oil and connected relative biodegradation rates of in the field to those measured in the laboratory. The literature which connects laboratory studies to field studies is sparse, and much more needs to done. The difference between closed vessel studies, which are usually performed in the laboratory, and open systems, which occur in the field, has been discussed by Oudot (1984).

Of primary concern here is the fact that the products of biodegradation often inhibit microbial activity and retard degradation, whereas these products can be removed by tidal action or large volume dispersion in field environments.

Advantages of bioremediation

It has been demonstrated by the "Exxon Valdez" project and the Delaware project cited above, that bioremediation is an effective and safe technology for beached petroleum hydrocarbons in appropriate areas. Areas appropriate for this technology include those areas in which it can be demonstrated that hydrocarbon degraders are present and active. It is also necessary to demonstrate that the selected area is nutrient limited. This is not always the case, as demonstrated by the "Amoco Cadiz" spill (cited above), in which fertilizer runoff from nearby farms enhanced nutriation on oil covered beaches. This also appears to be the case at the head of the Persian Gulf where the Tigris and Euphrates rivers enter the Gulf delivering nutrients washed from farm lands in Iraq. If the oil spill affected area receives a large input of nutrients, further addition maybe counterproductive.

Experience in the "Exxon Valdez" spill also indicated that bioremediation was perceived as a "natural technology", yielding higher public acceptance. Adding nutrients, especially familiar agricultural products gave the appearance of "helping nature". Thus, higher public acceptance resulted in more rapid attainment of required approvals for application by public regulatory agencies. Addition of non-indigenous organisms is ineffective and receives negative public scrutiny. Genetically modified organisms have not been demonstrated to yield any advantage at this point and are often viewed with alarm by segments of the public. It was found to be of great advantage in the "Exxon Valdez" project to respond to public concerns regarding bioremediation with "townhall meetings" to describe the process and to answer questions about the process. A frequently asked question was "what happens to the bacteria after they consume the oil"? The answer to this question is that the bacteria are consumed by higher organisms and safely converted to biomass, thus increasing "food" available to local species. This question is illustrative of the concerns expressed by the public.

Bioremediation is a relatively inexpensive technology using available nutrient additives. Application involves low impact application techniques and a low degree of labour intensity. Low impact application techniques are an advantage in addressing the problem of beached hydrocarbon remediation with minimal disturbance of the beaches themselves. Nutrient enhancement bioremediation techniques are not only applicable to oil spills but also to soil remediation of weathered hydrocarbons which widely occur in refineries and transportation terminals, though the case studies are not common in the open literature and mostly occur within the affected companies.

Obstacles to application of bioremediation

A major obstacle to the application of bioremediation techniques during an oil spill has been a lack of agreement of among all appropriate international, national and local regulating agencies regarding the use of bioremediation techniques after the occurrence of an oil spill. This delays or in some recent cases eliminates the use of bioremediation as a tool for oil spill clean-up. Since there has not been agreement regarding the application of nutrient enhanced bioremediation techniques, there also has not been a general effort to prepare for oil spills. Agreement followed by preparedness could greatly mitigate the effect of oil spill which will inevitably occur, especially in high traffic areas.

Application of nutrient enhanced bioremediation techniques to problems of soils contaminated with weathered hydrocarbons has been hampered by the existence of published examples demonstrating success. This makes it difficult for regulators to approve the process for clean-up without extensive demonstration projects. The connection between published bioremediation projects which have been accomplished on "freshly spilled oil" and bioremediation of "weathered oil" has not been demonstrated. A particularly difficult point is the determination of an acceptable endpoint for the application. It is clear that bioremediation never fully mineralizes the oil leaving a residue. Therefore, what is an acceptable endpoint for bioremediation? One answer appears to be that after limitations such as nutrients and oxygen are removed, the bacteria will proceed until no more hydrocarbons are "available". By definition, if the hydrocarbons are not available to the bacteria, they are probably now in "safe" condition, and this may be an acceptable end point.

Technological limitations requiring more research and field demonstration need to be demonstrated. Foremost among these limitations is analytical determination of the agreed endpoint of bioremediation. Currently, established techniques such as the determination of "biomarker" molecules such as hopanes which occur in small quantities require costly mass spectrometry/gas chromatography. These techniques are very expensive and time consuming. It is estimated that in the case of the "Exxon Valdez", the cost of application of these analytical techniques required to demonstrate effectiveness was of the same order as the nutrient enhanced application itself. This fact is probably a major factor in preventing the occurrence of more bioremediation projects and the concurrent publication of scientifically sound results. Additionally, the application of these techniques require removal of the samples to remote facilities, elaborate chemical separation and ensuing time delays. It would seem that development of a rapid technique based on molecular genetics that could be applied in the field would represent a major advance.

Many scientific questions exist which are not only important scientifically but if answered will greatly advance the usefulness and applicability of bioremediation. These research directions are outlined in the next section.

Research directions

Insight into fundamental processes of bioremediation will be gained by pursuing the objectives described below:

1) *Development of strong laboratory/field connections*: Fundamental understanding of microbial transformations which can be studied in the laboratory are often difficult to connect to field bioremediation studies because the connection is weak between them, and expensive and time consuming field experiments are required. When laboratory, macrocosm and field experiments are strongly connected via proper statistical design of each, the time and cost required to demonstrate field effectiveness is greatly reduced (Bragg *et al.*, 1994). Additionally, controlled laboratory experiments can be performed in the laboratory having direct meaning to real environmental situations.

2) *Development of strong connection between biological growth rates and rates of molecular transformation*: The work of microbiologists studying hydrocarbon degrading organisms and bioremediation scientists studying rates of molecular destruction, are often only weakly connected. Development of rates of transformation which are strongly connected to biological dynamics would greatly assist scientists and engineers design, accelerate and control bioremediation processes. For example, hydrocarbon destruction is most rapid when microbes are in the growth phase. Developing rate models which directly relate the number of microbes in the growth phase with the rate of destruction of hydrocarbons would greatly enhance our understanding of the process and how to accelerate and control it. This knowledge would allow application of rate limiting nutrients more effectively and with proper timing. It would also allow greater control by providing biological assays (perhaps based on RNA/DNA), which would more easily and rapidly indicate the state of the bioremediating population, rather than traditional time consuming chemical analyses, allowing quicker intervention to adjust unfavourable conditions.

3) *Development of an understanding of the microbial ecology of soil micro-organisms:* Bioremediation experiments often slow down or accelerate in the field for reasons which are not well understood and are not related to the chemical composition of the substrate. Understanding of the relation of the degrading organisms and their relation to the physical surroundings is an important next step in controlling the delivery of nutrients and the removal of products. Such understanding can lead to new treatments which will greatly accelerate the rate of destruction and the completeness of the remediation.

4) *Understanding predator/prey relations in bioremediation experiments:* As biomass accumulates during hydrocarbon destruction, higher organisms begin to feed on the microbes. If this occurs too rapidly, bioremediation may be slow. If it occurs too slowly, bioremediation may also slow because the microbes may fill available niches and stop growing. Understanding and control of predation could lead to new ways of accelerating bioremediation by removing biomass from the available niches, allowing more organisms to grow.

5) *Understanding of chemical endpoints of bioremediation processes*: In the destruction of spilled and/or aged hydrocarbons by microbial degradation, not all the hydrocarbon is mineralized. The reason for this is not completely understood, although it is usually attributed to "bioavailability". However, there is evidence that microbes may participate in cross-linking heavier fractions of hydrocarbons via oxygen addition. This process has been termed "bioasphalting" by analogy to chemical asphalting. If this can be demonstrated, it is good news for the bioremediation community because then a natural endpoint is determined for the process. The bioasphalting renders that which is not mineralized safe and inert.

Policy recommendations

The following recommendations are made to the OECD for consideration:

- Identify high risk tanker routes where accidents are likely to occur. This process should include historical data regarding previous accidents and consideration of overall tanker volume and particularly difficult weather conditions. For example, tanker accidents in the northern hemisphere often occur during the winter months. Of course, preventative strategies should be reviewed.

- Determine applicability of nutrient enhanced bioremediation in high risk regions. Local authorities should be encouraged to measure hydrocarbon degrading microbial activity and existing nutrient levels.

- International agreement on rules for application of nutrient enhanced bioremediation should be developed with clear lines of authority established among international, national, regional and local authorities.

- High risk areas should be encouraged to develop plans for rapid implementation of bioremediation techniques when a spill occurs. This would include availability of nutrients, application equipment and personnel. Ecologically vulnerable or economically important tourist areas should receive top priority. Where possible, controlled spill experiments should be undertaken to determine optimum treatments and techniques for particular areas.

- Research recommendations for specific scientific questions have been outlined above. However, several bioremediation applications need to be further developed and demonstrated. The advantages of applying nutrients to open water spills needs to be demonstrated. Evidence suggests that rapid application of oleophilic nutrients to open water oil spills may prevent oil from reaching sensitive coastline areas. Similarly, evidence suggests that application of nutrient packages in wetlands and swamps where vegetation occurs can also help recovery of plant life, but techniques need to be developed and demonstrated.

- To achieve the above, improved nutrient packages are necessary and methodology needs to be developed which allows continuous delivery of the optimum nutrient levels until completion of the bioremediation is accomplished.

- Interdisciplinary teams should be formed to address research and development problems. Experience has clearly indicated that truly interdisciplinary teams make the most rapid and effective progress in developing and applying bioremediation techniques because of the complex nature of the problems which involve microbiology, chemistry, toxicology, hydrology, engineering and other disciplines.

- Corporations should be encouraged to publicise successes in applying bioremediation techniques to clean-up of hydrocarbon spills within their facilities. Ways should be found to foster co-operation between regulators and corporations to reduce fear of liability and make available information to the scientific community.

- Examples of successful bioremediation projects should be publicised and provided to the general public.

Questions, comments and answers

Q: What happens to the microbe population after degradation is complete?

A: This is what we call the slime monster question. People worry about the build-up of biomass and fear that it will come out and attack something else. It's just biomass and part of the food chain. It is consumed by the next higher organism, and there are many indications that fish populations boom as a consequence.

Q: Twelve hundred miles of beach got the spill but only 70 were sprayed -- what happened to the rest?

A: Don't take the mileage too seriously -- there's something here about fractal geometry. Most of the 1 200 miles were very lightly oiled, and clean-up occurred naturally. The 70 miles were heavily impacted.

Q: Did you add bacteria or was it the natural consortia?

A: We didn't add any bacteria. These are indigenous organisms from the area. As an aside, we were under pressure from vendors, and we set up a separate group to evaluate their treatments but didn't find any that were significant. The indigenous organisms were excellent. Seeding and augmentation had been tried on the beaches of Alaska but without success, so there seemed no need to add organisms.

Q: Did any of the added nutrients enter the sea?

A: There was extensive toxicology done both in laboratory and field -- on baskets of mussels, etc. -- and measurements of run-off. These showed that there was no nitrogen run-off that would be injurious to local species. Avoiding this was a prime objective.

Questions, comments and answers cont'd.

Q: I presume there was also oil in the water. How did you remove it?

A: The beaches were boomed and the oil off the beach was washed into skimmers and collected.

Q: Can you comment on the usefulness of laboratory research when applied to a real case?

A: Its usefulness lies in the ability to run control experiments. The problem is the heterogeneity and variability of the samples collected. It is very difficult to design statistically, but what we did in the lab really did predict the effectiveness of the treatment. We did a variety of experiments from small flasks to microcosms. The latter reasonably well simulated conditions on the beach, except that it was easier to deliver oxygen on the beach!

Q: Does this kind of micro-organism live in the presence of lead?

A: In soils containing tetraethyl lead, these organisms do just fine.

C: It is important to point out that the hopane analysis was developed late because branched and straight chain hydrocarbons were degraded at much the same rate.

A: We found out that the local organisms were so active that the original ratios were not useful. The hopane analysis biomarker was developed for this situation.

REFERENCES

ATLAS, R.M. (1990), personal communication.

BRAGG, J.R., R.C. PRINCE, E.J. HARNER, and R.M. ATLAS (1994), "Effectiveness of Bioremediation for the Exxon Valdez Oil Spill", *Science* 368, pp. 413-418.

CHIANELLI, R.R. *et al.* (1991), "Bioremediation technology development and application to the Alaskan spill", in *Proc. 1991 Int. Oil Spill Conf.*, pp. 549-558, 4-7 March, San Diego, California, Am. Petrol. Inst., Washington, DC.

CHIANELLI, R.R. (1994), "Bioremediation: Helping Nature's Microbial Scavengers", *Proceedings of the Royal Institution* 65, pp. 105-126.

GUNDLACH, E.R., P.D. BOEHM, M. MARCHAND, R.M. ATLAS, D.M. WARD, and D.A. WOLFE (1983), "The Fate of Amoco Cadiz Oil", *Science* 221, pp. 122-129.

HALMO, G. (1985), "Enhanced Biodegradation of Oil", in *Proc. 1985 Oil Spill Conf.*, pp. 531-537, 25-28 February, Los Angeles, California, Am. Petrol. Inst, Washington, DC.

LEE, K. and E.M. LEVY (1989), "Biodegradation of Petroleum in the Marine Environment and its Enhancement", in J.O. Nrigau and J.S.S. Lakshminarayana (eds.), *Aquatic Toxicology and Water Quality Management*, p. 221, John Wiley & Sons, New York.

LEE, K. and E.M. LEVY (1991), "Bioremediation: Waxy crude oils stranded on low-energy shorelines", in *Proc. 1991 Oil Spill Conf.*, pp. 541-547, 4-7 March, San Diego, California, Am. Petrol. Inst., Washington, DC.

LEE, K., G.H. TREMBLAY, and E.M. LEVY (1991), "Bioremediation: Column application of slow release fertilizers on low energy shorelines, in *Proc. 1991 Oil Spill Conf.*, pp. 449-454, Tampa, Florida, Am. Petrol. Inst., Washington, DC.

OUDOT, J. (1984), "Rates of Microbial Degradation of Petroleum Components as Determined by the Computerized Gas Chromatography and Computerized Mass Spectrometry", *Marine Environmental Research* 13, pp. 277-302.

PRINCE, R.C., J.R. CLARK, and J.E. LINDSTROM (1990), "Bioremediation Monitoring Program", report to US Coast Guard, Exxon Co. USA, US EPA, Alaska Dept. of Environmental Conservation, Anchorage.

PRINCE, R.C., S.M. HINTON, J.R. BRAGG, D.E. ELMENDORF, J.R. LUTE, M.J. GROSSMAN, W.K. ROBBINS, C.S. HSU, G.S. DOUGLAS, R.E. BARE, C.E. HAITH, J.D. SENIUS, V. MINAK-BERNERO, S.J. MCMILLEN, J.C. ROFFALL, and R.R. CHIANELLI (1993),

"Laboratory studies of oil spill bioremediation; toward understanding field behavior", *Preprints of the Division of Petroleum Chemistry, American Chemical Society* 38, pp. 240-244, Washington, DC.

PRITCHARD, P.H. and C.F. COSTA (1991), "EPA's Alaska oil spill bioremedation report", *Envir. Sci. Technol.* 25, pp. 372-379.

PRITCHARD, P.H., C.F. COSTA, and L. SUIT (1991), *Alaska Oil Spill Bioremediation Project*, Report No. EPA/600/9.91/O.46a, b, US Environmental Protection Agency, Gulf Breeze, Florida.

SENDSTAD, E. *et al.* (1982), "Enhanced Oil Biodegradation on an Arctic Shoreline", in *Proc. 5th Arctic Marine Oil Spill Program Tech. Semin.*, pp. 331-340, 15-17 June, Edmonton, Alberta, Environmental Protection Service, Ottawa, Ontario.

SENDSTAD, E., P. SVEUM, L.J. ENDAL, Y. BRANBAKK, and O.L. RONNING (1984), "Studies on a Seven Years Old Seashore Crude Oil Spill on Spitsbergen", in *Proc. 7th Arctic Marine Oil pill Program Tech. Semin.*, Edmonton, Alberta, 60-74, 12-14 June, Environmental Protection Service, Ottawa, Ontario.

STONE, R. (1992), "Oil Clean-up Method Questioned", *Science* 257, p. 320.

SVEUM, P. and A. LADOUSSE (1989), "Biodegradation of Oil in the Arctic: Enhancement by Oil-Soluble Fertilizer Application", in *Proc. 1989 Oil Spill Conf.*, pp. 439-446, 13-16 February, San Antonio, Texas, Am. Petrol. Inst., Washington, DC.

TELLIER, J., A. SIRVINS, J.C. GAUTIER, and B. TRAMIER (1984), *US Patent No.* 4460692.

VENOSA, A.D., M.T. SUIDAN, B.A. WRENN, K.L. STROHMEIER, J.R. HAINES, EBERHART, D. KING, and E. HOLDER (1996), "Bioremediation of an Experimental Oil Spill on the Shoreline of Delaware Bay", *Environ. Sci. & Tech.* 30:5, pp. 1 764-1 775.

ZOBELL, C.E. (1973), "Bacterial Degradation of Mineral Oils at Low Temperatures", in D.G. Ahearn and S.P. Meyers (eds.), *The Microbial Degradation of Oil Pollutants*, pp. 153-161, Publication No. LSU-SG-73-01, Center for Wetland Resources, Louisiana State University, Baton Rouge.

Parallel Session II

II-A. PREVENTION OF WATER POLLUTION: MUNICIPAL SOURCES

SUSTAINABLE DEVELOPMENT IN THE URBAN WATER CYCLE

by

M.M.A. Ferdinandy - van Vlerken and A.H. Dirkzwager
Ministry of Transport, Public Works and Water Management
Institute for Inland Water Management and Waste Water Treatment (RIZA)
Lelystad, the Netherlands

Introduction

The RIZA (Institute for Inland Water Management and Waste Water Treatment) is a national advisory and research body of the Dutch Ministry of Transport, Public Works and Water Management. Amongst others, one of the tasks of RIZA is to supply decision-makers with information needed to formulate policy in the field of inland water management. The urban water cycle is part of this field.

In the Netherlands, an extended system for collecting, transport and treatment of urban wastewater has been installed since the beginning of this century. Although some negative aspects have been recognised, in general the system is operating properly. Nevertheless, in line with the relatively young principles of 'sustainable development', new approaches to the urban water management are under consideration at the moment. 'Sustainable development' implies a broad and fresh look at processes and activities. Sustainable development is based on an integral and interdisciplinary evaluation of possible measures, directed to future desires and needs. Careful use of space, energy and raw materials, as well as minimisation of emissions and waste production, are some of the starting points of sustainable development.

This paper tries to explain the reasons and methods used in the Netherlands to develop a sustainable technology strategy for the urban water cycle. Some of the principles and background of sustainable development will be presented, followed by a method used for the definition of possible new approaches for the urban water cycle. A few examples of current research on technology and assessment will be presented, ending with the possible role of biotechnology within the future urban water cycle. The paper starts with a brief history, leading to the present Dutch situation.

The present urban water cycle in the Netherlands

Brief history

The Netherlands are located at the downstream part of large European rivers (Rhine, Sheldt, Meuse). The country has always been active in the field of water and water management, which is and one third of the country is situated below mean sea level. Half of the surface is protected by dikes against flooding by either sea or rivers. The Netherlands with its 15 million people can be considered as a densely populated country.

While protection from flooding was the first activity (Middle Ages), the supply of water and drainage were important items in a later stage. Water quality became of interest during the last decades. In the beginning of this century, wastewater from most houses was discharged directly into open water, resulting in public health problems. Therefore, public water supply systems were built, while sewer systems were constructed to transport the wastewater to large rivers or lakes outside the cities. The industrialisation process after the Second World War, combined with agricultural growth, resulted in pollution of the surface waters. In 1970, the Dutch Pollution of Surface Waters Act (Wvo) came into force, leading to an adequate water management. This water Act requires permits and fees for every discharge of wastewater into surface water. Furthermore, the permit conditions include emission standards, based on targets for quality of surface waters. After installation of the Act, an intensive program for building wastewater treatment plants was started.

Present situation

Nowadays, the water supply system reaches almost all houses, and 97 per cent of the houses are connected to sewer systems. Besides domestic wastewater and rainwater, (pre-treated) industrial wastewater is also collected in the central sewage system. The sewage system leads the water to wastewater treatment plants (WWTPs). The effluents of WWTPs are discharged into surface waters. In total, about 440 WWTPs are operational, with a design capacity of around 24 million population equivalents (p.e.'s). In Figure 1, a schematic view of the present situation in the Netherlands with regard to the urban water cycle is presented.

Figure 1. **The present urban water cycle in the Netherlands**

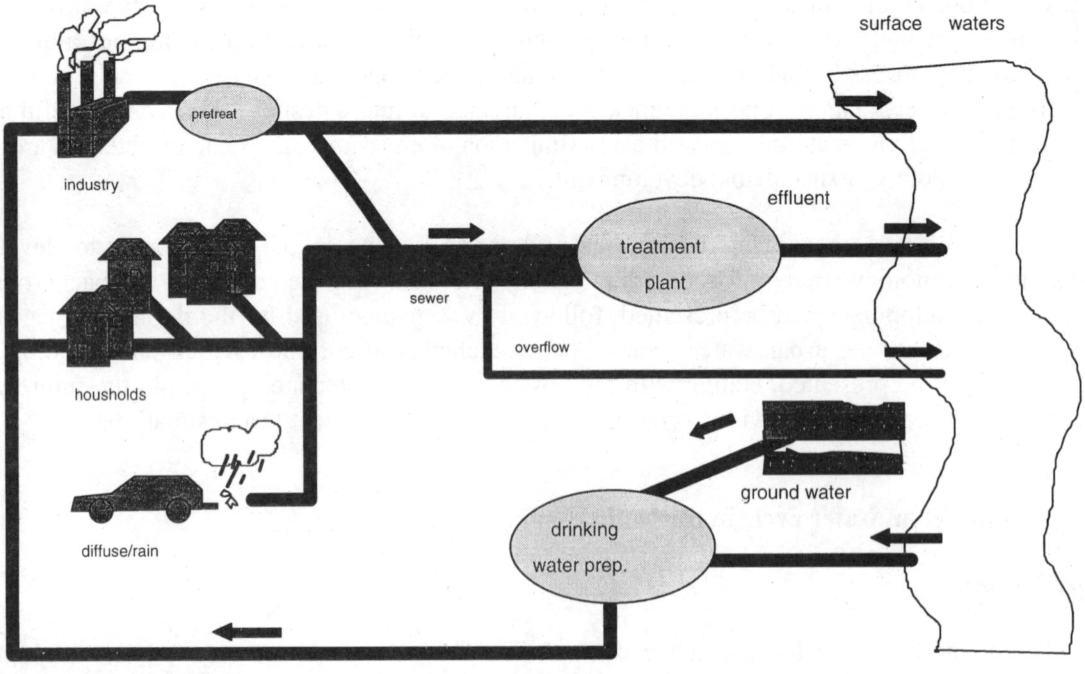

Source: Author.

The cycle starts at the preparation of drinking water from ground- or surface-water. The drinking water is then used for (some) industrial and domestic purposes.

Domestic wastewater is directly collected by the sewer system. Most industrial wastewater is pre-treated at location and then transported to either the sewer system or the surface water. Rainwater is collected in industrial and housing areas where impervious surfaces are applied. Rainwater can contain pollution from diffuse sources (air, corrosion from materials, traffic), whose origin often is hard to localise.

The total length of the sewers is 70 000 kilometres, and two types of sewer systems are currently operated. Around 78 per cent of the sewers are 'combined systems', in which wastewater and storm water are collected together. During times of heavy rainfall, overflow structures in the system enable direct discharge of the storm water into the open waters. The remaining sewer systems (22 per cent) are 'separate systems', in which wastewater and storm water are collected separately. The wastewater is transported to the WWTP, and the rainwater to the surface water.

Although both composition and volume of the three types of urban wastewater (domestic, industrial and diffuse) differ, all types of wastewater are collected and treated together. The treatment in the WWTP (see Figure 2) has always been based on biodegradation. The treatment originally was meant to remove BOD and COD (Biological and Chemical Oxygen Demand). At present, the national legislation on effluent quality also contains legal standards for nitrogen and phosphorus content, in order to prevent eutrophication of the surface waters.

The treatment has known a rather pragmatical development, resulting in different technological concepts (e.g. oxidation tanks or carrousels) for biological remediation in active sludge systems. For P-removal, biological and chemical processes are applied. The biodegradation results in an increase of the biomass; surplus sludge must be removed after consolidation. The surplus sludge used to be applied as manure in agriculture. However, due to the presence of heavy metals and organic pollutants, legislation does not allow this application any more. Therefore, the sludge must be incinerated, treated in another way, or disposed of in landfills (Dirkzwager *et al.*, in press). In some cases the sludge is fermentated, resulting in biogas (H_2S, NH_4) for self-supplying the required energy of the treatment plant. In Figure 2 a general flowsheet of the WWTPs is depicted.

Present problems

Although the present approach of water management in the urban water cycle (see Figure 1) is functioning properly, some negative aspects have been identified.

As already mentioned, the surplus sludge produced during the treatment is polluted. A solution (prevention or treatment) for this problem of creating a waste stream must be found.

Secondly, the overflow from the sewage system during times of heavy rain results in unnecessary pollution of surface waters. The overflow is also partly responsible for the Dutch problem of high water in the waterways, leading to flooding (e.g. 1994 and 1995, Meuse). At the same time, the present system also is one of the reasons of desiccation, because around 66 per cent of the drinking water is prepared from groundwater. Finally, targets for surface water quality often are not met, partly due to emission of oxygen binding compounds (overflow) and heavy metals (effluents).

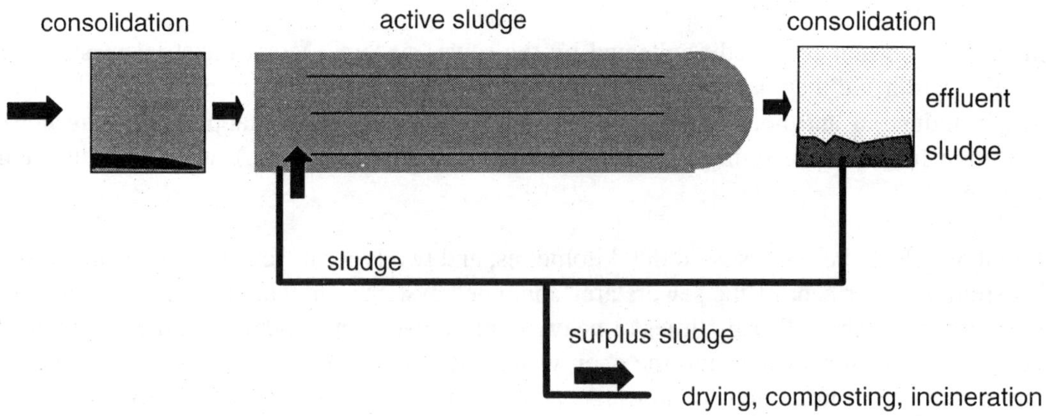

Source: Author.

Sustainable development

The Dutch environmental policy for the medium-long term, as described in the Dutch National Environmental Policy Plan (NMP), is directed at *sustainable development*. The Policy Plan distinguishes three basic points: 1) integral life cycle management (closing of substance cycles), 2) decrease of energy consumption, and 3) quality improvement. When following the principles of *sustainable technology development*, setting up a development strategy will include more aspects than the solution of the above-mentioned problems with the current system. Not only the direct results of a process (effluent quality, sludge production), but also the process-related features (like use of energy, chemicals, space, building materials and emissions), become important.

When looking at the urban water cycle from a larger distance, some questions on efficiency come to mind, like:

- why use high quality drinking water for purposes for which a lower quality is sufficient?
- why the same treatment of different types of wastewater?
- why an end-of-pipe approach, instead of prevention of pollution or source-tackling?
- why introduce long transport distances and dependency on a central system?
- why have national effluent standards when variations in quality of the receiving surface waters exist?

Discussion on these kinds of new and drastic reflections is opened in the framework of development towards a sustainable future.

Although 'sustainability' is one of the topics in most disciplines all over the world, many different perceptions of this rather abstract concept exist. The original definition of sustainable development from the Brundtland Commission (Brundtland, 1987) is as follows.

> Sustainable development is a process of change in which the exploitation of resources, the direction of investments, the orientation of technological development, and institutional change are all in harmony and enhance both current and future potential to meet human needs and aspirations.

This definition implies a different, broad approach in both policy and technology development. Sustainable development requires an equilibrium between technological, economic, social, institutional and cultural aspects, now and in the future. Development in different fields (industry, agriculture, transportation and environment) should be interlinked. Strategies of development within a certain field should be based on:

- a look into the future:
  meeting possible future needs and desires;
  fitting in future infrastructure.

- a look over the borders of the field:
  without interference of performance of other activities outside the field;

- a look over the borders of the environmental compartment:
  without shifting environmental problems from air to water or soil.

Not only development in other sectors, but also the interpretation of the term *sustainability* is subject to changes in time. Therefore, sustainable development is an ongoing process, in which visions will have to be updated continuously.

Finally, sustainability is a subjective term, for which new standards and values must be developed. The first step in this development is the definition of criteria or factors of sustainability. Discussions and evaluation on which aspects must be considered as sustainability are still going on. Some examples of possible sustainability factors are 'long-lasting', 'good environmental impact', 'little use of chemicals, energy, primary material' and 'little nuisance'. Discussions within both scientific and political fields will have to lead to a general agreement on criteria and strategies of sustainability. Also, tools to assess these criteria must be developed.

Since so much abstraction surrounds the concept of sustainability, a two-track approach could be applied in strategies for technology development (see Table 1).

Table 1. **Steps within the traject of sustainable development**

Track A: Assessment	Track B: Technology development
1. Perception of the term 'sustainability'	1. Picturing future, sustainable situations (alternative approaches)
2. Definition of sustainability criteria	2. Identification of required technology development
3. Development of assessment tools	
4. Assessment	
selection -------------->	3. Feasibility studies on the proposed technological solutions
Updating assessment	
selection -------------->	4. Technology development
selection -------------->	5. Implementation

Source: Author.

The two tracks should be followed simultaneously and be interactive. Track A will lead to an agreement of sustainability factors and tools to assess sustainability of technologies and processes. Track B is set up to develop technologies, starting with the definition of possible, sustainable situations in the future. Then an inventory is made of the technology required to reach these situations. Various techniques are hence submitted to feasibility and development studies. Finally, Track B will lead to full-scale technologies: concrete solutions. During the different steps within the technology development track, sustainability assessment will take place regularly. The assessment might lead to abandoning ideas as well as to development of new ideas. The two tracks together will result in a choice of the optimal solution. A continuous verification of the perception of sustainability, the criteria and the pictured future situations is essential, since these matters are subject to change over time.

Developing tools for assessment

Criteria of sustainability

The first step towards assessment of sustainability, is to identify and achieve general agreement on the aspects that must be considered as affecting the sustainability. Since the term is subjective, many variations in the meanings as to which factors belong to sustainability occur, for instance, whether or not costs belong to the sustainability factors. In Table 2, some criteria are summarised.

Table 2. **Some criteria affecting sustainable development**

storage	energy raw materials building materials chemicals space (and time)	social	nuisance (smell, noise) public health, hygiene acceptance NIMBY (not in my backyard)
nature	biodiversity desiccation pollution (water, air, soil) creation of waste streams	infrastructure	fulfilling future needs and desires fitting in future infrastructure
		technology	technically feasible flexible (fitting in changing situations) process control reliability
		economics	costs

Source: Derived from Graaf van der *et al.*, 1996.

Tools for assessment

Assessment of the sustainability criteria mentioned in Table 2 is an important factor within sustainable development. The assessment will always contain subjective elements.

The assessment of techno-economical aspects will be relatively simple, although as to the costs, discussion on both political and public level must take place on the question: 'how much money do we want to invest in a sustainable future?' The assessment of the other criteria is more complicated. Some starts are made with the development of tools for sustainability assessment.

One of those tools is the Life Cycle Analysis (LCA), with which the environmental impact (part of the sustainability criteria) is investigated. The LCA method is a tool originating from ecology, in which the complete consequences for the environment of a process or product are quantified. In 1995, an LCA analysis was carried out on WWTPs in the Netherlands (Roeleveld *et al.*, 1996).

The LCA method starts with the definition of the 'working unit': the borders of the system to be analysed are set (only WWTP, or entire water chain). From this working-unit, all phases (the life cycle) are analysed: all required materials (and their production), the production process (with energy and chemical demands, emissions to air/water/soil), as well as the (waste) products are subject of study.

All aspects are translated (*'classification'*) into ten environmental effects: mineral depletion, energy depletion, global warming, ozone layer depletion, human toxicity, aquatic ecotoxicity, terrestrial ecotoxicity, photochemical ozone creation, nutrification potential and acidification potential.

This classification results in a quantification of the environmental effects of the working unit. The next step is to assess the gravity or unsustainability of these effects. This is done by means of *'normalisation'*: the contribution of the working unit to each of the ten environmental effects is related to the total input on these effects of a larger unit (a country, the world).

The working unit from the LCA analysis on WWTPs in the Netherlands was set at a plant for 100 000 population equivalents. All events from the arrival of the wastewater at the plant, up to effects in the surface waters, were classified.

Different 'scenarios' were analysed: no treatment (zero-option), BOD removal, BOD and nutrient removal, and BOD/nutrient removal with effluent polishing. The first problem occurred with the zero-option: the LCA methodology has no translation for the effects of oxygen-binding substances on the quality of surface water. Furthermore, the LCA does not account for the production of normal, toxic and nuclear final waste. For these impacts, no classification into the general environmental effects could be performed. However, the effects were quantitatively analysed, in order to compare the different treatments.

The normalisation was performed with the Netherlands as total input area. The effects of the WWTP on each of the ten environmental problems were related to the total effect of the Dutch society. This resulted in ten relative contributions of the WWTP, expressed as percentage of the total (100 per cent) input of the Netherlands. The analysis showed that the contributions of the WWTP to the ten environmental effects were relatively low (this means that other activities than wastewater treatment are mainly responsible for the effects).

Logically, the highest contributions of the WWTP were found for the zero-option, on the effects of nutrification (38 per cent), aquatic toxicity (9 per cent) and the production of final waste (around 5 per cent). To all other effects, the contribution of the WWTP was smaller than one per cent. Especially for the effect of energy depletion, this was a surprising result.

The effects were reduced through treatment of the wastewater. For instance, nutrient removal resulted in a decrease of the effect nutrification to less than 20 per cent of the total Dutch nutrification. The effect on the aquatic ecotoxity was reduced to 3 per cent with BOD removal and mainly caused by emission of mercury (1.7 per cent), whereas copper, cadmium and zinc had smaller contributions (<1 per cent). A summary of the results is presented in Figure 3.

The conclusions of this study were directed towards the need for development. To improve the sustainability of the WWTP, most attention should be paid to minimisation of discharge from pollutions with the effluent and minimisation of sludge production. Development of energy-saving treatment technology was regarded as less important.

Figure 3. **Results of Life Cycle Analysis of wastewater treatment plants**

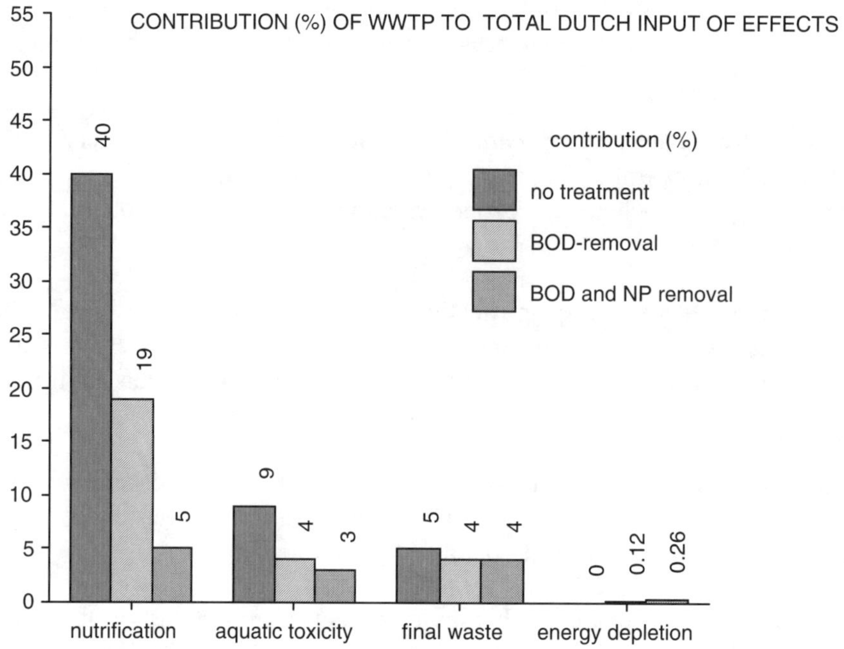

Source: Derived from Roeleveld *et al.*, 1996.

However, the interpretation of the results must be regarded with some caution. The normalisation method strongly influences the results. In this case, the impact of the WWTP was compared to the total input from the Netherlands. When compared to bigger or smaller units (the world, or a region), the contribution of the WWTP could be different.

Besides the working unit (world, country) of the normalisation set, the reliability of these data is also of importance. The data used for the total Dutch input were defined in 1993 and are currently being verified (IVAM, 1995) and will be related to a European level.

A second remark concerns the interpretation of 'low' contributions (e.g. one per cent) as being 'not important'. When the total contribution of the effect (100 per cent) is built up out of many small contributions, each of them is relevant. A more detailed insight in the partitioning of all contributions is required as a basis for conclusions.

Another way for the evaluation of the classification is to avoid any normalisation. Judgement of the seriousness of the effects would then be based on an internal comparison within the working unit.

The totals of the working unit to each of the ten environmental effects would be seen as 100 per cent. An inventory within the life cycle could reveal phases during which relatively high impacts on a certain environmental effect occur. For instance, the use of a specific building material could represent 75 per cent of the effect on ozone depletion of the working unit. In this case, replacement of this material would be a significant improvement.

This might be a more conscious approach, in which we do not point the finger to other processes and activities.

Finally, the LCA method considers all ten environmental effects to be equally important. It is possible to apply weighing factors for the effects. Weighing factors might be based on national or global agreement on the seriousness of environmental problems. This might have a large effect on the results of an LCA test. For instance, if energy depletion were regarded as extremely important, the contribution of one per cent from the WWTP could increase. Weighing factors might be adapted to local situations, like the quality of receiving surface water, available space or energy supply system.

If and how weighing factors must be defined is a matter of intense political and scientific discussions. Developments within the ecotoxicological science is important in this field.

Increased knowledge of the behaviour of various compounds might lead to increased importance of (aquatic) toxicity.

A search for new, sustainable approaches within the urban water cycle

The search for sustainable solutions for the future, started in the Netherlands in the early 90s. An interdepartmental research programme was set up: Sustainable Technological Development (STD). The program aims at defining future production and consumption patterns, in order to *reduce the pressure on the environment* with a factor 20. The challenge of this task is to define how to reach this decrease and which technology development would be necessary.

In June 1993, a workshop on the subject of water and sustainable management was organised. In 1994, the STD started a pilot study in which the possibilities for the urban water cycle would be examined. This STD water study was carried out by a consulting agency (Witteveen en Bos) and a project team of experts in all fields of water management (Graaf van der *et al.*, 1996).

The working unit chosen was a city of 100 000 inhabitants, for which the reduction of 'unsustainability' would have to be in the year 2040.

First the (un)sustainability of the present urban water cycle was determined. For this purpose, several factors of unsustainability were defined, like emission into surface water, production of solid waste, use of chemicals, energy, building materials, water balance, space and nature, hygiene and reliability. Hence several ideas or solutions to improve sustainability were developed, some of which have already been mentioned in this paper. Since a single idea can not solve all problems, the ideas were combined into five scenarios. Some main starting points of the scenarios will be described below.

Scenario 0: Autonomous development

This scenario is based on a continuation of the current policy, combined with several optimisations. This could lead to a decrease of groundwater use for drinking water preparation (50 per cent), renovation of the sewer system, optimisation of the WWTP and in some cases disconnection of rainwater from the sewage system, followed by infiltration.

Scenario 1: Zero emission

The starting point is extreme reduction of emissions to the surface water. All the water is collected in the sewage system, and no overflows occur. Treatment in the WWTP leads to effluent qualities corresponding with the quality of the surface water, for instance with effluent polishing techniques. Produced sewage and drinking water sludge is remediated for beneficial use. Drinking water is prepared from surface water only, in order to maintain a sufficiently high groundwater level.

Scenario 2: Flux restriction

Reduction of the quantity of water flows, realised by savings on water consumption in many places of the water cycle. Drinking water would only be prepared from surface water and supplied in bottles, combined with a second quality tap water. Water savings within domestic use could be vacuum or compost toilets, water-saving showers, collection and use of rainwater for secondary purposes. Overflows would hardly occur, and rainwater would be disconnected from the sewerage and infiltrated.

Scenario 3: Source reduction

Prevention of pollution and tackling at the source would lead to an improved quality of the wastewater and a decreased need for end-of-pipe techniques. To reduce emissions from diffuse sources, the following measures would be taken: replacement of copper pipes for drinking water distribution and street fixtures (traffic pools, signs, roof gutters) by non-corroding materials or coatings. A decrease in domestic emissions reached by an enforced product policy would result in replacement of many consumer products (soaps, cleaning materials, paints). Industrial emissions would be decreased by total provisions, realised by prevention (clean technology), closing substance cycles and sanitation at the source.

Scenario 4: Small scale approach

The starting point is a local approach, leading to a more conscious relation between man, environment and nature. Central provisions are no longer at stake. Also, some disadvantages (over-dimensioning, dependency and unclear cost structure) of central provisions in a large-scale approach are avoided. Drinking water is supplied in bottles, treatment of wastewater takes place in compact local plants, and rainwater is collected and (after natural treatment with helophytes) suitable for use. The water-saving measures from scenario 2 are applied.

In order to analyse possible results of the implication of these scenarios, the effects on sustainability as well as the economic consequences were analysed. The results are summarised in Figure 4.

For the evaluation of sustainability, the importance of several effects on the urban water cycle was weighed, based on the direct effects on the water chain. This resulted in four categories, with different importance, expressed in weighing factors of 1 000, 100, 10 and 1 points per category. Factors within the most important category were:

- emission of heavy metals into surface water;
- emission of organic micro-pollutants into surface water;

– sludge production;
– groundwater management;
– emission of oxygen-binding materials through overflows at peak discharges.

Figure 4. **Results from STD water study**

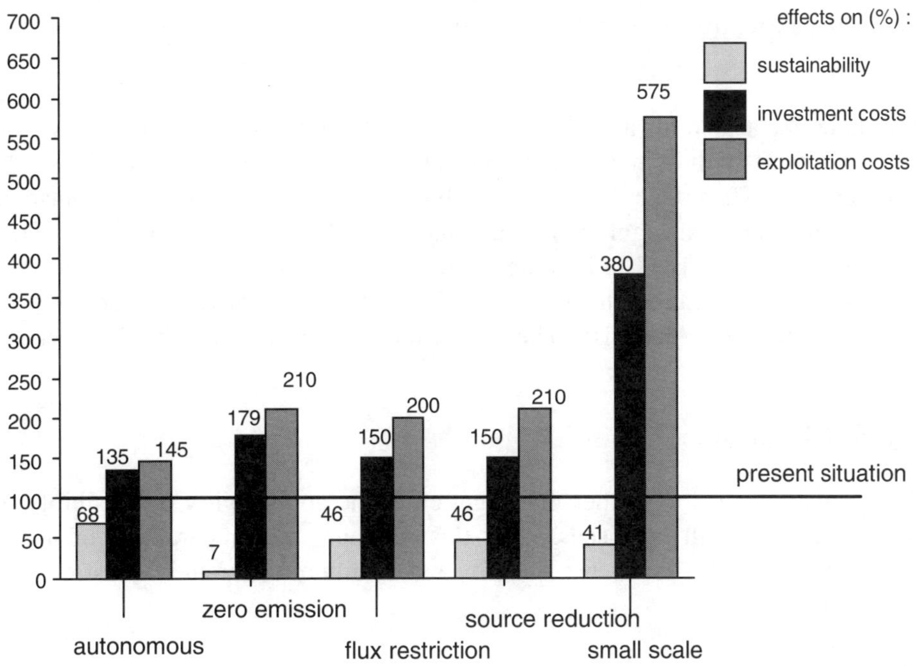

Source: Derived from Graaf van der *et al.*, 1996.

The 'zero emission' scenario would have the best environmental impact, followed by the 'source reduction'. In these scenarios, residual emissions of heavy metals and organic compounds would still occur. The costs for scenarios 1, 2 and 3 are higher (maximum with a factor 2) than for autonomous development. The small-scale approach would lead to much higher costs. Combinations of the different scenarios and a further development of the ideas most probably will lead to more optimal situations.

The study pointed out three major aspects of unsustainability in the present urban water cycle:

– discharge of (micro)-pollutants;
– production of sewage sludge;
– use of groundwater.

This result of course is related to the weight which was connected to the different aspects of unsustainability. In this case, the weighing was based on the direct effects on the water chain. Another approach is used in the LCA method (see section on developing tools for assessment above), in which the importance of environmental effects is weighed by comparing the contribution of the unit in question (e.g. the water chain) with all national or global activities.

Whether or not, and if so, 'how much' weight should be related to the aspects, is a question which needs an intense scientific and political discussion.

The STD study concluded with a general starting point for the development within the urban water cycle: a more separate and local collection and treatment of different types of water streams.

The project was one of the first steps on the route to sustainable development of the urban water chain. Some of the ideas developed in the STD project follow the technology development track (see Table 2) and are currently the subject of further study (see section on technology development below).

Furthermore, the search for new directions within the urban water cycle is a continuous process. Therefore, a new STD-study is now being conducted. In this second search, the working unit is an urban agglomeration. The project has started with the definition of an ideal, sustainable, image of the water management in an urban agglomeration in the Netherlands for the year 2040. The picture given is one of peace, where man lives in harmony with his environment. New ideas are being developed by experts, resulting in variations to reach the ideal situation. These solutions will be assessed by method of LCA (Life Cycle Analysis). The final results will be available by Summer 1997.

Technology development

The STD study has led to further investigation of the applicability or feasibility of some of the ideas. Some examples will be discussed in this section: the preparations of a residential area, feasibility of a second quality tap water, quantity and quality of different water streams, and advanced wastewater treatment.

Preparations for construction of a new residential area

Implementation of new approaches within the urban water cycle will have to take place gradually. An immediate and total replacement of existing provisions would be impossible from a practical point of view, and would have a negative economic result, due to investments already made. However, in view of building residential quarters, application of available technologies is being considered in several Dutch communities.

An example is the city of Ede, where an area-specific design ('tailor-made') for water management for a new quarter is being made. This requires an integral set-up, resulting in a working committee of urban developers, ecologists, hydrologists and civil engineers. All infrastructure and provisions required for the district are taken into account. Where possible, new technologies will be applied in a local approach. The following measures are being evaluated:

- use of sustainable materials (tested with e.g. LCA method);
- water saving measures (shower/toilet);
- supplying different water qualities;
- re-use of rainwater and separate collection of remaining rainwater and wastewater;
- infiltration of rainwater in the higher parts of the district;
- creation of non-pervious parts in streets;
- minimisation of energy consumption;
- instruction of inhabitants.

These measures should lead to closing the urban water cycle as much as possible, and to minimal burdens on other sites (surface water, WWTP of Ede).

Feasibility of separate collection and treatment

Separate, local collection and treatment of different water flows is one of the recommendations of the STD study. This could be an effective approach, since strong differences in both composition and fluxes exist within the urban water chain. The various flows are rainwater (lightly polluted), domestic wastewater (polluted) and industrial discharges (severely polluted).

The type of treatment depends on both the quality and the quantity of the wastewater. Remaining pollution in industrial wastewater might lead to specific requirements at the central WWTP in which all water is treated. Large volumes of lightly polluted rainwater have consequences for the dimensions of the WWTP. It seems more effective to keep the different streams separate and apply treatment technologies fit for each of the flows.

Besides a more effective treatment, disconnection of rainwater from the sewerage would result in several positive effects. The rainwater could be used for certain purposes, like irrigation and car-washing, resulting in a decrease in tap water use. Infiltration of the water would cope with the problem of drought, while the disconnection from the sewerage system would lead to avoiding overflow problems like pollution and high water in the waterways.

A possible disadvantage of this local and separate approach might be the creation of a less 'orderly' situation and higher costs, compared to that of a central system. It also brings along an increased dependence on the 'good behaviour' of the citizens. Nevertheless, this idea is worthy of further investigation.

Therefore, a feasibility study is being carried out by Tauw (Tauw, 1995). This study aims at a description of different urban water streams, including compositions and quantities and possible treatment technologies. In order to define the different streams, 'reference systems' will be defined. Reference systems reflect various infrastructural situations, a clayish polder with a high population density, or a sandy area with a low population. Also, some targets for sustainable situations will be described. In the second phase, scenarios for the water management in the different reference systems will be made. Several ideas will be combined within these scenarios:

- continuation of present technology;
- relatively clean rainwater is treated and used at the source;
- all rainwater is treated and used at the district, while other water is transported to the WWTP;
- grey water is treated and used close to the house;
- grey water for second quality tap water.

Ultimately, 25 scenarios for the different reference systems will be assessed in relation to the target situations.

The study will result in a strategy tool (a framework) for sustainable development within the urban water cycle. This strategy will include aspects of management, technology, jurisdiction, communication and finances. This should be a useful tool for further development.

Feasibility of a second quality water supply system

Reduction of the amount of high quality drinking water for secondary purposes is an idea which comes forward in almost all discussions and meetings. On a yearly basis in the Netherlands, 235 million cubic meters drinking water is used for toilet flushing and 135 million cubic meters for clothes-washing. This represents almost half of the drinking water consumption.

Besides water-saving measures in toilets and showers, a reduction could be realised by collecting and using rainwater around the house, or through the installation of a second quality tap water system.

These possibilities have been analysed for the residential area in Ede (NUON, 1995). Rainwater systems at very small scale at household level turned out to be less desirable. The costs would be high (Gld 20 per cubic meter), the water savings would be only 20 litres per person per day, the hygiene would be unreliable or biologically unstable, and consumer comfort would be low.

A second system for supply of a lesser quality tap water (e.g. surface water) was considered to be more efficient. The effect of the costs for installation and distribution would be balanced by lower costs for treatment in the city of Ede. The feasibility study concluded with a statement that a local second water system could contribute to a sustainable water regime. Implementation of this system will depend on local factors, such as availability of groundwater and surface water, on administrative and legal aspects.

Advanced wastewater treatment

Development within the urban water chain is directed towards prevention of pollution and minimisation of wastewater quantities. The consequence is a decreased need for central end-of-pipe technologies. However, successful and total implementation of prevention and source-tackling will take a long period. In the mean time, improved end-of-pipe treatment can offer a temporary solution. After this period, end-of-pipe technologies may still be applied at industrial sites for treatment of separate wastewater streams. Development within treatment technologies should therefore be directed towards two situations: the existing system and totally new approaches in the urban water chain (Dirkzwager *et al.*, 1994).

Medium-long term

The interim solution would be based on the existing sewage system and WWTP. The treatment plant should be optimised, in order to increase the effluent quality (heavy metals and organic micro-pollutions) and deal with the problem of polluted sludge.

For improved effluent quality, effluent polishing is one of the options. A feasibility study on effluent polishing was conducted by DHV consulting engineers (Meinema and Rienks, 1995). The target effluent quality was set at the national policy target for surface water quality (Ministry of Transport, Public Works and Water Management, 1991). The starting point was the highest concentrations of various compounds in the effluents of the Dutch WWTPs in the period between 1982-1989. The effect of different combinations of technologies like flocculation filtration, chemical oxidation or membrane technology was calculated, based on known removal performances.

could be reached by nanofiltration, combined with activated carbon. Depending on the techniques, costs would vary from Gld 45 per p.e. for flocculation filtration or cascade aeration, to Gld 1 188 per p.e. for microfiltration with hyperfiltration and cascade aeration. Since the costs for the present treatment are around Gld 82 per p.e., the extra costs were considered too high.

Effluent quality might also be realised by enhancement of the biological process within the aeration basin. Research in the field of environmental biotechnology has taught us that the potential for biodegradation is increasing. More and more organic pollutions can be degraded by bacteria (Keuning and Jansen, 1987). The disadvantage of specialised bacteria is that they generally are more sensitive to process conditions and competition with heterogenic organisms, and have low growth rates. In a suspended system, these bacteria would soon disappear from the basin (dilution). These problems can be overcome by immobilisation in biofilms or within gel-matrices. The carrier material allows retainment of the bacteria.

Figure 5. **The effect of immobilisation on sludge production**

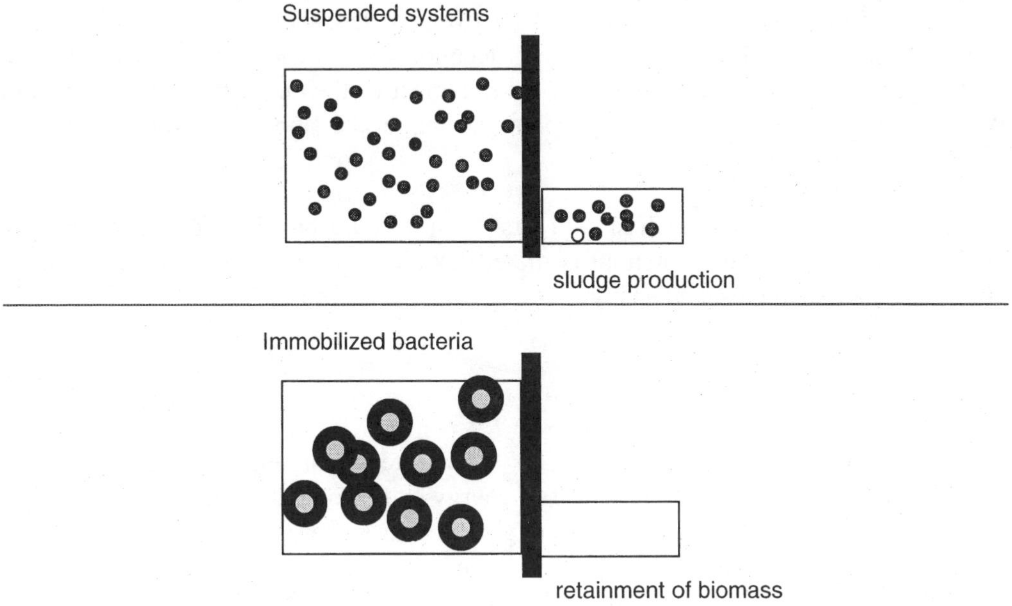

Source: Author.

Research on the agricultural university of Wageningen (Leenen *et al*., 1996*a,* 1996*b*) showed that nitrification within gel-matrices is possible. The bacteria are inside of small 'balls' of gel. This sort of immobilisation results in a kind of buffer capacity for nutrients and carbon sources, which helps in surviving times of low feeds. Comparable systems are being applied in Japan on a larger scale (Sumino *et al*., 1992; Tanaka *et al*., 1991). It might be possible in the (far) future, to use these gel-balls to make stocks of different kinds of bacteria, which can be used in biosystems whenever needed for a specific waste stream.

Nowadays, the problem of contaminated sludge is mostly dealt with by incineration in large scale plants. Biotechnology might also be effective here, by means of the technology of landfarming. This technology is applied for remediation of soils and sediments contaminated with organic pollutants (Ferdinandy *et al*., 1996). Heavy metals will have to be extracted from the sludge, possibly by means of phytoremediation, an extreme uptake of metals by plants. Phytoremediation has been successful for sediments in the north of France (Dubourguier, 1996). Also, other processes, like thermal immobilisation or flotation, might be applicable.

Long term

When thinking of a new approach of the urban water cycle, development of treatment technologies should focus on separate streams with specific compositions. This means that the existing system would be abandoned, offering more opportunity to consider all sustainability criteria during the design process. Aspects like saving energy, chemicals, space, little emissions to air, soil and water, flexible process control, and careful selection of materials, might be easier to realise when starting off with a totally new system.

Treatment can be more effective, since it will be applied on separate streams of specific composition. When effluents are destined to secondary use, more technologies will be effective.

Development of treatment technologies will be strongly dependent on the composition of the various water streams. For several compounds, technologies already exist (effluent polishing, drinking water preparation), but are not suitable for treatment of the entire wastewater stream of the present WWTP. It is possible that these techniques might be applicable for specific, smaller wastewater streams.

Research within the new approach should first focus on the physical, chemical or biological principles of processes. When the potential is there, flexible applications (for different streams, in large or small scale) should be reached easily.

The role of biotechnology

Biotechnology has always been applied in wastewater treatment. In general, biological techniques are environmental friendly (little use of energy and chemicals) and cheap. As mentioned in the previous section, the urban water cycle might shift to a more small scale approach, where treatment of specific waste streams are at hand. In the meanwhile, enhanced treatment at the WWTP could offer a (temporary) solution. For both applications, biological techniques are suitable.

As already mentioned, degradation of xenobiotics (in gel-balls) and sludge remediation (landfarming) are possible working areas for biotechnology. Another application can be found in the use of wetlands (helophytes) for treatment of rainwater and domestic wastewater. This technique is applied mostly for rural areas, where central provisions are absent.

Research within biotechnology should be directed to application of already known biological processes into technologically sound techniques. Some examples of subjects of study are:

- efficiency on different types of wastewater;
- different kinds of gel-matrices;
- efficiency of metal uptake in plants;
- effect of fungi in landfarming;
- process control.

The first step in this track is to know the (possible) composition of the different waste streams, as will be estimated in the feasibility study on 'separate treatment'.

Conclusions

The perception of 'sustainability' and criteria affecting this term are both of subjective character and subject to changed visions over time. Therefore, a strategy for sustainable development should continuously perform the following activities:

- multidisciplinary discussions on the meaning of 'sustainability';
- development of methods for assessment sustainability;
- definition of the sustainable future situation.

Developed systems will have to be flexible, in order to adjust to new visions of sustainability and to changes in infrastructure or compositions of wastewater streams. Prevention and clean technology will certainly lead to different compositions of streams.

Since techno-economic, social and sustainability aspects of measures are depending on local situations, a local approach (fit-to-situations, tailor-made) could be the best solution (Dirkzwager, 1997).

Questions, comments and answers

Q: How do government officials react to local solutions in the Netherlands?

A: We have invested in central systems so we want to use them, but in newly built areas local solutions are possible.

Q: In the United Kingdom we see pressures on the environment from agriculture, industry, transport and energy. Are the sources the same in the Netherlands, and have you calculated their respective contributions to water quality and environmental pollution?

A: When life cycle analyses were made, these calculations were done, especially for wastewater treatment.

REFERENCES

BRUNDTLAND, G.H. (1987), *Our Common Future*, World Commission on Environment and Development, Oxford University Press.

DIRKZWAGER, A. (1997), "Sustainable development: new ways of thinking about water in urban areas", *Eur. Water Pollut. Control* 7:1, pp. 28-40.

DIRKZWAGER, A.H., E. EGGERS, and M.M.A. FERDINANDY - VAN VLERKEN (1994), "Developments in waste water technologies and municipal waste water treatment systems in the future", *European Water Pollution Control* 4:1, pp. 9-19.

DIRKZWAGER, A., L. DUVOORT VAN ENGERS, and J. VAN DEN BERG (in press), "Production, treatment and disposal of sewage sludge in the Netherlands", *Eur. Water Pollut. Control.*

DUBOURGUIER, H. (1996), "Possibilités de traitement et de mise en dépot des sédiments toxiques", Conférence Européenne sur les boues et sédiments toxiques, Salon Ecotop96, 7 June, Lille, France.

FERDINANDY - VAN VLERKEN, M., W. BRUGGEMAN, and G. STOKMAN (1996), "Sediment remediations: technology development and pilot projects", presented at the OECD Workshop Mexico '96 on Biotechnology for Water Use and Conservation, 20-23 October, Cocoyoc, Morelos, Mexico.

GRAAF VAN DER, J., H. MEESTER-BROERTJES, W. BRUGGEMAN, and E. VLES (1996), "Sustainable technology development for urban water cycles", Proceedings of the Congress on Advanced Wastewater Treatment, 23-27 September, AquaTech Amsterdam, pp. 321-329.

IVAM ENVIRONMENTAL RESEARCH (1995), "Verbeterde normalisatie ten behoeve van de LCA-methodiek", Project proposal, intern. RIZA.

KEUNING and JANSEN (1987), "Microbiologische afbraak van zware en prioritaire stoffen voor het milieubeleid", in *Bioclear milieubiotechnologie*, Bioclear, Groningen, The Netherlands (update ready in 1997).

LEENEN, E., A. BOOGERT, A. VAN LAMMEREN, J. TRAMPER, and R. WIJFFELS (1996*a*), "Quantitative characterization of viability and growth dynamics of immobilized nitrifying cells", in R. Wijffels, R. Buitelaar, C. Bucke, and J. Tramper (eds.), *Immobilized cells: basics and applications*, pp. 341-348, Elsevier Science, Amsterdam, the Netherlands.

LEENEN, E. V. DOS SANTOS, J. TRAMPER, and R. WIJFFELS (1996*b*), "Characteristics and selection criteria of support materials for immobilization of nitrifying bacteria", in R. Wijffels,

R. Buitelaar, C. Bucke, J. Tramper (eds.), *Immobilized cells: basics and applications.* pp. 202-212, Elsevier Science, Amsterdam, the Netherlands.

MEINEMA, K. and J. RIENKS (1995), "Inventarisatie en evaluatie van technieken voor effluent polijsten van rioolwaterzuiveringsinstallaties", RIZA nota no. 95.033.

MINISTRY OF TRANSPORT, PUBLIC WORKS AND WATER MANAGEMENT (1991), "National policy document on water management, Water in the Netherlands: a time for action".

NUON VNB, RIZA, MUNICIPALITY OF EDE (1995), "Feasibility of a local water system for the Doesburg district in Ede", DHV Water bv.

ROELEVELD, P., A. KLAPWIJK, P. EGGELS, W. RULKENS. and W. VAN STARKENBURG (1996), "Sustainability of municipal wastewater treatment", Proceedings of the Congress on Advanced Wastewater Treatment, 23-27 September, AquaTech Amsterdam, pp. 331-338.

SUMINO, T., H. NAKAMURA, N. MORI, and Y. KAWAGUCHI (1992), "Immobilization of nitrifying bacteria in porous pellets of urethane gel for removal of ammonium nitrogen from wastewater", *Appl. Microbiol. Biotechnol.* 36, pp. 556-560.

TANAKA, K., M. TADA, T. KIMATA, S. HARADA, Y. FUJII, T. MIZUGUCHI, N. MORI, and H. EMORI (1991), "Development of new nitrogen removal system using nitrifying bacteria immobilized in synthetic resin pellets", *Wat. Sci. Tech.* 23, pp. 681-690.

TAUW and RIZA (1995), "Gescheiden waterstromen", internal project proposal.

SANITATION IN PERI-URBAN AREAS

by

Roland Schertenleib
Water & Sanitation in Developing Countries (SANDEC)
Swiss Federal Institute for Environmental Science & Technology (EAWAG)
Duebendorf, Switzerland

Introduction/background

The dramatic and unprecedented urban transition is one of the biggest challenges local governments and societies are facing today and in the years to come. By the year 2025, almost two thirds of the world's population, an estimated five billion people, will be living in urban areas. In 1950, this figure did not even amount to one third, and the number of urban dwellers was smaller than 750 000. The most rapid change is occurring in the developing world, where urban populations experiencing an average growth of 3.5 per cent per year (Figure 1). The population of many of the cities quadrupled between 1950 and 1990; some even increased sevenfold, and have reached unprecedented sizes (e.g. Sao Paulo: 16.4 million; Mexico City: 15.6; Bombay: 15.1; Shanghai: 15.1). Figure 2 indicates that in 1960 there were world-wide just four megacities with a population exceeding 10 million, and only one of the four was located in the developing world (Shanghai). By 1980, there were eight megacities, five of them in a developing country, and by the year 2000, there will be 22 megacities, 17 of them in the developing world.

Apart from benefits (e.g. driving forces in economic and social development), urbanisation also involves unparalleled environmental and social problems. These include lack of access to clean drinking water and safe sanitation services. Especially where urban growth is rapid, provision of even the basic needs to all citizens, such as clean drinking water and sanitation, is mostly beyond the means of most local governments. The people who are most affected are the people living in the densely populated peri-urban areas.

General sanitation situation in urban areas of the developing world

Current situation

One of the greatest threats to human health in the urban areas of the developing world is the lack of adequate sanitation services. In 1994, only about 60 per cent of the urban population in Africa, Asia and Latin America had access to save sanitation facilities. Despite significant efforts by local governments and national and international professionals and agencies, the overall situation has only been deteriorating. In 1980, at the beginning of the International Drinking Water Supply and

Sanitation Decade, roughly 375 million people in urban areas of developing countries had no access to safe sanitation systems. By 1994, this number had increased to almost 600 million. The numbers differ from one region to another and even from one country to another; however, the trend is the same everywhere (Table 1).

Figure 1. **Development of the world population**

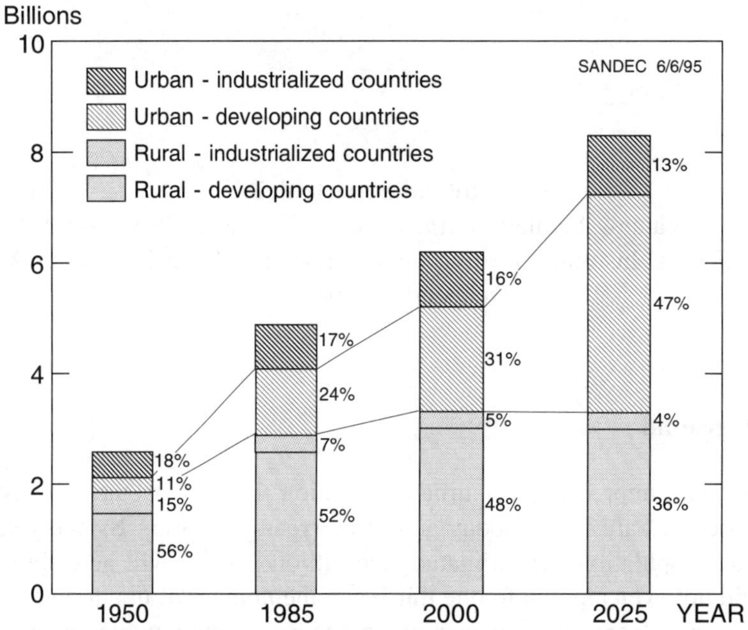

Source: United Nations, 1984.

Figure 2. **Development of urban agglomerations**

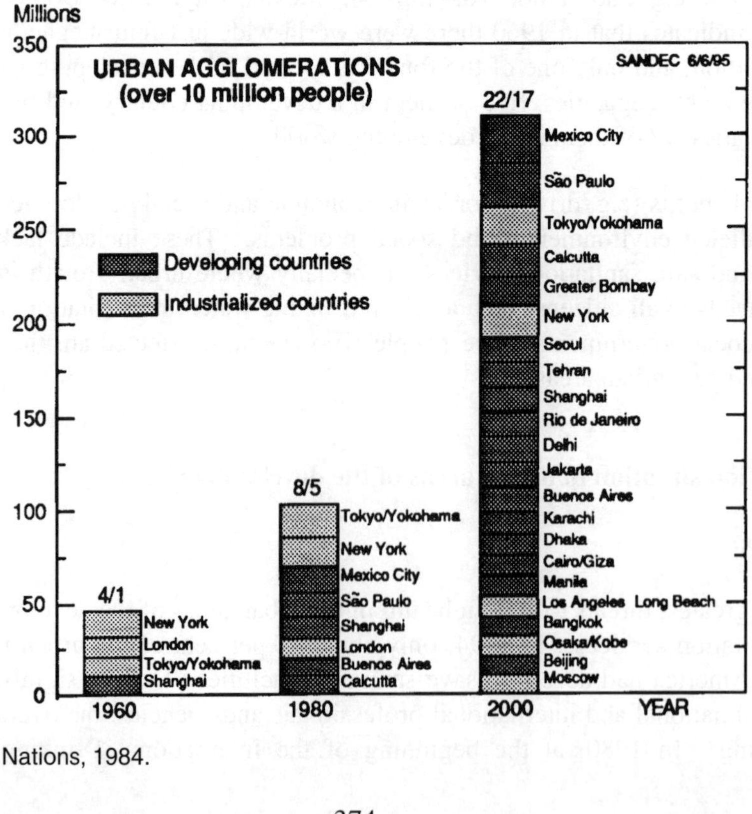

Source: United Nations, 1984.

Table 1. **Urban sanitation coverage by region (1994)**

	Population (in millions)			Coverage
	Total	Served	Unserved	%
Africa	239	131	108	55
Latin America and the Caribbean	348	254	94	73
Asia and the Pacific	955	584	371	61
Western Asia	52	36	16	69
Global	1 594	1 005	589	63

Source: World Health Organisation, 1996.

This alarming situation calls for the development of alternative approaches that the hard-pressed officials responsible for urban services can adopt (or adapt) to solve their particular problems. This should replace the "philosophy" commonly encountered in the "development community" of the North, to export and use the conventional sanitation model which was developed under different circumstances and which is, in the long run, neither economically nor environmentally sustainable.

Conventional sanitation approach and its problems

The "conventional" sanitation model of collecting the human waste of an entire urban area in a waterborne sewer system and treating it in centralised mechanical-biological wastewater treatment plants was first developed in ancient times: the urban civilisations of the Indus Valley, some 6 000 years ago, had centralised water supply and wastewater disposal systems, and we also know of the existence of the sophisticated aqueducts and sewers in ancient Rome. However, these systems also mainly served the well-to-do. The model was reintroduced and improved in Europe during the middle of the nineteenth century, when the industrial revolution resulted in high urban densities and consequent unsanitary conditions, which led to the outbreak of cholera and other public health hazards. The model was basically developed in countries which were (a) located in the temperate zones and normally blessed or even cursed by abundance of rain, and (b) of relative prosperity, mainly derived from new industries and, to some extent, colonisation, thus allowing massive investments in public health schemes.

At the time the model was adopted, scientific knowledge was limited and did not foresee the ultimate effect which the ever increasing water consumption and waste discharge could have on the environment, and no one appreciated that there might be financial and technical limits to the application of this approach.

It is indisputable that the sanitation model developed and optimised by industrialised countries in the course of time has, by and large, served them well. However, an increasing number of professionals are aware that even in industrialised countries, the model should be revised substantially in order to continue to provide adequate services to the entire population and to protect the environment. The model needs to be revised to reduce (a) capital, operating and maintenance costs, and (b) limit the enormous use of energy and natural resources, to render the services economically and ecologically more sustainable. Because water resources are often more limited, populations grow faster, and finances and human resources are even scarcer, the need for adjusting the model is greatest, more urgent and affects more people in developing countries.

Specific situation in peri-urban areas

The problems described above are the most serious and have the most dramatic consequences in the densely populated and fast growing peri-urban areas. The crowding of large numbers of people without adequate sanitation facilities creates condition very favourable to the rapid spread of a variety of infectious diseases, often in the form of disastrous epidemics, such as the recent cholera outbreak in Latin America. In contrast to higher-income urban dwellers and some rural populations, the urban poor have a lower life expectancy at birth and a higher infant mortality rate (Bradley *et al.*, 1992). Therefore, the rest of this paper will focus on the specific problems and issues in peri-urban areas.

Characteristics of peri-urban areas

Peri-urban settlements have a number of unique characteristics that distinguish them from formal urban and rural areas and which have to be considered carefully when designing and implementing a sanitation improvement project. They can be summarised as follows (Hogrewe *et al.*, 1993):

- Families settle in peri-urban areas primarily because land prices or rents are low. Sites which are affordable for poor families are usually those located on steep slopes, along gullies and ravines, on soil that is too rocky to excavate easily, in desert lands, or in areas prone to flooding.

- Most peri-urban settlements do not have piped water and depend often on water vendors. The cost of water sold by the water vendors is significantly higher than what families pay in higher income areas. The water is often of poor quality, and due to its high price, families can only afford a limited amount. Therefore, peri-urban settlers are usually not prepared to purchase water to flush a toilet. Even if families have access to a standpipe nearby or to water piped into the house, this water is usually also limited in quantity and of erratic flow. This leads to low levels of personal and domestic hygiene and thus favours the transmission of excreta-related diseases.

- Peri-urban settlements have typically high population densities which are often greater than 400 people per hectare. Such high densities lacking a basic infrastructure lead to serious health and environmental risks.

- Since residents of peri-urban settlements have migrated from various parts of a region or country, these settlements are generally not homogeneous with respect to ethnic background, language, income levels, and social norms and habits. Consequently, a widespread sense of community cannot be expected. Nevertheless, community organisations can and do exist and are often formed around an issue of common interest, such as school construction and night security.

- In most countries, regulations do not include the legal development of land and buildings in peri-urban areas, and their residents do not have legal tenure of a site which is in itself not legally urbanised. Therefore, governments generally do not validate the presence of these settlements.

- Since peri-urban settlements are generally not recognised as legal areas, their community leaders and residents have limited political influence. However, since peri-urban residents do represent an enormous electoral block and can have a major influence on elections, a political change might occur along with the general trend of democratisation.

- Poor families in peri-urban areas operate in a cash economy, and their workers rely mainly on an informal source of revenue. As cash in the informal economy is unsteady and unreliable, residents are not deemed credit-worthy and will not obtain bank loans.

The peri-urban sanitation challenge

The present sanitation situation of peri-urban communities varies from one country to another and even from one city to another. The most common sanitation practices can be described as follows (Hogrewe *et al.*, 1993):

- *No system:* Defecation occurs in open areas within the settlement, on the perimeter of the settlement, or in drainage ditches. Absence of a planned excreta disposal system is characteristic of most peri-urban areas.

- *Latrines:* Use of a wide range of latrines (e.g. bucket latrines, pit latrines, ventilated improved pit latrines) is the second most common sanitation practice in peri-urban areas, especially in Africa and Latin America. However, these latrines are often poorly designed and maintained and may not be used by all family members.

- *Pour-flush toilets/septic tanks:* In peri-urban settlements experiencing regular or even irregular water supplies, pour-flush toilets with household or community septic tanks and/or soakaways may exist. However, the septic tanks are often poorly maintained and/or undersized.

A major difference between urban and peri-urban programme design is the fact that the former is a "supply-driven" and the latter a "demand-driven" process. "Supply-driven" is used here to denote the traditional/conventional approach whereby a supply of housing stock is developed and then offered to the public for sale. This situation greatly differs in the more common urbanisation processes taking place in peri-urban areas. In this "demand-driven" process, people move in prior to the provision the water supply, sanitation, and other infrastructure services. Once on the land and after construction of a house, their "demand" for infrastructure evolves.

Non-technical issues to improve the sanitation situation in peri-urban areas

Improving the public health situation in peri-urban areas is a very complex and difficult endeavour. Choosing and applying the right technology is only one of the issues that must be addressed in this process. Other concerns which must be considered include health benefits, and environmental issues, as well as social, financial, legal and institutional considerations.

Sanitation-related health benefits

The most compelling justification for improving community sanitation is the achievement of a better public health standard. Important in this context is the question of the relationship between the stages of improved sanitation and the degrees of improved health. According to a study conducted in Guatemala, sanitation coverage of 75 per cent or more of a densely populated community is required before a health impact becomes apparent. Therefore, the project approach to sanitation improvement in peri-urban areas should focus on community-wide improvements, and not on individual households. In contrast, in rural areas where houses are further apart, the approach to sanitation improvement has largely been towards households.

Environmental issues

The lack of adequate peri-urban sanitation provisions leads to serious environmental consequences. Peri-urban areas are the largest non-point sources of faecal contamination in a given city. Inadequate human waste disposal contaminates soil, surface water and groundwater. Reducing such environmental contamination via improved sanitation technologies can actually be more compelling to decision-makers than improving general health (Whittington *et al.*, 1992).

Social considerations

Citizen involvement has been found to be a key ingredient in the success of peri-urban sanitation projects. Community participation in all stages of the project can lead to cost reductions, increased cost recovery, and more effective operation and maintenance of the systems. However, in order to organise residents effectively, project planners need to explore opportunities for bridging cultural and other differences within peri-urban settlements early on in the planning of sanitation improvement activities.

When planning improved sanitation services, other major social considerations should centre around the recipient's hygiene behaviour. Hygiene education and modifications in people's hygiene behaviour might be the first step in improving the public health situation. It is often also a prerequisite for the appropriate use and maintenance of improved facilities (Yacoob *et al.*, 1992).

Financial, legal and institutional considerations

Key elements among the various factors limiting the range of acceptable and appropriate sanitation interventions are people's priorities and their willingness to pay, the institutional capabilities, and the legal constraints.

To help determine the appropriate level of service for a given peri-urban community, a willingness-to-pay survey must be conducted. Such surveys have been used successfully with water supply projects, but very few have been applied to peri-urban situations (Whittington *et al.*, 1992). Willingness-to-pay surveys not only help identify technologies that are acceptable and affordable to a community, they can also help determine feasible methods of financing (Table 2).

The underlying legal issue is the aforementioned absence of legally recognised and approved areas for peri-urban settlements. Households may be reluctant to invest time and money in infrastructure improvements that may not benefit them in the long run. Therefore, specific legal issues that may need to be assessed when designing a sanitation improvement project include land tenure, plot registration with municipal authorities, and the applicability and appropriateness of building codes, design and construction standards, as well as environmental regulations.

Table 2. **Important components in a willingness-to-pay study**

A well-designed willingness-to-pay study will achieve the following:

- Assess household decision-making and resource allocation, given users' limited time and money.
  - What types of technologies do people want?
  - What are they willing to pay to build, operate, and maintain a system?
  - Who within the household makes the decisions?
  - Who controls the resources?
- Determine the financial resources available from the user group, and for what use they are intended.
- Clarify cost recovery issues.
  - How are costs to be divided among users?
  - What is the acceptable time frame for cost recovery?
  - What subsidies, if any, are needed?
- Define the appropriate level of service based on technical and institutional options, recognising that the design process takes an requires user participation.

Source: Whittington *et al.*, 1992.

A small number of formal institutions -- government and private -- have the motivation, mandate, experience, or capability to implement peri-urban sanitation programs. As mentioned earlier, governments do not want to condone the existence of informal communities for several reasons. Furthermore, formal institutions have been set up to provide conventional, "supply-driven" urbanisation infrastructure. They are not structured and prepared to assist or support the "demand-driven" urbanisation process that normally takes place in peri-urban areas. On the other hand, NGOs often play a key role not only in initiating the organisation of the community, but also as intermediary organisations.

Technical issues to improve the sanitation situation in peri-urban areas

The use of the most appropriate technology or combination of technologies in peri-urban sanitation intervention is probably the most challenging task for project designers. Sanitation technologies for either rural areas or formal urban areas have been found inappropriate for use in peri-urban settlements. In recent years, new and more appropriate technologies for peri-urban sites have been developed by engineers in close collaboration with social scientists, however, the current experience with these technologies is still limited. Perhaps the most important point for project planners is to be aware that there is no "magic bullet" technology which fits the conditions of every peri-urban settlement (e.g. combination of harsh physical conditions, high population densities, extreme poverty, and the inability or unwillingness of governments to provide assistance). This awareness could free project planners to explore different technical and non-technical approaches to address and improve peri-urban sanitation.

Design standards

Project planners should be careful when choosing design guidelines for a sanitation technology. Existing selection and design guidelines were mostly developed for rural or formal urban situations and are often not appropriate for peri-urban areas. They often do not include, for instance, alternative technologies such as condominial, shallow, and simplified sewer systems, nor do they consider some of the constraints (such as lack of legal tenure) unique to peri-urban areas.

Sanitation technologies

On-site technologies: latrines and septic tanks

In on-site systems, the excreta are disposed at the point of defecation, i.e. the material is not removed from the site, although liquids may leach into the ground. On-site systems present offer both advantages and disadvantages for peri-urban areas. Their lower costs and minimal institutional requirements appropriately meet the needs of the lower incomes and weak institutional capacities often found in informal settlements. However, peri-urban population densities are frequently too high to accommodate latrines, and site conditions might be also inappropriate and may lead to pit flooding or groundwater contamination.

Communal sanitation facilities can be an appropriate interim solution for peri-urban areas. In general, communal latrines address the earlier discussed constraints commonly found in peri-urban areas. However, one should use caution when considering this option. Because of the ambiguity as to who owns the latrine and who is responsible for its maintenance, communal latrines often fall into disrepair and/or become unsanitary.

It is beyond the scope of this paper to discuss the specific advantages, disadvantages and design criteria of the different types of latrines and septic tanks. The reader is referred to the extensive literature on the subject.

Off-site technologies: sewerage systems and latrines with cartage systems

In off-site systems, the waste is transported away from the user. A simple bucket latrine is one example of an off-site system. The more complex waterborne sewage collection system is another example. Off-site sanitation interventions often solve excreta-related health problems by isolating the excreta from the user but, at the same time, exposing downstream populations to health and environmental problems.

Conventional sewage collection systems have a high initial construction cost. These systems often include pump stations and screening equipment, which call for trained maintenance personnel and considerable operation and maintenance expenditure. Furthermore, they require a significant supply of water to minimise clogging of sewage pipes. As discussed earlier, the majority of peri-urban settlements around the world do not have the required amount nor the reliable supply of water for the proper operation of conventional sanitation systems. The most significant constraint, however, is the inability of most peri-urban areas to afford conventional sewage systems.

In recent years, the following alternatives to conventional sewage systems have been developed:

- *Simplified sewerage*: design standards are altered so that sewers can be built with smaller pipes, shallower trenches, and fewer appurtenances;

- *Condominial sewers:* can be built and maintained by users and located between houses, rather than in streets or other rights of way;

- *Small bore sewers:* septic tanks are connected to small-diameter pipes that carry the solids-free effluent.

All these alternatives not only have significantly lower initial and operating costs and lower requirements with regard to the quantity and reliability of piped water supply, but also provide the advantages of waterborne sewage systems.

Wastewater treatment

One of the disadvantages of any waterborne sewer system is the requirement for wastewater treatment in order to achieve full health and environmental benefits. It is obvious that conventional, highly sophisticated, wastewater treatment systems are usually too expensive to build and operate for peri-urban areas. Some of the lower-cost wastewater treatment alternatives worth considering are:

- communal septic tanks combined with anaerobic filters;
- oxidation ditches;
- stabilisation ponds;
- duckweed ponds;
- constructed wetlands.

It is again beyond the scope of this paper to discuss the specific advantages, disadvantages and design criteria of the different types treatment systems. The reader is referred to the literature on the subject.

Summary and key principles (see also Hogrewe *et al.*, 1993)

- Improving the health of the rapidly growing number of families living in peri-urban areas and protecting the urban environment are urgent needs.

- To improve health in densely populated peri-urban areas, sanitation programs must target the community rather than individual households.

- The current planning paradigm for formal urbanisation, which begins with the installation of basic urban services, does not coincide with the actual peri-urbanisation process, which begins with the informal and/or illegal settlement by poor urban families.

- To improve health, changing individual and community behaviours that cause faecal-oral transmission of diseases is at least as important as constructing new sanitation facilities.

– Improving peri-urban community sanitation is a complex process and requires the involvement of the many institutional actors that influence or have responsibility for peri-urban sanitation.

– Peri-urban sanitation should not be solely technology driven. Successful sanitation interventions should also consider health, economics, social, legal, and institutional factors.

– The conceptualisation, design and construction of peri-urban sanitation systems pose extremely complicated engineering challenges that require skilled, experienced and innovative engineers working in interdisciplinary teams, including planners, social scientists, environmentalists, lawyers and economists.

– Citizen involvement and community participation are critical to successful peri-urban sanitation programs.

– Institutions providing peri-urban sanitation services should seek to recover as much of their costs as possible, in order to reach some level of financial sustainability and be able to expand services to other peri-urban areas.

REFERENCES

BRADLEY, D., S. CAIRNCROSS, T. HARPHAM, and C. STEPHENS (1992), *A Review of Environmental Health Impacts in Developing Country Cities*, The World Bank, Washington, D.C.

HOGREWE, W., S.D. JOICE, and E.A. PEREZ (1993), "The Unique Challenges of Improving Peri-Urban Sanitation", *WASH Technical Report No. 86*, Washington, D.C.

UNITED NATIONS (1984), *Estimates and Projections of Urban, Rural, and City Populations 1950-2025: The 1982 Assessment*, United Nations, New York.

WHITTINGTON, D., A.M. LAURIA, A.M. WRIGHT, C. KYEONGAE, J.A. HUGHES, and V. SWARNA (1992), *Willingness to Pay for Improved Sanitation Services in Kumasi, Ghana*, The World Bank, Washington, D.C.

WORLD HEALTH ORGANISATION (1996), *Water Supply and Sanitation Sector Monitoring Report 1996*, World Health Organisation, Geneva.

YACOOB, M., B. BRADDY, and L. EDWARDS (1992), "Rethinking Sanitation: Adding Behavioral Change to the Project Mix", WASH Project, Arlington, Virginia.

NEW ADVANCED SEWAGE WASTEWATER TREATMENT PROCESSES

by

Masayoshi Kitagawa* and Koji Mishima**
*EBARA Research Co., Ltd.; **EBARA Corporation, Japan

Introduction

The usage of sewage wastewater treatment systems is now becoming widespread in Japan (average expansion ratio of over 50 per cent in 1995), and most major cities have municipal sewage treatment plants. The official policy for a sewage treatment systems is now directed to construct small-scale sewage treatment plants, to remove organic pollutants such as biochemical oxygen demand (BOD) and suspended solids (SS), in rural cities and towns. Usage of sewage treatment systems has proven to be vital for improving the surface water quality in rivers and lakes susceptible to pollution by organic matter. Despite this, other types of pollution, such as red tide or waterbloom by eutrophication in closed water bodies, have appeared. They are causing problems such as malodour of drinking water and the nonsustenance of fish and other water creatures due to insufficient oxygen. Accordingly, wastewater treatment now requires removal of not only organic matter but also nutrients, such as nitrogen and phosphorus to prevent eutrophication. This has led to a demand in compact, advanced sewage wastewater treatment plants because of the limitation of space for such plant sites in Japan.

The following introduces two advanced sewage wastewater treatment processes developed recently in Japan, the media-anaerobic-anoxic-oxic process (M-AAO) and the intermittent aeration process, both of which feature high nitrogen and phosphorus removal performance.

The Media-Anaerobic-Anoxic-Oxic Process (M-AAO)

Outline of the M-AAO

A biological suspended solid process well known as an anaerobic-anoxic-oxic process, which can simultaneously remove nutrients in wastewater, requires a reactor with a volume of more than twice that used in a conventional activated sludge process. Moreover, its operation requires long solids retention time (SRT) and low BOD-SS loading to prevent nitrifier wash-out from the reactor due to a slow growth rate at low temperature.

The M-AAO process is a simultaneous denitrification and phosphorus removal process. Nitrifiers adhere to the surface of fluidized media which are charged into an oxic tank. Figure 1 shows a flowchart of a M-AAO process which was constructed as a pilot plant. This process consists of an anaerobic tank, an anoxic tank, an oxic tank and a sedimentation tank. A screen was equipped at the outlet of the oxic tank to separate the media from suspended mixed liquor and keep it in the oxic tank. If nitrifiers could be adhered on the surface of fluidized media and kept steadily in an oxic tank, the volume of the reactor could be made smaller, as a high nitrification rate can be attained even in the winter season. However, if the media were covered with BOD oxidising bacteria, which have a higher growth rate than that of the nitrifiers, nitrification rate would drop.

Therefore, it is important to reduce most of the BOD components in wastewater upstream an oxic tank. In this process, nitrifiers can easily attach to the surface of the media and grow densely by becoming dominant over BOD oxidising bacteria in the oxic tank, as most of the influent soluble BOD components are removed in an anaerobic and an anoxic tank.

Figure 1. **Flowchart of the M-AAO pilot plant**

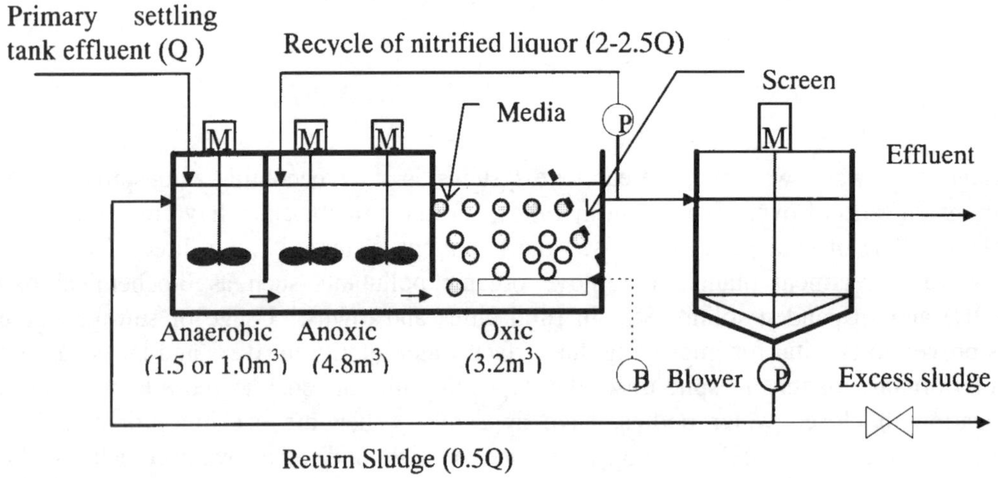

Source: Mishima *et al.*, 1996.

Media

Nitrification rates of four kinds of media (Table 1), selected by characteristic of fluidity, density and endurance, were studied previously by performing bench-scale denitrification plant tests (600 *l*) using a domestic sewage. The plant had a common anoxic tank and five separate oxic tanks in parallel, and each media was added into an oxic tank running on the same BOD-loading. The volume concentration of the media was 10-20 v/v per cent. One of the tanks was run without any media (only activated sludge) to constitute a control system. Nitrification rates in each oxic tank were sequentially compared.

Table 1. **Support media tested in bench-scale experiment**

Media name	Appearance (mm)	Media charge (%)
1. Polyurethane foam (PUFa; 30 mesh)	Rectangular block: 10x12.5x12.5	20
2. Polyurethane foam (PUFb; 7 mesh)	Rectangular block: 10x12.5x12.5	20
3. Polyethylene glycol gel (PEG)	Sphere: Mean diameter of 3	10
4. Polyvinyl alcohol gel (PVA)	Sphere: Mean diameter of 4.7	10

Source: Mishima *et al.*, 1994.

The result of the test indicated that polyethylene glycol gels (PEGs) had the highest nitrification rate, next to polyvinyl alcohol gels (PVAs), and that polyurethane foam had the lowest nitrification rate under a steady state condition (Table 2). However, nitrification rates indicated a proportional relation among media surface areas but no relation among media materials. Hydrophilic PEG gels were selected as the fluidized medium in a follow-up study because, compared to PVAs, they demonstrated more physical strength, higher nonbiodegrability, and were easier to handle due to their uniform particle size.

Table 2. **Nitrification performance in a steady state**

	PUFa	PUFb	PEG	PVA	No media
Nitrification rate (kg/m^3-tank/d)	0.05	0.07	0.21	0.15	<0.01
Nitrification rate of media (kg/l-media/l)	10.1	13.9	85.4	63.3	-
Immobilized biomass (mg/l)	4 450	4 320	6 920	1 620	-
Water temperature (°C)	15	14	13	13	13
DO in nitrification tank (mg/l)	5.9	6.1	2.8	6.0	7.1

Source: Mishima *et al.*, 1994.

Pilot plant

Operation conditions and analysis

The pilot plant shown in Figure 1 had been operated for a few years using the effluent sewage from the first sedimentation tank in the Aso Sewage Treatment Center, Kawasaki City. Table 3 shows operational conditions of the pilot plant from May 1993 to March 1995. PEG gels media, each medium with a diameter of five mm, were initially used under a volumetric concentration of 20 per cent (in the oxic tank). After May 1994, however, this was replaced with 4.2 mm in diameter media under a volumetric concentration of 16 per cent.

Because the PEG gels had a light specific gravity of 1.02, it was easy to fluidize them using only process air (by aeration).

Composite samples of influent and effluent wastewater were collected and measured using an auto-sampler. Grab samples of each tank were also collected and measured. Samples containing suspended solid were immediately filtered using glass fibre filters (wattman GF/B).

Table 3. **Operation conditions of the pilot plant**

	May 1993 - July 1993	August 1993 - April 1994	May 1994 - March 1995
Influent flow rate (m^3/d)	25.0	25.0	28.6
HRT (h)	8.7	9.0	8.0
Anaerobic tank (m^3)	1.0	1.5	1.5
SRT (d)	16	11-14	8-11
A-SRT (d)	5.7	3.7-4.7	2.7-3.7
Media diameter (mm)	5.0	5.0	4.2
Media charge (v/v %)	20	20	16
DO (mg/l)	1.8-5.0	2.0-5.8	2.8-7.8
Influent BOD (mg/l)	49-151	45-110	48-118
T-N (mg/l)	14-47	13-47	16-45
T-P (mg/l)	1.2-2.9	2.0-5.8	2.8-7.8

Source: Mishima *et al.*, 1996.

Results

Nutrient removal performance of the pilot plant

Figure 2 shows daily changes in performance from 1993 to 1995. As most of the soluble BOD was removed in the anaerobic and anoxic tanks, it was possible to maintain a high-density adherence of nitrifiers on the surface of the media. At a retention time of eight to nine hours in the reactor, the average T-N (total nitrogen) in the influent was 32 mg/l, while that in the effluent was 8.3 mg/l. The average T-P (total phosphorus) in the influent was 2.6 mg/l, while that in the effluent was 0.6 mg/l. Our target effluent water quality, T-N 10 mg/l and T-P 1.0 mg/l, was sufficiently achieved, even under low temperatures during winter.

Figure 2. **Nutrient removal performance of the M-AAO pilot plant**

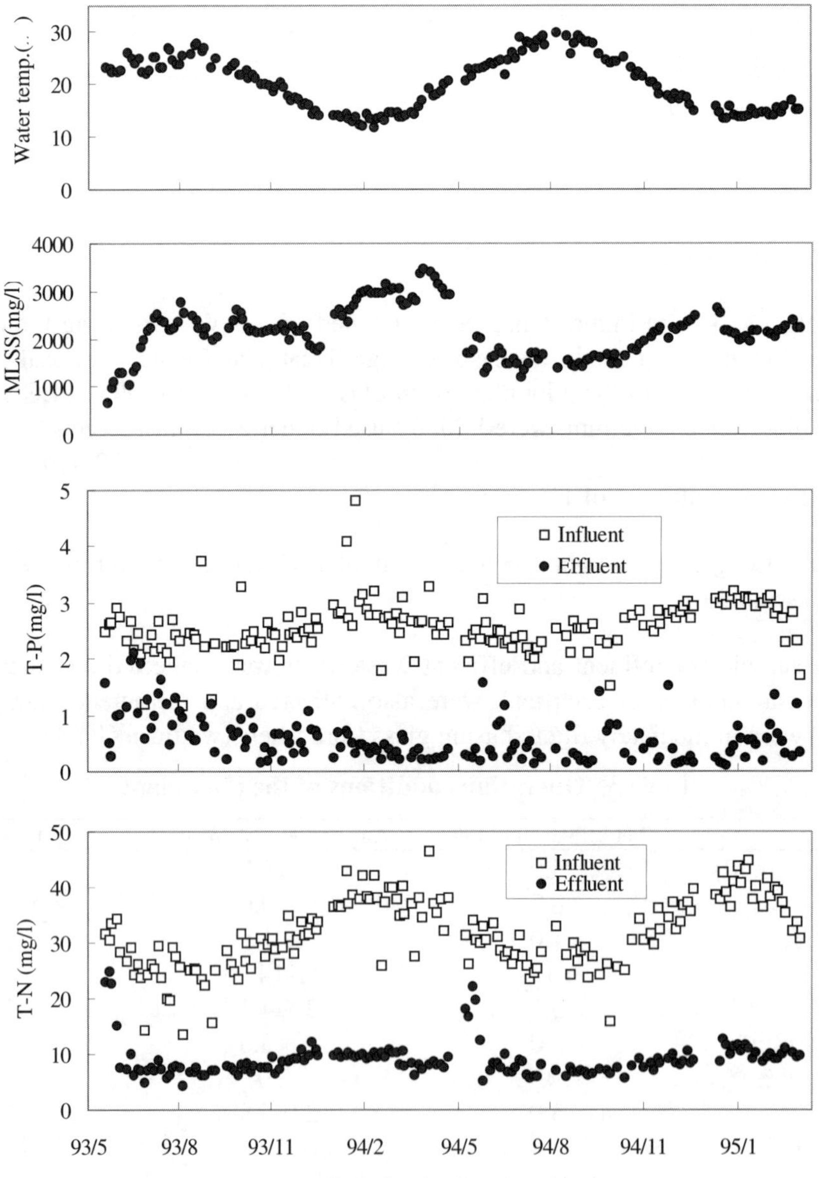

Note: Water temperature in °C.

Source: Mishima *et al*., 1996.

Nitrification rate and denitrification rate

Activated sludge and media were sampled from the pilot plant to study the nitrification and denitrification rates. This included a study on the reaction characteristics of nitrification and denitrification, for which a small 2 L batch reactor was used. The effect on the nitrification rate by DO (dissolved oxygen) was also studied by measuring its rate at constant dissolved oxygen concentrations in the reactor. Figure 3 shows the relationship between the DO and media nitrification rate coefficient (Kn). It was apparent that the Kn largely depended on the DO, while the Kn of the activated sludge in a conventional denitrification process only depended on the same up to DO concentration of about two mg/l. This high dependency suggested that the DO diffusion rate limited the Kn, due to high respiration rate by dense biofilm on the media. Figure 4 shows the relationship between BOD-SS loading and the denitrification rate coefficient (Kd) of the pilot plant in a steady state. Because the BOD-SS loading of the M-AAO was higher than those in a conventional denitrification process, it was determined that M-AAO denitrification tank could be made smaller than that used in a conventional process.

Figure 3. **Relationship between DO and the media nitrification coefficient Kn**
(batch test at 14 °C)

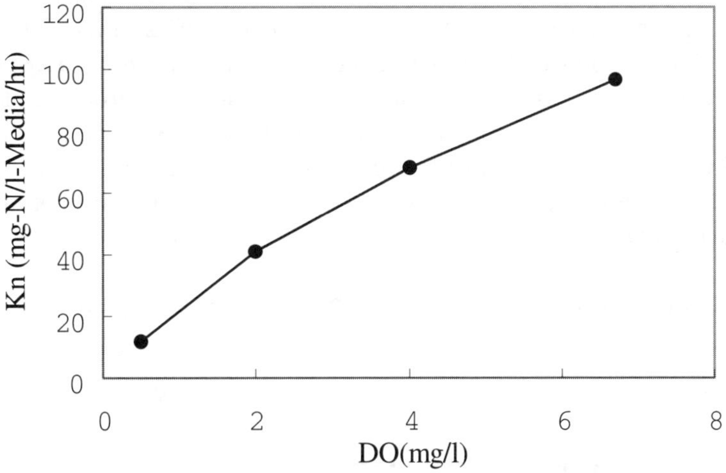

Source: Mishima *et al.*, 1996.

Figure 4. **Relationship between BOD-SS loading and the sludge denitrification rate coefficient Kd in an anoxic tank at 13-15 °C**

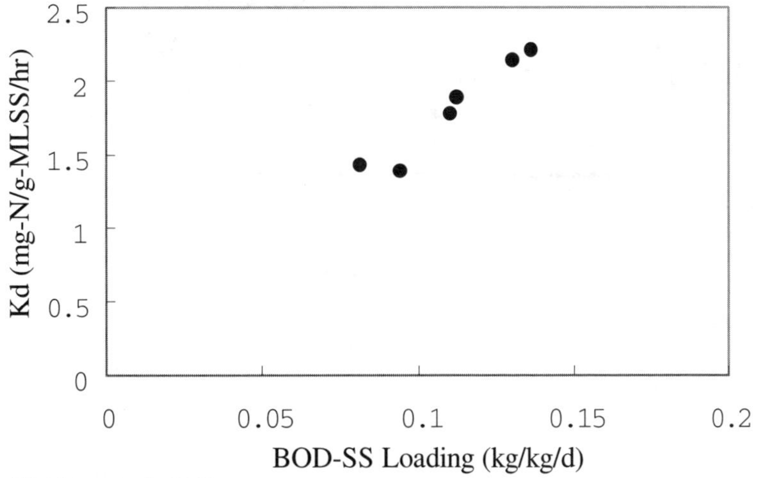

Source: Mishima *et al.*, 1996.

The intermittent aeration process of a single tank

Outline of the process

It is known that oxidation-ditch and sequencing batch activated sludge processes used in small-scale sewage treatment help yield high quality effluent water. However, they require considerable installation space and are difficult to operate and maintain. The authors had previously confirmed that a single-tank, continuous inflow/intermittent aeration process was capable of removing not only BOD and SS but also nutrients, in spite of the simplicity of its operation. To make it more compact, an integrated configuration was applied, by which the sedimentation section was located at around the upper part of the system, while the aeration section was located at the bottom part (see Figure 5). Raw sewage was made to flow directly into the bottom of the aeration section, where it was treated continuously by intermittent aeration by submerged mechanical aerators. The mixed liquor was circulated between aeration and circulation sections, part of which ascended to sedimentation section. The sludge blanket zone, formed at the foot of the sedimentation section, filtered SS particles and thus improved the clarity of the effluent. The sludge in the aeration section settled down to the bottom when aeration was stopped. The inflow sewage made a countercurrent contact with the settled sludge in the aeration bottom. The soluble BOD in the influent sewage was decomposed by denitrification reaction under anoxic conditions, or uptaken by phosphorus storage bacteria accompanying the released phosphorus from the cells under anaerobic conditions. The luxury uptaking of phosphorus was done during aeration time. Other raw sewage SS compartments were able to be captured by the filtration effect of the settled sludge in the aeration section, as well as by that of the sludge blanket zone in the sedimentation section. Adjusting the intermittent aeration time cycle to timely and seasonal changes in BOD loading was found to be easy.

Figure 5. **Schematic drawing of intermittent aeration process**

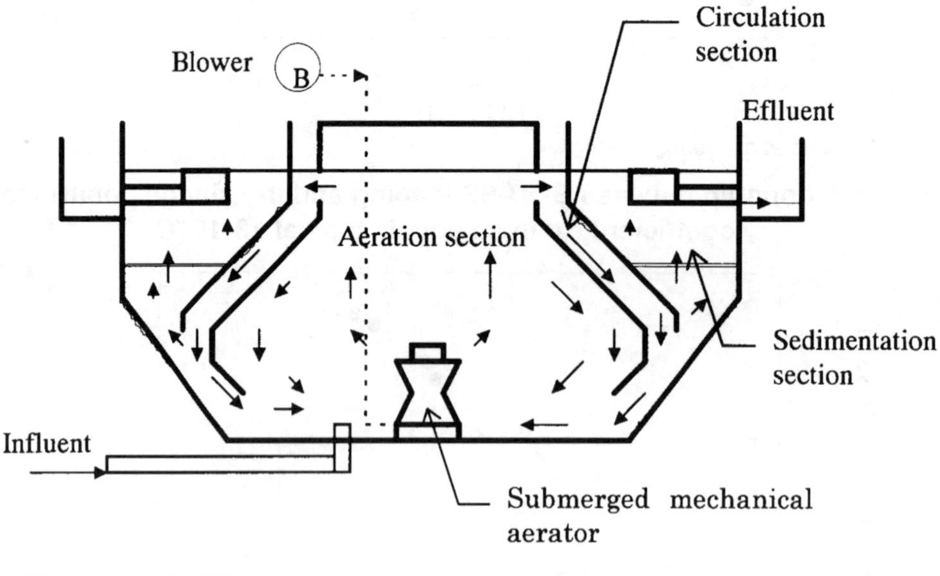

Source: Kitagawa *et al.*, 1992.

Performance at a community plant

Design specifications

The said process was applied to modify a small sewage treatment plant. This plant had been operated for near 30 years at a hot spring resort village. Table 4 shows the design criteria and equipment specifications of this plant.

Table 4. **Design criteria and equipment specifications**

Design criteria	Design flow	476 m^3/d	
	Water quality	Influent	Effluent
	SS (mg/l)	200	30
	BOD (mg/l)	200	20
Equipment specifications	Grid chamber	1.0 m (W) x 1.0 x (L) x 0.3 m (H) 2 vessels	
	Equalisation tank	4.0 m (W) x 10 m (L) x 2.0 m (H)	
	Reactor	8.6 m (W) x 10 m (L) x 3.5 m (H)	
	Aeration section volume	200 m^3	
	Sedimentation section volume	86 m^3	
	Sedimentation section area	46 m^2	
	Submerged aerator	3.7 kW (2 sets)	
	Root blower	2.5 m^3/min. (2 sets)	
	Thickener	2.0 m (W) x 2.0 m (L) x 2.75 m (H)	
	Sludge reservoir	2.0 m (W) x 2.0 m (L) x 3.0 m (H)	

Source: Kitagawa *et al.*, 1992.

Performance of the plant

Figure 6 shows daily changes in performance of the operation from January 1990 to April 1994. The sewage temperature rose to 25 °C in summer and fell to 8 °C in winter. The average ratio of on- and off-time of the aeration per day was around 1:2, though the on-time was set from 30 to 60 minutes, while the off-time was set from 45 to 75 minutes, timely corresponding with changes in the BOD load. This ratio was also adjusted according to seasonal BOD loading. The concentration of MLSS in aeration section was kept between approximately 2 000 and 3 000 mg/l. The sludge level of the blanket zone in the sedimentation section was kept submerged within 1-2.5 m of the surface water, although there was some fluctuation in the level caused by hourly and daily flow rate changes.

Removal performance of BOD and SS

The plant was found to be capable of treating sewage wastewater; i.e. the effluent water quality almost satisfied design flow conditions. The average BOD and SS concentrations in the influent were 130 mg/l and 90 mg/l, while those in effluent were 7.3 mg/l and 5.5 mg/l respectively. The favourable quality of the effluent was achieved even at the lowest temperature in winter. There was almost no difference between the effluent quality during the on-time of the aeration vs. off-time of the same.

Figure 6. **Variation of temperature, MLSS and BOD removal performance of the small sewage treatment plant**

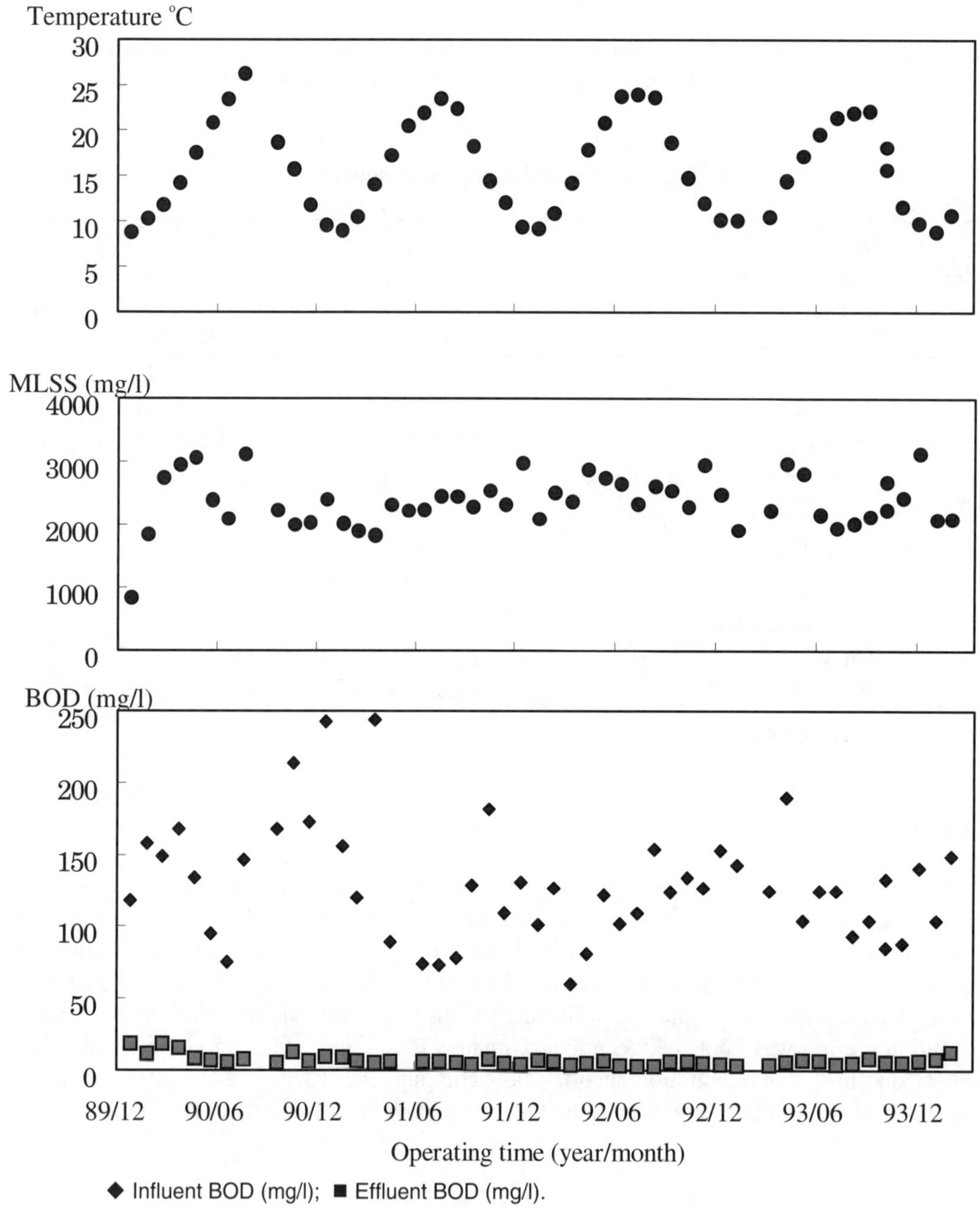

Source: Japan Sewage Works Agency, 1995.

Removal performance of nitrogen and phosphorus

Figure 7 shows daily changes in performance of the nutrients removal from April 1990 to March 1994.

Figure 7. **Nitrogen and phosphorus removal performance**

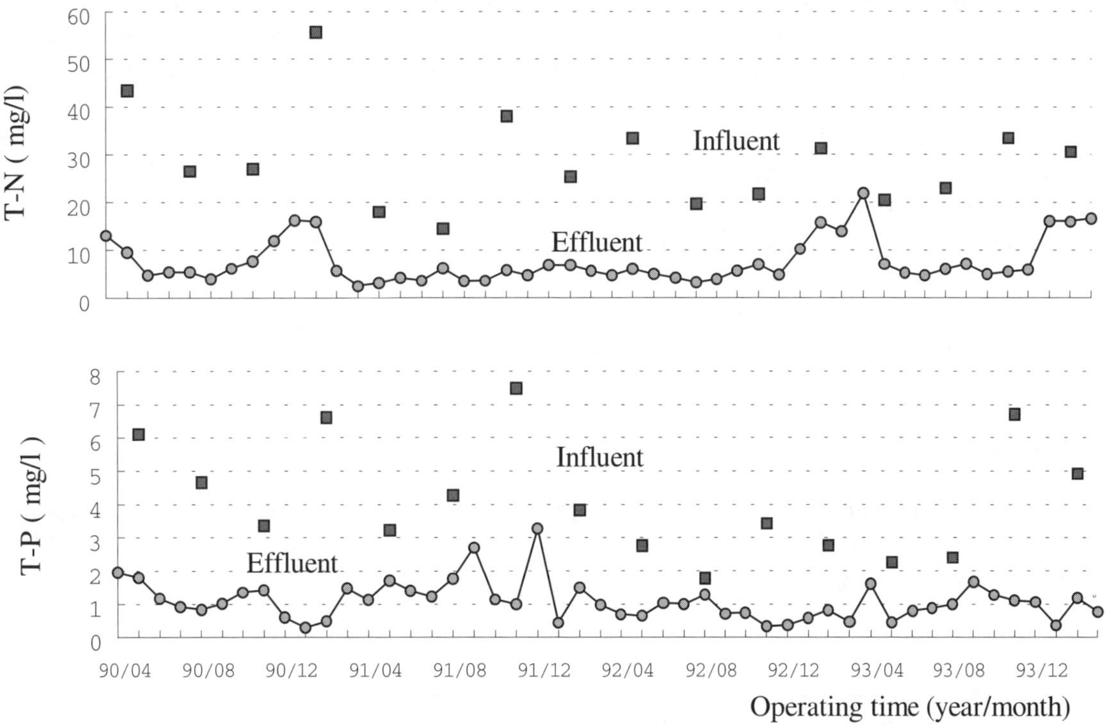

Source: Japan Sewage Works Agency, 1995.

The nitrogen removal rate decreased during the winter season, due to a lower nitrification rate. This was because the actual aeration time was only four to six hours, although the retention time of the aeration section was about 12 hours. However, during other seasons, the nitrogen concentration in the effluent was less than 10 mg/l and high removal rates (65-85 per cent) were attained. In contrast, the phosphorus removal rate decreased to less than 50 per cent during summer, while it was between 60 and 93 per cent during winter. The average annual phosphorus removal rate was 70 per cent. Although the process was not designed with emphasis put on stable and high nutrient removal throughout the year, it was demonstrated that using a simple intermittent aeration process made it possible to achieve such a removal.

As this plant was located in a remote area, far away from the central sewage plant in the city, maintenance personnel from the municipal sewage department had to visit the plant once a day for maintenance. Maintenance was also carried out electronically by remote control from the central sewage plant using telephone lines. A vacuum truck was used for the daily transport of excess sludge from the thickener to municipal sewage treatment facilities. Although the plant featured a modification under which the aeration and sedimentation sections were integrated, no significant problems occurred, and operation was stable.

Conclusion

The authors studied the performances of two newly developed processes for advanced sewage treatment. One is the media-anaerobic-anoxic-oxic process (M-AAO) which features the use of fluidized media (PEG gel beads) in the oxic tank. Nitrifiers adhere to the surface of the fluidized media. Results of a performance study using a pilot plant indicated that this process was capable of a high and stable removal of both nitrogen and phosphorus from municipal sewage, using the same retention time as the conventional activated sludge processes. Accordingly, it is possible to remove both organic pollutants and nutrients by modifying a conventional activated sludge tank.

The other process is the intermittent aeration process, which features an integration of the aeration and sedimentation sections. The performance data obtained by studying this process applied in a small-scale sewage treatment plant showed that high removal rates of not only BOD and SS, but also nutrients on average, were achieved. This was the case in spite of the simplicity of the operation. Note must be taken here that there was a bit of fluctuation in the removal rate of nutrients. This process is ideal for use in a small-scale sewage treatment plant, especially in one where it is difficult to hire specialists and technicians for advanced treatment.

Acknowledgement

This study was supported by Sewage Works Bureau, Municipal Government of Kawasaki City, Japan, as well as Sewage Works Division, Nagato City, Japan, both to whom we would like to express our sincerest appreciation.

Questions, comments and answers

Q: Do you characterise influent in terms of volatile fatty acids? Our studies show that for consistent removal of phosphorus you need rapidly degradable organics.

A: Our sewage contains 70-80 mg/l of volatile fatty acids.

REFERENCES

JAPAN SEWAGE WORKS AGENCY (1995), "The activated sludge process combined an intermittent aeration tank with a sedimentation tank containing sludge blanket zone into one body - SAF process", the estimating report of a newly developed industrial technology for sewage wastewater 1995, Japan Sewage Works Agency, No. 603.

KITAGAWA, M., E. OHKUBO, and S. OGATA (1992), "Performance of the newly developed single-tank continuous inflow and intermittent aeration sewage treatment system", WEF Asia/Pacific Rim Conference on Water Pollution Control, Proceedings, pp. 439-446.

MISHIMA, K., T. YOSHIZAWA *et al.* (1994), "Nitrogen and Phosphorus Removal by Anaerobic-Anoxic Process with Supporting Media", *EBARA-INFILCO Engineering Review* 111, pp. 17-25.

MISHIMA, K., T. NISHIMURA *et al.* (1996), "Characteristics of nitrification and denitrification of the media-anaerobic-anoxic-oxic process", Water Quality International '96, 18th IAWQ Biennial International Conference and Exhibition, Singapore, Proceedings, pp. 118-124.

BIOLOGICAL TREATMENT OF RESIDUAL MUNICIPAL SLUDGE

by

Gabriela Moeller
Mexican Institute of Water Technology, Jiutepec, Morelos, Mexico

Introduction

Mexico, a country of around two million of square kilometres (Encyclopedia Hispanica, 1993) and with approximately 91.6 million inhabitants by 1995, does not have a uniform distribution of its population; around 70 per cent is concentrated in urban areas, and 22 per cent in the Valley of Mexico. The remaining 29 per cent lives in approximately 153 813 rural localities, most of them of less than 100 inhabitants. This distribution increases the difficulties of providing them with drinking water services, sewerage and sanitation in these rural areas (Poder Ejecutivo Federal, 1995).

It is estimated that the country consumes approximately 240 m^3/s of potable water. This volume generates around 170 m^3/s of wastewater. Not all of it is treated, and sometimes is discharged in a raw way to water bodies, a lot of which have been polluted (Secretaria de Desarrollo Social, 1994).

By 1995, it was estimated that the coverage of drinking water was 87 per cent, while sewerage was around 67 per cent. At the end of the same year, the national inventory of wastewater treatment plants reported 684 plants with a total installed capacity of 53.9 m^3/s. Not all of them operate correctly. Around 67 per cent were reported to operate satisfactorily, with an approximate flow of 43 m^3/s. With these installations, just 25 per cent of the total wastewater flow is treated. Forty-nine per cent of these plants are waste stabilization ponds systems, and 24 per cent activated sludge systems. The remaining 27 per cent are other kind of systems. All these systems produce different quantities of sludge.

The handling and disposal of sludge, a by-product of wastewater treatment processes, is a big problem; most of the sludge that is produced is discharged or disposed without any treatment in body waters, sewers or in the soil, without considering its quality and nuisance characteristics. In the past, when a wastewater treatment plant was constructed, the sludge treatment system was not considered because of its high costs. It is well known that almost 50 per cent of the wastewater treatment plant operation costs are due to sludge. A recent study by the National Water Commission (CNA) reported that 90 per cent of the sludge produced in Mexico is discharged in this manner.

Sludge characteristics

The sludge generated in wastewater treatment has to be safely disposed or used in a better way. Sludge management begins with its generation, treatment and its use or disposal. These operations are the most complex and expensive part of wastewater management. Municipalities currently generate approximately 26 kg per person per year (EPA, 1989). This quantity is expected to increase as the population increases, and also if more sophisticated wastewater treatment systems are developed or installed. Also, sludge generated by more advanced treatment processes is more difficult to handle.

Sludge usually contains 93 to 99.5 per cent water as well as solids and dissolved substances that were present in the wastewater. These wastewater solids have to be treated prior ultimate use or disposal to improve their characteristics. The characteristics of a sludge depends on both the initial wastewater composition and the subsequent wastewater and sludge treatment processes used. Different treatment processes generate radically different types and volumes of sludge. Sludge is named depending upon the operation unit that generates it. Table 1 describes the different types of sludge produced, and Figure 1 presents the generation, treatment and disposal of wastewater sludge.

Table 1. **Types of sludge produced in a wastewater treatment plant**

Sludge	Characteristics
Primary	Generated during primary wastewater treatment, which removes the solids that settle out readily. Primary sludge contains 3 to 7 per cent solids; usually its water content can be easily reduced by thickening or dewatering.
Secondary	Also called biological sludge, because it is generated by secondary biological treatment processes, including activated sludge systems and attached growth systems. Secondary sludge has a low solids content (0.5 to 2 per cent), and is more difficult to thicken and dewater than primary sludge.
Tertiary	Produced by advanced wastewater treatment processes, such as chemical precipitation and filtration. The characteristics of tertiary sludge depend on the wastewater treatment process that produced it. Chemical sludges result from treatment processes that add chemicals, such as lime, organic polymers, and aluminum and iron salts, to wastewater. Generally, lime and polymers improve the thickening and dewatering characteristics of a sludge, whereas aluminum and iron salts usually reduce its dewatering and thickening capacity by producing very hydrous sludges which bind water.

Source: EPA, 1989.

The mass of sludge produced depends upon the degree of wastewater treatment. The production of sludge measured as dry solids is approximately 80 grams per person per day for a primary treatment plant, compared to 115 grams per person per day for secondary treatment, and 145 grams per person per day for a secondary treatment plant with chemical addition for phosphorous removal (tertiary treatment) (Environmental Protection Programs Directorate, 1984). Figure 2 presents typical sludge quantities generated by various wastewater treatment processes.

Figure 1. **Sludge production in wastewater treatment**

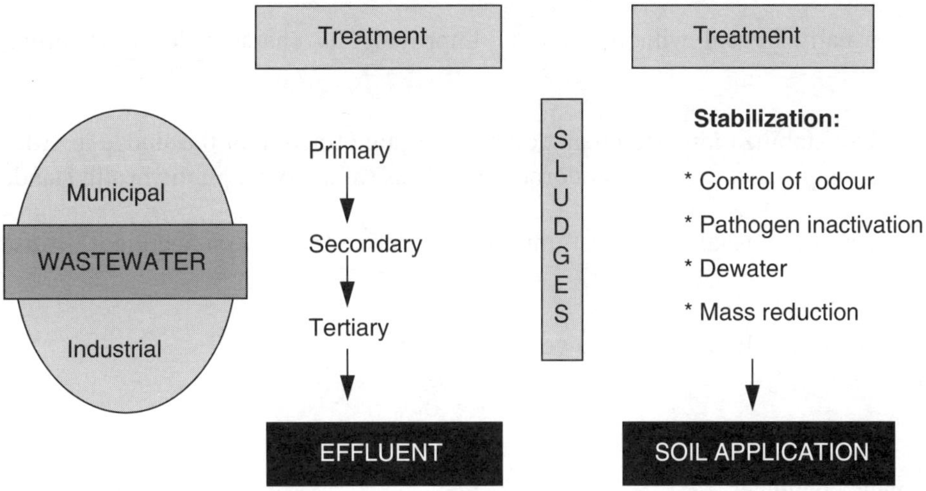

Source: Adapted from EPA, 1989.

Figure 2. **Typical sludge quantities generated by various treatment processes**

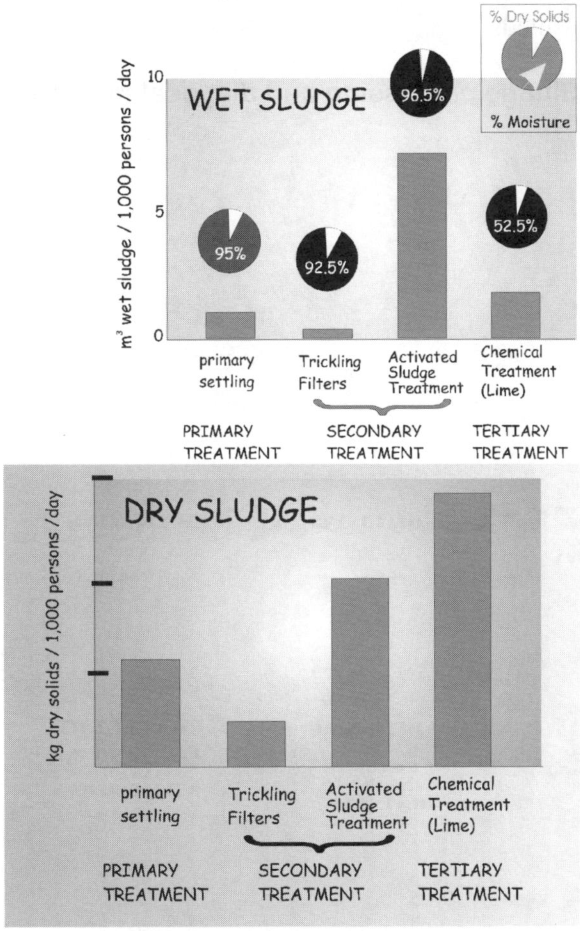

Note: 1 m^3 = 264.2 gal; 1 kg = 2.205 lb.

Source: EPA, 1989.

Stabilization of sludge, a biotechnological application

Sludge is usually treated with the aim of improving its characteristics for ultimate use or disposal.

The purpose of stabilization is to break down the organic fraction of the sludge in order to reduce its mass and obtain a product that is less odorous as well as safer from a public health standpoint.

Five constituents are usually the most important to define the way of treating it, its use and final disposal:

- organic content (usually measured as volatile solids);
- nutrients;
- pathogens;
- metals;
- toxic organic chemicals.

Stabilization refers to a number of processes that reduce the potential for odour, pathogen level and volatile solid content (EPA, 1992). The most common methods for stabilization include anaerobic digestion, aerobic digestion, lime stabilization and composting. Figure 3 presents the sludge processing and disposal methods.

Figure 3. **Sludge processing and disposal methods**

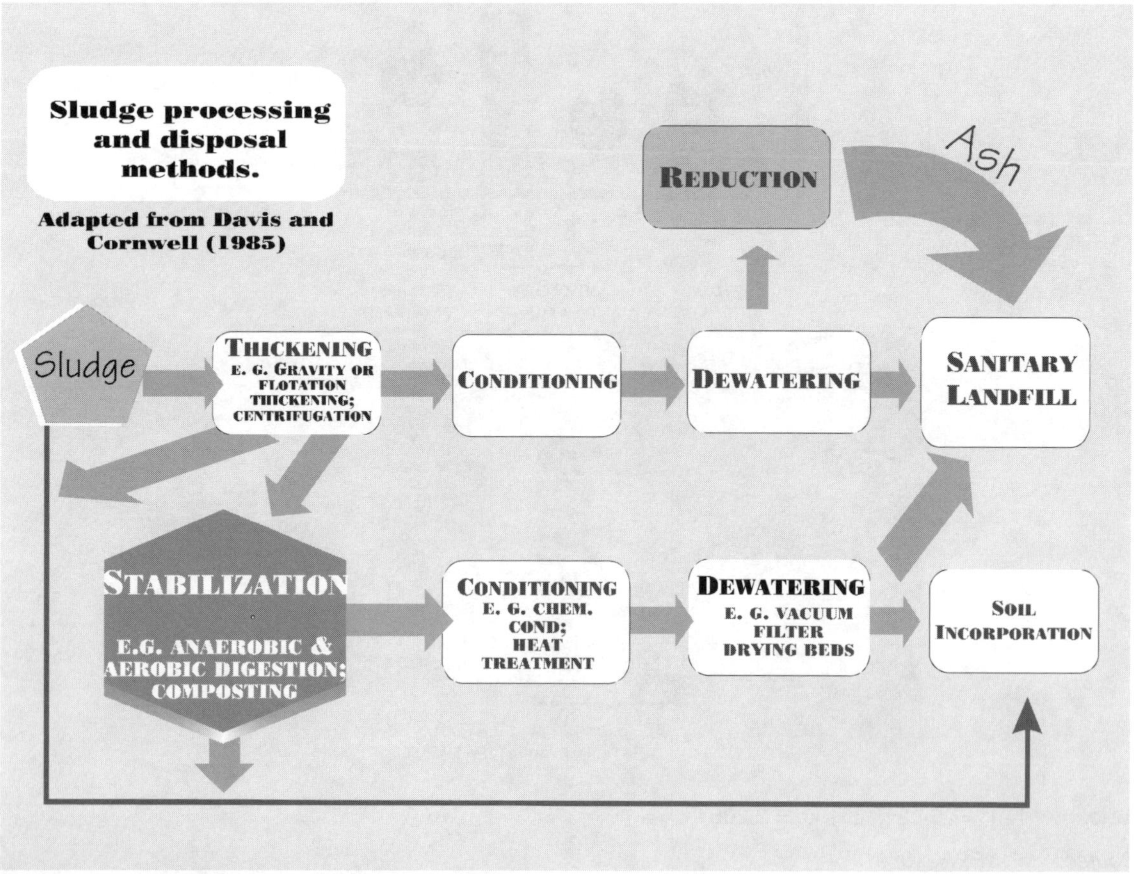

Source: Adapted from Davis and Cornwell, 1985.

Anaerobic and aerobic digestion, as well as composting, are biotechnological applications for the stabilization of sludge.

Anaerobic digestion

The stabilization of putrescible organic wastes by anaerobic digestion represents one of the longest established and most successful applications of biotechnology in the field of pollution control (Bruce and Loll, 1981).

A well organised series of microbiological processes that convert organic matter to methane is the so-called anaerobic digestion. The anaerobic process is driven mostly by bacteria. For a long time, anaerobic digestion has been used for the stabilization of wastewater sludges, and in recent years, it has also been used for the treatment of industrial and even municipal wastewaters.

A better understanding of the microbiology of the process and the improvement in reactor designs has made it possible to improve the process.

A well organised consortia of micro-organisms, mostly bacteria, are involved in the transformation of complex high molecular weight organic compounds to methane. There are synergistic interactions between the various groups of bacteria implicated in anaerobic digestion of wastes. The overall reaction is the following:

$$\text{organic matter} \Rightarrow CH_4 + CO_2 + H_2 + NH_3 + H_2S$$

Large numbers of strict and facultative anaerobic bacteria are involved in the hydrolysis and fermentation of the organic compounds. There are four categories of bacteria that are involved in the transformations of complex materials into simple molecules (methane and carbon dioxide). These groups are the following: the hydrolytic bacteria are a consortia of anaerobic bacteria that break down complex organic molecules into soluble monomer molecules. These monomers are directly available to the next group of bacteria. This hydrolysis is catalysed by extracellular enzymes. The acidogenic bacteria convert sugars, amino acids and fatty acids to organic acids, alcohols and ketones, acetate, CO_2 and H_2. Acetate is the main product of carbohydrate fermentation. Acetogenic bacteria convert fatty acids and alcohols into acetate, hydrogen, and carbon dioxide, which are used by methanogens. Ethanol, propionic acid and butyric acid are converted to acetic acid by acetogenic bacteria according to the following reactions:

$$\underset{\text{ethanol}}{CH_3CH_2OH} + CO_2 \Rightarrow \underset{\text{acetic acid}}{CH_3COOH} + 2H_2$$

$$\underset{\text{propionic acid}}{CH_3CH_2COOH} + 2H_2O \Rightarrow \underset{\text{acetic acid}}{CH_3COOH}$$

$$\underset{\text{butyric acid}}{CH_3CH_2CH_2COOH} + 2H_2O \Rightarrow \underset{\text{acetic acid}}{CH_3COOH} + 2H_2$$

Acetogenic bacteria grow much faster than methanogenic bacteria.

The methanogenic group is composed of both gram positive and gram negative bacteria with a wide variety of shapes. They grow slowly in wastewater and sludge, and their generation time ranges from three days at 35 °C to as high as 50 days at 10 °C. There are two subcategories, including hydrogenotrophic methanogens that convert hydrogen and carbon dioxide into methane:

$$CO_2 + 4H_2 \Rightarrow CH_4 + 2H_2O$$

The hydrogen utilising methanogens help maintain the very low level partial pressures necessary for the conversion of volatile acids and alcohols to acetate.

The acetotrophic methanogens, also called acetoclastic bacteria, convert acetate into methane and CO_2:

$$CH_3COOH \Rightarrow CH_4 + CO_2$$

These bacteria grow much more slowly than the acid forming bacteria. About two thirds of the methane produced in anaerobic digestion is derived from acetate conversion by acetotrophic methanogens. The other third is the result of carbon dioxide reduction by hydrogen.

From a taxonomic viewpoint, methanogens belong to a separate kingdom, *The Archaebacteria*, and differ from prokaryotes in several ways.

Aerobic digestion

This process consists of adding air or oxygen to sludge that is contained in an open tank three to seven metres deep. Oxygen concentration must be maintained above 1 mg/L to avoid production of foul odours. The detention time in the digester is 12 to 30 days, depending on the prevailing temperature. Micro-organisms degrade aerobically available organic substrates. They enter into the endogenous phase of growth when the biodegradable organic matter is depleted. The prevailing conditions in the tank also promote nitrification, which converts NH_4 to NO_3. The end result is a reduction in sludge solids. The stabilized sludge is allowed to settle, and the supernatant is generally recycled to the head of the plant because it has a high BOD and high N and P levels.

An important innovation in aerobic digestion of sludges has been the development and implementation of autoheated thermophilic aerobic digestion. This process generates heat as a result of free energy released by sludge micro-organisms. Much of the energy that results from the oxidation of organic compounds is released as heat. The rise in temperature is a function of the level of organic matter. Most of the heat generated during biological oxidation is lost during sludge aeration. Therefore to achieve autoheating, the heat resulting from organic matter oxidation can be conserved by providing enough biodegradable organic matter, by insulating the bioreactor, and increasing oxygen transfer efficiency to levels exceeding 10 per cent (Steinle, 1993).

The advantages of aerobic digestion are low capital cost, easy operation, and production of odourless stabilized sludge. Some disadvantages are the high consumption of energy necessary for supplying oxygen, dependence upon weather conditions, and an end product with a relatively low dewatering capacity.

Composting

The main objectives of composting are to produce stabilized organic matter, accompanied by reduction of odours, and to destroy pathogens and parasites.

Composting consists essentially of mixing sludge with a bulking agent, stabilizing the mixture in the presence of air, curing, screening to recover the bulking, and storing the resulting compost material. There are three main types of composting systems: aerated static pile process, windrow process and enclosed systems.

The composting process is dominated by the degradation of organic matter by micro-organisms under aerobic, moist, and warm conditions. Composting results in the production of a stable product that is used as a soil conditioner or a fertilizer, or as feed for fish in aquaculture. Furthermore, this process inactivates pathogens and transforms organic forms of nitrogen and phosphorus into inorganic forms, which are more bioavailable for uptake by agricultural crops.

Composting consists of several temperature controlled phases, each of which is driven by specific groups of micro-organisms:

- latent phase, for the acclimatisation of microbial populations to the compost environment;
- mesophilic phase, dominated by bacteria that raise the temperature following decomposition of organic matter;
- thermophilic phase, which is characterised by the growth of thermophilic bacteria, fungi, and actinomycetes (organic matter is degraded at high rates during this phase);
- cooling phase, during which temperature decreases again to the mesophilic range (thermophilic micro-organisms are replaced by mesophilic ones);
- maturation phase, in which the temperature drops to ambient levels, allowing the establishment of biota from other trophic levels (protozoa, rotifers, beetles, mites, nematodes), as well as other important processes such as nitrification, which is sensitive to high temperatures.

The composting mass may be regarded as a microbial ecosystem. Microbial succession during composting is relatively rapid, since it may take only a few days to reach 55 °C. Community diversity during the thermophilic stage is relatively low. The thermophilic microflora of the compost are mostly bacteria of the genus Bacillus, a spore-forming bacterium. Thermophilic bacteria play a role in the degradation of carbohydrates and proteins, whereas actinomycetes (e.g. *Streptomyces* spp. and *Thermoactinomyces* spp.) and fungi (e.g. *Aspergilus fumigatus*) play a role in the degradation of more complex organics such as cellulose and lignin (Bitton, 1994).

Several physical and chemical parameters control the activity of micro-organisms during composting: temperature, aeration, moisture, pH, and carbon and nitrogen content, as well as the type of material being composted and the composting system used. Composting produces a stable end product that is a good source of nutrients and serves as a useful soil additive in agriculture.

Current status of sludge treatment in Mexico

It is estimated that just 25 per cent of the wastewater generated is treated, and the sludge produced as a result of this treatment is in a lesser extent treated. The production of sludge in the country is estimated to be around 2.58 million kilograms of dry solids per day, on the basis of the population benefited by the wastewater that is treated.

Biotechnologies for the stabilization of sludge in Mexico are not widespread; few wastewater treatment plants use them.

Aerobic digestion is the system more often preferred, in spite of its high operational costs. To a lesser extent, composting is the biotechnology selected in second term. Just a few municipal wastewater treatment plants utilise anaerobic digestion for the stabilization of sludge.

An intensive training program for wastewater treatment plant operators is necessary to support the correct application of these biotechnologies.

Case studies

Some particular case studies are presented, focusing the attention on the type of biotechnology used for stabilizing the sludge; they were selected in different parts of the country.

The five states selected, located across Mexico, one in the north, one in the southern part, one in the eastern part and two in the central part of the country, were Morelos, Guanajuato, Guerrero, Quintana Roo and Nuevo Leon.

Table 2 presents the plants and biotechnologies used to stabilize sludge, and Figure 4 the location of the states selected.

Table 2. **Case studies**

State	City	Biotechnology for stabilization
Morelos	Cuernavaca	Composting
Guanajuato	Irapuato	Aerobic digestion
Quintana Roo	Cancún	Aerobic digestion
Nuevo Leon	Monterrey	Anaerobic digestion
Guerrero	Acapulco	Anaerobic and aerobic digestion, composting

Source: Author.

State of Morelos: Cuernavaca wastewater treatment plant

The Cuernavaca wastewater treatment plant is located near Temixco, Morelos, and serves a part of the population of Cuernavaca city. It treats 35 000 m^3/d using the Rotating Biological Contactors (RBC) system. It produces 6.5 tons per day of sludge that is stabilized by composting using cane bagasse as the bulking agent. The compost is produced using the static pile process. In the earlier research stages of this project, The Mexican Institute of Water Technology (IMTA) contributed during the laboratory and pilot research stages, in the selection of the best bulking agent, the best way of aeration and the optimum moisture level. The design parameters were obtained. This technology was transferred in 1994 to US Filter Corporation, Mexico, for the Cuernavaca wastewater treatment plant mentioned here. This plant was constructed and operated by US Filter. The final product named "Nutriplante" is sold as a fertilizer to the agriculturists and horticulturists of the area.

Figure 4. **Selected case study areas**

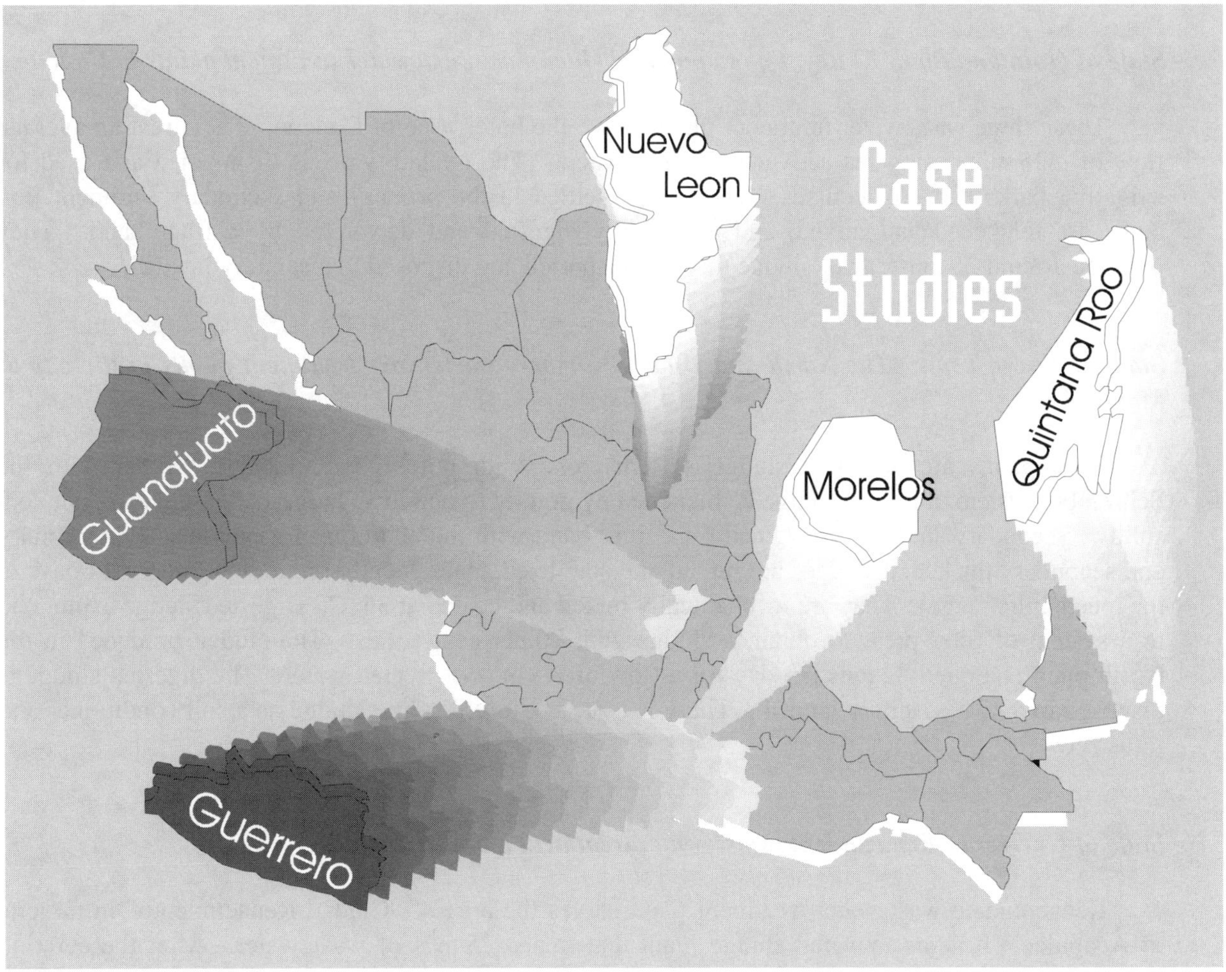

Source: Author.

State of Guanajuato: Irapuato wastewater treatment plant

The Irapuato wastewater treatment plant serves the population of Irapuato. It has a primary treatment and a baffled waste stabilization pond. Sludge produced is aerobically digested and dewatered. It is disposed in a sanitary landfill. Since 1995, the Institute of Agricultural Sciences of the University of Guanajuato, together with the Wastewater Technology Center in Canada, the State Commission of Water and Sanitation of Guanajuato (CEASG) and the Mexican Institute of Water Technology (IMTA), have been working in the area of beneficial use of wastewater sludge, better named as biosolids. These biosolids can increase the productivity of the soil through the addition of moisture, organic content and nutrients. The Institute of Agronomical Sciences was the leader of this project. Its objective was to demonstrate the safe use of the biosolids being applied in agricultural lands. The location of sites was determined, and the characterisation of the soil and sludge made. The selected crops were corn, beans, wheat and sorghum. Biosolids were added in three different concentrations and compared with conventional fertilizers. The results obtained demonstrate that yields for sludge treatments which supplied enough nitrogen quantities for crop growth were as high or higher than for a commercial fertilizer treatment (Diaz and Romero, 1996; Webber and Mac Donald, 1995). Experiments continue, and it is expected that in the near future, there will be enough information to develop national guidelines for the application of sludge in agricultural lands.

State of Quintana Roo: El Rey, Pok ta pok and Gucumatz wastewater treatment plants in Cancun

These three wastewater treatment plants serve the hotel zone of Cancun. They treat an average flow of .325 m^3/s, using the activated sludge system. The treated water is disinfected and used for irrigating parks and golf courses. The sludge settled in the primary and secondary sedimentation tanks are thickened and mixed, and aerobically digested and dewatered by a filter band. They produce around 25 tons/day of sludge that is transported and disposed in a sanitary landfill.

State of Nuevo Leon: The North and Dulces Nombres wastewater treatment plants in the city of Monterrey

Both plants are activated sludge systems, one with pure oxygen, and the other with the Schreibber system. Primary sludge is thickened by gravity to obtain a sludge with around six per cent solids. Secondary sludge is thickened by a filter band with the addition of a polymer. The primary and secondary thickened sludges are mixed and stabilized anaerobically (detention time of 14 days in the mesophilic range). They are mechanically mixed and heated at 35 $^{\circ}C$. Digested sludge is pumped to a system of filter press to obtain a sludge with 20 per cent solids. The sludge produced in the North plant is around 82 tons per day for a flow of 2.5 m^3/s of treated water. The digested sludge is disposed in a 47 ha sanitary landfill. There is a project to utilise the sludge as a soil conditioner and fertilizer.

State of Guerrero: Renacimiento wastewater treatment plant in Acapulco

Renacimiento wastewater treatment plant serves the area of "Ciudad Renacimiento" in the city of Acapulco. It is an activated sludge plant that treats .25 m^3/s of wastewater. After the aeration basin, the sludge passes to a secondary settler. The water is disinfected and discharged to La Sabana river. The sludge is thickened in a gravity thickener and is then mixed with polymers and calcium hydroxide, and dewatered by a filter press. The sludge is partially stabilized and carried to a sludge dumping.

The Mexican Institute of Water Technology (IMTA), together with the National Water Commission (CNA), are carrying out a research project to select the best stabilization system for that particular plant. Pilot installations have been constructed to compare which stabilization technique is the best for the same quality of sludge. There are aerobic and anaerobic digesters and a static pile composting system. Experiments are underway, and it is expected that at the end of this year, there will be enough information to recommend the best technology for the stabilization of sludge in this particular wastewater treatment plant. Experiments for land application of sludge are being carried out to contribute to the database for the development of national guidelines for the application of sludge in agricultural lands.

Research activities and trends related with the biotechnologies for the stabilization of sludge

Several groups in Mexico are related with environmental biotechnology. Most of them are concentrated at the National Autonomous University (UNAM), the Metropolitan Autonomous University (UAM), The National Polytechnic Institute (IPN), the Autonomous University of the State of Morelos (UAEM), and the Mexican Institute of Water Technology (IMTA).

Dealing with anaerobic technologies, the Engineering Institute, UNAM and the UAM Iztapalapa have consolidated research groups dedicated to bioprocess engineering, reactor design and microbiology. IMTA also has a research group on anaerobic digestion of wastewater and sludge. Efforts have been made to enhance the anaerobic digestion of sludge by adding growth stimulants, improving mixing and reactor design, and studying the microbiology of anaerobic sludge.

Sludge is an unavoidable product of wastewater treatment, and as effluent quality objectives increase, so does the quantity of resulting sludge. Contaminants which are not destroyed by the treatment process tend to be concentrated in the sludge, and the public is becoming increasingly concerned about the impacts of the sludge on the environment and human health. As treatment plants are constructed to eliminate the discharge of pollutants into aquatic environment, they will produce large quantities of sewage sludge which must be managed in an environmentally acceptable manner. Addressing the issue in the initial planning stages will ensure cost-effective and environmentally acceptable solutions.

Related with the biotechnological processes that are used to stabilize the sludge, research on the micro-organisms' dynamics and metabolic pathways has to be pushed. Prospective epidemiological studies have to be conducted to assess the potential health risks associated with sludge treatment and disposal, for example, methodological problems in determining the number of pathogens in sludge, high variability of pathogen densities in sludges, and lack or fragmentary nature of information about the minimum ineffective doses for pathogens, which may vary in a very wide range.

Odour control at composting facilities is absolutely essential for public acceptance. This will push the technology to develop the biological treatment of the gases and odours.

Autothermal aerobic digestion has to be deeply studied, because it seems to be a promising technology for the stabilization of sludge.

Earlier surveys have shown that anaerobic digestion is by far the most widely used method for sewage sludge stabilization in most countries (Bruce, 1987; Saabye *et al.*, 1994). The successful use of anaerobic digestion depends on the detailed knowledge of the basic biochemistry and microbiology of the process. Research and development is needed to reduce costs and improve efficiency. *Black boxes* must be analysed to reveal individual components and processes that can be understood, optimised and controlled.

Land application of municipal sludge is likely to remain a major option for the future. It may represent the major cost-effective option, particularly for small plants, and it poses little environmental risk. The question of the impact of organics from sludges is not fully answered but if sludge is applied according to generally accepted scientific guidelines, the risk appears to be minimal. There is an ongoing need for a proactive public education program to ensure that affected communities are aware of the benefits of land application.

There is a definite role for new or emerging technologies that offer the potential of reduced cost, increased flexibility and increased environmental protection. The definition of new technology is very dependent on perspective, and could range from processes that are unique and still at a fundamental level of development, to those which might be considered as only a minor modification of existing technology.

Predicting the future of biosolids management is extremely difficult because of the dynamic nature of the wastewater business. Diversification to multiple management strategies is desirable (Campbell and Webber, 1994).

The current high profile of biosolids management can be viewed as a mixed blessing: the number of stakeholders that must be satisfied has never been greater, but at the same time, this provides a powerful incentive to develop new and innovative solutions.

Biotechnology is one of the available technologies for the maintenance of environmental quality and must be viewed within the large spectrum of scientific and engineering disciplines.

Innovative sludge technologies must harmonize with the global material cycle in the biosphere and should have the following characteristics: renewability of resources, mildness of production processes, compatibility of the resultant products with humans and the environment, and the recyclability of wastes.

Research has to be encouraged in the following topics:

- microbial ecology and dynamics (adaptation, identification of micro-organisms in sludge);
- bioprocess engineering (scale-up problems, modelling, kinetics, process design and optimisation);
- removal of specific pollutants and recovery of materials;
- on-line monitoring (OECD, 1993).

Micro-organisms are the tools that we have in our hands for a better and cleaner tomorrow. The work that we do today with them, and the knowledge and understanding of their way of living, will allow us to develop innovative technologies that help us to eliminate the wastes we have created. The beginning of life was due to micro-organisms, and we have to work with them (Aldama, 1995).

Questions, comments and answers

Q: With the building of new municipal wastewater plants, what is the trend for sludge management in Mexico? What process will be used most?

A: Priorities have not yet been established, and there are differing views on the alternatives. My personal view is that biological treatment should be promoted, as it makes recycling easier. When chemicals are added, there are more recycling problems and compounds accumulate.

Questions, comments and answers cont'd.

Q: Do you have problems with heavy metals in sludges? In the Netherlands, we have difficulties spreading sludges for this reason.

A: If the origin of the sludge is only municipal, there is no problem, but if it comes from industrial areas, it cannot be used freely. At present, Mexico has no limits or guidelines.

C: If you have lead pipes, there is lead in the sludge from municipal wastewater.

A: We need to understand wastewater characteristics. A problem in the water is multiplied in the sludge.

REFERENCES

ALDAMA, A.A. (1995), "The role of biotechnology in environmental remediation", in *Wider Application and Diffusion of Bioremediation Technologies, the Amsterdam '95 Workshop*, pp. 433-438, OECD Documents, OECD, Paris.

BITTON, G. (1994), *Wastewater Microbiology*, John Wiley, New York.

BRUCE, A.M. (1987), "Progress on anaerobic digestion within the framework of the EC concerted action cost 681", in M.P. Ferranti, G.L. Ferrero, and P.L. Hermite (eds.), *Anaerobic digestion: Results of research and demonstration projects*, Elsevier Applied Science, Essex, England.

BRUCE, A.M. and U. LOLL (1981), "A review of methods for stabilizing sewage sludges", P. de l'Hermite and H. Ott (eds.), *Characterization, treatment and use of sewage sludge*, Reidel Publishing Co., Dodrecht, the Netherlands.

CAMPBELL, H.W. and M.D. WEBBER (1994), "Biosolids management in Canada: current practice and future trends.", WEAO Seminar "Biosolids treatment and utilization: innovative technologies and changing regulations", 20-23 September, Missisagua.

DAVIS, M.L. and D.A. CORNWELL (1985), *Introduction to Environmental Engineering*, PWS Engineering, Boston, Massachusetts.

DIAZ, S.F. and V.M. ROMERO (1996), "Una guía para la aplicación de lodos residuales en agricultura, una experiencia en Guanajuato", in *Plantas de tratamiento de aguas residuales, operación y economía*, Gto. Secretaría de Ecología del Edo., Guanajuato, and WTI, Canada.

ENCYCLOPEDIA HISPANICA (1993), Vol. 10, p. 97, Encyclopaedia Brittanica Publishers, Inc., Rand McNally, United States.

ENVIRONMENTAL PROTECTION AGENCY (EPA) (1989), "Use and disposal of municipal wastewater sludge", Environmental Regulations and Technology, WH-595.

ENVIRONMENTAL PROTECTION AGENCY (EPA) (1992), "Control of pathogens and vector attraction in sewage sludge", Environmental Regulations and Technology, 625/R92/013.

ENVIRONMENTAL PROTECTION PROGRAMS DIRECTORATE (1984), "Land application of sludge", Manual EPS6-EP-84-1, Canada.

ORGANISATION FOR ECONOMIC CO-OPERATION AND DEVELOPMENT (OECD) (1993), *Biotechnology for a clean environment*, OECD, Paris.

PODER EJECUTIVO FEDERAL (1995), "Programa hidraùlico 1995-2000", Secretaría del Medio Ambiente Recursos Naturales y Pesca.

SAABYE, A., A.S. KRUGER, and H.C. SCHWINNING (1994), "Treatment and beneficial use of sewage sludge in the European Union", *Times* 3, ISWA, pp. 1-6.

SECRETARIA DE DESARROLLO SOCIAL (1994), "Informe de la Situación General en Materia del Equilibrio Ecológico y Protección al Ambiente, 1993-1994", Instituto Nacional de Ecología.

STEINLE, E. (1993), "Sludge treatment and disposal systems for rural areas in Germany", *Wat. Sci. & Tech.* 27:9, pp. 159-170.

WEBBER, M.D. and R. MAC DONALD (1995), "Feasibility demonstrations for sludge application on agricultural land in Guanajuato State, México", Final report WTC, Burlington, Canada.

SOURCE WATER PROTECTION THROUGH PLANT AND PLANT-MICROBE INTERACTIONS*

by

John A. Glaser, Steven A. Rock, and Donald S. Brown
U.S. Environmental Protection Agency
National Risk Management Research Laboratory, Cincinnati, Ohio, United States

Introduction

A major threat to water quality in the United States arises from non-point source contamination. Agricultural activities, urban runoff, atmospheric deposition, land disposal, construction work and mining operations contribute to the non-point source pollution. Due to the inherent absence of source definition, it is difficult to design effective, low-cost solutions for this pollution category (Committee on Ground Water Alternatives, 1994). Selected plant species and plant communities have been shown to have unexpected abilities to remove, treat, and/or sequester organic and inorganic toxic chemicals. Vegetation has been shown to enhance microbial activity in soils under a wide range of environmental conditions (Cunningham and Ow, 1996; Cunningham *et al.*, 1995). The diversity and density of soil micro-organisms has been shown to increase in vegetated areas in contrast to poorly vegetated regions. Strategic placement of selected plants can offer means to interdict water flows contaminated with pollutant chemicals occurring as part of runoff or contaminated subsurface water. Naturally occurring and constructed wetlands have offered significant treatment capacity as interceptors of contaminated water flows whereby degradation/removal of contaminants is sufficient to achieve permitted discharge specifications. Wetlands can function as very useful buffers between land and water to improve water quality through purification mechanisms attributable to both biological and abiotic processes. Judicious selection and use of single plant species can offer significant advantages for the treatment of non-point source pollution. Examples of wetland application and selected plant species will be used to portray the potential of phytoremediation to diverse pollution problems.

Vegetation has been used in a number of settings to assist the remediation of both organic and inorganic pollutants in soil and groundwater (Tables 1 and 2). This new potential treatment technique has been named phytoremediation (*phyto* = plant and *remedium* = correct evil) and is focused on the use of green plants to "remove, contain, or render harmless environmental contaminants" (Cunningham and Berti, 1993). Plants can naturally remove chemical contaminants from soil and water by various mechanisms. Certain plants degrade organic pollutants directly, or the degradation mechanism may rely on associated and symbiotic microbial communities (Kingsley *et al.*, 1993).

Selected inorganic contaminants in soil or water can be removed by plant species known for their ability to concentrate the contaminants in plant tissue. This sequestration of the inorganic contaminant(s) permits the design of useful treatment strategies where the contaminant can be removed and disposed of separately, leaving the soil clean. Plants are possibly the only cheap renewable resource available to this civilisation.

Table 1. **Phytoremediation trials (organic contamination)**

Contaminant	Site	Media	Plants	Location	Status
Crude oil	Pipeline	Soil	Sorghum, rye, cowpeas	Louisiana	Planted 1995
Leachate	Landfill	Groundwater, surface water	Poplar trees	Oregon	Planted 1994
Nitrates, pesticides	Agricultural fields	Groundwater, surface water	Poplar trees	Iowa	Planted 1994
PAHs (250 ppm), PCP (160 ppm)	Wood trating site	Soil	Ryegrass	Oregon	Planted 1996
Petroleum	Pipeline	Soil	Grass, clover	Virginia	Planted 1996
Petroleum (1 200 ppm)	Refinery	Groundwater, soil	Poplars, alfalfa	Utah	Planted 1996
TCE, TCA	Army base	Groundwater	Wetland	Tennessee	Installed 1996
TCE (850 ppm)	Air force base	Groundwater	Cottonwood trees	Texas	Planted 1996
TNT	Army ammunition plant	Groundwater	Wetlands	Tennessee	Installed 1996
Organics, BOD	Sewage	Surface water	Wetlands	Widespread	Ongoing

Source: Author.

Table 2. **Phytoremediation trials (inorganic contamination)**

Contaminant	Site	Media	Plants	Location	Status
Pb, Cd, Cr	Industrial sites	Soil	Indian mustard	Widespread	Planted 1996
Uranium	Chernobyl	Surface pond	Sunflower	Ukraine	Completed 1995
Acid mine drainage	Mine sites	Surface water	Wetland	US mountains	Ongoing

Source: Author.

Background

Phytoremediation has been shown to reduce concentrations of hydrocarbons, polychlorinated biphenyls (PCBs), wood preserving chemicals, and agricultural runoff contaminants (nitrates, pesticides and herbicides) (Schnoor *et al.*, 1995). Some plants can extract heavy metals such as lead, chromium and uranium (Raskin, 1996). Phytoremediation is best suited for clean-up operations over a wide area, with contaminants in low to medium concentrations.

The mechanisms of phytoremediation can be classified into four broad categories: enhanced rhizosphere biodegradation, phytodegradation, accumulation, and physical effects. Enhanced rhizosphere biodegradation refers to a series of effects that plants have on the microbial population in the rhizosphere, or the immediate area surrounding the root (Bolton *et al.*, 1993). Microbial populations have been reported to be two orders of magnitude higher in the soil of the root zone than in adjacent unplanted soil. The roots of many plants provide growth requirements to the soil microbiota as a by-product of normal plant growth, and there can be simple and symbiotic associations possible between a plant and the soil biota. As roots penetrate the soil, there is passive aeration as the roots loosen the soil, and positive aeration as the roots release oxygen as part of respiration. Phytodegradation refers to biochemical processes within a plant, leading to changes in the pollutant chemical's structure.

Parts of tree roots die during seasonal water and temperature fluctuations. These sloughed roots become a nutrient source to the rhizobial community. This resource may provide cometabolites which sustain microbes which only incidentally degrade contamination. Plants also draw water into the near surface root zone, sometimes from great depth. Leaf fall and root death add to the organic matter content of the soil, which adds to the soil's ability to retain water.

Inorganic contaminants can be sequestered and stored in large quantities by plants. This process of accumulation can take different forms in different plant species (Ryan *et al.*, 1988). Sequestration is the process of immobilising a contaminant. The humic acid fraction in soil has been found to chemically bind some organic contaminants through the action of micro-organisms (Bollag, 1992). Poplar trees have been shown to take metals into the roots between the root cells. These intracellular metal contaminant concentrations can exceed the expected toxic levels for the plant (Hinchman and Negri, 1994). Sunflowers have been found to accumulate large quantities of uranium in their root structure (Raskin, 1996; Salt *et al.*, 1995).

Certain plants have been identified that not only accumulate metals in the plant structure, but also translocate the accumulated metals from the root to the leaf and shoot (Baker and Brooks, 1989; Baker and Walker, 1990; Reeves *et al.*, 1983). While many plants have this function, some plants, known as hyperaccumulators, can concentrate contaminant inorganic pollutants to final concentrations of several parts per hundred in plant biomass on a dry weight basis. The plant can then be harvested, dried, and the heavy metals recovered or disposed of, cleaning the soil of the contamination. Often, the plants that naturally hyperaccumulate are native to remote and unique locations, and are not suitable to temperate climates which include most of the United States. Some of the hyperaccumulators are small in size, so that even though they accumulate significant percentages of heavy metals, their biomass accumulates in so low a mass that the total amount of contaminant removed from the ground is not acceptable as a treatment. To overcome these limitations, standard or slightly modified crops and agriculture practices are being evaluated. Indian mustard has been used as a plant that can both accumulate significant quantities of metals, and be grown easily in many parts of the United States (Cunningham and Lee, 1995).

Plants also can influence the movement of contaminants, especially in groundwater. These physical effects include transpiration of volatile contaminants (Simonich and Hites, 1995), and desiccation of a groundwater flow which in effect forms a hydrologic barrier. Trees of the willow family have been shown to draw as much as 200 gallons or 760 litres of water per day (Dickman and Stuart, 1983). The willow family (*Salicaceae*) provides several useful tree species, including poplars, cottonwood, and aspens, which have wide distribution throughout the northern hemisphere. Trees of this family can be propagated from cuttings and have vigorous growth habits (10 feet or 3.2 metres of vertical growth per year is common).

Wetlands constructed with reeds and cattails are used to prevent acid mine drainage from polluting streams; poplar, cottonwood, and willow trees have been planted as interceptor barriers to remediate groundwater contamination (Mitsch and Gosselink, 1993). Common crop plants like mustard are used for contaminant extraction, and alfalfa and ryegrass are used as *in situ* soil remediation applications. Planted areas strategies can be used in conjunction with other remedial technologies, for example, following a removal action of high concentration hot spots.

Source water applications of phytoremediation can be conceptualised as plant mediated pump and treat for contaminated groundwater to control contaminant plumes and interception of non-point source pollution by riparian zone plantings. Treatment wetlands offer considerable potential as interception/treatment techniques applied to wastewater and non-point sources and mine drainage (Dushenkov *et al.*, 1995; Nada Kumar *et al.*, 1995).

Phytoremediation mechanisms

Uptake and transport of chemicals within plants

The conductive systems implicated in the pollutant uptake by plants are roots, stems, leaves, and vascular tissue (xylem and phloem) (McFarlane, 1995). The mechanisms associated with contaminant uptake may limit contaminant uptake to the root zone or other parts of the conductive system (Schroll and Scheunert, 1992; Trapp, 1995). Physiochemical properties of contaminants important to understanding uptake mechanisms are vapour pressure, lipophilicity, water solubility, acid strength, and transport across membranes (Walker, 1972; Reiderer, 1995; Bromilow and Chamberlain, 1995).

Transformation of organic chemicals by plant metabolic processes

Plants have been characterised as functioning as "green livers" acting as global sinks for environmental pollutants and their degradation (Sandermann *et al.*, 1977). The plant enzymes responsible for this activity are constitutive. There exists a wide variety of enzymes available from plant sources which are useful in the destruction of contaminants. Nitroreductase, dehalogenase, peroxidase, and other enzymes have been found to be produced by plants (Cunningham *et al.*, 1996). These enzymes can either detoxify a contaminant, or render it vulnerable to microbial consumption. Transformation of organic chemicals (Figure 1) by the metabolic processes of plant systems has been organised as Phase 1) transformation; Phase 2) conjugation; and Phase 3) internal compartmentalisation and storage (Sandermann, 1992; Komossa *et al.*, 1995).

Figure 1. **Example of plant metabolic phases**

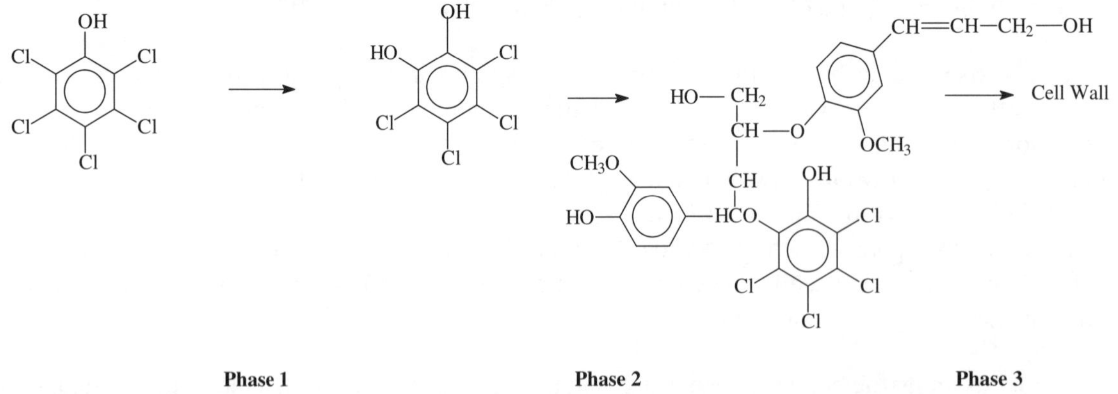

Source: Sandermann *et al.*, 1977.

Transformation of organic chemicals in the rhizosphere

The rhizosphere or root zone has been identified as a source of desirable pollutant degrading activity (Anderson *et al.*, 1993; Keyser *et al.*, 1993). Microbial populations of the rhizosphere are typically larger by an order of magnitude than non-vegetated soil. This population difference is referred to as the "rhizosphere effect" and expressed as the R/S ratio, i.e. ratio of micro-organisms found in the rhizosphere to non-rhizosphere soil. The ratio ranges generally between five and 20, but values of 100 can be encountered. Rhizosphere soil slurries have been shown to degrade trichloroethylene (TCE), whereas soil from non-vegetated areas exhibited no activity towards TCE (Walton and Anderson, 1990). Mineralization of TCE was found to account for 25 per cent loss of labelled TCE in contrast with 15 per cent in non-vegetated soil by different grass species (Anderson and Walton, 1995). Plant tissue incorporation accounted for 1-21 per cent of the TCE mass. The depletion of PAH concentrations in contaminated soil has been attributed to similar rhizosphere activity (Reilley *et al.*, 1996). Root exudates may be responsible for part of the activity by providing carbon substrates which support the cometabolic microbial degradation of TCE (Rovina, 1969; Rao, 1980).

Qualitative and quantitative shifts in soil microbial population have been observed with the introduction of mycorrhizae (Allen and Allen, 1992; Lindermann, 1988). These organisms have been shown to increase plant tolerance to heavy metals (Bradley *et al.*, 1981). Heavy metal translocation to new plant shoots can be reduced by plant root mycorrhizal colonisation which leads to the binding of heavy metals to cell walls of fungal hyphae in the plant roots. Inoculation of soil with mycorrhizal fungi can significantly improve the growth and establishment of certain plants in contaminated soil, whereas inoculation with other micro-organisms had negligible effects (Shetty *et al.*, 1994). The extent to which plants rely on microbial amendments for establishment on moderately contaminated soil is greatly dependent on plant species and type of microbial inoculum. The interaction of plant species with soil micro-organisms must be considered for its importance to revegetation strategies. Ectomycorrhizal fungi have been shown to degrade PCBs (one to five attached chlorines) in liquid culture (Donnelly and Fletcher, 1994*a* and 1994*b*; Donnelly *et al.*, 1994; Fletcher *et al.*, 1995). However, the observed activity was much more limited than selected "super PCB-degrading" bacterial isolates.

Wetlands

Ecologists have often referred to wetlands as "nature's kidneys" for their water purification characteristics. The microbial activity of wetlands is known to be rich and diverse. Aerobic and anaerobic processes within most wetlands lead to a wide range of biological activity potentially useful to the degradation of pollutant chemicals. Wetland plant species and plant microbe interactions offer potential approaches for the enhancement of wetland treatment performance.

Treatment wetlands can be categorised by their origin, flow regime and vegetation type. Treatment wetlands can be either natural or constructed. A natural treatment wetland is a pre-existing wetland that has been modified by adding flow control structures, such as flow dispersion or collection devices, and perimeter and/or interior dikes. A constructed wetland is completely manmade. Two flow regimes are generally used to categorise treatment wetlands. Free water surface wetlands have also been called surface flow wetlands. Subsurface flow is a generic term which includes "root zone method", "rock reed filters", and "vegetated submerged beds" systems. Vegetation categories are descriptive of the predominant type of vegetation in the wetland.

Categories include marsh (with emergent species such as cattails or bulrush), forest (with trees such as cypress or red maple), and floating aquatic (with plants such as duckweed or water hyacinth).

Wetlands have been used to treat wastewater for many years. Originally, wetlands were just convenient places to discharge wastewater, with little thought given to either the treatment capabilities of the wetland or the potential impact of the wastewater on the wetland. However, over time, more attention has been paid to the use of wetlands as not just receiving waters for discharges, but as the treatment system for the wastewater. During the past decade, wetlands have been receiving more and more attention around the world as treatment systems for a variety of wastewaters (USEPA, 1993). In the USEPA's database compilation of information about existing treatment wetlands in the United States, 178 wetland treatment systems were identified (USEPA, 1994). Historical perspectives on the use of wetlands for wastewater treatment, including a history of the important early research efforts on wetlands treatment, are well documented (Kadlec and Knight, 1996; Hartmann, 1975). Table 3 highlights the historical utilisation of treatment wetlands.

Table 3. **Selected wastewater treatment wetlands**

Date	Location	Description
1912	Concord, Massachusetts	Natural wetland
1939	Waldo, Florida	Natural cypress swamp
1955	Wildwood, Florida	Natural forested wetland
1973	Mt. View, California	Constructed free water surface wetland
1974	Othfresen, Germany	Constructed soil-based subsurface flow wetland
1980	Show Low, Arizona	Constructed wetland for wastewater and wildlife
1984	Fremont, California	Constructed wetland for urban stormwater
1985	Pflugerville, Texas	Constructed gravel-based subsurface flow wetland

Source: USEPA, 1994; Kadlec and Knight, 1996.

Removal of conventional pollutants

The primary removal mechanisms for BOD and TSS are sedimentation and filtration, with subsequent microbial degradation of biodegradable materials. Since wetlands typically have detention times ranging from days to weeks, and wastewater flows through either a medium or a mass of stems, leaves and plant detritus, conditions are excellent for sedimentation and filtration. While emergent aquatic plants can supply a limited amount of oxygen to their roots, due to the primarily anoxic or anaerobic conditions within the medium or water column, the degradation appears to be accomplished by facultative or anaerobic microbes. Effluent concentrations of 5 mg/L of BOD and TSS have been achieved in treatment wetlands.

The primary mechanism for nitrogen removal is microbial nitrification followed by denitrification. For wetlands receiving wastewater with less than secondary pre-treatment, plant uptake of nitrogen is minimal compared to the amount in the wastewater. The ability of wetlands to remove nitrogen is still a matter of debate, with some wetlands removing over 60 per cent of influent nitrogen, and other wetlands removing no nitrogen at all. A key factor in the debate is the ability of the plants to supply sufficient oxygen to their roots to maintain an active population of nitrifying bacteria.

The primary mechanism for phosphorus removal is the reaction with the medium or bottom soils via adsorption, complexation and precipitation (Reed *et al.*, 1995). As with nitrogen, plant uptake does occur; however, uptake is minimal compared to influent wastewater concentrations in wetlands used only for tertiary polishing. Once the adsorption capacity of the medium or soils is reached, phosphorus removal will be minimal.

Removal of organics and metals

The removal of organic compounds in wetlands has not received the same attention as the removal of metals. The two primary removal mechanisms for organic compounds are thought to be 1) volatilisation, and 2) adsorption to organic matter followed by microbial biodegradation. The ability of wetland plants to extract and metabolise or immobilise organic compounds is suspected, but not documented. Early work indicated that bulrush continued to grow when fed only tap water and phenol, and that the plants appeared to metabolise the phenol for the production of amino acids (Siedel, 1976). Investigation of the reduction of a variety of aromatic and aliphatic compounds in a pair of wetlands -- one with plants (reeds) and one without plants -- exhibited removals ranging from 49 per cent to 99 per cent (Wolverton and McDonald-McCaleb, 1986). The removal of organic pollutants may be attributed to soil microbial activity surrounding the plant roots. Furthermore, the planted wetland always outperformed the unplanted wetland. Remarkably little research has been conducted since these early studies to further investigate these early findings. Wetlands are primarily anaerobic reactors with hydraulic retention times ranging from days to weeks; hence, contaminant chemicals known to be anaerobically biodegradable should be removed by a wetland.

Metals are primarily removed by accumulation in the wetland sediments, with little or no uptake or sequestering by either plants or microbes. The one exception to this primary mechanism is the formation of metal precipitates in anaerobic mine drainage wetlands; this process is mediated by sulfate reducing bacteria. Removal of metals in wetlands is not well understood, and a wide range of removal efficiencies has been reported, as shown in Table 4 (Kadlec and Knight, 1996).

Table 4. **Metal removal by treatment wetlands**

Metal	Influent concentrations (ug/L)	% removal	Contributing studies
Aluminum	30-300	-33-0	3
Cadmium	10-70	0-99.7	7
Chromium	<2-160	0-88	5
Copper	1-1 510	36-96	8
Iron	105-205 000	-217-98	7
Lead	2-300	-181-98	8
Manganese	3-5 900	40-98	6
Mercury	1-6	17	2
Nickel	3-300	25-90	7
Silver	0.4-4	-49-76	2
Zinc	21-6 900	33-98	9

Source: Kadlec and Knight, 1996.

There are two major concerns regarding metals removal in wetlands. First, because metal removal is primarily a physical/chemical process, the capacity of the sediments to retain metals will eventually be exceeded. The duration of the wetland's metal retention capacity depends upon the type of sediments and the metals loading rate. The designer must consider ways to revitalise or close the wetland when the metals capacity is reached. Second, there is some concern that accumulating metals in wetlands may pose a threat to biota living in or using the wetland. While there are no known cases of adverse effects from biota exposed to a treatment wetland, ways to minimise exposure should be considered by the designer if high metals concentrations are expected.

Ecological engineering

The use of ecosystems to treat pollutants has been widely recognised as a desirable effort, but loss of desired treatment performance is poorly understood (Mitsch, 1993). Existing wetland databases offer little information relating poor performance to controllable causes. There are instances where decreased pollutant removal is observed with time. For specific pollutant chemicals, wetland treatment system overloading is a major problem. Finally, treatment wetland failures can be partially attributed to poor design. Wetland restoration and implementation requires a more thorough understanding of the complex ecology found in a wetland. Whether human intervention can successfully duplicate environmental conditions to support the complex functioning of natural wetlands (bogs, fens, swamps or marshes) requires extensive investigation. Wetland restoration is a new science in terms of understanding the complex ecology found in wetlands.

Current examples of vegetation based treatment

Carswell AFB demonstration

A shallow dissolved TCE plume near at Carswell AFB, near Fort Worth, Texas, was selected to demonstrate the ability of eastern cottonwood trees to remediate shallow groundwater in an arid climate (Figure 2). The site is a gently sloping buffer for a golf course and is vegetated with short lawn grass species. The demonstration entails the planting and cultivation of eastern cottonwood trees over a dissolved TCE plume in a shallow (six to 11 feet or 1.8 to 3.3 metres below grade) alluvial aquifer. The trees are expected to enhance bioremediation of the contaminated groundwater and any potentially contaminated soil by the following mechanisms: biochemical transformations of contaminants within the soils due to microbial activity in the rhizosphere that is stimulated by the release of root exudates and enzymes; metabolism or mineralization within the vegetative tissues of contaminants from water taken up by the roots; and transpiration of water by the leaves.

Groundwater levels and TCE concentrations in the aquifer will be monitored to establish baseline conditions and to map changes within the aquifer throughout the life of the demonstration. Changes in the flow field and the position of the dissolved plume will be modelled. In general, data will be interpreted to identify the overall affect of the planted trees on the dissolved TCE plume in the aquifer.

Contaminant concentrations will also be monitored in the soil rhizosphere and tree tissue. Ratios of daughter and parent compounds will be computed for groundwater, soil, and tissue samples throughout the demonstration to assay the effect of phytodegradation. Microbial activity in the rhizosphere will also be monitored, and transpiration rates will be measured and plantation plot transpiration effects modelled. These data will be used to determine the fate of the TCE by the phytoremediation process, including abiotic processes that affect its fate. Tissue from a mature cottonwood tree currently growing near the site may be sampled near the beginning of the demonstration to provide early evaluation of the potential fate of the TCE. The trees are planted in two stands. A stand of whips (cuttings) and a stand of 5-gallon (or 19 L) bucket trees will be included in the study. Each stand has seven rows, and the stands are separated by an open control area. The total planted area is approximately one acre. The demonstration includes 29 monitoring wells and piezometers, some of which have continuous groundwater level measurements. These stands will be monitored in a similar fashion. The length of time it takes trees of each age to affect the dissolved plume will be determined. Root development and tree mortality rates within the two tree stands will be monitored. Costs associated with the planting and cultivation of each tree stand will be compared. Information relating the practicability and performance of the technology will be developed.

Figure 2. **Treatment layout for the Carswell site**

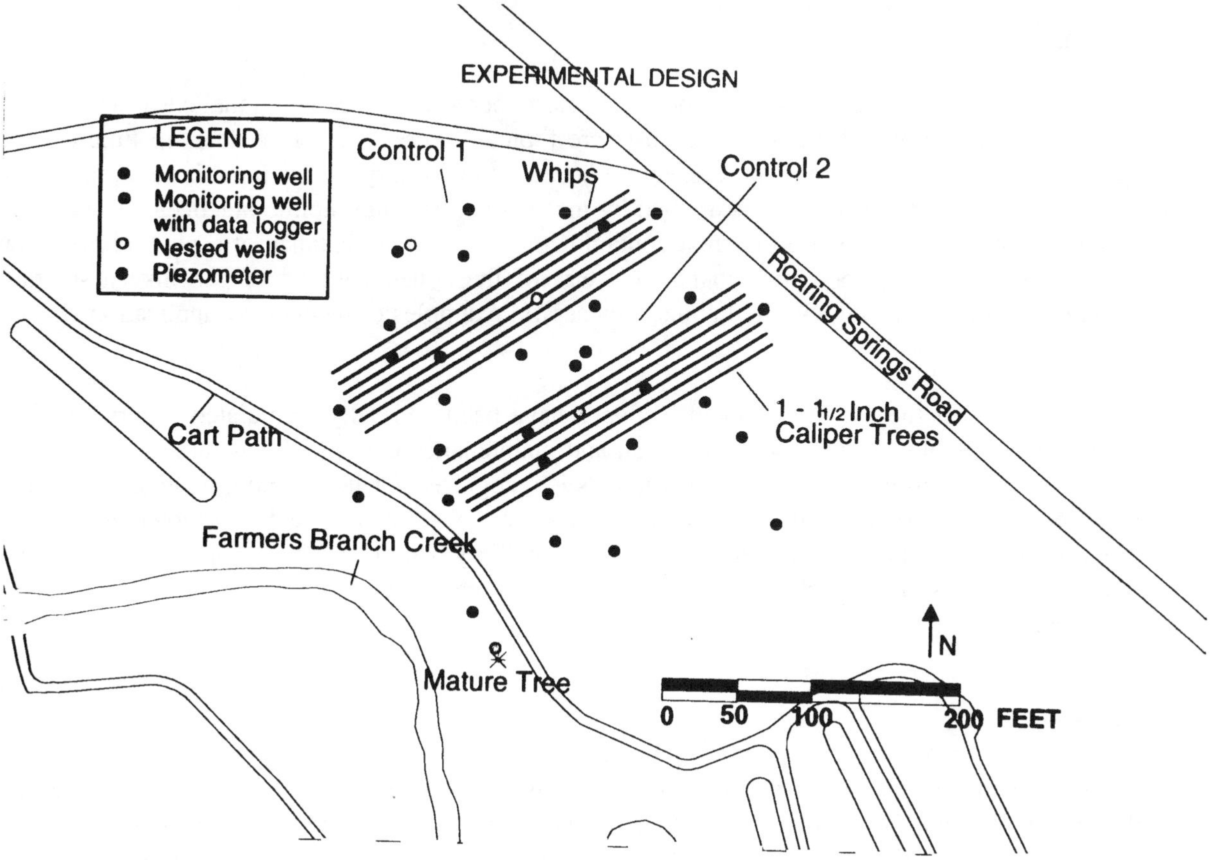

Source: Author.

The results of this study will be used evaluate the affects of planted trees as an active remediation of the groundwater TCE plume. The fate of TCE within the area will also be determined. Planting, irrigation techniques and site specific conditions under which the technology was applied will be documented. Finally, the technology will be evaluated in terms of its overall practicability. The trees and groundwater will be monitored for three years.

Everglades nutrient removal demonstration

In response to legislation and a lawsuit, the South Florida Water Management District has undertaken the evaluation of wetlands as part of a comprehensive effort to re-establish the landscape of the Everglades to the extent possible. This demonstration is designed to interdict the flow of agricultural runoff, which is rich in phosphorous, from the highly productive farmland surrounding

Lake Okeechobee. A full-scale prototype wetland of 3 681 acres has been constructed and evaluated for its performance. The project was designed to reduce the phosphorous load by 75 per cent to 50 mg/L. With an influent phosphorous concentration of 170 mg/L, the first year operation showed phosphorous outflow concentrations of 20 to 40 mg/L at a removal of 83 per cent, which surpassed the original design expectations (Abtew *et al.*, 1995; Guardo *et al.*, 1995; South Florida Water Management District, 1995). Expansion of the project to provide treatment needs is expected to require the construction of some 43 000 acres of wetlands.

Advantages

Using plants to remediate a site can be much less expensive than conventional clean-up options, because the installation and maintenance costs are typically very low. The clean-up time can be longer than with other remediation processes, but because the primary energy input is solar, operating costs are also low. The technology is applied by means of simple agronomic practices. The natural ability of the environment is used to treat contamination. Public acceptance of a phytoremediation project on a site can be very high, in part because of the added benefits of park-like aesthetics, including providing bird and wildlife habitat. Successful examples of technology application already exist.

Advantages for the use of treatment wetlands have been generally formulated in terms of low capital costs for implementation, low operation and maintenance costs, and general aesthetic values realised. The high biological diversity of wetlands is recognised for its potential treatment capacity, but is poorly understood for its intricate microbiology. Wetlands have been shown to be less susceptible to influent pulses of targeted pollutants. Ancillary benefits such as property improvement to recreational areas, and development of supporting circumstances to enhance educational values, can be realised.

Limitations

The use of vegetation for pollutant treatment is not universally applicable, or fully tested. Pollutant availability to the plant from long distances may be limited. The treatment of deep contaminant zones will be limited mainly by the growth habit of the roots of selected plants. Treatment is expected to be optimal during the growing season, which will constrain the treatment performance in temperate and greater latitude zones. The adaptability of individual plant species to unnatural growth conditions may be limiting, but only important when native plant species options are exhausted. Pollutant phytotoxicity can be treatment limiting, but several strategies are available to protect plant species for toxic effects. The treatment end points may not be acceptable to the regulatory community.

Wetlands generally require large land areas for desired treatment. The optimal design and operation of wetlands are poorly understood. Insect pestilence associated with wetlands may be difficult to control. Impacts of wetlands to wildlife are unknown. Existing wetland performance databases offer little information relating poor performance to causes. There are instances where wetlands have been observed to lose their ability to treat influent contamination with time. In the case of specific pollutants, wetland system overloading can be a problem. Failures of treatment wetlands can be partially attributed to poor design of the wetland to accommodate the influent loading.

Future developments

The use of plants as described is an emerging technology which requires significant investigation into components of fundamental and applied research. There are basic issues requiring thorough research, such as limiting factors controlling pollutant uptake, metabolism, and translocation within the plant. The understanding of plant tolerance to soil contamination must be developed further.

Within the rhizosphere, extensive investigation of taxonomy and biodiversity of root-associated populations related to soil type, composition, and plant type is necessary to fully develop the understanding of the contaminant treatment contribution associated with this highly active environmental component. Determination of mechanisms at the micro level involved in the adhesion of bacteria on host plant roots requires further elucidation. An understanding of the carbon and nitrogen mass balance in soil-plant systems is necessary to aid the control of the various mechanisms for carbon and nitrogen transfer and the mineralization of soil organic carbon.

Field scale investigation of the mechanisms and control factors relating to optimal use of phytoremediation present significant analytical hurdles. Attempts to determine the mass balance of contaminant are often incomplete and are complicated by the level of understanding of fate and metabolic processes implicated in the degradation of the targeted contaminants. The heterogeneity of the contaminated matrix in composition and pollutant significantly complicates the resolution of treatment fate components. Incorporation of the pollutant within components of the rhizosphere is difficult to assess even under the best conditions using radio-labelled pollutants.

Additional understanding of plant-microbe-toxicant interaction in soil will permit a more complete utilisation of phytoremediation as a treatment technology. Species-specific properties of selected plant species, root morphology, plant physiology, ecological and physiological characteristics of the microbial communities associated with the plant root systems, and the role of plant root exudates in the selection of microbial plant root communities. The role of non-bacterial plant associations, such as mycorrhizae, and the effects of abiotic soil management processes (aeration, nutrient addition, or chemical stresses associated with toxic pollutant mixtures), must be assessed for their contribution to the treatment process. The more complete understanding of the mechanistic interactions between plant roots and the soil microbial communities will supply information to assist the successful field demonstration of phytoremediation and the effective selection and management of vegetation to achieve *in situ* bioremediation.

Recommendations

The United Nations / Economic Commission for Europe's Industry and Technology has rated phytoremediation in contrast with other available technologies (United Nations/Economic Commission for Europe's Industry and Technology Division, 1996). This analysis shows that phytoremediation is expected to be less than US$ 100/metric ton in overall cost, with a clean-up time greater than one year. At its current level of development, this technology is judged to have a low degree of reliability, a high maintenance component, and to be limited to minimum achievable contaminant concentrations of 5-50 mg/kg, but with high community acceptance. The technology's major advantages are low projected costs and simple techniques for implementation. Not all the functioning features of this emerging technology are well understood, but they represent a remarkable opportunity for developing countries to respond to their pollution control needs with a technology that can be implemented by simple application technology techniques in a cost-effective manner. Demonstrated on the greenhouse and pilot scale, phytoremediation is too new to have widespread acceptance among site managers, owners and responsible parties. More field demonstrations and applications of the technology will be needed to verify the information indicated on the lab and greenhouse scale.

Questions, comments and answers

Q: When you do a planting, do you consider root zone drainage or do you just plant as close together as possible?

A: In the designed study (Environmental Protection Agency and Air Force), planned planting was used, but I am not sure to what extent they planned for rhizosphere effects.

Q: What happens to the quality of the plants, and what do you do with them after they have been used?

A: This has yet to be determined. In some instances, for example, incorporation of PCPs may lead to poor growth characteristics. We are not totally informed about where contaminants are ultimately located in the plants.

Q: What about uptake by plants of cadmium and chromium? What happens to them? Is this a temporary solution?

A: If a metal species is sequestered (that is, hyperaccumulated) and if the concentration is high in the stem area, the plant species can be harvested. It can be used on a dry weight basis as a potential metal ore. We need to know where the pollutants end up.

C: In the United Kingdom, it has been shown that some plants take up up to seven per cent of their dry weight in zinc and cadmium.

REFERENCES

ABTEW, W., M.J. CHIMNEY, T. KOSIER, M. GUARDO, S. NEWMAN, and J. OBEYSEKERA (1995), "The Everglades Nutrient Removal Project: a constructed wetland designed to treat agriculture runoff/drainage", Amer. Soc. Agric. Eng. Versatility in the Agricultural Landscape Conf. Proc., 17-20 September, Tampa, Florida, pp. 45-57.

ALLEN, M.F. and E.B. ALLEN (1992), "Mycorrhizae and plant community development: mechanisms and patterns", in G.C. Carroll and D.T. Wicklow (eds.), *The Fungal Community*, pp. 455-479, Marcel Dekker, New York.

ANDERSON, T.A., E.A. GUTHRIE, and B.T. WALTON (1993), "Bioremediation in the rhizosphere", *Environ. Sci. Technol.* 27, pp. 2 630-2 636.

ANDERSON, T.A. and B.Y. WALTON (1995), "Comparative fate of [C-14]trichloroethylene in the root zone of plants from a former solvent disposal site", *Environ. Toxicol. Chem.* 14, pp. 2 041-2 047.

BAKER, A.J.M. and R.R. BROOKS (1989), "Terrestrial higher plants which hyperaccumulate metal elements - a review of their distribution, ecology, and phytochemistry", *Biorecovery* 1, pp. 81-126.

BAKER, A.J.M. and P.L. WALKER (1990), "Ecophysiology of metal uptake by tolerant plants", in A.J. Shaw (ed.), *Heavy Metal Tolerance in Plants: Evolutionary Aspects*, CRC Press Inc., Boca Raton, Florida.

BOLLAG, J.-M. (1992), "Decontaminating soil with enzymes", *Environ. Sci. Technol.* 26, pp. 1 876-1 881.

BOLTON, Jr., H., J.K. FREDRICKSON, and L.F. ELLIOT (1993), "Microbial ecology of the rhizosphere", in F.B. Manning Jr. (ed.), *Soil Microbial Ecology*, pp. 27-63, Marcel Dekker, New York.

BRADLEY, R., A.J. BURT, and D.J. READ (1981), "Mycorrhizal infection and resistance to heavy metal toxicity in *Calluna vulgaris*", *Nature. Lond.* 292, pp. 335-337.

BROMILOW, R.H. and K. CHAMBERLAIN (1995), "Principles governing uptake and transport of chemicals", in S. Trapp and J.C. McFarlane (eds.), *Plant Contamination: Modeling and Simulation of Organic Chemical Processes*, pp. 37-68, Lewis Publishers, Boca Raton, Florida.

COMMITTEE ON GROUND WATER CLEANUP ALTERNATIVES (1994), *Alternatives for Ground Water Cleanup*, National Academy of Science, Washington, DC.

CUNNINGHAM, S.D. and W.R. BERTI (1993), "Remediation of contaminated soils with green plant: an overview", *In Vitro Cell Dev. Biol.* 29P, pp. 207-212.

CUNNINGHAM, S.D. and C.R. LEE (1995), "Phytoremediation: Plant-based remediation of contaminated soils and sediments", in H.D. Skipper and R.F. Turco (eds.), *Bioremediation: science and applications*, pp. 145-156, Soil Science Society of America, Spec. Publ. 43, Madison, Wisconsin.

CUNNINGHAM, S.D. and D. OW (1996), "Promise and prospects of phytoremediation", *Plant Physiol.* 110, pp. 715-719.

CUNNINGHAM, S.D., W.R. BERTI, and J.W. HUANG (1995), "Phytoremediation of contaminated soils", *Tends Biotechnol.* 13, pp. 393-397.

CUNNINGHAM, S.D., T.A. ANDERSON, A.P. SCHWAB, and F.C. HSU (1996), "Phytoremediation of soils contaminated with organic pollutants", *Adv. Agron.* 56, pp. 55-114.

DICKMAN, D. and K.W. STUART (1983), *The Culture of Poplars in the Eastern United States*, Michigan State University Press, East Lansing, Michigan.

DONNELLY, P.K. and J.S. FLETCHER (1994*a*), "PCB metabolism by mycorrhizal fungi," *Bull. Environ. Contam. Toxicol.* 53, pp. 507-513.

DONNELLY, P.K. and J.S. FLETCHER (1994*b*), "Potential use of mycorrhizal fungi as bioremediation agents," in T. Anderson and J. Coats (eds.), *Bioremediation Through Rhizosphere Technology*, pp. 93-99, American Chemical Society, Symposium Series 563, Washington, DC.

DONNELLY, P.K., R.S. HEGDE, and J.S. FLETCHER (1994), "Growth of PCB-degrading bacteria on compounds from photosynthetic plants," *Chemosphere* 28, pp. 981-988.

DUSHENKOV, V., P.B.A. NADA KUMAR, H. MOTTO, and I. RASKIN (1995), "Rhizofiltration: the use of plants to remove heavy metals from aqueous streams", *Envir. Sci. Technol.* 29, pp. 1 239-1 245.

FLETCHER, J.S., P.K. DONNELLY, and R.S. HEGDE (1995), "Biostimulation of PCB-degrading bacteria by compounds released from plant roots," in *Bioremediation of Recalcitrant Organics,* pp. 131-136, Battelle Press, Columbus, Ohio.

GUARDO, M., L. FINK, T.D. FONTAINE, S. NEWMAN, M. CHIMNEY, R. BEARZOTTI, and F. GOFORTH (1995), "Large-scale constructed wetlands for nutrient removal from stormwater runoff: an Everglades Restoration Project", *Environ. Manag.* 19, pp. 879-890.

HARTMAN, W.J., Jr. (1975), "An evlauation of land treatment of municipal wastewater and physical siting of facility installations", U.S. Department of Army, Washington, DC.

HINCHMAN, R. and C. NEGRI (1994), "The grass can be greener on the other side of the fence", *Logos* 12, pp. 8-12.

KADLEC, R.H. and R.L. KNIGHT (1996), *Treatment Wetlands*, CRC Press, Boca Raton, Florida.

KEYSER, H.H., P. SOMASEGARAN, and B.B. BOHOLL (1993), "Rhizobial ecology and technology", in F.B. Manning Jr. (ed.), *Soil Microbial Ecology*, pp. 205-226, Marcel Dekker, New York.

KINGSLEY, M.T., F.B. METTING, J.K. FREDRICKSON, and R.J. SEIDLER (1993), "In-situ stimulation vs bioaugmentation: can plant inoculation enhance biodegradation of organic compounds", Proceedings of Air and Waste Management Association's 86th Annual Meeting and Conference, 14-18 July, Denver CO-WA-89:04, AWMA Pittsburgh, Pennsylvania.

KOMOSSA, D., C. LANGEBARTELS, and H. SANDERMANN, Jr. (1995), "Metabolic processes for organic chemicals in plants", in S. Trapp and J.C. McFarlane (eds.), *Plant Contamination: Modelling and Simulation of Organic Chemical Processes*, pp. 69-103, Lewis Publishers, Boca Raton, Florida.

LINDERMANN, R.G. (1988), "Mycorrhizal interactions with the rhizosphere microflora: the mycorhizosphere effect", *Phytopathol.* 78, pp. 366-371.

MCFARLANE, J.C. (1995), "Anatomy and physiology of plant conductive systems", in S. Trapp and J.C. McFarlane (eds.), *Plant Contamination: Modeling and Simulation of Organic Chemical Processes*, pp. 13-34, Lewis Publishers, Boca Raton, Florida.

MITSCH, W.J. (1993), "Ecological Engineering", *Environ. Sci. Technol.* 27, pp. 438-445.

MITSCH, W.J. and J.G. GOSSELINK (1993), *Wetlands*, 2nd Edition, Van Nostrand Reinhold, New York.

NADA KUMAR, P.M.A., V. DUSHENKOV, H. MOTTO, and I. RASKIN (1995), "Phytoextraction: the use of plants to remove heavy metals from soils", *Environ. Sci. Technol.* 29, pp. 1 232-1 238.

RAO, A.S. (1980), "Root flavonoids", *Bot. Rev.* 56, pp. 1-84.

RASKIN, I. (1996), "Plant genetic engineering may help with environmental cleanup", *Proc. Natl. Acad. Sci.* 93, pp. 3 164-3 166.

REED, S.C., R.W. CRITES, and E.J. MIDDLEBROOKS (1995), *Natural Systems for Waste Management and Treatment,* 2nd Edition, McGraw-Hill, Inc., New York.

REEVES, R.D., R.R. BROOKS, and T.R. DUDLEY (1983), "Uptake of nickel by species of *Alyssum, Bornmuellera* and by other genera of old world tribus Alyssae", *Taxon* 32, pp. 184-192.

REILLEY, K.A., M.K. BANKS, and A.P. SCHWAB (1996), "Dissipation of polycyclic aromatic hydrocarbons in the rhizosphere", *J. Environ. Qual.* 25, pp. 212-219.

RIEDERER, M. (1995), "Partitioning and transport of organic chemicals between the atmospheric environment and leaves", in S. Trapp and J.C. McFarlane (eds.), *Plant Contamination: Modelling and Simulation of Organic Chemical Processes*, pp. 153-190, Lewis Publishers, Boca Raton, Florida.

ROVINA, A. (1969), "Plant root exudates", *Bot. Rev.* 35, pp. 35-59.

RYAN, J.A., R.M. BELL, J.M. DAVIDSON, and G.A. O'CONNOR (1988), "Plant uptake of non-ionic organic chemicals from soils", *Chemosphere* 17, pp. 2 299-2 323.

SALT, D.E., M. BLAYLOCK, N.P.B.A. KUMAR, V. DUSHENKOV, B.D. ENSLEY, I. CHET, and I. RASKIN (1995), "Phytoremediation: a novel strategy for the removal of toxic metals from the environment using plants", *Bio-Technology* 13, pp. 469-474.

SANDERMANN, H. (1992), "Plant metabolism of xenobiotics" *Trends Biol. Sci.* 17, pp. 82-84.

SANDERMANN, H., H. DIESPERGER, and P. SCHEEL (1977), "Metabolism of xenobiotics by plant culture" in W. Barz, E. Reinhard, and M.H. Zenk (eds.), *Plant Tissue Culture and Its Bio-technological Application*, pp. 178-196, Springer-Verlag.

SCHNOOR, J.L., L.A. LICHT, S.C. MCCUTCHEON, N.L. WOLFE, and L.H. CARREIRA (1995), "Phytoremediation of contaminated soils and sediments", *Environ. Sci. Technol.* 29, pp. 318-323.

SCHROLL, R. and I. SCHEUNERT (1992), "A laboratory system to determine separately the uptake of organic chemicals from soil by plant roots and by leaves after vaporization", *Chemosphere* 24, pp. 97-108.

SEIDEL, K. (1976), "Macrophytes and Water Purification", in J. Tourbier and R.W. Pierson, Jr. (eds.), *Biological Control of Water Pollution*, pp. 109-121, University of Pennsylvania Press, Philadelphia, Pennsylvania.

SHETTY, K.G., B.A.D. HETRICK, D.A.H. FIGGE, and A.P. SCHWAB (1994), "Effects of mucorrhizae and other soil microbes on revegetation of heavy metal contaminated mine spoil", *Environ. Poll.* 86, pp. 181-188.

SHIMP, J.F., J.C. TRACY, L.C. DAVIS, E. LEE, W. HUANG, and L.E. ERICKSON (1993), "Beneficial effects of plants in the remediation of soil and groundwater contaminated with organic materials", *Crit. Revs. Environ. Sci. Technol.* 23, pp. 41-77.

SIMONICH, S. and R.A. HITES (1995), "Organic pollutant accumulation in vegetation", *Envir. Sci. Technol.* 25, pp. 2 905-2 914.

SOUTH FLORIDA WATER MANAGEMENT DISTRICT (1995), "Everglades Nutrient Removal Project: Year 1 Synopsis", December 1995, West Palm Beach, Florida.

TRAPP, S. (1995), "Model for uptake of xenobiotics into plants", in S. Trapp and J.C. McFarlane (eds.), *Plant Contamination: Modelling and Simulation of Organic Chemical Processes*, pp. 107-151, Lewis Publishers, Boca Raton, Florida.

WALKER, A. (1972), "Availability of atrazine to plants in different soils", *Pestic. Sci.* 3, p. 139.

WALTON, B.T. and T.A. ANDERSON (1990), "Microbial degradation of trichloroethylene in the rhizosphere: potential application to biological remediation of waste sites, *Appl. Environ. Microbiol.* 56, pp. 1 012-1 016.

WOLVERTON, B.C. and R.C. MCDONALD-MCCALEB (1986), "Biotransformation of Priority Pollutants Using Biofilms and Vascular Plants", *Journal of the Mississippi Academy of Sciences* 31, pp. 79-89.

UNITED NATIONS/ECONOMIC COMMISSION FOR EUROPE'S INDUSTRY AND TECHNOLOGY DIVISION (1996), EIEET's Soil-Remediation Technology Summary Page, http//gnew.gn.apc.org/eieet/techdesc.html.

U.S. ENVIRONMENTAL PROTECTION AGENCY (USEPA) (1993), *Constructed wetlands for wastewater treatment and wildlife habitat, 17 case studies*, EPA 832-R-93-005.

U.S. ENVIRONMENTAL PROTECTION AGENCY (USEPA) (1994), *North American Wetlands for Water Quality Treatment Database*, USEPA, ORD, Cincinnati, Ohio.

APPLICATION OF BIOTECHNOLOGY FOR THE CONTROL OF NUTRIENT RELEASE TO RECEIVING WATERS

by

Samuel S. Jeyanayagam
Stanley Technology Group Ltd., Vancouver, British Columbia, Canada

Introduction

The aquatic environment (lakes, rivers, streams, estuarine or marine) is the most common receiving environment for the disposal of wastewater effluents. Traditionally, effluent discharge limits are established based on an examination of the receiving water quality, and its assimilative capacity. The appropriate level of treatment is then selected to achieve the desired effluent limits.

Secondary treatment, which typically achieves better than 85 per cent removal of Biochemical Oxygen Demand (BOD) and total suspended solids, is commonly used for the treatment of many community wastewaters. However, due to population growth and associated urban development, the receiving environment experiences increased wastewater and nutrient loadings. As a result, in many instances, treatment levels once considered adequate are not sufficient anymore. Existing secondary treatment facilities are required to be upgraded to provide nutrient removal to ensure environmental protection. This is evidenced by an upsurgence of interest in nutrient removal in Canada, the United States, Europe and Australia.

Impact of nutrient loading

Nitrogen or phosphorus is often the limiting nutrient in environmentally sensitive receiving waters. Hence, the discharge of wastewater effluents containing these nutrients is likely to stimulate growth of algae and other photosynthetic aquatic life. This leads to:

- the depletion of dissolved oxygen (DO) caused by the biodegradation of algal biomass;
- DO depletion caused by effluent ammonia-nitrogen;
- fish kills due to reduced DO levels;
- upper layers of the water becoming turbid due to the presence of algae, blocking sunlight to lower layers. A shift in aquatic ecology towards undesirable organisms may be encountered.

The above events could lead to accelerated eutrophication of the receiving streams. In an effort to ensure environmental protection, regulatory agencies are imposing increasingly stringent effluent standards for nitrogen and phosphorus. For example, the Bellair Research Institute in Ontario, Canada has established total nitrogen and total phosphorus limits of 9.8 and 2.4 microgram per litre in coastal waters to protect coral reef organisms. Such low levels can be attained by a combination of both old and new technologies.

Old technology -- effluent dilution

The old adage "dilution is the solution" is still a commonly used concept to reduce effluent pollutant concentrations to harmless levels. However, proper outfall design based on detailed oceanographic studies is required to achieve the desired dilution in the receiving waters. For example, by strategically locating an ocean outfall in the island of Barbados, it was possible to achieve dilutions of up to 250 times in the rising plume at a depth of 60 m. An additional dilution of 3:1 was achieved by utilising surface currents. Back calculations reveal that with such dilutions, wastewater effluent containing 7.4 mg/L total nitrogen and 1.8 mg/L total phosphorus can be safely discharge to coastal waters without exceeding the limits established by the Bellair Research Institute.

New technology -- biological nutrient removal

Introduction

Biological Nutrient Removal (BNR) utilises naturally occurring micro-organisms to remove nitrogen and phosphorus from wastewaters. While by biological means has become the predominant process for nitrogen control, there are two basic phosphorus removal processes -- chemical and biological. Chemical precipitation methods for phosphorus removal have become costly because of increasing chemical costs and the high cost of disposal of the large quantities of sludge generated. Over the last two decades, considerable progress has been made in optimising the BNR process. As a result, it is now possible to eliminate the use of chemicals and achieve reliable nutrient removal at lower cycle costs. In short, *BNR is a proven biotechnology for nutrient control.*

The remainder of the paper focuses on the historical development of the BNR technology, process description and present status.

Historical development

The development of the single-sludge biological nitrogen removal process was an improvement over the two and three sludge processes of the early 1960s. This took place when it was recognised that a single sludge mass could perform carbonaceous removal as well as nitrification and denitrification when subjected to sequential aerobic and anoxic conditions. The two principal single-sludge processes noted for total nitrogen removal were the post-denitrification Wuhrmann process and the pre-denitrification Ludzack-Ettinger process. In 1974, the 4-stage Bardenpho process was developed to combine the positive aspects of the two processes.

In the late 1950s and early 1960s, enhanced biological phosphorus removal was reported to occur on an erratic basis in various parts of the world. Levin and Shapiro (1965) and Vacker *et al.* (1967) observed up to 80 per cent and 96 per cent biological phosphorous removal, respectively, at plants in Washington, DC, and San Antonio, Texas. However, very little was understood about the nature of the biological phosphorus removal mechanisms and the conditions necessary to reliably trigger it in the activated sludge process.

In the 1970s, a number of researchers (Barnard, 1975; Fuhs and Chen, 1975) attributed the "luxury uptake" mechanism for phosphorous removal to the consequence of subjecting the sludge to an anaerobic stress condition (defined as the absence of both nitrate and dissolved oxygen) of such an intensity at some point in the process ahead of the aerobic zone that phosphorus is released by the organisms into solution. The "luxury uptake" of phosphorus then occurs in the aerobic zone. The 5-stage Modified Bardenpho (Phoredox) process was developed out of this hypothesis. In this process, the required anaerobic stress condition is intended to be created in the anaerobic zone at the head end of the plant. Levin and Topol (1972) utilised the anaerobic phosphorus release phenomenon by introducing a sidestream "P stripper" reactor into the sludge recycle line of their combined biological-chemical Phostrip phosphorus removal process. Phosphorous released into the supernatant of the reactor is precipitated using lime and the stream returned to the head end of the process.

In the late 1970s and early 1980s, researchers began to explore the biochemical nature of the phosphorus (P) removal mechanism, and the reason why the anaerobic-aerobic sequence was essential for the mechanism to operate. Research workers of the City of Johannesburg, South Africa proposed that simple carbonaceous substrates (principally volatile fatty acids, VFAs) were stored by *Acinetobacter*, the organism involved in enhanced P removal. The stored substrates were used by these organisms to generate the required energy for phosphorus uptake in the subsequent aerobic zone. It was suggested that increasing the availability of VFAs in the anaerobic zone would improve the phosphorus removal characteristics, and that this may be achieved by anaerobic fermentation of the primary sludge.

The research group at the University of Cape Town (UCT) demonstrated conclusively that the P removal capabilities of a particular process were more closely a function of the availability of simple carbonaceous substrates in the anaerobic zone, as opposed to a certain "minimum degree of anaerobic stress", as postulated previously. It was also shown that nitrate entering the anaerobic zone had a detrimental effect on phosphorus removal because the substrate required in the P removal mechanism is utilised preferentially in the presence of nitrate (denitrification reaction). Subsequently, the UCT group proposed a simple modification to the Bardenpho process that introduced a large degree of operational flexibility and significantly reduced the risk of returning nitrates to the anaerobic zone under a wide variety of influent sewage characteristics. This modification relocated the RAS recycle discharge from the anaerobic zone to the anoxic zone and introduced a separate recycle from the anoxic zone to the anaerobic zone. The configuration became known as the UCT process.

The UCT process proved to be capable of consistent P removal from wastewaters with a wide variation in influent characteristics, without a major detrimental effect on nitrogen removal. In a further refinement of the UCT process, the anoxic zone was split in two. The first anoxic zone is used for denitrification of the RAS recycle and the second anoxic zone provides denitrification of the nitrates recycled from the aerobic zone. This modification facilitated an even greater degree of operational flexibility and protection of the anaerobic zone by separating the high nitrogen recycle from the phosphorus removal cycle of the process. This process is commonly referred to as the Modified UCT Process.

The Virginia Initiative Plant (VIP) process was developed in the United States in the mid 1980s. The VIP process is based on the same process configuration as the UCT, but at much shorter retention times. The detention times for the UCT process in South Africa are similar to an extended aeration process, and the VIP process more closely simulates the detention times of a high rate activated sludge process.

The Johannesburg process was developed in the late 1980s and is similar in purpose to the UCT process but is less flexible. Denitrification of the recycle streams occurs in an anoxic zone ahead of the anaerobic zone to prevent nitrates from entering the anaerobic zone.

The following is a brief description of the advantages and disadvantages of various enhanced phosphorus removal process configurations:

1) Modified Ludzack-Ettinger process (A/O)

Advantages: High denitrification capacity.

Disadvantages: No enhanced P removal; cannot remove N to low concentrations much lower than 10 mg/L.

2) Wuhrmann process

Advantages: In theory, can achieve complete denitrification.
Disadvantages: No enhanced P removal, slow denitrification rates.

3) 4-Stage Bardenpho process

Advantages: Good N removal from suitable wastewaters.
Disadvantages: No enhanced P removal.

4) 5-Stage Modified Bardenpho process (Phoredox)

Advantages: Achieves good N and P removal under ideal influent and operating conditions.

Disadvantages: Inflexible -- very limited operational optimisation possible; loss of P removal if nitrate enters anaerobic zone; problems with diurnal load variations

5) 3-Stage Modified Phoredox process (A^2/O)

Advantages: Achieves a high level of nitrogen removal from high strength wastewaters.

Disadvantages: Will not produce an effluent that consistently meets stringent nutrient criteria; loss of P removal if nitrate enters anaerobic zone; problems with diurnal load variations. When operated as a high-rate A^2/O process, produces an effluent with a high ammonia concentration.

6) UCT process

Advantages: High flexible operation; P removal is independent of effluent nitrate concentration; good P removal under non-ideal influent characteristics; great potential for automated process control.

Disadvantages: Additional recycle and flow equalisation required; problems with diurnal load variations.

7) Modified UCT process

Advantages: Same as for UCT process but more flexible in operation. No problems with nitrate entering the anaerobic zone.

Disadvantages: Same as for UCT process.

8) VIP process

Advantages: Small bioreactor volumetric requirements.

Disadvantages: Same as UCT process, but nitrogen removal may be low depending on waste characteristics.

9) Johannesburg process

Advantages: Process simplicity. Nitrate removal prior to anaerobic zone. Small bioreactor volumetric requirement.

Disadvantages: Good phosphorus removal but nitrogen removal limited significantly less N removal than modified UCT process. RAS recycle rate affects HRT of anaerobic zone.

The above review shows that a number of BNR process configurations are available. All configurations include anaerobic, anoxic and aerobic zones. The primary differences between the various BNR processes are zone sequences and recycle stream locations. Because each plant is unique, an evaluation of the multiple nutrient removal alternatives is required to compare them quantitatively and qualitatively.

Primary sludge fermentation

Primary Sludge Fermentation (PSF) for the generation of volatile fatty acids (VFAs) may be coupled with any of the BNR processes to improve phosphorus removal. Variations of the PSF system have been developed to maximise VFA production and to enhance operational flexibility in order to meet stringent effluent phosphorus limits. Use of PSF is especially critical with organically weak waste. For example, the addition of a PSF to a UCT process has been shown to improve phosphorus removal by 50 per cent and achieve consistent overall phosphorus removal efficiency exceeding 90 per cent. With organically weak waste, the use of a PSF resulted in a median effluent total phosphorus concentration of 0.3 mg/L over a one year study period without chemical addition or effluent filtration. The presence of VFAs has also been found to enhance denitrification rates in the anoxic zone. With the inclusion of PSF, high costs associated with chemical polishing, such as the costs of chemicals, and sludge handling and disposal, can be eliminated.

There are three PSF systems in full scale application at the present time. The advantages and disadvantages of these fermenters are listed below:

- Complete mix fermenter

  Advantages: Full scale operation optimises VFA production, single reactor.

  Disadvantages: Does not allow addition of VFA direct to bioreactor, some VFA lost in recycle through primary clarifier.

- Static fermenter

  Advantages: Allows direct addition of VFAs to anaerobic zone of bioreactor, minimises volume of waste primary sludge to be handled.

  Disadvantages: No SRT control, potential for anaerobic projects with long SRT, large reactor.

- Separate complete mix/thickener fermenter

  Advantages: Optimises VFA production, allows addition of VFA direct to anaerobic zone of bioreactor.

  Disadvantages: High cost of two stage reactor, complex operation.

Process design

In a full-scale advanced wastewater treatment facility, biological nutrient removal is usually carried out in a series of continuous flow complete-mix reaction tanks in which a mixed population of micro-organisms is conditioned for phosphorus uptake. In the anoxic reactor, recycled nitrates from the aerobic zone are converted to elemental nitrogen gas by denitrifying bacteria using carbon in the wastewater as an energy source. Some phosphorus uptake also occurs in the anoxic zone. Organic carbon removal, conversion of organic nitrogen and ammonia to nitrates, and phosphorous storage in the sludge (luxury uptake) are achieved in the aeration section. Phosphorus is removed by wasting sludge directly from the process. An integral part of a good BNR facility is the PSF system. The VFAs generated in the PSF system are channelled to the anaerobic zone of the BNR bioreactor. In order to achieve significant phosphorus removal, approximately 5 mg/L VFA is required for each mg/L of influent phosphorus. If an adequate supply of rapidly biodegradable substrate is not available, year-round phosphorus removal is difficult to achieve.

Table 1 below summarises Stanley's BNR experience to date. The five levels of project scope indicated are:

- Engineering study: Modelling and evaluation of the various BNR process configurations.

- Pilot study: Small scale testing of BNR process configurations. Normally conducted on-site in one of Stanley's transportable pilot units.

– Demonstration study: Large scale testing of BNR process configurations by retrofitting part of the treatment plant.

– Design: Actual design of all of the components of the BNR facility. This could be modification and/or expansion of an existing facility or a brand new plant.

– Operation: Involves start-up, BNR process optimisation, and training of plant operators, as well as long-term operational assistance.

Table 1. **Biological nutrient removal experience**

BNR project	Plant capacity m^3/d	Project scope				
		Engineering study	Pilot study	Demonstration study	Design	Operation
Friederiksvaerk, Denmark	12 000	x			x	x
Solrod, Denmark	15 000	x			x	x
Tarnby, Denmark	15 000	x			x	x
Aostorp, Sweden		x	x			
Bonnybrook, Canada	552 000	x	x	x	x	
Penticton, Canada	19 000	x	x	x	x	x
Saskatoon, Canada	120 000	x			x	x
Edmonton, Canada	318 000	x	x	x		
Jasper, Canada	9 500	x				
Kalispell, US	12 000				x	x
Howard County, US	37 000	x	x	x		
Hagerstown, US	30 000	x				
Brendale, Australia	28 000	x			x	
Murrumba Downs, Australia	204 000	x			x	
Dalby, Australia	3 000	x			x	
Bridgeton, Barbados	9 500		x			

Source: Stanley Consulting Group Ltd.

The North American BNR plants indicated above have typical influent wastewater characteristics of 200 mg/L BOD_5, 250 mg/L total suspended solids, 6-8 mg/L total phosphorus and 45-50 mg/L total nitrogen. In contrast, Australian wastewaters tend to be stronger (300 mg/L BOD_5, 300 mg/L total suspended solids, 12-20 mg/L total phosphorus and 50-70 mg/L total nitrogen). Typical effluent limits that are being achieved by these BNR facilities (with effluent filtration) are: less than 5 mg/L BOD_5, less than 5 mg/L total suspended solids, 0.2 mg/L total phosphorus and 5 mg/L total nitrogen.

Conclusion

A properly designed BNR system can achieve effluent total nitrogen of less than 5 mg/L and total phosphorus of less than 0.2 mg/L on a consistent basis, even in cold weather with wastewater temperatures as low as 8 °C. This is possible due to the presence of rapidly biodegradable organic substrate in the form of volatile fatty acids (VFAs). In full scale applications, VFAs are produced from anaerobic fermentation of primary sludge.

BNR is a proven biotechnology for nutrient control in wastewater effluent discharges. It allows the complete elimination of the use of chemicals.

Questions, comments and answers

Q: You state that BNR is always the cheapest option for phosphorus removal? Surely this depends on the nature of the sewage and, in the context of the developing world, on the ability to operate these complicated plants? How appropriate are these plants in areas where training is limited?

A: We are finding this out in North America. The use of chemicals is the most straightforward means, and therefore the decision involves location, type of water, etc., and is very site-specific. When I say cost-effective, I am speaking of what we have found in British Columbia.

REFERENCES

BARNARD, J.L. (1975), "Nutrient removal in biological systems", *Water Pollution Control* 74:2, pp. 143-154.

FUHS, G.W. and M. CHEN (1975), "Microbial basis of phosphate removal in the activated slude process for the treatment of wastewater", *Microbial Ecology* 2, pp. 119-138.

LEVIN, G.V. and J. SHAPIRO (1965), "Metabolic uptake of phosphorus by wastewater organisms", *Journal of Water Pollution Control Federation* 37:6, pp. 800-821.

LEVIN, G.V. and G.J. TOPOL (1972), "Pilot plant test of a phosphate removal process", *Journal of Water Pollution Control Federation* 44:12, pp. 1 940-1 944.

VACKER, D., C.H. CONNELL, and W.N. WELLS (1967), "Phosphate removal through municipal wastewater treatment at San Antonio, Texas", *Journal of Water Pollution Control Federation* 39:5, pp. 750-771.

ECOPARQUE: A SUSTAINABLE DEVELOPMENT PROJECT, BASED ON A WASTEWATER TREATMENT PLANT

by

Oscar Romo
Ecoparque, El Colegio de la Frontera Norte, Tijuana, Baja California, Mexico

Introduction

Mexico has approximately 320 water basins distributed randomly throughout the country. Half of these resources are located in the south-eastern region of the country, within an area that represents less than 20 per cent of the nation's total territory. The northern part of the country has only three per cent of the total amount of such resources (García, 1982). As a result, a regulated growth planning system geared towards increasing productivity and water optimisation, and making it more efficient for agriculture, the industrial-urban sector, the generation of electric power, leisure activities and the like is sorely needed. Such a system must also avoid adversely impacting the environment.

The activities undertaken in Mexico's northern region[1] are basically dependent on the Colorado River and the Rio Grande, and the amounts allotted pursuant to international agreements and treaties signed with the United States. In the near future, these resources will be insufficient to meet the needs of the accelerated urban growth that the border region has experienced over the last several years[2].

In Baja California, due to the climate and minimal rainfall[3], there are almost no rivers or permanent surface currents to speak of. Annual rainfall in the region comes to only 100 mm [INEGI, 1995 (National Statistics, Geography and Data Processing Institute)]. The Colorado River Basin generates a small amount of runoff during the course of the year, which represents 88 per cent of the water supply available to the entire state. This volume is regulated by an international treaty signed by Mexico and the United States in 1944.

---

1 In the 1983 agreement, the border zone was defined as up to the first 100 kilometres of territory on both sides of the international land and sea boundary lines between Mexico and the United States.

2 According to the UN Population Fund (UNPF, 1991), the high demographic growth, together with the lack of socioeconomic infrastructure has contributed in large measure to accelerating the crisis of the demand for water.

3 The State of Baja California is considered to be Mexico's driest region. The region's climate is an important factor influencing its natural endowment of water resources.

The state of Baja California is located in the north-western part of Mexico. It shares approximately 253 km of border with the United States. The eleventh largest state, it occupies an area of 71 576.26 km^2, which is 3.6 per cent of Mexico's total land extension (INEGI, 1995). Its border region with the United States has been the focus of intensive demographic and economic growth[4]. For its water supply, it depends on two sources, the Colorado River Basin and the Tijuana River Basin. Both are shared with the United States, a situation requiring bilateral agreements, enabling the states to obtain their appropriate share.

Most of the Colorado River Basin is located in the United States, with only a small part in Mexico. The part that lies in the United States is in Utah, Wyoming, Colorado, New Mexico, California and Arizona. The Colorado River delta in Mexico is located between the states of Baja California and Sonora, in the Mexicali Valley and San Luis Rio Colorado, respectively, until it flows into the Gulf of California (also known as the Sea of Cortes).

On 3 February 1944, the United States and Mexico signed the "International Water Distribution" Treaty regulating the use of the Colorado River water supply. This treaty guaranteed Mexico a volume of 1 850 234 000 m^3 yearly. When there is a surplus, the United States is committed to provide Mexico with additional amounts up to a total annual volume of 2 096 931 000 m^3. At times of drought or other misfortune that might hinder yearly water distribution, Mexico's allotted water volume is reduced in the same proportion as US consumption.

The Tijuana River Basin is located in the corner where the international border between the United States and Mexico joins the Pacific Ocean. The Basin comprises a total 4481 km^2, of which three quarters (approximately 3 233 km^2) are located in Mexico and the remainder in the United States (1 251 km^2).

The source of the Tijuana river current is the Hechicera stream which flows from the Sierra de Juárez in a north-westerly direction until it reaches rancho La Tortuga, where it is joined by an important tributary. From this confluence, it changes direction and heads west, to be joined by another tributary from the right bank, which is the junction of the currents originating in the Las Palmas River, and continues until reaching the Abelardo L. Rodríguez Dam. Downstream, the river is called the Tijuana River. After crossing the city, it heads into the United States, modifies its course slightly to head west and flow into the Pacific Ocean 1.5 km north of the international border. It spans a total distance of 128 km (INEGI, 1995).

In US territory, the Tijuana River Estuary was decreed a natural reserve and a research centre. The estuary is located in the outermost coastal portion of the Tijuana River, along the border. It encompasses a surface area of 2 500 acres. The importance of this international river lies in the fact that it is the only hydrological current in the region[5]. Its capacity to serve as a sure source of water for the community has been overtaken by rapid demographic growth, resulting in pollution, which in

---

[4] In recent decades, Mexico's northern border has mushroomed both economically and demographically. For example, Tijuana's population growth rate between 1970 and 1990 was 5.06 per cent, while that of the whole country was 2.5 per cent. Tijuana's economic growth rate (measured as a function of Gross Domestic Product – GDP) was 3.8 per cent between 1980 and 1990, while that of the whole country was 1.7 per cent (Clement and Zepeda, 1993).

[5] The Tijuana river basin was declared a restricted zone as of 13 November 1956, according to data from the INEGI in 1995. The decree hoped to achieve efficient water resource management through the regulation of its development and use.

turn has been a source of controversy between the two nations, mainly because of damage resulting from a series of unusual events, plus the discharge of sewage[6] from the city of Tijuana, which has seriously affected the Tijuana River Estuary's ecological reserve.

At the beginning of the last decade, Tijuana consumed a little over 21 million cubic meters per year; in 1989, its water consumption came to 64 million cubic meters, of which 54 million came from the Colorado River water. Some 4 000 litres per second (lps) of this vital liquid are extracted from the Colorado River in order to meet this demand. Although there is currently a surplus, in view of the fact that Tijuana's water consumption at present is 3 023 lps, and the water supply has been guaranteed through the year 2005 (SAHOPE, 1996), the demand for water will continue to grow due to demographic growth and the partially unserved current demand. Of an estimated 174 567 homes in the municipality of Tijuana, the area's water system serves only 161 338.

Tecate, Tijuana and Ensenada have the State's highest growth rates. It is estimated that by the year 2010, demand for water in these cities will be approximately 240 million m^3/year, while the Mexicali Valley will require 110 million m^3/year.

Baja California's supply of potable water was found to be deficient. In some areas, only 37 per cent of the community had direct access to drinking water. To overcome this situation, the global water potabilization capacity was by increased 136 per cent state-wide. That is to say, the initial capacity of 3 550 lps grew to 8 380 lps.

Sewage management in Tijuana

The problem of how to manage sewage began in the 1930s. The issue has been the focus of an ongoing negotiation process between both countries. Finally in 1990, a *Plan for an International Solution to the Border Problem of Environmental Rehabilitation in Tijuana* took shape in IWBC Regulation 283, which proposed creation of a binational sewage treatment plant. After a series of studies and proposals, it was decided that the plant should be built in the United States along its southern border, and that both governments would share construction and operating costs. According to the agreement, the plant would be up and running by the end of 1996. The treatment plant's capacity was planned at 1 500 lps. However, this appears to be only a technical solution to the problems generated by inefficient sewage treatment in the basin and coastal areas, not a clear and balanced solution for overall management of sewage and sludge.

In the 80s, sewage generated by the urban area was channelled through an international discharge facility which received 68 per cent of the total volume of sewage generated by the City of Tijuana. In 1980, mechanical failures in pumping stations 1 and 2, operated by Mexico's Federal

---

[6] The problems caused by the discharge of sewage both in the river basin (occasioned by the low level of coverage of the sewage system) and in the coastal area (because of the lack of a treatment plant for urban generated sewage) gave rise to the need for bilateral negotiations between the two nations, in light of the expansion of the environmental conflict and the lack of any corrective action. As a response to the set of environmental problems which transcend political and economic borders, a bilateral accord was signed in 1983, called the Peace Accord, whose purpose was to bring the institutions from each country into contact for co-ordinated efforts to resolve and prevent the environmental impact caused by the border area's diverse economic activities. Among the principal problems requiring action was pollution caused by sewage generated in Tijuana.

Department of Hydraulic Resources (SARH), hindered efficient operation of the sewage removal system.

On the other hand, Tecate lacked the necessary infrastructure to manage the service, which led to serious problems due to the elevated volumes of soda discharged by the brewery there. The inability to control discharges and insufficient capacity to collect and treat sewage generated serious problems in the Tijuana River Estuary reserve and the beach areas used by swimmers.

In 1987, the San Antonio de los Buenos Treatment Plant began operating with a design capacity for treatment of 1 100 lps, but it currently operates at a capacity of 750 lps, meaning it treats only 53 per cent of the total amount of sewage generated by the City.[7]

Decentralised sewage treatment system

Unfortunately, official reaction to sewage management has been to collect or treat it and then discharge it into the ocean. Treated sewage should be considered a valuable urban resource, not a problem which requires solving. It should be utilised to meet the city's needs, the main one being the water supply for the community, given the scarcity of water in the region.

Tijuana's sewage problem is caused by a set of factors, including extremely rapid and unregulated urban and demographic growth, insufficient and deficient urban infrastructure, and the topographical features inherent to its location. All of the above factors have hindered the ability of the drainage system and sewage treatment to provide adequate service.

The reasons behind the demographic explosion[8] can be attributed largely to the evolution of the US economy, and particularly to that of neighbouring California. This demographic growth is closely linked to the migratory flow from the interior of Mexico, which steers a large floating population into Tijuana. 50.8 per cent of the undocumented flow into the United States is headed for California.

Tijuana's environment features a dry steppe climate, scarce rainfall (270 mm), relatively poor soil conditions and minimal desert vegetation. Topographically, it is basically a city of steep hills, a circumstance that has made urban growth arduous. Only 13 per cent of the urban area is usable. Mesas and hills have had to be transformed into residential or industrial areas, resulting in the destruction of already scarce vegetation and erosion problems caused by rain run-off and wind. Sewage discharge into areas lacking appropriate sewage systems has resulted in subsoil filtration, polluting ground and surface water supplies.

A study of the impact of the 1993 floods in Tijuana undertaken by Bocco *et al.* (1993), indicates that Tijuana's vulnerability to these types of natural phenomena is due, among other things, to "*... the urbanization of the natural waterways and the loss of vegetation, and soil erosion resulting from urban activities*". This gives us a clear idea of Tijuana's need for green areas, not only to prevent soil erosion, but also to meet other important ecological requirements: to act as stabilizing agents in the urban micro-climate, and in noise control, to fulfil an aesthetic and recreational function, to aid in oxygenation of the environmental, and to filter pollutants generated by the urban environment (Ojeda,

[7] Figures compiled by CESPT (State Public Service Commission in Tijuana) show that Tijuana's total sewage production in 1995 came to 1 416 lps.

[8] With a growth rate of 1.8 per cent, greater than that of the interior of the country (INEGI, 1995).

1992). Unfortunately, Tijuana has few parks and green areas, and those that exist are in deplorable condition. In 1990, total green space area was 1 469 023.90 m^2 for a population of 698 752 inhabitants, which represents 2.10 m^2 of green area per inhabitant. This is far below established international and national norms of 9-10 m^2 and 8.5 m^2 per inhabitant, respectively.

Tijuana's predominant technological sewage treatment model is conventional, limited to collecting sewage by means of the drainage system, pumping it to be treated far from the city, and discharging it into the ocean. The area's steep hills make installation of a network of drainage systems very costly, because so many pumping stations are needed. Only 65.1 per cent of all homes have a drainage system (INEGI, 1995). The deficit in collection and sewage treatment systems has led to the gradual increase of contamination levels.

In Tijuana, the model called *Decentralized System of Sewage Treatment and Reuse in Urban Areas* (SIDETRAN, Sistema Descentralizado de Tratamiento y Reuso de Aguas Negras en Zonas Urbanas) was developed as Ecoparque. Its basic premise is that sewage is a resource that must be utilised, not an unsolvable problem. Therein lies the importance of conceptualising a decentralised sewage collection and treatment system that would manage the city's various flows of sewage separately. This system is an alternative treatment option to the currently used traditional sewage management systems. It is comprised of compact treatment plants with modular growth capacity, is easy to operate, and has a minimal number of moving parts, ensuring efficient and effective operation. It proposes using technologies that are within reach of the community, which in turn would improve urban resource management.

SIDETRAN is an alternative to traditional and centralised sewage treatment systems. The proposal was conceived as a decentralisation of the primary collection system (primary network) of the sanitary sewage and sewage treatment systems in order to manage the city's sewage streams separately, and manage the resource in urban areas. It is not a question of changing the sanitary drainage system's existing urban infrastructure, but rather, as Carlos de la Parra (1995) stated, "an educational revolution is needed", which would change the ingrained idea that the primary collection system is simply a cog in the system of getting rid of sewage so as to prevent disease (a system that requires requiring a pumping network). Implementation of this system is subject to the following sub-criteria (de la Parra, 1995):

1) Availability of sewage: preferably there should be an existing sewage system, or a flow of polluted water.

2) A market for selling or using sewage for environmental restoration or protection is required. This requires a certain level of population density.

3) A drop between the elevation of inflows and effluents; terrain slope should be 30 per cent or greater.

4) Available cost-effective land to treat and make use of the sewage.

5) Access and viability are important, mainly because of the "social function" the system represents for the community, i.e. Ecoparque.

SIDETRAN's main objective is to optimise the manner in which the cities use their water resources. Using appropriate technology, it proposes a sewage treatment model that offers, among other things:

- compactness and capacity for modular growth;

- ease of operation and few moving parts, ensuring reliable and efficient operation;

- minimal need for pumping to transport the sewage;

- under the umbrella of an integrated water resources management system, a sewage system that channels sewage flows separately so as to make use of them in the area of their origin.

The elements and processes that make up this modular treatment system are available in the market. The sequence of stages in the treatment process are the same as those in any treatment plant. The difference lies in the above four points, adapted to local conditions.

Ecoparque

In 1986, the Colegio de la Frontera Norte in Tijuana began to study decentralised sewage treatment models. The project was titled Sistema Descentralizado de Tratamiento y Reuso de Aguas Negras (SIDETRAN). On 19 October 1992, it came into being with the name Ecoparque.

The treatment system underway at Ecoparque is considered an exemplary form of modular treatment that may be duplicated in order to build a decentralised sewage treatment system. Ecoparque's treatment module occupies an area of 2.5 hectares, while the entire complex encompasses 6.2 hectares. It is located on the Otay Buena Vista ramp, on the southern slope of the Mesa de Otay near the Universidad Autónoma de Baja California. The topography of the terrain is broken, and its slope of over 30 per cent favours gravity-based treatment.

Ecoparque is an alternative technology for optimum sewage management; it permits urban growth, and at the same time meets the region's basic needs, resources and development potential. Its basic premise is that sewage, far from being a problem, should be considered a resource to be utilised and managed within the urban area.

At present, the treatment system generates water that is suitable for green-area irrigation. This is known as "secondary quality" water; secondary treatment is available for up to four lps. Upgrading of the infrastructure for treatment is planned to give Ecoparque sufficient capacity to treat 12 lps of sewage using simple, modular, passive and natural technology, encompassing preliminary, primary, secondary and tertiary treatment. The tertiary treatment phase consists of an artificial swamp that will be covered with rush and tulle, and will have an underground impermeable membrane, allowing for recovery of deposited water. Here the water's remaining BOD will be eliminated, and the nitrification and denitrification required to reduce the levels of nitrogen to acceptable levels for the receiving bodies of water will be achieved. The resulting renovated water (or treated water) will be used for green-area irrigation.

In addition to being an alternative sewage treatment system, Ecoparque is a green area that contributes to improving the city's environment and a place where educational and recreational activities[9] may be enjoyed. Its general services include environmental education, visitor services, and community-oriented activities (ecotechniques).

Ecotechniques are activities developed by applying regionally appropriate technologies in order to optimise the use of natural resources and energy. To this end, Ecoparque plans on experimenting with solar energy, waste separation, composting and efficient water management. Solar-energy experiments will be based on solar cell technology; waste separation will promote community participation in recycling solid wastes such as glass and aluminium. Residents will take such trash to Ecoparque's recycling deposit center[10] , where trash processing companies having agreements with Ecoparque will pick it up when a sufficient amount has been gathered. In exchange, the individuals will be given a plant from the Ecoparque nursery to use for reforestation of their community. Composting will be carried out by using dead vegetation and solid wastes eliminated by the microcribs in the primary treatment process.

The preliminary treatment phase of the process is effected in the inflow or de-sanding channel which uses grills to separate objects and larger waste materials from the rest of the waste. The primary treatment phase uses microcribs with fine grills, removes thick solids (floating and sedimentary), and controls odours by increasing the level of oxygen dissolved in the water to approximately three mg/l. This passive aeration initiates the flocculation of colloidal material for sedimentation. The secondary treatment phase uses biofilters and a sedimentation tank. The first increases particulate flocculation and reduces the soluble portion of BOD through bacterial activity. The sedimentation tank is a circular tank with a peripheral intake, a central deflector, and a cone in the lower regions to avoid traces of solid waste concentrations.

Final comments

The secondary treatment process of sewage by means of biofilters, such as those used at Ecoparque, has generated a great deal of interest over the last several years, mainly because the process is extremely simple and does not require much space, and because the BOD effluents produced are of a quality similar to that obtained in the activated sludge process.

Integrated urban waste management involves a balanced use of community resources and opportune management of urban services, whose goal is to attain sustainable development. Integrated water management via decentralised sewage system systems thus becomes very important, in order to use treated sewage and the products and by-products derived from the treatment process for diverse various activities, such as irrigation of green areas and compost production. Moreover, the use of environmentally "friendly" technology with minimal energy requirements, functioning through compatible processes actually improves the quality of the city's environment.

---

[9] The recreational activities refer to relaxation and contemplation of the surrounding scenery, and the educational activities consist of lectures on ecological awareness, the importance of re-using solid wastes, and development of urban resources, information concerning appropriate and economical techniques, and the treatment of sewage, etc.

[10] Actual separation of the glass and aluminium will take place in the homes of the persons resident in the community. Once it is separated, they will take it to Ecoparque's Recycling Deposit Center. This activity will promote awareness of environmental improvement.

Within the framework of integrated urban resource management, the alternative sewage treatment technology proposed by Ecoparque offers greater advantages than the City of Tijuana's Pumping and Sewage Treatment System, according to a recent master's degree thesis at the Colegio de la Frontera Norte (Gómez, 1996). The main reason lies in its multiplicity of functions. Not only is it an alternative sewage treatment system, it is a space that provides green areas for relaxation and environmental education, and activities that promote the use or re-use the community's refuse.

Questions, comments and answers

Q: What is quality of water before irrigation in terms of biological oxygen demand (BOD), suspended solids, coliforms, etc.?

A: It is fairly good because it has been pumped from a distant source which serves as treatment, but it is very high in organics. Compared with other facilities in the same area, it is within the same parameters.

Q: What proportion of the wastewater flow of Tijuana is treated?

A: So far this is only a pilot plant, but the topography will allow several Ecoparques around Tijuana and with those we can cover the whole effluent.

REFERENCES

BOCCO, G., R. SÁNCHEZ, and H. RIEMANN (1993), "Evaluación del Impacto de las Inundaciones en Tijuana (enero de 1993). Uso Integrado de Percepción Remota y Sistemas de Información Geográfica", *Revista Frontera Norte* 10:5, pp. 53-81, July-December.

CLEMENT, N. and E. ZEPEDA (1993), *San Diego-Tijuana in transition, a regional analysis*, Institute for Regional Studies of the Californias, United States.

COMISIÓN ESTATAL DE SERVICIOS PÚBLICOS DE TIJUANA (CESPT) (1995), *Agua en Tijuana*, Government of the State of Baja California, Mexico.

DE LA PARRA, C. (1995), "Manejo integral de recursos y desarrollo urbano sustentable", in Mendoza Berrueto (ed.), *Reunión de Alcaldes Fronterizos sobre Desarrollo y Medio Ambiente*, Colección memorias 2, El Colef, Mexico.

GARCÍA, M. (1982), "Los recursos hidráulicos", in M. López Portillo (ed.), *El medio ambiente en México: temas, problemas, y alternativas*, Fondo de Cultura Económica, México.

GÓMEZ, I. (1996), "Aplicación de técnicas multicriterio para el analisis de alternativas de tratamiento y reuso de aguas residuales: el caso de ecoparque en Tijuana" (Application of multicriterial techniques for analysis of alternatives in the treatment and re-use of sewage: the case of Ecoparque in Tijuana), Masters Thesis on Integrated Environmental Management, Colegio de la Frontera Norte, unpublished.

INEGI (1995), *Cuaderno Estadistico Municipal. Tijuana Estado de Baja California*, 1993 edition, México.

OJEDA, L. (1992), "Areas verdes en ciudades de la frontera norte", *Ciudades* 16, pp. 48-53, October-December.

SECRETARIA DE ASENTAMIENTOS HUMANOS Y OBRAS PÚBLICAS DEL ESTADO (SAHOPE) (1996), *Baja California, obra pública 1989-1995*, Government of the State of Baja California.

UNITED NATIONS POPULATION FUND (UNPF) (1991), *La población, los recursos y el medio ambiente, los desafios críticos.*

Parallel Session II

II-B. PREVENTION OF WATER POLLUTION: INDUSTRIAL SOURCES

INDUSTRIAL WATER MANAGEMENT

by

Jacques van de Worp and Wim Harder
Commission of the European Communities, DGXII
TNO Institute of Environmental Sciences, Energy Research and Process Innovation
Apeldoorn, the Netherlands

Introduction

The past decades have witnessed an increasing awareness that human activities, in particular intensive agriculture and industrial technologies, must be brought in harmony with the global material cycles in the biosphere. In other words, we shall have to make a transition from exploitation of our natural resources towards a partnership with the global ecosystem (Harder, 1995).

At the end of the 80s, after the publication of the Brundtland report *Our Common Future* (1987), sustainable development became a key issue (Brundtland, 1987). In the years following the publication of this report, several attempts have been made to translate its basic philosophy and recommendations into an operational approach for the immediate future (Jansen and Vergragt, 1995). In the case of industrial technologies, a working group of the World Business Council for Sustainable Development (WBCSD, 1995) considered the following elements to be essential:

- dematerialise: reduce the amount of raw materials used;
- increase energy efficiency;
- eliminate negative environmental impact of processes and products;
- close material cycles: design for recyclability, but not at any cost;
- borrow from natural cycles, particularly where renewable resources and recycling are concerned;
- extend the durability and service life of products.

Water is an essential resource in three of the six elements mentioned, but so far it has not been identified as a key issue by decision-makers. As water becomes more important in the next decades, particularly in several regions on this planet (western United States, Middle-East, India, China, Australia), the time has come to put the issue on the political and economic agenda (Donkers, 1994).

Water is a most important natural resource. On a global scale, there is no shortage of water, since more than 70 per cent of our planet is covered with it. Since there is essentially no exchange of materials between the earth and outer space, the total amount of water on the planet is constant. On earth, water is only found in a thin layer of approximately 60 kilometres at the outer surface (Donkers, 1994).

Liquid water is essential for life. Many organisms consist for more than 90 per cent water. The human body contains some 65 per cent water, and man can only survive for approximately three days without drinking. The problem with regard to the requirement for water by terrestrial ecosystems (including humans) is not only related to its availability, but more in particular to its quality, notably its chemical (and bacteriological) composition. On earth, there is a relationship between water quality and the type and composition of ecosystems that it can support. Thus, water quality always has to be related to the required purpose for use. Water fulfils many different functions, such as:

- essential reactant for organisms;
- environment for aquatic organisms;
- drinking water for animals and human beings;
- utility in household and industry;
- power supply (steam and waterpower; coolant).

In order to fulfil the different functions, water has to be "fit-for-use".

Water cycles

Global water cycle

According to the latest calculations, the earth contains some 1 386 million km^3 ($1.4 * 10^{18}$ m^3) of water. However, 97.5 per cent of this water amount is salty seawater. From the total freshwater reserves on earth, only 0.26 per cent (93 000 km^3) is available to terrestrial lifeforms (humans, animals, vegetation, lower organisms) (Figure 1) (Donkers, 1994).

Figure 1. **The earth's waterstock**

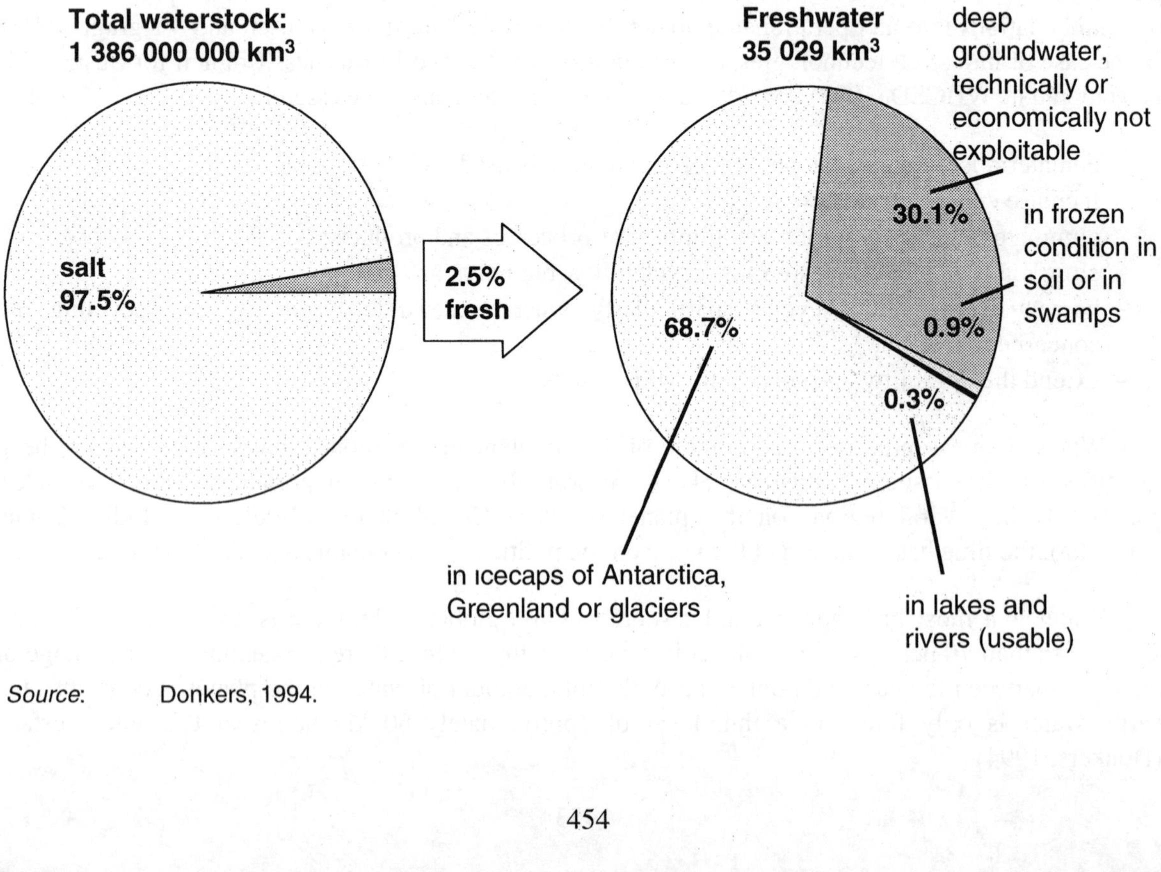

Source: Donkers, 1994.

The water on earth is not in rest, but participates in a cycle maintained by solar energy and the rotation of the earth (Figure 2). At sea, the evaporation exceeds precipitation, which leads to the building up of clouds. The clouds are transported from the oceans to the continents by wind, and since the temperature over continents is higher than over sea, the airmass is forced to rise. The resulting drop in temperature causes the vapour in the clouds to condense, and precipitation occurs. This precipitation meets the land surface. In humid areas there is a surplus situation (precipitation exceeds evaporation). The surplus water flows to rivers and aquifers, or directly to the sea. In arid areas, however, there is no flow of surplus water, since evaporation equals precipitation. The residence time of the water in the atmosphere mostly is short, some 10 days. In acquifers, the residence time on average is 600 years (range <1 - >> 10 000 years). In rivers, the residence time amounts to 10 to 20 days (average), and in oceans 3 000 years (average) (Donkers, 1994).

Figure 2. **The natural water cycle**

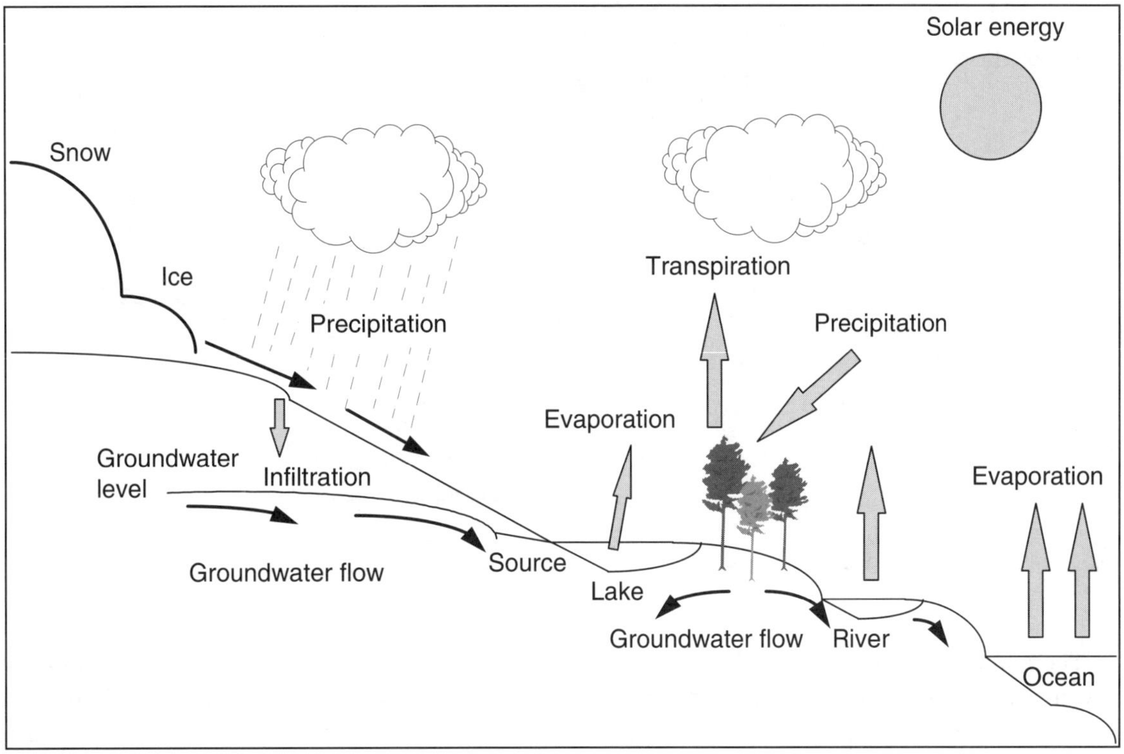

Source: de Jong, 1995.

The amount of water available for terrestrial activities on an annual basis depends on the precipitation on the continents. Over two-thirds of the precipitation eventually evaporates, while the rest (47 000 km^3) is supplied to the groundwater stocks, rivers and lakes (Figure 3). Man is not capable of using all the water before it flows back into the oceans. From the available 47 000 km^3, man can use only 9 000 km^3, which translates to some 1 600 m^3 per person. However, this amount of water is not equally divided over the world: in some parts it never or seldom rains, in other parts it rains excessively. Unfortunately, in the dry areas, more water evaporates, and agriculture is only possible when the land is irrigated. In some cases, nature offers some help by transporting water from wet to dry areas. Where nature fails, man intervenes and transports water in an artificial manner (Donkers, 1994).

Water is not only unequally divided geographically, but also in time. In some regions, it rains very intensively in a short period of time. In other regions, precipitation is more or less equally divided over the year (Donkers, 1994).

Figure 3. **Renewable amount of freshwater on earth**

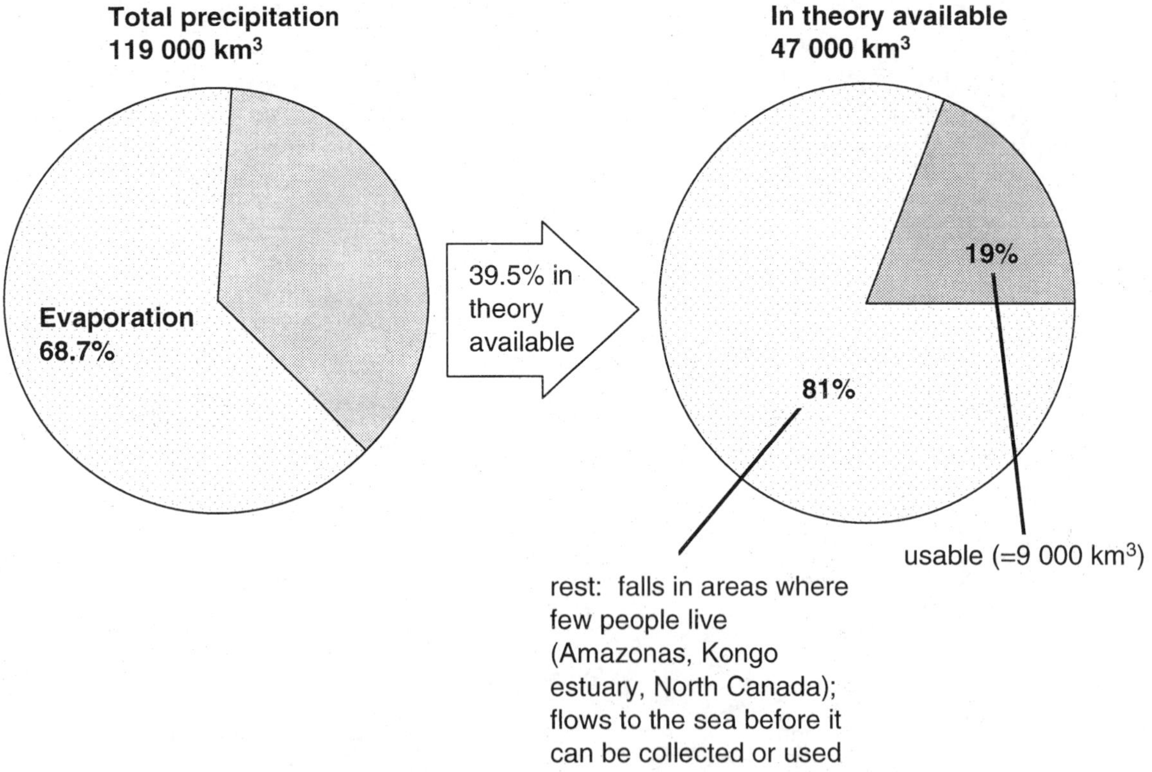

Source: Donkers, 1994.

The annual amount also can change from year to year, as in the south-east of Spain. Moreover, the water consumers are unequally divided over the continents. In some regions (Africa, the Middle-East) relatively many people have very little water available, while in other regions (Canada, Iceland, the Amazon, the Congo delta) few people have access to excessive amounts of water (Donkers, 1994)

The local water situation not only depends on the water supply, but also on the water demand, which usually is subdivided into domestic, industrial and agricultural demand. Because of the growth of both the domestic demand per capita and the world population, the domestic water demand increased from 100 billion m^3 in 1977 to 260 billion m^3 in 1987. The predicted figure for the year 2000 is 920 billion m^3. The industrial demand increased over the same period (1977-1987) from 445 to 745 billion m^3. The fastest growing sector, however, is agriculture, due to extensive irrigation programmes. Between 1960 and 1990, the earth's irrigated area increased from 90 to 234 million hectares. In Israel, 65 per cent of the water supply is used for agriculture. Figures for Egypt (88 per cent), Iraq (92 per cent) and Sudan (99 per cent) are even higher (Donkers, 1994).

Urban water cycles

It is not a coincidence that the first settlements and industries were erected on the shores of brooks, rivers and lakes, since in these areas fertile soil, water for consumption and water for use in various production processes is found. After use, when the water quality has decreased, it can be disposed of readily by discharge into the surface water. In our modern age, a city has a very complex water cycle and infrastructure.

Even in the Netherlands for instance, where most of the urban and industrial wastewater is treated continuously, it is necessary to draw attention to the prevention of pollution and the purification of wastewater (Figure 4). This is the case because the urban and industrial water cycles have a significant impact on the greater water cycle, both in quantitative and qualitative terms. Some recent issues are listed below (DTO, 1995; van der Graaf *et al.*, 1995):

- modern urban society urges a non-disturbed functioning; this requires sufficient drainage;
- due to ageing, many sewer systems show leakages, leading to groundwater pollution;
- due to water extraction, groundwater levels have dropped, leading to withering;
- due to human activities including agriculture and industry, ground and surface water have become polluted with various substances, both from natural and manmade origin.

Figure 4. **The urban water cycle**

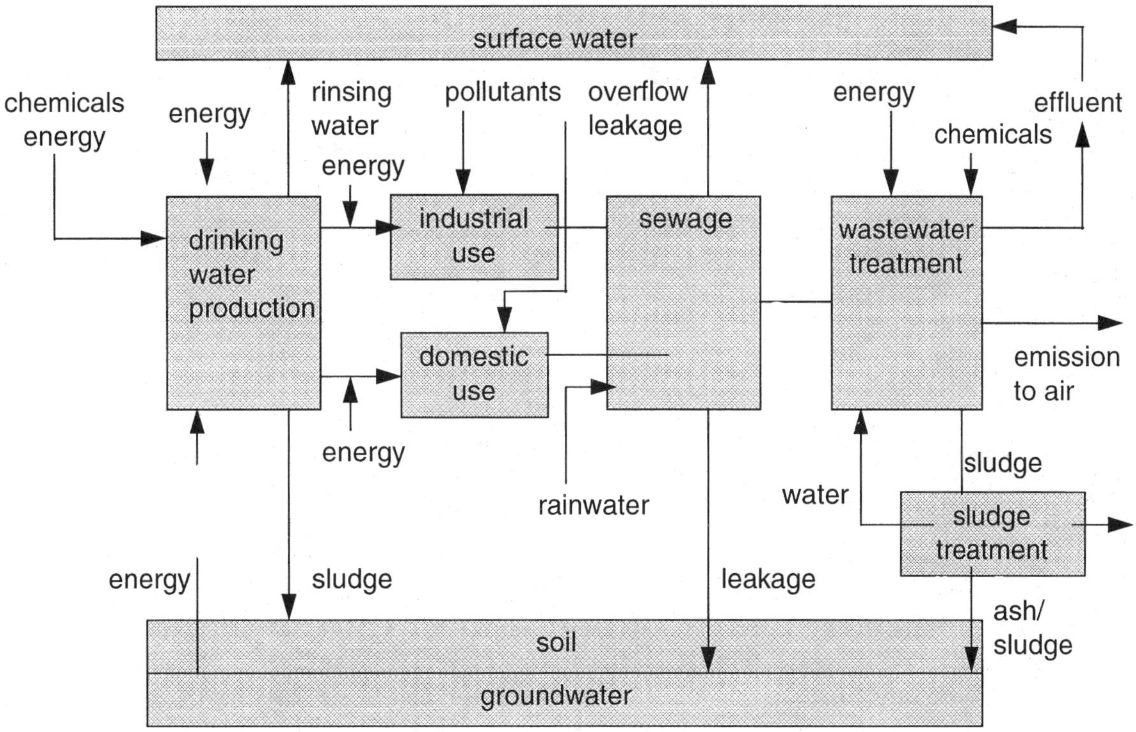

Source: DTO-Water Workgroup, 1994.

Industrial water cycles

In current practice, an industrial water cycle is generally not closed. The industry draws water from acquifers, from surface water or the public drinking water net. Used water is disposed of. In industry, water is often seen as a utility and is used for various purposes, such as (Assink *et al.*, 1996):

Function of process water:	Examples:
- product, reactant	production of beverages, hydrolysis
- solvent, absorption	gasscrubber, pickling
- washing, adsorption	textile finishing
- (energy) transport	cooling, steam circuits, solid wastes, sugar canes
- washing and rinsing	cleaning of equipment, installations and piping

In the first application (product and reactant), water is a raw material which cannot be replaced by any other component. In the other applications, however, water is a utility. In these cases, water is qualified as irreplaceable, due to the absence of adequate alternatives. Water is an attractive substance because of its physico-chemical properties, its relatively low price and its abundant availability in many industrialised parts of the world (Assink *et al.*, 1996).

Especially for cooling, washing and rinsing purposes, industry uses relatively large quantities of water. Cooling accounts for some 90 per cent of the total industrial water consumption. The main water pollution occurs by extensive conditioning and cleaning activities, and in processes where water is in direct contact with water-soluble components (Assink *et al.*, 1996).

Groundwater, surface water and even potable water cannot always be used without pretreatment, in order to satisfy the quality requirements. Examples are the removal of suspended solids (SS), water softening (removal of Ca, Mg) and removal of iron (Fe) and manganese (Mn). More specific examples are water demineralisation and sterilisation (Assink *et al.*, 1996).

After use in a process, water generally contains components which deteriorate its quality in such a way that the water cannot be re-used in the process without treatment, because it would lead to negative effects on product quality or production costs. For these reasons, in many cases, water is used only once, and after use is discharged in a sewer system for treatment on-site or off-site or disposal (Assink *et al.*, 1996).

Sustainable water use

General

An important lesson we can learn from nature is to endeavour to close the water cycles (Figure 5). In the water supply, the renewability of the water is essential, either by internal sources (precipitation) or external sources (surface water from rivers) (Donkers, 1994). When the amount of available water is limited, recirculation at a local or regional scale will have to be increased. Recirculation is only feasible when the water to be re-used is made fit-for-use.

To implement sustainable use of water, the use and quality of the water will have to be integrated into an economic context. This way, the prevention of water pollution may be promoted and funds for implementing measures can be set aside (Cramer *et al.*, 1992). At present, companies often do not know what the total costs for using water are, since these costs are presented by different actors. These may cover:

- intake of water;
- processing of raw water;
- water conditioning;
- wastewater treatment, inclusive waste disposal (sludge);
- wastewater disposal.

In Dutch industry, the use of one m^3 of water may cost between US$ 3.00 to 10.00 per m^3. Even looking at total costs, the costs of water are relatively low compared to other production costs (lower than 1-2 per cent of the product-unit-price). In cases where there is no economic drive to take measures aimed at water reduction, legislative measures will have to be taken (van der Graaf *et al.*, 1995).

Figure 5. **The urban water cycle**

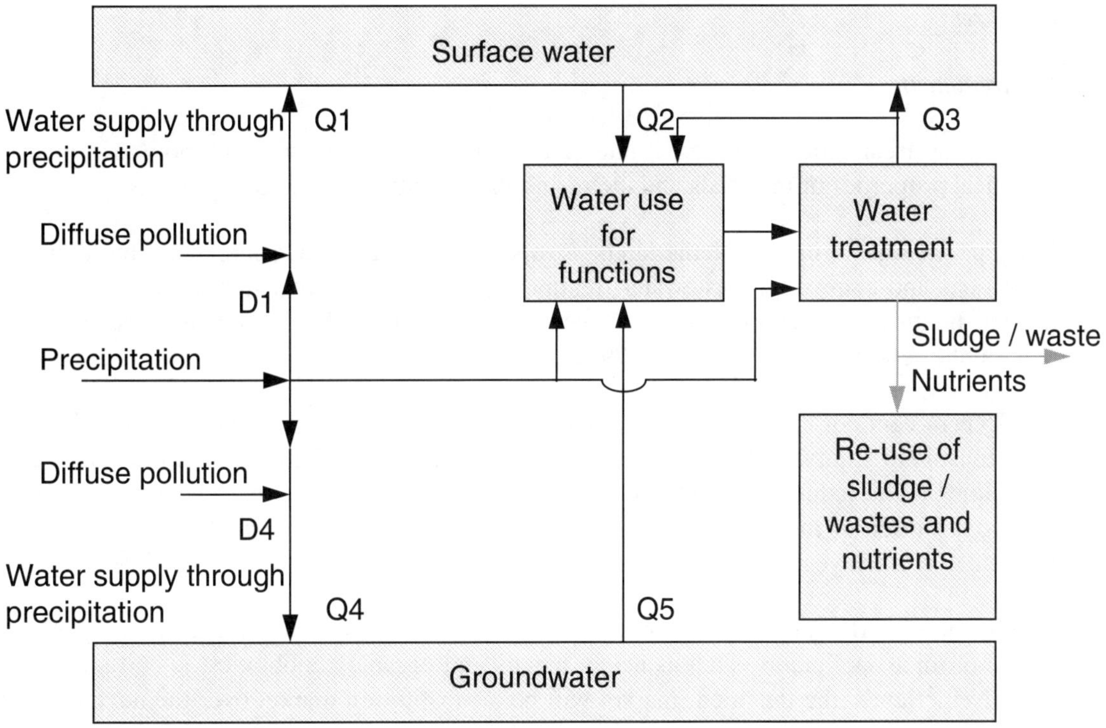

Source: TNO-MEP.

In the Netherlands, the government authorities see two pathways to reduce the use of water and emissions, both striving at zero-discharge (Senhorst and de Wit, 1996).

The first approach is a step-by-step method of implementing *add-on measures*, starting at the existing situation (historic pathway). The add-on measures can be put on a prevention ladder as follows (Senhorst, 1995; Senhorst and de Wit, 1996):

- end-of-pipe measures (wastewater treatment);
- good housekeeping measures;

- decentral water treatment for re-use purposes;
- process optimisation;
- process modifications;
- application of alternative raw and aid materials.

These add-on measures can be implemented in a time period of one to five years.

The second approach is the introduction of *clean technology*, the highest step on the prevention ladder. The concept of clean technology aims at achieving changes in product and process designs that in a period of 10-20 years will lead to a significant reduction of the environmental impact (Senhorst, 1995).

The add-on measures and the clean technology concept will be discussed in the next two sections.

Add-on measures

End-of-pipe treatment

In developing treatment technology, man has learned from nature and now exploits the physico-chemical properties of materials and the metabolic potential of (micro)organisms.

End-of-pipe treatment in the Netherlands is operated on a large scale, both by industrial treatment facilities and communal wastewater treatment plants (mostly two-step activated sludge or oxidation ditch treatment systems). Some 97 per cent of the households and industrial sites are connected to public sewer systems (de Jong, 1995).

Some 75 per cent of the produced chemical oxygen demand (COD) from domestic (15.3 million Inhabitant Equivalents -- I.E.) and industrial sources (10.2 million I.E.) is removed in treatment facilities. The industrial COD load was reduced by 65 per cent in 25 years (1970-1995). The relative contribution of industry to the total emission of priority pollutants has decreased by at least 50 per cent over a ten year period (1985-1995) (de Jong, 1995).

The application of effluent polishing by using rapid sand filters or membrane technology, aiming at the removal of micro-pollutants, such as heavy metals and organic xenobiotics, is currently under study. In the Netherlands, the treatment market will be a substitution market over the next 10 years. In substituting existing installations, sustainable development issues (treatment efficiency, energy consumption, space-saving, cost-saving) are being taken into account (de Jong, 1995).

Good housekeeping measures

Good housekeeping refers to measures to be taken after appropriate brainstorming and using common sense. For instance, a key question can be whether it is necessary and efficient to use the same (high) water quality for various purposes.

It might be more efficient and profitable to produce and use, for instance, five different water qualities in a certain production process than only one (high) standard quality. Introduction of good housekeeping can take place in a short period of time without having to change production processes much. It starts with training of the personnel, aiming at awakening awareness (Jansen and Vergragt, 1995). Good housekeeping is one of the main items in Company Integral Quality systems.

Decentral water treatment for re-use purposes

In many production processes, water is used in once-through systems, since water is both relatively cheap and not fit-for-use for a second time. Making the water fit-for-use again requires treatment. Treatment costs money and often is more expensive than discharge. However, when looking at rising costs for water use, treatment and disposal, as well as limitations for use by law, the industry is looking more and more for re-use and recirculation options. Re-using water and closing water cycles is only possible when the water quality can be brought back to the required specifications. Universities and technological institutes have been working on new technical solutions, such as compact biological processes, advanced oxidation systems, membrane technology, combined with extraction or adsorption processes, etc. With due regard to total costs (water, sludge, emissions, loss of raw materials and product), economically feasible solutions can be found. Important in this respect is the fact that decentral treatment is often more effective and efficient, since the facility has to deal with lower quantities of water, higher concentration levels and less variation in quality (lower number of different pollutants). Furthermore, decentral treatment often makes re-use of raw materials possible, thus leading to an added (economic) value of materials in the water. This aspect contrasts to the end-of-pipe treatment, where water and substances have a negative economic value.

Taking the above set of aspects into consideration, biotechnological processes in combination with physico-chemical working principles offer new interesting prospects. The main advantages are:

- overall treatment costs are relatively low;
- proven technology;
- can be used for a wide range of substances, even for recalcitrant compounds (in combination processes);
- relatively insensitive to toxic substances (in combination processes);
- compact in using sludge-on-carrier technology (biofilm reactors);
- high sludge residence time and low in sludge production;
- closed reactors, avoiding emission of volatile substances, including odour;
- in-process application opportunities.

A main point of concern is the reliability of in-process treatment facilities, such as bioreactors. In a production process, the (economic) consequences of a process breakdown due to malfunctioning or disruption of the treatment facility are by far greater than in end-of-pipe applications.

Examples of new biotechnology applications in process water treatment are:

- three phase airlift reactor;
- upflow anaerobic sludge blanket reactor;
- membrane bioreactor;
- combination of advanced oxidation (O_3, H_2O_2, UV) and biotechnology. The newest development is the electrochemical *in situ* generation of H_2O_2 in a bioreactor.

Process optimisation

Process optimisation aims at increasing the efficiency of the process, and reducing unnecessary spillage and the risk of failures or calamities. This is possible by using adequate on-line measurement equipment and process control. In measuring it is important to know the critical process parameters and to set an optimum range. In controlling the process, it is important to be able to adjust the process as quickly as possible whenever the setpoints are sub- or exceeded. An example is the discharge of cooling water in a closed system at the time when the water quality is below the set-point, instead of discharging at a pre-set time.

Process modifications

Many production processes are frequently modified in the course of time. These modifications often address a specific process step instead of an integral solution. Consequently, the modification may give an improvement in a certain part of the production process, while causing problems in other parts, leading to a suboptimal overall production process. When regarding the production process in a global manner, process modifications may lead to more optimal situations. Examples are counter-current washing and rinsing, the application of heat pumps, and the use of alternative chemicals that may increase the water recirculation time.

Application of alternative raw and support materials

In many production processes, materials and chemicals are used that have a great impact on the environment (heavy metals, xenobiotics). In some cases, it is possible to use less toxic substances which are biodegradable. Examples are the use of higher alcohols instead of dichloromethane in degreasing, and the replacement of the herbicide diuron by glyfosate (van de Worp, 1996).

Clean technology

In applying clean technology, historic production aspects must be set aside. The existing situation is no longer the starting point for improving the process with regard to environmental pollution. A set-point is chosen in the near future (for instance 2010 or 2040), and by careful evaluation of process and product characteristics while focusing on a zero-discharge, the pathway to achieve the desired results can be drawn by backcasting. Important characteristics of a clean technology approach are (Senhorst, 1995):

- prevention is better than cure: deal with the cause of pollution, and not its results (size of emissions);
- the time horizon ensures that there is sufficient conceptual room in order not to be curtailed by today's restrictions, while at the same time it remains necessary to come up with solutions that can be implemented;
- it requires sufficient creative thinking about long-term solutions so that the future also offers challenges.

The concept of clean technology is not just a product of wild imagination. Several clean technology processes are under investigation or can already be applied, such as (Senhorst, 1995):

- supercritical carbon dioxide textile-dyeing;
- alternatives for hot-dip zinc coating;
- dry paper production;
- biotechnological paper bleaching.

An example of a clean sustainable technology with respect to freshwater supply may be the desalination of seawater, making use of the ultimate sustainable energy source: the sun. The technology already is available, namely evaporation and reversed osmosis. In this way, man is able to influence the world's water cycle.

The clean technology concept has not yet matured. There are still problems and bottlenecks to overcome, both technological and economic in nature. The challenge is to solve them and create possibilities for sustainable development.

Sustainable industrial water management

Sustainable development is necessary if mankind wants to survive in the long run. One important aspect is sustainable water use and management. There is enough water available on this planet, but water quality is crucial for ecosystems and production processes alike. However, this does not mean we only have to focus on water; renewable energy is necessary as a driving force for water cycles (recirculation), and the formation of solid wastes and off-gases must be prevented (closed material cycles) (Figure 6) (Palsma *et al.*, 1993).

Figure 6. **Example of an ideal sustainable material cycle**

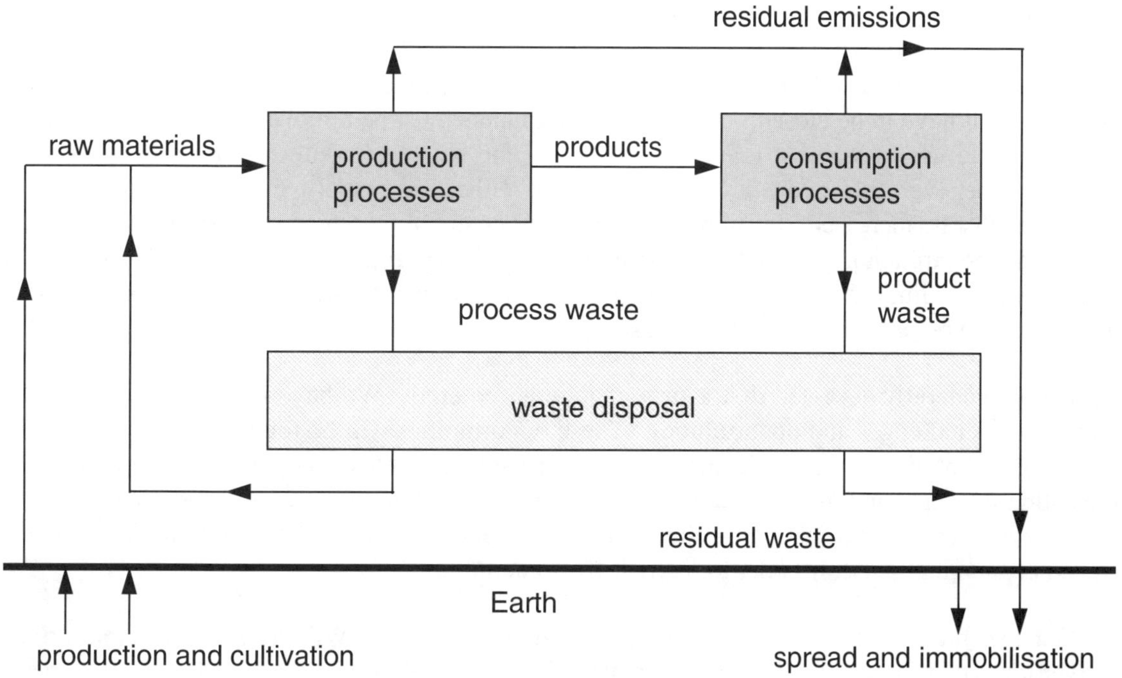

Source: Palsma *et al.*, 1993.

A company in general has two main goals: economic profit and continuity. From the viewpoint of a company, industrial water management must contribute to these goals to be considered seriously. The current practice in many companies is to strive for optimisation of the water use and emissions, rather than minimisation. Key issues are the effects on:

- product quality;
- cost minimisation;
- technical feasibility;
- the environment (mainly legislation-driven).

Unfortunately, this approach gives a limited and unpractical scope of the possibilities that have to be considered.

The use of water has to be related to the water quality, not just the water availability and the economic aspects. The goal of technological innovation should be to preserve the value (quality) of process flows (water and substances). Why should we clean surface water to drinking water quality standards for the flushing of toilets? Why should we pump up large amounts of groundwater only for the addition of low-value energy (heat) to it in cooling processes, and spill this water into the surface water?

Conclusions and recommendations

There is no lack of water on a global scale. However, there are many differences in the availability of water, in geographic terms, in time and in quality. Water quality is a key factor for the development of ecosystems and in production processes alike. Terrestrial life can make use only of freshwater, which encompasses only 0.26 per cent of the global water reserves. The world water cycle is closed. The non-biodegradable pollution we discharge in the water system therefore builds up, deteriorating water quality and having an impact on the world's ecosystems. From a sustainability point of view, this situation is not desirable.

The main lesson we can learn from nature is to strive for closing local or regional water cycles. These cycles will have to be closed; consequently pollutants must be removed at source. In order to create sustainable use of water it is essential to cover the net use by either water supply or water recycling. Whenever this cycle is out of balance, either the supply has to be increased, or recirculation must be increased. In supplying water, it is essential that this water be renewable, either from internal (precipitation) or external sources (surface water from rivers). When the amount of renewable water is limited, recirculation will have to be increased. Recirculation is only feasible when the water to be re-used is made fit-for-use.

It is now generally accepted that a firm change is required. We have reached a turning point which sets new challenges and opportunities. These opportunities will be found not only in the field of technologies, but also in the administrative and organisational fields. We must be willing to depart from existing concepts and give new concepts a chance. This requires an active participation of the players. Experiments require proper preparation, implementation and monitoring in a technological, social and economical sense, in order to really learn something.

Which incentives are required for taking the right measures? Will they be principle-driven, law-driven or driven by economics? What approach will be followed: taking add-on measures or developing clean, sustainable technology based processes? Will the attention focus on problems and bottlenecks, or are we willing to respond to the challenge and generate new opportunities? We strongly hope the latter, in order to not only continue industrial activities at desired profits, but also give our children and the earth's ecosystems a future.

REFERENCES

ASSINK, J.W. and A. WEENK (1996), *STEPS: a Systematic Approach for Integral Industrial Water management*, TNO, Apeldoorn, the Netherlands.

BRUNDTLAND, G.H. (1987), *Our Common Future*, World Commission on Environment and Development, Oxford University Press.

CRAMER, J. *et al.* (1992), *Sustainable Development: Closing the Material Cycles in Industrial Production*, TNO, Apeldoorn, the Netherlands.

DONKERS, H. (1994), *The White Oil - Water, Peace and Sustainable Development in the Middle-East*, the Netherlands Organisation for International Development Co-operation (NOVIB), Jan van Arkel, Utrecht, the Netherlands.

DTO-WATER WORKGROUP (1994), *Sustainable Urban Watercycle: Scouting study DTO-Water*, Institute for Inland Water Management and Waste Water Treatment (RIZA), the Netherlands.

GRAAF VAN DER, J.H.J.M. *et al.* (1995), "Possibilities of sustainable watermanagement in the cities", H_2O 28:25, p. 754.

HARDER, W. (1995), "The Microbe's Contribution to Sustainable Development", *Wider Application and Diffusion of Bioremediation Technologies, The Amsterdam '95 Workshop*, OECD, Paris.

JANSEN, L. and P. VERGRAGT (1995), "Towards Sustainable Development with Technology: Challenge in Programmatic Perspective", in *Practices in sustainability - perspectives towards 2040*, Jan van Arkel, Utrecht, the Netherlands.

JONG, J. DE (1995), "Optimization of Wastewater Systems: a Turning-point", Proceedings of the annual NVA-fallmeeting (Netherlands Association for Water Management), 24 November, Rhenen, the Netherlands.

PALSMA, A.J. *et al.* (1993), *Sustainable Soil Use, a TNO-concept*, a TNO-Environmental and Energy Research Publication, Apeldoorn, the Netherlands.

SENHORST, H. (1995), *Clean Technology at RIZA*, RIZA, Lelystad, the Netherlands.

SENHORST, H. and S. DE WIT (1996), "Zero-emission as Strategy for the Future", H_2O 29:1, p. 17.

WORLD BUSINESS COUNCIL FOR SUSTAINABLE DEVELOPMENT (WBCSD) (1995), "The Six Dimensions of Eco-Efficiency", *Tomorrow* 5, p. 40.

WORP, J.J.M. VAN DE (1996), "Physico-chemical Processes", Proceedings of the Euroforum-conference on "Industrial Wastewater Treatment", 30-31 January, Leidschendam, the Netherlands.

THE DETECTION OF POLLUTION - SOME NOVEL APPROACHES

by

Issy Caffoor and David Fearnside
Yorkshire Water Services Ltd., United Kingdom

Historical review

Sewage treatment works are normally designed to treat sewages of 'domestic' character, i.e. those that do not inhibit the biological treatment processes as defined in the UK Water Industry Act (1991) Chapter 56, Section 111. The Act specified that Water Companies are required to accept such wastes, unless they will not adversely affect the treatment processes, etc. However, it has not been technically possible to quantify and enforce the meaning of this part of the 1991 law. As a consequence, many sewage works today receive inhibitory wastes and do not perform at 100 per cent of design capability. This represents an expense in extra facilities to achieve the required effluent standard, which is met by the general service charge. The current charging system in the United Kingdom has few incentives for traders to produce less toxic wastes. Many of the UK companies use a charging formula based on the following principle:

$$\text{Unit Charge} = R + P + (Ot / Os)B + (St / Ss)S,$$

where
- R = Reception Charge
- P = Primary Treatment Charge
- B = Biological Treatment Charge
- S = Sludge Treatment Charge
- Os = Biological strength of combined sewage (mg/litre COD)
- Ot = Biological strength of trade effluent (mg/litre COD)
- Ss = Sludge strength of combined sewage (mg/litre Settled Solids)
- St = Sludge strength of trade effluent (mg/l Settled Solids).

As toxicity does not always correlate directly with COD or solids, this charging formula does not reflect the cost of treating inhibitory wastes. Pre-treatment plants on a trader's premises may remove easily degradable substances leaving the toxics which then cause inhibition at the sewage works. The BOD test, a measure of the strength of a waste, can give a false reading if the waste contains toxics. Yorkshire Water has decided to try and control toxicity problems at source rather than accept them and cope with the consequences.

The words 'inhibition' and 'toxicity' describe different characteristics of the same substance(s). Inhibition is what we observe on a sewage works -- it fails to achieve 100 per cent performance. Toxicity is what we observe in the tests -- the organisms are damaged in some way. If toxicity or inhibition is eliminated during sewage treatment, we say it is treatable, but of course it may affect the process while it is being treated.

Yorkshire Water has engaged in a number of technology and analytical method developments in order to secure the optimum performance of Yorkshire Water sewage treatment plants and sludge disposal operations. The aim is to reduce harmful and inhibitory properties of trade effluents, which result in toxic or inhibitory conditions in sewage treatment processes, including incident (emergency) management, investigation of drinking water, etc.

Figure 1 contains data reflecting the level of inhibition to carbonaceous treatment experienced at ten of Yorkshire's worst affected waste water treatment plants (WWTP's). Works 1, on average, receives effluent of relatively low toxicty, but on occasion very high levels of inhibition are observed, so high as to put the treatment plant at risk of failure in terms of quality of effluent discharged to the environment. Works 7 demonstrates relatively high levels of inhibition to biological treatment most of the time. This reflects the need for increased treatment capacity at this particular works in order to meet regulatory standards, in terms of BOD/COD removal.

Figure 1. **Levels of inhibition to activated sludge organisms measured at ten of Yorkshire's WWTP's**

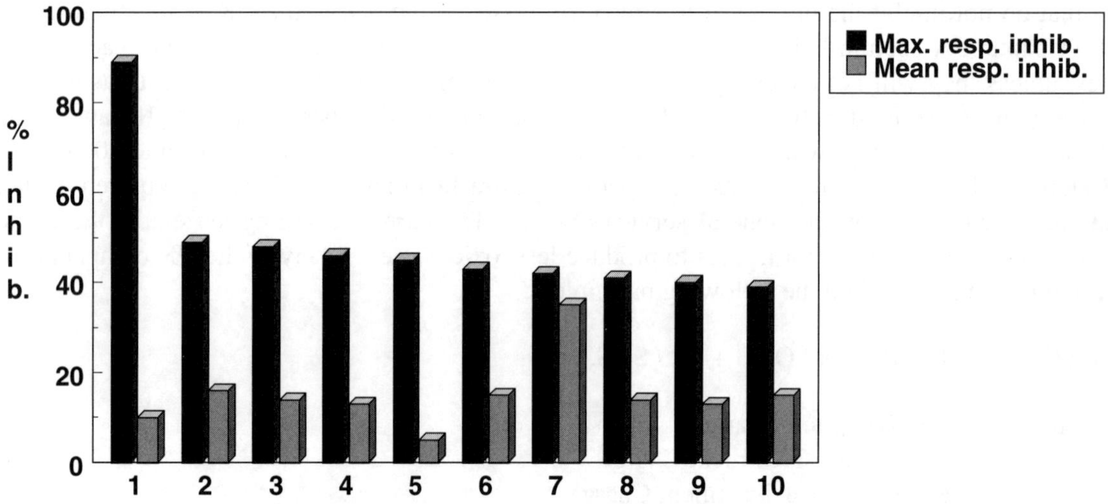

Source: Yorkshire Water, 1996*a*.

Figure 2 demonstrates very high levels of nitrification inhibition encountered at some of Yorkshire's WWTP's that are required or will require, in the near future, to nitrify in order to reduce the level of ammonia in discharges to the environment.

Current activities

Yorkshire Water has engaged in a programme of test and technology development to assist in the identification of inhibitory discharges. The credibility of the tests used by Yorkshire Water has been established by a programme of evaluation, involving wide ranging experimentation involving tests of reproducibility and sensitivity. Concurrently, a programme of R&D was established to improve the efficiency and reproducibility of existing tests, as well as to develop other tests which expand the range of organism available.

Figure 2. **The level of nitrification inhibition measured at ten of Yorkshire's WWTP's**

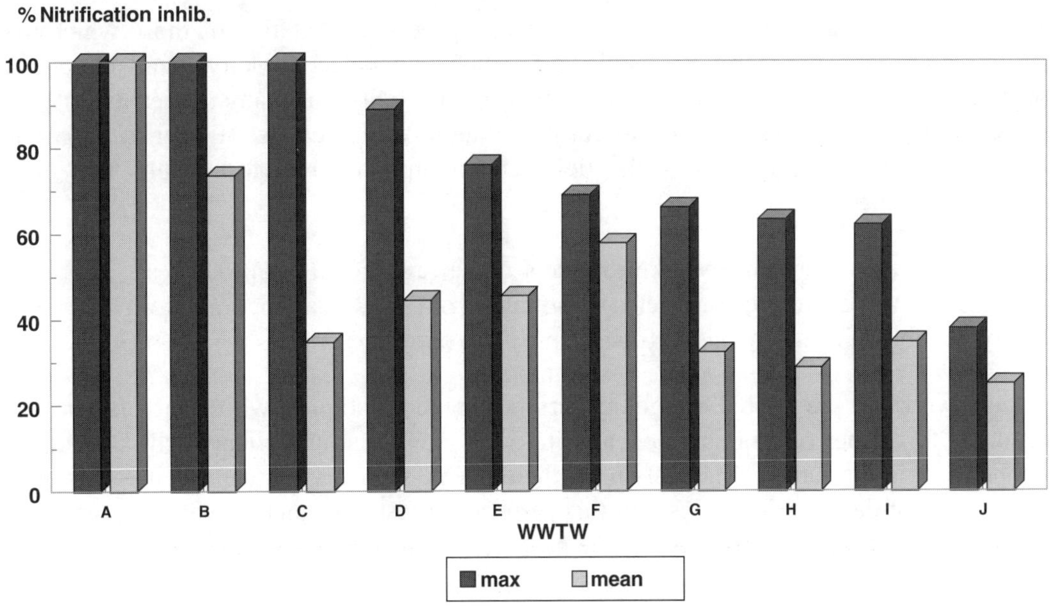

Source: Yorkshire Water, 1996*a*.

Application of multiple-test toxicity assays

Yorkshire Water has evaluated and developed a suite of tests using the following organisms: *Photobacterium phosphoreum, Nitrosomonas, Nitrobacter* and respiration inhibition of a mixed culture of domestic activated sludge.

No single test or even a small set of tests would be appropriate in all cases. For example, to assay the effects of a biodegradable mothproofer would require tests on sensitive invertebrates from the system likely to be impacted (e.g. percolating filter grazers and receiving river invertebrates) linked to tests on a standard organism likely to show a fast and sensitive response (e.g. *Daphnia*). The mothproofer may show little effect on *P.phosphoreum* and other bacterial systems. In contrast, a bactericide would be unlikely to affect the invertebrates, and a herbicide may be so specific that in order to guard against possible damage to river weeds and sludge disposal to agriculture operations, tests with algae and plants would be essential. Vasseur and Ferrard (1984) describe how the results of Microtox and Daphnia tests can differ according to the source of the industrial effluents.

Specific sub-sets of the tests can be used at the outset if the toxicants are known. However, this is not normally the case, since wastes comprise a range of unknown compounds. In order to make the most efficient use of the tests, a stepwise procedure is employed. It extends to chemical analyses, bench-scale sewage treatment and plant growth trials also. No battery of tests can ever provide 100 per cent certainty of reaching the correct decision, and the procedure acknowledges this in the requirement for skilled judgement.

Such tests must fulfill the following requirements to be of operational use.

Laboratory tests - test requirements

1) A range of organisms and ecosystems is needed, to be representative of those which may be harmed. There may never be a single test which will detect toxicity in all circumstances (Roesler, 1984), because of the high organism-specificity of many toxicants. We must recognise (Hellawell, 1988) that no range of tests will ever be 100 per cent certain of representing what will happen in the field, and interpretation of test results will therefore always be required.

2) The tests have to represent the likely effects on sewage treatment processes. The full-scale processes are subject to such a variety of variables from day to day that monitoring of inhibition is only possible in medium to long timescales (up to one year).

3) Sewage works in the United Kingdom currently need to comply with their effluent consent standards for 95 per cent of samples, and it seems likely that compliance will be required for 100 per cent in the future. The tests must therefore be secure against unusual events, e.g. loss of acclimatisation to toxic wastes. In this respect, all tests use unacclimatised organisms and ecosystems in the first instance to achieve no adverse effect in the 'worst case'. Some harmless substances may give a 'toxic' reading on a test, requiring further testing to establish the cause. This is a normal consequence of using any form of screening test.

4) The tests need to have public, scientific and legal credibility. A high degree of reproducibility is therefore important in the application of a test to, for example, legally enforceable effluent discharge concentrations. Standardised and reliable tests strengthen the scientific basis of environmental impact assessments (Robinson, 1989).

5) The tests need to produce relatively quickly, and some tests should be both 'instant' and portable to aid in incident management.

Test application procedure

At each step the question is "Do we know enough to pass or fail now or should we test further?", and, to facilitate this, a set of pass/fail criteria is required for each test in each type of situation. The normal criteria for a 'pass' is if the waste would be at a no-effect concentration during sewage treatment, and for a 'fail' if the waste would cause more than five per cent inhibition. The results of all tests, especially the initial 'screening' tests, require skilled judgement to decide whether or not a 'fail' merits further testing. For example, wastes which are toxic but might be acceptably diluted and/or treated could be tested for degradation of toxicity and failed if they are less than 95 per cent degradable in a three-hour test (Fearnside and Hiley, 1991).

Sewage treatment may acclimatise to some inhibitory wastes without loss of performance, as shown by a test later in the procedure, and these may then 'pass'. The criteria can, of course, be altered to suit the needs of individual circumstances. Yorkshire Water methods for calculating no-effect concentrations from test data are given by Fearnside and Hiley (1991), who also give examples of the application of the tests. Eagleson *et al.* (1986) asserted that acute testing such as this will identify a large proportion of problems without the need to perform long-term chronic tests. Synergism and/or antagonism are assessed by re-testing separate components where these are known

Subsequent monitoring

One or two simplified tests are chosen from those most sensitive to the critical characteristics/components of the waste, and repeated to establish an acceptable range of variation, which is then applied as a standard. Monitoring results outside the range are taken to mean a change has occurred without specifying its nature [e.g., just because the (Microtox?) value is low do not assume the waste has 'got better'].

The degree of inhibition at a sewage works may be identified and controlled conservatively by equating inhibition with organic load, assuming no acclimatisation and no breakdown of inhibitors. Thus process failures can be anticipated and/or prevented. For example, assume a waste with an inhibition to respiration of 20 per cent in the test would have a maximum inhibition of five per cent if discharged into three times its volume of sewage. The effect of this would be the same as increasing the organic load on the works by five per cent. As a sewage works which is just meeting its performance target, any increase in load or inhibition may cause the works to fail. Because of the inherent variability of sewages and sewage works, this cause and effect link would not be discovered in the normal course of things for many months. For example, inhibitors may break down during treatment, reducing the overall effect. While a sewage works biological stage may acclimatise so that it treats certain toxic substances, it may never acclimatise to others. It may not remain acclimatised under all conditions, and in its acclimatised state may not be as efficient as expected. Acclimatisation, therefore, involves a risk to our legal obligation to meet effluent consent concentrations.

Case study

In the early 90s, WWTW's *A* was redesigned in order to deal with increasing organic load and to reduce the level of ammonia in its final effluent discharge. An ammonia level of one mg/l was to be in consent by April 1994. A capital scheme costing £ five million was required in order to achieve the new performance objectives.

Soon after commissioning, it was realised that the plant would not nitrify. Operational and design problems were investigated and eliminated; the only source of the problem was thought to be toxic sewage effluent. A programme of screening selected sample points was initiated.

The mixed crude sewage entering the works via two major sewerage systems (Figure 3) was screened first.

Nitrification Inhibition analysis indicated the works was being subjected to very high levels of inhibition (Figure 4).

The two receiving sewers were then individually screened. Both were found to contain toxic material which would inhibit the nitrification process. Unusually, there was a significant "additive" effect when the two sewers mixed. In theory, the resultant inhibition after mixing should have been no greater than the average of the sum, as the volumes of sewage within the sewers are very similar. Obviously some form of chemical interaction resulted in the mixed sewage being more toxic.

The next stage was to identify and screen the trade discharges that were suspected of contributing to the toxicity measured (Figure 4). Six major traders were identified for screening.

Figure 3. **Nitrification inhibition survey of trade effluents into Works A**

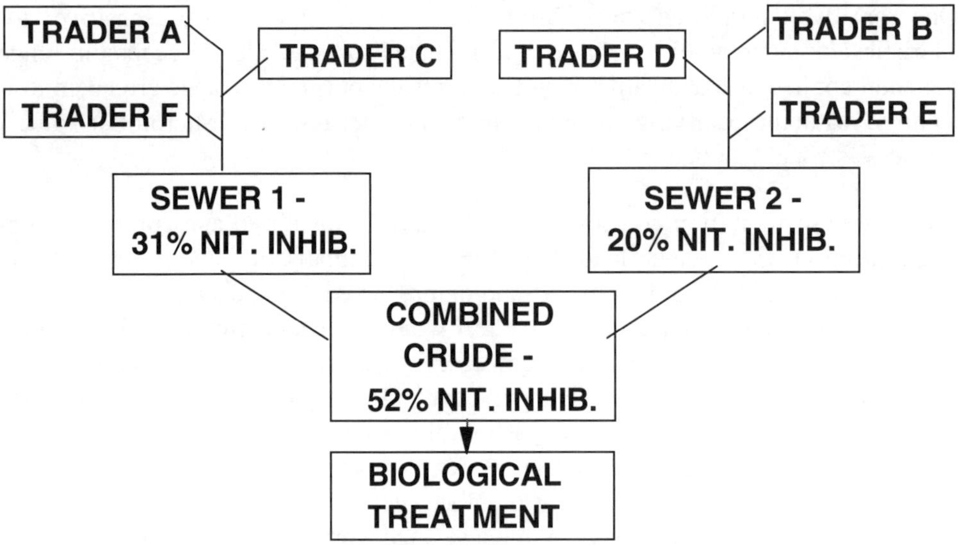

Source: Yorkshire Water, 1994.

Figure 4. **The level of nitrification inhibition of traders effluents entering works A**

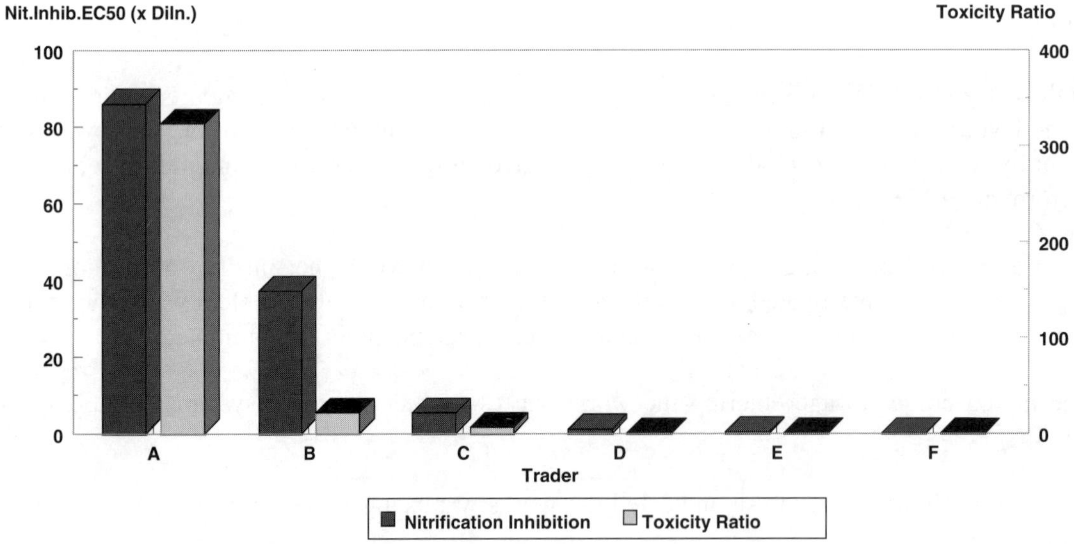

Source: Yorkshire Water, 1994.

The types of discharges ranged from waste from metal processing, chemical manufacturing, waste recovery (oils, etc.), metal plating and a dairy and creamery. Inhibition analysis identified three traders with measurable inhibition values. When the analysis was converted to toxicity ratios, only two were confirmed as the source of the problem.

Both effluents were from metal plating processes; each was approached by our Trade Effluent Control Department in order to establish whether, with co-operation, the toxicity within their discharge could be identified and eliminated.

Analysis of their internal process effluents (Figures 5 and 6) identified a discharge from the "Waste spill dump" at trader *A* and the "cyanide rinse" process effluent at trader *B*, as the source of problem. Trader *A* was able to very quickly alter the chemical process used, resulting in a drastic reduction in toxicity. Trader *B* has not yet implemented any changes to processes which will reduce the toxicity of their effluent. Trader *C*, though not thought to be a significant contributor to the problem, has also reduced the toxicity of its effluent.

Figure 5. **The toxicity of trade effluent A**

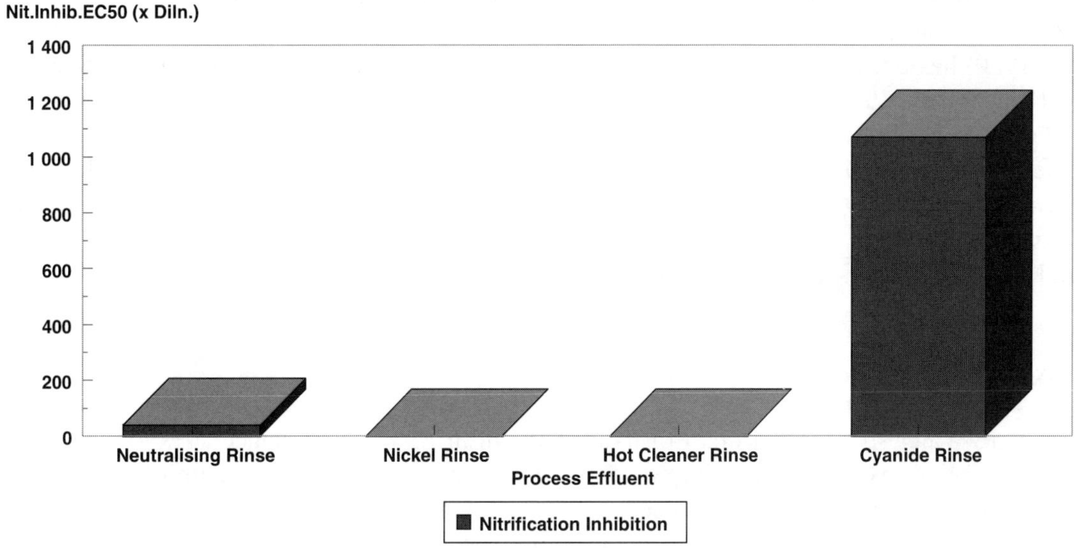

Source: Yorkshire Water, 1994.

Figure 6. **Toxicity of effluent B**

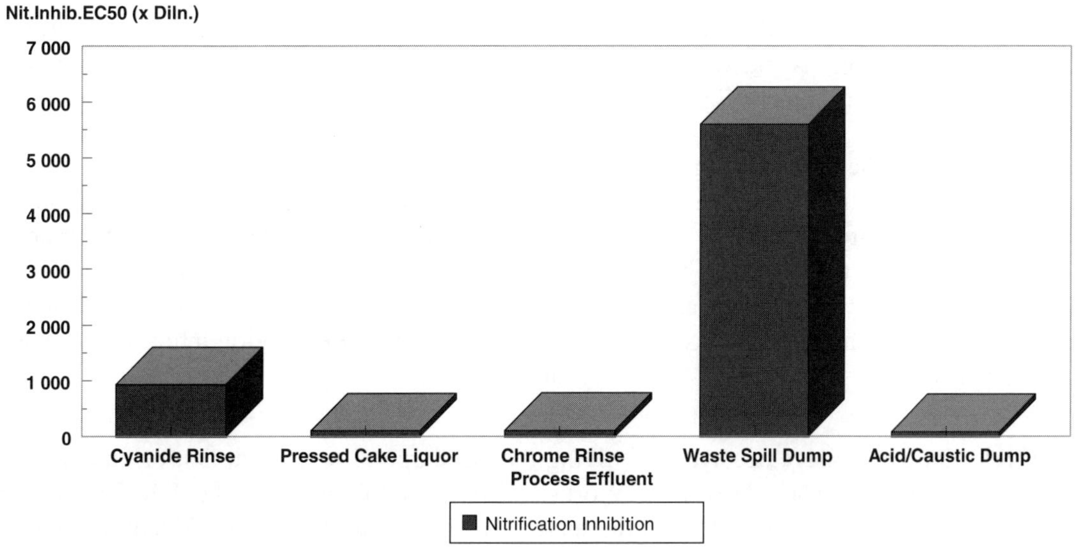

Source: Yorkshire Water, 1994.

Since going through the process of "toxicity reduction evaluation" there has been a measurable improvement at the WWTW *A*. On the whole, the sewage received is now domestic in nature; only occasionally does the works receive a "shock toxic discharge" resulting in measurable levels of nitrification inhibition at the works. This was the situation in early 1994.

Likely developments in the near future

Although UK discharge consent legislation is currently based on chemical composition, there is a growing trend to adopt microbial biosensors in routine analysis, and the UK Environment Agency has recently issued a discussion document to gauge industry opinion on the imposition of such types of consent within a legal framework. Microbial biosensors have been shown to be rapid, reproducible and sensitive indicators of organic and inorganic pollutants (Bitton and Koopman, 1986). The bioluminescence based biosensor Microtox®, using the marine bacterium *Vibrio fischeri* (Bulich and Isenberg, 1980) has been adopted as a screening procedure in Canada, France, Germany, Spain and Sweden (Richardson, 1995) for water quality testing. Paton *et al.* (1995) demonstrated the use of a bioluminescence based biosensor using a genetically modified (lux-marked) terrestrial bacterium. The bacterium *Pseudomonas fluorescens* was found to be more sensitive to metals than the marine *Photobacterium phosphoreum.* Yorkshire Water has engaged on two biosensors developments (Microtox-os, NEWT) in collaboration with major partners Siemens Environmental Systems Ltd. (UK) and The Microbics Corporation (US).

Microtox-os

Yorkshire Water sponsored, as part of its investment in Analytical Science at UMIST, a project to develop the Microtox® bench test into a continuous toxicity monitor. The main objective was to produce a toxicity monitor for use within sewage and effluent treatment, though the technique which has been developed can also be applied to the monitoring of raw water sources in potable water treatment.

The research phase undertaken at UMIST resulted in the identification of a technique of measurement and not an actual working instrument. To transform this technique into a useable instrument, the involvement of a company actually engaged in the business of instrument manufacture was essential. A process was initiated to find a suitable collaborator, which resulted eventually in the choice of Siemens Environmental Systems (SES) as collaborator.

Recent events in the United Kingdom, following the unknown discharge of a toxic substance from a sewage works which was subsequently picked up in a raw water intake, demonstrate one aspect of the vulnerability of a water company. Figure 7 demonstrates the ability of Microtox to identify toxic discharges to WWTP's, hence its potential as a treatment protection device. Though based in a laboratory, the system was able to detect a problem within hours of a suspected incident occuring, but this relies on rapid response and rapid sample transfer. Respiration inhibition tests were used to verify positive responses from Microtox analysis.

Other relevant events have occurred, involving the discharge of toxic material, originating from industrial sources, into Sewage Treatment Works. This was followed by the UK Environment Agency prosecuting a water company when the materials passed into the works effluent discharge to the environment. The detection, and hence management, of such events is impossible without the provision of appropriate instrumentation to detect such events as they occur.

Figure 7. **Examples of toxic incidents where Microtox detected potential**

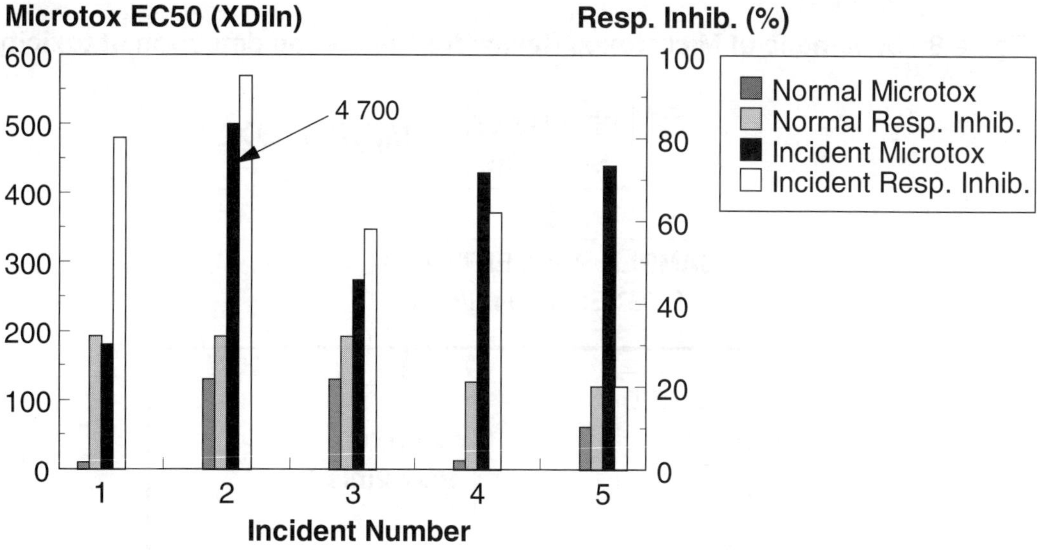

Source: Yorkshire Water, 1996*b*.

The evidence within Yorkshire Water Services from off-line analysis is that a large number of sewage treatment plants experience an ongoing level of chronic toxicity interspaced with acute events. This means that the treatment capacity of many large works in practice is not fully utilised, and that if toxicity could be identified, controlled, or removed, then additional capacity could be made available without the need for additional capital expenditure.

The objective of the project is to develop a robust and reliable on-line toxicity monitor that can determine and record a measurement of toxicity at least every 15 minutes (Figure 8). If a toxic substance is detected, then the instrument will generate an alarm. The alarms produced by the instrument will be capable of initiating an on-site audible alarm and a sampler will take a sub-sample. The facility will also be provided for the instrument to communicate with other computer/telemetry systems. The instrument will be capable of measuring the toxicity levels in samples of raw water, sewage and trade effluent.

Lux response in genetically modified organisms (GMOs) - NEWT Project

One of the criticisms of the Microtox technology is that it utilises a bacterium which is not endemic to sewage treatment or freshwater ecosystems, and therefore will not provide comparable responses to indigenous species. Genetic engineering now makes it possible to engineer the luminescence response into bacteria. Bacteria indigenous to the sewage treatment and aquatic environments have been chromosomally marked with *lux* AB genes, which only contains the genes encoding the luciferase enzyme. This construct requires the addition of exogenous aldehyde. In addition, bacteria were marked with lux genes containing all the necessary genes for light production (*lux* CDABE), which in this format does not require exogenous aldehyde. This was achieved by introducing a multi-copy broad host range plasmid. A method for freezing the bioassay organisms was developed for storage to optimise reproducibility of the assay. Project work is now targeted at assessing the response of the bioassay to toxic compounds, in immobilisation techniques and in the development of on-line instrumentation and test kits for toxicity detection. Table 1 compares the response of GMOs, Microtox and Respirometry to various metals and organics.

Figure 8. **Schematic of Microtoxos device for the on-line detection of toxicity**

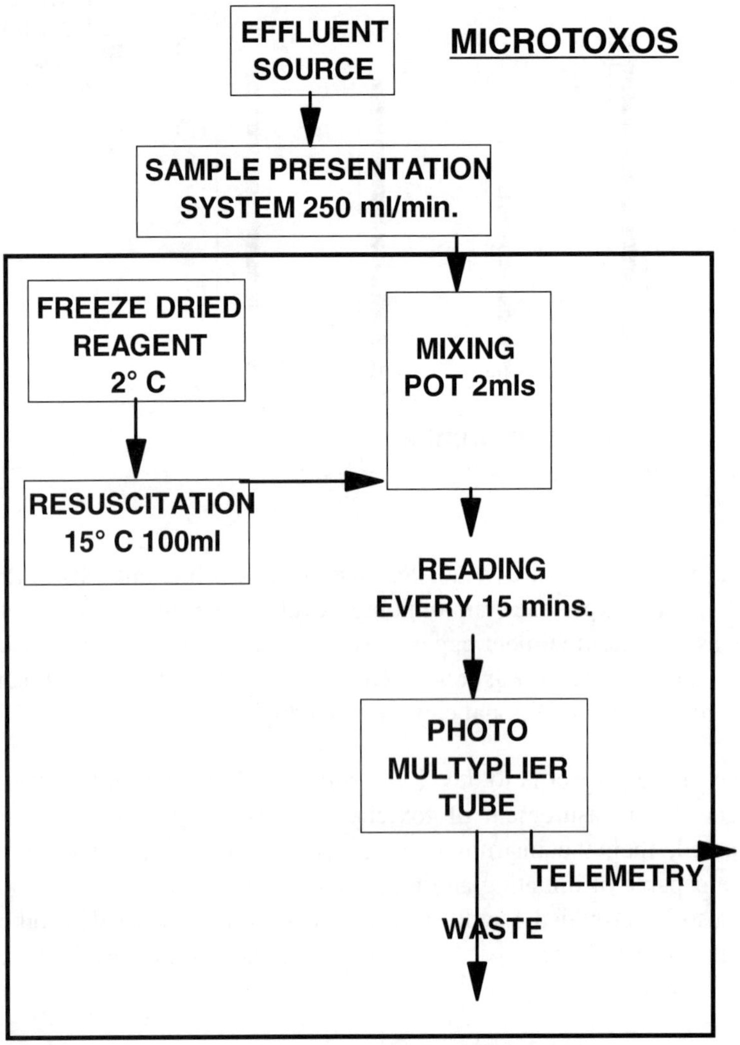

Source: Yorkshire Water, 1996*c*.

Table 1. **EC_{50} values ($mg\ l^{-1}$) for a range of pollutants as assessed by the *lux* - GMM assay and respirometric measurement of Ps. fluorescens at pH 7.0 in spiked double deionised water and compared with the Microtox assay**

	Ps. fluorescens 10586s and pUCD607			
Pollutant	Unwashed	Washed KCL	Respirometric	Microtox
Hg	0.14	0.13	0.70	-
Cd	0.20	0.20	4.32	2.35
DCP	35.1	4.51	34.2	1.63
PCP	27.5	0.80	52.7	1.80
CoV	6.8	10.3	11.2	10-15

Source: Brown *et al.*, 1996.

Respirometry

The respiration rate of an organism may be depressed by certain chemicals found in effluents. The extent to which this occurs can be used as a measurement of the toxicity of that effluent. Figure 9 shows data which indicates that BOD and Respiration Inhibition are not necessarily correlated.

Figure 9. **Diurnal pattern of BOD/Inhibition of a trade effluent**

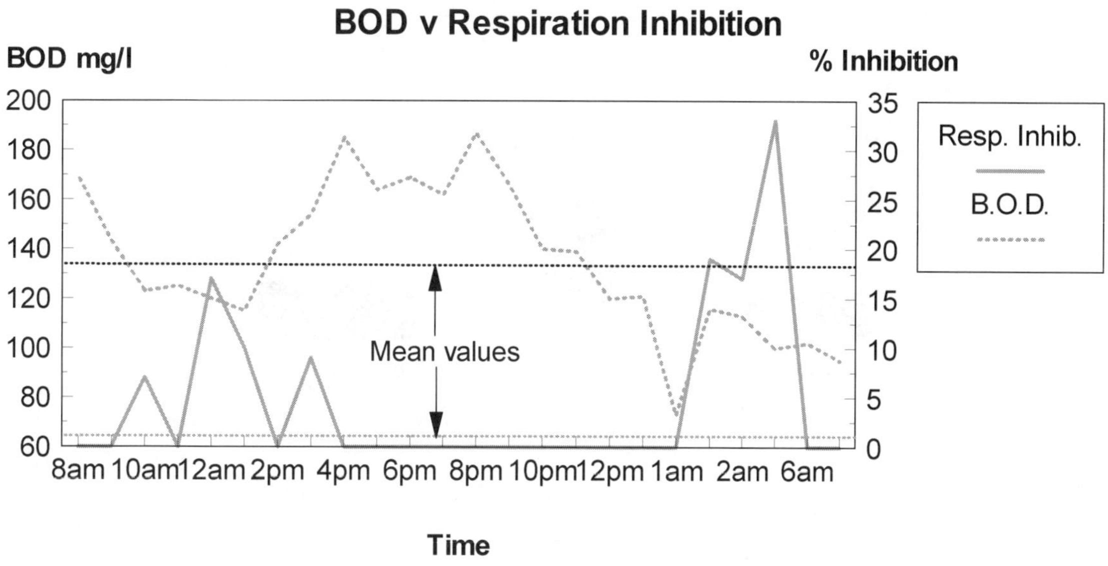

Source: Yorkshire Water, 1996*d*.

Respiration rate measurements can be made in two ways. Firstly, the sample is allowed to respire in a confined atmosphere, absorbing oxygen with either the change in volume with consumption of oxygen being measured on a calibrated scale or with the atmosphere being recharged with oxygen to maintain a constant volume.

The second method relies on the observed decrease in the concentration of dissolved oxygen in a sample which is confined so that no further replenishment of the dissolved oxygen can take place. Yorkshire Water has helped to design a new respirometer the Meritox 20.

The Meritox 20 consists of 20 identical measuring chambers which contain the respiring sample and the compensator (a similar non-respiring sample) (Figure 10).

The chambers utilise oxygen generating electrolytic cells which are driven by differential pressure transducers. Figure 10 shows a schematic and diagram of the chambers and associated electrolytic cell. The transducer is attached to both the reaction and compensation chambers, which eliminates errors due to barometric pressure changes. The respiring material reduces the amount of oxygen present and carbon dioxide is emitted. A differential pressure is produced by the absorption of the carbon dioxide by the presence of soda lime granules. The pressure differential transducer detects a pressure change (less than one Pascal) and gives a linear electrical output, increasing with decreasing pressure in the reaction vessel. This output is connected to a comparitor which compares the incoming voltage to a reference voltage. When the two are identical, the comparitor switches on the electrolytic cell and starts the timer in the micro-processor. The comparitor remains on until the electrolytic cell has produced enough oxygen to increase the pressure differential above a certain fixed level. Measurement of time taken to re-establish equilibrium is proportional to the amount of oxygen absorbed by the biological process. A measure of the oxygen uptake over time provides a respiration rate. The effect of toxicity generally decreases respiration (Figure 11).

Figure 10. **Schematic of Meritox Respirometer - (single vessel)**

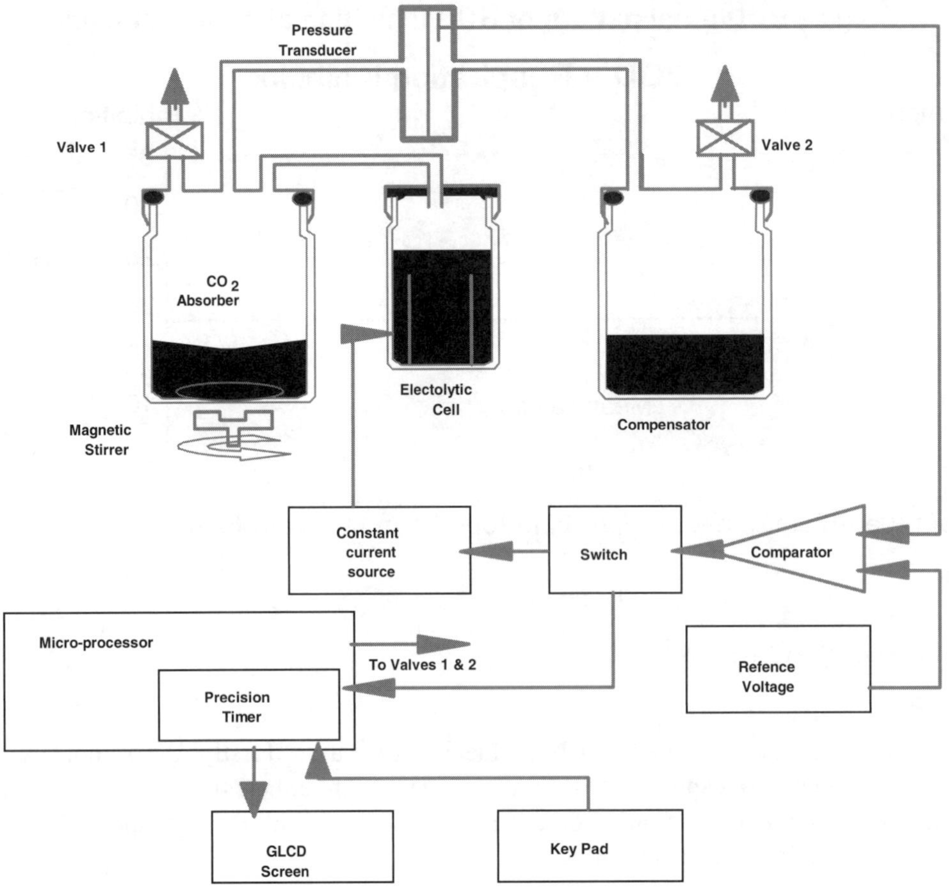

Source: MERITOX, 1995.

This relationship allows the interpretation of effective concentration values (EC). In the case of respiration inhibition by pollutants, the EC_{50} is often expressed as the effective concentration at which the respiration rate is inhibited by 50 per cent.

Table 2 presents results of the response of Psudomonas fluorescens in the Meritox 20, to trade effluent from a paper works.

Table 2. **EC values for a papermill effluent as determined by the Meritox 20 Respirometer**

Paperworks effluent	EC values determined by Meritox 20 Respirometer	
	EC_{10}	EC_{50}
Treated	11.2	>100
Untreated	14.6	42.3

Note: Results are expressed as a percentage of effluent in a distilled water diluent to reach the stated EC value.

Source: Yorkshire Water, 1995.

Figure 11. **Respiration measured by Meritox**

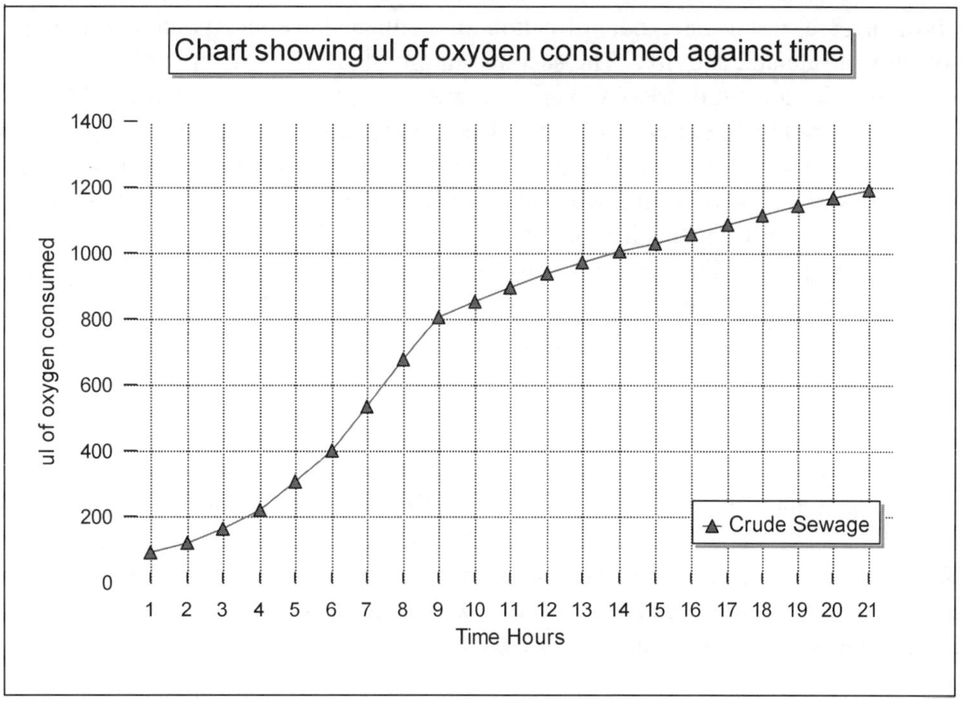

Source: Yorkshire Water, 1995.

The results from the respirometer compared favourably with other standard toxicity tests. Furthermore, respirometry proved to be sufficiently sensitive to enable the toxicity of individual components of the effluent to be assessed. The heavy metals, mercury and cadmium, for example, which are commonly associated with paper processing were found to have EC_{50} values of 0.67 and 2.50 mg l^{-1}, respectively. Use of modern electrolytic respirometers such as the Meritox offers rapid and sensitive toxicity testing of environmental samples. The ease of use of such machines and their flexibility in terms of sample type and size will prove to be a powerful combination in this rapidly growing field.

rRNA16 probes

Ammonia consents are likely to become tighter in the future. There is a need therefore to develop technologies to optimise and monitor the nitrification process. As a result of slow growth and low biomass yield, nitrifying bacteria do not readily form visible colonies on solid media. Traditional enumeration depends on the Most Probable Number (MPN) technique. This requires inoculation, serial dilution and incubation for at least 28 days. The limitations of traditional techniques prevents rapid assessment of the nitrification potential of sewage sludge or filter beds. This delays, and often prevent, identification of conditions which are unfavourable or inhibitory to nitrification, and limits the ability to assess the efficiency of remedial measures. It also prevents identification of conditions which are optimal for nitrification within particular plants. As a result of the extended doubling times of nitrifiers, cf. carbonaceous bacteria, the level of mixed liquor suspended solids (MLSS) does not provide reliable indication of the levels of nitrifying bacteria present. rRNA-based techniques provide an alternative approach for both enumeration and activity measurements of ammonia oxidisers which are potentially rapid and specific to ammonia oxidising bacteria, the organisms which determine the overall rate of nitrification. Individual cells contain in the order of 50 000 ribosomes, the number increasing with growth rate. As faster growing cells synthesise protein at a faster rate, quantification of cellular ribosome content therefore enables estimation of growth rate and activity. The second feature of rRNA is its specificity. The sequence of nucleotides which constitute rRNA varies between different organisms. Most sequence information comes from the 16S rRNA sub unit. Certain regions of 16S rNA are highly conserved , i.e. they are identical, or vary little in all bacteria. Others show intermediate conservation and may be similar in organisms of the same genus will vary in species belonging to that genus. It is possible therefore to design and construct specific probes, consisting of these common sequences, which are unique to ammonia oxidisers. These specific probes can be labelled with fluorescent dyes. Numbers of ammonia oxidisers can be determined by hybridisation of samples with fluorescently labelled probes followed by UV microscope examination.

Recommendations

Many previous attempts at establishing toxicity testing in the water industry have failed, or met with very limited success. They were characterised by the use of only one or two tests, in limited circumstances, and with restrictions on the involvement of other disciplines. The UK Environment Agency has now committed to the introduction of toxicity based consents for discharge to rivers and are currently in the process of consultation

In order to obtain optimum performance out of treatment works assets, it is essential that toxicity is removed from or prevented from entering biological treatment processes. In order to achieve this, a range of test protocols and detection technologies need to be in place. The following issues need to be considered:

1) No single test is appropriate for the range of effluents that could be received. Each test should be relevent to the environment under test, and therefore a range of tests needs to be developed.

2) Any test protocol should be reproducible from part of an accepted methodology which would stand up to legal scrutiny if necessary. The methodologies and technlogies used should be simple and relatively low cost in order to encourage uptake by trade effluent producers. This will encourage self assessment.

3) On-line and test-kit technologies should be developed to facilitate continuous monitoring by both trade effluent producers and treatment centres. This will facilitate operational control of toxic discharges.

4) Charges for treatment of effluents should reflect the effects of inhibition to the biological treatment process.

5) Treatment authorities should consider the imposition of consents to discharge based on the above test protocols and technologies.

6) Control technologies and knowledge based systems need to be developed in order to respond to data and alarms generated by such detection technologies.

Questions, comments and answers

Q: What toxicity are you measuring, toxicity of the activated sludge plant or toxicity of the discharged effluent? Is there any correlation between plant acclimatisation and toxicity?

A: We measure toxicity to the biological process. By removing it we save money on capital expenditure. Some plants will acclimate but there are fluctuations. Acclimatisation is difficult when there is no steady state of pollution.

Q: Can you comment on the sensitivity limits for the different methods of detection of pollutants?

A: Some have been used in raw water intake and will work, but we wouldn't expect to use them under these circumstances. Others are certainly sensitive enough and reproducible enough for sewage treatment applications.

Q: Can you force traders to build pre-treatment facilities?

A: It is very difficult to enforce treatability, but we can charge them in ways that make it beneficial for them to put in treatment systems and thereby save themselves money.

REFERENCES

BITTON, G. and B. KOOPMAN (1986), "Biochemical tests for toxicity screening" in G. Bitton and B.J. Dutka (eds.), *Toxicity Testing Using Microorganisms,* Volume 1, pp. 32-41, CRC Press Inc., Boca Raton, Florida.

BROWN, J.S., E.A.S. RATTRAY, G.I. PATON, G. REID, I. CAFFOOR, and K. KILLHAM (1996), "Comparative assessment of the toxicity of a papermill effluent by respirometry and a luminescence based bacterial assay", *Chemosphere* 32:8, pp. 1 553-1 561.

BULICH, A.A. and D.L. ISENBERG (1980), "Use of luminescent bacterial system for the rapid assessment of aquatic toxicity", *ISA* 20:35-4-, MRL 72-80.

EAGLESON, K.W., S.W. TEDDER, and L.W. AUSLEY (1986), "Strategy for Whole Effluent Toxicity Evaluations in North Carolina", in T.M. Poston and R. Purdy (eds.), *Aquatic Toxicology and Environmental Fate*, pp. 154-160, ASTM, STP 921.

FEARNSIDE, D. and P.D. HILEY (1991), "The Use of Toxicity Testing Methods to Identify and Trace Treatability Problems on Sewage Treatment Works", presented at Conference on Toxicity Assessment, 11 May, Kurashiki, Japan, Environment Canada.

HELLAWELL, J.M. (1988), "Toxic Substances in Rivers and Streams", *Environ. Pollut.* 50:61, p. 85.

MERITOX (1995), *MERITOX Operator's Manual*, E.A. Addingtons.

PATON, G.I., C.D. CAMPBELL, M.S CRESSER, L.A. GLOVER, E.A.S. RATTRAY, and K. KILLHAM (1995), "Bioluminescence-based ecotoxicity testing of soil and water", in *Bioremediation: the Tokyo '94 Workshop*, pp. 547-552, OECD Documents, OECD, Paris.

RICHARDSON, M. (ed.) (1995), *Environmental Toxicology Assessment,* Taylor and Francis Ltd., London.

ROBINSON, R.M. (1989), "Environmental Impact Assessment: The Growing Importance of Science in Government Decision Making", *Hydrobiologia* 188/189, pp. 137-142.

ROESLER, J.F. (1984), "Potential Analyzers of Toxic Materials for On-Line Use", *Govt. Reports Announcements & Index* (GRA+1), issue 10.

VASSEUR, P. and J.F. FERRARD (1984), "Comparison des Tests Microtox et Daphnia pour L'Evaluation de la Toxicite Aigure d'Effluents Industriels, *Environmental Pollution (Series A)* 34, pp. 225-35.

YORKSHIRE WATER (1994) "Aldwarke Toxicity Investigation", internal report.

YORKSHIRE WATER (1995), "MERITOX Evaluation", internal report.

YORKSHIRE WATER (1996*a*), "Regional Toxicity Survey 1993-1996", internal document.

YORKSHIRE WATER (1996*b*), "Regional Toxicity Investigation", internal report.

YORKSHIRE WATER (1996*c*), "Microtox-OS Development", internal report.

YORKSHIRE WATER (1996*d*), "Knostrop On-line Instrument Evaluation", internal report.

STABLE ENZYMES FOR INDUSTRIAL APPLICATIONS

by

Barry L. Marrs
Arres Enterprises, Kennett Square, Pennsylvania, United States

Enzymes, nature's most highly evolved catalysts, excel in the two most important features that are required of an industrial catalyst: selectivity and specificity. Little or no waste is generated by the typical enzyme functioning in a living system. This translates into minimising water use for waste treatment in an industrial setting. Since enzymes are naturally compatible with renewable resources, their expanded use as industrial catalysts would lead toward cheaper, easier wastewater treatment as compared to the burdens imposed by petrochemical feedstocks and the catalysts currently used with them. On the other hand, the enzymes that have been studied most thoroughly, and that are most readily available in commercial quantities, evolved to function the narrow range of conditions found in mammals, a few microbes and a few plants. Furthermore, enzymes in living systems did not evolve to be especially stable, since they could be replaced when necessary. If enzymes are to gain widespread use in industrial processes, they must be more stable than those that are currently commercially available. It would also facilitate matters if a wider variety of enzymes, which could function under a wider range of conditions, were available in large quantity and at low cost. Methods for providing better industrial enzymes include: searching biodiversity, especially among extremophiles, engineering improved enzymes via either rational design or Edisonian approaches, cross-linking, and immobilisation strategies and recombinant production technologies.

The universal Phylogenetic tree shows three main branches: Eucarya, including plants, animals and protozoans; Bacteria, including Gram-negative and Gram-positives, cyanobacteria (blue-green algae), Green non-sulfur bacteria, and others; and Archaea, including methanogens, halophiles and hyperthermophiles (Figure 1; Woese, 1987). From this analysis, it becomes clear that most of the evolutionary sequence diversity among living things is to be found among micro-organisms. It therefore follows that if one wishes to examine the greatest diversity of enzymes, in hopes of finding those most useful for industrial applications, one should examine microbial enzymes. Harvesting microbial enzymes has traditionally started with growing the microbes that produce them, and therein lies a problem. Most microbes cannot be cultivated using today's know-how and equipment. Most microbiologists agree that less than one per cent of the various types of microbes in nature are currently culturable (Amann *et al.*, 1995; see Table 1). If one cannot cultivate an organism, one cannot obtain enough biomass to perform analytical enzyme assays, let alone produce the tons of enzymes that are used in large scale industrial processes.

Figure 1. **A universal Phylogenetic tree**

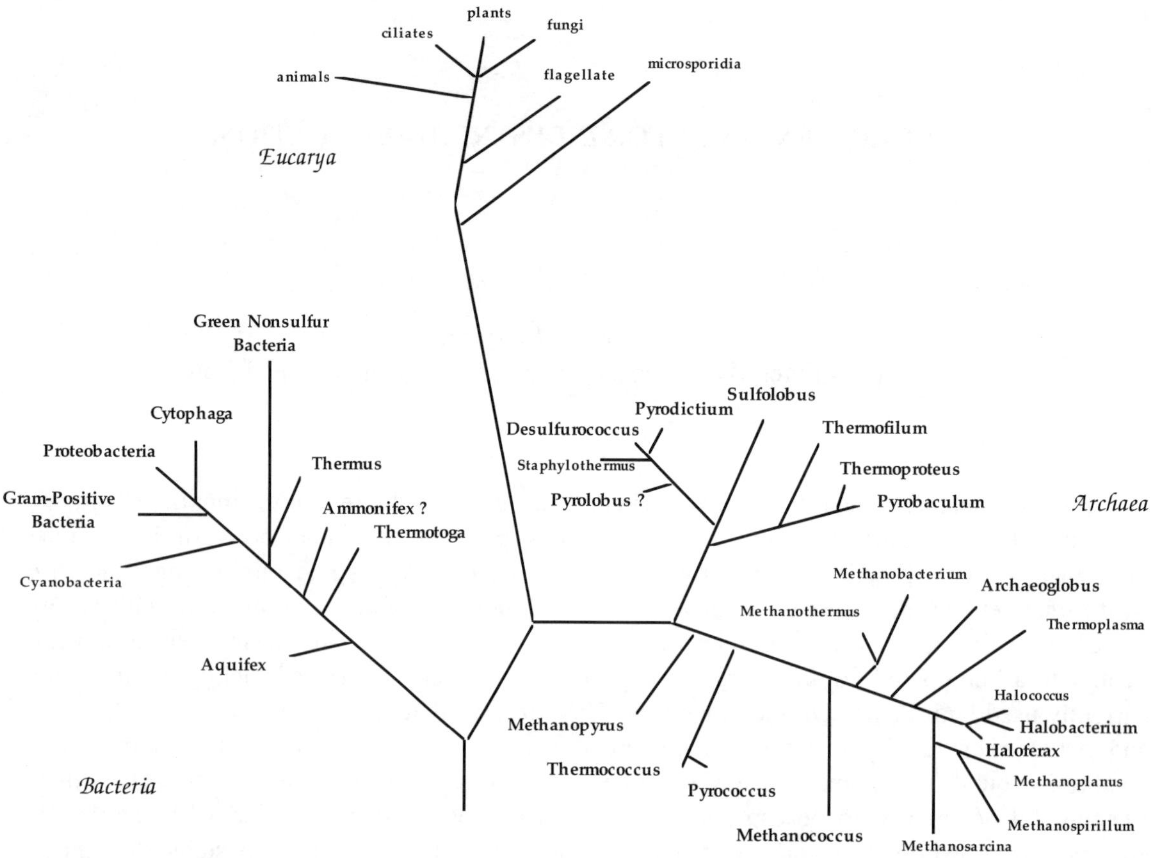

Note: The relatedness or diversity of all living things can be portrayed by comparing their DNA sequences which encode ribosomal RNA. The lengths of the branches on this tree are proportional to the sequence differences among the different organisms depicted.

Source: Modified from Woese *et al.* (1990) by Karl Rusterholtz of Recombinant BioCatalysis, Inc.

Table 1. **Micro-organisms in the environment**

Habitat	Culturability(%)
Seawater	0.001 - 0.1
Freshwater	0.25
Mesotrophic lake	0.01 - 1.0
Unpolluted esturine waters	0.1 - 3.0
Activated sludge	1.0 - 15.0
Sediments	0.25
Soil	0.3

Note: Culturability (%) determined as per cent of culturable bacteria in comparison to total cell counts.

Source: Amann *et al.*, 1995.

Industrial enzyme companies have responded to this challenge in two different ways. The more widespread and traditional approach is to become better at cultivating microbes. This is a fertile area of microbial research. A second approach is to learn how to isolate useful genes from microbes without ever growing them in pure cultures in the laboratory. The latter approach is relatively new and involves randomly cloning genes from environmental samples into domesticated expression hosts. Each expression host clone must then be assayed for each enzyme of interest. Since this involves large numbers of enzyme assays, automation is critical for the success of this approach (Robertson *et al.*, 1996).

Although nature contains a truly vast array of enzymes, it has not exhausted the types of protein-based catalysts that can be produced. Enzymes have evolved to function under that set of environmental conditions compatible with living things. The needs of industrial process developers frequently go beyond anything found in nature, but it is possible to produce enzymes with properties significantly different from any found in nature. The sorts of changes that are most often sought are either to increase an enzyme's catalytic rate or its stability or both under a certain set of conditions. There are two major ways of achieving improved process compatibility for enzymes: the chemical modification of naturally occurring enzymes, and the generation of enzymes based on new sequences of amino acids.

There is a large body of literature on the stabilization of enzymes through chemical modification (e.g. Laskin, 1985). These processes usually involve bifunctional chemical reagents, which can attach to the surface of an enzyme, without inactivating it, and can also then attach either to another molecule of enzyme or to a solid support. The result in either case is an enzyme that is less mobile, and that results in increased stability. An added benefit of this approach is that immobilised enzymes are often easier to handle and recover from the process stream for disposal or regeneration. New approaches to the chemical modification of enzymes continue to emerge.

More recently, success has been reported for the genetic engineering of enzymes with new properties based on new amino acid sequences. Here also, there are two fundamentally different approaches being applied. The first is that of rational design, wherein one attempts to form an improved catalyst based upon a thorough understanding of its structure and function. The fruits of this approach are currently used as enzymes added to laundry products that function in the presence bleach, detergents and high temperatures. The second approach has be called "Edisonian", since it does not depend upon understanding first principles, but rather on large scale trial and error. This process has also been termed "directed evolution", since it mimics natural evolution with the difference that the experimenter rather than nature controls the direction of the changes. Briefly, directed evolution is carried out by mutagenizing the gene for a particular enzyme, and then screening the resulting mutant genes for ones that perform slightly better under a given set of conditions. The process is repeated using the best of each generation of enzymes as the starting point for subsequent generations. This process is rapid and cost-effective.

It relies upon high-throughput enzyme assays, because many clones of each enzyme must be screened at each generation. Enzymes capable of functioning in high concentrations of powerful organic solvents have been produced in this way (Arnold, 1996).

The major obstacle to the application of these technologies within the chemical process industry is in the area of training. Process designers and operators are largely trained as chemical engineers, and they are not knowledgeable in enzyme catalysis. Enzyme manufacturers are attempting to make their products so like conventional chemical catalysts that chemists and chemical engineers do not need to learn new skills, but the optimal rate of penetration of biocatalysis into the chemical process industry would be achieved if chemical engineers were trained in the use of enzymes.

Questions, comments and answers

Q: Can you give an example of how this technology can help in water conservation?

A: If you have difficulty degrading materials, there are some excellent hydrolytic enzymes that can be used to degrade them. We feed enzymes to cattle to improve their digestion, and that constitutes waste treatment. Also, nature uses enzymes as anti-microbials, in tears, for example. We could have a similar process which uses a more natural solution to destroy bacteria or viruses.

Q: How might enzymes be applied?

A: As immobilised reactants in contained systems.

Q: In most cases of mutagenesis, activity usually decreases. What are the chances of generating a better enzyme in this way?

A: If you screen a few thousand random mutants, you will always have two or three with higher activity in the desired dimension and thus have additive effects. For example, an enzyme has been selected which works in 80 per cent dimethyl formamide. Nature never had an enzyme that did that.

REFERENCES

AMANN, R., W. LUDWIG, and K-H. SCHLEIFER (1995), "Phylogenetic identification and *in situ* detection of individual microbial cells without cultivation", *Microbiol. Rev.* 59, p. 143.

ARNOLD, F.H. (1996), "Directed Evolution: Creating Biocatalysts for the Future", *Chemical Engineering Science* 51, pp. 5 091-5 102.

LASKIN, A.I. (ed.) (1985), *Enzymes and Immobilized Cells in Biotechnology*, The Benjamin/Cummings Publishing Company, Inc., Menlo Park, California.

ROBERTSON, D.E., E..J. MATHUR, R.V. SWANSON, B.L. MARRS, and J.M. SHORT (1996), "The discovery of new biocatalysts from microbial diversity", *SIM News* 46, pp. 3-8.

WOESE, C.R. (1987), "Bacterial Evolution", *Microbiol. Rev.* 51, pp. 221-271.

WOESE, C.R., O. KANDLER, and M.L. WHEELIS (1990), "Towards a natural system of organisms: Proposal for the domains Archaea, Bacteria, and Eucarya", *Proc. Nat. Acad. Sci. USA* 87, pp. 4 576-4 579.

MICROBIAL METAL ACCUMULATION AND MICROSOLID SEPARATION FOR TREATING AND RECOVERING METALS FROM INDUSTRIAL RINSEWATERS

by

Bradley M. Tebo*, Rajagopalan Ganesh, William K. Tolley**, and Ann E. Grow****
*Marine Biology Research Division and Center for Marine Biotechnology and Biomedicine
Scripps Institution of Oceanography, University of California-San Diego, California, United States
**Biopraxis, Inc., San Diego, California, United States

Introduction

Micro-organisms interact with metals in a variety of ways that lead to the accumulation of the metal on or in the cell and the removal of the metal from solution. These interactions can be either passive in which the sorption of metals or biosorption (Tsezos, 1990) is mediated by various functional groups that can be present on either live or dead microbial cells, or active, in which living cells or "active" cell materials (e.g. enzymes) take up, precipitate or transform metals from a dissolved to solid form.

Extracellular metal precipitation results from the production of a metabolite that combines with a metal ion, metal complex, or other metal-containing compound forming a solid phase (i.e. insoluble precipitate). This precipitation reaction is dependent on the solubility product (K_{sp}) of the precipitated salt: when the product of the concentrations of the dissolved species exceeds the K_{sp} of the solid then the formation of the solid is thermodynamically favoured. Two of the major classes of these precipitates are metal sulfides and metal phosphates.

A variety of oxidation and reduction reactions are catalysed by living micro-organisms or, in some cases, by reactive sites on their cell surfaces (Beveridge and Murray, 1980), and often these redox transformations bring about the precipitation of solid phases because the new species has reduced solubility. Reduction of elements such as chromium (Cr), selenium (Se) and uranium (U), or oxidation of iron (Fe) and manganese (Mn) can lead to insoluble or low solubility materials.

Micro-organisms can also accumulate and precipitate metals within cells as a result of specific enzymes or matrices. The classic example is the formation of magnetite within magnetosomes in magnetotactic bacteria (Blakemore, 1982). Intracellular accumulation can also occur without mineral formation. Micro-organisms require certain metals as nutrients and therefore have mechanisms to take them up from the environment. In addition, non-essential metals are often taken up, usually non-specifically through energy-dependent transport systems intended for other metals. Cadmium, aluminum, and zinc have all been reported to be accumulated as phosphate or polyphosphate granules or bodies in prokaryotes (Higham *et al.*, 1984; Higham *et al.*, 1986; Gadd, 1988; Sakurai *et al.*, 1990).

Biological based technologies for the detoxification and removal of the variety of metals and radionuclides that exist have received great interest in relation to the treatment of wastes and for the remediation of pollution (Tebo, 1995). Most biological processes to date have employed sorption onto a variety of different types of biomass, including bacteria, yeast, fungi, algae and their biopolymers, to remove the metals and radionuclides from solution. More recently, living systems exploiting active metal precipitation processes are receiving more attention for metal removal, particularly in the decontamination of wastes where metals (and metalloids) are not at levels toxic to the desirable organism. For example, different bacteria including sulfate reducing and dissimilatory iron reducing bacteria are able to reduce Cr(VI) to Cr(III) or U(VI) to U(IV), forms that are much less soluble and hence less toxic. Thus this process may be exploited for the remediation of contaminated waters and soils (Lovley and Phillips, 1992; Lovley, 1993; Mehlhorn *et al.*, 1994).

The high metal content of many industrial rinsewaters poses a particularly challenging problem for treatment. In California in 1987, hazardous heavy metal wastes were produced by industries at a rate of greater than 127 million kg per year. A relatively small number of industries generate most of California's metal-bearing wastes, such as metalworking foundry activities, surface cleaning and stripping, surface treatment (e.g. chromating, anodising, and passivating), electroplating and electroless plating, draining and rinsing, and coating operations; printed circuit board and semiconductor manufacturing; and photofinishing and printing. Metals frequently found in California's waste streams include cadmium, chromium, lead, copper, nickel, silver, and cobalt (Table 1).

Table 1. **California industries that use listed metals**

Industry	Number of businesses in California	Examples of metals used
Metal mining	123	As, Ba, Be, Cd, Cr, Cu, Pb, Ni, Se, Ag, V, Zn
Oil and gas extraction	1 038	Ba, V
Nonmetallic minerals, except fuels	325	V
General contractors/builders	14 583	Pb, V
Heavy construction, except buildings	2 533	V
Contractors, special trade	28 536	V
Lumber & wood products	2 406	As, Cr, Cu, V
Publishing & printing	6 585	Ag
Chemicals & allied production	1 440	Sb, As, Ba, Cd, Cr, Co, Pb, Hg, Ni, Se, Ag, V, Zn
Petroleum refining	249	Pb, Ni, V
Rubber & misc. plastic products	1 931	Sb, Zn
Stone, clay & glass products	1 658	Sb, As, Cr, Pb, Se
Primary metals industry	782	As
Fabricated metal products	4 522	Cd, Cr, Cu, Pb, Ni, Zn
Machinery, except electric	7 389	Sb, Cr, Co, Cu, Ni, Ag, V, Zn
Electric & electronic equipment	3 509	Sb, As, Be, Cd, Cr, Co, Cu, Pb, Hg, Ni, Se, Ag, Zn
Transportation equipment	1 625	Be, Co, Cu, Pb, Ni, V, Zn
Instruments & related products	1 623	Hg, Ag
Miscellaneous manufacturing	2 033	Ag
Pipelines, except natural gas	44	V
Electric, gas & sanitary services	1 370	Be
Miscellaneous repair services	6 140	Cr
ALL ESTABLISHMENTS	110 550	

Source: U.S. Department of Commerce, 1985; U.S. Department of the Interior, 1985.

Some of the most diverse and complex wastes arise in electric and electronic equipment industries. For example, the wastes generated in the semiconductor, printed wiring board (PWB), and cathode ray tube (CRT) manufacturing industries are composed of mixtures of a variety of acids, organic compounds, anions, metals and other inorganic compounds (Table 2), that are used at different steps in the manufacturing process. These wastes tend to be acidic, which further complicates treatment strategies.

Table 2. **Typical waste constituents from the electric and electronic equipment manufacturing industries**

Class of pollutant	Specific pollutant
Semiconductor industries	
Acids	sulfuric, hydrochloric, hydrofluoric, phosphoric and nitric acids
Organics	acetone, xylene, ethylene glycol, methanol, freon, 1,1,1-trichloroethane, methylethylketone, phenol, toluene, trichloroethylene, ethylbenzene, di- and tri-chlorobenzene, nitriloacetic acid
Anions	sulfate, chloride, fluoride, phosphate, nitrate
Metals	lead, copper, antimony, cobalt, nickel, zinc
Others	ammonium sulfate, ammonium nitrate, chlorine dioxide
Printed wiring board industries	
Acids	sulfuric, hydrochloric, hydrofluoric, phosphoric and nitric acids
Organics	glycol ethers, formaldehyde, acetone, methanol, trichloromethane, ethanol, toluene, ethylene glycol, phenol, trichloroethane, xylene, trichloroethylene
Anions	sulfate, chloride, nitrate
Metals	copper, lead, nickel, zinc
Others	ammonia, ammonium sulfate, ammonium nitrate, barium carbonate
Cathode Ray Tube (CRT) manufacturing industries	
Acids	sulfuric, hydrochloric, hydrofluoric, and nitric acids
Organics	acetone, methanol, 1,1,1-trichloroethane, toluene, methylketone, xylene
Anions	chloride, nitrate, sulfate
Metals	lead, barium, zinc, copper, arsenic, chromium, nickel

Source: U.S. Environmental Protection Agency, 1995.

The primary waste management concern of many metal fabricators and electronics and computer manufacturers appears to be meeting wastewater effluent requirements for discharges (Meltzer *et al.*, 1990). Limitations on the chemical concentrations that can be discharged to sanitary sewers have increased the demands on the plant's industrial waste treatment system. In addition, the potential fines and other penalties associated with violating these discharge requirements have become more severe in recent years. As a result, electronics industries commit significant resources to maintaining their industrial waste treatment systems.

Conventional treatment strategies available for process solution wastes generated by the electronics industry include a complicated treatment train such as pH adjustment, contaminant oxidation or reduction, metal precipitation, flocculation, dewatering, and sludge drying. These conventional processes require that many wastes, such as chromium, cyanide, electroless plating, and printed wiring board wastes, be carefully segregated from other streams in order to reduce waste volumes and avoid the chance that some complexed metals may escape the treatment system.

The electronics industry, especially the high-tech telecommunications and computer industry, is one on which California is basing its hopes for significantly expanding business, increased revenues, and greatly enlarged job markets to replace the rapidly disappearing defense industry. Yet the high costs, liabilities and burdensome paperwork associated with complying with environmental regulations threaten to drive away the very business California most wants to encourage. Clearly, development of economical technologies for hazardous metal waste treatment that would reduce processing cost, minimise the amount of sludge requiring disposal or permit on-site recycling, would be a substantial benefit to industry.

Biopraxis, Inc. of San Diego and the University of California-San Diego/Scripps Institution of Oceanography (UCSD/SIO) have teamed together to develop a family of technologies, referred to as MOP-UP™, that employ the accumulation of metals by micro-organisms from a variety of different media coupled with separation and recovery of those metals. One of the technologies, MicroSolids Separation (MSS™)-MOP-UP™, may provide a simple, cost-effective technology for the recovery and recycling of heavy metals and the simultaneous purification and recycling of rinsewaters and spent plating baths.

MicroSolids Separation-MOP-UP™ (MSS™-MOP-UP™)

MSS™-MOP-UP™ is composed of three steps: 1) removal of metals from solution by micro-organisms; 2) metal separation; and 3) metal recovery.

Metal removal

To remove metals from solution, the bioaccumulation properties of diverse micro-organisms are exploited. Although biosorptive mechanisms enhance the process, the strategy for MSS™-MOP-UP™ primarily exploits active processes that micro-organisms can catalyse: precipitation and redox transformations. But whether passive or active metal removal mechanisms are used, there are a number of factors that are necessary for the technology to be successful (Brierley *et al.*, 1985; Hancock, 1986; Gadd, 1988):

1) The microbes must be able to remove large quantities of metal.

2) Metal removal must be selective enough to remove specific metals or mixtures of metals from solutions which often contain other ions or chemicals that compete with or inhibit metal removal or complex the targeted metal(s).

3) Metal removal must occur over a range of environmental conditions (e.g. pH, temperature, metal concentration).

4) The microbes must be able to do at least as well as other physicochemical procedures, such as chemical precipitation, ion exchange, solvent extraction, and adsorption onto activated charcoal.

5) It must be cost-effective relative to physicochemical treatments.

6) The microbes should be resistant to physical and chemical stresses and biological effects that might affect them in the metal removal process.

In addition, there are a several other factors that would be desirable for such a microbial process and would have economic benefits (Tebo, 1995):

7) The microbes should be re-usable.

8) The value of the metal recovered or the value of the problem being solved must be more than the cost of producing or re-using the organisms.

9) Microbial growth should not have to be sustained.

10) The microbial process should be able to be engineered to improve or "tailor" the process for specific problems.

To date, biosorption has achieved the most interest for applications in metal removal because it does not require live or growing microbes. Thus, one does not have to worry about maintaining activity or growth of cells which can be drastically affected by various environmental parameters, including physical parameters (e.g. temperature), the chemical environment (e.g. pH, redox potential, and the presence of toxic or competitive substances), as well as biological effects (e.g. competition for growth substrates by other microbes, attack by viruses, or predation). Biosorption is also very rapid and efficient and often inexpensive, employing waste biomass from other biotechnologies (e.g. fermentation). Frequently the loaded biomass can be treated so that the valuable elements can be recovered for recycling or waste minimisation. Thus, biosorption has been shown to be a feasible alternative to physical and chemical treatments for removing metal and radionuclide pollutants from liquid waste (Gadd, 1992).

Active metal removal processes have a number of different advantages depending on the application. Frequently, active metal removal processes can reduce metal concentrations to levels lower than those achieved through biosorption and may have a higher degree of metal specificity or metal affinity. Since metals are transformed into particulate phases they are "fixed" and less susceptible to desorption that could be induced by changes in pH, chemical complexation, or competition with other metals. Because the organisms are alive, there is more versatility for the development of tailored processes. For example, there is the potential for the cells to metabolise organometallic compounds or for multiple organisms and processes to be combined into a "synergistic" process for waste treatment. Furthermore, there is the potential for living systems to be used in continuous processes for extended time periods without the need to regenerate the biomass, and living systems may provide additional capabilities for bioremediation of organic pollutants and nitrate (Gadd, 1992). Finally, and perhaps one of the greatest advantages, is that these active microbial metal removal processes are good systems for genetic manipulation. Those that may lend themselves most readily to genetic improvement are those in which the activity is catalysed by proteins that may be modified or hyperexpressed to enhance metal removal processes.

The different active processes, i.e. redox transformations, precipitation reactions, and intracellular accumulation, have different advantages depending on the application. Redox transformations allow the alteration of metal speciation and permit the process to be tailored to those metals whose solubilities vary with redox state. Metal precipitation reactions have applications to a broad spectrum of metal pollutants and mixed wastes. The use of intracellular accumulation is somewhat more limited: since the metals are taken up within the cells, cell disruption is required in order to recover them, and re-use is not possible.

Probably the most critical issue for potential application of an active microbial metal removal process is to evaluate whether it will require growth or just the activity of the microbes. Clearly, the problem will be more complex if growth of the organisms has to be sustained (Brierley *et al.*, 1985). To sustain growth, the treatment system essentially becomes either a batch or a continuous culture. In either case, a variety of different factors can affect growth, including the availability of the necessary nutrients (including metal nutrients) required for growth, the concentration in the medium of the target metal(s) to be removed, the potential for components in the medium to be toxic, to complex, or to compete for uptake of either metals required for growth or the target metal(s). Some of these problems are diminished in continuous flow-through systems because the input of the various chemicals will be relatively constant and the organisms will have a chance to adapt. However, the flow rate will be limited to that with which the organisms can physiologically keep up. For wastes with extremely high levels of substances that potentially could interfere in the bioprocess, a pre-treatment step could be performed and the biological process used as a secondary treatment to further reduce metal concentrations. One advantage of a growing system, however, is that because the organisms are constantly growing and making new cells, the system is more self-sustaining and thus potentially can have a greater capacity for metal removal.

Any treatment system will be substantially simplified if non-growing active microbes, either immobilised or in suspension, can be used. Although some of the same considerations are necessary, as for treatment systems involving growth, not having to sustain growth makes this approach much more attractive for possible application. The active microbial biomass can be prepared under controlled conditions, presumably enabling the production of a product that has high capacity and specificity for target metals. Cells attached to supports can be employed in batch or column systems if they possess good mechanical properties. If the metals can be stripped or separated from the biomass such that the biomass can be re-used and the metal recovered, then the technology becomes even more feasible.

In MSS™-MOP-UP™ a variety of different groups of organisms that use different metal accumulating processes are employed. For example, metals can be precipitated using sulfate-reducing bacteria (SRB) or organo-phosphate utilising bacteria by a process in which the production of a metabolite combines with a metal ion, metal complex, or other metal-containing compound forming a solid phase (insoluble precipitate). Alternatively, bacteria that oxidise or reduce the metal and thereby create a metal species with reduced solubility, such as iron and manganese oxidising bacteria or chromium reducing bacteria, can be used.

Metal separation

Conventional clarifier systems typically include chemical flocculation, coagulation, sedimentation, and media filtration. In MSS™-MOP-UP™, however, no additives are required; the microbes encrusted with their metal precipitates are separated from wastewaters using filtration. A large number of durable micro and ultra filtration membranes have been developed over the last three decades. They are used in disposable cartridges and in crossflow filtration systems. The Biopraxis team has recently invented a new micro-solids separator (MSS™). The MSS™ is capable of separating particulates as small as 0.01 μm at a higher flow rate and lower energy consumption than those typical for conventional crossflow micro or ultra filters, producing purified water and a moist solids fraction in a single step. Thus, both solid particles or emulsions can be highly concentrated, facilitating their recovery and recycling or disposal. MSS™ is ideally suited for applications where fine filtration is required and the concentrate is to be recovered. Even if the metals are never recycled, MSS™-MOP-UP™ will significantly reduce the amount of sludge that is generated and that which requires disposal.

Resource recovery

If the value of the metals being treated is high, then their recovery and recycling from the concentrated "microbial sludge" will further decrease overall treatment and disposal costs. For example, nickel and silver are precious metals that are discarded in significant amounts by the electronics industry and photoprocessing operators. Cost savings can be achieved either by recirculating the recovered metals back into process baths or plating processes or by selling the recovered metal to a metals reclaimer. Recovery of metals even from conventional treatment sludges by off-site commercial plants is already employed to some extent in California, and will likely see rapid growth in the near future. Off-site recovery services include those offered by smelters, hydrometallurgical plants, and ion exchange/electrolytic recovery plants.

On a volume basis, contaminated rinsewater accounts for the majority of the waste from electronic parts, printed circuit boards, and semiconductor manufacturing, plating, and other metal finishing processes. These industrial processes require many clean water rinses after each treatment, to remove plating chemicals from parts prior to the next treatment. The water is usually discarded after one use. In California, water itself is a precious resource. Even if the metals themselves were not recovered and recycled, simply purifying rinse and wastewaters for re-use would save a substantial amount of money by reducing costs for water and sewage, and at the same time, conserve water supplies.

Waste minimisation programs can benefit the environment by preventing the generation of wastes, residues, and contaminants that, if released, could pose a threat. The high costs associated with hazardous waste disposal are making it more attractive for manufacturers to implement measures that prevent the generation of those wastes, or recycle the wastes once they are generated, rather than seek ways to treat and dispose of them. Technologies that permit both recycling of process chemicals and reduction of wastes that have to be disposed will gain great popularity for being a cost-effective way to manage hazardous wastes and reduce the volume of sludge that must be disposed.

Possible bioaccumulation processes for treatment of industrial rinsewaters

Sulfate reducing bacteria have been receiving increased attention in recent years for precipitating heavy metals in a variety of different types of samples and media. Laboratory scale experiments of SRB with wastewaters from the precious metal industry have demonstrated that they can effectively reduce metal concentrations in complex mixed waste media to very low levels (see Table 3) (Ellwood *et al.*, 1992). As one would expect, metals that form insoluble metal sulfide phases, such as silver, mercury, lead, copper and zinc, are removed from solution. However, other metals that do not form insoluble sulfide phases, such as chromium, gold, and ruthenium, were also removed. Apparently other metals can be co-precipitated or adsorbed on the metal sulfide phases (Ellwood *et al.*, 1992).

Table 3. **Results of the treatment of precious metal effluents with *Desulfovibrio***

	Solution 1			Solution 2			Solution 3			Solution 4		
	0d (mg l^{-1})	5d (mg l^{-1})	% removed	0d (mg l^{-1})	5d (mg l^{-1})	% removed	0d (mg l^{-1})	5d (mg l^{-1})	% removed	0d (mg l^{-1})	5d (mg l^{-1})	% removed
Rh	0.6	0	100	23	1.5	93	0.2	0.1	50	2	0.1	95
Ag	15	0.5	97	0.9	0	100	0.6	0	100	0.6	0	100
Ir	0	0		1.9	1	47	0	0		0.5	0	100
Au	1	0.1	90	0.5	0	100	0	0		0.5	0	100
Ru	25	0.8	97	25	11	56	32	0.1	100	2	0.4	80
Pd	2.2	0.7	68	43	15	65	165	50	70	2.2	0.2	91
Os	0.1	0	100	0.5	0	100	0	0		0	0	
Pr	5	1.5	70	61	30	51	32	0.2	99	23	0.2	99
Hg	2.5	0.2	92	1.6	0.8	50	12	0.3	98	1.2	0.2	83
Pb	30	0	100	1	0	100	1.2	0	100	3.5	0	100
Si	4 020	1 545	62	4 515	1 785	60	4 008	1 562	61	2 114	1 325	37
Cr	10	1	90	13	3	77	6	0.8	87	10	1	90
Fe	250	82	67	245	86	65	204	102	50	250	64	74
Ni	32	2	94	29	2	93	13	1	92	72	20	72
Cu	118	0	100	72	2	97	76	0	100	59	0	100
Zn	33	0.5	98	89	23	74	20	1.5	93	1 004	0	100
Sb	2	0.4	80	34	13	62	0.5	0.5	0	0	0	
P	861	238	72	795	265	67	855	242	72	531	50	91
Mn	5	0.5	90	5	0.5	90	3	0.5	83	10	0.5	95
As	182	0.5	100	532	222	58	192	35	82	201	0.5	100
Sn	5	0.5	90	181	60	67	8	0.5	94	5	3	40
Al	14	2	86	17	9	47	5	8	-60	15 208	40	100
Mg	7 053	7 429	-5	7 203	7 400	-3	7 116	7 100	0	14 148	1 350	90
Sr	19	19	0	20	20	0	20	18	10	15	14	7

Source: Modified from Ellwood *et al.*, 1992.

Microbial metal oxidation processes may be another appropriate treatment strategy for certain industrial rinsewaters. One organism that may be particularly appropriate is the marine manganese oxidising marine *Bacillus* sp. strain SG–1. In SG-1, it is the mature (dormant) spores that oxidise Mn(II) (Nealson and Tebo, 1980; Rosson and Nealson, 1982; de Vrind *et al.*, 1986). SG–1 catalyses Mn oxidation under conditions where oxidation is thermodynamically favoured, but kinetically slow; thus no coupling to metabolic energy is involved. SG–1 spores are capable of binding and/or oxidising a variety of different metals, and thus may have applications for solving problems where mixed metals occur. SG-1 is capable of binding or oxidising metals both directly on the spore surface or through the adsorption or oxidation on the Mn oxides that accumulate on the spore surface. Thus, SG-1 spores have metal removal capabilities that exploit active metal precipitation processes as well as passive sorption. To date, we have demonstrated that SG-1 can oxidise cobalt(II) (Tebo and Lee, 1993; Lee and Tebo, 1994) and iron(II) (Tebo and Edwards, unpublished) in addition to Mn(II). The spores also bind copper, cadmium and zinc. From studies of the adsorption of metals onto Mn oxides, we know that a variety of other metals and radionuclides will be removed including uranium, lead, thorium, radon, cadmium, zinc, mercury, chromium, nickel and silver.

SG–1 spores have both a high affinity for metals and a high capacity for metal precipitation (He and Tebo, in press). SG-1 spores oxidise Mn(II) and Co(II) over a wide range of concentrations, from less than nanomolar to greater than millimolar (Rosson and Nealson, 1982; Tebo and Lee, 1993; Lee and Tebo, 1994). Manganese oxides can accumulate on the surface of the spores up to approximately six times their own weight under ideal conditions. Using radioactive tracers of Mn(II) (^{54}Mn) and Co(II) (^{57}Co), we have been able to demonstrate that SG-1 is able to bind and oxidise Mn(II) and Co(II) even at low tracer only additions of radioisotope to HEPES buffer. Both Mn(II) and Co(II) oxidation follow Michaelis-Menten kinetics. The kinetics of Co(II) oxidation as a function of Co(II) concentration are shown in Figure 1. An Eadie-Hofstee plot of the data suggests that SG-1 spores have two oxidation systems for Co(II), a high-affinity–low-rate system (K_M, 3.3×10^{-8N} M; V_{max}, 1.7×10^{-15} M·spore^{-1}·h^{-1}) and a low-affinity–high-rate system (K_M, 5.2×10^{-6} M; V_{max}, 8.9×10^{-15} M·spore^{-1}·h^{-1}). The K_M for the high affinity system (33 nM) translates to an affinity on the order of two parts per billion (ppb) suggesting that SG-1 spores can remove metals (at least Mn and Co) to much lower levels than achieved by other chemical or biological procedures. In fact, we have demonstrated that SG-1 can remove Co(II) to levels less than parts per trillion!

SG-1 spores have a high specificity for catalysing metal precipitation. SG-1 catalyses the oxidation of Mn(II) almost equally well in seawater, with its abundance of potentially competing divalent cations as in freshwater or buffered distilled water. In addition, the oxidation of Mn(II) is not significantly inhibited by mixtures of other metals such as Cu, Cd, Ni, and Zn.

There are a variety of other useful properties of SG-1 spores that offer advantages for applications. It is the inherent nature of bacterial spores to be resistant to physical and chemical agents and stresses. Spores are resistant to heat, UV light, toxic chemicals, detergents, denaturants, reductants and proteolytic enzymes. The fact that spores are a dormant resting stage means that growth of the organisms does not have to be maintained in order to have metal removal activity. In fact, SG-1 spores rendered nonviable (so they don't germinate) still oxidise Mn(II) (Rosson and Nealson, 1982). Thus, if SG-1 were to be applied in the environment, one would not have to worry about their uncontrolled growth, because nonviable spores could be employed. SG-1 is also active over a wide range of conditions of pH (>6), temperature (2-80 °C), and osmotic strength (Rosson and Nealson, 1982; Tebo and Lee, 1993; Lee and Tebo, 1994), so the organism is environmentally versatile. Finally, the spores can be treated to remove the metal oxides, and they maintain their metal oxidising capability (de Vrind *et al.*, 1986). Thus, there is a potential for the spores to be recycled.

Current activities

Biotechnologies employing the active metal accumulating properties of micro-organisms (i.e. bioprecipitation and redox transformations) for metal removal applications are still very much in their research and development phase. Biopraxis/UCSD has recently received funding from the California Office of Strategic Technology (OST) to evaluate and further develop MSS™-MOP-UP™ for treatment of industrial rinsewaters. Rinsewater "simulants" from semiconductor, printed wiring board, and cathode ray tube manufacturing industries are being employed to demonstrate the minimum level of residual heavy metal contamination that can be achieved through microbial metal accumulation. The focus is on the heavy metals, lead, cadmium, silver and chromium. Hexavalent chromium is considered to be especially difficult to treat using conventional techniques, and waste streams containing this pollutant are usually segregated from other waste streams and treated separately.

Figure 1. **Kinetics of cobalt oxidation by SG-1 spores in artificial seawater**

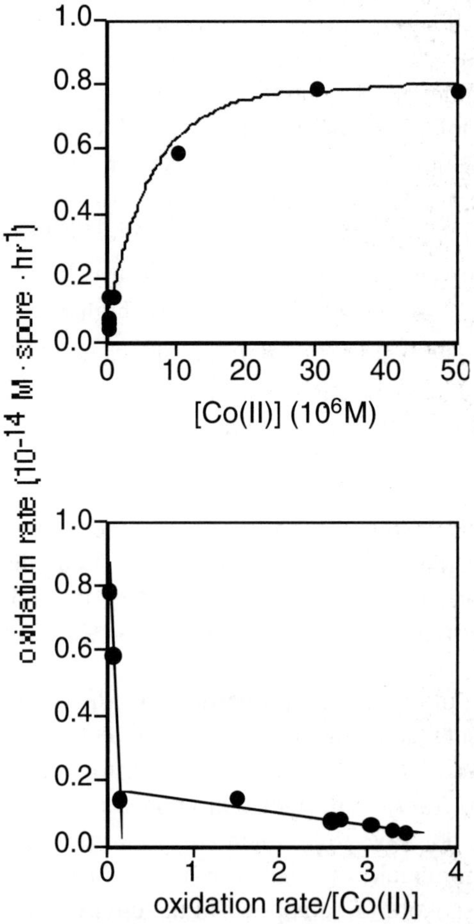

Note: Kinetics of cobalt oxidation by SG–1 spores in artificial seawater. Top, saturation of cobalt oxidation by high concentrations of Co(II) typical of Michaelis-Menten kinetics. Cobalt oxidation occurs even with tracer addition of the radioisotope ^{57}Co(II). Bottom, Eadie–Hofstee plot of the data shown in the top panel indicates that there are two cobalt oxidation sites on the spores, one with high affinity and one with low affinity.

Source: Tebo, 1995.

Initial studies are focusing on screening and selecting microbial metal accumulating processes that will be effective in the rinsewater simulant. In order to evaluate the spores of *Bacillus* sp. strain SG–1 which have been shown to effectively bind a variety of metals in mixed metal solutions, metal removal will be tested at various pHs. Sulfate reducing bacteria, like those reported on by Ellwood *et al.* (1992), will also be tested under anaerobic conditions. Other pure and enrichment cultures of metal accumulating micro-organisms will also be screened. The project will then test the effect of different growth parameters, environmental conditions, and potentially competing ions on metal accumulation. The results of the studies will be used to develop operating system concepts, process flow diagrams, material balances, and operating costs for application of MSS™-MOP-UP™ for treating rinsewaters from the electronic/computer industry.

Future opportunities and recommendations

Microbial biotechnology has an important role to play in water use and conservation. The traditional biotechnological approach for metal removal has employed biosorption, and while this can be effective for a variety of applications (Gadd, 1992), the fact that the metals are passively adsorbed to biomass limits its application. Active processes that are catalysed by specific biological processes offer tremendous opportunity for application. For example, many of these processes are catalysed by specific proteins (enzymes) that function as a part of a metabolic pathway. In some cases, the enzyme alone may be useful, whereas in other cases, the metabolic pathway within a living organism might be exploited. However, in most cases of metal-microbe interactions, fundamental knowledge concerning these processes is lacking. We clearly need a better understanding of the diversity, physiology, biochemistry and molecular genetics of the organisms that accumulate metals.

When one considers the diversity of bacteria, few metal-accumulating organisms have been isolated and characterised. We need to develop a better understanding of the mechanisms involved in metal accumulation, particularly those that involve active processes. Only then will it become more feasible to manipulate the metal-precipitating properties through physiological and genetic means. A better basic understanding will permit molecular geneticists and biochemists to engineer the enzymes and pathways so to better develop tailored treatment processes for metal removal and recovery for a variety of applications. One of the greatest advantages of the active metal accumulation process is the potential to use genetics for strain improvement, because the mechanism of metal removal involves a property of the microbe rather than a product. Since enzymes are often involved in active metal accumulation, a variety of molecular approaches can be taken to enhance the process, for example, over-expression or modification of the proteins/pathways. Depending on the goal, it may be desirable to engineer properties related to the metal accumulation process, such as rate, efficiency, stability, capacity, affinity and specificity, or to combine compatible processes within a single organism. The ultimate goal, the design of an efficient and effective metal removal process, will require exploiting an organism's natural ability, as well as engineering improvements.

The application of microbial biotechnology to water use and conservation also requires new engineering efforts. Thus, communication between biotechnologists/microbiologists and engineers is essential for developing satisfactory technologies that can compete with traditional water treatment strategies. One tremendous advantage of microbial biotechnology is its flexibility. Engineered operating systems have to become equally flexible -- such as those based on MicroSolids Separation.

Acknowledgements

BMT thanks the following agencies for providing research support: The National Science Foundation, the University of California Toxic Substances Research and Teaching Program, the National Sea Grant College Program (National Oceanic and Atmospheric Administration, U.S. Department of Commerce grant NA36RG0537; project R/CZ-123 of the California Sea Grant College), and the California State Resources Agency. Biopraxis, Inc. would like to acknowledge the Department of Energy, Morgantown Energy Technology Center and the State of California Office of Strategic Technology for funding. The views expressed herein are those of the authors and do not necessarily reflect the views of the funding agencies. The U.S. Government is authorised to reproduce and distribute for governmental purposes.

Questions, comments and answers

Q: You showed sulphate-reducing bacteria and then a precipitate of noble metals. How do the bacteria add something beyond the normal solubility product of the metal sulphide they produce?

A: There are a variety of things. One, the precipitate forms at the cell surface, so the cells may have bioadsorptive effects and entrap the metals. Second, the precipitates are very amorphous and are highly disordered, with vacancies in the mineral lattice, and metals may be trapped in the mineral. This is co-precipitation.

REFERENCES

BEVERIDGE, T.J. and R.G.E. MURRAY (1980), "Sites of metal deposition in the cell wall of *Bacillus subtilis*", *J. Bacteriol.* 141, pp. 876-887.

BLAKEMORE, R.P. (1982), "Magnetotactic bacteria", *Ann. Rev. Microbiol.* 36, pp. 217-238.

BRIERLEY, C.L., D.P. KELLY, K.J. SEAL, and D.J. BEST (1985), "Materials and biotechnology", in I.J. Higgins, D.J. Best and J. Jones (eds.), *Biotechnology Principles and Applications*, pp. 163-213, Blackwell Scientific Publications, Oxford.

DE VRIND, J.P.M., E.W. DE VRIND-DE JONG, J.-W.H. DE VOOGT, P. WESTBROEK, F.C. BOOGERD, and R.A. ROSSON (1986), "Manganese oxidation by spores and spore coats of a marine *Bacillus* species", *Appl. Environ. Microbiol.* 52, pp. 1 096-1 100.

ELLWOOD, D.C., M.J. HILL, and J.H.P. WATSON (1992), "Pollution control using microorganisms and magnetic separation", in J.C. Fry, G.M. Gadd, R.A. Herbert, C.W. Jones, and I.A. Watson-Craik (eds.), *Microbial Control of Pollution* 48, pp. 89-112, Cambridge University Press.

GADD, G.M. (1988), "Accumulation of metals by microorganisms and algae", in H.-J. Rehm (ed.) *Biotechnology, A Comprehensive Treatise in 8 Volumes*, pp. 401-433, Volume 6b, VCH Verlagsgesellschaft, Weinheim.

GADD, G.M. (1992), "Microbial control of heavy metal pollution", in J.C. Fry, G.M. Gadd, R.A. Herbert, C.W. Jones and I.A. Watson-Craik (eds.), *Microbial Control of Pollution* 48, pp. 59-88, Cambridge University Press.

HANCOCK, I.C. (1986), "The use of Gram-positive bacteria for the removal of metals from aqueous solution", in R. Thompson (ed.), *Trace Metal Removal from Aqueous Solution*, pp. 25-43, Special Publication No. 61, The Royal Society of Chemistry, London.

HE, L.M. and B.M. TEBO (in press), "Surface characterization of and Cu(II) adsorption by spores of a marine *Bacillus*", *Environ. Sci. Technol.*

HIGHAM, D.P., D.P. SADLER and M.D. SCAWEN (1984), "Cadmium-resistant *pseudomonas putida* synthesizes novel cadmium proteins", *Science.* 225, pp. 1 043-1 046.

HIGHAM, D.P., P.J. SADLER and M.D. SCAWEN (1986), "Cadmium-binding proteins in *pseudomonas putida*: Pseudothioneins", *Environmental Health Perspectives.* 65, pp. 5-11.

LEE, Y. and B.M. TEBO (1994), "Cobalt oxidation by the marine manganese(II)-oxidizing *Bacillus* sp. strain SG-1", *Appl. Environ. Microbiol.* 60, pp. 2 949-2 957.

LOVLEY, D.R. (1993), "Dissimilatory metal reduction", *Ann. Rev. Microbiol.* 47, pp. 263-290.

LOVLEY, D.R. and E.J.P. PHILLIPS (1992), "Bioremediaton of uranium contamination with enzymatic uranium reduction", *Environ. Sci. Technol.* 26, pp. 2 228-2 234.

MEHLHORN, R.J., B.B. BUCHANON, and T. LEIGHTON (1994), "Bacterial chromate reduction and product characterization", in J.L. Means and R.E. Hinchee (eds.), *Emergining Technology for Bioremediation of Metals*, pp. 26-37, Lewis Publishers, Boca Raton.

MELTZER, M., M. CALLAHAN, T. JENSEN, CA DEPARTMENT OF HEALTH SERVICES, and U.S. ENVIRONMENTAL PROTECTION AGENCY (1990), "Metal-bearing Waste Streams: Minimizing, Recycling and Treatment", Noyes Data Corporation, Park Ridge, NJ, *Pollution Technology Review* 196.

NEALSON, K.H. and B. TEBO (1980), "Structural features of manganese precipitating bacteria", *Origins of Life* 10, pp. 117-126.

ROSSON, R.A. and K.H. NEALSON (1982), "Manganese binding and oxidation by spores of a marine bacillus", *J. Bacteriol.* 151, pp. 1 027-1 034.

SAKURAI, I., Y. KAWAMURE, H. KOIKE, Y. INOUE, Y. KOSAKO, T. NAKASE, Y. KONDOU, and S. SAKURAI (1990), "Bacterial accumulation of metallic compounds", *Appl. Environ. Microbiol.* 56, pp. 2 580-2 583.

TEBO, B.M. (1995), "Metal precipitation by marine bacteria: potential for biotechnological applications", in J.K. Setlow (ed.), *Genetic Engineering–Principles and Methods*, pp. 231-263, Vol. 17, Plenum Press, New York.

TEBO, B.M. and Y. LEE (1993), "Microbial oxidation of cobalt", in A.E. Torma, J.E. Wey, and V.L. Lakshmanan (eds.), *Biohydrometallurgical Technologies* I, pp. 695-704, The Minerals, Metals, & Materials Society, Warrendale, PA.

TSEZOS, M. (1990), "Engineering aspects of metal binding by biomass", in H.L. Ehrlich and C.L. Brierley (eds.), *Microbial Mineral Recovery*, pp. 325-339, McGraw-Hill Publishing Co., New York.

U.S. DEPARTMENT OF COMMERCE (1985), "County Business Patters 1983 – California", Bureau of the Census.

U.S. DEPARTMENT OF THE INTERIOR (1985), "Mineral Facts and Problems", Bureau of Mines, Bulletin 675.

U.S. ENVIRONMENTAL PROTECTION AGENCY (1995), EPA Office of Compliance Sector Notebook Project, Profile of the Electronics and Computer Industry, U.S. Environmental Protection Agency, EPA/310-R-95-002.

METAL REMOVAL FROM LIQUID EFFLUENTS -- WHY SELECT A BIOLOGICAL PROCESS?

by

Harry Eccles
Remediation Technologies Group, British Nuclear Fuels, plc, Preston, United Kingdom

Introduction

The removal/recovery of metals from aqueous solution has been practised for more than five thousand years, since Moses sweetened water using wood. Since Moses' demonstration, scientists and engineers have developed an arsenal of separation technologies for both process and environmental applications for a diversity of industries, such as petrochemicals, pharmaceuticals, food, metal and nuclear.

Environmental protection is now one of the key considerations by governments, industry and the general public. It is predicted that the environmental technologies market will increase by about five per cent per annum, from US$ 200 billion to US$ 300 billion by the end of this century.

It was estimated that environmental expenditure in the United Kingdom in 1994, by the mining/quarrying, energy and water supply and manufacturing industries, was a total of £ 2.34 billion. Further analysis of this trend and expenditure are provided.

Nowadays research attention has focused on two key strategies, namely the development of cleaner technologies and "end-of-pipe" solutions. Interest in the latter has been stimulated by increasing, more stringent legislation now being required by EC Directives. The application of metal biosorptive and bioaccumulative techniques have found little favour in resolving some of industry's waste management problems; why is this? In response to the question, this paper will briefly review the favoured separation technologies currently used by industry and those under development. An explanation of biosorption and bioaccumulation of metals is presented, and the criteria which should be considered when selecting a liquid effluent clean-up process are described.

Some cost comparisons of metal removal processes are presented. The potential for biological processes to capture a greater market share is explained.

Liquid waste management in the United Kingdom

The most significant sector in terms of environmental expenditure in the United Kingdom in 1994 was the chemical sector (HMSO, 1996). This industry accounts for 21 per cent of gross environmental expenditure. Other significant sectoral expenditure is in the paper and pulp industry (13 per cent) and food processing (14 per cent). Sectoral expenditure is summarised in Table 1. Of the £ 2 340 million environmental expenditure, £ 1 010 million (43 per cent) was accredited to water protection, while air pollution control accounted for £ 670 million (28 per cent of total).

Table 1. **Sectoral environmental expenditure, 1994**

Sector	Capital expenditure		Current expenditure		Total expenditure	
	(£m)	(%)	(£m)	(%)	(£m)	(%)
Chemicals	260	22	240	21	500	21
Food processing	80	7	250	22	330	14
Paper & pulp	100	8	200	18	300	13
Minerals processing	160	13	30	2	190	8
Energy supply	80	7	70	6	150	7
Metals manufacture	80	7	50	4	130	6
Rubber/plastics	70	5	20	2	90	4
Other sectors	370	31	280	25	650	28
Total expenditure	1 200	100	1 140	100	2 340	100

Source: HMSO, 1996.

It is not surprising that water protection accounts for a significant proportion of the total environmental expenditure, as there are some 250 000 liquid effluent treatment facilities in the United Kingdom, operated by industry, public utilities and other organisations. Expenditure in other environmental media is presented in Table 2.

Table 2. **Environmental expenditure by media, 1994**

Environmental media	Capital expenditure		Current expenditure		Total expenditure	
	(£m)	(%)	(£m)	(%)	(£m)	(%)
Water protection	330	28	680	60	1 010	43
Air pollution control	590	49	80	7	670	28
Waste management	150	12	370	32	520	22
Noise pollution control	30	2	0	0	30	1
Other (non-attributed)	110	10	10	10	110	5
Total expenditure	1 200	100	1 140	100	2 340	100

Source: HMSO, 1996.

These facilities have to treat a diversity of pollutants, ranging from complex organic molecules to inorganic materials, or a combination of both. It is the composition of these effluents which largely influences the selection of the effluent treatment process.

Process criteria

By definition, the end-of-pipe solution has to be compatible with existing operations, and therefore has to comply with several criteria. The major criteria are listed below, and those that contribute directly to cost effectiveness and environmental considerations will have by far the most significant influence on the selection of the clean-up process. The installation of an effluent process which generates significant quantities of secondary waste, or smaller quantities of a waste which is more toxic and less amenable to handling/disposal, has not resolved the problem.

The effluent clean-up process should be:

- compatible with existing operations, in particular with upstream processes;
- cost-effective, as environmental processes add to overall production costs;
- efficient, to meet existing and future legislation;
- flexible, to handle fluctuations in quality and quantity of effluent feed;
- reliable, as most effluent processes will operate continuously;
- robust, to minimise supervision and maintenance;
- selective, to remove only the contaminants (metals) under consideration;
- simple, to minimise automation and the need for skilled operators.

Invariably, effluent processes have to handle a variable feed stream both in quantity and quality. Industrial and household liquid wastes often show a great variability in chemical and physical parameters: pH values ranging from strongly acidic to basic values; simultaneous presence of inorganic and organic components; dissolved and volatile species, colloids, emulsions and particles (Table 3 presents some typical industrial wastewater characteristics).

Processes which are incapable of dealing with such perturbations tend to rely on upstream buffer storage facilities to blend effluent arisings, smoothing such perturbations. Provision of such buffer storage facilities will add significantly to the cost of effluent clean-up. With solid waste disposal costs increasing, it is now more important to attempt the recycling of the recovered metals. This may require the selective removal of the metal(s), and hence selectivity may automatically be required of the clean-up process.

Table 3. **Typical wastewater characteristics**

Parameter	Effluent A	Effluent B	Effluent C
Temperature °C	Ambient	Up to 40	Ambient
Salinity (% w/v)	3 to 5	1.5 to 5	0.20
pH value	3 to 4	5 to 6	2
Major substrates	Solvents (e.g. methanol, ethanol, acetone, dimethyl formamide), acetic acid TOC 1 000 to 5 000 mg/l	Volatile fatty acids (acetate, proprionate) TOC approximately 1 000 mg/l	Solvents (methanol pyridine, dimethyl formamide) TOC 500 to 600 mg/l
Difficult compounds	Chlorinated, sulphonated and nitrated aromatics, colours	-	Brominated and chlorinated organics
Toxic substances	Phenolics, aromatic amines, agrochemicals, biocides, sulphides	Amines, biocides, mercaptans, sulphides, oil	Aromatic amines, agrochemicals, biocides/bactericides
Nutrients	P deficient	N and P deficient	-
Metals	Various (Cu, Ni, Fe)	Many	Cu, Ni, Zn, etc.
Other inorganics	Sulphate	Sulphates, boron	Sulphates

Source: Private communication.

The use of best environmental option techniques are now being employed to minimise the costs and timescales in the development of effluent clean-up processes. These techniques generally employ mathematical models and other computer predictions to select the more appropriate solution(s) against process and environmental criteria. They will not, however, eliminate the need for both laboratory and upscale demonstration studies.

Separation technologies for metal removal

Metals are a class of pollutants, often toxic or dangerous, widely present in industrial and household wastewaters. Electro-plating and metal finishing operations, electronic circuit production, and steel and aluminium processes, to name but a few industries, produce a large quantity of wastewaters containing metals. Although metal precipitation using a cheap alkali such as lime (calcium hydroxide) has been the most favoured option, other separation technologies are now beginning to find favour. Precipitation, by adjusting the pH value, is not selective, and any iron (ferric ion) present in the liquid effluent will be precipitated initially, followed by other heavy metals, as illustrated in Figure 1. Consequently, precipitation produces large quantities of solid sludge for disposal; for example, precipitation as hydroxides of 100 mg/l of copper (II), cadmium (II), or mercury (II) produces as much as ten-, nine- and five-fold mg/l of sludges respectively.

Figure 1. **Precipitation of metals with caustic soda**

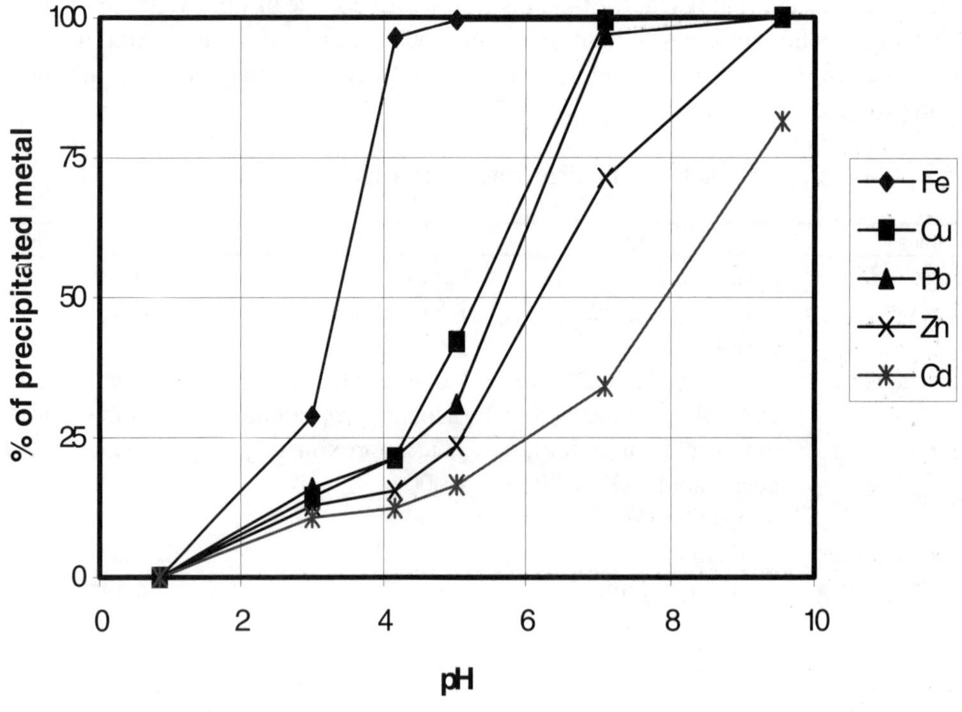

Source: Fabiani, 1992.

Nonetheless, precipitation processes can be highly efficient as initially they rely on solubility products and on the effectiveness of solid-liquid separation. The former, however, can be influenced by the presence of complexing agents such as cyanides or if the metal is capable of existing in an anionic form e.g. chromium as chromates. Solid-liquid separation can be improved by the use of polyelectrolytes or flocculants such as aluminium, but at the expense of sludge generation.

The performance characteristics of non-biological heavy metal wastewater treatment technologies are identified in Table 4. It should be emphasized that other parameters may require consideration during the selection process. The performance characteristics reported in Table 4 are comparative and indicative.

Table 4. **Performance characteristics of heavy metal removal/recovery technologies**

Technology	Performance characteristics				
	pH change	Metal selectivity	Influence of suspended solids	Tolerance of organic molecules	Working level for appropriate metal (mg/l)
Adsorption, eg. granulated activated carbon (GAC)	Limited tolerance	Moderate	Fouled	Can be poisoned	<10
Electrochemical	Tolerant	Moderate	Can be engineered to tolerate	Can be accommodated	<10
Ion exchange	Limited tolerance	Chelate-resins can be selective	Fouled	Can be poisoned	<100
Membrane	Limited tolerance	Moderate	Fouled	Intolerant	>10
Precipitation					
a. Hydroxide	Tolerant	Non-selective	Tolerant	Tolerant	>10
b. Sulphide	Limited tolerance	Limited selective-pH dependent	Tolerant	Tolerant	>10
Solvent extraction	Some systems pH tolerant	Metal selective extractants available	Fouled	Intolerant	>100

Source: Eccles, 1995.

The versatility, simplicity and other technology characteristics will contribute to the overall process costs, both capital and operational. Some typical values are presented in Table 5.

Table 5. **Evaluated capital and operating costs for industrial wastewater treatment processes**

Technology	Costs ($/m^3)		
	Capital		Operating
	A	B	
Precipitation*	12.5	8	0.003-0.013
Adsorption (GAC)	500	250	0.020-0.050
Membrane (microfiltration)	12.5	11	0.013-0.050
Ion exchange	100	75	0.050-0.250

Notes: Case A is for a plant with a process capability of up to 1 000 m^3/d, whereas Case B is for a 10-20 000 m^3/d facility; *Precipitation includes neutralisation, coagulation, flocculation and separation.

Source: Fabiani, 1992.

It is difficult, if not impossible, to obtain comparable cost data for different technologies for a given liquid effluent clean-up system. Consequently, the values quoted in Table 5 are indicative for comparable systems. The cost values quoted will also be dependent on other parameters, such as:

- concentration of metals in solution;
- operational mode of the equipment;
- secondary treatments needed, such as regeneration of GAC or ion exchange resins;
- selectivity of GAC's or ion exchange resin, coupled with their respective capacities for given metal(s);
- disposal of secondary wastes such as sludges.

At present, many of the above technologies such as ion exchange are firmly established, well understood and represent significant capital investment by industry. Biological systems are therefore most likely to succeed in areas where established competition does not exist or where a significant advantage can be identified, e.g. at low metal concentrations or where different selectivities are required.

Biosorption and bioaccumulation of metals

Biosorption/bioaccumulation of metals by micro-organisms is neither novel and/or new; nature has been using these fascinating and intricate interactions for several thousand years.

Scientists, however, have been attempting, and in many instances succeeding, to understand these subtleties for only five or six decades whilst the process technologists are still wondering how we can benefit by or put to use this ingenious science. So what is biosorption and bioaccumulation of metals? I have defined biosorption as "the passive sorption and complexation of metal ions by microbial biomass, or material derived from this", whereas bioaccumulation includes all "processes responsible for the uptake of metal ions by living cells", and thus includes biosorptive mechanisms, together with intracellular accumulation and bioprecipitation mechanisms. These mechanisms are responsible for the formation of ferric and manganese oxide deposits by organism including *Metalogenum* and *Gallionella*.

The mechanisms by which micro-organisms can remove heavy metals from solution have been reviewed extensively elsewhere (Gadd, 1988; Macaskie and Dean, 1989) and are covered in part by other authors in these proceedings. Consequently, this section will compare and contrast the characteristics of metal biosorptive and bioaccumulative processes only. The comparison of the two processes will hopefully assist in the selection of which one to use in a specific circumstance, although it should be emphasised that other factors could be equally or more important in the selection procedure.

The major characteristics of metal biosorption and bioaccumulation are presented in Table 6.

Table 6. **Metal biosorption and bioaccumulation characteristics**

Feature	Biosorption	Bioaccumulation
Metal affinity	high under favourable conditions	toxicity will affect metal uptake by living cells, but in some instances high metal accumulation
Rate of metal uptake	usually rapid, a few seconds for outer cell wall accumulation	usualy slower than biosorption
Selectivity	variety of ligands involved, hence poor	better than biosorption, but less than some chemical technologies
Temperature tolerance	within a modest range	inhibited by low temperatures
Versatility	metal uptake may be affected by anions or other molecules	requires an energy source
	extent of metal uptake usually pH dependent	dependent on plasma membrane ATP-ase activity
		frequently accompanied by efflux of another metal

Source: Eccles, 1995.

While the use of living organisms is often successful in the treatment of toxic organic contaminants, living organisms in conventional biological treatment systems generally have not been useful in the treatment of solutions containing heavy metal ions. Once the metal ion concentration becomes too high, or sufficient metal ions are absorbed by the micro-organism, the organism's metabolism is disrupted, thus causing the organism to die. This disadvantage does not exist if non-living organisms or biological materials derived from micro-organisms are used to absorb metal ions from solution. Biosorption is thus surrogate ion exchange, and the biomass is behaving as the exchange resin.

The biomass, however, unlike mono-functional ion exchange resins, will contain a number and variety of functional sites. These sites, contributed by the cell biopolymers, include carboxyl, imidazole, sulphydryl, amino, phosphate, sulphate, thioether, phenol, carbonyl, amide and hydroxyl moieties (Bedell and Darnall, 1990). Various algal species and cell preparations have quite different affinities for different metal ions (Darnall *et al.*, 1986). The unusual metal binding properties exhibited by individual algae species are explained by the fact that various genera of algae have different cell wall compositions. Thus, certain algal species may be much more effective and selective than others for removing particular metal ions from aqueous solution.

It is not unexpected therefore that biosorptive processes have received greater attention and aroused greater interest than bioaccumulative ones. The need to supply nutrients, accommodate increasing biomass within the chosen reactor, coupled with generally slower kinetics, adds to the burdens of process engineering. Although metal bioaccumulative processes have appeared to be at a disadvantage, there are one or two excellent examples whereby the above parameters have been overcome with some engineering and process control foresight. One such example is the Shell process using sulphate reducing bacteria (Barnes, 1991).

The way forward

Removal of metals from industrial liquid waste streams is currently achieved using largely physio-chemical and chemical processes. The selection is largely based on economics, but other factors such as compatibility and effectiveness will have been considered.

With few exceptions, existing effluent processes are not capable of simultaneously treating both organic pollutants and toxic heavy metals, but invariably industrial effluents will contain a diversity of pollutants, as indicated in Table 3.

As yet relatively untapped is the ability of micro-organisms to tackle simultaneously two problems, a solution not possible using conventional chemical processes. This is exemplified by reference to the Shell process (Barnes, 1991), in which both metal ions and sulphate contaminants are reduced/removed in a single operation. It is this versatility which is unique to biotechnological solutions.

Questions, comments and answers

Q: Can you give some examples of simultaneous removal of organics and heavy metal inorganics by biological systems?

A: Metals and sulphates, yes, and also nitrates and metals, but not yet organics and inorganics, although these are being developed. This is a very important area.

REFERENCES

BARNES, L.J. (1991), "Treatment of Aqueous Waste Streams", EP 0436, 254 A1, 10 July.

BEDELL, G.W. and D.W. DARNALL (1990), "Immobilisation of Non-Viable Biosorbent Algal Biomass for the Recovery of Metal Ions", in B. Volesky (ed.), *Biosorbents and Biosorption - Recovery of Heavy Metals*, pp. 313-326, CRC Press, Boca Raton, FL.

DARNALL, D.W., B. GREEN, M. HOSEA, R.A. McPHERSON, M. HENZL, and M.D. ALEXANDER (1986), "Recovery of Heavy Metal Ions by Immobilised Alga", in R. Thompson (ed.), *Trace Metal Removal from Aqueous Solution*, pp. 1-24, Special Publication No. 61, Royal Soc Chem, London.

ECCLES, H. (1995), "Removal of Heavy Metals from Effluent Streams - Why Select a Biological Process?", *Int. Biodeterioration and Biodegradation* 35:1-3, pp. 5-16.

FABIANI, C. (1992), "Recovery of Metal Ions from Waste and Sludges", ENEA, ISSN/1120-5555.

GADD, G.M. (1988), "Accumulation of metals by Mircoorganisms and Algae", in H.J. Rehm and G. Reed (eds.), *Biotech Vol 6B, Special Microbial Processes,* VCH, Verlagsgesllschaft, Weinheim.

HER MAJESTY STATIONERY OFFICE (HMSO) (1996), "Environmental Protection Expenditure by Industry", ISBN 0117533009.

MACASKIE, L.E. and A.C.R. DEAN (1989), "Microbial Metabolism, Desolubilisation and Deposition of Heavy Metals: Metal Uptake by immobilised cells and application to detoxification of liquid wastes", in A. Mizrahi (ed.), *Advances in Biotech Processes, Vol. 12, Biological Waste Treatment*, Alan R. Liss, New York.

UASB TREATMENT OF FOOD WASTEWATERS WITH BIOLOGICAL SULFUR REMOVAL

by

Sosuke Nishimura
Kurita Water Industries, Ltd., Atsugi, Japan

Introduction

Since Lettinga *et al.* (1980) introduced the concept of Upflow Anaerobic Sludge Blanket (UASB) process, a lot of successful applications of UASB have been reported, especially in treatment of food industry wastewaters, such as brewery wastewater and potato processing wastewater. The major advantages of the UASB are high organic loading rate, low operating cost, and capability of methane gas recovery. Even though recent awareness of global environment protection sometimes criticises that methane gas is one of the substances responsible for the greenhouse effect, it is obvious that recovery of energy from wastewater is by no means worse than using a conventional activated sludge process, that consumes huge amount of electricity and also emits carbon dioxide.

When the methane gas is utilised as a fuel, it is needed to remove hydrogen sulfide (H_2S), which is produced from sulfate in the wastewater by sulfur reducing bacteria. This is to prevent corrosion problems in boilers and reduce air pollution caused by SO_x in the exhaust gas. Conventional technology for H_2S removal is chemical absorption. However, chemical absorption requires high operating costs for disposal of waste chemicals, as well as the cost of the chemicals themselves. To overcome this limitation, the bioscrubber can be an effective and economical option.

Principles of biological H_2S removal

Sulfur oxidising bacteria can degrade H_2S to sulfate under aerobic conditions (ASM, 1974). This property is widely utilised in the biological odor removal process (Kanagawa and Mikami, 1989). In the deodorization of the air, two essential steps, i.e. the gas-absorption into the liquid phase and the oxidisation, may take place in one reactor. However, when the H_2S removal from the biogas is concerned, the oxidation step must be separated from the absorption process to avoid mixing air into the biogas. So to meet this demand, a two-reactor system, which consists of a gas-liquid contact tower (bioscrubber) and an aeration tank, is employed (Honda and Fukuyama, 1988). In the contact tower, H_2S in the gas is absorbed by the mix liquor of activated sludge, which is continuously fed to and withdrawn from the contact tower. The effluent sludge from the tower, which contains H_2S, is introduced to an aeration tank, where the H_2S is oxidised to sulfate by sulfur oxidising bacteria, such as *Thiobacillus* sp. This process can be efficiently applied to the biogas from the UASB process, because UASB is often followed by an activated sludge process, which can be used also for oxidising H_2S, as shown in Figure 1.

Figure 1. **Flow diagram of UASB process with biological sulfur removal**

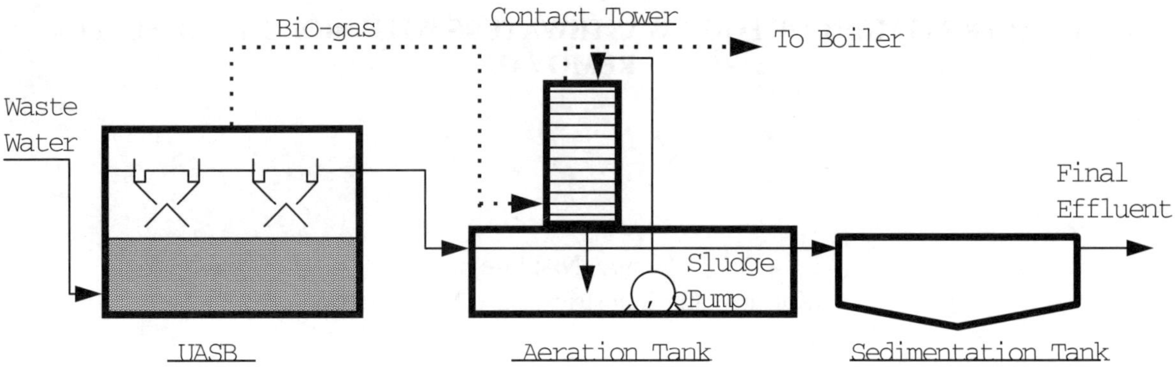

Source: Author.

Design criteria

When designing a bioscrubber, both absorption efficiency and prevention of clogging problems should be considered. Among several types of contact towers, the multiple-bubble-tray contact tower was chosen because it had fairly good absorption efficiency and less risk of clogging. Each tray keeps a sludge-liquid layer upon it, and the trays have many small holes through which the gas passes upward to make bubbles in the sludge layer. The sludge liquid on the tray flows downward through a pipe that connects one tray to the next. The size and the number of the trays were calculated to satisfy a desirable H_2S concentration in the off-gas. Practically, the H_2S concentration in the off-gas is set at 100 ppm, which is recommended to avoid corrosion problems in boilers. A computer simulation model which incorporated material-balances and gas/liquid transfer rate of H_2S was used for the design calculations. Although oxidation of H_2S is expected to take place in the aeration tank outside the contact tower, a considerable amount of partial oxidation of H_2S to elementary sulfur was observed in the contact tower in a pilot-scale experiment (Nishimura and Yoda, 1995). Thus, this in-tower oxidation capacity of the sludge was also taken into account for the simulation, using reaction constants obtained from the pilot experiment. A typical dimension of the contact tower is shown in Figure 2.

The sulfur oxidation tank was designed based on the oxidation rate constants measured in the pilot-scale experiment. H_2S is oxidised in the two steps as shown in Table 1.

Table 1. **Sulfur oxidation steps and rate constants**

	Reaction	Rate constant at 25 pH7
Step 1	$2HS^- + O_2 \rightarrow 2S^0 + 2OH^-$	2 400 mg-S/g-VS/day
Step 2	$2S^0 + 3O_2 + 4OH^- \rightarrow 2SO_4^{2-} + 2H_2O$	190 mg-S/g-VS/day

Note: Rate constants were obtained from a pilot-scale experiment.

Source: Nishimura and Yoda, 1995.

When the sulfur oxidation tank works also as a BOD remover for the effluent of UASB, the volume of the aeration tank should be larger than that required solely for sulfur oxidation. In this case, calculation factors should be determined by consideration of the sulfur/BOD ratio in the loading rate to the aeration tank. However, in many cases, the volume calculated simply based on the BOD loading rate to the aeration tank is sufficiently large for simultaneous sulfur oxidation.

Figure 2. **Typical dimension of full-scale bioscrubber**

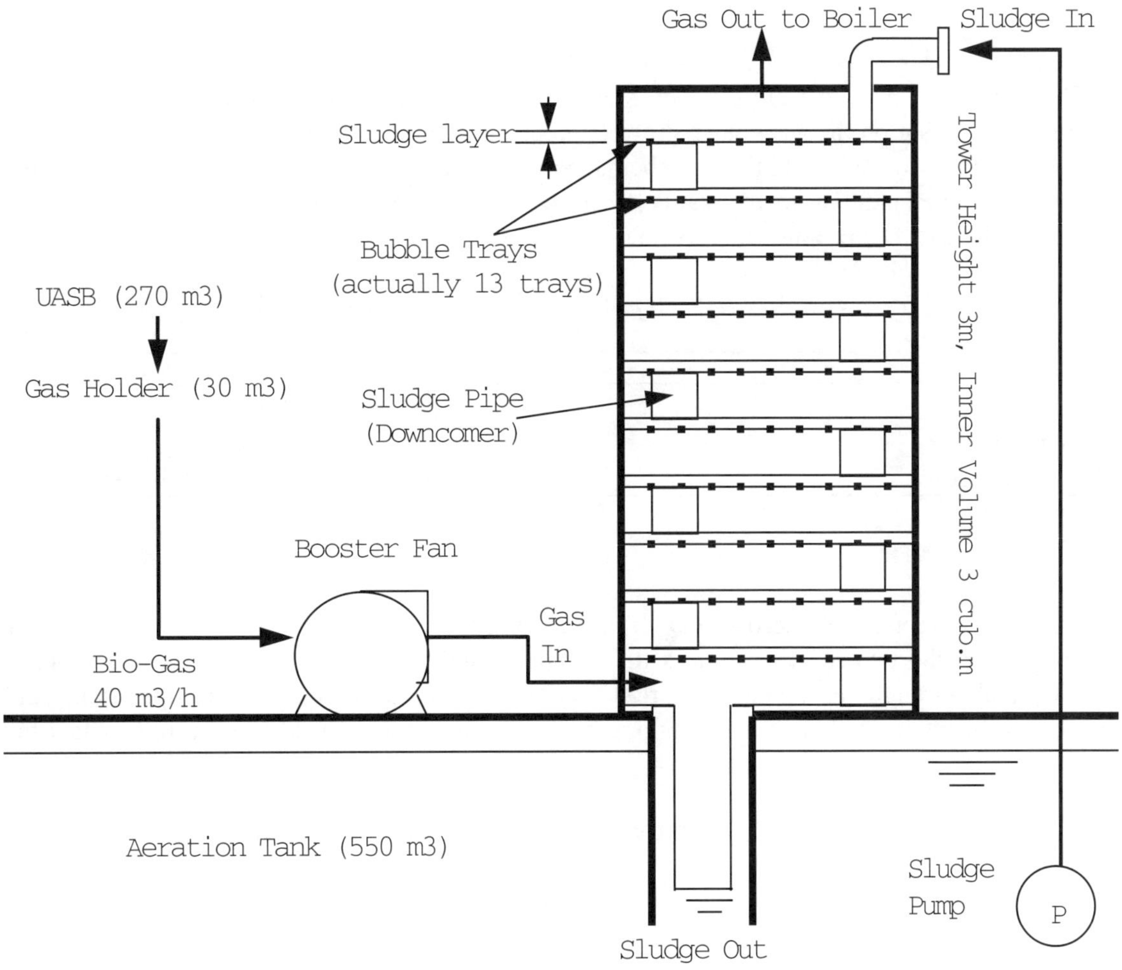

Source: Author.

Performance of a full-scale plant

Potato processing wastewater was treated using the UASB process with a biological H_2S scrubber. The wastewater characteristics along with the treatment steps are shown in Table 2. The removal of BOD and COD was satisfactory as intended.

Table 2. **Characteristics of wastewater from potato processing**

	Flow rate	T-COD	T-BOD	SO_4^{2-}
	m^3/d	mg/L	mg/L	mg/L
Raw wastewater	550	5 000	4 000	100
UASB treated	550	700	350	n.a.
Activated sludge treated (final effluent)	550	56	20	100

Source: Author.

The configuration of the bioscrubber was shown in Figure 2. The booster fan was turned on when the level sensor on the gas holder showed "high", and turned off when the level reached "low". Other equipment associated with the scrubber, such as the sludge pump and the boiler, was controlled automatically, corresponding to the status of the booster fan operation. The characteristics of the biogas from the UASB are shown in Table 3.

Table 3. **Characteristics of biogas from the UASB**

	Average flow rate 40 m^3/h
Methane	80 %
Carbon dioxide	20 %
Hydrogen sulfide	300 - 2 500 ppm (avg. 1 300 ppm)

Source: Author.

The H_2S concentration before and after the bioscrubber is shown in Figure 3. The H_2S concentration in the treated gas was less than 20 ppm, and the removal efficiency was more than 99 per cent all through the operation period. After six months of continuous operation, neither corrosion problems in the boiler, nor clogging problems (increase of head loss) in the contact tower were found.

Figure 3. **H_2S removal in the full-scale bioscrubber**

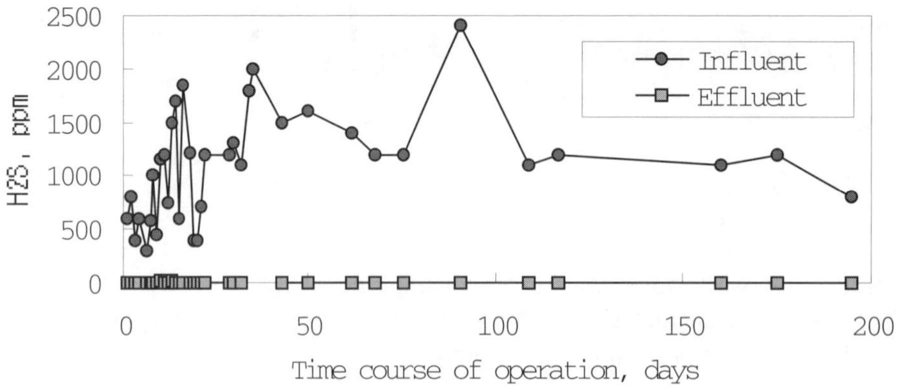

Source: Author.

In the aeration tank of the full-scale plant, the sulfur oxidation rate for step 1 and step 2 was 870 mg-S/g-VS/d and 50 mg-S/g-VS/d, respectively. Since the sludge in the full-scale plant contained BOD degrading bacteria besides sulfur oxidising bacteria, the rate constants were lower than those measured for the sludge prepared only for sulfur oxidation (Table 1). However, the actual size of the aeration tank was large enough to oxidise all the H_2S down to sulfate, even with the lower oxidising rate per unit sludge weight.

In this particular case, the operating the cost for the bioscrubber was ¥ 11 600/month for electricity (1 056 kWH/month) for the booster fan and sludge pump. On the other hand, the estimated cost for the chemical (ferric oxide pellets) was ¥ 36 000/month (200 kg/month) if a chemical scrubber was selected. This comparison showed the bioscrubber has brought ¥ 24 400/month of cost merit, which corresponds to ¥ 293 000/year (US$ 2 660/year). Besides the cost merit, it should be emphasized that the bioscrubber can eliminate the generation of waste chemicals and labour costs for replacing hard-handled chemical pellets.

Conclusion

The bioscrubber process, which utilised sulfur oxidising bacteria, was effective for removing hydrogen sulfide in the biogas from the UASB process. The process consisted of a gas/liquid contact tower and a sulfur oxidising aeration tank. The post-treatment aeration tank for UASB could be used as a sulfur oxidising tank, which supplied the gas-scrubbing liquid to the contact tower. A full-scale bioscrubber showed more than 99 per cent removal efficiency when treating 2 000 ppm of H_2S in the 40 m^3/hr of the biogas from the UASB. This full-scale case study showed that the bioscrubber could save in operating costs, and could eliminate the generation of waste chemicals, which are hard-handled and require a dumping area.

Questions, comments and answers

Q: In a bioscrubber, is it better to use a biofilm reactor to have a larger active surface area?

A: The basic difference between a biofilm reactor and a bioscrubber is growth. When treating fuel gas, we cannot allow organisms to grow in the scrubber. Also we cannot add oxygen, so no biofilm grows. We need an aeration tank to oxidise sulfides.

Q: Do you observe corrosion or water problems in the system?

A: No!

Q: In the potato effluent, what is the prime pollutant, chemical oxygen demand (COD), BOD, or sulfate or a combination of all three?

A: The first priority is COD removal; sulfate is not targeted.

Q: In what form was the carbon substrate? Did this influence the rate of sulfate conversion?

A: There is an acidification tank before the UASB and therefore volatile fatty acids are the COD source for sulfate reduction.

Questions, comments and answers cont'd.

Q: The scrubber is an aerobic process. Did you measure the composition of the gas emanating from the scrubber, and is there any loss in methane or any change other than hydrogen sulfide removal?

A: There was no obvious change at the percentage level, but we did not measure it very accurately.

Q: There was no growth in the scrubber?

A: We measured pressure loss in the scrubber and there was none over a period of 200 days. Therefore, we concluded there was no growth.

REFERENCES

AMERICAN SOCIETY OF MICROBIOLOGY (ASM) (1974), *Bergey's Manual of Determinative Bacteriology*, 8th edition, R.E. Buchanan (ed.), The Williams & Wilkins Company, Baltimore.

HONDA and FUKUYAMA (1988), *Japanese Patent*, No. 1432746 (in Japanese).

KANAGAWA, T. and E. MIKAMI (1989), "Removal of methanethiol, dimethyl disulfide, and hydrogen sulfide from contaminated air by *Thiobacillus thioparus* TK-m", *Appl. Environ. Microbiol.* 55:3, pp. 555-558.

LETTINGA, G., A.F.M. VAN VELSEN, S.W. HOMBA, W. DE ZEEUW, and A. KLAPWIJK (1980), "Use of the upflow sludge blanket (USB) reactor concept for biological wastewater treatment, specially for anaerobic treatment", *Biotech. Bioeng.* 22, pp. 699-734.

NISHIMURA, S. and M. YODA (1995), "Removal of hydrogen sulfide from anaerobic biogas using bioscrubber", *Proc. The 29th Conference of Japan Society on Water Environment*, p. 246 (in Japanese).

MONITORING BIOLOGICAL WASTEWATER TREATMENT PROCESSES

by

Freda R. Hawkes
School of Applied Sciences, University of Glamorgan, Mid Glamorgan, United Kingdom

Introduction

Biotreatment process performance has been commonly monitored by off-line analyses. The lack of robust on-line sensors is an obstacle to increased uptake of biotreatment technologies, which may be perceived as less reliable than physico-chemical treatments. A major disadvantage of the latter processes, which include the use of coagulants and flocculants, settling or floatation, and various forms of filtration, is that the organic content is only removed, not treated, and the organic sludge still requires disposal by incineration, landfill, land spreading, etc. The sludge often contains the chemicals added during processing.

Biological treatment, however, removes the degradable organic and nitrogen content of the wastewater, producing gases (CO_2, CH_4, N_2) by microbial action. The growth of excess microbial cells can be minimised by utilising anaerobic digestion as the first step in the process. Anaerobically, cell growth is much more restricted, and the excess cell sludge for disposal is approximately one-fifth or less of that from the aerobic treatment of the same amount of organic load (Mergaert *et al.*, 1992).

Biotreatment plants operating on industrial sites are subject to unpredictable variations in organic load, and are commonly designed with sufficient capacity to deal with these surges and still maintain effluent quality. Plants over-engineered in this way will frequently run under capacity, and capital costs will be higher. It is proposed that with additional information from reliable, continuously-operating (on-line) sensors, performance of existing biotreatment plants could be optimised in the face of these variations, and capital costs for new treatment plants reduced.

Typical biotreatment plant monitoring

The configuration of a typical biotreatment plant for the removal of organic material and nitrogen from industrial effluent is shown in Figure 1, together with the analyses and monitoring typically performed.

Parameters monitored in the influent are routinely flow rate, which is normally continuously recorded, and some measure of the organic content of the influent, measured off-line. This may be assayed as BOD_5 (taking five days for the results to be available), COD (taking about three hours minimum by Standard Methods) or Total and Volatile Solids for a particulate wastewater (taking about 18 hours). Total Organic Carbon (TOC) is the fastest measurement, if the equipment is available at the site.

Figure 1. **Typical biotreatment plant monitoring**

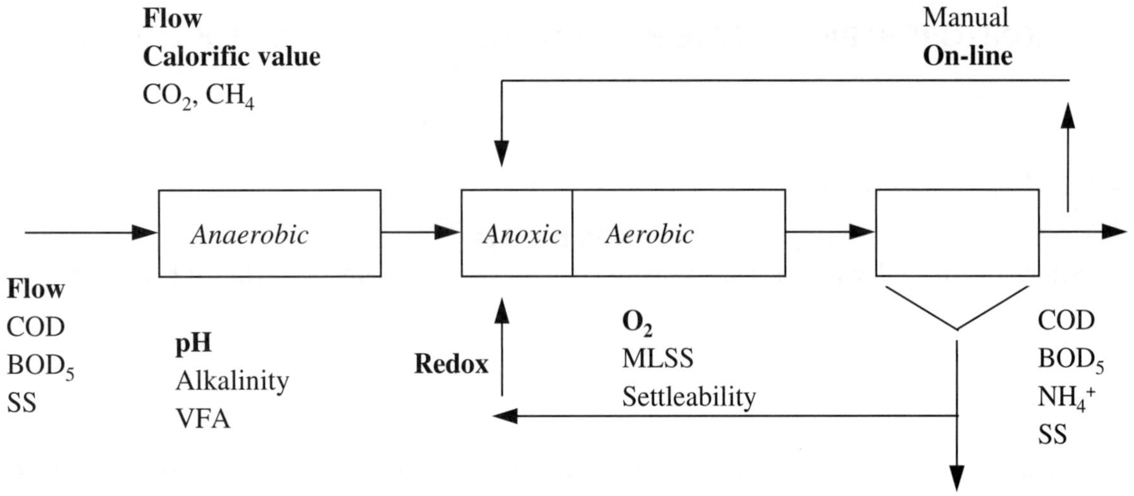

Source: Author.

Influent passes to an anaerobic stage, which is usually designed to retain biomass, for example an anaerobic upflow filter, fluidised bed or Upflow Anaerobic Sludge Blanket (UASB) reactor. From a world survey of over 900 commercially supplied anaerobic systems, 67 per cent were UASB reactors (Habets, 1996). The flow of biogas is constantly monitored, and the calorific value of the gas may be measured on-line, particularly if the biogas is used as an energy source. The gas composition (methane and CO_2 percentage) is commonly measured off-line using gas liquid chromatography (GLC) or cheaper methods depending on differential solubility. In the anaerobic digester liquor pH may be assayed on-line, although with many wastes, for example those with a high fat content, the pH electrode may rapidly foul. Sensor fouling is particularly a problem near the start of the treatment, where the incoming wastewater is untreated. Off-line analyses routinely performed on the effluent from this stage include a titration assay for alkalinity and a GLC assay for ethanoic (acetic), propionic, butyric and similar volatile fatty acids (VFA). These two assays are related in that alkalinity (measured by standard methods in units of mg/l $CaCO_3$) represents pH buffering capacity, which is destroyed by the production of VFA if the food chain of the microbial consortia becomes unbalanced.

Effluent then passes to an anoxic stage, without oxygen but with nitrate entering in the recycle line from the aerobic process. This compartment contains denitrifying bacteria which convert nitrate to nitrogen gas, using organic material not removed in the anaerobic stage as an energy source. Redox (oxidation-reduction potential) may be monitored on-line by means of an electrode in the liquor, providing information on the availability of nitrate and its competitor electron acceptor, oxygen, particularly useful where nitrification and denitrification are carried out in the same basin (Plisson-Saune *et al.*, 1996).

The final stages are an activated sludge process with oxygen supplied for the aerobic respiration of the remaining biodegradable organic compounds to CO_2, followed by a settler which removes microbial cells. Some cell sludge is recycled to maintain cell density, and the excess sludge must be disposed of. In the activated sludge stage, dissolved oxygen is routinely monitored on-line by oxygen electrodes, and the cell concentration measured off-line by the crude gravimetric measure of Mixed Liquor Suspended Solids (MLSS), which includes inert particles and non-viable cells. As a major cause of poor performance in activated sludge plants is bulking sludge, due for example to a developing growth of filamentous organisms, the settleability of the sludge is routinely measured.

The quality of the final effluent will be monitored not only by the plant operator but also by the regulatory body ensuring compliance with the agreed discharge standards. The regulatory body will monitor the discharge as often as finances allow, for example once per fortnight. The sample taken will be as a minimum subjected to off-line tests for its organic content (BOD_5, COD, or TOC), nitrogen content (NH_4^+-N and probably NO_3^--N) and suspended solids (SS) content.

If all the off-line assays shown in Figure 1 were done five days a week, and all the on-line sensors were operational, the plant would actually be very well monitored by most current standards. It should be noted that in this example the only on-line parameters monitored are influent flow rate, biogas flow rate, and pH, redox and dissolved oxygen monitored by electrodes. Another on-line sensor which may be utilised is a turbidimeter to provide a measure of the particle content of the activated sludge liquor, assumed related to the viable cell content, although operators do not always put confidence in this measurement.

Stricter environmental legislation has recently stimulated the development and marketing of a number of on-line sensors. There are now on-line monitoring procedures described in the literature or commercially available which give measurements related to BOD_5 COD, MLSS, and settleability, and can measure CO_2, CH_4 and H_2 content in biogas, alkalinity and VFA in the digester liquor, and NH_4^+-N and NO_3^--N in the final effluent. However, their use is not typical, and for a number of these sensors there are limited reports in the literature of their reliability and effectiveness.

Implementation of on-line monitoring

It can readily be appreciated that on-line monitoring of relevant parameters at the outflow of the treatment plant will provide a check on effluent quality. However, a major advantage of on-line monitoring at the inflow to the plant is that it warns of variations in incoming load. There are then remedial actions which could be taken by the operator to maintain effluent quality, for example increasing the alkalinity in the anaerobic stage, and the oxygen supply and the cell recycle in the aerobic stage, or diverting the influent temporarily to a buffer tank. On-line monitoring also allows for a saving in the addition of chemicals and power for aeration (Schlegel and Baumann, 1996). It provides accurate data to characterise variations in the influent during the design stage for new plants (Longdong and Wachtl, 1996). By making continuous measurements, information on the process and its behaviour as conditions vary is acquired, which when sufficient to model the process allows for the ultimate goal of automatic control. With reliable implementation of control actions, plant size at the design stage could be reduced, saving capital cost. Because the process can be optimised by automatic control procedures, the running costs too should be lower and the effluent quality consistently meet the discharge standards set.

Why then is on-line monitoring not more widely implemented? One obstacle to its use is instrument unreliability. A recent review states that sensors are the weakest part of the chain in real-time control of wastewater treatment plants, and that compared to the relevant computer technology, both hardware and software, the sensor technology is far behind (Lynggaard-Jensen *et al.*, 1996). Many wastewaters are harsh environments, microbial slime can develop, and probes can start to drift from the true reading within hours of calibration. Systems for cleaning in place involve liquid shear, brushes and ultrasound. Alternatively, the sensor can work on filtered liquor, the filtration apparatus itself then requiring maintenance. If a control system is to be based on sensors, both must be fail-safe, or there is the possibility of a catastrophic failure which could set back the use of automatic control in biotechnology.

Instrument cost is also a barrier. Industry is often reluctant to spend on its effluent treatment plant, which is often not perceived as part of the whole production process. Many of the sophisticated types of sensor, for example for organic content or nitrogen cost £ 30 000 (sterling) or more and weigh over 50 kg. It is perceived that the data provided does not have a sufficiently high information content to warrant such expenditure. Even for low-cost common sensors such as dissolved oxygen electrodes the information may not be highly valued unless (as may be the case) it is actually used to control the aeration system and save power. On-line sensors produce a mass of data; the way in which this data will be logged, accessed and used must be considered. Finally, on-line monitoring does not seem to be needed to meet current environmental standards. However, as these get tighter, it will be.

The features that on-line sensors need in order to be more useful are shown in Table 1. A multi-national company of instrument manufacturers who are our collaborators in instrument development set as a criterion that the sensor should require maintenance by a technician only once in every fortnight. An instrument user with whom we collaborate requires as a minimum that the instrument should run unattended at least over a weekend.

Table 1. **Desirable features of on-line sensors**

◆	Should measure significant parameters
◆	Should be low cost
◆	Should need infrequent maintenance (once per 14 days)
◆	Should not foul, e.g. should be self-cleaning, operate in filtered liquor or not be inserted into the liquor.

Source: Author.

Classes of on-line sensors

Fortunately, unlike other areas of biotechnology, wastewater treatment sensors do not need to be sterilizable. A general classification of types of sensors available is given in Table 2. The most commonly used types of sensors so far are in the first category, and use flow or potentiometric detectors (pH, redox and dissolved O_2 electrodes, and electrodes selective for ammonium or nitrate ions). Turbidimeters and auto-analysers using photometric detectors are routinely used in some circumstances for measurement of MLSS, ammonium-N, phosphate, etc. (Thomsen and Nielsen 1992; Nyberg *et al*, 1996; Lynggaard-Jensen *et al.*, 1996). Selective ion electrodes and colourimetric on-line assays are particularly used where local environmental restraints justify the use of equipment which is relatively expensive to purchase and operate (Schlegel and Baumann, 1996; Thomsen and Kisbye, 1996). Biosensors, such as variants of respirometers (Vanrolleghem *et al.*, 1994), utilise the same detectors as the direct physical/chemical sensors, but generally have added complexity because they also utilise a microbial sub-culture, often representative of the main reactor culture. They may be used to monitor organic content, toxicity or NH_4^+ or $N0_3^-$ concentrations (see for example Spanjers *et al.*, 1994; Massone *et al.*, 1996). Software sensors (sensors implemented through software) interpret the meaning of parameters measured by conventional sensors through modelling events (Montague *et al.*, 1992).

Table 2. **Classes of on-line sensors**

◆	*Direct physical/ chemical sensors* using, e.g. pressure, flow, potentiometric and photometric detectors
◆	*Biosensors* using the same detectors with a microbial sub-culture or enzymes
◆	*Software sensors* interpret parameters by software

Source: Author.

Sensors based on gas measurements

Sensors based on gas measurements avoid the problem of fouling, as they are not inserted into the liquor. A bicarbonate alkalinity monitor developed in our laboratory (Hawkes *et al.,* 1993; Guwy *et al.,* 1994) is an example of such a sensor. The detector for this instrument is a low flow gas meter based on a solenoid valve and pressure transducer (Guwy *et al.*, 1995). Alkalinity is a measure of pH buffering capacity in anaerobic digesters. In liquid under biogas, the main buffering agent is bicarbonate. A schematic representation of the monitor is shown in Figure 2. The sample of anaerobic digester liquid contents (about 10 ml per minute) enters a CO_2 saturation chamber to ensure there is a constant background level of dissolved CO_2. It proceeds to an acidification chamber where acid is added constantly at a low rate, sufficient to ensure all of the bicarbonate is converted to CO_2. The CO_2 evolved passes to the low flow gas meter and the signal to the computer can be simply related to the bicarbonate content. The detector and the equipment associated with it are low cost.

Figure 2. **Schematic diagram of the bicarbonate alkalinity monitor**

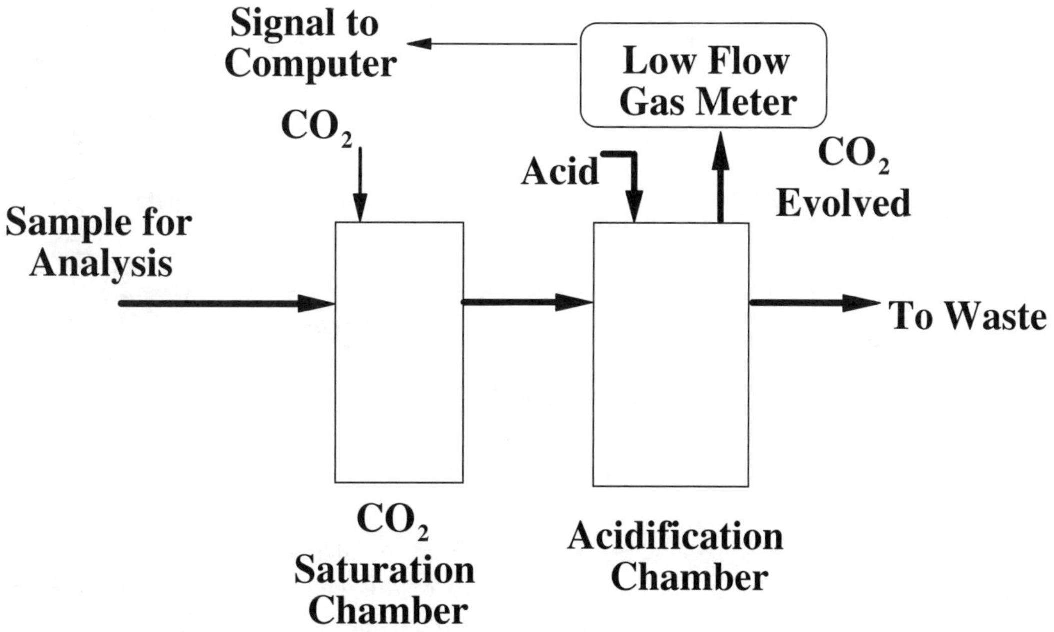

Source: Author.

An imbalance in the microbial food chain in the anaerobic digester destroys bicarbonate because volatile fatty acids are produced. To protect the bacteria, particularly the methanogens, against a sudden unfavourable drop in pH, a safe minimum level of alkalinity agreed by operators is 1 000 mg/l. Many industrial effluents have very low levels of natural alkalinity, and additions of bicarbonate or lime are needed in steady state to ensure stable operation. The bicarbonate alkalinity monitor could be used in a simple on-off control loop with a dosing pump for bicarbonate to ensure minimum consumption (Wilcox *et al.*, 1994). As bicarbonate levels drop as a result of increased organic load, a control mechanism could protect against a fall in pH, and warn indirectly of an accumulation of VFA (Hawkes *et al.*, 1994).

A typical response of the bicarbonate alkalinity monitor to an increase in load is shown in Figure 3. The results are from a fluidised bed reactor with high levels of retained bacteria on a sintered glass support (Guwy *et al.*, in press). The feed simulated a bakers' yeast wastewater with a COD of 6 600 mg/l COD. In the step load experiment shown in Figure 3, the organic content of the feed (indicated by the output of the pump delivering concentrated feed to the dilution stream) was increased to 24 300 mg/l COD for 8.3 hours, keeping the hydraulic retention time constant at eight hours. The organic loading rate was thus increased from 18 kg COD/m^3/day to 71 kg COD/m^3/day. The bicarbonate alkalinity decreased from around 1 800 mg/l $CaCO_3$ to near the danger level of 1 000 mg/l, when the step load was terminated and the feed organic content returned to the initial level. The bicarbonate alkalinity then increased, corresponding to a decrease in VFA concentration.

Figure 3. **On-line measurement of bicarbonate alkalinity during a step load indicated by the output of the feed pump delivering concentrates to the dilution stream**

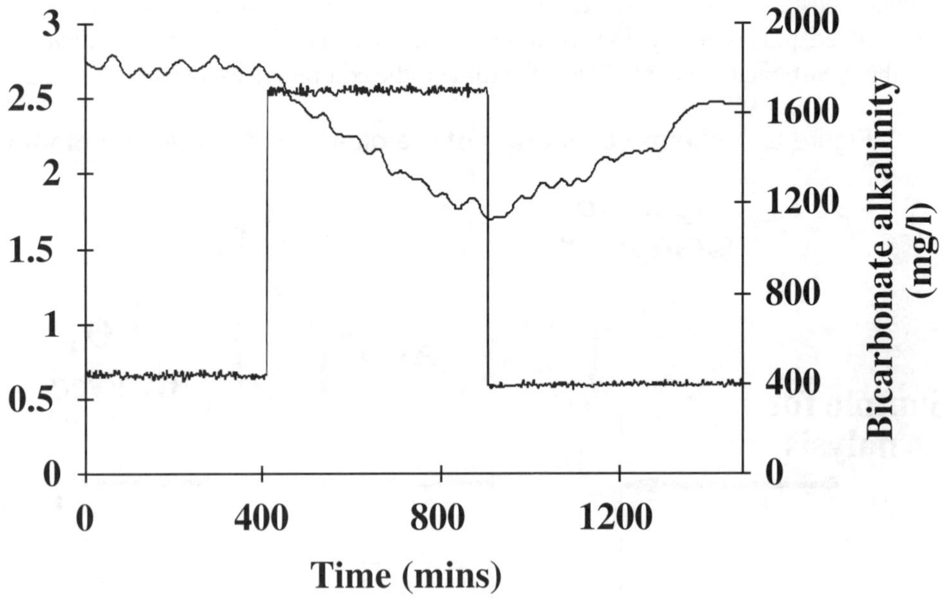

Source: Author.

Other instruments utilising the low flow gas meter to avoid the insertion of detectors when measuring liquor parameters are also being developed. A readily degradable organics monitor takes in a sample for analysis into a reaction chamber. The oxidising agent peroxide is added. In the catalytic chamber, unused peroxide is broken down to oxygen which is measured by the low flow gas meter. We are now establishing the relationship of this signal to the readily degradable organics content of the sample. A monitor for viable aerobic cells also involves the low flow gas meter. Both of these monitors are patented with a route to manufacture by a major company, and both use common, low-cost components, leave no environmentally-unfriendly chemicals in the waste stream and avoid the use of potentially harmful reagents such as ozone.

In anaerobic digestion, the amount of biogas is proportional to organic material destroyed, with only a small proportion used as a carbon source for bacterial growth. The rate of biogas production can therefore indicate the degree of treatment of the waste, and the organic load applied to the reactor. If the reactor was on the verge of failure, the biogas flow rate would decrease; thus, a decrease in gas flow could signal either serious problems or just a decrease in influent strength. Because of this uncertainty in interpretation, information on biogas flow rate should be combined with other parameters to ascertain the state of the digestion process.

Information on the performance of the digester can also be obtained from the composition of the gas phase, since the content of carbon dioxide and hydrogen increases as fermentative metabolism outstrips methanogenesis. Biogas CO_2 content can be measured on-line using commercially available infra-red based detectors, and biogas hydrogen monitored using an instrument originally developed to monitor hydrogen levels in human breath, and modified in the laboratory.

Both parameters are difficult to interpret if considered alone. Physical factors affect the carbon dioxide content of the gas because it has a considerable solubility in water. The absolute value of hydrogen in the biogas from a stable digester operating on simulated bakers' yeast production wastewater can be, in our experience, 50 ppm or 800 ppm with no deterioration in performance. One variable affecting biogas hydrogen levels other than the organic load is the degree of acidification of the feed, so that prior fermentation in the pipelines and storage tank before the digester reduces biogas H_2 content. Hydrogen levels in the biogas during a step overload can have significantly different absolute values, e.g. peaks of 300 ppm or 1 500 ppm, with no difference in the COD reduction achieved (Guwy *et al.*, in press). Thus, in order to monitor the operating condition of an anaerobic digester using biogas flow and composition measurements, it is recommended that all three parameters be considered by the control system in parallel.

Approaches to automatic control

The simplest application of automatic control on a wastewater treatment plant is on-off control. This can be used to maintain a set value of a single parameter, providing the value necessary for good operation has been defined by years of off-line monitoring. An example is alkalinity in anaerobic digesters, dissolved oxygen in aerobic treatment or Mixed Liquor Suspended Solids in activated sludge plants. With reliable on-line instruments in place, this could be implemented in the immediate future, and control of the two latter parameters is already practised on selected plants.

In the longer term, automatic control can give optimisation of one reactor, or the whole plant. This requires multi-parameter control, taking in information from a number of sensors, such as those discussed above for anaerobic digestion. Biological systems are non-linear and time-variant, and it is difficult to obtain a mechanistic model with enough detail to account for the responses of real biotreatment plants (Côté *et al.*, 1995). Their control requires adaptive control using systems which can learn and predict, so the appropriate control action can be taken. One such system is based on neural networks (Montague and Morris 1994; Wilcox *et al.,* 1995).

In our laboratory, data on the parameters described for the anaerobic digester has been used to train and then to test a neural network. Input parameters were pump output metering influent strength, bicarbonate alkalinity, pH, biogas flow and biogas H_2 and CO_2 composition. The training data and the test data were taken one month apart in a period of constant operation. From the six inputs the neural network predicts each of the five variables. At present we are able to get a good match between the predicted and the actual values for each of the five variables ten minutes in advance. Even this short time span would allow a plant operator to take appropriate remedial action, an improvement on the typical situation where the running of the plant depends on yesterday's results. Using this approach with more training data, we anticipate a longer prediction period will be possible. By incorporating this into a control strategy, it is possible a system to take over control and optimisation of the anaerobic digester stage could soon be developed. It must be understood that adjustments made to the operation of the anaerobic digester would affect stages downstream, so the ultimate, longer term goal is integrated control of the whole plant. This could be applied to existing plants, with the only modification required being the provision of controllable equipment (particularly pumps with variable speeds). The control technologies exist -- the obstacle to their application is poor sensors.

Recommendations

Recommendations to bodies influencing research and development policy are that strategic R&D leading to reliable, low-cost on-line sensors be encouraged and the interaction of control engineering with biotechnology facilitated.

Acknowledgements

The readily degradable organics monitor and viable aerobic cell monitor are being developed under a grant from the UK Biotechnology and Biological Sciences Research Council through the LINK Biological Treatment of Soil and Water programme. Work on control is supported by EU funding ENV4-CT95-0064. This presentation draws on the work of others in the Wastewater Treatment Laboratory of the University of Glamorgan, including Mr. G. Premier, Prof. D.L. Hawkes, Dr. S. Wilcox, Dr. A. Guwy, and Mr. R. Dinsdale.

Questions, comments and answers

Q: In your monitoring scheme, what do you do to accommodate the failure of one or more of the monitoring components?

A: Any control system must be fail-safe to allow for sensor failure. Sensor redundancy is the easiest way, and control engineers have this all worked out.

C: Waste treatment systems often have detection devices which are not maintained and therefore fail.

A: We need robust sensors. This may mean that those in the inflow should not be put directly into the liquid.

Questions, comments and answers cont'd.

Q: How good is oxidation/reduction potential for predicting oxygen concentration?

A: We haven't used this. I was referring to other people's work.

Q: In terms of the organics monitor, were you measuring chemical oxidation but relating it to biodegradability?

A: We oxidise the organics with peroxide and relate the remaining peroxide to the chemical (or biological) oxygen demand.

Q: Do you measure or predict for a real plant or only at lab scale?

A: At lab scale. We are working on a 14 litre fluidised bed anaerobic digestor.

Q: What do you do with the answers from data measurement, for example of alkalinity?

A: Off-line work has given a minimum level of alkalinity to be maintained for optimum degradation. You need a model system, or neural net trained on previous data.

Q: Is a ten-minute prediction of values useful?

A: It would be better if it were longer, but it is a lot better than the hours or days of delay involved in measurements of BOD or COD.

REFERENCES

COTE, M., B.P.A. GRANDJEAN, P. LESSARD, and J. THIBAULT (1995), "Dynamic modelling of the activated sludge process: improving prediction using neural networks", *Water Research* 29:4, pp. 995-1004.

GUWY, A.J., D.L. HAWKES, F.R. HAWKES, and A.G. ROZZI (1994), "Characterisation of a prototype industrial on-line analyser for bicarbonate/carbonate monitoring", *Biotechnol. Bioeng*. 44, pp. 1 325-1 330.

GUWY, A.J., D.L. HAWKES, and F.R. HAWKES (1995), "On-line low flow high-precision gas metering systems", *Water Research* 29:3, pp. 977-979.

GUWY, A.J., F.R. HAWKES, D.L. HAWKES, and A.G. ROZZI (in press), "Hydrogen production in a high rate fluidised bed anaerobic digester", *Water Research*.

HABETS, L. (1996), Paques Water Systems BV, PO Box 52, 8560 AB, Balk, The Netherlands, Communication at Society for Chemical Industry Conference Industrial Anaerobic Waste Water Treatment, 18 September, SCI, London.

HAWKES, F.R, A.J. GUWY, A.G. ROZZI, and D.L. HAWKES (1993), "A new instrument for on-line measurement of bicarbonate alkalinity", *Water Research* 27:1, pp. 167-170.

HAWKES, F.R., A.J. GUWY, D.L. HAWKES, and A.G. ROZZI, (1994), "On-line monitoring of anaerobic digestion: application of a device for continuous measurement of bicarbonate alkalinity", *Wat. Sci. Technol*. 30:12, pp. 1-10.

LONGDONG, J. and P. WACHTL (1996), "Six years of practical experience with the operation of on-line analysers", *Wat. Sci. Tech*. 33:1, pp. 159-164.

LYNGGAARD-JENSEN, A., N.H. EISUM, I. RASMUSSEN, H.S. JACOBSEN, and T. STENSTROM (1996), "Description and test of a new generation of nutrient sensors", *Wat. Sci. Tech*. 33:1, pp. 25-35.

MASSONE, A.G., K. GERNAEY, H. BOGAERT, A. VANDERHASSELT, A. ROZZI, and W. VERSTRAETE (1996), "Biosensors for nitrogen control in wastewaters", *Wat. Sci. Tech*. 34:1, pp. 213-220.

MERGAERT, K., B. VANDERHAEGEN, and W. VERSTRAETE (1992), "Applicability and trends of anaerobic pre-treatment of municipal wastewater", *Water Research* 26:8, pp. 1 025-1 033.

MONTAGUE, G.A., A.J. MORRIS, and M.T. THAM (1992), "Enhancing bioprocess operability with generic software sensors", *J Biotechnol*. 25, pp. 183-201.

MONTAGUE, G.A. and A.J. MORRIS (1994), "Neural-network contributions in biotechnology", *TIBTECH* 12, pp. 312-324.

NYBERG, U., B. ANDERSSON, and H. ASPEGREN (1996), "Experiences with on-line measurements at a wastewater treatment plant for extended nitrogen removal", *Wat. Sci. Tech.* 33:1, pp. 175-182.

PLISSON-SAUNE, S., B. CAPDEVILLE, M. MAURET, A. DEGUIN, and P. BAPTISTE (1996), "Real-time control of nitrogen removal using three ORP bending points: signification, control strategy and results", *Wat. Sci. Tech.* 33:1, pp. 275-280.

SCHLEGEL, S. and P. BAUMANN (1996), "Requirements with respect to on-line analysers for N and P", *Wat. Sci. Tech.* 33:1, pp. 139-146.

SPANJERS, H., G. OLSSON, and A. KLAPWIJK (1994), "Determining short term Biochemical Oxygen Demand and respiration rate in an aeration tank by using respirometry and estimation", *Water Research* 28:7, pp. 1 571-1 583.

THOMSEN, H.A. and K. KISBYE (1996), "N and P on-line meters: requirements, maintenance and stability", *Wat. Sci. Tech.* 33:1, pp. 147-157.

THOMSEN, H.A. and M.K. NIELSEN (1992), "Practical experience with on-line measurements of NH_4, NO_3, PO_4, Redox, MLSS and SS in advanced activated sludge plants", Proceedings HYDROTOP Conference, The City and Water, vol. 2, 8-10 April, Marseilles, France, pp. 378-388.

VANROLLEGHEM, P.A., Z. KONG, G. ROMBOUTS, and W. VERSTRATE (1994), "An on-line respirographic biosensor for the characterisation of load and toxicity of wastewaters", *J. Chem. Tech. Biotechnol.* 59, pp. 321-333.

WILCOX, S.J., A.J. GUWY, D.L. HAWKES, and F.R. HAWKES (1994), "The control of anaerobic digestion for agro-industrial wastewaters", Proceedings 4th FAO/SREN Workshop, 14-17 June, MIGAL, Israel, pp. 46-58, FAO/UN Rome, ISSN 1024-2368.

WILCOX, S.J., D.L. HAWKES, F.R. HAWKES, and A.J. GUWY (1995), "A neural network, based on bicarbonate monitoring, to control anaerobic digestion", *Water Research* 29:6, pp. 1 465-1 470.

Parallel Session II

II-C. PREVENTION OF WATER POLLUTION: AGRICULTURAL SOURCES

WATER TREATMENT IN RURAL AREAS

by

Martin Wegelin
Water & Sanitation in Developing Countries (SANDEC)
Swiss Federal Institute for Environmental Science & Technology (EAWAG)
Duebendorf, Switzerland

Introduction

In developing countries, a large part of mankind is forced to use surface water, water drawn from polluted rivers, irrigation canals, ponds, and lakes. This surface water, however, is a carrier of many infectious and tropical diseases and hence, must usually be treated prior to consumption. The main target of any water treatment is the removal or inactivation of disease-causing organisms (pathogens), such as harmful bacteria, viruses, protozoal cysts, and worm eggs. Disinfection -- usually by application of chlorine -- and slow sand filters are the two most widely used treatment processes for bacteriological water quality improvement.

The use of chemicals in rural water supply schemes is often bound to fail. Hence, a reliable and adequate application of water treatment processes such as flocculation and chlorination is generally beyond the capacity of local skill and resources. As presented hereafter, slow sand filters preceded by roughing filters and solar water disinfection are simple, efficient and sustainable water treatment processes adequate for rural water supplies.

The crux in rural water treatment

Disinfection efficiency and slow sand filter performance are strongly influenced by the turbidity of the water to be treated. Turbidity mainly reflects the amount of solids and colloids present in the water. A large part of the micro-organisms are attached to the surface of these solids, which acts as shelter. Hence, chemical disinfection will be impaired in turbid water. Organic solid or dissolved matter using the disinfectant for oxidation processes will further reduce the efficiency of chemical disinfection. Consequently, adequate water disinfection is only possible with water of low turbidity or virtually free of solid and organic matter.

Slow sand filters also require relatively clear water. Reasonable filter operation can be expected with raw water turbidities below 20-30 NTU. The solid matter of water with higher turbidities will rapidly clog the top of the sand bed and interfere with the biological processes located in this part of the filter. The filter resistance will increase drastically, calling for filter cleanings at short intervals of a few days or weeks. Such an operation will considerably impair the biological filter activities and greatly reduce the bacteriological quality of the treated water. Therefore, only raw water of low turbidity will enable slow sand filters to produce a good water quality which is safe for consumption.

Pre-treatment of surface water for the reduction of turbidity or solid matter concentration is, therefore, required for both discussed treatment processes, i.e. chlorination and slow sand filtration. Sedimentation, possibly preceded by chemical flocculation, is applied as solid removal process. Plain sedimentation will remove the settleable and part of the suspended solids. Particles smaller than 20 µm will hardly be separated, although this fraction might represent the bulk of the solids in a surface water. In such situations, sedimentation is enhanced by the addition of chemicals which destabilize the suspension and induce flocculation of the solids.

A reliable use of chemicals in rural water supplies in developing countries is, however, extremely difficult and often bound to fail, as is illustrated in Figure 1. The chemicals must frequently be imported and therefore require foreign currency often scarce in developing countries. Apart from the availability of chemicals, transportation poses another problem, generally pertinent to developing countries. The adequate supply of chemicals to remote rural treatment plants might therefore be difficult and unreliable. The dosage of chemicals depends on the raw water quality and, hence, calls for careful supervision. The sensible dosing equipment is exposed to the aggressivity of the chemicals which can attack and damage the installations. Finally, chemical water treatment requires skilled personnel, often not available in rural areas. These factors, therefore, very much question a reliable and successful application of chemicals in rural water supply schemes of developing countries. Numerous malfunctioning or abandoned water treatment plants supposed to use chemicals endorse this statement.

Figure 1. **Operational problems in conventional water treatment plants**

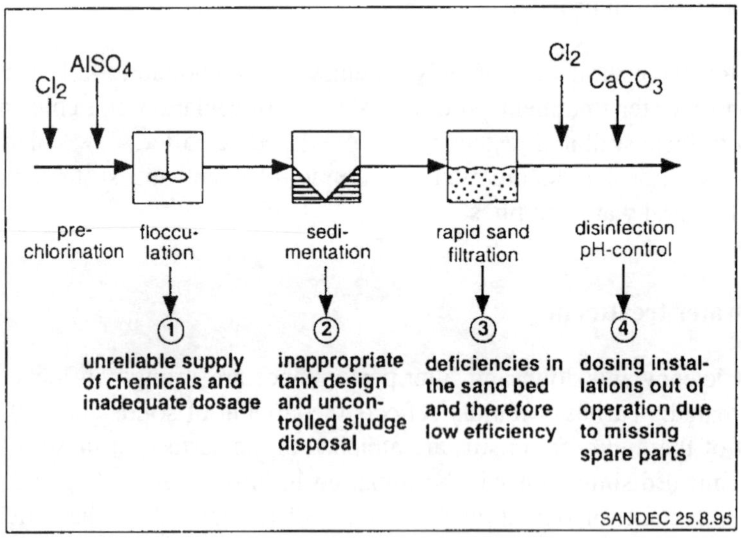

Source: SANDEC, 1995.

Consequently, chemical flocculation, a process reacting sensitively to water quality changes and therefore difficult to operate, generally constitutes an inappropriate technology for the considered situation. Similarly, a safe and reliable chlorination of the water remains a target difficult to attain under rural conditions in developing countries.

The multiple filter approach

Particle removal is far more efficient by filtration as compared to plain sedimentation. This is illustrated in Figure 2 by the comparison of a sedimentation tank with a roughing filter. In a sedimentation tank, each particle has to overcome a settling distance equal to the tank's depth, e.g. around one to three m, in order to be separated. Suspended solids with a small settling velocity might not reach the tank's bottom and, hence, be flushed out of the sedimentation tank. However, the same suspended solids moving through a filter will fortunately touch the gravel surface already after a few millimetres of settling distance. Thus, filtration becomes a more effective process for solids removal, since the settling distance is extremely reduced by the filter material. The presence of a small pore system and large internal filter surface area enhances sedimentation, adsorption, as well as chemical and biological activities.

Figure 2. **Particle removal in a sedimentation tank and a roughing filter**

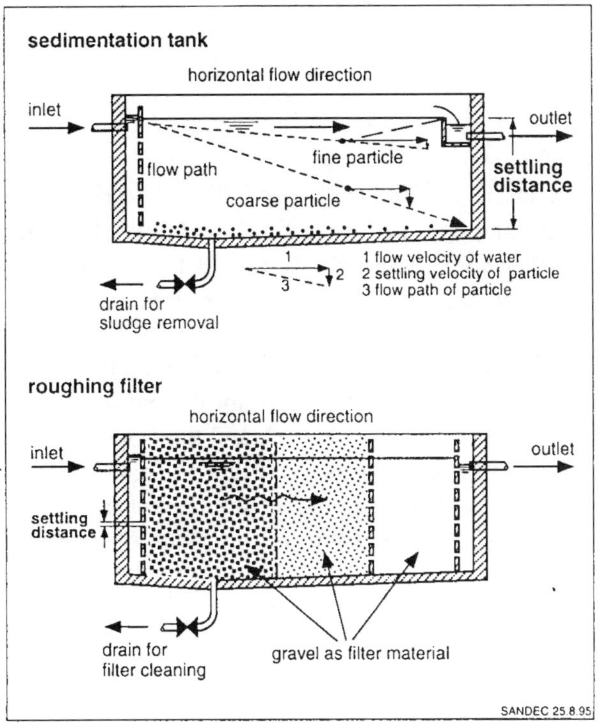

Source: SANDEC, 1995.

Sound water treatment concepts start with the separation of coarse matter, usually easy to achieve, and end with the removal or inactivation of small solids and micro-organisms, generally more difficult to separate. Figure 3 illustrates the step-by-step treatment approach. In a first step, coarse matter, possibly present in surface water, will be removed by screens, and settleable solids retained by sedimentation tanks. In a second stage, prefilters, known as roughing filters, will separate the fine and light particles classified as suspended solids. And finally, slow sand filtration as the main treatment stage will substantially reduce the number of micro-organisms present in the water, retain the last traces of solid matter and oxidise organic compounds dissolved in the water.

Figure 3. **Multiple barrier water treatment concept**

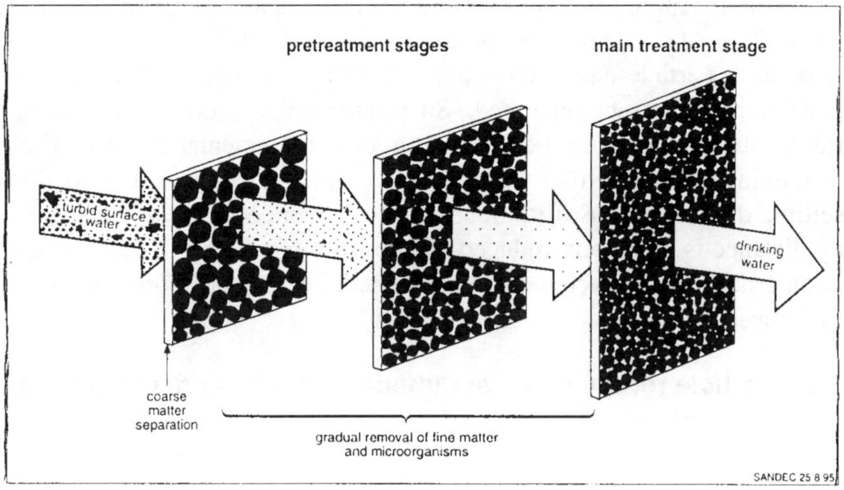

Source: SANDEC, 1995.

Roughing filters usually consist of differently-sized filter material which successively decrease in size. The bulk of solids is separated by the coarse filter medium, while the subsequent medium and fine filter media have a more polishing function. Roughing filters are able to reduce turbidity to a level which allows a sound and efficient slow sand filter operation.

In other words, roughing filters mainly act as physical filters and are applied to retain the solid matter. Slow sand filters are biological filters used to improve particularly the bacteriological water quality. Both filter types are of an equal technical level, and their operation is characterised by a high process stability. The treatment combination shown in Figure 4 makes full use of natural purification processes, and does not require chemicals to support or supplement the treatment scheme. Therefore, roughing filtration combined with slow sand filtration presents an appropriate process scheme for rural water treatment in developing countries.

Roughing filters

The filter medium of roughing filters is composed of relatively coarse (rough) material ranging from 25 mm to four mm in size, installed in layers of different fractions. Gravel is usually used as filter material. The design and mode of application of roughing filters vary considerably. The different filter types can be classified according to their location within the water supply scheme, and with respect to the flow direction. One therefore distinguishes between intake and dynamic filters, which form part of the water intake structure, and the actual roughing filters which are integrated in the water treatment plant. Roughing filters are further subdivided into down, up and horizontal-flow filters. Figure 5 illustrates the layout of intake and dynamic filters, and presents the main features of the three roughing filters types.

Figure 4. **Water treatment by roughing and slow sand filters**

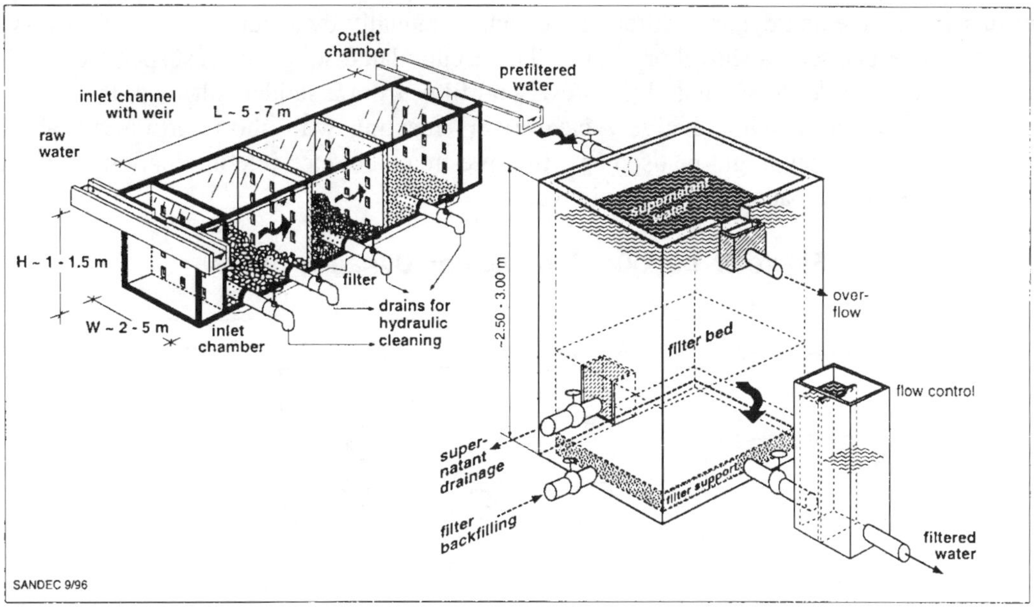

Source: SANDEC, 1996.

Figure 5. **Layout of pre- and roughing filters**

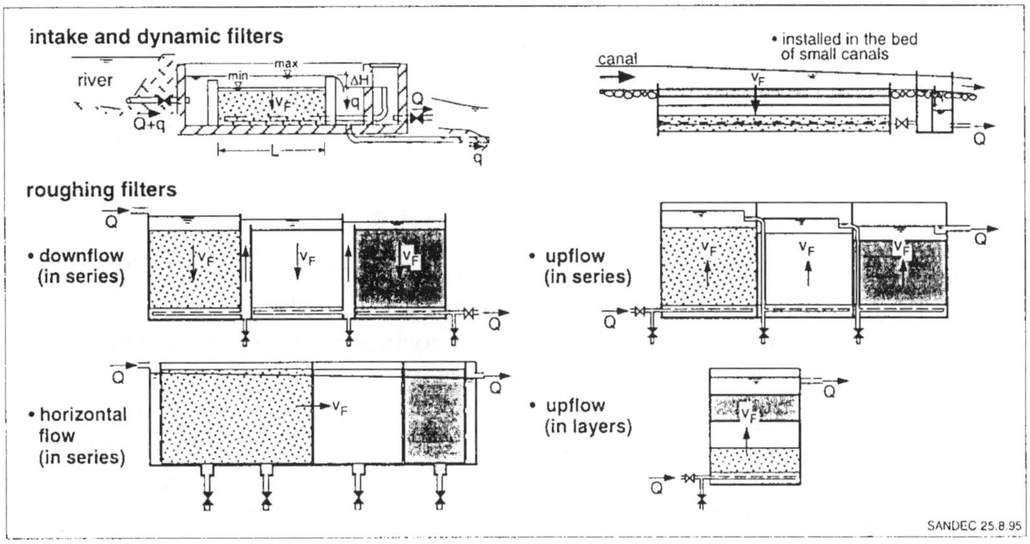

Source: SANDEC, 1995.

Intake and dynamic filters are installed next to small and narrow river beds as shown in Figure 6. Part of the river water, impounded by a small weir, is diverted either into a filter box or into a small canal where a subsurface filter is installed. The river water is filtered through different gravel layers, the top layer being of finest size. Therefore, these two filters act mainly as surface filters.

Intake and dynamic filters are not only designed for continuous solids separation, but also to protect the treatment plants from shock loads of high solid concentration. Such peaks rapidly clog the filter and thereby interrupt the flow of highly turbid water to the treatment plant, especially when high filtration rates of approximately 10 m/h are applied to a top layer of relatively fine gravel of about 1-2 mm in size. Intake and dynamic filters are cleaned manually by scouring, normally every week, the top of the filter bed with a shovel or a rake. The accumulated solids are resuspended and washed back to the river. In order to safeguard the slow sand filter from a sudden silt load, these two filters should preferably be used with relatively clear rivers reacting with short turbidity peaks during rainfall. Alternatively, they might be used as a first pre-treatment step in combination with roughing filters to reduce the solid load of highly turbid water.

Figure 6. **Layout of intake and dynamic filters**

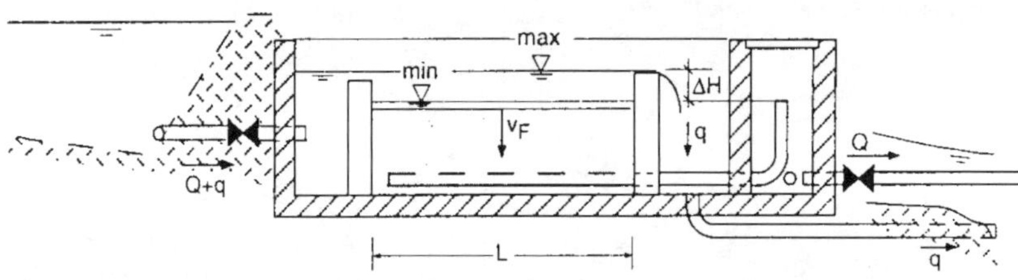

list of symbols

d_g	(mm)	gravel size
L	(m)	filter length
W	(m)	filter width
A	(m^2)	filter area
ΔH	(cm)	headloss
Q	(m^3/h)	flow rate
v_F	(m/h)	filtration rate
q	(m^3/h)	surplus flow rate

design guidelines	intake filter	dynamic filter
filtration rate $v_F = \frac{Q}{L \cdot W} = \frac{Q}{A}$	0.3 - 2 m/h	> 5 m/h
max. headloss (operation) ΔH	~20 - 40 cm	~20 - 40 cm
gravel size	gravel layer height	
d_g = 2 - 4 mm	—	20 - 30 cm
d_g = 4 - 8 mm	30 - 40 cm	10 cm
d_g = 8 - 12 mm	(10 - 20 cm)	10 cm

SANDEC 25 8 95

Source: SANDEC, 1995.

Roughing filters are designed to treat surface water of high turbidity over prolonged periods. The water is filtered by a sequence of normally three filter fractions. Due to the deep penetration of the solids into the filter medium, roughing filters act, in contrast to intake and dynamic filters, as space filters and therefore possess a large silt storage capacity. In vertical-flow roughing filters, the available height of the filter medium is limited to about 1.0-1.5 m due to structural constraints. Hence, the total length of a three-stage vertical-flow filter will amount to maximum 4.5 m. Alternatively, the filter length of horizontal-flow roughing filters is theoretically unlimited, but varies in general between five and nine metres. Roughing filters are operated at filtration rates ranging from 0.3-1.5 m/h. Up and down-flow roughing filters as shown in Figure 7 can cope with raw water turbidities of 50-150 NTU, whereas horizontal-flow roughing filters can handle even short turbidity peaks of 500-1 000 NTU on account of their comparatively long filter length. The filters are cleaned periodically by a fast filter drainage and, if necessary, manually by removing, washing and reinstalling the filter material.

Figure 7. **Layout of roughing filters**

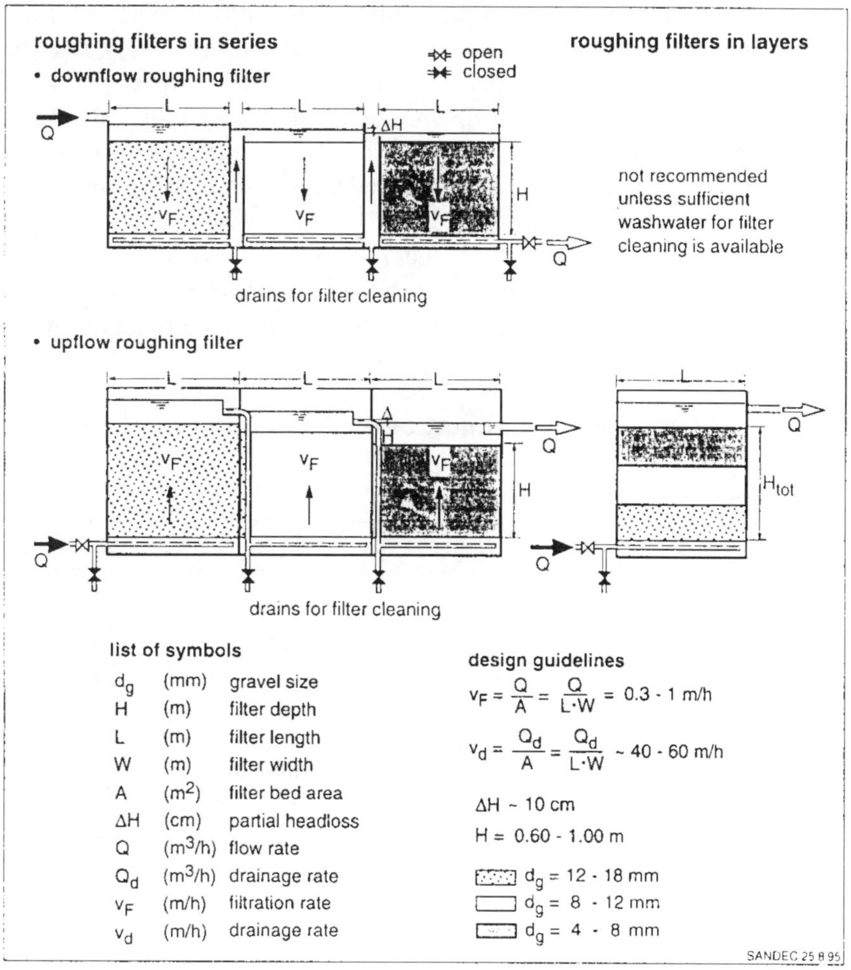

Source: SANDEC, 1995.

Field experience

In the past decade, horizontal-flow roughing filtration received greater attention than any other prefiltration techniques. The comprehensive development of the horizontal-flow roughing filter technology through research, laboratory and field tests, and the successful application of the treatment process also enhanced the development of other types of roughing filters. SANDEC has been strongly involved in the development and promotion of the horizontal-flow roughing filter technology since 1982 (Wegelin *et al.*, 1991). This long-term engagement enabled a world-wide dissemination of the treatment process. Field experience, process limitation, and promotion of the horizontal-flow roughing filter technology is presented hereafter.

Blue Nile Health Project, Sudan

The rural population of the Gezira/Managil irrigation draws its water from small irrigation canals with a high bacteriological contamination. Besides malaria, diarrhoeal diseases and bilharzia are predominant in this region. The surface water supplied by the Blue Nile is further characterised by high turbidities reaching 1 000 and more turbidity units during the rainy season. Small water supply schemes, operated by gravity and consisting of a horizontal-flow roughing filter and a slow sand filter unit, a clear water tank and a handpump, as illustrated in Figure 8, were constructed to supply villages of 200-500 inhabitants. Part of the horizontal-flow roughing filter medium has been replaced by broken burnt bricks, since gravel and stones are scarce in the project area located in an alluvial zone. Turbidity is reduced to approximately 5-10 NTU, and a substantial improvement of the bacteriological water quality is recorded. In the raw water, *E. coli* counts of 200 to several 1 000/100 ml are reduced to 10-30/100 ml in the treated water (Basit and Brown, 1986). The achieved water quality improvement is a step towards better health conditions for the farmers in these irrigation schemes.

Figure 8. **Water treatment plant of the Blue Nile Health Project**

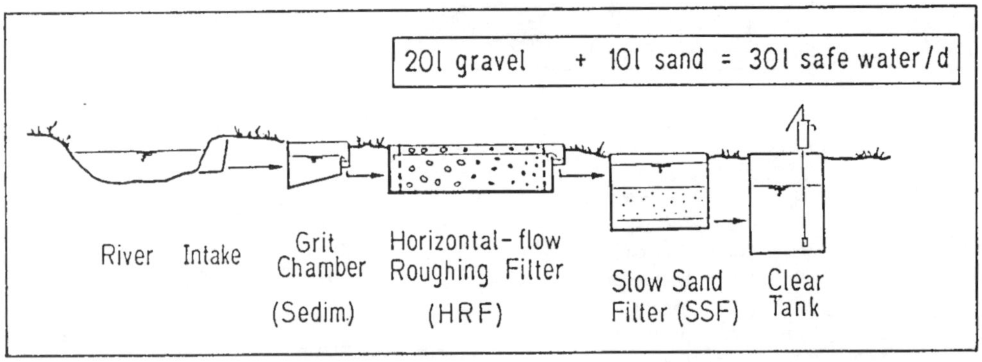

Source: SANDEC, 1992.

Jinxing, China

In co-operation with the Zhejiang Health and the Anti-Epidemic Station, two pilot plants were constructed in the Zhejiang Province in 1989. Jinxing, which is one of them, treats water from a canal grossly contaminated by small-scale industry and heavy navigation. The treatment plant of 240 m^3/d capacity has two lines, each consisting of a small tilted plate settling tank, a horizontal-flow roughing filter and a slow sand filter unit. Turbidity of the canal water is relatively low, ranging between 20 and 90 NTU. The horizontal-flow roughing filters operated at a high filtration rate of 1.7 m/h still drastically reduce turbidity, which is further decreased by slow sand filtration, as shown in Figure 9. The two horizontal-flow roughing filter units are cleaned by drainage every 40 days. After five months of operation, headloss in the slow sand filters was recorded to be only 31 cm. The remarkable operational experience with the first horizontal-flow roughing filter/slow sand filter schemes in China has been presented and discussed in a workshop (Xu, 1994) and is now attracting the interest of the local authorities.

Figure 9. **Operational conditions and turbidity reduction of the Jinxing treatment plant**

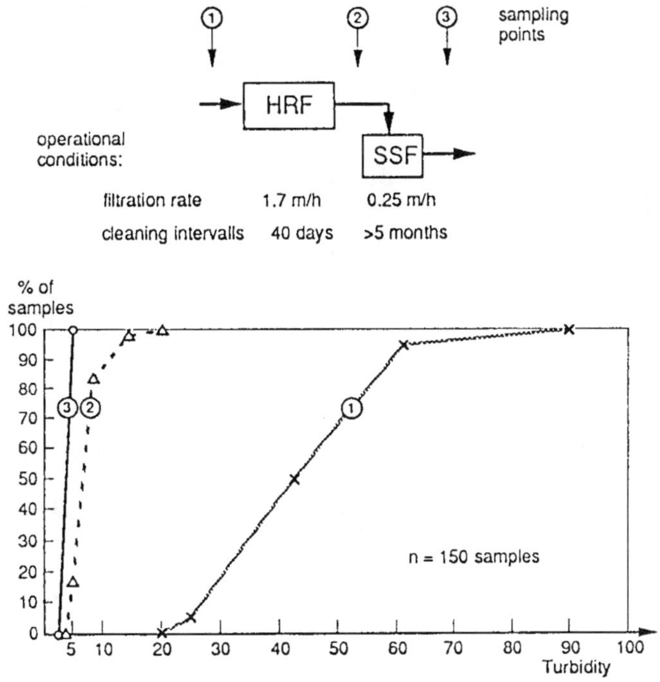

Source: Xu, 1994.

Mafi Kumase, Ghana

Slow sand filter operation is not only impaired by high turbidities originating from inorganic solid matter, but also by high algae concentrations. The field tests of Mafi Kumase also demonstrate the potential of horizontal-flow roughing filter for algal removal. The population of this Ghanaian village has suffered under guinea worm and bilharzia. These diseases originated from a shallow lake used as a water supply source. Under a local expertise (Dorcoo, 1992), the community constructed a horizontal-flow roughing filter/slow sand filter treatment plant in a self-help project. The lake virtually acts as an efficient sedimentation tank and also as a potential reactor for biomass production. In this case, the horizontal-flow roughing filter technology is used to reduce the algae concentration. A grab sample illustrated in Figure 10 indicates the efficiency of horizontal-flow roughing filters and slow sand filters for algae separation. The algae counts found in the slow sand filter effluent are partly explained by the organotrophic regrowth in the treated water. Nevertheless, practical experience shows that the slow sand filter units have to be cleaned every four to six months only. The horizontal-flow roughing filter/slow sand filter process combination also reveals its potential for the treatment of algae loaded surface water.

Figure 10. **Algae removal by Mafi Kumase treatment plant**

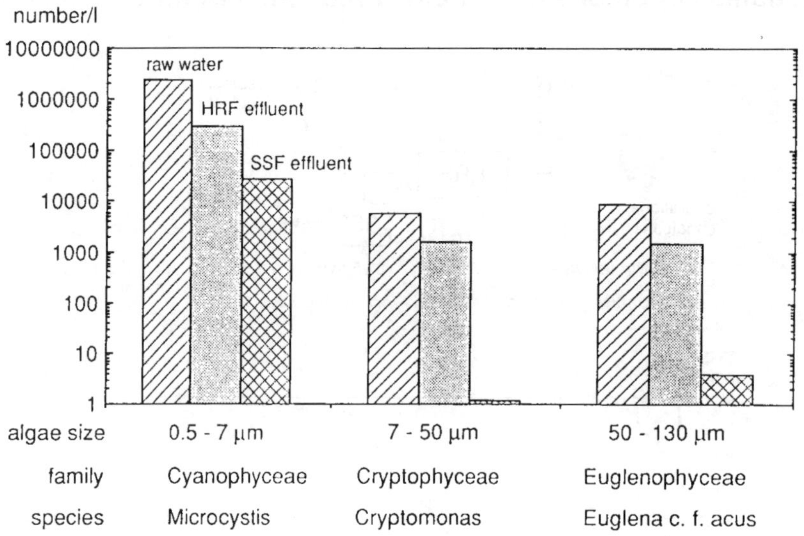

Note: Filtration rates: horizontal-flow roughing filter 1.5 m/h, slow sand filter 0.25 m/h.

Source: SANDEC, 1992.

La Javeriana, Columbia

The treatment plant of La Javeriana in Cali, Columbia, is a good example of the multiple filter approach. The treatment process scheme consists of an intake filter, two horizontal-flow roughing filter and two slow sand filter units. The respective filtration rates amount to 4.7, 1.2 and 0.15 m/h. The performance of the treatment plant with a capacity of 260 m^3/d has been extensively monitored by the Centro Inter-Regional de Abastecimiento y Remoción de Agua (CINARA) (1990). The variation of the different water quality parameters and their respective reduction by the different treatment stages is summarised in Table 1. Figure 11 illustrates the gradual reduction of turbidity, apparent colour and faecal coliforms through the different treatment stages. The gradual improvement of these three parameters is an indication for the development of biochemical processes in the different filters. The applied natural purification processes improve the bacteriological water quality by an order of three to four magnitudes, and also reduce significantly the intense colour of the raw water.

Table 1. **Water quality improvement by the different treatment stages for the treatment plant La Javeriana, Cali, Columbia**

		Raw water	Intake filter	Effluent of horizontal-flow roughing filter	Effluent of slow sand filter
Turbidity (NTU)	Average	22	12	4.4	2.6
	Stand. deviation	17.2	9.5	3.0	1.9
Suspended solids (mg/l)	Average	49	27	11	7.5
	Stand. deviation	41	19	7	4.9
Apparent colour (CU)	Average	132	79	34	22
	Stand. deviation	113	53	20	17
Faecal coliforms(/100 ml)	Average	4 800	2 400	295	3.9
	Stand. deviation	2 400	1 600	300	6.6

Source: CINARA, 1990.

Figure 11. **Turbidity, apparent colour and faecal coliform reduction at the treatment plant La Javeriana, Cali, Columbia**

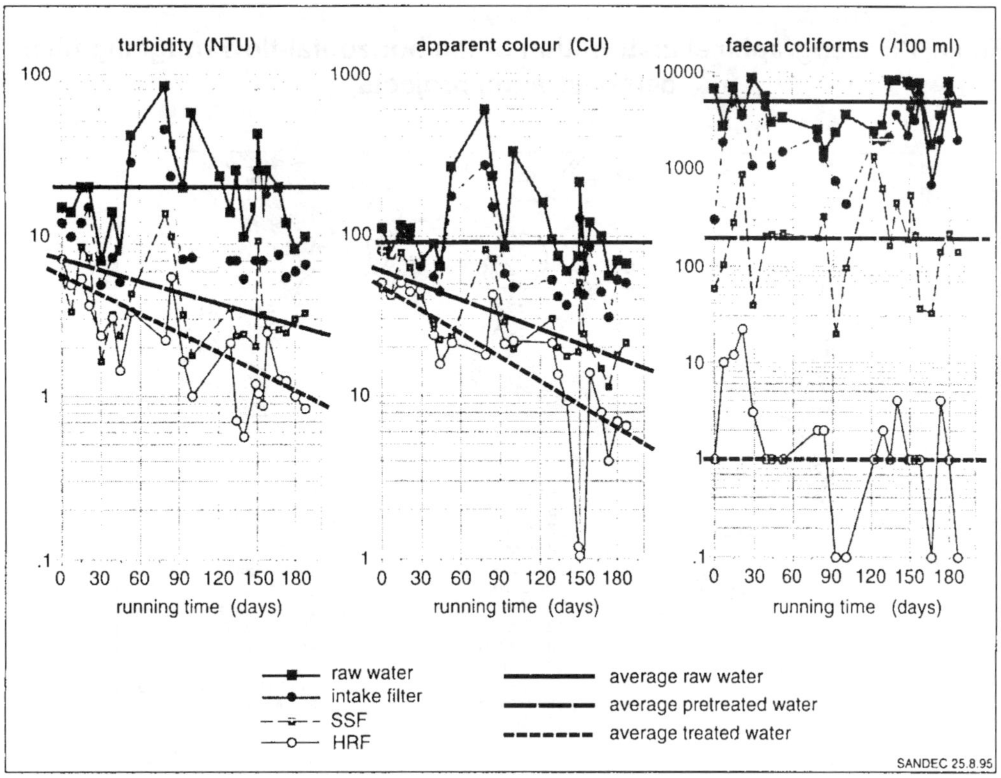

Source: SANDEC, 1995.

Limits of the roughing technology

The roughing filter technology has also limitations. First of all, roughing filters are predominantly designed for rural water supplies. The treatment process requires relatively large installations and can therefore hardly be applied in large urban water supply schemes. Roughing filters operate at relatively small filtration rates requiring larger installations and consequently higher specific construction costs per cubic metre design capacity. The initial costs will be higher than those required for the construction of flocculation/sedimentation tanks. However, the operation and maintenance costs will be lower for the roughing filter option, which is an important factor with respect to the economical sustainability. Construction costs can be saved by the use of the multi-stage treatment concept, which generally minimises the total required filter volume.

The treatment efficiency of roughing filters is also a limiting factor. They are not able to treat any kind of raw water, i.e. raw water with a large amount of colloidal matter and a high suspension stability can not efficiently be treated by roughing filters. Destabilisation of the suspension with chemicals and separation of the formed flocs by direct filtration using roughing filters has been proved to be an efficient treatment method with a high process stability.

The horizontal-flow roughing filter technology has spread to more than 25 countries (see Figure 12) in the past 10 years. According to SANDEC's information, over 80 horizontal-flow roughing filter plants were constructed during this period. This not only enabled the development of local horizontal-flow roughing filter expertise, but positive experience with the treatment process convinced the co-operation partners, who now strongly identify with the horizontal-flow roughing filter technology.

Figure 12. **Geographical distribution of the horizontal-flow roughing filter demonstration projects**

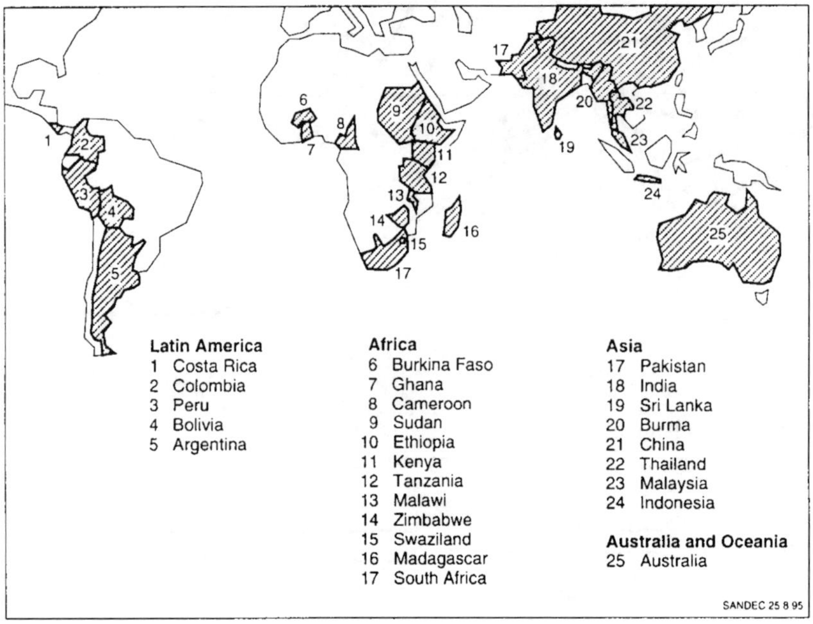

Source: SANDEC, 1995.

An overview of the pre-treatment methods and current experience is presented in a report by the International Water and Sanitation Centre, The Hague (IRC) (1989). Comprehensive design details and guidelines on roughing filters are summarised by Wegelin (1996). The different pre-treatment methods are currently field tested in parallel in Cali, Columbia, where CINARA investigates, in collaboration with IRC and different other international technical institutions and supporting agencies, possibilities to optimise and simplify the pre-treatment processes (CINARA, 1989).

Solar Water Disinfection (SODIS)

Micro-organisms are vulnerable to light and heat. The use of solar energy, which is universally available and free of charge, is the basis for a water treatment process called solar water disinfection (SODIS). The treatment basically consists of filling transparent containers with water and exposing them to full sunlight for several hours as shown in Figure 13. SODIS may be used as a batch process at household level to treat small quantities of drinking water in bottles or plastic bags. However, the daily capacity of the batch process is limited to the water volume stored in the bottles and exposed to the sun. Therefore, SODIS can also be used in continuous-flow systems. Such installations consist of solar collectors and heat exchangers and significantly increase the daily output. Hence, SODIS continuous-flow reactors might be used to provide disinfected water for institutions (e.g. hospitals or schools).

Figure 13. **Application of solar water disinfection**

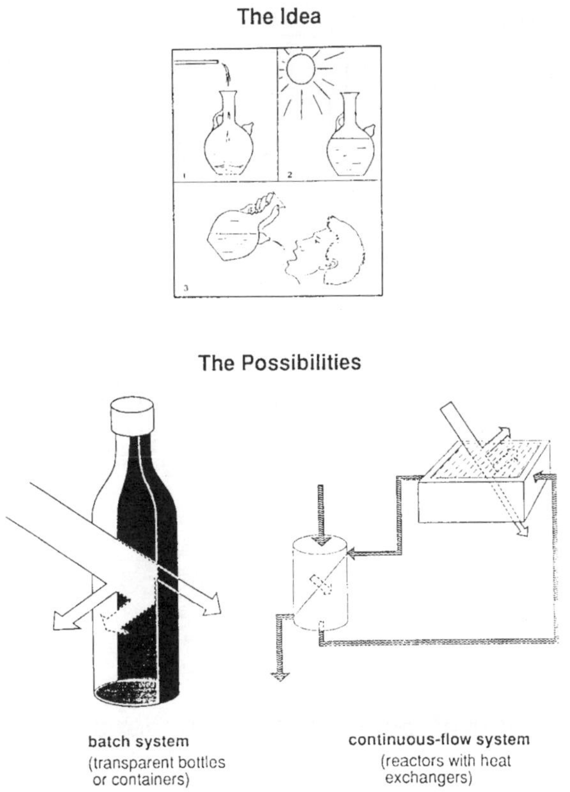

Source: SANDEC, 1994.

EAWAG/SANDEC embarked on extensive laboratory and field tests in 1991 in order to assess the potential of SODIS and to develop an effective, sustainable and low-cost water treatment method. At that time, two different processes using solar energy for water treatment were propagated. The first focused on solar water treatment by radiation, and the second applied solar thermal water treatment. However, the research conducted at EAWAG (Wegelin *et al.*, 1994) and its partners in developing countries (Wegelin *et al.*, 1997) revealed that synergies induced by the combined application of radiation and thermal treatment have a significant effect on the inactivation rate of the micro-organisms. Figure 14 illustrates this phenomenon: two different faecal coliform inactivation tests were carried with constant water temperatures of 30 °C and 50 °C, respectively. The dark control (dc) samples wrapped in aluminium foil show the inactivation effect of temperature only. The faecal coliform inactivation rate for the combined treatment (sunlight and water temperature) is moderate at 30 °C; however, it increases significantly at 50 °C. Parameter tests with different water temperatures ranging from 20-55 °C revealed that for the inactivation of bacteria (*E. coli, Str. faecalis, Enterococci*) the threshold temperature for synergetic effects is 50 °C. The die-off rate of faecal coliforms under different stress factors is also shown in Figure 15 and clearly illustrates the synergetic effect. Viruses (Rotavirus, encephaloyocarditis virus and bacteriophage f2), however, are more temperature sensitive as their die-off rate steadily increases with temperatures in the range of 20-50 °C. Hence, the recorded synergetic effects of solar radiation -- in which the UV-A light is the most germicidal range -- and thermal water treatment favour a combined use of these two water treatment processes.

Figure 14. **Influence of the water temperature on the inactivation of faecal coliforms**

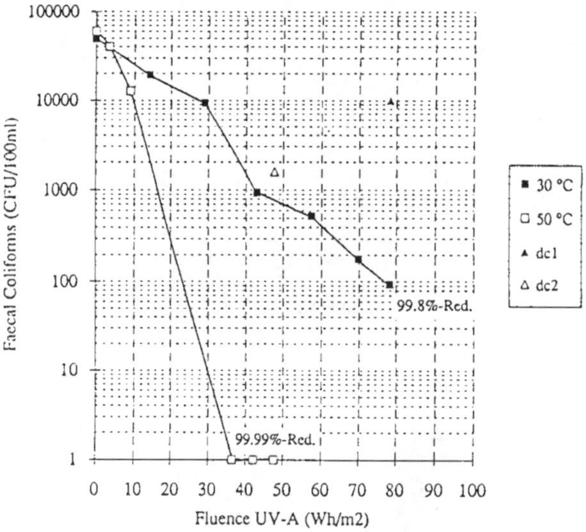

Source: SANDEC, 1994.

Figure 15. **Die-off rates of faecal coliforms under different stress factors**

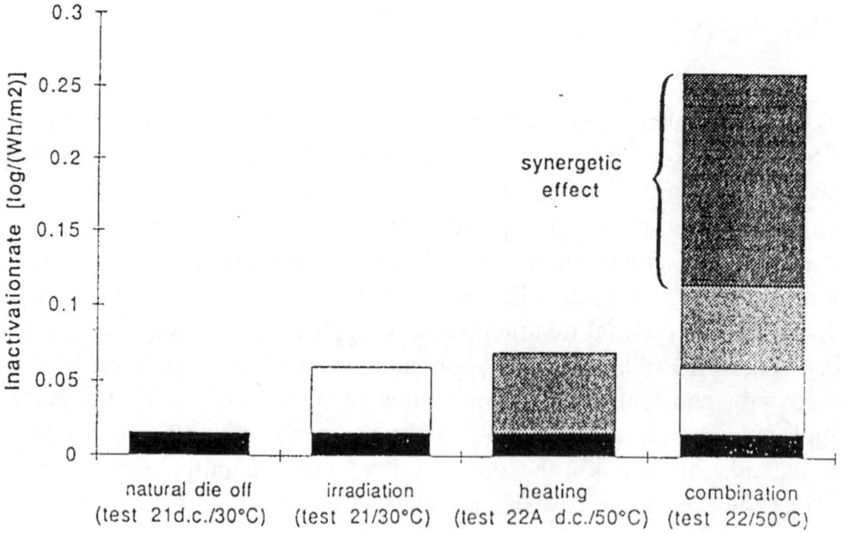

Source: SANDEC, 1996.

SODIS batch process

The SODIS batch process has a limited output capacity, and therefore will be used to treat small quantities of drinking water only, i.e. in the order of two to three litres per person and day. In order to achieve sustainability, locally available material should be used as much as possible. Soft drink glass or plastic bottles are generally available on the local market. Such bottles should be half blackened with a paint as illustrated in Figure 16 in order to increase the heat adsorption. The graph shows the increase of water temperature in a 1.5 litre plastic bottle exposed to the sunlight. The water temperature rose from 25 °C to 50 °C within three hours and attained a maximum of 57 °C whereas the initial faecal coliforms concentration of 50 000 CFU/100 ml was totally removed within four to five hours exposure time. The graph shows that during the first two hours (or 30 Wh/m²), inactivation was very small (only four per cent reduction) as the water temperature rose to 43 °C only. During this period, only the radiation effect was active. In the following, the water temperature increased to over 50 °C, and the inactivation efficiency improved considerably. After an additional 120 minutes, the inactivation of faecal coliforms achieved 100 per cent. Thus, a three to four log reduction of faecal coliforms found in raw water and stored in plastic bottles requires a UV-A fluence of approximately 75 Wh/m² or five hours of sunlight exposure on a sunny day.

Figure 16. **Water temperature increase and inactivation of faecal coliforms in a plastic bottle**

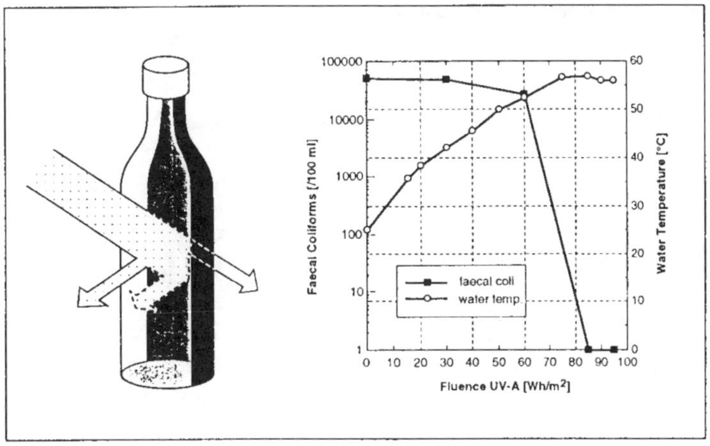

Source: SANDEC, 1995.

The use of plastic bags is another possibility for the batch process option. The used plastic bags were made of a 0.15 mm thin polyethylene foil, filled with water up to 10 litres volume to form a two to six cm thick water layer. The ratio of exposure area to water volume is two to three times larger for plastic bags as compared to bottles. Hence, the water temperature will increase more rapidly, provided the bag is placed on an adequate surface. The small water depth of two to four cm enhances the inactivation of micro-organisms by radiation. The graph of Figure 17 illustrates the inactivation of faecal coliforms and *Vibrio cholerae* which have a similar die-off characteristic. A three to four log reduction could be achieved within an exposure time of 140 minutes at a UV-A dose of 54 Wh/m². Hence, plastic bags allow the most effective heating and irradiation, and therefore, the most efficient inactivation of micro-organisms. The release of taste and odour to the water might pose a problem by the use of plastic bags. However, EAWAG/SANDEC has developed in co-operation with a Swiss plastic manufacturer a special SODIS plastic bag. This bag is made of two different polyethylene foils, a transparent one on top and a black one at the bottom, and is equipped with a handle and screw plug. This bag is now undergoing extensive field tests to assess its durability and acceptance by the consumers.

Figure 17. **Water temperature increase and inactivation of faecal coliforms and *Vibrio cholerae* in plastic bags**

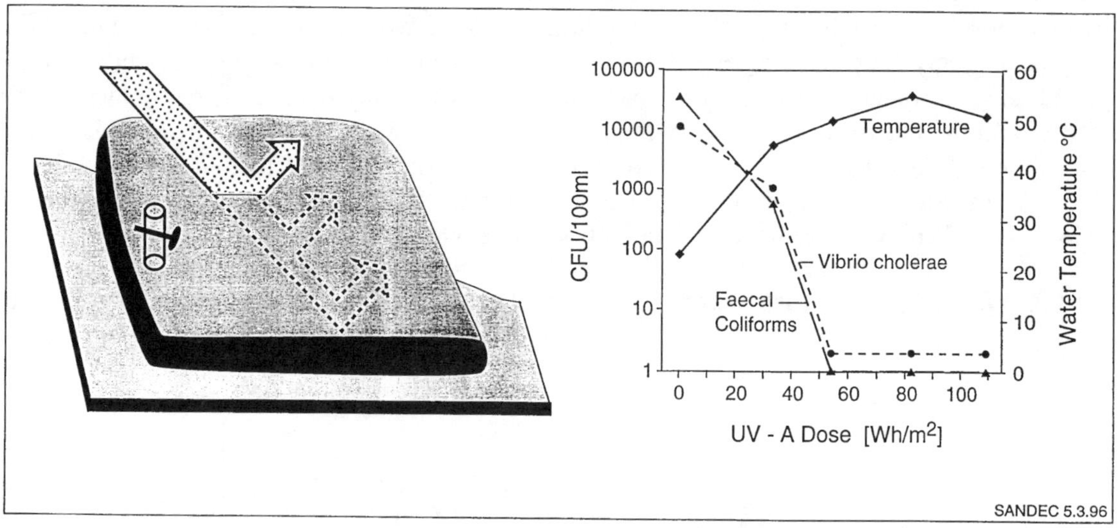

Source: SANDEC, 1996.

SODIS continuous-flow process

The daily output of the SODIS process can be considerably increased by applying continuous-flow systems. The raw water is stored in an elevated tank, flows from there by gravity through an heat exchanger and a reactor for sunlight exposure into the clean water tank. As illustrated in Figure 18, the raw water is preheated in the heat exchanger by the 50 °C hot and treated water, runs from there through the SODIS reactor, where it is heated up and exposed to sunlight for one hour, and finally is cooled down in the heat exchanger before it is stored in the clean water tank. Hence, the available solar energy is constantly recycled which significantly increases the output. The daily capacity of such a system amounts to about 100 litres of disinfected water per square metre solar collector.

Figure 18 also shows the water temperatures at different places of the system. Flow through the installations starts after a two hour heating up period when the water temperature has reached the necessary 50 °C. At that time, a thermovalve used for flow control opens. With increasing operation time, the water temperature in the SODIS reactor increases to over 60 °C which enables a very efficient inactivation of the micro-organisms. The faecal coliforms found in the raw water at an initial concentration of more 30 000 CFU/100 ml are totally inactivated by the SODIS reactor. Field tests carried out in co-operation with Swisscontact and AyA (Instituto Costarricense de Acueductos y Alacantarillados) in San José, Costa Rica revealed that the SODIS plant works at full capacity even on days when the sky is 50 per cent overcast with clouds. On very cloudy days, the flow through the system was unsteady and caused problems which could be partly overcome by reducing the flow rate.

Figure 18. **Water temperature increase and inactivation of faecal coliforms in a SODIS reactor**

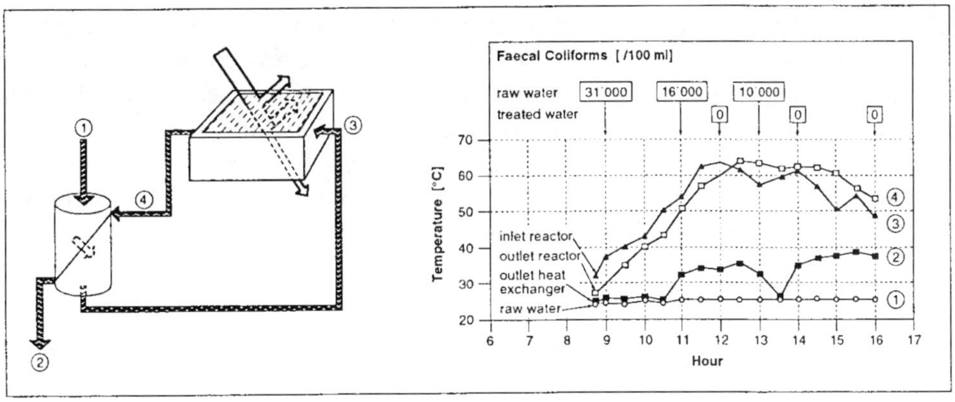

Source: SANDEC, 1996.

SODIS demonstration projects

EAWAG/SANDEC has conducted a market analysis on SODIS to assess the potential and local conditions required to develop a demand-oriented and sustainable technology. Four different target groups might benefit from the new water treatment process: the unserved rural population dependent on individual water supply, institutions in rural areas not served by a public water supply, people in urban low-income areas supplied by poor quality water, and finally, people affected by catastrophes and war, i.e. migrants and refugees. The potential and limitation of SODIS has been assessed by numerous field tests carried out in Switzerland, Columbia, Costa Rica, Jordan and Thailand. The project is now focusing on technical, socio-cultural and economic aspects. In a demonstration project programme co-ordinated by EAWAG/SANDEC, over 700 households will participate using the batch process at various sites in South America (Columbia and Bolivia), Africa (Burkina Faso and Togo) and in Asia (Indonesia, Thailand and China). The addressed implementations aspects will be studied in these demonstration projects prior to applying SODIS on a large scale.

EAWAG/SANDEC in co-operation with Swisscontact is also in the process of modifying a solar collector to be used as a SODIS reactor. This system will be field tested in ten demonstration plants in Honduras and Costa Rica. The aim of this co-operation is the development of a full-proved SODIS continuous-flow system fit for mass production.

In order to enhance information exchange on the use of SODIS, EAWAG/SANDEC is building up an information network between its co-operation partners. Any other institution interested in the application of SODIS is invited to join this network and to share with us their practical experience.

Final remarks

The provision of safe water remains a challenge, particularly in developing countries facing increasing economic, institutional and socio-cultural problems. Under such conditions, self-reliant, sustainable and community-based water supplies will be less affected. Slow sand filtration meets these criteria and is therefore recognised as a particularly appropriate technology for developing countries. In the past, however, slow sand filter operation was often hampered by inadequately pre-treated turbid surface water. Over the last decade, roughing filtration was rediscovered and is now used for its simple and efficient process.

The research on SODIS is driven by the motivation to develop a reliable and inexpensive water treatment method for developing countries. SODIS as a batch process could cover the demand for safe drinking water of single families in rural and urban low income areas. Hospitals, schools and neighbourhood groups could use this treatment method in continuous-flow systems. Finally, SODIS could be used in refugee camps or as temporary water treatment in emergency situations (e.g. after earthquakes or hurricanes). SODIS has the potential to contribute to public health improvement; a controlled dissemination of the methods is now required.

Questions, comments and answers

Q: Have you monitored for other pathogens, such as *Giardia*? What about growth of biofilms in pipelines?

A: Viruses and bacteria are the main problem, and we can remove these and we concentrate on them. Pipelines are not a problem because the treated water is used immediately.

Q: Does solar disinfection depend on global radiation or total sunshine hours?

A: It is not visible light that is important but UVA radiation in combination with visible light.

C: Protozoa can be removed by slow sand filtration, and *Cryptosporidium* is susceptible to temperature and would be inactivated by heat and sunlight.

REFERENCES

BASIT, S.E. and D. BROWN (1986),"Slow sand filter for the Blue Nile Health Project", *Waterlines* 5, pp. 29-31.

CENTRO INTER-REGIONAL DE ABASTECIMIENTO Y REMOCIÓN DE AGUA (CINARA) (1989), Proyecto Integrado de Investigación y Demonstración de Métodos de Pretratamiento para Sistemas de Abastecimiento de Agua, CINARA/IRC.

CENTRO INTER-REGIONAL DE ABASTECIMIENTO Y REMOCIÓN DE AGUA (CINARA) (1990), Proyecto Filtración Gruesa Horizontal, Informe Final.

DORCOO, K. (1992), "The case of Mafi Kumase, Ghana", International Workshop on Roughing Filters for Water Treatment, 25-27 June, Zurich.

INTERNATIONAL WATER AND SANITATION CENTRE (IRC) (1989), "Pre-Treatment Methods for Community Water Supply", IRC, The Hague.

WEGELIN, M. (1996), *Surface Water Treatment by Roughing Filters*, SANDEC Report No. 2/96, Intermediate Technology Publications, London.

WEGELIN, M., R. SCHERTENLEIB, and M. BOLLER (1991), "The decade of roughing filters - development of a rural water treatment process for developing countries", *J. Water SRT-Aqua* 40, pp. 304-316.

WEGELIN, M. *et al.* (1994), "Solar water disinfection: scope of the process and analysis of radiation experiments", *J. Water SRT-Aqua* 43, pp. 154-169.

WEGELIN, M. *et al.* (1997), "SODIS - an emerging water treatment process", *J. Water SRT-Aqua* 3.

XU, X-K. (1994), "Roughing filtration combined with slow sand filtration", National Workshop on Rural Water Treatment, 9-13 May, Hangzhou, China.

SUTRANE AND ZENOGEM®: BIOLOGICAL WASTEWATER TREATMENT SYSTEMS FOR WATER RE-USE APPLICATIONS

by

Bruce Jank
Water Technology International Corporation, Burlington, Ontario, Canada

Introduction

A shortage of water supplies in many countries has become the driving force for the re-use of treated wastewater for numerous industrial and municipal applications. The development, installation and verification of innovative biological treatment systems capable of providing a product which meets specific use criteria, will assist in reducing this water supply problem. This paper describes two biotechnology-based treatment options which can be used in water re-use systems. They are the SUTRANE system developed for the re-use of water, nutrients and energy at the domestic level, and the ZenoGem® system which combines biological treatment with membrane filtration. The SUTRANE system incorporates a low-cost version of a constructed wetland which functions best in warm climates and is ideally suited for developing countries. The ZenoGem® technology is a modification of the activated sludge process with the secondary clarifier replaced by a membrane filtration module. This system produces a high quality product ideally suited for a complete range of re-use options. The assessment includes a description of the principles of operation, system performance, the status of full scale utilisation, operation and maintenance requirements and the range of system applicability for both technologies.

SUTRANE system

Process description

The SUTRANE Integrated Wastewater Treatment System was developed by Professor Jesús Arias Chavez of the University of Chapingo in Texcoco, the State of Mexico. The technology has been further developed and marketed by the Xochicalli Eco-Development Foundation, A.C.

The English translation for SUTRANE is the Unit Treatment System for the Re-use of Water, Nutrients and Energy at the domestic level. As illustrated in Figure 1, the system provides primary and secondary treatment. The primary system includes an anaerobic digester for the treatment of black water and a two-stage reactor for the treatment of grey water, a pre-oxygenator (a box filled with stone and gravel) followed by a grease trap. Both primary effluents flow into a channel with aquatic plants. These effluents sub-irrigate a secondary filtration field constructed of stone, gravel, and sand, with the entire bed placed on an impermeable film. Selected plants are grown on the filtration bed. A multi-purpose greenhouse can be used to provide optimal growth for the plants in both stages of secondary treatment.

For larger systems, the SUTRANE system concepts have been incorporated into a design referred to as the Dual Microplant system. The components of this system are presented in Figure 2. Blackwater combined with the biodegradable organic fraction of solid wastes are treated in a three-stage anaerobic reactor followed by solid/liquid separation and effluent polishing. The selection of the effluent polishing tertiary treatment technology is based on the water quality re-use requirements.

Figure 1. **Schematic of SUTRANE system**

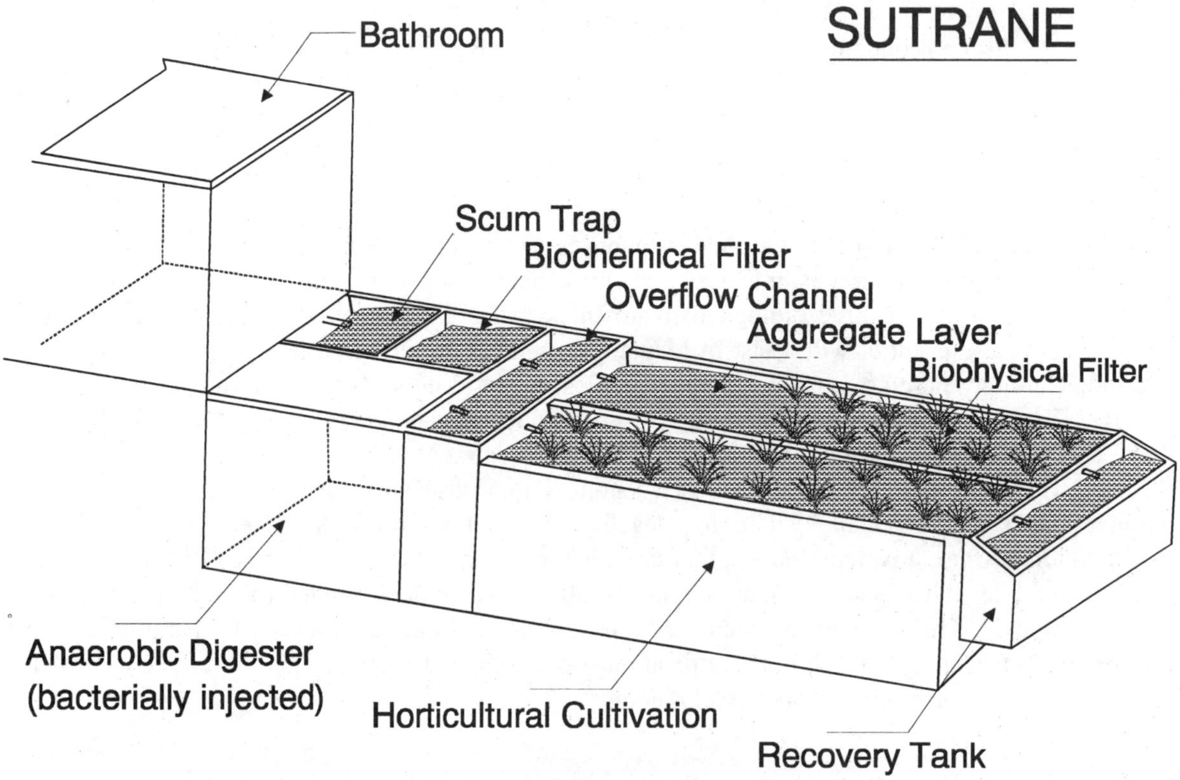

Source: Jank, 1995.

The anaerobic digesters decompose complex organic material, thereby generating methane gas and liberating essential nutrients for plant growth in the secondary treatment system. The methane gas is used as a fuel source for cooking or heating.

Pre-treatment, including the pre-oxygenator, provides film flow on the surface of the rock media, absorbing oxygen necessary to counteract the harmful effect of the detergents. This effluent flows to the grease trap where the oil and grease floats to the surface; the grease is re-used for soap production or placed in the anaerobic digesters to enhance digester loading and performance.

The plants in the secondary process consume the available nutrients and, with the assistance of the soil micro-organisms in the filtration bed, provide a relatively high degree of treatment.

Figure 2. **Schematic of Dual Microplant system**

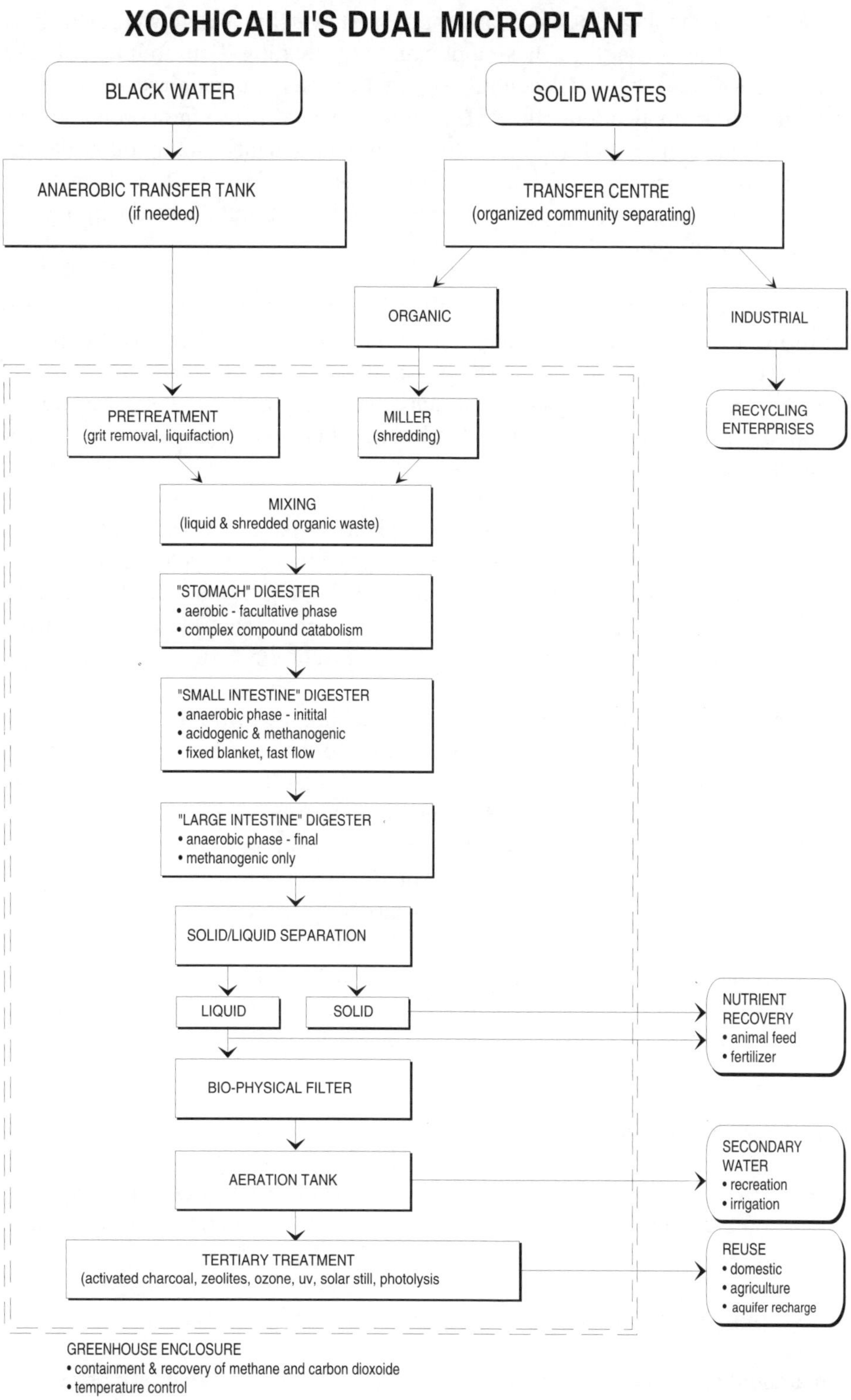

Source: Jank, 1995.

Status of application of the technology

The 'SUTRANE' technology for the treatment of wastewater was reviewed by a Canada-Mexico team (Jank, 1995). The technology has application in treating both municipal and industrial wastewater under conditions where decentralised applications are required. Compared to other technologies with similar applications, the 'SUTRANE' system offers numerous advantages. The process includes simple primary and secondary treatment components. Its modular design allows the system to be tailored to a variety of operating environments. The SUTRANE technology is best applied in warm climates and is ideal for developing countries. Capital and operating costs are very low in comparison to other systems.

The initial system was developed and installed in early 1970 in Mexico. There is a limited amount of performance data available, mainly due to the simplicity and size of the process. Design data for the larger Dual Microplant systems should be developed with the assistance of Professor Jesús Arias Chavez. For systems of ten dwellings or less, design could be based on technical descriptions provided by individuals that have participated in the construction and operation of systems of such size.

An entire SUTRANE system and Dual Microplant systems can be constructed using local material and labour. An organisation similar to the Xochicalli Eco-Development Foundation, A.C., could be established to market, design and construct the technology, or a local firm could provide the same services on a fee for service basis. In all cases, an agreement should be reached with Xochicalli Eco-Development Foundation, A.C., with respect to the utilisation/marketing of technology.

Emphasis has been placed on marketing the technology within Mexico, while research, development, and full-scale performance assessment have not been a high priority. Since the systems are very effective in warm weather conditions, and resources are not available for research and development, efforts have been focused on the commercialisation of the technology. This has been highly successful, as the developers estimate that there are now thousands of systems operating in Mexico and other countries in Central and South America and the Caribbean. While the majority of these systems are for single dwellings, a number of larger systems provide treatment for industries, institutional facilities, and residential subdivisions. The two largest systems are Dual Microplants, located at the Iberoamericana University in Puebla, and at the University of Chapingo in Texcoco. The former serves a population of approximately 2 000, and the latter a population of approximately 8 000.

Assessment of SUTRANE performance

The effluent quality for the three litres/second. Dual Microplant system treating the wastewater from approximately 2 000 inhabitants at the Universidad Iberoamericana in Puebla is given in Table 1.

This data is representative of the performance of the SUTRANE systems.

A limited amount of heavy metals and bacteriological data is available. The results indicate that the quality is relatively comparable to an effluent from a conventional secondary activated sludge process. The limited data makes it difficult to conduct a statistically valid comparison with other technologies. However, the effluent quality appears to be much better than what is generally obtained from low cost technologies.

Table 1. **Performance - SUTRANE system**

Parameter	Influent (mg/L)	Effluent (mg/L)
BOD_5	260	20
Oil and grease	430	1
Organic N	84	0.5
TSS	223	48

Source: Jank, 1995.

Operation of the SUTRANE system

The operational requirements for the SUTRANE system are clearly less than 25 per cent of those for a conventional mechanical system, and maybe under ten per cent. For the larger Dual Microplant systems serving 8 000 inhabitants, the operator would spend two hours per shift. This time would be spent on cleaning and housekeeping responsibilities in addition to operational functions.

For systems for individual dwellings, the following maintenance is required: skim grease from the grease trap three times per year, clean the pre-oxygenator screen once per month, and remove sludge from the digester once every ten years. This does not include the care and harvesting of plants grown on filtration beds.

For the larger Dual Microplant systems, the aquatic plants are harvested and shredded in a mechanical device developed, manufactured and marketed by the Xochicalli Eco-Development Foundation. The plants are fed to the first stage of the three-stage anaerobic reactor in order to increase methane gas production, a valuable energy by-product of the treatment system.

The SUTRANE system operates best in warm climate situations. The complete SUTRANE system could only be utilised in colder climates if the primary grey water treatment unit and the secondary treatment units were enclosed in a greenhouse. Energy requirements would likely invalidate this approach. However, if methane generation from the anaerobic reactor was optimised, the energy costs may be substantially reduced.

The anaerobic digester concept and design is directly applicable in rural areas and could be considered as a replacement for a septic tank or similar pre-treatment systems. Operating the digester at loadings which ensure methane gas production (and gas utilisation) is essential for optimal process efficiency and reduced operating costs.

Economic assessment of SUTRANE technology

The SUTRANE technology is definitely lower in capital and operating cost than competing technologies generating a similar quality effluent. SUTRANE systems have been designed so that the individual homeowners, multiple dwelling owners, or communities, can construct the plants using their own labour. The capital cost of construction is the cost of building materials, and every effort has been made to reduce this cost through optimal process design. Capital costs estimates of \$0.06-0.12/m^3[*] have been provided by Prof. Arias Chavez for communities of 10 000 inhabitants. Data from the Comisión Nacional del Agua would indicate that comparable costs for conventional

[*] These cost estimates are in Mexican pesos and were developed in January 1995.

technologies providing a similar effluent quality would range from $0.30 to 0.60/m^{3*} or five times the cost of SUTRANE systems: SUTRANE operational costs are a small fraction of the cost of the conventional mechanical systems.

Xochicalli Eco-Development Foundation has expended considerable effort on cost reduction initiatives. Reactor design is based on the efficient use of re-used waste products.

As with the performance data, there is a limited amount of cost data available. Since cost is specific to the individual sites, it is difficult to make cost comparisons using similar population numbers for different sites. However, there is no question that trends presented and cost estimates indicate that these technologies will cost significantly less than a competitive technology providing a similar effluent quality.

A comparison of SUTRANE and competing technologies

The Wastewater Technology Centre completed a critical assessment of appropriate technologies for wastewater treatment and disposal for rural communities. The study evaluated both on-site and off-site wastewater treatment technologies. All appropriate technologies were considered and evaluated against the following criteria: principles of operation, status of the technology, suitability to local conditions, operation and maintenance requirements, capital and installation costs, effluent quality from full scale plants, and utilisation of local resources for construction and operation.

The study (Wastewater Technology Centre and Davis Engineering & Associates, 1995*a* and *b*), carried out for the Province of Newfoundland and Environment Canada, recommended the use of peat filters, the Waterloo biofilter (a packed-bed absorbent biofilter), the EcoFinn trickling filter, the New Hamburg process, and constructed wetlands, as technologies with the greatest potential for meeting assessment criteria.

The SUTRANE system was one of the technologies evaluated. It was not selected as an appropriate wastewater treatment system for the Province of Newfoundland only because it would be adversely affected by the low ambient air temperatures encountered during the Canadian winters.

Socio-cultural impact of technology implementation

The Xochicalli Eco-Development Foundation is an organisation which promotes a holistic approach involving environmentally sustainable economic development and the conservation of natural resources. The Foundation understands these ecological principles, and is able to provide integrated solutions to environmental issues related to air, water, solid waste, environmental hygiene and environmental health.

Another unique attribute of Xochicalli Eco-Development Foundation is its marketing philosophy. Each participant in the foundation's program agrees to transfer his new knowledge to at least four new participants. This concept allows for effective dissemination of the Foundation's ecological principles. The promotion and the implementation of the principles of ecological sustainability have created a powerful marketing approach. This has resulted in the widespread acceptance of the treatment systems, in both rural and urban communities.

---

[*] These cost estimates are in Mexican pesos and were developed in January 1995.

Summary

- The SUTRANE and Dual Microplant systems represent examples of biotechnology used extensively throughout Mexico, Central and South America and the Caribbean, providing a high quality effluent at reduced capital and operating costs.

- The Xochicalli Eco-Development Foundation has developed an integrated ecological approach to the management of environmental issues. The approach has resulted in the installation of thousands of SUTRANE systems in Mexico and other countries.

- The SUTRANE technology represents an ideal technology for warm climates. Because of its simplicity and reduced cost, it should be identified as an appropriate technology for developing countries.

The ZenoGem® process

Process description

The membrane bioreactor (MBR) process is a highly efficient activated sludge process (ASP) which has been operated at full scale in industrial and municipal applications for the last decade. The process consists of an aerobic biological reactor integrated with a membrane system which replaces the secondary clarifier. The membrane system provides an ultimate barrier for biomass control.

The original MBR systems shown in Figure 3 used crossflow membranes to replace the clarifier and post filtration process found in most traditional treatment processes. The membrane crossflow separation stage provided a simple and reliable positive barrier to soluble organics and micro-organisms. Its separation performance was independent of the quality or condition of the biological process fluid. In addition, the membranes completely eliminated the maintenance and process limitations associated with a passive clarifier design. The positive membrane filtration process dramatically simplified the entire treatment process by eliminating clarifier maintenance, routine process adjustments including chemical additions for enhancing settleability, and regular sludge management. Since process upset problems associated with sludge bulking and difficult mixed liquor floc conditions were eliminated, a very stable and efficient biological process was maintained. In addition, since regular process adjustments, manual cleaning and maintenance operations were not necessary, the membrane/biological process could be highly automated using industry standard programmable controllers to operate the process.

The most significant benefits of integrating biological treatment with membrane technology, however, occur with the biological reactor. Membrane technology allows reactors to be operated at substantially higher mixed liquor suspended solids than previously possible with conventional technology. They enable long solids retention times (SRT) while operating at short hydraulic retention times (HRT). This makes integrated membrane biological systems excellent processes for treating high strength complex wastes and also allows for significantly reduced amounts of sludge. It is not unusual for integrated membrane biological systems to operate at reactor mixed liquor suspended solids (MLSS) of over two per cent -- which is two to ten times higher than MLSS levels for conventional systems.

Figure 3. **Membrane bioreactor using crossflow membrane separation**

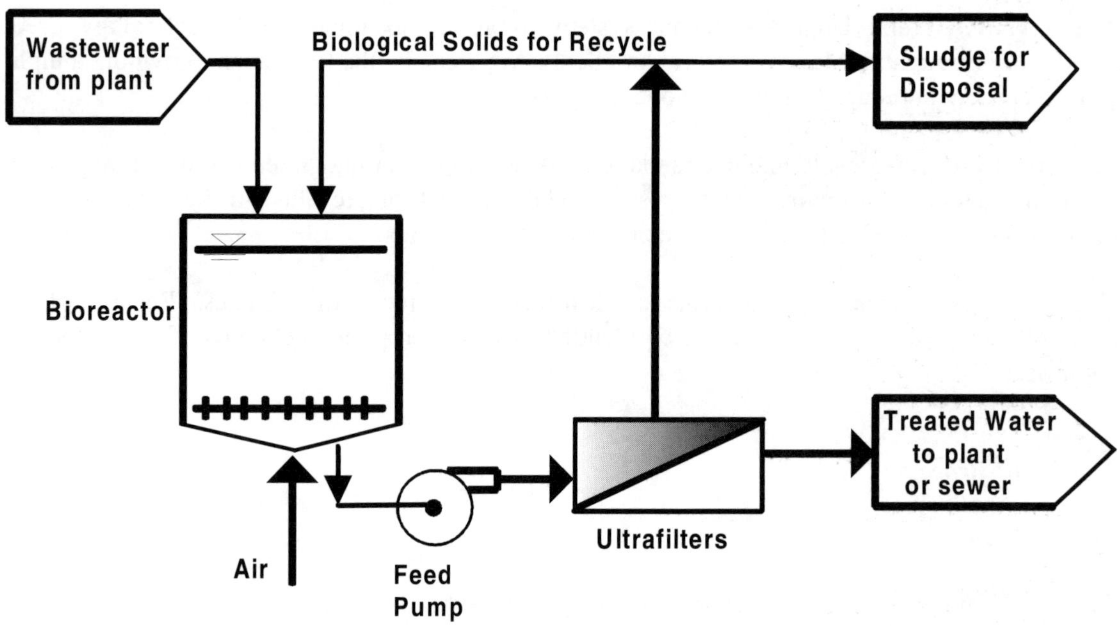

Source: Jordan and Clifton, 1994.

Zenon Environmental commercialised two product lines using this advanced MBR technology. The ZenoGem® technology used for the treatment of complex and oily industrial wastewaters and their Cycle-Let® technology developed for commercial, private and institutional wastewater treatment applications.

The Cycle-Let® technology has been successfully incorporated in a wide variety of applications since 1979. The Cycle-Let® system includes the bioreactor and membrane system shown in Figure 3 and may include activated carbon and ultraviolet disinfection for specific re-use applications.

In commercial, private and public buildings, i.e. office buildings, shopping centres, hotels, multifamily residential dwellings, etc., a major potential exists for Cycle-Let® to reduce water use through on-site wastewater treatment and reclamation. On-site reclamation will include recycling of treated effluent for irrigation, toilet flushing, evaporative makeup and other non-human contact uses. On-site wastewater treatment simply refers to the treatment and discharge of wastewater from a facility on the site in which it is located rather than discharging to a public sewer or a water course where the effluent is taken away from the site. Recycling in such applications can have a major impact on water conservation and can help to solve both water supply and wastewater disposal problems.

Descriptions of specific applications of ZenoGem® and Cycle-Let® have been presented by Jordan and Clifton (1994). A ZenoGem® system, combined with reverse osmosis for recycle has been installed at a new automotive manufacturing plant in Saltillo, Mexico. As shown in Figure 4, the process treats the oily wastewater using an ultrafiltration membrane system integrated with the aerobic biological reactor. The system is followed with reverse osmosis to further polish the processed wastewater. The concentrate is returned to the ZenoGem® process for further oxidation, and the permeate will be re-used for in-plant process water.

Figure 4. **Flow diagram for ZenoGem® oily wastewater treatment and recycle system**

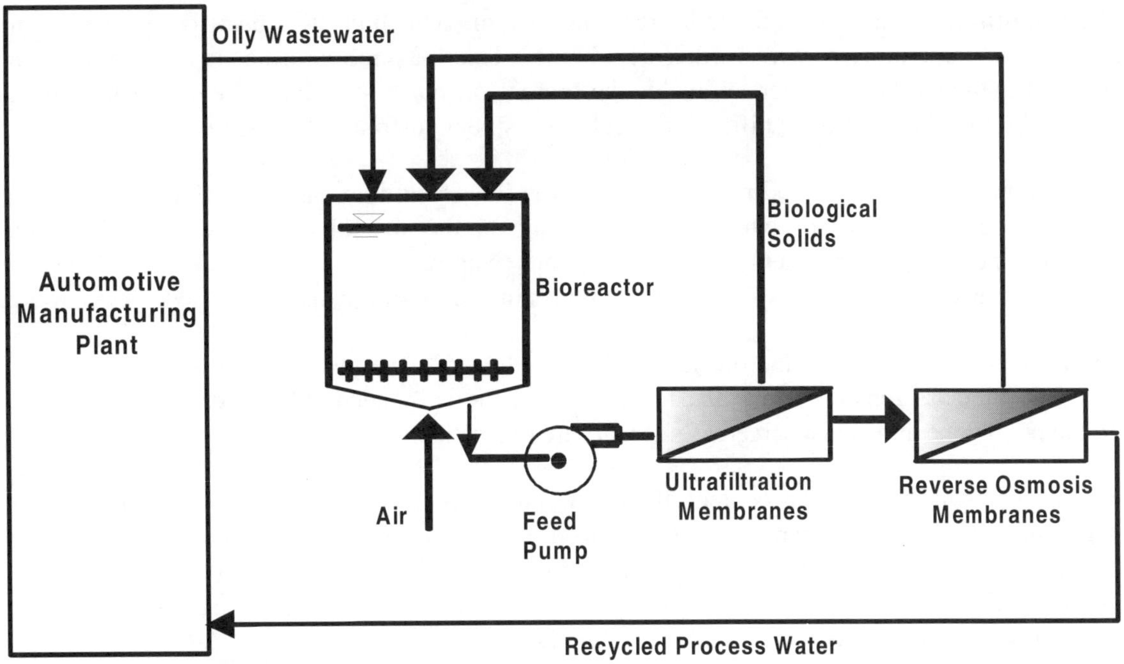

Source: Jordan and Clifton, 1994.

The ZenoGem® system has been refined by replacing the external ultrafiltration membrane module with a hollow-fiber microfiltration membrane module immersed in the aeration tank of the activated sludge process. This modification has substantially reduced the capital and the operating costs of the MBR system.

As with the initial design, the main operating advantages of this system are that the quality of the effluent is independent of the setteability of the mixed liquor, and no operating failure can result in the presence of suspended solids in the effluent. The lack of sludge settling allows operation of the bioreactor in extreme conditions such as high biomass levels (15 000-20 000 mg/L) and high sludge retention times (SRT > 50 days) which cannot be achieved in a conventional activated sludge process.

Operating at high biomass levels means that the membrane bioreactor is four to seven times smaller than a conventional bioreactor, or that an existing bioreactor can be upgraded to treat four to seven times more wastewater without the need of infrastructure construction. Operating at a high SRT allows for sludge digestion to occur within the bioreactor, generating digested sludge volumes 50 to 80 per cent lower than conventional activated sludge plants, and in small to medium plants it allows both functions to be combined in a single tank, further reducing the plant's footprint and ease of operation.

Finally, the process is an efficient tertiary treatment system, reducing suspended solids and coliforms to a minimal level and producing an oocyst and coliform free effluent characterised by its low residual ammonia and phosphorus, even in winter conditions. The plant's effluent meets the most strict effluent criteria and can be discharged directly or fed directly into a reverse osmosis plant for aquifer recharge.

ZeeWeed® the microfiltration membrane

The microfiltration membrane is the heart of the system. This high flow plant is designed with an innovative, low operating cost microfilter membrane. This 0.2 µm microfilter is revolutionary in terms of its design and direct immersion inside the bioreactor, but has a membrane chemistry which has been used for the last decade in difficult to treat, corrosive industrial effluents.

The membrane is an immersed membrane which is integrated into cassettes of modules which are then inserted into frames or into tanks, depending on the size of the plant. Based on the municipality's needs, the ZenoGem® ZeeWeed® plants can be designed as packaged plants or full-scale plants, using frames to suspend the membranes inside existing tanks.

The flow is generated from the outside of the membrane to the inside of the fiber by a small vacuum. The biomass remains inside the tank, whereas the suspended solids free effluent travels through the membrane and is discharged as a tertiary treated effluent.

Because of its higher biomass content, the membrane bioreactor process can efficiently operate at high organic loadings which translates into the need for significantly smaller bioreactors. Typical carbon and nitrogen loadings for the design of ZenoGem® and the conventional activated sludge process are presented in Table 2. The ZenoGem® loadings are five to ten times higher than the typical loadings for conventional municipal activated sludge plants.

The MBR plant for sewage treatment is simple in its construction and operation. With no clarifiers, no sludge return pumps, and the increased biomass concentrations, the ZenoGem® plant is an easy to operate and maintain plant. The entire operation is now based on MLSS control instead of on SVI, sludge return ratios and strict SRTs.

Table 2. **Typical design parameters**

	ZenoGem™ design Typical design	Activated sludge design Typical design
MLVSS	1.0-2.0%	0.1-0.3%
COD	2.6-5.0 kg/m^3/day	< 1.0 kg/m^3/day
BOD	1.5-2.5 kg/m^3/day	< 0.5 kg/m^3/day
NH_3	0.20-0.40 kg/m^3/day	< 0.07 kg/m^3/day

Source: Mourato and Marshall, 1996.

Performance of ZenoGem® process

A comparison of the performance of a conventional activated sludge process and a ZenoGem® system is presented in Table 3. For a typical domestic sewage averaging 200 mg/L BOD, the MBR effluent BOD ranges from non-detectable to two mg/L representing BOD removal from 98 per cent to greater than 99 per cent. These values have been confirmed during comparative field trials conducted in St. John, New Brunswick (Mourato and Marshall, 1996; Mourato *et al.*, 1996).

Table 3. **Comparison of Conventional Activated Sludge Process and ZenoGem® performance**

	Secondary treatment	ZenoGem® treatment
BOD	10-12 mg/L	< 2 mg/L
TSS	10-15 mg/L	n.d.
NH_3	1-10 mg/L	$\leq$ 0.3 mg/L
Total phosphorus	> 1 mg/L	$\leq$ 0.1 mg/L
Total coliforms	> 1 000 cfu/100 mL	< 100 cfu/100 mL
Fecal coliforms	> 100 cfu/100 mL	< 10 cfu/100 mL
Oocysts	> 10 counts/mL	n.d.
Sludge yield	0.3-0.6 kg/kg BOD	0.10-0.3 kg/kg BOD

Notes: n.d.: non detectable; cfu: colony forming units.

Source: Zenon Environmental Inc., 1996.

Since the MBR effluent is filtered through a microfiltration membrane, the effluent suspended solids (TSS) are typically non-detectable. Even with feed streams containing very high suspended solids (>1 000 mg/l TSS), the membrane permeate remains crystal clear, with no suspended solids. Hence, 100 per cent removal of suspended solids is achieved, which is expected, since the nominal pore size of the membrane is 0.2 μm.

The ZenoGem® reactor configuration required to provide nutrient removal is presented in Figure 5. Nitrified mixed liquor is returned to the anoxic zone of the reactor. The quantity of nitrified mixed liquor returned to the denitrification stage will control the extent of nitrate removal and thus the total nitrogen removal. The MBR operates at a low HRT, with a high SRT and mixed liquor suspended solids concentration ensuring that the nitrifiers are retained in the system, even in less than optimal conditions. In winter with a HRT of two hours, the system typically achieves 96 per cent to greater than 99 per cent removal, meeting the strictest tertiary treatment criteria.

Figure 5. **ZenoGem® incorporating nitrogen removal**

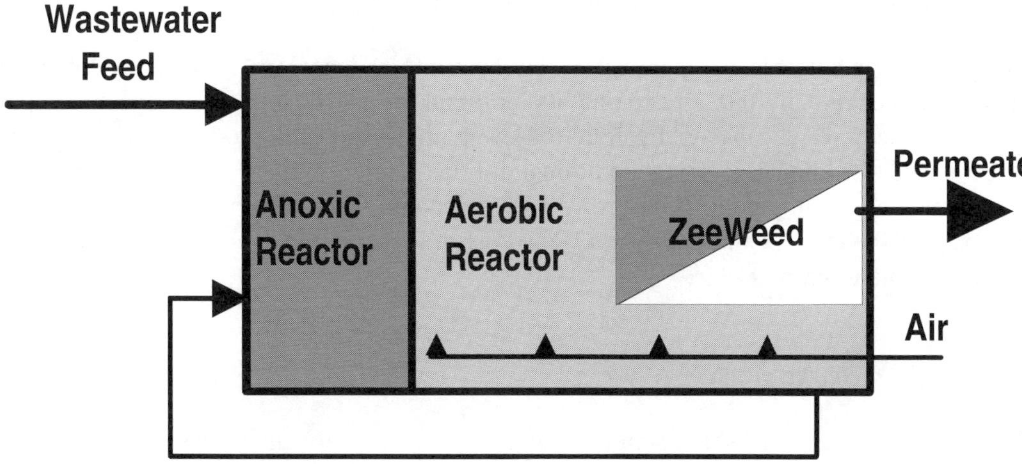

Source: Zenon Environmental Inc., 1996.

The microfiltration unit is an effective disinfection tool with the capability to remove oocysts and bacteria from the treated effluent, avoiding in many instances the need to further disinfect prior to discharge. Viruses are removed by adsorption to the high biomass content in the bioreactor. The phenomenon has been reported in the past by other authors (Fane, 1994).

The bioreactor sludge yield is an important factor in estimating the volume of sludge to be disposed and subsequently on the operating costs. The system can be operated on a primary clarified effluent and hence be very efficient in digesting the biomass within its bioreactor, or operate directly on non-clarified sludge with a premium to pay in terms of sludge yield.

The typical sludge yield in non-clarified sewage, with the system operating at 50 days SRT, is about 0.26 kg TSS/kg BOD consumed. If the primary sludge is removed prior to the bioreactor, higher SRT can be maintained in the bioreactor with subsequent lower sludge yields (<0.2 kg MLVSS/kg BOD consumed). This is a significantly lower sludge production rate than conventional biological processes that operate at an average of 0.6 kg MLVSS/kg BOD consumed. Furthermore, MBR sludge is already concentrated to one to two per cent solids, which represents a further sludge volume generation reduction of two to four times.

Applications of the ZenoGem® process

The most significant applications of the ZenoGem® process are the following:

- upgrading existing sewage treatment plants;
- treating for aquifer recharge or water re-use;
- treating complex high strength wastewaters of variable strength and flow.

Upgrading sewage treatment plants

Many municipalities have sewage treatment plants that require upgrading to increase hydraulic capacity, to provide biological nutrient removal or to produce better quality effluent. In an existing plant, meeting current and future discharge criteria forces municipalities to add more treatment steps to their conventional activated sludge plant. The result is complex sewage treatment plants which require large footprints and are expensive to operate and maintain.

Existing sewage plants can be upgraded at low cost into a ZenoGem® process by lowering the ZeeWeed® membranes directly into an existing aeration tank or clarifier. In this way, municipalities can increase treatment plant capacity by four to seven times and still produce a tertiary treated, microfiltered effluent without incurring additional infrastructure charges. Plant capacity can be increased incrementally as the population grows and as income from development charges becomes available to pay for the incremental expansion. The cost of this solution is considerably less than the cost of conventional alternatives.

Treating for aquifer recharge or water re-use

In areas where water is in short supply and/or expensive, water may need to be re-used for irrigation, industrial, recreational and other non-potable urban uses. Currently, aquifer recharge requires a multi-step treatment process (Figure 6) to bring the sewage to drinking water quality before re-injection into the aquifer. The basic building block for the ZenoGem® process for recycle/re-use is the ZeeWeed® microfiltration membrane module presented in Figure 7. After the ZenoGem® process, the water quality is frequently suitable for re-use; however, if salts need to be removed for water recharge, the ZenoGem® effluent can be polished by a reverse osmosis system. When this is required, the usual problem of reverse osmosis membrane fouling is completely eliminated, since the effluent of a ZenoGem® plant is low in bacteria and has a low silt density index. The combination of ZenoGem® and reverse osmosis produces effluents better than drinking water quality while significantly reducing capital and operational costs. Plant operation becomes both simple and reliable. Also, as an additional benefit, the sewage treatment plant can be upgraded incrementally at minimal capital cost.

Figure 6. **Conventional ASP upgrade for aquifer recharge or water re-use**

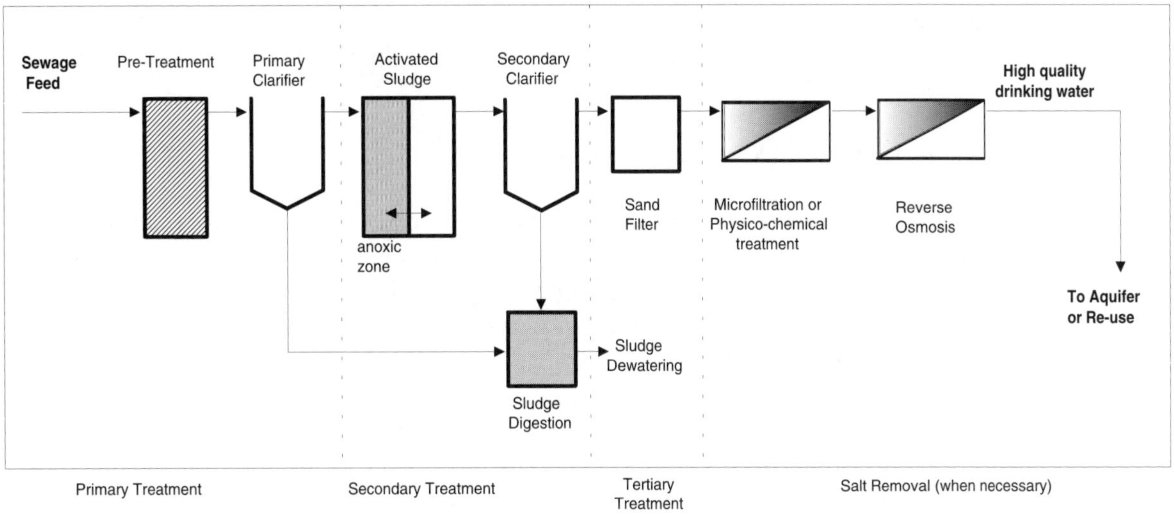

Source: Zenon Environmental Inc., 1996.

Figure 7. **ZenoGem® upgrade for aquifer recharge or water re-use**

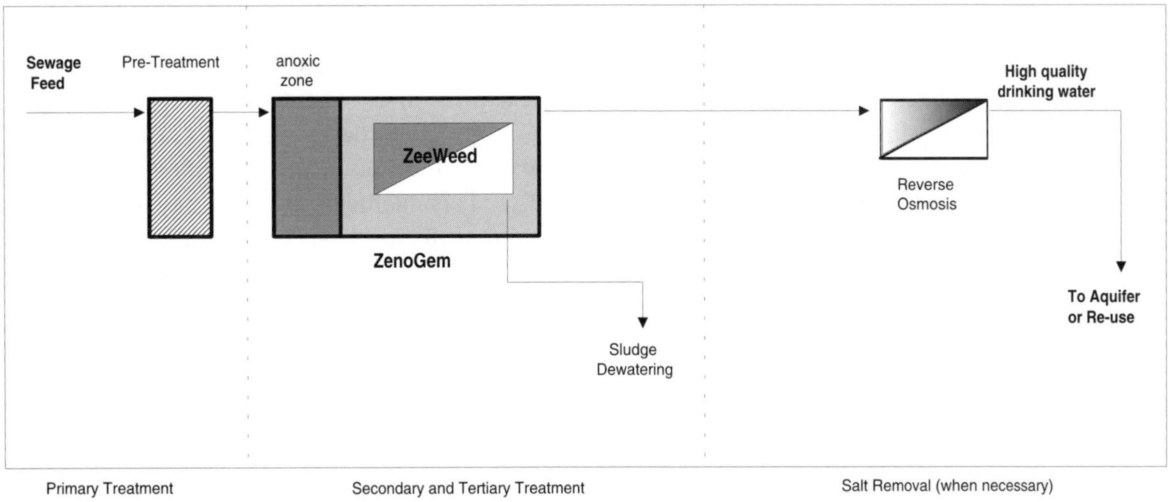

Source: Zenon Environmental Inc., 1996.

Treating high strength and variable feed wastewaters

The ZenoGem® process is an economical solution for municipal landfill leachates, certain combined municipal and industrial effluents, septage treatment and plants where the feed flow or concentration vary significantly.

Typically in landfill leachate applications, the ZenoGem® process is delivered in a mobile trailer -- mounted and ready-to-use. Experience has proven these systems to provide excellent quality effluents. These systems are reliable and easy-to-use even in remote locations.

In plants that are exposed to variable feed conditions, the ZenoGem® easily handles both low and high organic loading conditions, consistently producing high quality effluent. In high organic loading conditions, the ZenoGem® uses its high mixed liquor population to efficiently treat the effluent. During low organic loading periods, the ZenoGem® process functions effectively by further digesting its sludge. Regardless of feed conditions, the ZenoGem® process will not fail.

Cost comparison of conventional ASP versus ZenoGem®

A comparative cost estimate has been prepared by Mourato *et al.* (1996) for the expansion of the Saint John, New Brunswick ASP from 3 800 m^3/d to 20 000 m^3/d. In addition to the expansion of the volumetric capacity, the effluent quality was to be improved with the BOD to be reduced from the existing 20 mg/L to <10 mg/L. The plant which did not nitrify was to be expanded to provide a fully nitrified effluent, i.e. a NH_3 value <3 mg/L.

The cost estimate provided by the consulting firm for the ASP expansion was C$ 28 950 000. The cost estimate for retrofitting the existing 3 800 m^3/d plant with ZeeWeed™ to a capacity of 20 000 m^3/d was C$ 14 804 000. This resulted in a 50 per cent saving in the capital costs for the expansion. Operating costs, which included membrane replacement for the ZenoGem®, were essentially equal.

Summary

The ZenoGem® MBR process has exhibited significant advantages over the designs of conventional wastewater systems in a wide size range of applications. The advent of the submerged membrane now allows the designs to be not only cost-effective but energy efficient. The operation and maintenance costs have been proven to be dramatically reduced while providing an effluent quality that potentially meets drinking water standards. This includes nitrogen and phosphorus removal to well below required limits, combined with virtual disinfection to provide a system that meets or exceeds all the current and potential future discharge requirements of all the major regulation agencies.

Acknowledgement

The author would like to acknowledge the assistance of Professor Jesús Arias Chavez in the preparation of the SUTRANE assessment and the Zenon Environmental team headed by Dr. Andrew Benedek in the preparation of the review of the ZenoGem® technology.

Questions, comments and answers

Q: What is the microbial quality of the first process and the efficiency of virus removal by microfiltration in the second?

A: We don't have data but it would be similar to any wetland or flow-through system. There are indications of considerable absorption of viruses into the biomass. In other words, they are retained in the system.

Q: With high organic loads you would have problems with oxygen transfer. What systems do you use?

A: Coarse bubble aeration is used at present. We hope to use fine bubble aeration when we look at larger systems. We switch to a thermophilic mode of operation if the organic loading is high enough.

Q: Can you estimate the life of your membranes, and have you noticed a drop in efficiency over time?

A: The membrane is back-pulsed every 30 minutes for one second with clean water. The units have been operating for three to four years and the suppliers guarantee a ten-year life.

Q: What systems have been installed and what are costs?

A: The largest systems are 2 000 and 8 000 population equivalents. The estimated cost of a 10 000 population equivalent plant is 0.03-0.06 pesos per m^3 of reactor capacity.

REFERENCES

FANE, A.G. (1994), "An overview of the Use of Microfiltration of Drinking Water and Wastewater Treatment", presented at the Microfiltration Symposium for Water Treatment, Irvine, California.

JANK, B.E. (1995), "Evaluation of the Technical and Financial Viability of the 'SUTRANE' Integrated System for the Treatment of Domestic/Municipal Wastewaters," report for Department of Foreign Affairs and International Trade, Ottawa, Canada.

JORDAN, E. and R.W. CLIFTON (1994), "Alternatives for Recycling and Reuse of Industrial/commercial Wastewater", presented at PRO-ECO Conference, 17-19 May, Monterrey, Mexico.

MOURATO, D. and H. MARSHALL (1996), "Are Membranes the Future for Wastewater Treatment?", *Environmental Science and Engineering* 9:4, pp. 44-46.

MOURATO, D., H. BEHMAN, and G. MCGINNI (1996), "The ZenoGem® Process for Municipal Sewage Treatment Plant Upgrades", presented at the 69th Annual Water Environment Federation Conference, 5-9 October, Dallas, Texas.

WASTEWATER TECHNOLOGY CENTRE, BURLINGTON, ONTARIO, and DAVIS ENGINEERING & ASSOCIATES LIMITED, PORT BLANDFORD, NEWFOUNDLAND (1995*a*), *Assessment of Appropriate Technologies for Wastewater Treatment and Disposal for Rural Communities in Newfoundland, Volume I: Newfoundland Conditions and Technology Selection*, report for Government of Newfoundland and Labrador and Environment Canada, Halifax, Nova Scotia.

WASTEWATER TECHNOLOGY CENTRE, BURLINGTON, ONTARIO, and DAVIS ENGINEERING & ASSOCIATES LIMITED, PORT BLANFORD, NEWFOUNDLAND (1995*b*), *Assessment of Appropriate Technologies for Wastewater Treatment and Disposal for Rural Communities in Newfoundland, Volume II: Technology Assessment,* report for Government of Newfoundland and Labrador and Environment Canada, Halifax, Nova Scotia.

ZENON ENVIRONMENTAL INC. (1996), "Advanced Membrane Systems for Municipal Wastewater Treatment", promotional literature prepared by Zenon Environmental Inc., 845 Harrington Court, Burlington, Ontario, Canada.

COST EFFECTIVE SEWAGE TREATMENT FOR RURAL COMMUNITIES

by

John Upton
Severn Trent Water Ltd., Birmingham, England, United Kingdom

Background and introduction

Sewerage and sewage treatment in the United Kingdom is taken for granted by most of the population, with some 96 per cent of households connected to a public sewerage system.

The four per cent of households not connected to public sewers are nearly all in rural areas. These rural communities have been historically left till last in priority terms, as the more significant environmental effects of the larger urban discharges have been tackled by the UK water undertakings.

The UK water industry was privatised in 1989, and this removed some of the historical restrictions on investment, and the regional water companies were set demanding targets of capital infrastructure development to improve the existing assets, and to meet the stringent discharge standards of the EC Waste Water Directive.

This European Directive provides a convenient sizing criteria for small sewage treatment plants, defining baseline discharge consents for communities greater than 2 000 population equivalent (pe). For lesser populations, it requires that appropriate treatment is provided, but member states are free to set discharge consents which are more restrictive than the baselines established by the Directive.

These rural communities are served by combined or partially separate sewerage systems with often no overflow on the system other than emergency overflows at pumping stations. Many of these rural works therefore receive all flows through treatment, sometimes provided with storm tanks, but frequently the consequences of high rainfall are local flooding and potential pollution of the watercourse.

Severn Trent Water is one of the largest water companies formed following privatisation of the UK water industry in 1989, and it serves the water and wastewater needs of eight million people in the central part of England. The company inherited the responsibility for operation and maintenance of over 730 sewage works serving communities of less than 2 000 pe. Whilst this large number of works represented less than three per cent of its customer base, the regulation provided for tight controls on these discharges with fines, imprisonment or both, should the consents be compromised.

The performance of these small works had deteriorated prior to 1989 and Severn Trent was faced with a strategic requirement to replace many of these ageing assets, as they were proving a cost burden because of high maintenance and attendance.

A decision to replace 200 of these small rural works in the first five years of operation demanded a high pace of investment and construction. A streamlined approach was considered essential and the normal site by site evaluation and procurement procedures were superseded by a process design matrix (Crabtree and Rowell, 1993).

Process selection

This was established for works below 2 000 pe quite simply as a package type Rotating Biological Contactor (RBC) as the preferred treatment system. Where river water quality objectives required discharges to meet conditions more restrictive than 25 mg/l BOD and 35 mg/l TSS, constructed reed beds were chosen as the most appropriate form of tertiary treatment (Green and Upton, 1994.) The success of constructed reed beds for effluent polishing opened the way for their application for cleaning up combined sewer overflows at these rural works.

The decision was taken to limit the flow treated by the RBC Unit to 6 x dry weather flow (dwf) to protect its performance, and excess flows were diverted at the head of the works to storm treatment on reed bed systems (Green *et al.*, 1995).

At the very smallest sites treating sewage from communities less than 50 pe, Severn Trent's preferred approach is to install a septic tank arrangement followed by a reed bed. The performance of such a system has been described previously by Upton and Griffin (1990) and Green and Upton (1993), and some results are included later in this paper.

There is little doubt that this process selection has served the company well, with vast improvements in performance at these small rural works and robust, low energy operation. The one essential element of this strategy that has underpinned this performance has been the use of constructed reed beds which are low cost in construction and maintenance. Reed beds, when used to polish the variable effluent quality from secondary treatment at small sites, produce excellent quality of effluents that require little further treatment for re-use as irrigation or low grade water in the community. Severn Trent has now installed more than 130 reed bed treatment systems to date (1996) and examples of performance figures from such systems are included later in this paper.

Rotating biological contactors

These prefabricated treatment units are readily available in the United Kingdom from a number of suppliers, and Figure 1 illustrates the principals of the design which is enclosed under a removable cover. The primary zone receives raw sewage, usually unscreened, and returned secondary sludges from the clarification zone.

The secondary stage or biozone is suspended over the primary tank, and as the rotating discs alternatively submerge into the sewage and emerge into the air space, the biomass on the media oxidises the sewage much the same as more conventional processes.

Integral treatment units such as that described above occupy a relatively small area, and are free from odour and flies, and the modular size or treatment capacity of these proprietary systems is up to 250 pe. Multiple units can be installed but require the necessary flow distribution chambers to equalise the split of sewage to the multiple RBC units. Experience with RBCs in Severn Trent Water has determined the design loadings for these units to achieve reliable effluent qualities. At a disc or media loading of 4 g BOD per m^2 an effluent quality of 25 mg/l BOD 45 mg/l SS can be expected. For the reduction of ammonia to less than 10 mg/l NH_3^{-N} the organic loading to the media should be reduced to 2.5 g BOD per m^2.

Figure 1. **Typical RBC unit**

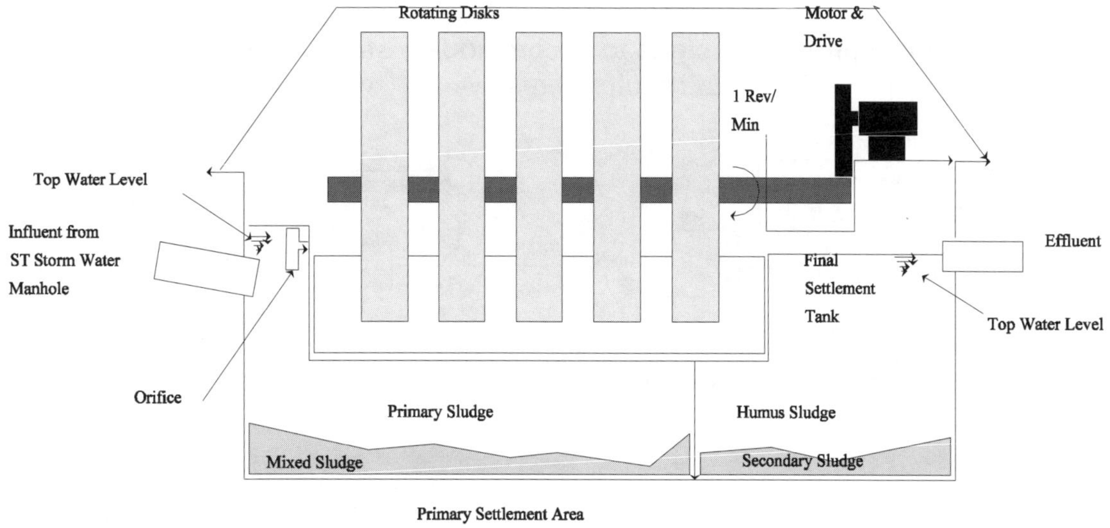

Source: Severn Trent Water plc internal report.

Historically, these proprietary RBC systems have exhibited a number of serious operational problems. One common fault, the collapse of the media and breakage of the rotational shafts and bearing assemblies, has led to unhappy experiences for the operators. Severn Trent has thoroughly researched these problems, and broadly concluded that the significant tensile stresses imposed by the rotational forces of the shaft assemblies were the principal cause of failure. Re-design of these essential components has resulted in reliable, robust RBC operation, and an improved design of RBC has been reported on by Findlay and Bannister (1993).

Another problem area with small treatment plants in general, not solely RBCs, has been the performance sensitivity to high flows. Excess flows do create problems quite understandably as essential retention times in the reactor required for adequate treatment can be dramatically reduced.

Treatment of storm flows

A strategic decision to limit the sewage flows to the RBCs to 6 x dwf has overcome this problem of storm flow treatment. Separation of flows in excess of 6 x dwf proved difficult to engineer satisfactorily, as the relatively small flows generated in these rural communities makes accurate storm separation over conventional weir arrangements very difficult. A workable alternative is to limit the flows into the RBC unit by a correctly sized orifice at the inlet to the biozone, and the excess flows are diverted to alternative treatment via an adjacent upstream chamber.

Where treatment is required, the diverted storm flows can be screened in a simple bag screen arrangement (COPASAC) and then fed into constructed reed beds.

These reed beds are designed to accommodate the intermittent nature of the storm discharges with overflow arrangements provided to relieve the beds, should the flows exceed 12 x dwf. At these flows, the reed beds still provide settlement and dilution, and Green (1993) calculates a retention period of about seven hours at 12 x dwf for a 1 000 pe rural works.

It has been possible in many instances to develop these reed bed systems into combined storm and tertiary treatment units illustrated in Figure 2. The effluent quality from these storm reed beds is unqualifyingly good, and examples of some results are reported later.

Figure 2. **Typical process designs to accommodate stormwater discharge requirements**

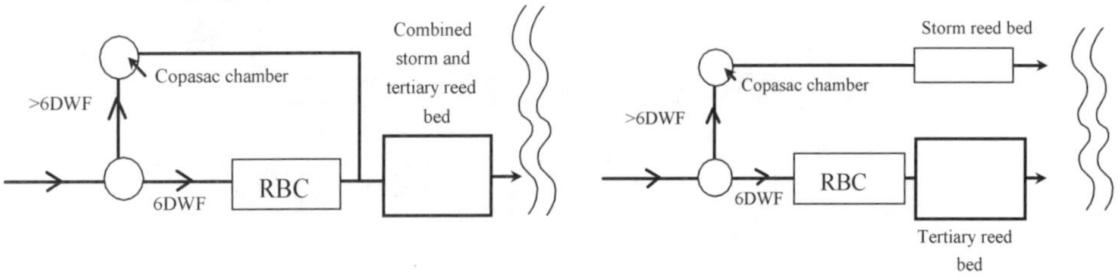

Source: Severn Trent Water plc internal report.

Reed beds

The design and operation of constructed reed beds have been described by Cooper (1990) and the specific applications for effluent polishing in Severn Trent has been described by (Green and Upton, 1994).

At their simplest (Figure 3), the reed beds are constructed by simple excavation lined with an impermeable membrane of bentonite or polyethylene, and provided with inlet and outlet collection zones.

Figure 3. **Longitudinal section through a tertiary reed bed**

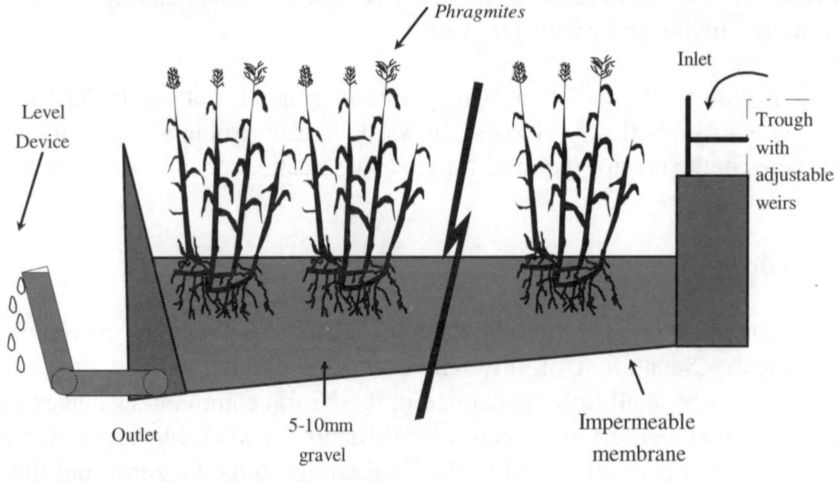

Source: Severn Trent Water plc internal report.

The bed is filled to 0.6 m with washed gravel usually 5-10 mm nominal size, and pot grown seedlings of the common reed mace (*Phragmites australis*) are planted at four plants per m^2.

The size of these reed bed systems, necessary to achieve a predicted effluent quality can be derived from the following equation:

$$A_h = Q_d(\ln C_o - \ln C_t)/K_{BOD}$$

where A_h = area (m^2)
Q_d = average flow (m^3/d)
C_o = inlet BOD_5 (mg/l)
C_t = outlet BOD_5 (mg/l)
K_{BOD} = BOD_5 reaction constant (m/d)

Cooper (1990) recommended using a value of 0.1 for K_{BOD} rather than the reaction constant of 0.19 originally proposed to the UK Reed Bed Treatment Group (Boon, 1986). This would predict a requirement of 4.6 m^2 to achieve an average BOD_5 of 20 mg/l from an average settled wastewater of 200 mg/l using 200 litres/pe per day. The recommendation was for 5 m^2/head per day where the settled wastewater (or septic tank effluent) strength varied between 150 and 300 mg BOD_5/l (Cooper, 1990).

The same equation is believed to be valid for tertiary treatment. Green and Upton (1994) predicted the need for 1 m^2/head to achieve a 95 percentile BOD_5 of 15 mg/l from an input with a 95 percentile of 25 mg/l. In practice, beds of this size have comfortably achieved 95 percentiles of less than 10 mg BOD_5/l.

Costs

Capital

Severn Trent has installed over 130 RBC + reed bed sites to date (1996) and an analysis of the installed costs of these sites reflects the range of costs associated with construction on new 'greenfield' sites or modification and asset refurbishment of existing works.

The mechanical and electrical components of this design of works represents at least 50 per cent of the scheme costs and for small community sites less than 200 pe, this proportion of the scheme cost related to the mechanical components can rise substantially. A cost range of US$ 800-1 200 per head for communities of 250 pe to 1 000 pe would seem to be typical, and these costs would include all infrastructure requirements, such as access road, sludge tanker access, security fencing, and electrical control and monitoring systems. These elements are considered an essential part of the specification for utility company assets, as these are required to have an expected life of 15-20 years for both the mechanical components and the reed bed system. Water utility operations in Europe (Belgium) confirm similar costs for small community installations (personal communication).

Operational costs

The revenue costs for these small treatment sites are difficult to accurately assess, but clearly they include sludge tankering, supervision, energy costs and maintenance. Severn Trent was able to compare the costs of operating the RBC package type treatment concept with the traditional trickling filter historically used in the United Kingdom for this size of installation. The RBC is visited less frequently (weekly minimum), and sludge is withdrawn at three monthly intervals; these costs are at least 30 per cent lower than those incurred with the traditional type of installation. The operational costs for RBC + reed bed in the range US$ 8-20 per person per annum.

The reed beds require a suggested input of one man day per 1 000 m^2 for weed control in years one and two, but the reeds are not harvested or cut back except where they impede access to the inlet distribution zone.

Performance

RBC + reed beds

The performance of the secondary treatment system can be variable dependent on the sewage characteristics and flow variation. The influence of industrial discharges into some of the sewer catchments can result in a poorly settling secondary sludge in the RBC clarifier zone, but this is unusual. Figure 4 illustrates how the combination of RBC + reed bed meets the challenge of discharge standards which are more stringent than 25 mg/l BOD and 25 mg/l S.solids.

Figure 4. **BOD and TSS influent and effluent average data for 10 sites**

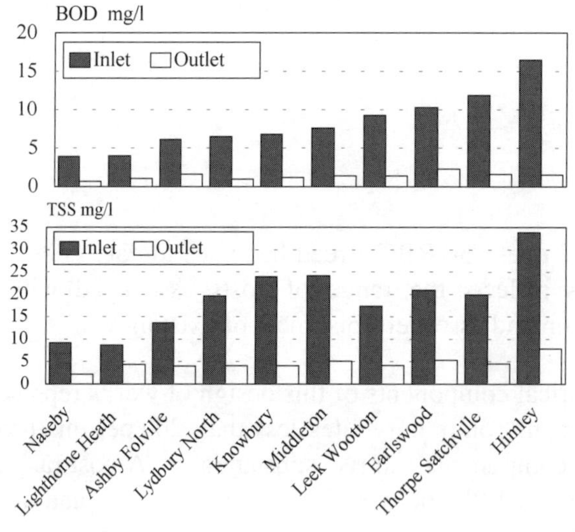

Source: Severn Trent Water plc internal data.

The consistent robust performance of reed beds as tertiary treatment systems is illustrated in Figure 5 using the regulatory data collected by the UK Environment Agency (EA) for a number of reed bed sites in Severn Trent Water.

Figure 5. **Average BOD_5 concentrations for the period Jan.-Dec. 1995 for works with tertiary treatment reed beds**

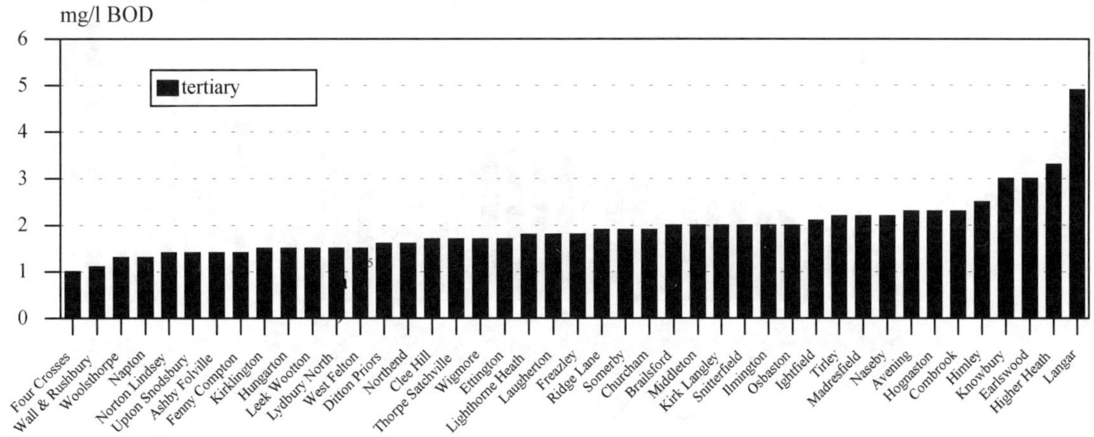

Source: UK Environment Agency data.

Storm reed beds

The anecdotal evidence from plant operators consistently endorses the benefit of storm flow treatment on reed beds, and the beneficial effluent to the watercourse from the cessation of errant storm discharges must be significant. The effluent quality following a storm event has been tracked on a number of occasions by Severn Trent, and Figure 6 illustrates two such storm events at Lighthorne Heath works (1 400 pe).

Figure 6. **Lighthorne Heath influent and effluent quality during storm conditions**

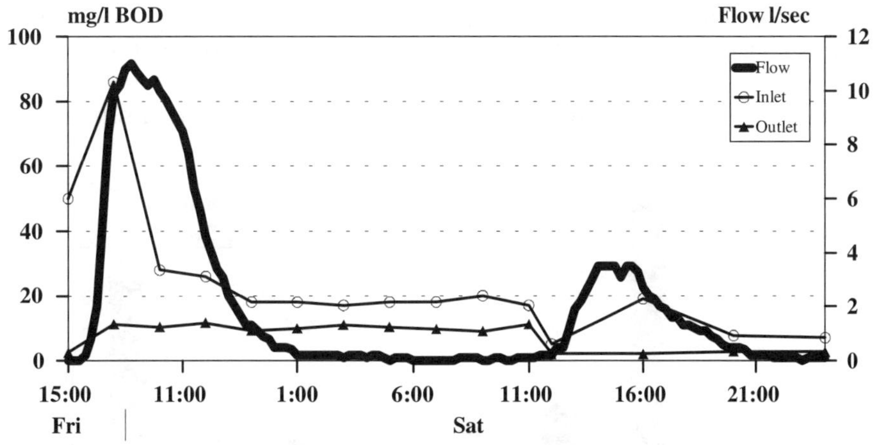

Source: Severn Trent Water plc internal report.

Similar performance characteristics can be expected at combined storm and tertiary treatment reed beds, and Figure 7 illustrates the BOD_5 effluent quality at a number of sites (Green *et al*., 1996*b*).

Figure 7. **Average BOD_5 concentration for the period Jan.-Dec. 1995 for works with combined storm and tertiary or remedial reed beds**

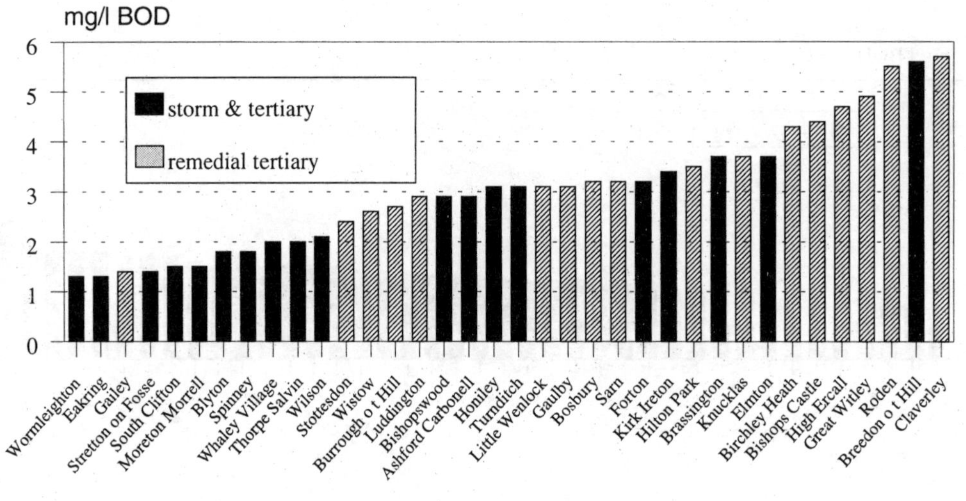

Source: UK Environment Agency data.

Secondary treatment by reed beds

At a number of sites, reed beds have been designed to provide the full secondary treatment requirement. This is the preferred treatment route for very small communities of less than 50 pe, and the sewage gravitates on to the reed bed system following settlement via a septic tank.

At Little Stretton, extensively reported by Upton and Griffin (1990) and Green and Upton (1993), the reed bed was constructed in a series of terraced beds to suit the topography of the site. Little Stretton is typical of the 'unsewered' villages in the United Kingdom. It comprises 15 dwellings and two farms, the latter contributing drainage from milk parlour and farmyard areas. In the nine years since commissioning in 1987, performance has steadily improved to the extent that BOD_5 and Amm N concentrations in 1993 were negligible at 1.7 and 0.2 mg/l respectively (Figure 8).

Figure 8. **Average influent and effluent analysis at Little Stretton since commissioning**

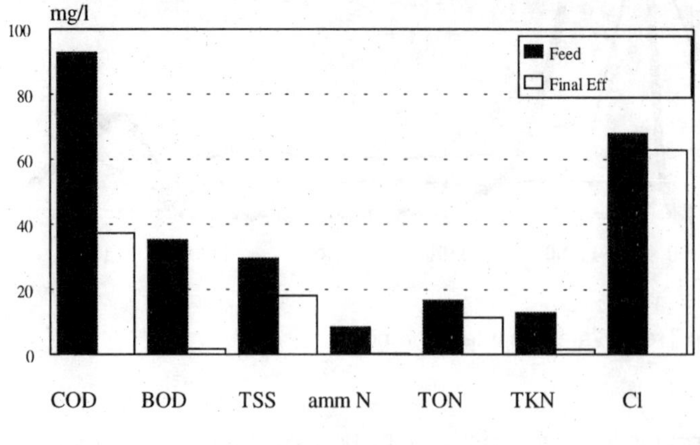

Source: Severn Trent Water plc internal data.

Bacteria removal and potential water re-use

Natural treatment systems such as reed beds offer a potential for reducing the numbers of pathenogenic bacteria in the wastewater. Detailed studies at both pilot and full scale reed beds by (Green *et al*., 1996*a*) show good removal of Coliform bacteria in beds with retention periods greater than 24 hours.

Removals of *E. Coli* of about 1.5 to 2.1 log were found in dry weather. Removal rates fell in wet weather, although no changes were detected in removal of BOD_5, TSS and Amm N.

Pilot studies (Figure 9) illustrate a trend of increasing removal as retention increases beyond 12 hours.

Figure 9. **Bacterial removal vs. retention time for pilot scale reed bed**

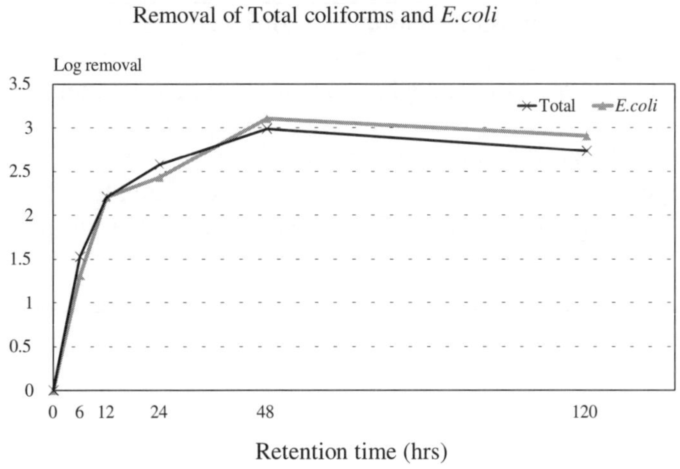

Source: Severn Trent Water plc internal report.

The WHO guidelines for the use of wastewater in agriculture and aquaculture (WHO, 1989) suggested microbiological standards for re-use for unrestricted irrigation of ≤1 000 faecal coliforms per 100 ml.

Severn Trent experience shows that this standard can be comfortably achieved with reed beds following secondary treatment.

Conclusions

– Package plant type RBC and constructed reed beds show competitive advantages for the treatment of wastewater from rural communities (less than 2 000 pe) over conventional treatment processes, such as trickling filters or activated sludge plants, where the manpower, energy and visit frequency makes these systems uneconomical.

- There are now over 350 reed bed treatment systems operating in the United Kingdom designed as secondary treatment units, tertiary treatment and storm flow treatment systems. The performance of these low cost systems is excellent.

- The future scenario for reed beds in the United Kingdom, Europe and elsewhere will expand with more systems designed to treat septic tank wastes from small populations. Reed beds will become most attractive for small private sites, or small entrepreneurs and increasingly reed beds will be used for the treatment of highway run-off and urban pollution management.

- Reed bed systems are effective at removing large numbers of pathenogenic faecal bacteria from sewage, and reported evidence from work in the United Kingdom indicates that this *E. Coli* removal can meet the WHO guidelines criteria for re-use of effluents to agriculture.

Questions, comments and answers

Q: What is the largest plant you have and what do you estimate is its lifetime?

A: We expect to renew the gravel and replace plants after ten years. This allows for a build-up of debris. The largest constructed wetland in the United Kingdom is for 20 000 population equivalents and is about two ha. In the United States, there are much larger wetlands in use, but they are not of the same type.

REFERENCES

BOON, A.G. (1986), "Report of a visit by members and staff of WRc to Germany (GFR) to investigate the root zone method of treatment for wastewaters", Water Research Centre, U.K. 376-S/1.

COOPER, P.F. (1990), "European Design and Operations Guidelines for Reed Bed Treatment Systems", *Report No. U117,* Water Research Centre, Swindon.

CRABTREE, H.E. and M.R. ROWELL (1993), "Standardization of small wastewater treatment plants for rapid design and implementation", *Wat. Sci. Tech* 28:10, p. 17.

GREEN, M.B. (1993), "Growing confidence in the use of constructed reed beds for polishing wastewater effluents", *Proc. Water Environ. Fed. 66th Annual Conf. & Exposition, General Topics* 9, p. 86.

GREEN, M.B. and J.E. UPTON (1993), "Reed-bed treatment for small communities; UK experience", in G.A. Moshiri (ed.), *Constructed Wetlands for Water Quality Improvement,* Lewis Publishers, Ann Arbor.

GREEN, M.B. and J.E. UPTON (1994), "Constructed reed beds: a cost-effective way to polish wastewater effluents for small communities", *Wat. Environ. Res.* 66:13, p. 188.

GREEN, M.B., J.R. MARTIN, and G.E. FINDLAY (1995), "Evaluation of Constructed Reed Beds treating combined sewer overflows on small wastewater treatment works", Natural and Constructed Wetlands for Wastewater Treatment and Re-use Conference, 15-17 October, Perugia, Italy.

GREEN, M.B., P. GRIFFIN, J.K. SEABRIDGE, and D. DHOBIE (1996*a*), "Removal of bacteria in subsurface flow wetlands", presented at the 5th Int. Conf. on Wetland Systems for Water Poll. Control, 15-19 September, Vienna.

GREEN, M.B., P.J. O'CONNELL, and P. GRIFFIN (1996*b*), "Uprating and rescuing small sewage treatment facilities by adding tertiary treatment reed beds", presented at WEFTEC Dallas, 5-9 October, Dallas, Texas.

FINDLAY, G.E. and R.H. BANNISTER (1993), "Designing for reliability. Quality requirements of rotating biological contactors and reed beds", paper presented to meeting of ICE Coventry, 28 October, Coventry.

UPTON, J. and P. GRIFFIN (1990), "Reed-bed treatment for sewer dykes", in P.F. Cooper and B.C. Findlater (eds.), *Constructed Wetlands in Water Pollution Control*, Pergamon Press, Oxford.

WORLD HEALTH ORGANIZATION (WHO) (1989), "Health Guidelines for the use of wastewater in agriculture and aquaculture", *WHO Tech. Report Series No. 778,* WHO, Geneva.

ENZYMATIC METHODS FOR WASTE MANAGEMENT

by

Gregory F. Payne* and Jeffrey S. Karns**

*Department of Chemical and Biochemical Engineering, Center for Agricultural Biotechnology
University of Maryland Baltimore County, Baltimore, Maryland, United States
**Soil Microbial Systems Laboratory, Beltsville Agricultural Research Center
United States Department of Agriculture, Beltsville, Maryland, United States

Biotechnology has gained considerable attention for the treatment and remediation of contaminated wastewaters. For instance, there has been substantial investment over the last couple of decades to characterise the biochemistry of the degradation of complex and recalcitrant organic pollutants such as polychlorinated biphenyls (Abramowicz *et al.*, 1993), chlorinated alkanes (McCarty, 1993) and nitroaromatics (Spain, 1995). Also, there are extensive research efforts to isolate individual microbes or microbial consortia and to characterise the enzymatic machinery responsible for biodegradation. Examples include studies on the peroxidase systems of white rot fungi (Barr and Aust, 1994) and cytochrome P-450 enzymes (Poulos, 1995).

In addition to understanding the relevant biochemistry, the application of biotechnology for waste management will also require a detailed understanding of the environmental problem being considered. This is especially true for environmental problems associated with agriculture, because often these problems involve large, non-point source problems (e.g. agricultural runoff) or small, point source wastes (e.g. pesticide-containing rinsates). Often solutions to these problems must employ simple, small scale approaches which have been individually tailored to the specific problem. In our work, we are employing a series of case studies to demonstrate the potential of such small-scale biotechnological approaches. Specifically, we are focused on the use of enzymes in waste management and have considered waste management in a broad sense to include not only treatment and remediation, but also waste minimisation. Although we have performed various studies with pesticide hydrolysing (Karns and Tomasek, 1991), phenol oxidising (Sun and Payne, 1996), and nitrate ester-degrading enzymes (Meng *et al.*, 1995), we will limit our discussion to a single case study involving the enzyme parathion hydrolase.

Historical review of the problem

During the late 1800s, the cattle industry in the Southern United States was experiencing substantial losses due to Texas cattle fever. By 1895, surveys had been conducted to determine the northern limit of this cattle fever, and the "Texas fever line" was established. In 1889, the US Department of Agriculture (USDA) restricted the northern movement of cattle from tick-infested areas, and in 1905, the US federal government quarantined cattle from large tick-infested regions.

When it became clear that cattle fever was transmitted by a tick, there was considerable pressure to develop a tick eradication program. In 1907, a state and federal program was initiated to eradicate the fever-causing tick. Early efforts of tick control involved spraying, mopping or brushing various solutions onto the cattle, while later efforts employed simpler "plunge" or "dip" vats. Over the years the active ingredients in the vats have changed from arsenic compounds to organochlorides (including DDT) and organophosphates. Currently, the organophosphate pesticide coumaphos is the most common active ingredient in dip vats to control cattle fever. For a more detailed review, see Graham and Hourrigan (1977).

Since the ticks which transmit cattle fever are endemic in Mexico, cattle in Mexico and at the Texas-Mexico border are routinely dipped in vats containing an aqueous suspension of the pesticide coumaphos (typical coumaphos concentrations of 1 500-3 000 mg/L). This dipping procedure requires a large number of relatively small vats (e.g. 3 500 gallons) distributed throughout rather remote regions. Although vats can sometimes be used for several months (and even years), a by-product of coumaphos occasionally accumulates, and this by-product is toxic to cattle. As illustrated in Figure 1, this potasan by-product appears to be produced from the anaerobic dechlorination of coumaphos (Shelton and Karns, 1988). In the United States, when the potasan level exceeds 300 mg/L, the dip vat must be taken out of service, and the liquid is disposed in a holding pond. Thus, even though a dip vat fluid may contain substantial amounts of the coumaphos pesticide, it must be disposed when relatively small amounts of potasan accumulate. This disposal not only results in the generation of a pesticide-containing wastewater, but recharging the vat requires the use of more pesticide which can be costly.

Figure 1. **Reactions involving the pesticide coumaphos**

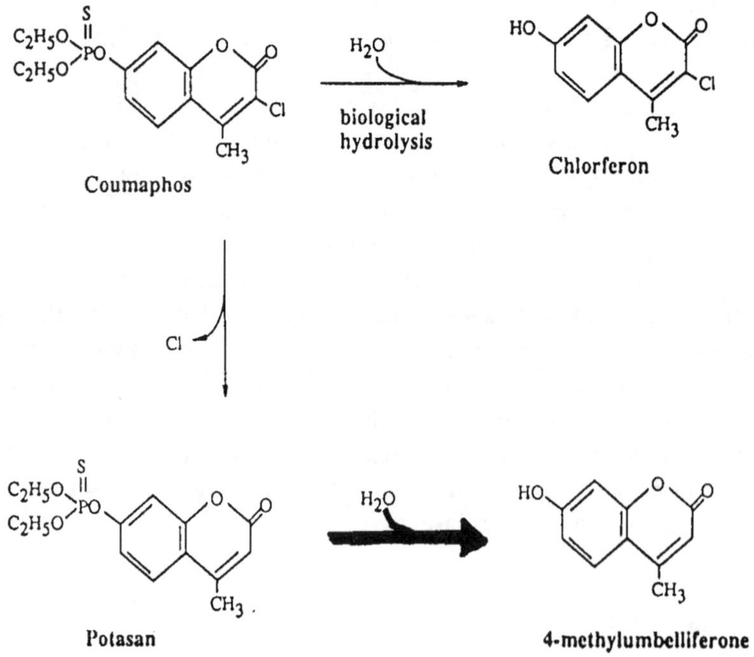

Note: Potasan is the undesired by-product formed in some vats. The horizontal lines indicate the enzyme-catalysed hydrolytic reactions.

Source: Adapted from Smith *et al.*, 1992.

Biotechnological approach for waste minimisation

The goal of our research was to determine if a biological catalyst (i.e. enzyme) could be used to selectively degrade the toxic potasan by-product, while minimising degradative losses of the desired coumaphos. Such a solution would allow the lifetime of dip vats to be extended and would reduce the total amount of wastes generated from the dip vat operation. Further, by reducing the need to re-charge dip vats, the cost associated with the addition of fresh coumaphos could be reduced. This approach is based on initial studies which demonstrated that the *Flavobacterium* enzyme (Brown, 1980), organophosphate phosphotriesterase, or parathion hydrolase, is capable of degrading both coumaphos (Kearney *et al.*, 1986) and the potasan by-product (Coppella *et al.*, 1990) in dip vat liquids, and that the rate of potasan hydrolysis was about an order of magnitude larger.

Results to date

Source of enzyme. Initial studies identified the *opd* gene from *Flavobacterium* which codes for the parathion hydrolase enzyme (Mulbry and Karns, 1989). For our study, the *opd* gene was cloned into *Streptomyces lividans* which can express and secrete the enzyme (Steiert *et al.*, 1989). Using this organism, it is possible to culture the cells under well-defined laboratory/factory conditions, harvest the fermentation broth and simply remove the cells by centrifugation or filtration (Payne *et al.*, 1990; DelaCruz *et al.*, 1992). The enzyme-containing supernatant or filtrate can then be stored or used directly without the need for extensive downstream recovery or purification.

Enzymatic hydrolysis. Laboratory studies were conducted to quantify the hydrolysis rates for the pesticide (coumaphos) and the undesired by-product (potasan). A careful kinetic study demonstrated that the intrinsic kinetics for enzymatic hydrolysis are identical for coumaphos and potasan, and that kinetic differences observed in dip vat liquids are likely the result of differences in mass transfer. Specifically, in dip vat liquids, the overall hydrolysis rates appear to be limited by how fast the coumaphos and potasan dissolve into the liquid (since both potasan and coumaphos have low water solubilities, they exist as a suspension in the dip vats). The eight-fold higher solubility of potasan relative to coumaphos appears to explain potasan's preferential hydrolysis in dip vat liquids (Smith *et al.*, 1992). Thus, parathion hydrolase is capable of hydrolysing both the undesired potasan by-product and the desired pesticide, coumaphos, but as illustrated by the thick line in Figure 1, potasan is preferentially hydrolysed.

Enzyme inactivation. In order to use parathion hydrolase to selectively hydrolyse potasan while minimising coumaphos losses, it is necessary to inactivate the enzyme after potasan hydrolysis is complete. Unless inactivated, the enzyme will continue to hydrolyse the desired pesticide coumaphos. Although various strategies could be envisioned to inactivate parathion hydrolase after potasan hydrolysis is complete, we chose to exploit the limited stability of the enzyme -- half-lives on the order of eight to 40 hours have been reported (Munnecke, 1980; Coppella *et al.*, 1990; Rowland *et al.*, 1991). To facilitate our quantitative characterisation, a model was developed which combined mass transfer and hydrolysis kinetics, and enzyme inactivation (Smith *et al.*, 1992; Grice *et al.*, 1996). This model permits estimation of the optimal amount of enzyme required for complete potasan hydrolysis while minimising hydrolytic losses of the desired pesticide, coumaphos.

Laboratory demonstration. Figure 2 from Grice *et al.* (1996) shows a laboratory test in which varying amounts of an enzyme solution were added to five ml of dip vat liquid containing both potasan and coumaphos. As shown in Figure 2, when high levels of enzyme were used (0.01 mg/L), both potasan and coumaphos were completely hydrolysed. Although hydrolysis of the undesired potasan was complete, this enzyme level was excessive and resulted in substantial loss of the coumaphos. Large losses preclude coumaphos' re-use in subsequent dipping operations. When the lowest enzyme level (0.001 mg/L) was studied with dip vat liquids, Figure 2 shows complete degradation of potasan and limited degradation of coumaphos. This level of enzyme is close to optimal. The lines in Figure 2 are generated from our mathematical model which show reasonable agreement with the experimental data.

Figure 2. **Laboratory experiment demonstrating the selective hydrolysis of potasan**

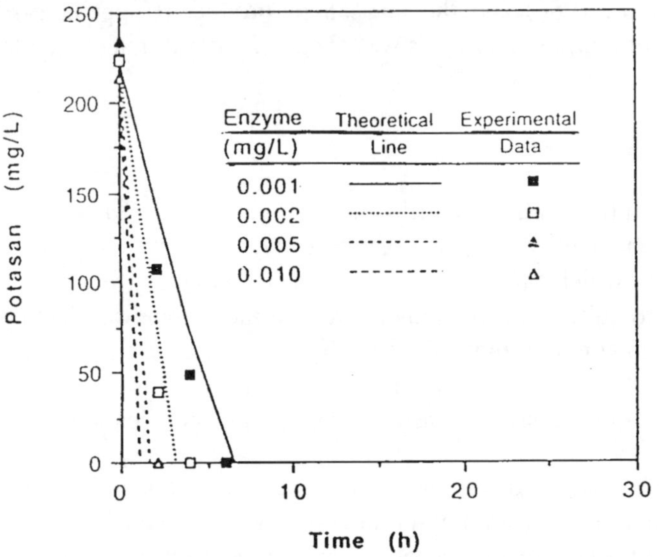

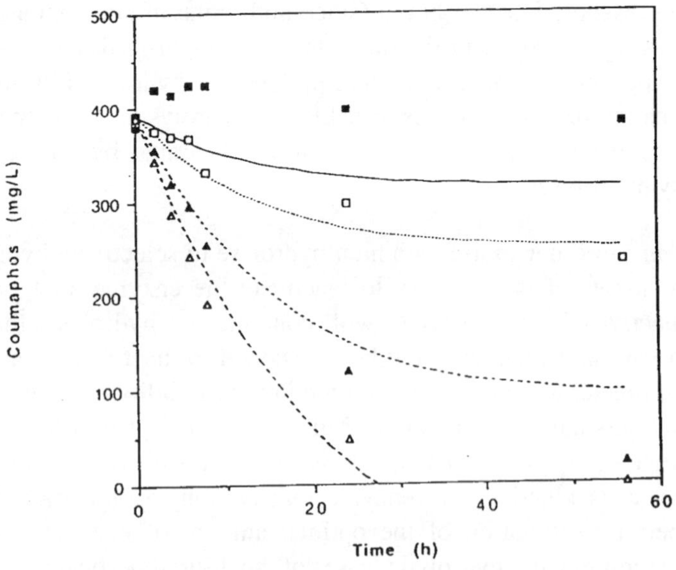

Note: For this study, varying amounts of enzyme were added to five ml of dip vat liquid. The enzyme was used as a crude, cell-free solution obtained by centrifuging the fermentation broth of *Streptomyces lividans* (i.e. the enzyme was not concentrated). The amount of cell-free broth added varied from five to 50 μl (estimated enzyme levels of 0.001 to 0.01 mg/L).

Source: Adapted from Grice *et al.*, 1996.

Field test and economic analysis. Studies conducted on a larger scale (2 100 gallon) demonstrated the technical feasibility for using parathion hydrolase to selectively degrade the potasan by-product (Grice *et al.*, 1996). Further, the economic analysis of Figure 3 shows that substantial savings would be realised if potasan were selectively degraded, and the lifetime of the dip vat extended. For comparison, the upper, horizontal line in Figure 3 shows the coumaphos cost when fresh vats are prepared. As indicated by this upper line, the coumaphos cost is high, and does not depend on the amount of potasan which may have been present in the previous vat. In contrast, the lower lines in Figure 3 are estimates of the two costs incurred if enzyme were added to selectively degrade potasan. Since the amount of enzyme required for complete potasan hydrolysis is small, the lower line in Figure 3 shows the enzyme cost to be small, but it increases with increased potasan levels. The second cost for employing the enzymatic approach to extend the lifetime of the dip vat is the cost to replace the small amount of coumaphos which would be hydrolysed by the enzyme. As can be seen, the sum of the two costs for adding enzyme are small relative to the cost to re-charge the dip vat with fresh coumaphos. Thus under most reasonable conditions, it would be less expensive to add enzyme to extend the lifetime of the vat, compared to emptying the vat and adding fresh coumaphos.

Figure 3. **Economic comparison of adding enzyme to selectively hydrolyse potasan and extending the lifetime of the dip vat versus disposal and re-charging the vat with fresh coumaphos**

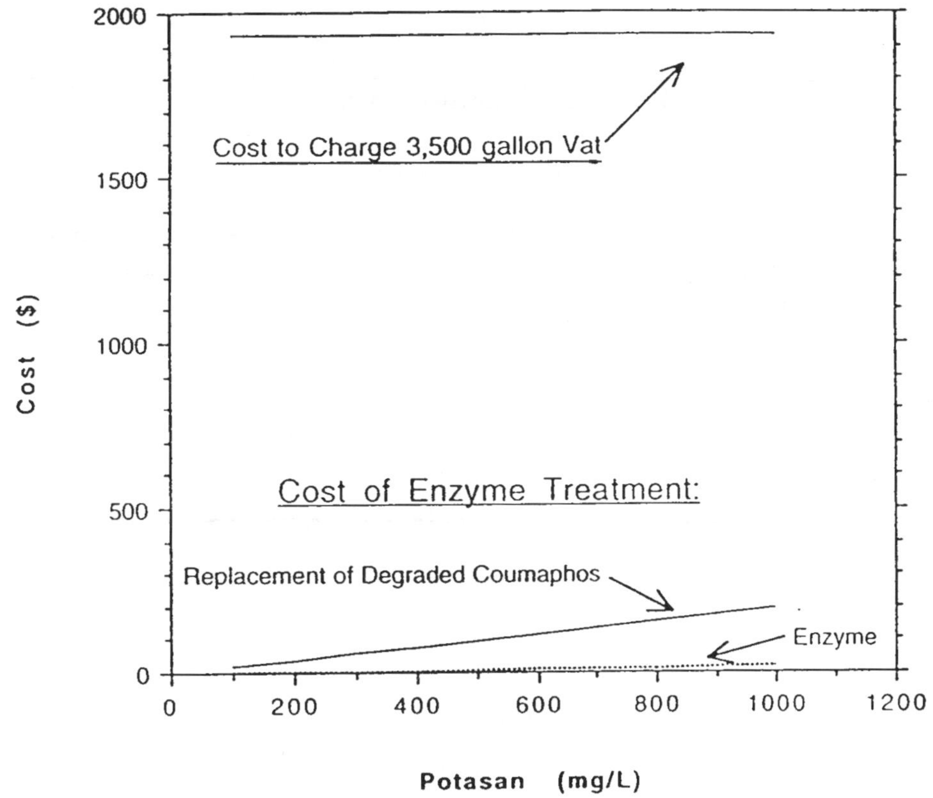

Note: For details, see text.

Source: Adapted from Grice *et al.*, 1996.

Conclusion

In this case study, we have demonstrated the technical and economic feasibility for using parathion hydrolase to selectively degrade the potasan by-product and to permit the lifetime of the dip vat to be extended. By extending the lifetime of the dip vat, the total amount of pesticide-containing water which must be disposed would be reduced. In addition, the total pesticide usage could be reduced, which would have significant environmental and economic ramifications. Finally, by understanding the individual features of the problem and matching these with the unique capabilities of the enzymatic hydrolysis, it was possible to employ a biotechnological approach to achieve waste minimisation. It is difficult to imagine any alternative technology which could achieve this waste minimisation as simply and cheaply. It is important to note that although developing such an enzymatic approach required rather sophisticated technologies (e.g. molecular biology and fermentation), the end-user does not need advanced skills to implement this approach. The end-user would simply need to add a specific amount of enzyme to a potasan-containing dip vat.

A general conclusion of this work is that by using a case study approach, innovative solutions can be found to solve specific environmental problems. A case study approach is particularly important in agriculture, because environmental problems associated with agriculture often involve small volume point source contamination. Appropriate solutions to these problems will generally be different than generic solutions developed for the routine treatment of industrial manufacturing effluents, or for the remediation of contaminated landfills.

Questions, comments and answers

Q: Is this approach used (commercialised) and is it accepted by farmers?

A: We have just completed our economic analysis and are working with US Department of Agriculture which is responsible for dips. They have not yet accepted it because they are exploring alternatives.

Q: How stable is the enzyme?

A: In the dip bath it has a half-life of about eight hours. It is important to emphasize that we are exploiting *instability* -- we want activity to cease.

REFERENCES

ABRAMOWICZ, D.A., M.J. BRENNAN, H.M. VAN DORT, and E.L. GALLAGHER (1993), "Factors influencing the rate of polychlorinated biphenyl dechlorination in Hudson River sediments", *Environ. Sci. Technol.* 27, pp. 1 125-1 131.

BARR, D.P. and S.D. AUST (1994), "Mechanisms white rot fungi use to degrade pollutants", *Environ. Sci. Technol.* 28, pp. 78A-87A.

BROWN, K.A. (1980), "Phosphotriesterases of *Flavobacterium* sp.", *Soil Biol. Biochem.* 12, pp. 105-112.

COPPELLA, S.J., N. DELACRUZ, G.F. PAYNE, B.M. POGELL, M.K. SPEEDIE, J.S. KARNS, E.M. SYBERT, and M.A. CONNOR (1990), "Genetic engineering approach to toxic waste management: Case study for organophosphate waste treatment", *Biotechnol. Prog.* 6, pp. 76-81.

DELACRUZ, N., G.F. PAYNE, J.M. SMITH, and S.J. COPPELLA (1992), "Bioprocess development to improve foreign protein production from recombinant *Streptomyces*", *Biotechnol. Prog.* 8, pp. 307-315.

GRAHAM, O.H. and J.L. HOURRIGAN (1977), "Eradication programs for the arthropod parasites of livestock", *J. Med Entomol.* 13, pp. 629-658.

GRICE, K.J., G.F. PAYNE, and J.S. KARNS (1996), "Enzymatic approach to waste minimization in a cattle dipping operations: Economic analysis", *J. Agric. Food Chem.* 44, pp. 351-357.

KARNS, J.S. and P.H. TOMASEK (1991), "Carbofuran hydrolase - purification and properties", *J. Agric. Food Chem.* 39 pp. 1 004-1 008.

KEARNEY, P.C., J.S. KARNS, M.T. MULDOON, and J.M. RUTH (1986), "Coumaphos disposal by combined microbial and UV-ozonation reactions", *J. Agric. Food Chem.* 34, pp. 702-706.

MCCARTY, P.L. (1993), "In situ bioremediation of chlorinated solvents", *Curr. Opin. Biotech.* 4, pp. 323-330.

MENG, M., W.-Q. SUN, L.A. GEELHAAR, G. KUMAR, G.F. PAYNE, M.K. SPEEDIE, and J.R. STACY (1995), "Denitration of glycerol trinitrate by resting cells and cell-free extracts of *Bacillus thuringiensis/cereus* and *Enterobacter agglomerans*", *Appl. Environ. Microbiol.* 61, pp. 2 548-2 553.

MULBRY, W.W. and J.S. KARNS (1989), "Parathion hydrolase specified by the *Flavobacterium opd* gene: Relationship between the gene and the protein", *J. Bacteriol.* 171, pp. 6 740-6 746.

MUNNECKE, D.M. (1980), "Enzymatic detoxification of waste organophosphate pesticides", *J. Agric. Food Chem.* 28, pp. 105-111.

PAYNE, G.F., N. DELACRUZ, and S.J. COPPELLA (1990), "Improved production of heterologous protein from *Streptomyces lividans*", *Appl. Microbiol. Biotechnol.* 33, pp. 395-400.

POULOS, T.L. (1995), "Cytochrome P-450", *Curr. Opin. Struct. Biol.* 5, pp. 767-774.

ROWLAND, S.S., M.K. SPEEDIE, and B.M. POGELL (1991), "Purification and characterization of a secreted recombinant phosphotriesterase (parathion hydrolase) from *Streptomyces lividans*", *Appl. Environ. Microbiol.* 57, pp. 440-444.

SHELTON, D.R. and J.S. KARNS (1988), "Coumaphos degradation in cattle-dipping vats", *J. Agric. Food Chem.* 36, pp. 831-834.

SMITH, J.M., G.F. PAYNE, J.A. LUMPKIN, and J.S. KARNS (1992), "Enzyme-based strategy for toxic waste treatment and waste minimization", *Biotechnol. Bioeng.* 39, pp. 741-752.

SPAIN, J.C. (1995), "Biodegradation of nitroaromatic compounds", *Annu. Rev. Microbiol.* 49, pp. 523-555.

STEIERT, J.G., B.M. POGELL, M.K. SPEEDIE, and J. LAREDO (1989), "A gene coding for a membrane-bound hydrolase is expressed as a secreted, soluble enzyme in *Streptomyces lividans*", *BioTechnol.* 7, pp. 65-68.

SUN, W.-Q. and G.F. PAYNE (1996), "Tyrosinase-containing chitosan gels: A combined catalyst and sorbent for selective phenol removal", *Biotechnol. Bioeng.* 51, pp. 79-86.

CLOSED CYCLE OPERATION IN MILK PRODUCTION UNITS FOR GROUNDWATER PROTECTION

by

H. Diestel*, H.J. Schwartz, U. Kühl* and H. Mieth*****
*Technical University of Berlin, **Humboldt University, Berlin, ***University of Rostock, Germany

Summary

For the analysis and the planning of animal production units in tropical zones, a pragmatic concept of "cycle closure" is applied. The objective is to achieve a minimum input of energy from fossil energy carriers into the unit under consideration, and a minimum output of contaminants into the landscape.

One of the principal constraints to milk production in the humid and sub-humid tropics is the low availability of protein in grasses and other forage plants which form the diet of dairy cattle. This needs to be supplemented, usually by costly concentrates, to achieve satisfactory performance levels. An integrated, low-technology management concept for dairy production units in tropical climates presented. It aims at replacing some of these concentrates by nitrogen recycled from animal manure into protein of aquatic plants, while simultaneously reducing contamination of surface and groundwaters by animal wastes. Central elements of the system are a biogas reactor and hydroponic ponds with the hydrophytes Azolla and Lemna. Here, the dissolved nutrients function as fertilizer to the aquatic plants, producing a plant biomass with high protein content which can be fed to the cattle. The discharge from the pond with its reduced nutrient content and the compost are both used for supplemental irrigation and for fertilization of pastures and crops, which in turn are fed to the dairy stock.

With reference to a pilot project in Cuba which is in its initial phase, the cycles of nutrition, energy and water as well as the in- and outputs of energy on a milk production unit are investigated. Management measures and technological means to reduce the output of contaminants from animal husbandry units and to achieve a reduction of the input of energy from fossil energy carriers are discussed. Basic quantifications were carried out. It is demonstrated that the energy potentials furnished by the wind, the solar radiation and the livestock excretions can be used to supply sufficient energy for operations like pumping, milking, water heating and cleaning, as well as for domestic uses. Proposals for future improvements are made, and some implications of the implementation of this concept are discussed.

Introduction

Outline of the problem and the task

Contamination of surface waters and of groundwaters from agricultural sources represents a serious world-wide problem. Crop production areas frequently represent diffuse sources of contamination of waters with fertilizer and pesticide compounds. Many animal husbandry units represent point sources for contaminations of waters with organic compounds. When the livestock grazes on the rangeland, or when the liquid manure is spread onto the land, milk and meat production units (frequently near larger settlements, which themselves contribute their share to contamination) also lead to diffuse contamination.

In view of the parallel processes of population increase and environmental degradation, it is essential to develop economically viable systems for milk and meat production which do not lead to contaminations of surface and groundwaters, and which do not require high fuel inputs. In addition, an important task to be achieved in the near future consists in the elaboration of objective, quantitative procedures for the identification and definition of the ecological and economic benefits which arise from such improved production systems.

Under a *temperate climate*, the untreated sewage from piggeries and dairy units is usually spread on fields and on rangeland, using an input of fuel. In most of the countries of this zone, the costs of treating liquid manure are prohibitive. A "purification effect" is achieved by applying the slurry on relatively large areas, thus leading to low contaminant concentrations. Precipitations of several hundreds of millimetres per year supply mostly sufficient water, but also represent a factor for surface and groundwater contamination. Milk production often takes place in hilly areas or in river plains, i.e. in zones with high surface runoff potential or with low distances to the groundwater. A realistic monetarization and internalisation of the contamination effect is difficult, due to the high subsidisation of agriculture in most of these countries.

In *arid regions*, water must frequently be obtained by investing energy from fossil energy carriers (directly in the case of pumping, or indirectly when water is diverted from reservoirs constructed with high fuel inputs). A biotechnological treatment of liquid manure with a purifying effect seems to be feasible, due to the high ambient radiation and temperatures. Due to the nature of water prices, the economic evaluation of external effects is problematic.

In zones with those *humid tropical climates* under which milk (and meat) production are feasible, the natural conditions favour the low-cost biotechnological treatment of slurry from piggeries and cattle corrals, precipitation rates are high, and the demand for milk and meat is high. In many countries of these areas, which are located roughly in the zones corresponding to sub-humid and semi-arid tropical regions, as well as in tropical highlands, a relatively realistic appraisal of the benefits which accrue from making use of regenerative energies and of biotechnological processes should be possible, due to the specific nature of the given social and economic structures.

In view of the fact that for the described pilot project, which is a *dairy* unit in a semi-humid area, the contamination of *groundwater* is the issue which represents the more serious problem; in the title of this paper, reference is made to the groundwater protection effect of closed cycle operation of *milk* production units. However, as in most cases liquid manure enters the soil as well as surface waters, there will be beneficial consequences for the quality of *surface waters* as well as groundwaters from managing animal production units with the concept discussed here. In the case of piggeries, which frequently constitute point contamination sources for surface waters, generally the benefits will primarily affect streams and lakes.

The notion of "closed" cycles

Interlinked cycles of successive substance and energy transformations may be considered as "closed", if, as the successive transformations take place, as much energy is taken up as is released, and/or if losses or gains of matter are accompanied by corresponding gains or losses of energy. Such a "closure" -- the proof of which frequently is unsatisfactory due to the lack of appropriate data and/or measurement techniques -- is difficult to achieve, even under controlled laboratory conditions. Much more so, a complete accounting for the sum of all components of such cycles is problematic if they take place in the open air and originate from piggeries or dairy units.

For the purposes of the work presented here, which is oriented towards the implementation of sustainable animal production systems, a different, pragmatic concept of "cycle closure" is applied. The objective is to achieve a *minimum input of energy from fossil energy carriers* into the unit under consideration, and a *minimum output of contaminants into the various elements of the landscape at, around and downstream* of this unit. The cycle would be considered to be "closed" if these two components of input and of output would be equal to zero. The basic unit for which it is attempted to achieve an optimum balance on a yearly, monthly or weekly balance is *the animal production unit including the rangeland* belonging to it. The lower boundary of this three-dimensional unit is considered to be above underlying aquifers. The energy input from the ambient radiation, the prevailing winds and the precipitation of the site are factors which enter into all the planning and implementation steps, but -- in the strict physical and economic sense -- these components are "mined". In order to achieve the mentioned objective, it is necessary to close to as high a degree as possible all subcycles of transformation within the system composed of the dairy units or pig-houses, and the land used in connection with the production processes.

A management concept for dairy production units in tropical climates

One of the principal constraints to milk production in the humid and sub-humid tropics is the low availability of protein in grasses and other forage plants which form the diet of dairy cattle. This needs to be supplemented, usually by costly concentrates such as soybean oilseed cakes, etc., to achieve satisfactory performance levels. The system that is suggested here aims at replacing some of these concentrates by nitrogen recycled from animal manure into protein of aquatic plants while simultaneously reducing contamination of surface and groundwaters by animal wastes.

Liquid manure from animal houses flows into a biogas digester where part of it is converted by fermentation into gas which can be used for domestic purposes in the farmer's household, or as the principal energy source for the refrigeration of milk, heating of water, or for operating the compressor of the milking machine (Figure 1). The effluent of the digester is separated into the solid phase which is composted, and the liquid phase which is channelled into a hydroponic pond. Here the dissolved nutrients, in particular nitrogen, function as fertilizer to aquatic plants like *Lemna* sp. and *Azolla* sp., producing a plant biomass with high protein content which can be harvested, air dried and fed to the cattle as protein concentrate. The discharge from the pond with its reduced nutrient content and the compost are both used for irrigating and fertilizing pastures and crops, which in turn are fed to the dairy stock. An alternative use for the hydroponic ponds discharge could be in a fish pond where it would stimulate algae growth, which would form the feed base for herbivorous fish.

Figure 1. **Schematic presentation of material and energy flows in an integrated livestock production system**

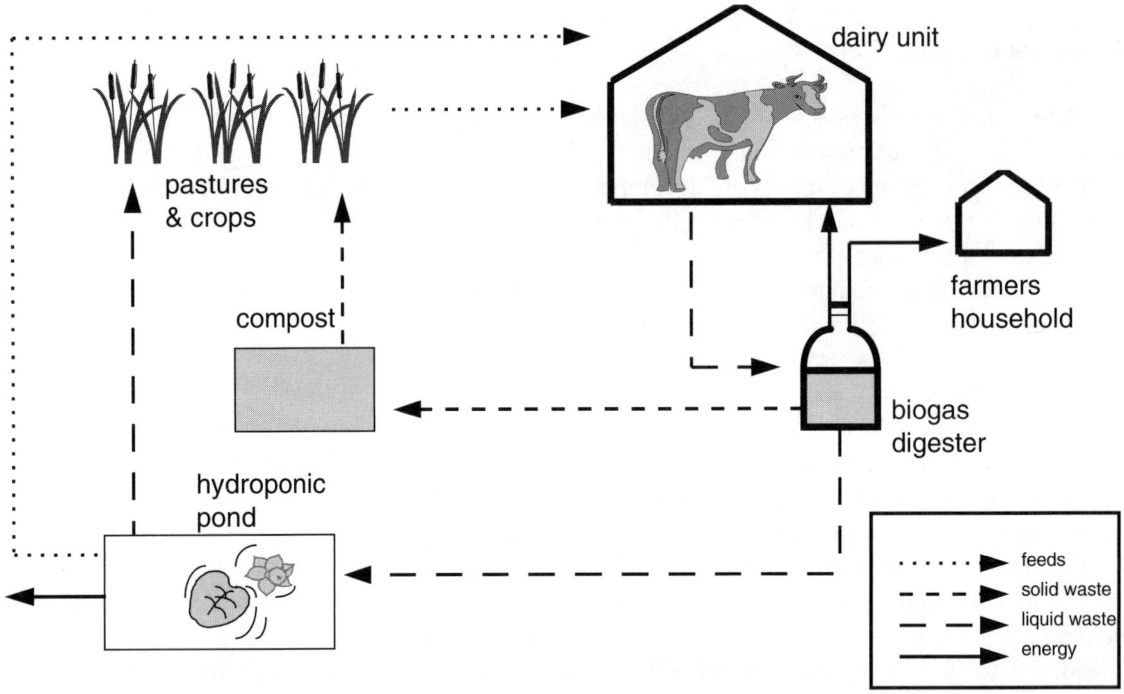

Source: Schwartz, 1995.

The individual components of such a system (Kühl *et al.*, 1995) are tried and known to function. However, considerable research still needs to go into optimising the various material flows, both with regard to quantities and to qualities, in order to achieve maximum efficiency. Of particular interest are the control of the effect of rainfall events and surface runoff, and the use of regenerative energies to facilitate the necessary material transport functions in the system.

In many cases, this concept will lead to direct additional income for the producers, *inter alia,* due to the savings in the purchases of fuel, fertilizers and feed, and as a result of the effects of the increase in the protein component of the nutrition of the livestock. The economically and ecologically beneficial effects from soil improvements due to the applications of organic compounds (not only on the rangeland, but also on the fields frequently cultivated with crops for the own consumption) are self-evident. The ecological benefits of applying such a management concept -- for the immediate vicinity of animal production units, for the more distant environment, and as a contribution to the reduction of atmospheric pollution with its negative consequences -- are obvious. They go beyond the effect of reducing the contamination of waters. Intuitively, economic benefits for the national societies and for the global community can be imputed which will result from the favourable ecological effects of a large scale implementation of this management concept. However, the instruments of economic analysis available at present do not allow objective quantifications of this kind.

The defined primary objective of introducing improved management on the pilot project described below was not "groundwater protection", but the implementation of an economically and ecologically sustainable concept for the production of milk in a country in which there is a serious shortage in the supply of this basic component of human nutrition. For the purposes of this paper, the main attention is directed to the aspects relevant to the subject defined by the title. As is the case for all truly sustainable and integrated systems, all components of the system must be taken into consideration in the attempt to implement the closed cycle operation concept. A farmer will not be stimulated primarily by ecological enthusiasm to contribute in the expected manner to low energy inputs from fossil energy carriers and to low contaminant contents in the dairy effluents. What will encourage him to modify his management and to install additional technology is the prospect of being able to sell good quality milk in sufficient quantities with satisfactory economic success.

Details on the implementation of the concept

The technological frame for a problem solution

In the Perú Valley in Cuba, a pilot project was initiated on the dairy unit "Vaquería No. 40" in 1993 to implement the described management concept (Schwartz *et al.*, 1995). Precipitation during the six summer months is approximately 1 000 mm and 200 mm during the six winter months. The mean annual temperature is 25 °C. The daily temperature amplitude does not exceed 4 °C. On the average, one can count on about eight hours of wind with windspeeds above four m/s per day. Global radiation varies between 5.2 and 8.0 kWh per square meter and day. There are two groundwater surfaces, one at about 13 meters and one at about 45 meters below the ground surface during the rainy season. The lateritic soils (Ferralsols/Oxisols) have derived from calcareous materials, and have a neutral reaction and rather high permeabilities for water. However, they can be sealed superficially quite well by mechanical action under wet conditions, a property which is of relevance for the planned hydroponic ponds.

Kühl (1996) has described the cycles of nutrition, energy and water as well as the resulting implications for the operations on the unit, providing data such as, *inter alia*, water quantities and qualities as well as soil properties. Mieth (1996) has analysed the in- and outputs of energy on this milk production unit and has made proposals to increase the self-sufficient supply with regenerative energies, providing detailed designs for some items. In Cuba -- as will be the case in general -- the corral area has been constructed with a gentle slope. A raised water tank (38 cubic meters capacity) within the cow-house area is filled once or twice daily with water pumped from a central electric pumping station which also supplies other dairy units. All raw effluents flow superficially on the concrete cover of the corral area and converge in two main concrete channels between the stables. The untreated slurry flows into the landscape from the downstream edge of the cow-house zone. A specific vegetation has developed along the surface channel which carries this effluent flow, a typical situation for such dairy units.

There are (status Spring 1996) approximately 55 lactating cows, 10 pregnant cows, 20 heifers and 10 to 15 calves on the unit (Holstein and Siboney) which was originally designed for 288 milk cows. The rangeland area is 79 hectares, there are two hectares of cropping area, and the livestock housing area -- which is nearly completely under concrete also between the stables -- amounts to 1 280 square meters. A minimum of 650 kg (0.71 cubic meters) of liquid manure are produced daily under the actual management. Milking is carried out with electric milking equipment. The cooling of the milk also consumes electric energy, as does the washing of the livestock, which is an indispensable procedure in tropical regions. Warm water is needed not only for domestic purposes, but also to clean the milking equipment.

Some of the components of a closed cycle operation system have already been implemented in the pilot project, e.g. the biogas fermentor[1]. Some (like the water heating unit) have been planned in detail but not yet implemented, some have been examined in other animal production units in Cuba, and some so far only exist as conceptions based on literature studies. The cycles of matter and energy on a dairy unit are very intricate and have quite complicated interlinkages. It is not possible to describe them in detail in the framework of this paper. Below, it is attempted to sketch out the main points of the implementation of the integrated, low-technology closed-cycle operation concept.

Reduction of the output of contaminants from animal husbandry units

Adaptation of management

The biogas reactor does not operate properly if quantities and qualities of the liquid manure vary too strongly. The cleaning operations in the stables and on the milking equipment must be adapted accordingly, especially the indiscriminate use of water to flush the stables. This operation must be modified, e.g. by employing scrapers with animal traction. The runoff in the corral area from the intensive rain showers must be diverted adequately. The microbes in the digester do not tolerate components like detergents or medications. The human excretions from the toilets must not be fed to the reactor haphazardly. Excessive dung should be piled on separate places, composted and applied -- using animal traction -- to the rangeland. As there is a tendency for cow dung to form a layer of "scum" on the decomposing material in the reactor, provisions should be made to turn and mix the contents of the reactor once per day. Inflow and outflow must be controlled and corrected as needed. Whenever it is feasible, "gray" water should be used for operations on the dairy unit.

Shade trees should be planted, and drinking water must be supplied to the pastures. This will lead to a higher elasticity in the time management of the grazing phases and thus to less erratic quantities in the production of excretions in the corral area. Besides, through their excretions, grazing milk cows return 75 to 80 per cent of the nitrogen and of the phosphorus which was taken up, 85 to 90 per cent of the potassium, 40 to 60 per cent of the organic matter, and 90 per cent of the calcium.

Technological measures

To ensure the proper functioning and operation of the complete integrated dairy production system, a regular, safe primary water supply must be secured by installing an independent pump next to the main raised tank above the unit.

By transformations of the raw liquid manure through biotechnological processes, it is avoided that raw liquid manure flows into surface streams and into the soil downstream of the dairy unit. The first step of these transformations is the passage of the livestock excretions -- in a 1:1 mixture with ("gray") water -- through a biogas reactor. The most common types -- used primarily in China and in India -- are the so-called "Chinese" type, with a rigid cupola (exit gas pressure not constant), and the "Indian" type, with a steel cupola floating on the fermentor chamber (exit gas pressure constant). In the pilot project described here, a rigid cupola biodigester was constructed (for more information on this installation see section on basic standard equipment below).

---

[1] The terms biogas reactor, biodigester, fermentor, biogas-digester, biogas installation, *inter alia,* are used as synonyms in this paper.

It was estimated that three to five hectares of total hydroponic pond area would be needed downstream of the corral area for an adequate slurry treatment and a sufficient dry matter production under the given natural conditions, operation possibilities and protein demands, based on what is known about the aquatic plants which are proposed. The lagoons should be laid out as rectangular, parallel elements, thus allowing proper harvest and management operations, as well as an adaptation to fluctuating effluent flows. It is not yet clear whether the lagoons should be operated with a fill-and-drain or a continuous-passage procedure. The harvested aquatic plants must be subjected to simple pre-processing and drying procedures before they are fed to the livestock. The exposure to ultraviolet radiation and the water loss will also eliminate pathogens in this feed to a large extent.

A strategic point for the introduction of modifications in the operation of animal husbandry units is the lower edge of the corral area. Here, a settling tank must be installed for the sediments contained in the slurry which leaves the biodigester. In this tank, the lignin material which tends to float on the sludge from dairy cattle excretions -- which would cause problems in the units of the system further downstream -- can also be removed. It can serve as organic fertilizer on the rangeland. In addition, to secure an adequate, regular supply of effluent under the erratic tropical rains and the fortuitousness of the operations in the stable area, a raised auxiliary reservoir which serves as a compensating tank will eventually become unavoidable.

From published work and experiences in Cuba, it follows that there would be a high probability of success if, for conditions like those in the Caribbean, the hydrophytes *Azolla* (a small water fern) and *Lemna* (often called "duckweed") were used. The Azolla hydrophyte species live in symbiosis with the blue-green alga *Anabaena azolla*, which supplies its host plant with nitrogen from the atmosphere. The biomass of Azolla -- the dry mass of which contains up to 30 per cent raw proteins -- can be expected to double within 2.5 days by growth and multiplication (Shiomi and Kitoh, 1987). The activity of these hydrophytes has quite high purification effects on the water in which they grow. The aquatic plants of the *Lemnaceae* show a quite comparable behaviour. However, as there is no symbiosis with an alga fixing nitrogen from the atmosphere, their protein production depends on the nitrogen content of the wastewater in which they thrive. They take up nutrients -- and contaminants -- in the absence of sunlight. Lemnaceae reduce the oxygen content of their substrate. Production rates of up to 53.8 kg of dry matter per hectare and day have been reported for *Lemna* species (Hillman and Culley, 1978).

The probable decontamination effect

The few, preliminary measurements of the wastewaters arising from various operation procedures in the dairy unit showed electrical conductivities between 0.88 and 3.00 mS/cm (groundwater: 0.56 mS/cm). Bacterial luminescence tests in the effluent channel downstream of the unit showed mostly high levels of toxicity. Bacterial counts showed colonies of *E. coli* and of coliphages in this channel, in which a "natural sewage treatment biotope" has developed which is subject to the natural and anthropogenic instabilities.

As only the biogas digester has been completed so far, it is not yet possible to substantiate with data the positive expectations inferred from the scarce available literature regarding the decontamination effect of the complete system. At present, the effluents which have passed the biogas reactor and those which have not passed this installation still flow in the old channels, which are filled with the organic sediments accumulated over approximately 30 years. However, it lies on hand that the water flowing in the surface channels which receive the raw effluents will be substantially less contaminated. It is problematic at present to quantify the positive effect which the decontamination of surface channels will have on the purity of the groundwater, especially on the groundwater in the higher aquifer (13 m depth). Data will become available when the lagoons have been constructed.

A satisfactory data basis does not yet exist, but the evidence is that the proposed system will lead to a substantial reduction of the contamination of the surface and groundwaters downstream of the dairy unit. It will remain an open question for quite a long time whether biotechnological measures and the implementation of this integrated low-technology system will lead to contaminant threshold values prescribed as limits for surface and groundwaters in national and international water laws. However, this is not yet the primary issue under most circumstances in tropical regions.

Reduction of the input of energy from fossil energy carriers

Adaptation of management

Maximum use must be made of the potentials furnished by the wind, the solar radiation and the livestock excretions to supply energy for the milking process, the refrigeration of the milk and the cleaning operations (cattle and milking equipment), as well as for the basic demands on the dairy unit (mainly domestic uses of the persons living there). Water (which has been raised with the input of energy) must be used economically, and the same applies to warm water heated with an energy input. Leakages must be repaired quickly. Operating times of energy consuming installations must be reduced by a rational organisation of work. Commonly, large potentials for such reductions lie in improvements of the milking procedure (avoidance of idle running times of the milking machine caused by inefficient conveyance of cows to the milking parlour). Analogous improvements will mostly be possible in the cattle washing operation. Animal traction should be used on the dairy unit as much as possible and feasible. Electrical appliances must be turned off when not used.

The supply of the biogas reactor should be adapted to the fermentation process. The contents of the fermentor should be circulated mechanically once per day. Whenever the outflow of the slurry and of the solid, fermented excretions does not function properly, corrective measures must be taken. The water/dung ratio of the biodigester charge should be adapted to its performance.

Technological measures

Basic standard equipment

The biogas reactor has already been discussed above when dealing with its other, complementary purpose. Its dimensions must be adapted to the expected quantities and qualities of excretions (some data for the biodigester on the pilot project include volume of the chamber: 16.25 cubic meters; daily charge: 300 kg of dung with 300 litres of water; retention time: 25 days; average temperature: 26 °C; available gas volume per day: 6.4 cubic meters). This unit proved to function adequately. Its economic analysis (again quite deficient due to inevitable, unsatisfactory assumptions) showed that its financial benefit for the dairy unit would be marginal. It should be noted that there is an unused additional potential energy reserve, as the daily production of dung is 650 kg.

The pipelines for the gas and the gas cookers (already installed in the pilot project) must be installed rigidly and safely. In most cases, the persons living on dairy units will be capable of installing and maintaining this equipment with very little external help.

Proposals for improvements and additions

A windmill-driven pump should be installed next to or on the main water tank to drive the pump which will serve to fill the reservoir (see above). For the pilot project it was estimated -- based on a locally available product and on some not very satisfactory but necessary economic assumptions -- that technically, such an installation would fulfil its objectives under the given conditions (pumping 15 cubic meters per day into the main water tank, the upper edge of which is 16 m above groundlevel). Financially, its benefit for the dairy unit would be marginal. Another windmill should be installed at the bottom edge of the dairy building area to drive the pump which will raise the water from the settling tank to the compensating tank mentioned above. This unit has not been planned in detail yet.

A combined solar energy/heat exchanger unit should be installed in order to secure water temperatures of 70 °C for cleaning the milking equipment. This temperature secures a dissolution of the milk fat which contains bacteria. Just as the improved diet with its effects on the quality of the milk, a high degree of purity of the milk -- in any case a prerequisite for a proper operation of the milk production process -- will secure a better price for the milk. The water heating unit consists of two elements: a thermo-syphon component in which the radiation of the sun is used to heat the water (2 x 50 litres per day) to 50 °C, and a heat exchanger which uses the compressed, hot air from the compressor of the milking machine to raise the temperature of the water by an additional 20 °C to 70 °C. Technically, such a unit would perform satisfactorily. It would also be financially beneficial for the dairy unit (again based on estimates with unsatisfactory economic data). A number of electric lamps should be replaced by biogas lamps.

The probable success of attempts to switch to regenerative energy sources

For the energy input component of the "closed cycle" balance, it is as difficult as for the other component (the output in contaminant loads) to give a reliable prognosis with some general validity. For the pilot project, the scarce data and experiences gained so far permit the statement that, under climatic conditions like those in the Caribbean, it will be technically feasible to attain a high degree of self-sufficiency with regards to energy inputs into dairy units. It can also be assumed that under most socio-economic conditions that can be expected in the sub-humid tropics, it will be profitable (to varying degrees) for milk production enterprises to invest management efforts and technology into the implementation of the concept proposed here. The use of regenerative energies will substantially increase the positive ecological effects and reduce the costs of biotechnological measures to reduce the contaminant loads of effluents from animal production units.

General appraisal, conclusions and recommendations

The pilot project

From the physical point of view, the choice of the pilot project conforms well with the objectives discussed here. After the breakdown of the Soviet Union, drastic problems with milk supply, as well as serious deficiencies in the supply with petroleum, machines and many other goods, arose in Cuba. This emergency situation led to the initiation of the assistance to the "Vaquería Nr. 40". The given economic background also makes this country a suitable place for this pilot project in the scientific and in the technological sense. The fact that Cuba has a centrally planned economy makes representative economic appraisals more difficult. The situation described earlier is rather representative of approximately 2 000 to 3 000 dairy units in Cuba. Regarding the implications of possible groundwater contamination, the situation is especially critical in the Perú Valley. Here, aquifers are located from which water for the supply of the city of La Habana is pumped.

The first results reported upon here on the one hand are offered as a contribution of data and experiences to a problem solution on a global scale for the mentioned tropical climatic zones. On the other hand, they should be seen as a first basis on which work in Cuba can be continued.

Suggestions for future improvements and refinements

Possibilities for modifications of the basic concept of the integrated system

From the experience gained, it seems advisable to locate the biodigester in future projects on the lower edge of the dairy building area, adjacent to the settling tank and the compensation tank.

As complementary additions to the concept presented above, fishponds and supplementary effluent treatment ponds with plants like "giant bulrushes" *Scripus validus* (Becker, 1993) should be taken into consideration.

Phototrophic cyanobacteria (formerly also classified as microalgae and commonly called "blue algae") can live in symbiosis with heterotrophic bacteria like *Bacillus subtilis* and *Bacillus thüringiensis* (Szigeti *et al.*, 1996). For effluents from piggeries and from households, a coupled purification and biomass production effect can be expected based on the following symbiotic system (O. Pulz, V. Ördög, and J. Szigety, 1996, personal communication): in a suspension containing the raw effluent and/or the slurry leaving a biodigester, the bacteria decompose organic substances, producing carbon dioxide which, in turn, is used for photosynthesis by the cyanobacteria. These organisms fix elementary nitrogen and produce oxygen. The oxygen is used for respiration by the bacteria. The suspension which develops from the metabolic activities of this symbiotic system contains the usable biomass of the cyanobacteria and of the heterotrophic bacteria, as well as valuable substances like proteins, vitamins and growth regulators. It should be investigated whether it would be feasible for piggeries and also for dairy units to inoculate, with a suspension from a central inoculate production plant (Pulz *et al.*, 1995), the sewage ponds of a number of animal production units from time to time. The ponds would have a function analogous to the hydroponic ponds with water fern and duckweed. As for the fishponds and/or the purification lagoons with wetland plants, it is not yet clear where their location would be downstream of the hydroponic ponds with hydrophytes, parallel to or combined with other components of the system. High technology equipment and management would be required at least to a certain degree, and -- as is known so far -- the lagoons would have to be shallow "raceway ponds" in which the suspension is kept moving with paddle wheels. Questions regarding the potential use of the biomass produced must be answered. It has been demonstrated that, for hogs and for fowl, the biomass represents a good addition to the diet.

Additional modifications in management to be considered

It might be beneficial for the operation of the fermentor to feed it with a 1.5-2:1 water/dung ratio instead of using a ratio of 1:1, and to use water which has already been used once for other purposes on the dairy unit ("gray" water).

The results and experiences gained in the framework of efforts to answer the open questions listed below should be incorporated into dairy unit operations as required.

Possible technical improvements

With unused biogas reserves (possibly stored in elastic balloons), modified engines could be driven and used for various operations. This would require the removal of sulfur from the gas. The design and testing of an integrated plant for pumping, storing water, producing electric power and washing of the livestock should be envisaged. There are windmill-driven pumps on the market which aerate the water body from which they pump water. This has the effect that the pumps are less sensitive to operation disturbances by sediments and other impurities. Such pumps should be tested.

Consequences regarding procedures of economic evaluation of ecological benefits and damages

Economists all over the world in international and national agencies as well as in scientific institutions struggle with the problem of how to assign an economic value to items such as the gain from reducing the driving factors for climatic change (such as burning fossil fuels, the future prices of which are open to question), the gain by not cutting wood for cooking purposes, or the damage from contaminants in surface and groundwaters. Fleischer (1995), in discussing the relevant environment standards and the ecological reference values, has pointed out the problems involved in executing environmental compatibility tests for animal husbandry plants. For the frame outlined in the sections above, the results of the work presented here may be a contribution to the narrowing in of ranges of values which can be used for economic and ecological evaluations.

In countries like China and India, governments promote the production and use of biogas. Other agencies encourage sequential water use or the involvement of the private sector in food processing industries (which also traditionally offer jobs to women in developing countries). There are projects to promote decentralised water supply and decentralised sewage treatment in rural areas or to recycle "waste". The studies presented here may also represent a usable data source for cost estimates for such programmes.

It might also be useful to submit the notion of a "closed cycle" as it is proposed here to a critical scrutiny with regards to the possibilities which it offers as a criterion to evaluate the worth of ecological measures.

Possible consequences for future work with the hydrotope concept

When agricultural production strategies as well as water and soil conservation measures are planned, the hydrologic suitability of landscape units must be specified. Diestel *et al.* (in press) have proposed to use the concept of the "hydrotope" for such purposes, and to apply it -- if conditions permit this -- to a planning level roughly corresponding to the administrative district. A hydrotope, as applied to such objectives, is a landscape unit with a *specified* hydrologic suitability for *defined* planning aim. A hydrotope may consist of aerial components designated as "smallest homogeneous units". Economic parameters (benefit/cost relationships), ecological factors (potential contamination of waters, *inter alia*) and sociological criteria (the capacity of the implementing agency to operate according to the objectives) enter into the hydrologic zonification procedure. If the low-energy-input/low contamination concept proposed here for the operation of animal husbandry units in some tropical regions proves to be successful, this will have the consequence that in many planning cases, the criteria to define the smallest homogeneous units and/or the hydrotopes must be modified. Landscape units which -- not taking into consideration this operation concept -- would be classified as "unsuited" for milk (or meat) production may then be defined as "suited" for that alternative of use.

Open questions

There are still many open questions to be answered which have not been touched upon in the above sections. Which are more precisely the tropical zones in which this concept could be implemented? Can hydrologic models be elaborated which will be applicable in different areas? Would other hydrophyte species be more adapted in specific cases? To which degree can cycles be "closed" in the sense defined here?

One block of questions refers to the acceptance of the dried hydrophyte plant material by the cattle, the digestibility of the diets, and the milk production during the rainy season and the dry season, the protein balance and the varying compositions of the dung with the resulting effects on the fermentation in the biogas reactor, and on the performance of the aquatic plants.

Will biogas production be increased by warming up the water entering the biodigester? Is it feasible to use "gray" water for irrigating the rangeland? Will it be better to operate the lagoons as a continuous flow system, or to use a fill-and-drain procedure? Can effluents from cane sugar factories be treated successfully with analogous systems? Can lateritic soils be utilised (due to their high iron content) as filter material to remove the sulfur from biogas to make it suitable as energy source for modified engines? These are some of the miscellaneous questions to which answers should be sought.

In countries like Germany, laws regulating the use and the disposal of wastes have strongly influenced the use of biogenic wastes (Weiland, 1995). A status study should be elaborated (if it does not yet exist) to investigate to which degree the implementation of integrated livestock production systems can be stimulated by laws in countries of tropical zones. In such a study, realistic threshold values should also be defined.

Questions, comments and answers

Q: What do you do with the plant material in the lagoon? How often does it have to be harvested, and what are the disposal options?

A: For about 100 cows we use a three to five ha hydroponic lagoon. This provides just enough feed. It is continuously harvested (every few days), then dried and processed as a protein feed.

REFERENCES

BECKER, H. (1993), "Constructed Wetlands Clean Up", *Agric. Res.*, December, p. 7.

DIESTEL, H., M. HAPE, and J.-M. HECKER (in press),"Specification of the hydrologic suitability of landscape units as a basis for planning water and soil conservation measures", *Proc. 9th ISCO (Intern. Soil Conserv. Org.) Conf.*, 26-30 August, Bonn, Germany.

FLEISCHER, E. (1995), "Zu einigen methodischenn Aspekten der Umweltverträglichkeitsprüfung UVP-pflichtiger Tierhaltungsanlagen", *Kühn-Archiv* 89:1, pp. 64-86.

HILLMANN, W. and D. CULLEY (1978), "The uses of duckweed", *Amer. Scientist* 66:4, pp. 442-451.

KÜHL, U. (1996), "Integriertes Nährstoffrecycling in einer Milchviehanlage auf Kuba mit Hilfe einer Biogasanlage und einer Hydrokulturlagune", Masters thesis, Faculty of Agriculture and Horticulture/Livestock Ecology Section, Humboldt University, Berlin.

KÜHL, H., H.J. SCHWARTZ, and H. DIESTEL (1995), "Proyecto lechero Mina Blanca, Valle del Perú", Seminar, Instituto de Ciencia Animal, 14-15 November, Cuba, posters and handouts.

MIETH, H. (1996), "Untersuchung zur autarken Energieversorgung eines landwirtschaftlichen Betriebes in Kuba unter Nutzung am Standort vorhandener Resourcen", Masters thesis, Faculty for Mechanical Engineering and Ship Technology, University of Rostock, Germany.

PULZ, O., N. GERBSCH, and R. BUCHHOLZ (1995), "Light energy supply in plate-type and light diffusing optical fiber bioreactors", *J. Applied Phycology* 7, pp. 145-149.

SCHWARTZ, H.J. (1995), "Lecture notes on Tropical Pastures and Feed Resources", Livestock Ecology Section, Humboldt University, Berlin.

SCHWARTZ, H.J., R. GARCÍA, and A. ÁLVAREZ (1995), "Proyecto lechero Mina Blanca, Valle del Perú, Habana. Milch für Kubas Kinder, Pilotprojekt Vaqueria No. 40, 1993/1994", Incidental publication, Asociacion Cubana de Produccion Animal/Cuba Sí/Humboldt University, La Habana/Berlin.

SHIOMI, N. and S. KITOH (1987), "The use of Azolla as a decontaminant in sewage treatment", in *Azolla Utilization,* Proceedings of the Workshop on Azolla Use, pp. 169-176, 31 March-5 April 1985, Fuzhou, Fujian, China, Intern. Rice Res. Inst., Manila, Philippines.

SZIGETY, J., V. ÖRDÖG, T. FÖLDES, and O. PULZ (1996), "Microbial groth stimulation and inhibition caused by cyanobateria". Abstracts of the Confer. on "Progress in Plant Sciences from Plant Breeding to Growth Regulation", 17-19 June, Pannon Agric. Univ., Mosonmagyaróvár, Hungary.

WEILAND, P. (1995), “Erfahrungen mit der Verwertung biogener Abfälle zur Biogaserzeugung in Deutschland”, in Internationale Erfahrungen mit der Verwertung biogener Abfälle zur Biogasproduktion, Umweltbundesamt, Vienna, Austria.

REDUCTION OF AGRICULTURAL NITRATE LOADING THROUGH MICROBIAL WETLAND PROCESSES

by

Torben Moth Iversen and Carl Chr. Hoffmann
National Environmental Research Institute, Silkeborg, Denmark

Introduction

Eutrophication of surface waters is caused by nutrient enrichment. The many consequences of eutrophication include increased algal biomass and production, changes in benthic macrophyte depth distribution and species composition, and in severe cases, oxygen depletion leading to mass death of fish and benthic fauna.

In Europe, eutrophication-related phenomena in coastal waters are more frequent and more serious than in the past (Stanners and Bourdeau, 1995), both in the North Sea and, not least, in the Baltic Sea. In the southern Kattegat, oxygen depletion has been observed in vast areas since the early 1980s, with its extent varying from year to year depending on nutrient input and wind speed and direction (Kronvang *et al.*, 1993). In 1988, 5 600 km^2 were affected, and fish and Norwegian lobster *Nephrops norvegicus* were absent or occurred in low numbers (HELCOM, 1991).

Marine eutrophication is mainly attributable to nitrogen leaching from agricultural areas. In Denmark and Sweden about 80 per cent and 40 per cent, respectively, of the nitrogen inputs to inland waters are derived from agriculture, even though farmland accounts for 65 per cent and six per cent of the land area, respectively (Kronvang *et al.*, 1993; Löfgren and Olsson, 1990).

Development in agriculture

Agricultural development in Europe this century has considerably increased production, especially since the 1950s, albeit that there has been a tendency towards stagnation since the late 1980s (Stanners and Bourdeau, 1995). This growth has been obtained by intensification of agriculture, a process that has significantly changed the landscape. Thus, in lowland areas such as the Netherlands, northern Germany, Denmark and southern Sweden, wetlands have been drained and rivers have been channelized and rigorously maintained to ensure efficient drainage of fields. In Denmark, for example, about 50 per cent of the agricultural area is drained and 80-98 per cent of all natural rivers have been channelized (Iversen *et al.*, 1993).

At the same time, nitrogen input in the form of commercial and animal fertilizer has increased, the net result being enhanced nitrogen loading of inland waters and consequently enhanced riverine loading of marine areas. In Denmark, riverine nitrogen transport increased until the late 1970s, stabilised in the 1980s, and now seems to be decreasing slightly due to measures implemented in the agricultural sector.

Nitrogen removal in riparian areas

Nitrogen removal through biological denitrification takes place under anaerobic conditions such as in waterlogged soils. There certain bacteria are able to respire by converting nitrate to atmospheric nitrogen in an energy-consuming process. In most cases, the energy required is derived from the breakdown of organic carbon sources such as peat. The process can be summarised as follows:

Organic matter + Nitrate + Hydrogen ions $\rightarrow$ Carbon dioxide + Atmospheric Nitrogen + Water

A common finding in many wetland studies is the very limited area in which denitrification takes place. Thus nitrate can disappear completely over a distance of only 5-20 metres along the groundwater flow line (Peterjohn and Correll, 1984; Lowrance *et al.*, 1984; Pinay and Decamps, 1988; Cooper, 1990; Fustec *et al.*, 1991; Haycock and Burt, 1993; Haycock and Pinay, 1993; Pinay *et al.*, 1993). An example from a fen in the river Gjern watershed in Denmark is illustrated in Figure 1. While the groundwater entering the fen at the hill slope is nitrate-rich, the concentration falls after a distance of only five metres to less than one mg nitrate-N l^{-1}. The deeper part of the groundwater penetrates further into the fen, but the nitrate-N concentration still declines to less than one mg nitrate-N l^{-1} within a distance of 13 metres, thereafter to fall below detection limit.

Figure 1. **Nitrate removal in a riparian fen at Voldby brook, Jutland, Denmark**

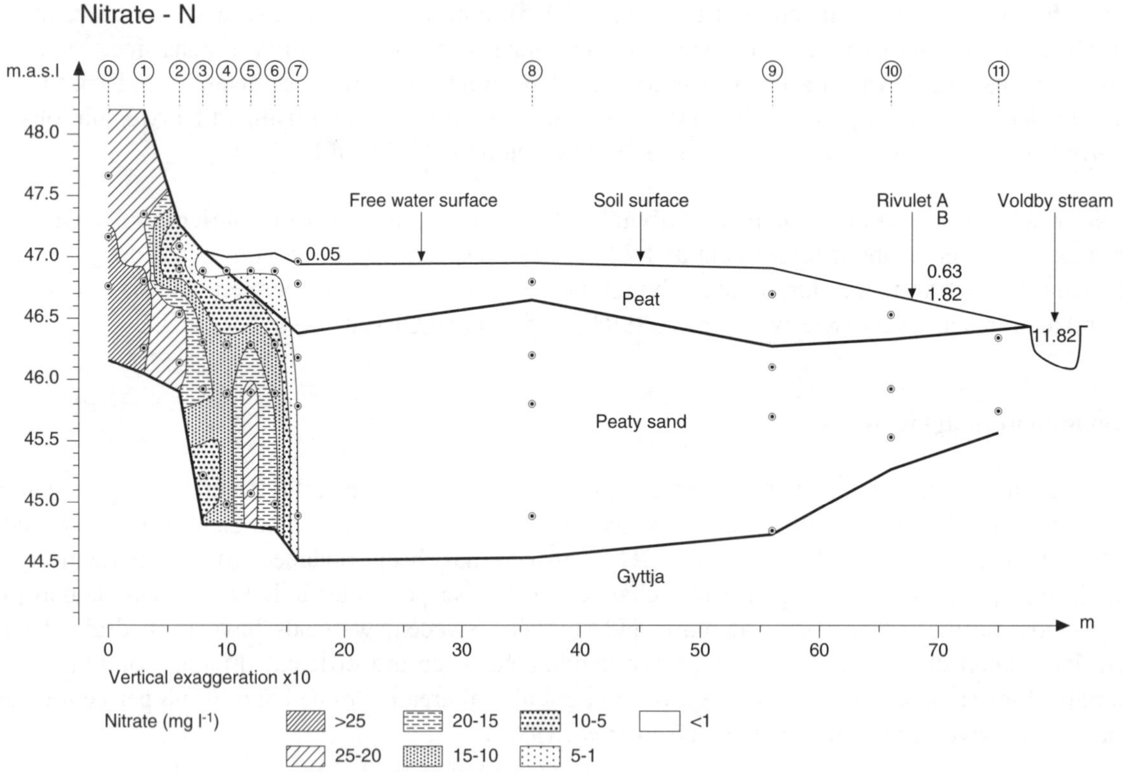

Note: At the hill slope, nitrate concentration rapidly decreases from 25 mg NO_3-N l^{-1} to below one mg NO_3-N l^{-1} over a distance of 16 m.

Source: Dahl, 1995.

Table 1 summarises the results of Danish studies of natural wetlands. The sites differ markedly with respect to the amount of nitrate reduced, mainly due to differences in hydrological properties between the wetlands and differences in nitrate inputs. In most cases, the relative effectiveness was very high and independent of the total amount of nitrate removed.

Table 1. **Nitrate removal and phosphorus retention in natural freshwater wet meadows and fens in Denmark**

Natural systems	Transformed kg nitrate-N ha^{-1} y^{-1}	Efficiency %	Phosphorus removal kg P ha^{-1} y^{-1}	Phosphorus removal efficiency %
River Stevns watershed; meadow	57	97	2.8	96
Rabis Brook watershed; meadow	398	56	0.4	100
River Gjern watershed				
A: meadow	140	67		
B: fen	2 169	97	-16.4	-410
C: meadow	590	99		
D: meadow	42	81		

Source: Dahl, 1995; Iversen *et al.*, 1995; Rebsdorf *et al.*, 1994.

Table 2 shows that the nitrogen removal capacity and efficiency of manipulated wetlands may be as high as that of natural systems.

Table 2. **Nitrate removal and phosphorus retention in manipulated freshwater wet meadows and reed swamps in Denmark**

Manipulated systems	Transformed kg nitrate-N ha^{-1} y^{-1}	Efficiency %	Phosphorus removal kg P ha^{-1} y^{-1}	Phosphorus removal efficiency %
Lake Glumsø; reed swamp	520	65	-49	-206
	975	62	-21	-41
	2 725	54	-365	-342
Lake Glumsø; full scale	569	94	-29.2	-389
River Stevns; meadow*	350	99	14.1	93
Syv Brook; meadow	300	72	0.1	1
River Stor; restored meadow	470	48	2	42
River Gjern; meadow*	34-200	88-98	2.2-4.8	

* Short-term experiments (one month).

Source: Hoffmann, 1986, 1991, in press; Hoffmann *et al.*, in press; Rebsdorf *et al.*, 1994.

In the case of the fen in the river Gjern watershed, both the relative (97 per cent) and total nitrate removal (2 169 kg nitrate-N ha^{-1} y^{-1}) were very high (Table 1). At the same time, though, 16.4 kg total P ha^{-1} y^{-1} leached into the stream as a result of decomposition of the organic matter used as the energy source for denitrification (Paludan and Hoffmann, 1995). This may have a detrimental effect on the down-stream lake and estuarine ecosystems. The situation is therefore non-sustainable as the system will eventually break down. The lack of sustainability is also stressed by changes in the plant community in favour of plants which can tolerate the very high nutrient concentrations.

Another example is the irrigation experiment in a reed swamp at lake Glumsø shown in Table 2. The high hydraulic and nitrate loads resulted in phosphorous export to the lake. During the three-year study period, the net loss of phosphorus to the lake ranged from 21 to 365 kg P ha^{-1} y^{-1}.

In the contrast to these two systems in which phosphorus was lost, the other systems studied -- both natural (nitrate removal rates 57-398 kg nitrate-N ha^{-1} y^{-1}; Table 1) and manipulated (nitrate removal rates 34-569 kg nitrate-N ha^{-1} y^{-1}; Table 2) -- actually retained phosphorus (Hoffmann, 1991; Hoffmann, in press).

Sustainable nitrate removal can be defined as a dynamic equilibrium in which the autochthonous plant production balances the consumption of organic matter for denitrification. The Danish studies thus suggest that the maximum sustainable nitrate removal rate is 300-400 kg nitrate-N ha^{-1} y^{-1}.

Restoration of rivers and riparian areas

River restoration has been on the political agenda in Denmark since 1982, where it was made possible by a revision of the Watercourse Act. The original goal was to improve the ecological quality of our watercourses, but the scope has since changed, and more and more projects now include the riparian wetland of the river valley (Iversen *et al.*, 1993).

The goal of river and riparian area restoration is to re-establish river valley hydrology. A number of methods of varying sophistication can be used, including:

- discontinuing drainage pipes at the river valley border;
- closure of ditches in the river valley;
- less rigorous or no stream maintenance;
- remeandering of previously channelized streams.

With each method, the groundwater level will increase, nitrogen removal capacity will improve, and the wetland flora and fauna will return. With the last method, ecological river quality will also improve significantly within one to two years (Kronvang *et al.*, 1994).

It is estimated that there are approximately 120 000 ha low-lying peat and loamy soils in Denmark presently under crop rotation, and where the natural nitrogen removal capacity could immediately be restored. At an average nitrogen removal rate of 200-400 kg nitrate-N ha^{-1} y^{-1}, total nitrogen removal would amount to 24 000-48 000 t nitrate-N y^{-1}, equivalent to 27-53 per cent of the average riverine nitrogen load from Denmark to marine areas.

The above mentioned findings clearly indicate that great care needs be taken to avoid an unsustainable nitrogen removal capacity. When planning to restore nitrogen removal capacity, future hydraulic and nutrient loads as well as soil type should therefore be evaluated in order to avoid unintended consequences.

In August 1994, a 3.0 km long straightened section of the river Brede Å in southern Denmark was remeandered into a 4.5 km section (Figure 2). During the period 1994-1996, nitrate removal in three transects of the riparian area ranged from -11 to 160 kg nitrate-N ha^{-1} y^{-1} (mean 92 kg nitrate-N ha^{-1} y^{-1}). This occurred despite the fact that 1995-1996 was the driest hydrological year this century. Average removal efficiency was 71 per cent of the nitrate load (Hoffmann *et al.*, in press). This case study clearly documents the large nitrate removal capacity of previously drained riparian areas.

Figure 2. **River Brede, Southern Jutland, Denmark**

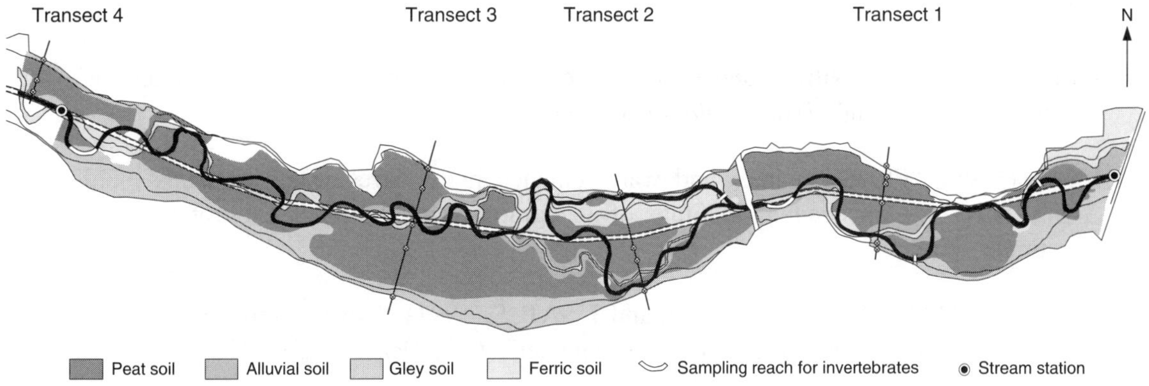

Note: A 3 km river channel was restored to 4.5 km meandering river. Soil type, investigating transects, old channel and new meandering river shown.

Source: Hoffmann *et al.*, in press.

In the European Environment Agency (EEA) report "Europe's Environment - The Dobríš Assessment" (Stanners and Bourdeau, 1995), eutrophication and the loss of biodiversity have been identified as major environmental issues in Europe. Similarly, the 5th Environmental Action Plan of the European Union, "Towards Sustainability", identifies integrated water management and stabilization/improvement of biodiversity as major goals. A significant measure to obtain these goals is restoration. The results presented here certainly document the potential of this measure.

Agricultural overproduction in the European Union has led to set-aside schemes with economic subsidies to reduce production. These same schemes could be used strategically with significant environmental advantage to promote large-scale restoration of wetlands.

In Eastern Europe, agriculture is generally extensive and many wetlands still exist. As future development of Eastern Europe will undoubtedly involve intensification of agriculture, it is extremely important to ensure that this does not lead to repetition of the mistakes made in Western Europe, and hence to the same undesirable ecological consequences.

A main environmental issue of agricultural policy should be to reduce nitrogen leaching from the root zone by avoiding excessive use and incorrect application of fertilizers, etc. However, nitrate loss from agricultural land will always be significantly greater than the background level, and an important additional measure would therefore be large-scale restoration of wetlands. There is certainly a great need to integrate environmental issues such as integrated water management and conservation/improvement of biodiversity into agricultural policy in both Eastern and Western Europe.

REFERENCES

COOPER, A.B. (1990), "Nitrate depletion in the riparian zone and stream channel of a small headwater catchment", *Hydrobiologia* 202, pp. 13-26.

DAHL, M. (1995), "Flow dynamics and water balance in two freshwater wetlands", PhD thesis, University of Copenhagen, Institute of Geography and National Environmental Research Institute.

FUSTEC, E., A. MARIOTTI, X. GRILLO, and J. SAJUS (1991), "Nitrate removal by denitrification in alluvial ground water: role of a former channel", *J. Hydrol.* 123, pp. 337-354.

HAYCOCK, N.E. and T.P. BURT (1993), "Role of floodplain sediments in reducing the nitrate concentration of subsurface run-off: a case study in the Cotswold, UK", *Hydrological Processes* 7, pp. 287-295.

HAYCOCK, N.E. and G. PINAY (1993), "Groundwater nitrate dynamics in grass and poplar vegetated riparian buffer strips during winter", *J. Environ. Qual.* 22, pp. 273-278.

HELCOM (1991), Baltic Marine Environment Protection Commision, Helsinki Commission 1991, "Interim report on the state of the coastal waters of the Baltic Sea", Baltic Sea Environment Proceedings No. 40.

HOFFMANN, C.C. (1986), "Nitrate reduction in a reedswamp receiving water from an agricultural watershed", Proceedings 13th Nordic Symposium on Sediments, 1985, Aneboda, Sweden, pp. 41-61.

HOFFMANN, C.C. (1991), "Water and nutrient balances for a flooded riparian wetland", in *Nitrogen and Phosphorus in Fresh and Marine Waters - Project Abstracts of the Danish NPo Research Programme*, C13b, pp. 203-220, NPo-forskning fra Miljøstyrelsen, C-Abstracts.

HOFFMANN, C.C. (in press), "Nitrate removal in a regularly flooded riparian meadow", *Verh. Internat. Verein. Limnol.* 26.

HOFFMANN, C.C., M.L. PEDERSEN, B.K. KRONVANG, and L. ØVIG (in press), "Implications of river restoration on nutrient retention", Proceedings River Restoration '96 - The physical dimension, International Conference, 9-13 September, Silkeborg, Denmark, *Aquatic Conservation*.

IVERSEN, T.M., B.K. KRONVANG, B.L. MADSEN, P. MARKMANN, and M.B. NIELSEN (1993), "Re-establishment of Danish streams, restoration and maintenance measures", *Aquatic Conservation* 3, pp. 1-20.

IVERSEN, T.M., K. KRONVANG, C.C. HOFFMANN, M. SØNDERGAARD, and H.O. HANSEN (1995), "Restoration of aquatic ecosystems and water quality", in H.S. Møller (ed.), *Nature Restoration in the European Union,* Proceedings of Seminar, pp. 63-69, 29-31 May, Copenhagen, Ministry of Environment and Energy, the National Forest and Nature Agency, Denmark.

KRONVANG, B., G. ÆRTEBJERG, R. GRANT, P. KRISTENSEN, M. HOVMAND, and J. KIRKE-GAARD (1993), "Nationwide monitoring of nitrients and their ecological effects, State of the Danish aquatic environment", *Ambio* 22, pp. 176-87.

KRONVANG, B., P. GRÆSBØLL, L.M. SVENDSEN, N. FRIBERG, A.B. HALD, G. KJELLSON, M.B. NIELSEN, B.D. PETERSEN, and O. OTTESEN (1994), "Restoration of River Gels at Bevtoft - Environmental effects in the stream and in the riparian areas", Technical Report No. 110, National Environmental Research Institute, Denmark.

LOWRANCE, R.R., R.L. TODD, J.Jr. FAIL, O.Jr. HENDRICKSON, R. LEONARD and L.E. ASMUS-SEN (1984), "Riparian forests as nutrient filters in agricultural watersheds", *BioScience* 34:6, pp. 374-377.

LÖFGREN, S. and H. OLSSON (1990), "Tilförsel av kväve och fosfor till vattendrag i Sveriges inland", Report No. 3692 from Naturvårdsverket (with English summary).

PALUDAN, C. and C.C. HOFFMANN (1995), "Fate of phosphorus in a Danish minerotrophic wetland", International Workshop, Sediment and Phosphorus, Erosion and delivery, transport and fate of sediments and sediment-associated nutrients in watersheds, 9-12 October, Silkeborg, Denmark.

PETERJOHN, W.T. and D.L. CORREL (1984), "Nutrient dynamics in an agricultural watershed: Observations on the role of a riparian forest", *Ecology* 65:5, pp. 1 466-1 475.

PINAY, G. and H. DECAMPS (1988), "The role of riparian Woods in regulating nitrogen fluxes between the alluvial aquifer and surface water: A conceptual model", *Regulated Rivers: Research and Management* 2, pp. 507-516.

PINAY, G., L. ROQUES, and A. FABRE (1993), "Spatial and temporal patterns of denitrification in a riparian forest", *Journal of Applied Ecology* 30, pp. 584-591.

REBSDORF, A., N. FRIBERG, C.C. HOFFMANN, and K. KRONVANG (1994), "Interactions between riparian areas and streams - state of art" (in Danish), Miljøprojekt No. 275, Danish Environmental Protection Agency.

STANNERS, D. and P. BOURDEAU (1995), "Europe's Environment - The Dobríš Assessment", European Environment Agency, Luxembourg.

Parallel Session II

II-D. DESERTIFICATION AND ENVIRONMENT MODIFICATION

MOLECULAR MECHANISMS OF THE DROUGHT RESPONSE IN HIGHER PLANTS

by

Alejandra A. Covarrubias, José Manuel Colmenero-Flores, Francisco Campos, Alejandro Garciarrubio and Adriana Garay-Arroyo
Departamento de Biología Molecular de Plantas, Instituto de Biotecnología, Universidad Nacional Autónoma de México (UNAM), Cuernavaca, Morelos, Mexico

Brief historical review

Water deficit is one of the most common environmental stress factors that soil plants are exposed to, since it is intrinsic to most abiotic forms of stress -- not only during drought, but also at low temperature and when the soil contains a high concentration of ions (Bartels and Nelson, 1994; Bohnert *et al.*, 1995). It interferes with normal development and growth and has a major adverse effect on their productivity. Among a diversity of responses, plants adapt to water deficit by the induction of specific genes (Ingram and Bartels, 1996). Some of these genes are also expressed during the normal embryogenesis program when seeds desiccate and embryos become dormant (Thomas *et al.*, 1991). During this developmental stage, plant tissues are viable for long periods of time in conditions of extremely high dehydration. Both environmental and developmental processes have in common the mediation of the phytohormone abscisic acid (ABA) (Bray, 1991; Bray, 1993; Thomas *et al.*, 1991). The application of ABA to non-stressed vegetative tissues can mimic many effects of drought on plants, including the induction of water deficit responsive genes (Bray, 1993; Thomas *et al.*, 1991). In recent years efforts have focused on the isolation of genes that are induced during water deficit or ABA treatments in order to study the function of their products. This approach has made possible the identification of stress proteins and the characterisation of their biochemical, cellular and adaptive roles in osmotic stressed plant cells. This is the case of proteins implicated in the biosynthesis of osmolytes (Bohnert *et al.*, 1995; Ingram and Bartels, 1996), in the uptake and compartmentation of ions (Molina and García-Olmedo, 1993; Niu *et al.*, 1993), in hydroxyl-radical scavenging (Bohnert *et al.*, 1995; Smirnoff and Cumbes, 1989), and in protein turnover (Borkird *et al.*, 1991; Bray, 1993; Kiyosue *et al.*, 1994; Koizumi *et al.*, 1993). Although other proteins have been identified whose levels are affected by water deficit, the lack of knowledge on their biochemical activities or *in vivo* functions in the cell, makes difficult to assign them a role in water stressed plants. Such is the case of some metabolic enzymes (Espartero *et al.*, 1994; Ludwig-Muller *et al.*, 1995; Umeda *et al.*, 1994); osmotin (Kononowicz *et al.*, 1993); lipid transfer proteins (LTP) (Kader, 1996; Torres-Schumann *et al.*, 1992); low molecular weight heat shock proteins (lmw-HSPs) (Almoguera and Jordano, 1992); water channels or aquaporins (Chrispeels and Maurel, 1994); and some cell wall structural proteins (Covarrubias *et al.*, 1995; Creelman and Mullet, 1991; Esaka *et al.*, 1992). In the case of the late embryogenesis abundant (LEA) proteins there is a strong circumstantial evidence for their involvement in the plant adaptation to water deficit.

The fact that *lea* transcripts and proteins accumulate in vegetative tissues of a number of drought-stressed plants, and that desiccation treatments can often induce their precocious expression in seeds, has led to the proposition that LEA proteins may play a protective role in specific cellular structures or ameliorate the effects of drought stress. This hypothesis is consistent with the properties predicted from their deduced amino acid sequence, such as their high hydrophilicity and randomly coiled moieties (Dure, 1993).

Some of the genes mentioned above have been characterised to respond not only to other environmental factors like heat shock, infection, wounding, light, etc., but also have been involved in the normal development process of non-stressed plants (Chrispeels and Maurel, 1994; Almoguera and Jordano, 1992; Espartero *et al.*, 1994; Kader, 1996; Showalter, 1993). This is not surprising given the multiple physiological and metabolic alterations induced in the plant by water deficit, as well as the fact that many of them belong to gene families. At this time, the role of some of these proteins during plant adaptation to water stress can not be defined, given the few data available regarding the functional characterisation of these proteins in response to this kind of stress. One of the reasons for this lack of progress is the multigenic nature of sensitive and tolerant plant phenotypes. In addition, it must be considered that different plant species have a variety of mechanisms that have evolved in a family-specific or order-specific manner to confer tolerance.

Conventional breeding approaches alone are valuable, but it would be more beneficial to include as molecular markers the genes for the mechanisms associated with stress tolerance. This kind of approach is supported by the observation that there are differences in the expression of specific genes between stress-sensitive and stress-tolerant plants, which indicates that tolerance is conferred by genetically encoded mechanisms (Basta, 1994; Bray, 1993). Recently, it has been reported that the overexpression of specific proteins in transgenic plants has some potential for reducing the effects of stress. Some of these proteins correspond to enzymes involved in the synthesis of osmolytes (mannitol, fructans, proline and glycine betaine) or enzymes whose overexpression increases oxygen-radical scavenging. Also, overexpression of some LEA proteins has led to the generation of transgenic plants with some tolerance to water stress (Delauney and Verma, 1993; Pilon-Smits *et al.*, 1995; Tarczynski *et al.*, 1993).

To learn about the biochemical and molecular mechanisms by which plants tolerate water stress, different strategies and model systems have been used. One approach has been to use specific structures or species that can withstand severe dessication. In this category are considered certain seeds (McCarty, 1995) and dessication-tolerant species, such as resurrection plants, mosses and ferns (Bartels *et al.*, 1990; Ingram and Bartels, 1996). Regardless of the evolutionary history of extremely tolerant species, they may be used to understand tolerance mechanisms, and as a source of genes able to enhance stress tolerance in other species (Bohnert *et al.*, 1995). In the case of seeds, it is well known that the final maturation stage of their development is characterised by the loss of 90 per cent of the original water. This dessicated state allows survival under extreme environmental conditions and contributes to a successful dispersal. Actually, the embryo acquires dehydration tolerance well before maturation drying, but is lost as germination progresses. It is during this maturation stage, at the onset of dessication, when LEA transcripts first appear to dominate the mRNA population in dehydrated tissues (Roberts *et al.*, 1993).

Another approach has been the analysis of dehydration tolerance in genetic model systems which take advantage of detailed genetic information, a wide range of mutants, and the feasibility of positional gene cloning. This is the case of *Arabidopsis thaliana* where different mutants affected in production or sensitivity to ABA, one of the mediators of the plant response to water stress, have been extensively characterised (Gosti *et al.*, 1995; Parcy *et al.*, 1994).

The use of species important for agriculture has also contributed to the knowledge of the plant response during and after drought stress. The studies carried out in these species have been very useful, since the transient and moderate drought stress represented in those analyses probably describe the most common form of dehydration that most plants may confront. Additionally, the intensive breeding has made available lines with different degrees of tolerance that may be used to obtain correlative evidences for the involvement of different genes in drought tolerance (White and Singh, 1991; Ribaut *et al.*, 1996; Boyer, 1982).

In an effort to contribute to understanding some of the mechanisms involved in the plant response to water deficit, we have followed three main research approaches: (a) the isolation and characterisation of genes, and their products, whose expression is induced by water deficit; (b) the study of the participation of cell wall proteins during this response and during the stress perception and signalling; and (c) a functional approach using the yeast *Saccharomyces cerevisiae*, one of the most powerful model systems for the molecular biology of eucaryotic cells, to isolate plant genes involved in halo- and osmo-tolerance, and to characterise plant proteins in regard to their cellular function during osmotic stress conditions. The first one has led us to the identification and characterisation of several genes in common bean (*Phaseolus vulgaris*) and yeast (*Saccharomyces cerevisiae*) whose expression is affected by water limiting environments (Colmenero-Flores *et al.*, in press; Garay-Arroyo and Covarrubias, data not shown). Among these genes, we have focused our attention on the study of the LEA proteins based on the correlation between their accumulation and environmental stresses, as well as on their particular structural properties consistent with a protective role of cellular structures and/or different macromolecules during adverse situations. The second approach has allowed us to identify two new cell wall glycoproteins (p36 and p33) whose levels are increased during the bean response to water deficit (Covarrubias *et al.*, 1995). The particular properties of these proteins allowed us to hypothesise that p36 and p33 participate as transducers of the cell water status and/or to maintain the cellular integrity by keeping the contact between the plasma membrane and the cell wall during water stress. Testing anticipated contributions of particular genes to stress tolerance does not necessarily have to be done in a plant. Based on similarities in biochemical pathways and response to stress between *Saccharomyces cerevisiae* and plants, we are using osmo- or halo-sensitive mutants to express plant cDNA libraries in order to isolate those plant genes able to confer osmo- or halo-tolerance to yeast strains (Serrano and Gaxiola, 1994). In this manuscript, most attention will be given to the results regarding the characterisation of the different plant cDNAs and yeast genes, and their role during water stress will be discussed. We will also address future developments and strategies for designing tolerance to water deficit in plants.

Current activities

In this manuscript we describe six cDNA clones from *Phaseolus vulgaris*, whose expression is induced by water deficit and ABA treatments. These cDNA clones were isolated from a complementary DNA (cDNA) library constructed from poly(A)$^+$ RNA isolated from 21-day-old bean plants exposed to water deficit treatment as described elsewhere (Covarrubias *et al.*, 1993). Of 27 clones which showed differential expression with regard to water deficit, after three screens (Covarrubias and Garciarrubio, 1993). All the cDNA clones that showed a higher expression in response to water deficit conditions were named with the **rsP** initials for the words in Spanish: **r**espuesta a **s**equía en **P**haseolus. Data obtained from their nucleotide and deduced amino acid sequences, as well as from the expression analysis of the corresponding genes, allowed us to postulate their identity (Table 1).

Table 1. **Characteristics of cDNA clones and predicted proteins**

cDNA clone	cDNA length	Estimated[1] mRNA length (bp)	Homology	Reference	aa[2] identity	Putative[3] product	Name[4] of the protein	Accession number
rsP12	350	1 180	SbPRP-1 and 2	27, 28	70[a]	proline rich protein	PvPRP-12	U72769
rsP18	525	550	--	--	--	Lea protein	PvLEA-18	U72764
rsP19	445	800	PsHsp17.7	36	81[b]	lmw-HSP	PvHSP17-19	U72766
rsP24	644	785	MZEPLTP	49	56[c]	lipid transfer	PvLTP-24	U72765
rsP25	443	850	pGmPM1	10	86[d]	group 4-LEA	PvLEA4-25	U72767
rsP37	1 100	1 180	SbPRP2	27	98[e]	proline rich protein	PvPRP2-37	U72768

Notes:

1. Estimated from electrophoretic mobility on northern hybridisations.
2. Compared to the proteins that present the highest amino acid (aa) sequence homology in fasta and blast analyses: **a** against aa 225-aa 257 of SbPRP1 and against aa 195-aa 230 of SbPRP2; **b** against aa 81-aa 155 of PsHsp17.9; **c** against whole polypeptide MZEPLTP; **d** against aa 80-aa 152 of pGmPM1; **e** against aa 66-aa 146 of SbPRP2.
3. Putative identity of the rsP gene products.
4. "Pv" indicates *Phaseolus vulgaris*; PRP, LEA, HSP and LTP define the protein families to which the gene products belong; the number after the dash corresponds to the cDNA clone number.

Source: Colmenero-Flores *et al.*, in press.

Clone rsP25 encodes an incomplete polypeptide which shows an 86 per cent identity to the carboxy-half of the 18 Kd soybean maturation protein GmPM1 that belongs to the group IV (family D-113) of LEA proteins (Chen *et al.*, 1992). This data strongly suggest that rsP25 contains the 3' half of a *lea* gene encoding a group 4-LEA protein (PvLEA4-25). The rsP19 nucleotide sequence predicts an incomplete 228 bp ORF that encodes a 77 amino acid-long polypeptide which shows 81 per cent identity with a *Pisum sativum* class 1-lmwHSP protein (HSP17.9) (Lauzon *et al.*, 1990). As other lmw-HSPs, the predicted polypeptide presents the GVLTV motif contained in a hydrophobic domain of its carboxy terminal region (Raschke *et al.*, 1988), indicating that rsP19 codifies a lmw-HSP (PvHSP17-19). The analysis of the rsP24 sequence showed a complete ORF that encodes for a 116 amino acid polypeptide that presents 57 per cent identity with a maize LTP (MZEPLTP) (Tchang *et al.*, 1988). The properties of the protein deduced from the rsP24 nucleotide sequence are common to plant LTPs: low molecular weight, basic isoelectric point, similar hydropathic profile, relative position among conserved cysteine residues and presence of the conserved signal peptide in the amino terminal region (Kader, 1996). These evidences indicate that rsP24 encodes for an LTP in bean (PvLTP-24). In the case of the rsp37 cDNA clone, the sequence analysis shows an incomplete ORF that lacks the initiation codon and encodes for an 89 amino acid polypeptide that is 98 per cent identical to the soybean PRP2 protein (SbPRP2) (Hong *et al.*, 1990). This sequence consists of almost perfect reiterations of the decamer PPVEKPPVYK that characterises PRP2 proteins. The 3' non-coding region in rsP37 also shows a high identity with the 3' non-coding region of different *prp* genes (Hong *et al.*, 1990). These structural evidences strongly suggest that rsP37 encodes for a bean PRP2 protein (PvPRP2-37). A putative PRP is also encoded by the rsP12 cDNA clone which contains a partial ORF corresponding to a 29 amino acid polypeptide that shows 70 per cent identity with both soybean PRP1- and PRP2- carboxy ends. Although this partial polypeptide presents the decamer PPVEKPPVYK, the decamers reiterations typical in PRP2 were not found. Additionally, as in the case of the rsP37 clone, the 3' non-coding region contained in rsP12 presents a high conservation with the 3' non-coding regions from different Soybean *prp* genes. However, given the partial length of the hypothetical polypeptide, at this point it is not possible to distinguish the type of PRP encoded by rsP12. In contrast to the other rsP clones, rsP18 did not show significant homology with any nucleotide or amino acid sequences deposited in data banks. However, the deduced protein sequence analysis showed striking characteristics of LEA proteins: a high hydrophilicity, the lack of cysteins and tryptophanes, a high glycine content and a putative predominance of random coil in the polypeptide (Dure, 1993). These characteristics, together with the high accumulation of the *rsP18* transcript in dry embryos (Figure 1), strongly indicate that rsP18 encodes a LEA protein. However, the lack of a significant homology and the absence of the conserved domains described for the known LEA proteins (Dure, 1993) suggest that this predicted protein corresponds to a novel LEA protein from bean (PvLEA-18).

Figure 1. **Expression patterns of the rsP genes during water deficit treatment, late embryogenesis and germination**

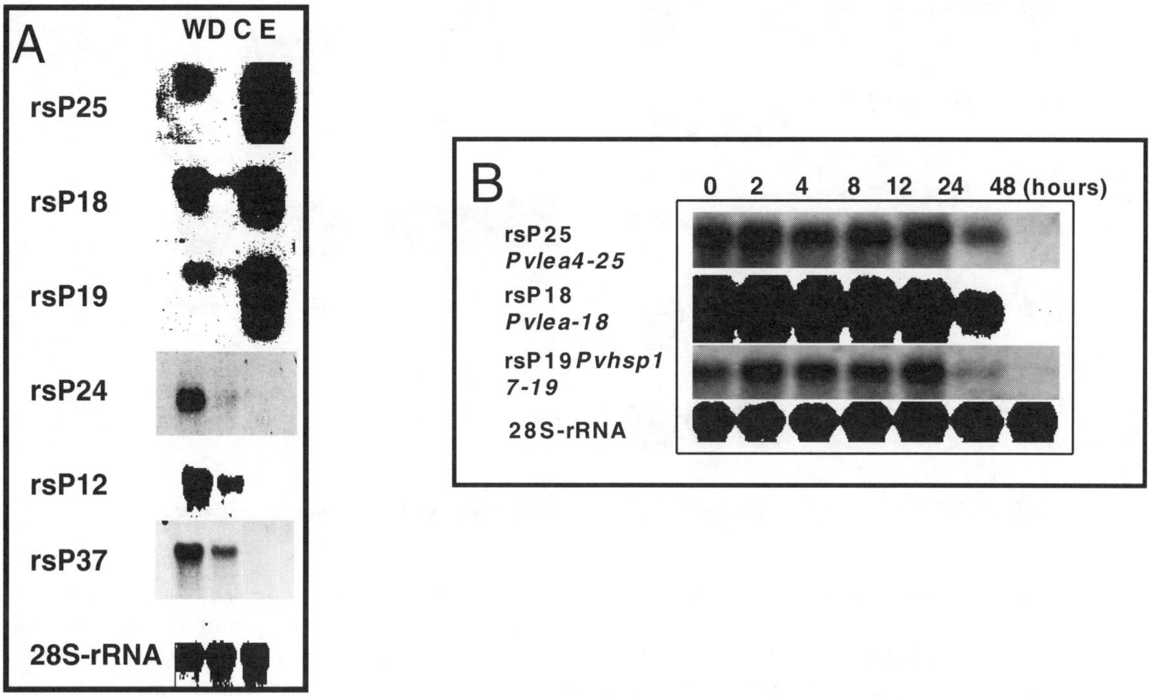

Note: Panel A shows the accumulation patterns of the rsP transcripts during water deficit treatment and dry bean embryos. Total RNAs were obtained from seedlings grown in the dark for 4d and transplanted to water-stressed vermiculite (Ψ_w=-0.35 MPa) (**WD**) or to well watered vermiculite (Ψ_w=-0.074 MPa) (**C**) during 24h. Total RNA was also obtained from embryonary axis excised from bean dry seeds (**E**). Total RNAs (5 µg) were electrophoresed, blotted to nylon membranes and hybridised against the indicated probes containing the same cpm. Hybridisation against a 28S-rRNA probe was used as a RNA loading control. Panel B shows the expression patterns of rsP18, rsP19 and rsP25 transcripts during germination. In this case, total RNAs (5 µg) obtained from dry seeds (time 0) and from germinating seeds 2, 4, 8, 12, 24 and 48h after imbibition were loaded on the gel. The results shown in this figure were reproducible in three independent experiments.

Source: Author.

Under water deficit and ABA treatments, the highest levels of expression for most of the genes occur in the root, except for the *Pvltp-24* gene whose maximum expression levels are found in the aerial regions of the plant (Figure 2), in agreement with its possible role in the cuticle formation (Kader, 1996). The fact that some rsP transcripts are induced by ABA and water deficit in roots, and that some others are induced by the same treatments in stems and leaves, not only supports the involvement of ABA in the induction of these genes by water deficit, but it also suggests a mechanism that correlates the ABA response with organ specific expression. The prevalence of a significant higher expression of the *rsP* genes in the root suggests a higher responsiveness of this organ to the water limiting environments, which may allow the plant to contend efficiently against this type of stress (Frensh and Hsiao, 1995; Jones *et al.*, 1987; Taiz and Zeiger, 1991).

Figure 2. **ABA and water-deficit modulation of rsP clones in different plant organs**

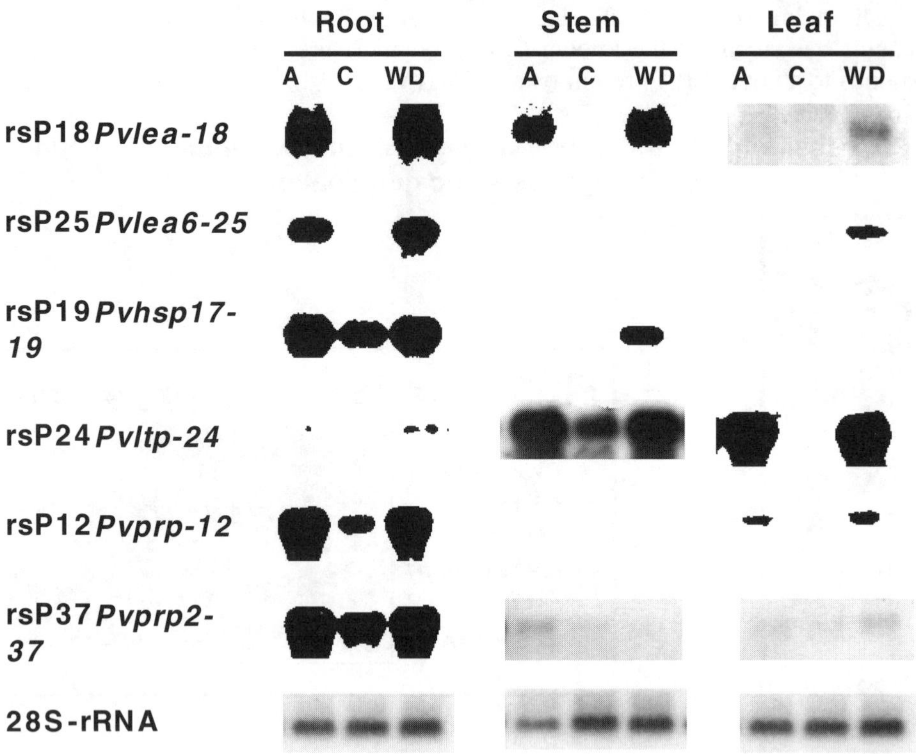

Note: Total RNAs were extracted from root, stem and fully expanded leaves of 21-day-old illuminated bean plants which were subjected to 24h water deficit (WD) (Ψ_w=-0.35 MPa) and 0.1 mM ABA (**A**) treatments. Control plants (**C**) were grown in well irrigated vermiculite (Ψ_w=-0.074 MPa). Total RNAs (5 µg) were hybridised against the indicated probes. Hybridisation against a 28S-rRNA probe was used as a RNA loading control. The results shown in this figure were reproducible in three independent experiments.

Source: Author.

Kinetics of their transcripts accumulation indicated that the transcripts of all these genes reach their highest accumulation around 16 to 24 hours after the induction of a mild water deficit (Ψ_w=-0.35 MPa) and decline thereafter. Expression analysis of the *rsp* genes after rehydration revealed that those encoding PRPs and the LTP maintain or transiently re-induce their expression when water is added to the soil after a dehydration period (Figure 3). A similar pattern was observed for the *chs1* gene. Although we ignore which could be the inductor implicated in this situation, the responsiveness of these genes to additional environmental or developmental stimuli that could be present during the rehydration treatment should be considered (Showalter, 1993; Kader, 1996). The accumulation of the *chs1* and *prp* transcripts has been observed to occur in response to ethylene (Ecker and Davis, 1987) or jasmonate (Colmenero, unpublished observations). The possibility that the rehydration treatment could induce a transient raise in such phytohormones may be taken into account. Gómez-Cadenas *et al.* have detected the sharp increases of ACC (1-aminocycloprpane-1-carboxylic acid, a precursor of ethylene biosynthesis) and ethylene after rewatering *Citrus reshni* (Cleopatra mandarin) plants that were subjected to water stress (Gómez-Cadenas *et al.*, 1996). This transient re-induction during rehydration was not observed for the *lea* genes whose transcripts disappear few hours (four hours) after rehydration, suggesting a more direct modulation by the plant water status (Figure 3).

Figure 3. **Accumulation of the rsP mRNAs after rehydration**

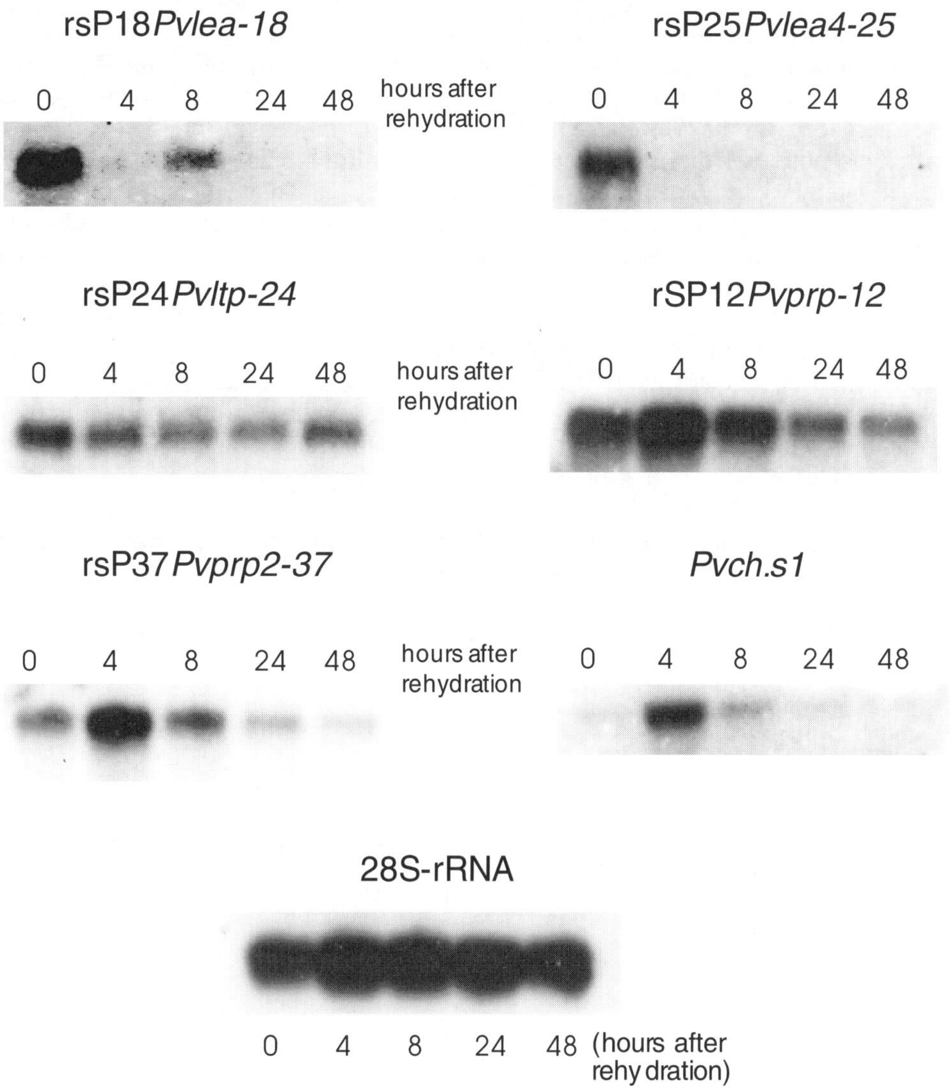

Note: Total RNAs were extracted from bean seedlings subjected to water deficit for 48h (t=0), and from seedlings after 4, 8, 24, and 48 hours of rehydration. The obtained RNA (5 µg/lane) was electrophoresed, blotted and hybridised against the probes indicated in the figure. The results shown in this figure were reproducible in three independent experiments.

Source: Author.

As in the case of other *lea* genes, their transcripts corresponding to the cDNA clones rsP25 (*Pvlea4-25*) and rsP18 (*Pvlea-18*) are highly accumulated in dry seeds, and they disappear during germination (Figure 1). In addition, they are able to respond to water deficit and ABA treatments in the plant vegetative tissues (Figure 2), and return to their basal levels after rehydration (Figure 3). The novel *lea* gene *Pvlea-18* shows the fastest induction and the highest accumulation of transcripts among the *rsp* genes. The involvement of the water status of the cell in the modulation of the *Pvlea-18* expression is also supported by the observation that the Pv*lea*-18 transcript is present in the growing regions but not in the non-growing regions of bean seedlings grown under well irrigated conditions (Colmenero-Flores *et al.*, data not shown). Although the function of the LEA proteins remains unknown, it is tempting to hypothesise that the high hydrophilicity and random coiling of these proteins allow them to interact with macromolecules to provide stability and/or a solvatation environment. It is worth mentioning that LEA proteins are commonly found in a wide variety of plant species in response to water deficit supporting their active role in the plant response to stress. Interestingly, we have found that proteins with structural characteristics similar to the LEA proteins are present in yeast (*Saccharomyces cerevisiae*). We have isolated and characterised three different genes from *Saccharomyces cerevisiae* whose expression is induced by high osmoticum: *OSR1, OSR2* and *OSR3* (Garay-Arroyo and Covarrubias, data not shown). As shown in Figure 4, the predicted polypeptide encoded by the *OSR*1 gene (OSR=**os**motic **r**esponse) presents all the structural characteristics distinctive of the LEA proteins: a high hydrophilicity and the lack of cysteines and tryptophanes (Dure, 1993). As in other LEA proteins, the 18 Kd predicted polypeptide is rich in charged residues and glycine, and the hypothetical secondary structure shows a predominant random-coil. The analysis of the *OSR1* expression patterns indicated that this gene not only responds to hyper-osmotic conditions (water deficit), but it is also induced by ionic stress imposed by adding LiCl (100 mM) or NaCl (300 mM) to the growth media (data not shown). This finding reiterates the importance of this type of proteins as possible osmoprotectors, that is to say, molecules that under a limiting water environment are able to function as compatible solutes that help to maintain the integrity of cellular structures. The random coiling and hydroxylated groups present in these proteins would permit their interaction with other structures, creating a solvatation environment that may provide a stable cohesive layer with proteins and other cellular components.

Figure 4. **Hydrophilic proteins from yeast and plants whose synthesis is induced by osmotic stress**

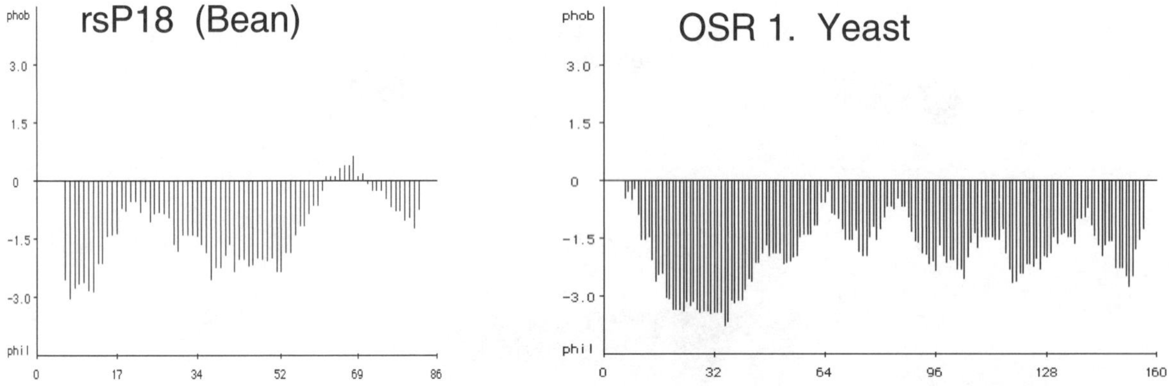

	LEA proteins					
	Gr. I Em Wheat	Gr. IV D-113 Cotton	Gr. II pcC2704 Crateros.	Gr. III D-7 Cotton	LEA 18 Bean	OSR1 Yeast
Size (aa)	93	165	117	136	86	160
% Gly	18	19	16	8	13	14
% Gly + Thr	23	28	23	19	22	22
% Glu or Asp	13	6.5	11	13	10.5	13
% Trp	0	0	0	0	0	1
% Cys	0	0	1	0	0	0
% Lys	6.5	5.5	12	13	14	3
% Lys + Arg	16.5	12.5	14.5	17.5	16	10

Note: The superior part of the figure shows the hydropathic patterns deduced from the amino acid sequences of two hydrophilic proteins from plants (LEA-18) and from yeast (OSR-1) that are accumulated in response to osmotic stress. The table shows the amino acid composition of different LEA proteins from plants, and that corresponding to the OSR-1 protein from yeast.

Source: Author.

In agreement with the idea of a more general role for some of the LEA proteins, the PvleaIV-25 gene, and not *Pvlea-18*, is able to respond to heat shock treatment (Colmenero-Flores *et al.*, in press). A similar expression pattern has been reported for another member of this family of proteins in sunflower (Almoguera and Jordano, 1992), suggesting a chaperone activity for these molecules.

As could be expected, besides water deficit and embryo development, the *Pvhsp17-19* gene expression is also induced by heat shock (Colmenero-Flores *et al.*, in press). The fact that a similar gene expression pattern has been observed for two lmw-hsp genes in sunflower (Almoguera and Jordano, 1992; Almoguera *et al.*, 1993; Coca *et al.*, 1994) reinforces a role for these proteins in the plant response, not only during thermal stress, but also during water deficit conditions. Accordingly, the *Pvlmw-hsp* gene expression is induced by ABA treatment (0.1 mM) in vegetative tissues (Figure 2). The concurrent and high accumulation of the lmw-HSPs and some LEA proteins observed in bean and sunflower, particularly in the dry embryo, as well as the nature of both kind of proteins, supports the hypothesis that they interact with macromolecules to protect them from the damage consequence of stressant conditions. Interestingly, we have observed that the *Pvlmw-hsp* gene is expressed in roots, not only in response to water deficit, but also under well irrigated conditions supporting again a contribution to some of the mechanisms that the root may use for a fast adaptation of the plant to changing soil environments (Creelman *et al.,* 1990; Frensh and Hsiao, 1995; Jones *et al.,* 1987; Taiz and Zeiger, 1991). This idea concords with data reported by Alamillo *et al.* that shows that the expression of an homologous gene from *Craterostigma plantagineum,* a dessication tolerant plant, is present at high levels in non-dessicated and dessicated plants (Alamillo *et al.*, 1995).

The possible role of LTP proteins has been related with the deposition of lipophilic compounds in the cell walls, such as cuticle precursors, being the most abundant proteins in the cuticular wax (Kader, 1996). This observation correlates with the presence of high amounts of *ltp* gene transcripts in the plant epidermal regions (Kader, 1996). Under water deficit conditions, we have shown that the *Pvltp-24* transcript accumulates mainly in the aerial regions of the plant (stems and leaves), while its levels are very low in stressed roots (Figure 1). This organ-specific expression can also be observed in well irrigated plants where their mRNA levels can be detected in stems and leaves, but not in roots. Such specific expression can be thought for a gene whose product may respond to the need of a higher impermeabilization of the plant surface (epidermal regions) in order to decrease water loss, particularly under water limiting conditions. Other reports describe the induction of *ltp* gene expression in response to additional environmental stimuli such as cold (White *et al.,* 1994), salinity (Torres-Schumann *et al.,* 1992), and pathogens attack (Molina and García-Olmedo, 1993). In particular, the *Pvltp-24* gene does not show any detectable induction neither under cold or heat shock treatments (Colmenero-Flores *et al.*, in press); rather it seems to respond in a specific manner to osmotic stress situations. Although we have observed that the *Pvltp-24* gene expression is also induced by wounding, since a transient expression was detected after the transplanting procedure, we have shown that the induction we detect under osmotic stress conditions is not the result of a possible tissue damage during the stress treatment (Colmenero-Flores *et al.*, in press); rather the *Pvltp-24* gene seems to be responding to the water deficit condition imposed.

It is known that the composition of the cell wall is affected under water stress (Botella *et al.,* 1994; Covarrubias *et al.*, 1995; Creelman and Mullet, 1991; Iraki *et al.*, 1989); however, little information exists on the nature of the cell wall proteins involved in the response. In this work we identified two cDNA clones (rsP12 and rsP37) that encode two different PRPs whose corresponding genes are induced by water deficit and ABA treatments (Figures 1 and 2). The putative product of the rsP37 corresponds to a PvPRP2, while rsP12 encodes a different PvPRP whose type has not been defined.

The characterisation of the PRPs and their genes indicates that their abundance is modulated by different regulatory systems (Keller, 1993; Showalter, 1993). The *prp* genes respond to a number of stimuli such as developmental programs, pathogens attack, wounding and environmental stress conditions (Keller, 1993; Showalter, 1993). A mechanism involving the crosslinking between some PRPs amino acid residues has been proposed to occur by an oxidative burst that may be taking place during wounding or pathogen attack (Bradley *et al.*, 1992). It has been suggested that this process may be inducing hardening of the plant cell walls, consequently affecting cell elongation or protecting against pathogen attack. Given the participation of the PRP proteins in multiple processes, it is difficult to assign them a function during water deficit conditions. Creelman and Mullet have reported that the *Sbprp1* transcript accumulates in the hypocotyl elongating and mature regions of soybean seedlings subjected to water deficit treatment. The accumulation of the *Sbprp1* transcript in the elongating regions allowed them to hypothesise a possible role for these proteins in the growth detention phenomena observed under water limiting conditions (Creelman and Mullet, 1991). A similar expression pattern has been observed for both *Pvprp* genes under stress situations where the hypocotyl elongation rate of bean seedlings significantly decreases. Since one of the *Pvprp* genes corresponds to *prp2*, both *prp1* and *prp2* genes seem to respond similarly under stress situations (Colmenero-Flores, unpublished results). Both observations could be related with the need under water stress conditions of cell walls with different properties. Under stress conditions, the crosslinking between the accumulated PRP proteins could contribute to a decrease in the cell wall extensibility and/or hydraulic conductivity. Further experimentation is needed to clear up the role of these proteins during the growth process and water limiting conditions.

Although the data described here supports a role for the proteins encoded by the rsP cDNA clones during the plant response to water deficit situations, much investigation needs to be done that will allow us to understand their function in this adverse condition.

Likely developments in the near future

Although most of the molecular studies on drought stress in plants, in its initial phase, have by necessity been largely descriptive, a large number of genes with a potential role in drought tolerance have been described, and major themes in the molecular response have been established.

The different nature of the proteins encoded by the cDNA clones isolated from bean in response to water deficit supports the involvement of multiple processes during this stress situation. Even though the biochemical functions and the physiological roles of these proteins in plants under water stress conditions remain unknown, the initial characterisation of the expression patterns of the corresponding genes, some of which have been included here, suggests some hypotheses that have to be tested. The plant seems to respond by synthesising proteins that help it avoid the loss of water when it needs to save it, such as the LTP possibly involved in the production of cuticle by the aerial organs. Also, the plant shows a compromise with growth detention under adverse conditions by synthesising cell wall proteins, PRPs, whose abundance could contribute to a decrease in the cell wall extensibility frequently associated with non-growing regions. Additionally, molecular chaperones, lmw-HSP, that protect proteins from denaturation or maintain a functional protein structure are also needed under water limiting conditions. Others, such as the LEA proteins, are predicted to protect or preserve protein and cellular structures from the effects of water loss. Recent experiments in our laboratory show that the PvLEA18 protein is able to protect yeast cells from oxidative stress. They may act as scavengers of oxygen radicals or as osmoprotectors of enzymes involved in the scavenging process for the hydroxy radicals generated during water deficit conditions. This is the first example of an *in vivo* protection against oxidative stress for this kind of molecule. A similar phenomenon has been observed *in vitro* for certain types of osmolites such as myo-inositol, sorbitol, mannitol and proline. We think that this type of protein may play an important function during dehydration tolerance, and this may be true not only for plants, but also for other organisms such as yeast. We consider that studies on the LEA proteins function(s) constitute an area of importance for the future.

Other important topics regarding the plant molecular response that should be considered for further studies are those related to (a) sugar metabolism (pathways of synthesis for osmotically active metabolites and their regulation); (b) the synthesis of specific proteins that control ion and water flux; (c) methylation reactions and photorespiration; and (d) the molecular mechanisms involved in the regulation of water deficit response genes: identification of *cis*- and *trans*- active elements and signal transduction elements. Modern genetic engineering requires the selection of "relevant genes", which makes a basic understanding of the phenomena essential.

Although much of the data generated during this period has been largely descriptive, the knowledge accumulated has been very useful to establish new strategies and future developments.

The use of novel approaches combining genetic, biochemical and molecular techniques should provide new insights to this complex phenomena. Among these should be considered the development of better plant models for molecular genetics, such as *Arabidopsis thaliana*, in order to generate tools that could be used for the selection of osmosensitive mutants and subsequent selection of relevant genes. Also, the possibility of controlling an entire water stress-response pathway by manipulating the expression of a "master" gene that could enhance the plant stress response in a directed manner might be analysed. The use of specific promoters that allow to modulate the expression of particular genes when they are really needed would be important to induce the expression and synthesis of specific proteins during a drought-sensitive developmental stage in the plant.

We should have in mind that the objective should be the increase of drought tolerance, rather than to modify crops to make them resistant to stress, since the acquisition of stress resistance inevitably results in yield penalties, a highly conserved plant strategy that has been selected during evolution. Nevertheless, it is important to consider that the design of drought tolerance in plants must target multiple mechanisms; thus the engineering should consider the transfer of multiple genes that can be achieved with present technology. Finally, advantage it should be taken of approaches that intrinsically consider the involvement of multiple genes for drought tolerance, such as the identification of quantitative-trait-loci (QTLs) which may well be an effective analytical tool.

Acknowledgements

We thank Rosa M. Solórzano, Claudia Smith and Marcela Castillo for technical assistance. This work was supported by grants from Consejo Nacional de Ciencia y Tecnología (México) N9106-0054, DGAPA-UNAM IN207592 and Programa de las Naciones Unidas para el Desarrollo PNUD/MEX/93/019 to A.A.C.

Questions, comments and answers

Q: Have you considered the risk for the environment?

A: There is not much risk. Most work has been done by conventional breeding. The strategy is the same, but we can do it in a more directed and controlled way, rather than randomly.

Q: Do you get accumulation of trehalose in seeds of transgenic plants?

A: We get high levels with plant (rather than yeast) trehalose genes because we are using a constitutive promoter.

Q: Do you expect the genes to be universal or restricted to a number of plants?

A: They are not universal, and we have to be very careful in designing a strategy for more tolerant plants. We have to look at which plants need what and get to the basics of the phenomenon.

REFERENCES

ALAMILLO, J, C. ALMOGUERA, D. BRTELS, and J. JORDANO (1995), "Constitutive expression of small heat shock proteins in vegetative tissues of the resurrection plant *Craterostigma plantagineum*", *Plant Mol. Biol.* 29, pp. 1 093-1 099.

ALMOGUERA, C., M.A. COCA, and J. JORDANO (1993), "Tissue specific expression of sunflower heat shock proteins in response to water stress", *Plant J.* 4, pp. 947-958.

ALMOGUERA, C. and J. JORDANO (1992), "Developmental and enviromental concurrent expression of sunflower dry-seed store low-molecular weight heat-shock-protein and lea mRNAs", *Plant Mol. Biol.* 19, pp. 781-792.

BARTELS, D., K. SCHNEIDER, G. TERSTAPPEN, D. PIATKOWSKI, and F. SALAMINI (1990), "Molecular cloning of abscisic acid-modulated genes which are induced during desiccation of the resurrection plant *Craterostigma plantagineum*", *Planta 181*, pp. 27-34.

BARTELS, D. and D.E. NELSON (1994), "Approaches to improve stress tolerance using molecular genetics", *Plant Cell Environ* 17, pp. 659-667.

BASTA, A.S. (ed). (1994), *Stress induced gene expression of specific proteins in plants*, Series Harwood Academic, New York.

BOHNERT, H.J., D.E. NELSON, R.G. JENSEN (1995), "Adaptation to environmental stesses", *Plant Cell* 7, pp. 1 099-1 111.

BORKIRD, C, C. SIMOENS, R. VILLARROEL, and M. VAN-MONTAGU (1991), "Gene expression associated with water stress adaptation of rice cells and identification of two genes as hsp70 and ubiquitin", *Physiol. Plant.* 82, pp. 449-457.

BOTELLA, M.A., M.A. QUESADA, A.K. KONOWICZ, R.A. BRESSAN, F. PLIEGO, P.M. HASEGAWA, and V. VALPUESTA (1994), "Characterization and *in situ* localization of a salt-induced tomato peroxidase mRNA", *Plant Mol. Biol.* 25, pp. 105-114.

BOYER, J.S. (1982), "Plant productivity and environment", *Science* 218, pp. 443-448.

BRADLEY, D.J., P. KJELLBOM, and C.J. LAMB (1992), "Elicitor and wound-induced oxidative cross-linking of a proline-rich cell wall protein: a novel, rapid defense response", *Cell* 70, pp. 21-30.

BRAY, E.A. (1991), "Regulation of gene expression by endogenous ABA during drought stress," in W.J. Davies and H.G. Jones (eds.), *Abcisic Acid, physiology and biochemistry*, Bios Scientific Publisher, Lancaster, United Kingdom.

BRAY, E.A. (1993), "Molecular responses to water deficit," *Plant Physiol.* 103, pp. 1 035-1 040.

CHEN, Z., Y.C. HSING, P. LEE, and T. CHOW (1992), "Nucleotide sequence of a soybean cDNA encoding a 18 kilodalton late embryogenesis abundant protein", *Plant Physiol.* 99, pp. 773-774.

CHRISPEELS, M.J. and C. MAUREL (1994), "Aquaporins: the molecular basis of facilitated water movement through living plant cells?", *Plant Physiol.* 105, pp. 9-13.

COCA, M.A., C. ALMOGUERA, and J. JORDANO (1994), "Expression of sunflower low-molecular-weight heat-shock proteins during embryogenesis and persistance after germination: localization and possible functional implications", *Plant Mol. Biol.* 25, pp. 479-492.

COLMENERO-FLORES, J.M., F. CAMPOS, A. GARCIARRUBIO, and A.A. COVARRUBIAS (in press), "Characterization of cDNA clones responsive to water deficit from *Phaseolus vulgaris*: a new late embryogenesis abundant protein", *Plant Mol. Biol.*

COVARRUBIAS, A.A., J.W. AYALA, J.L. REYES, M. HERNANDEZ, and A. GARCIARRUBIO (1995), "Cell-wall proteins induced by water deficit in bean (*Phaseolus vulgaris* L.) seedlings", *Plant Physiol.* 107, pp. 1 119-1 128.

COVARRUBIAS, A.A. and A. GARCIARRUBIO (1993), "Molecular characterization of the response to water deficit in *Phaseolus vulgaris* L", in W.M. Roca, J.E. Mayer, M.A. Pastor-Corrales, and J. Tohme (eds.), *Phaseolus Beans Advanced Biotechnology Research Network (BARN)*, CIAT, Cali, Columbia.

CREELMAN, R.A., H.S. MASON, R.J. BENSEN, J.S. BOYER, and J.E. MULLET (1990), "Water deficit and abcisic acid cause differential inhibition of shoot versus root growth in soybean seedlings", *Plant Physiol.* 92, pp. 205-214.

CREELMAN, R.A. and J. MULLET (1991), "Water deficit modulates gene expression in growing zones of soybean seedlings. Analysis of differentially expressed cDNAs, a new β-Tubulin gene, and expression of genes enconding cell wall proteins", *Plant Mol. Biol.* 17, pp. 591-608.

DELAUNEY, A.J. and D.P.S. VERMA (1993), "Proline biosynthesis and osmoregulation in plants", *Plant J.* 4, pp. 215-223.

DURE, III L. (1993), "Structural motifs in Lea proteins", in T.J. Close and E.A. Bray (eds.), *Plant responses to cellular dehydration during environental stress,* American Society of Plant Physiologists, Rockville, Maryland.

ECKER, J.R. and R.W. DAVIS (1987), "Plant defense genes are regulated by ethylene", *Proc. Natl. Acad. Sci. USA* 84, pp. 5 202-5 206.

ESAKA, M., H. HAYAKAWA, M. HASHIMOTO, and N. MATSUBARA (1992), "Specific and abundant secretion of a novel hydroxyproline-rich glycoprotein from salt-adapted winged bean cells", *Plant Phys.* 100, pp. 1 339-1 345.

ESPARTERO, J., J.A. PINTOR-TORO, and J.M. PARDO (1994), "Differential accumulation of S-adenosylmethionine synthetase transcrips in reponse to salt stress", *Plant Mol. Biol.* 25, pp. 217-227.

FRENSCH, J. and T.C. HSIAO (1995), "Rapid response of the yield threshold and turgor regulation during adjustment of root growth to water stress in *Zea mays*", *Plant Physiol.* 108, pp. 303-312.

GÓMEZ-CADENAS, A., F.R. TADEO, M. TALON, and E. PRIMO-MILLO (1996), "Leaf abscission induced by ethylene in water-stressed intact seedlings of Cleopatra mandarin requires previous abscisic acid accumulation in roots", *Plant Physiol.* 112, pp. 401-408.

GOSTI, F., N. BERTAUCHE, N. VARTANIAN, and J. GIRAUDAT (1995) "Abscisic acid-dependent and -independent regulation of gene expression by progressive drought in *Arabidopsis thaliana*", *Mol. Gen. Genet.* 246, pp. 10-18.

HONG, J.C., R.T. NAGAO, and J.L. KEY (1990), "Characterization of a proline-rich cell wall protein gene family of soybean", *J. Biol. Chem.* 265, pp. 2 470-2 475.

INGRAM, J. and D. BARTELS (1996), "The molecular basis of dehydration tolerance in plants", *Annu. Rev. Plant Physiol. Plant Mol. Biol.* 47, pp. 377-403.

IRAKI, N.M., N. SINGH, R.A. BRESSAN, and N.C. CARPITA (1989), "Cell walls of tobacco cells and changes in composition associated with reduced growth upon adaptation to water and saline stress", *Plant Physiol.* 91, pp. 48-53.

JONES, H., R.A. LEIGH, A.D. TOMOS, and R.G. WYN JONES (1987), "The effect of abscisic acid on cell turgor pressures, solute content and growth of wheat roots", *Planta* 170, pp. 257-262.

KADER, J.C. (1996), "Lipid-transfer proteins in plants", *Ann. Rev. Plant Physiol. Plant. Mol. Biol.* 47, pp. 627-654.

KELLER, B. (1993), "Structural cell wall proteins", *Plant Physiol.* 101, pp. 1 127-1 130.

KIYOSUE, T., K. YAMAGUCHI-SHINOZAKI, and K. SHINOZAKI (1994), "Cloning of cDNAs for genes that are early-responsive to dehydration stress (ERDs) in *Arabidopsis thaliana L*: Identification of three ERDs as HSPs cognate genes", *Plant Mol. Biol.* 25, pp. 791-798.

KOIZUMI, M., K. YAMAGUCHI-SHINOZAKI, H. TSUJI, and K. SHINOZAKI (1993), "Structure and expression of two genes that encode distinct drought-inducible cysteine proteinases in *Arabidopsis thaliana*", *Gene* 129, pp. 175-182.

KONONOWICZ, A.J., K.G. RAGHOTHAMA, A.M. CASAS, M. REUVENI, A-E.A. WATAD, D. LIU, R.A. BRESSAN, and P.M. HASEGAWA (1993), "Osmotin: regulation of expression and function", in T.J. Close and E.A. Bray (eds.), *Plant responses to cellular dehidration during environmental stress,* American Society of Plant Physiologists, Rockville, Maryland.

LAUZON, L.M., K. HELM, and E. VIERLING (1990), "A cDNA clone from *Pisum sativum* encoding a low molecular weight heat shock protein", *Nucleic Acid Res.* 18, pp. 4 274-4 274.

LUDWIG-MULLER, J., B. SCHUBERT, and K. PIEPER (1995), "Regulation of IBA synthetase from maize (*Zea mays L.*) by drought stress and ABA", *J. Exp. Bot.* 46:285, pp. 423-432.

McCARTY, D.R. (1995), "Genetic control and integration of maturation and germination pathways in seed development", *Annu. Rev. Plant Physiol. Plant Mol. Biol.* 46, pp. 71-93.

MOLINA, A. and F. GARCIA-OLMEDO (1993), "Developmental and pathogen induced expression of three barley genes enconding lipid tranfer proteins", *Plant Journal* 4:6, pp. 983-991.

NIU, X., R.A. BRESSAN, P.M. HASEGAWA, and J.M. PARDO (1995), "Ion homeostasis in NaCl stress environment", *Plant Physiol.* 109, pp. 735-742.

NIU, X., J.K. ZHU, M.L. NARASIMHAN, R.A. BRESSAN, and P.M. HASEGAWA (1993), "Plasma-membrane H+-ATPase gene expression is regulated by NaCl in cells of the halophyte *Atriplex nummularia L*", *Planta* 190, pp. 433-438.

PARCY, F., C. VALON, M. RAYNAL, P. GAUBIER-COMELLA, M. DELSENY, and J. GIRAUDAT (1994), "Regulation of gene expression programs during *Arabidopsis* seed development: roles of the *ABI3* locus and endogenous abscisic acid", *Plant Cell* 6, pp. 1 567-1 582.

PILON-SMITS, E.A.H., M.J.M. EBSKAMP, M.J. PAUL, M.J.W. JEUKEN, P.J. WEISBECK, and S.C.M. SMEEKENS, S.C.M. (1995), "Improved performance of transgenic fructan-accumulating tobacco under drought stress", *Plant Physiol.* 107, pp. 125-130.

RASCHKE, E., G. BAUMANN, and F. SCHOEFFEL (1988), "Nucleotide sequence analysis of soybean small heat shock protein genes belonging to two different multigene families", *J. Mol. Biol.* 199, pp. 549-557.

RIBAUT, J., D.A. HOISINGTON, J.A. DEUTSCH, C. JIANG, D. GONZALEZ-DE LEÓN (1996), "Identification of quantitative trait loci under drought conditions in tropical maize: I. Flowering parameters and anthesis-silking interval", *Theor. Appl. Genet.*, in press.

ROBERTS, J.K., N.A. DESIMONE, W.L. LINGE, and L. DURE III (1993), "Cellular concentrations and uniformity of cell-type accumulation of two Lea proteins in cotton embryos", *Plant Cell* 5, pp. 769-780.

SERRANO, R. and R.A. GAXIOLA (1994), "Microbial models and salt stress tolerance in plants", *Crit. Rev. Plant Sci.* 13, pp. 121-138.

SHOWALTER, A.M. (1993), "Structure and function of plant cell wall proteins", *The Plant Cell* 5:1, pp. 9-23.

SMIRNOFF, N. and Q.J. CUMBES (1989), "Hydroxyl radical scavenging activity of compatible solutes", *Phytochemistry* 28, pp. 1 057-1 060.

TAIZ, L. and E. ZEIGER (1991), *Stress Physiology. Plant Physiology*, Chapter 14, The Benjamin/Cummings Publishing Company, Inc., Redwood City, California, United States.

TARCKZYNSKI, M.C., R.G. JENSEN, and H. BOHNERT (1993), "Stress protection of transgenic tobacco by production of the osmolyte mannitol", *Science* 259, pp. 508-510.

TCHANG, F., P. THIS, V. STIEFEL, V. ARONDEL, and M.D. MORCH (1988), "Phospholipid transfer protein: full-length cDNA and amino acid sequence in Maize", *J. Biol. Chem.* 263:32, pp. 16 849-16 855.

THOMAS, T.L., J. VIVEKANANDA, and M.A. BOGUE (1991), "ABA regulation of gene expression in embyos and mature plants", in W.J. Davies and H.G. Jones (eds.), *Abcisic Acid, physiology and biochemistry,* Bios Scientific Publisher, Lancaster, United Kingdom.

TORRES-SCHUMANN, S., J.A. GODOY, J.A. PINTOR-TORO (1992), "A probable lipid transfer protein gene is induced by NaCl in stems of tomato plants", *Plant Mol. Biol.* 18, pp. 749-757.

UMEDA, M., C. HARA, Y. MATSUBAYASI, H.H. LI, Q. LIU, F. TADO KORO, S. AOTSUKA, and H. UCHIMILLA (1994), "Expressed sequence tags from cultured cells of rice (*Oryza sativa L.*) under stressed conditions: analysis of transcripts of genes engaged in ATP-generating pathways", *Plant Mol. Biol.* 25:3, pp. 469-478.

WHITE, A.J., M.A. DUNN, K. BROWN, and M.A. HUGHES (1994), "Comparative analysis of genomic sequence and expression of a lipid transfer protein gene family in winter barley", *J. Exp. Bot*. 45:281, pp. 1 885-1 892.

WHITE, J.W. and S.P. SINGH (1991), "Breeding for adaptation to drought", in *Common Beans*, A. van Schoonhoven and O. Voysest (eds.), *Research for crop improvement,* C.A.B. International, Colombia.

PLANT BIOTECHNOLOGY: A TOOL TO IMPROVE PLANTS AGAINST DROUGHT STRESS

by

Ramon Messeguer, Anna Campalans, and Montserrat Pagès
Dept. Genètica Molecular, Centre d'Investigació i Desenvolupament, Consejo Superior de Investigaciones Científicas, Barcelona, Spain

Introduction

Crop production in the world is influenced by biotic and abiotic stresses. Abiotic stress, such as drought, cold, and salinity represent the most limiting factor, being the main cause of yield reduction (Boyer, 1982). Moreover, water resources are reduced in large areas of the world, and in temperate and humid regions, drought periods tend to be more critical for the species adapted to this environment. In some areas, like the Mediterranean region and tropical areas in Africa, this situation is becoming more dramatic due to many factors, such as large periods of drought or increase of salinity by exhausting land resources. All these stresses have in common the change of water relation in plant tissues.

According to Levitt (1980) and Turner (1986) there are different strategies used by plants species to respond at limited water resources:

1) Escape. Plants with plasticity and phenological development, in which the vegetative and reproductive cycle is adapted to environmental resources or climate changes. This cannot be considered as a resistant response.

2) Avoidance. Plants can avoid dehydration by enhancing water uptake, increasing root biomass, or by reducing water losses by strict control of stomata closure and leaf movements, changing its form and architecture.

3) Tolerance. Plants are able to survive in a low tissue water status by osmotic adjustments -- synthesis of solutes or osmolytes and/or changes in elasticity of the cell wall.

Avoidance and tolerance represent strategies developed by plants to resist drought, and they involve complex response. Once plants have a loss of water, a cascade of events happens at the biochemical and genetic level. This response is very complex, and it is difficult to discriminate between the responses that allow plants to resist, and the ones which represent a consequence. Changes in membrane characteristics (saturation of fatty acids), protein turnover, osmolyte biosynthesis, metabolite balance, ion uptake, partitioning and sequestration, hormone sensitivity and balance, mRNA stability and transcription control, are examples (Bonhert *et al.*, 1995). Thus, one of the challenges of plant biotechnology is to elucidate this variety of responses to improve the capacity of plant survival and to give high yields under unfavourable conditions. In this review, we will give a picture of the studies carried out by plant molecular breeders and molecular biologists for understanding the physiological and genetic basis of water stress response, and the way that plants can be manipulated to improve their response.

Molecular breeding

The development of molecular markers in plant species has opened the possibility to construct dense genetic maps in any plant species. These markers are represented basically by the RFLPs (**R**estriction **F**ragment **L**ength **P**olymorphism) (Botstein *et al.*, 1980), that take advantage of the presence of insertion/deletion or point mutations in fragments of DNA generated by digestion with restriction enzymes, and RAPDs (**R**andom **A**mplified **P**olymorphic **DNA**s), that are PCR (**P**olymerase **C**hain **R**eaction) derived markers using random primers for amplifying genome sequences (Williams *et al.*, 1990). Recently, a new kind of markers, called AFLPs (**A**mplified **F**ragment **L**ength **P**olymorphism) has been described, which combine the use of restriction enzymes with a high fidelity in the DNA amplification by PCR (Vos *et al.*, 1995).

The main advantage of these maps, in terms of plant breeding, is the coverage of the whole genome. This allows us to screen genomes in order to find markers linked to any trait, and to select for or against the trait using the marker (marker assisted selection) in any tissue or developmental stage (Paterson *et al.*, 1991; Arús and Moreno-González, 1993).

As it was mentioned, the resistance to abiotic stresses, and particularly to drought stress, is very complex and it is not possible to measure it as a discrete response. This means that it is necessary to develop new systems to allow the genetic study of these complex traits by using quantitative genetics. Then, it will be possible to estimate of approximate number of loci affecting the character in a specific mating, the average gene action and the degree of gene interaction (Tanksley, 1993). Selection of the trait is one of the critical steps in this analysis. It is necessary to follow several physiological and biochemical responses to find a correlation between them and yield in different environmental situations. Examples are betaine content (McCue and Hanson, 1990), root morphology (Moons *et al.*, 1995; Champoux *et al.*, 1995) and ABA content (Moons *et al.*, 1995; Quarrie *et al.*, 1994; Janowiak and Dörffling, 1996). Once the trait is selected, each individual in a segregating population is characterised, on the one hand for the value of the quantitative trait of interest, on the other hand for its genotype at all available molecular marker loci (Paterson *et al.*, 1988). Then, biometrical techniques allow the correlation of all the possible association x marker traits and Quantitative Trait Loci (QTLs) that can be mapped and the estimation of their individual effects (Paterson *et al.*, 1988; Lander and Botstein, 1989; Jansen, 1996). Such an approach has been shown to be useful in the study of drought resistance in maize (Quarrie *et al.*, 1994; Quarrie *et al.*, 1996) and rice (Champoux *et al.*, 1995).

Molecular genetics

Osmotic stress

Osmotic stress is caused by drought, salinity and temperature, reducing the water level in the cell. Then, plants exposed to reduced water levels accumulate organic molecules to keep the cell osmotically balanced, called compatible (non toxic) solutes [reviewed by Skriver and Mundy (1990), Bartels and Nelson (1994) and Bonhert *et al.* (1995)].

It has been suggested that sugar alcohols, due to the presence of water-like hydroxyl groups, may mimic the structure of water and scavenge activated oxygen species, avoiding peroxidation of lipids. Mannitol, pinitol, sorbitol, proline, glycine-betaine are the best known compounds that accumulate under stress. Their osmoprotectant role leads to the hypothesis that controlling and inducing their synthesis by modifying the metabolic pathways would result in a better performance under abiotic stresses. Following this criteria, molecules that act as osmoprotectants have been extensively studied in many different organisms. McCue and Hanson (1990) proposed glycine betaine as a candidate because it was the only one that provided haloprotection to cianobacteria. Glycine betaine does not accumulate in many important crops, but some relatives are accumulators. Then, transforming those plants with an enzyme of its metabolic pathway, as choline monooxygenase, allows transgenic plants to synthesise glycine betaine. They noted that the biomass yield of maize that naturally accumulates betaine was 12 per cent higher than that of a betaine deficient maize mutant in dry site. Lilius *et al.* (1996) reported the transformation of tobacco plants with choline dehydrogenase, and by measuring dried weights, there was an 80 per cent increase in salt tolerance between tobacco transgenic and wild type plants. Similar results have been shown for mannitol (Bonhert *et al.*, 1995), fructan (Pilo-Smiths *et al.*, 1995) and proline (Kishor *et al.*, 1995). An example of genes that drive the synthesis of osmoprotectants in different organisms is the one presented by Hölstrom *et al.* (1996). In this work, tobacco plants were transformed with the gene coding for the trehalose-6-phosphate syntahase subunit of the yeast trehalose synthase, allowing the synthesis of trehalose by the transformants. It should be noted that trehalose is an osmoprotectant found in anhydroniotic organisms and some resurrection plants. In conclusion, a promising way of increasing abiotic stress resistance has been opened, although it will be necessary to closely follow the evolution of such transformants in order to verify if there is a real gain in terms of production and performing, and check the side effects, if any, in crop plants.

ABA signalling

ABA was first described as a fitohormone related to leaf abscission and bud latency. Later, it was found during embryo development and in response to water stress. This relation to embryo development has been assessed in mutants insensitive and deficient to ABA, where embryos escape from the dehydration process, leading to an early germination. Its effect is antagonist to giberellic acid, and has been suggested that the balance between both hormones mediates plant responses to environment. Thus, ABA must be considered the key hormone for the response to drought stress.

The biosynthesis pathway in plants has been elucidated (reviewed by Walton and Li, 1995). The precursor is mevalonate, and the synthesis proceeds from the carotenoid pathway, via xanthophylls. As proposed by Li and Walton (1990), zeaxanthin is transformed to violaxanthin by epoxidation catalysed by zeaxanthin epoxidase, the first step in ABA biosynthesis. Then, violaxanthin is transformed to neoxanthin, xanthoxin, ABA-aldehyde, that is finally oxidated to give ABA. Several mutants that show, at biochemical level, alterations in some of the biosynthetic steps have been described. But very little is known at the genetic level. The first report on cloning genes codifying for enzymes driving the synthesis of ABA has been described recently (Marin *et al.,* 1996). The authors cloned the gene codifying for the zeaxanthin epoxidase, using an *aba* mutant (ABA-deficient *aba2* mutant) from *Nicotiana plumbaginifolia*, and they open a door for understanding the regulatory signals for ABA biosynthesis and searching for the way to modify it in order to modulate a response to abiotic stresses.

ABA mediates two main responses to water deficit. First, there is an induction of stomatal closure by modifying the ion channels in guard cells. At this time, the mechanisms from which ABA induces potassium efflux from guard cells and triggers the closure of stomatal pores is still controversial. It seems that there is an effect upon the cytosolic Ca+, which is increased during water stress, leading to a modification of ion channels. However, an ABA-independent increase of cytosolic Ca+ has also been shown (Giraudat *et al.*, 1994; Giraudat, 1995; MacRobbie, 1995).

The second effect of ABA was observed during embryo development and vegetative tissues under water deficit, in which ABA modulates the transcription of some genes either by a positive or negative regulation. However, the signal transduction pathway remains unclear. As it was mentioned before, ABA induces an increase of cytosolic Ca+, and it has been suggested that there is an activation of Ca+ protein kinases, that would be involved in the kinase cascade that mediates the induction/repression of specific genes. One piece of evidence is the isolation of the gene responsible for the Arabidopsis ABA insensitive mutants *abi1* (Leung *et al.,* 1994; Meyer *et al*, 1994). The gene *abi1* is a phosphatase that could be implicated in the control of the activation of kinases by phosphorylation/dephosphorylation events. However, more information is needed in order to understand how the ABA signal is transmitted at the transcriptional level.

Gene induction

Three main strategies have been developed for studying the genes induced under water dehydration and ABA. The first one is based on changes in water status during seed development. In this process, mature seeds sustain a complete desiccation and must therefore develop a mechanism to tolerate this status until germination proceeds. Then, several genes related to this process were detected in vegetative tissues under desiccation (Gómez *et al.*, 1988; Mundy and Chua, 1988). There are examples of genes that mediate this response, the most important being the gene families of Rab (responsive to ABA), LEA (late embryogenesis) or dehydrins, as well as lipoxygenases and phopholipases.

The second strategy uses resurrection plants, as *Craterostigma plantagineum* (Bartels *et al.*, 1990). Such plants have developed an escaping strategy to resist unfavourable water conditions. Vegetative tissues lose their water content, up to 90 per cent, and when rewatering, they recover their normal water status. This is also true for callus tissues, that may withstand dehydration if they are treated with ABA prior to the drying treatment (Bartels *et al.*, 1990). By comparing the genes induced during water stress with the genes expressed in normal watered plants, it is possible to discover differential gene action that encompasses stress and non-stress induction.

Finally, the genetic approach takes advantage of the previously described mutants that affect plant responses to stresses or hormone induction. Examples of such mutants are the ABA deficient mutants *aba* from Arabidopsis, *viviparous* from maize, *flacca, sitiens* and *notabilis* from tomato, or the ABA insensitive mutants, such as *abi* from Arabidopsis or *vp1* from maize.

Except for genes where is it possible to find homology with previously characterised genes, such as the above mentioned lipoxygenases, kinases, phosphatases or phospholipases, and subsequently, to suggest a possible function, the genes classified as Rab/LEA/Dehydrins (Table 1) represent a novelty, and no homology has been found with previously cloned genes. Therefore, it is necessary to conduct very extensive research in order to elucidate its function and regulation. Two examples are the genes *Rab-17* and *Rab-28* from maize, both induced by water deficit and ABA. Both proteins are accumulated during embryo desiccation and in vegetative tissues under stress conditions; Rab-17 is located in the cytoplasm and in the nucleus, and it is phosphorylated by casein kinase 2, and Rab-28 is located in the nucleolus, but still the function of both proteins remains unknown (Vilardell *et al.*, 1990; Pla *et al.*, 1991; Goday *et al.*, 1994; Niogret *et al.*, 1996). The strategies used to elucidate the function of those proteins involve the use of homologous and heterologous systems to overexpress or repress by antisense those genes, and to study the phenotypic effects. These experiments are in course, and some promising results will be obtained by different groups.

Table 1. **Genes homologous to maize Rab17 cloned from different plant species**

Organism	Protein	Reference
Arabidopsis	dehydrin	Rouse, D. *et al.* (1992), *Plant Mol. Biol.* 19, pp. 531-532
	RAB18	Lang, V. and E.T. Palva (1992), *Plant Mol. Biol.* 20, pp. 951-962
	dehydrins: lti29, lti30 and lti45	Welin, B.V. *et al.* (1994), *Plant Mol. Biol.* 26, pp. 131-144 and (1995), *Plant Mol. Biol.* 29, pp. 391-395
Barley	dehydrins (8, 9, 17, 18, 5)	Close, T.J. *et al.* (1989), *Plant Mol. Biol.* 13, pp. 95-108
	dehydrin 7	Robertson, M. (1995), *Physiol. Plantarum* 94, pp. 470-478
	dehydrin (ABA3)	Gulli, M. *et al.* (1995), *Life Sci. Adv. Plant Physiol.* 14, pp. 89-96
	Lea like	Grossi, M. (1995), *Plant Sci.* 105, pp. 71-80
Blueberry	Dehydrin-like	Muthalif, M. *et al.* (1994), *Plant Physiol.* 104, pp. 1 439-1 447
Citrus	Lea protein	Naot, D. *et al.* (1995), *Plant Mol. Biol.* 27, pp. 619-622
Cotton	D11	Baker, J. *et al.* (1988), *Plant Mol. Biol.* 11, pp. 277-291
Craterostigma	rab	Piatkowski, D. (1990), *Plant Physiol.* 94, pp. 1 682-1 688
	dehydrin like	Michel, D. *et al.* (1994), *Plant Mol. Biol.* 24, pp. 549-560
Lophopyrum	salt-responsive protein	Gulick, P.G. and J. Dvorak (1992), *Plant Physiol.* 100, pp. 1 384-1 388
Maize	dehydrin3	Close, T.J. *et al.* (1989), *Plant Mol. Biol.* 13, pp. 95-108
Pea	dehydrin	Robertson, M. (1994), *Plant Mol. Biol.* 26, pp. 805-816
Radish	dehydrin-like	Raynal, M. *et al.* (1990), *Nucleic Acids Res.* 18, p. 6 132
Rice	rab21	Mundy, J.W. *et al.* (1988), *EMBO Journal* 7, pp. 2 279-2 286
	rab16D, rab16C and rab16B	Yamaguchi-Shinozaki, K. (1989), *Plant Mol. Biol.* 14, pp. 29-39
	rab25	Kusano, T. *et al.* (1992), *Plant Mol. Biol.* 18, pp. 127-129
Spinach	dehydrin	Neven, L.G. *et al.* (1993), *Plant Mol. Biol.* 21, pp. 291-305
Stellaria	dehydrin like	Zhang Xing-Hai *et al.* (1993), *Plant Physiol.* 103, pp. 1 029-1 030
Tomato	rab TAS14	Godoy, J.A. *et al.* (1990), *Plant. Mol. Biol.* 15, pp. 695-705

Another important step to elucidate the regulatory factors that control the expression of *Rab* genes is the study of *cis-* and *trans*-acting elements of these genes, that is, DNA sequences that are recognised by proteins -- transcription factors -- that induce or block the transcription of the gene. *Cis* elements may be located upstream of the transcription initiation site of the genes, or downstream, and their relative position will modulate their expression. Several *cis*-acting elements have been described. The most important are the ABRE (**AB**A **R**esponsive **E**lement) (Marcotte *et al.*, 1988; Mundy *et al.*, 1990), present in genes induced by ABA, the Sph element recognised by the transcriptional activator Vp1 and ABA in the C1 promoter and present in Rab/LEA/dehydrins genes (Hattori *et al.,* 1992, Kao *et al.,* 1996), and the DRE element (**D**rought **R**esponse **E**lement) involved in drought and osmotic stress but not in ABA induction (Yamaguchi-Shinozaki and Shinozaki, 1994). These sequences may be present several times in the promoter regions, and results have shown that these genes are differentially regulated in front of different stresses and/or different tissues. Therefore, the elucidation of the *cis*-acting elements in a particular response may help in the isolation of *trans*-acting factors that recognise such sequences. Despite the fact that many transcription factors have been cloned, until now none has unambiguously been shown to mediate induction of those genes *in vivo*. Thus, more information is needed to elucidate the molecular basis of the transcription factors involved in this response, because as pointed out by C. Martin (1996), the modulation of the expression of transcription factors will provide a tool for inducing changes that will improve the metabolic response of plant species.

Concluding remarks

In conclusion, there are several steps where plant biotechnology may act to improve the response to abiotic stresses by plant species, and particularly, crop plants. However, in some areas the amount of information and knowledge is low and far away from applied purposes. Nevertheless, the strategies used and the generation of plant transformants may give some insights into the effects at large scale of the modification of individual genes or, by extension, the metabolic pathway affected. The improvement made by molecular breeders in order to dissect these quantitative traits, and measure the effect of the loci identified by QTL analysis will be crucial. Mapping these QTLs, together with the increased information at genetic and molecular level, will help to improve the adaptation of plants to adverse environmental situations.

Acknowledgements

This work was funded in part by the European Communities BIOTECH Programme, as part of the Project of Technological Priority 1993-1996, and in part by grant BIO-94-0750 from Plan Nacional de Investigación y Desarrollo Tecnológico.

Questions, comments and answers

Q: In irrigation practice you can compensate to a degree for salinity -- you don't need new plants. There would be a dramatic change with such new plants if you could go into areas of higher salinity. Can you achieve such a jump in salt tolerance?

A: I don't really don't know yet, but we have the opportunity to try.

Q: When you over-express the osmolyte genes, does the water content increase?

A: No, they retain water but total content is lower.

REFERENCES

ARÚS, P. and J. MORENO-GONZALEZ (1993), "Marker-assisted selection", in M.D. Hayward, N.O. Bosemark, and I. Romagosa (eds.), *Plant Breeding: Principles and prospects,* pp. 314-331, Champan & Hall, London.

BARTELS, D., K. SCHNEIDER, G. TERSTAPPEN, D. PIATKOWSKI, F. SALAMINI (1990), "Molecular cloning of ABA-modulated genes from the resurrection plan *Craterostigma plantagineum* which are induced during dessication", *Planta* 181, pp. 27-34.

BARTELS, D. and D. NELSON (1994), "Approaches to improve stress tolerance using molecular genetics", *Plant, Cell and Environment* 17, pp. 659-667.

BONHERT, H.J., D.E. NELSON, and R.G. JENSEN (1995), "Adaptations to environmental stresses. *Plant Cell* 7, pp. 1 099-1 111.

BOTSTEIN, D., R.L. WHITE, M. SKOLNICK, and R.W. DAVIES (1980), "Construction of a genetic linkage map in man using restriction fragmento lenght polymorphisms", *Am. J. Hum. Genet.* 32, pp. 314-331.

BOYER, J.S. (1982), "Plant productivity and environment", *Science* 218, pp. 443-448.

CHAMPOUX, M.C., G. WANG, S. SARKARUNG, D.J. MACKILL, J.C. O'TOOLE, N. HUANG, and S.R. McCOUCH (1995), "Locating genes associated with root morphology and drought avoidance in rice via linkage to molecular markers", *Theor. Applied Genet.* 90, pp. 969-981.

GIRAUDAT, J. (1995), "Abscisic acid signaling", *Current Opinion and Cell Biology* 7, pp. 232-238.

GIRAUDAT, J., F. PARCY, N. BERTAUCHE, F. GOSTI, J. LEUNG, P-C. MORRIS, M. BOUVEIR-DURAND, and N. VARTANIAN (1994), "Current advances in acid abscisic action and signaling", *Plant Mol. Biol.* 26, pp. 1 557-1 577.

GODAY, A., A.B. JENSEN, F.A. CUALIÑEZ-MACIA, M.A. ALBA, J. SERRATOSA, M. TORRENT, and M. PAGES (1994), "The maize abscisic-acid-responsive protein Rab-17 is located in the nucleus and interacts with nuclear localization signals", *Plant Cell* 6, pp. 351-360.

GOMEZ, J., D. SANCHEZ-MARTINEZ, V. STIEFEL, J. RIGAU, and P. PUIGDOMENECH (1988), "A gene iduced by the plant hormone abscisic acid in response to water stress encodes a glycine-rich protein", *Nature* 334, pp. 262-264.

HATTORI, T., V. VASIL, L. ROSENKRAS, L.C. HANNAH, D.R. McCARTY, I.K. VASIL (1992), "The viviparous-1 gene and abcisic acid activate the C1 regulatory gene for anthocyanin biosynthesis during seed maturation in maize", *Gene Dev.* 6, pp. 609-618.

HOLMSTRÖM, K-O., E. MÄNTYLÄ, B. WELIN, A. MANDAL, and E.T. PALVA (1996), "Drought tolerance in tobacco", *Nature* 379, pp. 683-684.

JANOWIAK, F. and K. DÖRFFLING (1996), "Chilling of maize seedlings: changes in water status and abscisic acid content in ten genotypes differing in chilling tolerance", *J. Plant Physiol.* 147, pp. 582-588.

JANSEN, R.C. (1996), "Complex plant traits: time for polygenic analysis", *Trends Plant Science* 1, pp. 89-94.

KAO, C.-Y., S.M. COCCIOLONE, I.K. VASIL, and D.R. McCARTY (1996), "Localization and interaction of the *cis*-acting elements for abscisic acid, VIVIPAROUS1, and light activation of the c1 gene of maize", *Plant Cell* 8, pp. 1 171-1 179.

KISHOR, P.B.K., Z. HONG, G-H- MIAO, C-A.A. HU, and D.P.S. VERMA (1995), "Overexpression of d-pyrroline-5-carboxylate synthetase increases proline production and confers osmotolerance in transgenic plants", *Plant Physiol.* 108, pp. 1 387-1 394.

LANDER, E.S. and D. BOTSTEIN (1989), "Mapping mendelian factors underlying quantitative traits using RFLP linkage maps", *Genetics* 121, pp. 185-199.

LEUNG, J., M. BOUVIER-DURAND, P-C. MORRIS, D. GUERRIER, F. CHEFDOR, and J. GIRAUDAT (1994), "*Arabidopsis* ABA-response gene ABI1: features of a calcium-modulated protein phosphatase", *Science* 264, pp. 1 448-1 452.

LEVITT, J. (1980), *Responses of plants to environmental stresses, Vol. 1,* T.T. Kozlowski (ed.), Academic Press, New York.

LI, Y. and D.C. WALTON (1990), "Violaxhantin is an abscisic acid precursor in water-stressed dark-grown leaves", *Plant Physiol.* 92, pp. 551-559.

LILIUS, G., N. HOLMEBERG, and L. BÜLOW (1996), "Enhanced NaCl stress tolerance in transgenic tobacco expressing bacterial choline dehydrogenase", *Biotechnology* 14, pp. 177-180.

MacROBBIE, E.A.C. (1995), "ABA-induced ion efflux in stomatal guard cells: multiple actions of ABA inside and outside the cell", *Plant J.* 7, pp. 565-576.

MARCOTTE, W.R., CH.C. BAYLEY, and R.S. QUATRANO (1988), "Regulation of a wheat promoter by abscisic acid in rice protoplasts", *Nature* 335, pp. 454-457.

MARIN, E., L. NUSSAUME, A. QUESADA, M. GONNEAU, B. SOTTA, P. HUGUENEY, A. FREY, and A. MARION-POLL (1996), "Molecular identification of zeaxanthin epoxidase of *Nicotiana plumbaginifolia*, a gene involved in abscisic acid biosynthesis and corresponding to the *ABA* locus of *Arabidopsis thaliana*", *EMBO J.* 15, pp. 2 331-2 342.

MARTIN, C. (1996), "Transcription factors and the manipulation of plant traits", *Current Opinion in Biotechnology* 7, pp. 130-138.

McCUE, K.F. and A.D. HANSON (1990), "Drought and salt tolerance: towards understanding and application", *Tibtech.* 8, pp. 359-363.

MEYER, K., M.P. LEUBE, and E. GRILL (1994), "A protein phosphatase 2C involved in ABA signal transduction in *Arabidopsis thaliana*", *Science* 264, pp. 1 452-1 455.

MOONS, A., G. BAUW, E. PRINSEN, M. VAN MONTAGU, D. VAN DER STRAETEN (1995), "Molecular and Physiological response to abscisic acid and salts in roots of salt-sensitive and salt-tolerant indica rice varieties", *Plant Physiol.* 107, pp. 177-186.

MUNDY, J. and N.H. CHUA (1988), "Anscisic acid and water-stress induce the expression of a novel rice gene", *EMBO J.* 7, pp. 2 279-2 286.

MUNDY, J., K. YAMAGUCHI-SHINOZAKI, and N.H. CHUA (1990), "Nuclear proteins bind conserved elements in the abscisic acid-responsive promoter of a rice RAB gene", *Proc. Natl. Acad. Sci. USA* 87, pp. 1 406-1 410.

NIOGRET, M.F., F.A. CULIAÑEZ-MACIA, A. GODAY, M. MAR ALBA, and M. PAGES (1996), "Expression and cellular localization of *rab28* mRNA and Rab28 protein during maize embryogenesis", *The Plant Journal* 9, pp. 549-557.

PATERSON, A.H., E.S. LANEDR, J.D. HEWITT, S. PETERSON, S.E. LINCOLN, and S.D. TANKSLEY (1988), "Resolution of quantitative traits in to mendelian factors using a complete linkage map of restriction lenght polymorphism", *Knature* 335, pp. 721-726.

PATERSON, A.H., S.D. TANKSLEY, and M.E. SORRELS (1991), "DNA markers in plant improvement", *Adv. in Agronomy* 46, pp. 39-90.

PILO-SMITS, E.A.H., M.J.M. EBSKAMP, M.J. PAUL, M.J.W. JEUKEN, P.J. WEISBEEK, and S.C.M. SMEEKENS (1995), "Improved performance of transgenic fructan-accumulating tobacco under drought stress", *Plant Physiol.* 107, pp. 125-130.

PLA, M., J. GOMEZ, A. GODAY, and M. PAGES (1991), "Regulation of the abscisic acid-responsive gene rab28 in maize viviparous mutants", *Mol. Gen. Genet.* 230, pp. 394-400.

QUARRIE, S.A., A. STEED, V. LAZIC-JANCIC, and D. KOVACEVIC (1994), "Genetic variation in ABA production in maize determines the extent of drought-induced gene transcription", in J.H. Cherry (ed.), *Regulation of gene expression in response to drought and osmotic shock,* NATO-ASI Series, Vol. H 86, Biochemical and Cellular Mechanim of Stress Tolerance, pp. 3 222-3 328, Springer-Verlag, Berlin, Heilderberg.

QUARRIE, S., A. HEYL, A. STEED, C. LEBRETON, and V. LAZIC-JANCIC (1996), "QTL analysis of stress responses as a method to study the importance of stress-induced genes", in S. Grillo and A. Leone (eds.), *Physical Stresses in Plants. Genes and Their Products for Tolerance*, pp. 141-150, Springer, Berlin.

SKRIVER, K. and J. MUNDY (1990), "Gene expression in response to abscisic acid and osmotic stress," *Plant Cell* 2, pp. 503-512.

TANKSLEY, S.D. (1993), "Mapping polygenes", *Ann. Rev. Genet.* 27, pp. 205-233.

TURNER, A.M. (1986), "Adaptation to water deficits: a changing in perspective", *Aust. J. Plant Physiol.* 13, pp. 175-190.

VILARDELL, J., A. GODAY, M.A. FREIRE, C. MARTINEZ, J.M. TORNE, and M. PAGES (1990), "Gene sequence, developmental expression, and protein phosphorilation of RAB-17 in maize", *Plant Mol. Biol.* 14, pp. 423-432.

VOS, P., R. HOGERS, M. BLEEKER, M. REIJANS, T. VAN DE LEE, M. HORNES, A. FRITJERS, J. POT, J. PELEMAN, M. KUIPER, and M. ZABEAU (1995), "AFLP: a new technique for DNA fingerprinting", *Nucleic Acid Res.* 23, pp. 4 407-4 414.

WALTON, D.C. and Y. LI (1995), "Abscisic acid biosynthesis and metabolism", in P.J. Davis (ed.), *Plant hormones, Physiology, Biochemistry and Molecular Biology*, pp. 140-157, Kluwer Academic Publishers, Dordrecht, Netherlands.

WILLIAMS, J.G.K., A.R. KUBELIK, K.J. LIVAK, J.A. RAFALSKI, and S.V. TINGEY (1990), "DNA polymorphism amplified by arbitrary primers are useful as genetic markers", *Nucleic Acid Res.* 18, pp. 6 531-6 535.

YAMAGUCHI-SHINOZAKI, K. and K. SHINOZAKI (1994), "A novel *cis*-acting element in an Arabidopsis gene is involved in responsiveness to drought, low-temperature, or high-salt stress", *Plant Cell* 6, pp. 251-264.

PHYTOREMEDIATION: GREEN AND CLEAN

by

Ilya Raskin
AgBiotech Center, Rutgers University, Cook College
New Brunswick, New Jersey, United States

Soils and waters contaminated with toxic metals pose a major environmental and human health problem which is still in need of an effective and affordable technological solution. The partial success of microbial bioremediation has been limited to the degradation of some organic contaminants and has been ineffective at addressing the challenge of toxic metal contamination, particularly in soils. Current methodologies for remediating toxic metal polluted soils rely mainly on excavation and burial at a hazardous waste site at an average cost of $ 1 000 000/acre.

In our laboratory at Rutgers University, we helped to develop a cost-effective "green" technology based on the use of specially selected metal-accumulating plants to remove toxic metals, including radionuclides, from soils and water. We termed this technology phytoremediation. Phytoremediation takes advantage of the fact that a living plant can be compared to a solar driven pump, which can extract and concentrate particular elements from the environment. We have exploited this property to develop a method for enhancing the ability of specially selected and/or engineered plants to remove toxic metals from soil and water and to concentrate these metals in harvestable parts. To make phytoremediation possible, we have assembled a diverse group of scientists composed of plant biochemists, molecular biologists, soil chemists, agronomists, and environmental engineers. We have also formed close collaborations with other faculty members at Rutgers and scientists at other universities in the United States and abroad.

The basic idea that plants can be used for environmental remediation is, certainly, very old and can not be traced to any particular source. However, a series of fascinating scientific discoveries combined with interdisciplinary research approaches allowed the development of this idea into a promising environmental technology. Laboratory and greenhouse work on phytoremediation started in 1991, as a result of funding received from the United States Environmental Protection Agency (EPA) and the State of New Jersey. Specifically, two subsets of phytoremediation are approaching commercialisation.

1) *Phytoextraction*, in which high biomass metal-accumulating plants and appropriate soil amendments are used to transport and concentrate metals from the soil into above-ground shoots, which are harvested by conventional agricultural methods;

2) *Rhizofiltration*, in which plant roots grown in water, precipitate and concentrate toxic metals from polluted effluents. Two patents covering phytoremediation technology have been issued, with six more still pending.

The metals targeted for phytoremediation include lead, cadmium, chromium, arsenic and various radionuclides. The market is estimated to be billions of dollars, with government and private responsible parties being major clients. The harvested plant tissue, rich in accumulated contaminant, can be processed by drying, ashing or composting. The volume of toxic waste produced as a result is generally a fraction of that of many current, more invasive remediation technologies, and the associated costs are much less. Some metals can be reclaimed from the ash, further reducing hazardous waste and generating recycling revenues.

Phytoremediation effort received a real boost after Phytotech Inc. was formed in New Jersey in 1993. Phytotech is a developmental stage environmental biotechnology company that was founded to discover, develop and market phytoremediation products and services for the treatment of soil and water. Productive collaboration with Phytotech's research and business teams, under the leadership of company president Dr. Burt Ensley, has accelerated the development of phytoremediation and targeted this development towards the most pressing needs of the "real world". For the last two summers, Phytotech, in collaboration with Rutgers scientists, undertook an extensive field demonstration program that focused primarily on lead contaminated soils in New Jersey and on the radionuclide contaminated soils in the Ukraine. In one of these trials, specially selected, cultivated and fertilized *Brassica juncea* (Indian mustard) plants were able to accumulate over one per cent lead on the dry weight basis from the soils containing 1 200 ppm of lead. Special soil treatments developed by Phytotech and Rutgers were used in these trials to increase metal uptake by plants. Indian mustard can produce at least three crops in New Jersey with each crop yielding eight tons of dry biomass per acre. This translates into an annual lead phytoextraction rate of 240 kg/acre/year. The results from Ukrainian soils contaminated with Sr-90 were equally encouraging. We believe that these rates of metal removal will be substantially increased as a result of future research.

Phytotech Inc. also successfully tested rhizofiltration in the summer of 1995 at two locations: Ashtabula, Ohio, where site water was contaminated with 100-400 ppb uranium, and in a small pond within one kilometre of the Chernobyl nuclear power plant in the Ukraine. The field results demonstrated that rhizofiltration is a practical way to treat radionuclide contaminated water. As a part of the DOE-funded project in Ashtabula, groundwater was purified to below the regulatory standards (below five ppb of uranium) using a pilot-scale rhizofiltration system. In the Ukraine, Cs-137 and Sr-90 concentrations in water were reduced several hundred times to below 10 Bq/L for Cs-137 and below 200 Bq/L for Sr-90. In both cases, the estimated cost of the rhizofiltration process was much lower than the state-of-the-art ion exchange methods.

Phytoremediation research is rapidly gaining momentum in companies and universities. For example, DuPont Co., headquartered in Wilmington, Delaware, has its own phytoremediation program directed by Dr. Scott Cunningham. This program is targeted to the removal and stabilization of lead in soils using plants. Recent results from DuPont, presented at scientific meetings, are encouraging and demonstrate DuPont's interest in phytoremediation as a solution to their environmental problems.

Successful transfer of phytoremediation from laboratory to the field is a crucial step in the development of this technology. While Phytotech Inc. is making progress in commercialising phytoremediation, our team at Rutgers is directing its effort to understanding the physiological, biochemical and molecular mechanisms involved in metal uptake accumulation and resistance in plants. Rutgers and Phytotech teams are also developing molecular genetic approaches to improving phytoremediation and starting a program in the remediation of organics by plant roots and associated micro-organisms. This collaboration between a company and a public university is an example of successful technology transfer and demonstrates the value of properly structured interactions between university scientists and the private sector.

Plants have an enviable position of being the only cheap, renewable resource available to our civilisation. Gradual depletion of natural resources pushes scientists to explore the use of plants and their products as a substitute for oil and mining based commodities, and as alternatives for costly and energy intensive engineering processes. It may be the right time to utilise the potential of using plants for the remediation of metal and organic pollutants and to develop engineering and biological approaches that further improve this process. From the classical works of Dittmer published in 1930s, it can be calculated that a healthy rye field has 140 million miles of roots per acre, a number close to the diameter of Earth's orbit around the sun. Scientists should try to employ this hidden bioengineering marvel in the battle for a greener and cleaner environment.

Questions, comments and answers

Q: Why do you use radioactive species: why not use depleted uranium?

A: Our analytical ability and visualisation are much better. Detection is simple and we can process many samples a day.

Q: You spoke about the kinetic profile of uptake but what about pH?

A: It is very hard to generalise. The effect of pH is specific to individual metals. For example, lead becomes more soluble below pH 4.

Q: How does the speciation of the metal affect its uptake? For example, uranium as a carbonate?

A: We haven't really addressed this, but we do know it works.

C: The problem of accumulation of radionuclides is how to handle contaminated rhizospheres.

A: The goal is to reduce how much secondary waste is generated.

BIOPREVENTION OF WATER SHORTAGE: APPLICATION OF NEW SUPER-BIOABSORBENTS TO GREENING OF ARID SOIL

by

Ryuichiro Kurane
National Institute of Bioscience & Human-Technology (NIBH)
AIST/MITI, Tokyo, Japan

Introduction

Environmental impacts occur at all stages of a Product Life Cycle, such as material extraction, material processing, manufacturing product use, and waste management. Bioprevention technology has great potential for preserving our environmental quality and coping with these problems.

The concept of "design" can be employed to reduce these threatening impacts by changing the amount and type of materials used in the product, by creating more efficient manufacturing operations and processes, by reducing the energy and materials consumed during use, and by improving the recovery of energy and materials during waste management.

"Green products" and "green processes", which mean environmentally friendly products and processes, can be expected to grow in scale and to encourage, promote and stimulate further development of bioprevention technology to preserve our earth's environmental quality.

In a world where population growth and economic growth put increasing pressures on natural resources and ecosystems, the dominant paradigm upon which environmental policies are based can be expected to evolve from environmental protection involving resource management and eco-development.

How should one view the significance of green design as a competitive and environmental strategy? As a competitive strategy, green design can help manufacturers generate less waste and reduce production costs at the same time. As waste disposal costs and regulatory compliance costs increase, the environmental attributes of products will necessarily become more important to consumers and investors. Some OECD Member countries are already moving aggressively to integrate "clean" products into their industrial strategies for future competitiveness, and international trade will increasingly be influenced by environmental concerns. All of these trends suggest that having an environmental dimension to one's design capabilities will be an important competitive asset in the future.

Many organic synthetic high polymers have been frequently used as powerful and economical agents. However, among them, some of these high polymer agents are harmful to the environment and are a dangerous source of pollution that can adversely affect future generations. Thus, a safe biodegradable high polymer, that is, a biopolymer, that will minimise environmental and health risks, is both urgently needed and is attracting wide research interest.

As an environmental strategy, green product design offers a new way of addressing environmental problems.

For example, organic synthetic high polymer flocculating agents have been used for a wide range of applications, which include wastewater treatment, dredging and industrial processes. In particular, polyacrylamide derivatives have frequently been used because they are economical and effective agents.

Another example is a water absorbent. Due to life style modernisation, the use of various kinds of sanitary products and disposable baby diapers is increasing annually, with most of these water absorbent articles currently being made of synthetic high polymer materials such as polyacrylate and polyacrylamide derivatives. These organic synthetic high polymer absorbers are both cost-effective and high-activity agents. However, many of these products do not easily biodegrade and are disposed, discharged, and remain in the environment; they pose a significant environmental hazard. Studies indicate that the monomer of acrylamide is both neurotoxic (as indicated in the Merck Index) and a strong carcinogen in the human body (Vanhorick and Moens, 1983). It has been reported that the synthetic high polymer agent monomer, acrylic acid, is a strong irritant to the human body (as indicated in the Merck Index).

Another area of ecological global concern is the accelerated expansion of the world's deserts, which is a serious environmental protection problem because their area is reportedly expanding by about 60 000 km^2 each year (FAO/UNEP, 1984) (UNEP, 1990) (Touyama, 1990*a*). In some countries, soil aridity prevention and desert greening experiments are being done using synthetic high polymer absorbent materials. One major disadvantage of this technique is apprehension that these non-biodegradable synthetic high polymer materials will remain in the soil for a long period or a fine powder of the polymer and residual monomers will move from the soil to the rivers, thus contributing to environmental pollution. After plants begin to grow, the bioabsorbent will be expected to cause no damage to the soil, because it is composed of natural materials, as a polysaccharide, and is expected to be more easily decomposed.

To solve these ecological problems, the development of "environmentally friendly products", i.e. "green products" which are more compatible with the environment are urgently needed. For example, these new biopolymers may be expected to overcome the problems associated with conventional synthetic high polymer agents because they are polysaccharide bioabsorbents produced by a micro-organism and therefore do not pose a secondary pollution environmental hazard.

A new super polysaccharide bioabsorbent from *Alcaligenes latus*

A. latus B-16 bioabsorbent production (Kurane and Nohata, 1991) (Kurane and Nohata, 1994)

Micro-organism. Approximately 500 cultures were screened for biopolymer production; the B-16 bacterial strain isolated from soil produced a microbial bioabsorbent in the culture broth. This strain was identified as *Alcaligenes latus* as reported in our previous paper (Kurane and Nohata, 1991).

Bioabsorbent culture and production methods. Either fructose or sucrose (15 g), KH_2PO_4 (6.8 g), K_2HPO_4 (8.8 g), $MgSO_4 \cdot 7H_2O$ (0.2 g), NaCl (0.1 g), urea (0.5 g), and yeast extract (0.5 g) were dissolved in distilled water (1 000 ml), and the medium adjusted to pH 7.0. The fructose/sucrose was separately sterilised and the urea was filtered using a Millipore filter (pore size: 0.22 μ m). After cultivation on an agar slant of the same basal production medium, one loop of the cells was inoculated into 300 ml or 500 ml flasks containing the above basal medium (100 ml), and then cultivated with a rotary shaker (180 rpm) at 30 °C for five days.

Bioabsorbent samples were produced using different culture conditions, that is, by changing the culture carbon sources in the basal medium's composition. That is, fructose, sucrose, starch, and olive oil were used to study the effects of carbon sources on bioabsorbent formation. Urea was used as an inorganic nitrogen source, and yeast extract as also added as an organic nitrogen source.

By the recovering step described below, dried bioabsorbent biopolymer at the end stage of five days of incubation was obtained and weighed.

After four to six days of cultivation in the basal production mediums, 20-25 g of crude bioabsorbent was produced per litre of culture broth with either fructose and sucrose the most effective for bioabsorbent formation. Both urea, added as an inorganic nitrogen source, and yeast extract (0.05 per cent), as an organic nitrogen source, appeared to enhance bioabsorbent production.

Water absorption capacity (Kurane and Nohata, 1994)

Water absorption capacity. A method (Kurane and Nohata, 1994) generally called the "tea bag method" was used, i.e. a container (20 ml capacity) made from a non-woven fabric (Kitchen Tauper, Tokai Pulp Co., 100 per cent natural pulp) was filled with an accurately weighed sample, immersed in either pure distilled water or 0.9 per cent NaCl water for two hours, removed and left to drain for one hour to drain off excess water. This percolated sample was then put into a known weight beaker (10 ml) and weighed, then dried at 105 °C for 15 hours to completely evaporate the water, then weighed. The water absorption capacity (g) per gram of dried sample was then calculated by:

$$\text{Water absorption capacity} = \frac{\text{Sample weight after absorption (g)} - \text{Sample weight before absorption (g)}}{\text{Dried sample weight (g)}}$$

The bioabsorbent's water absorption capacity was measured using two different purified bioabsorbent (SP and FP) samples. Six control samples were also tested, i.e. pulp, silica gel (Kanto Chemicals Co., Ltd.), ion exchange resin (Dow Chemical), a high-grade water-absorbing synthetic high polymer (Sumika Gel S-50, Sumitomo Chemical Co., Ltd.), poly(vinyl) alcohol (PVA) (UP-100G, Unitika Co., Ltd.), and an anionic synthetic high polymer absorbent (Sumifloc FA-70, Sumitomo Chemical).

Table 1 clearly shows that the presented bioabsorbent absorbs the most water (up to 1 000 times its own weight), even more than the control group's synthetic high polymer water absorbents.

It should be noted that this bioabsorbent water absorption capacity is more than three to five times greater than that of the currently used synthetic high-polymer water absorbents.

Table 1. **Water absorption capacities of various absorbents**

	Sample description	Water absorption capacity per gram of dried sample (g)
Test Group	SP[1]	1 349.0
	FP[1]	1 295.4
Control group	Pulp	3.8
	Silica gel	1.4
	Ion-exchange resin	2.5
	PVA	4.6
	High-grade synthetic high-polymer absorbent[2]	249.4
	Anionic synthetic high-polymer absorbent[3]	363.6

Notes:

1. Bioabsorbent samples were produced using different culture conditions, i.e. by changing the culture medium's carbon sources; SP: sucrose, FP: fructose.
2. High grade synthetic high-polymer absorbent: polyacrylate/PVA derivative (copolymer of acrylate and vinyl alcohol).
3. Anionic synthetic high polymer absorbent: polyacrylamide derivative, MW 350 x 10^4.

Source: Kurane and Nohata, 1994.

Moisture absorption capacity (Kurane and Nohata, 1994)

Moisture absorption capacity. Three different desiccators containing saturated solutions of potassium nitrate, sodium nitrate, and magnesium chloride were used at the respective humidities of 91 per cent, 61.8 per cent and 31.9 per cent while being stored at 37 °C in a thermostatic chamber (Kurane and Nohata, 1994). Dried samples (100 mg) were weighed in plastic cups, and then left in the presence of the desiccators. After being stored two, four, six and 24 hours, the samples were weighed again, with their resultant moisture absorption capacity (per cent) being calculated by:

$$\text{Moisture absorption capacity (per cent)} = \frac{Wt - Wo}{Wo} \times 100$$

Wt: sample weight at given times

Wo: initial sample weight

The moisture absorption capacities of the SP bioabsorbent and a mixture of SP and FP (1:1) bioabsorbent (MIX) were measured and compared with several control samples, with the results showing a high moisture absorption capacity, as shown in Table 2.

Moisture retention capacity (Kurane and Nohata, 1994)

Moisture retention capacity. Three different desiccators containing saturated solutions of sodium nitrate, magnesium chloride, and phosphorus pentoxide (P_2O_5) were used at respective relative humidities of 64.8 per cent, 33 per cent and 34 per cent, while the samples were stored at 20 °C in a thermostatic with 20 μ l of water, weighed again and then left in the desiccators for different times. After standing two, four, six, eight, 10, 24 and 48 hours, they were weighed, with their moisture retention capacity being calculated by:

$$\text{Moisture retention capacity (per cent)} = 1 - \frac{Wo - Wt}{20} \times 100$$

Wo: initial sample weight

Wt: sample weight at given times

Table 2. **Moisture absorption capacities of various absorbents**

Relative humidity	Sample	Moisture absorption (%) at given interval (hr)						
		2	4	6	8	10	24	48
91%	SP_2	42	53	61	65	69	93	105
	MIX	38	50	58	62	66	87	99
	Silica gel	31	32	32	32	32	31	31
	PVP	29	34	36	37	39	46	48
	Urea	12	23	35	45	57	114	154
	PVA	8	11	13	13	14	15	16
	Glycerin	33	49	61	68	76	113	140
	PEG 200	27	38	45	50	55	78	92
	Anionic polymer	17	28	35	41	46	64	75
	Hyaluronic acid						35	
61.8%	SP_2	31	33	34	33	34	33	35
	MIX	28	34	31	32	32	31	30
	Silica gel	27	29	29	29	29	29	29
	PVP	18	19	19	19	19	18	19
	Urea	0	0	0	0	0	0	0
	PVA	3	3	4	5	5	7	7
	Glycerin	23	30	33	35	36	37	38
	PEG 200	18	20	20	20	20	22	21
	Anionic polymer	9	13	16	21	21	22	21
	Hyaluronic acid						17	
31.9%	SP_2	7	9	6	11	12	12	11
	MIX	9	10	7	14	14	14	14
	Silica gel	13	16	12	18	18	17	16
	PVP	8	9	9	9	9	9	10
	Urea	0	0	0	0	0	0	0
	PVA	0	1	1	1	2	2	2
	Glycerin	6	9	11	12	13	12	12
	PEG 200	3	7	7	7	7	6	7
	Anionic polymer	1	1	1	2	1	3	4

Source: Kurane and Nohata, 1994.

The moisture retention capacities of the SP bioabsorbent and a mixture of SP and FP bioabsorbents (MIX) were measured and compared with the same control samples using three different desiccators containing a saturated solution of sodium nitrate/phosphorus pentoxide (data not shown), and a saturated solution of magnesium chloride (Figure 1) at the respective relative humidities (20 °C) of 64.8 per cent, 34 per cent and 33 per cent.

Figure 1 clearly indicates that the new bioabsorbent has a high moisture retention capacity, i.e. in a dry environment (34 per cent and 33 per cent relative humidity), the SP bioabsorbent retained 82 per cent of its moisture after 24 hours.

Figure 1. **Moisture retention capacities of various absorbents**

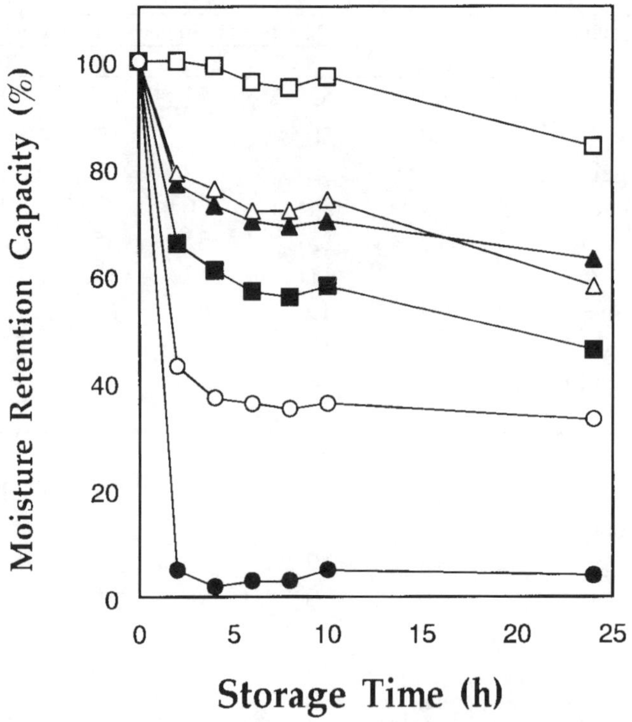

Notes: Condition: 33% relative humidity, 20 °C, in dessicator containing a saturated solution of magnesium chloride. □ bioabsorbent (SP sample); ● urea; ∆ glycerin; ♉ PVP; ■ PEG 200; O anionic polymer.

Source: Kurane and Nohata, 1994.

Water absorption capacity in the presence of NaCl (Kurane and Nohata, 1994)

The new bioabsorbent was placed in saline solution of varying NaCl concentrations to confirm its water absorption effectiveness, with the results being shown in Figure 2.

The water absorption capacity decreased in the presence of NaCl; however, this decrease was significantly less when compared to that of the synthetic high polymer absorbents (Sumika Gel S-50). This water absorption capacity in the presence of 0.9 per cent NaCl enabled the bioabsorbent to absorb 450-550 times its own weight, which is 20 times greater than that of a currently used synthetic high polymer absorbent (22 times, our test data).

As shown in Figure 2, the bioabsorbent was capable of absorbing water at 450 and 360 times its own weight at respective NaCl concentrations of 1 per cent and 2.5 per cent. It should be noted that these values are still slightly higher than those of conventional synthetic super water-absorbent high polymers for absorption of pure distilled water.

Figure 2. **Bioabsorbent and synthetic high polymer absorbent water absorption capacities in the presence of various NaCl concentrations**

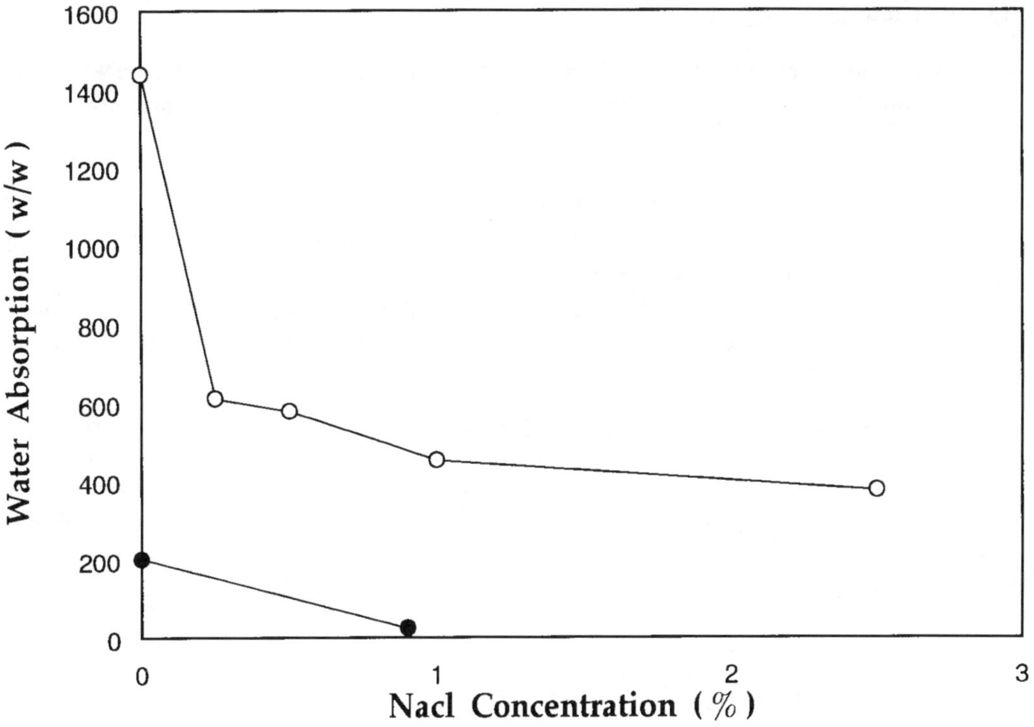

Notes: ○ bioabsorbent (SP sample); ● high-grade absorbing synthetic high-polymer (Sumika Gel S-50).

Source: Kurane and Nohata, 1994.

Retention of supplied water in sand with the bioabsorbent (Kurane and Nohata, 1994) (Nohata and Kurane, 1996)

Water retention capacity through the sand. The sand was obtained from the Hamaoka sand hill. The sand wads were rinsed with water to remove salt and dust, and then only the 50-80 mesh size grains were collected by filtering through a mesh.

A 150 g sample of the collected sand was put into a column-type holder (50 mn high), the bottom of which was an acrylic resin net (200 mesh). These holders were stored in a thermostatic chamber at 20, 45 and 70°C in 50-60 per cent relative humidity.

Each polymer was dispersed in pure distilled water so its apparent viscosity was 50 cps (no. 2 spindle, 20 °C, 30 rpm). Each polymer solution (80 ml) was poured over the sand in the holder, and then the holder containing the sand and the solution was weighed at regular intervals to determine the water retention rate.

$$\text{Water retention rate (per cent)} = \frac{\text{drained water of Blank - drained water of Sample}}{\text{drained water of Blank}} \times 100$$

Blank: no addition of polymer (pure distilled water only)
Sample: addition of each polymer solution

Expansion of desert areas has been recognised as a serious global environmental problem, and large amounts of water have been used to irrigate desert soil for greening the desert. However, use of large amounts of water causes serious salt damage to soil, and salt damaged soil in general can not be used for a long period.

In particular, a huge amount of water is needed for irrigation of the greening desert, and most irrigated water immediately disappears into the earth, causing injury from salt, and also disappears into the air at high temperatures.

Therefore, the identification of environmentally friendly agents to retain a water supply is required. We found that this acid biopolymer (bioabsorbent) has good water retention capacity characteristics through sand, as shown in Table 3 and Figure 3. The water retention capacity of polymers tested decreased when the temperature was increased (20 to 45 to 70 °C), but that of the bioabsorbent was the best. Moreover, less that one tenth of this bioabsorbent had the same effect when it was compared with sodium alginate.

Table 3. **Keeping capacity of supplied water in the sand with bioabsorbent**

	Keeping capacity (%)		
	5 minutes	1 hour	24 hours
Bioabsorbent	100	86	42
Sodium alginate	13	7	8
Xanthangum	8	0	0
Synthetic acrylamide high polymer[1]	8	0	0
Synthetic high polymer absorbent[2]	8	7	7

Notes: 1. Deafloc Inc., P-MP; 2. Sumika Co., Gel S-50.

Source: Kurane and Nohata, 1994.

Figure 3. **Effect of temperature on reducing drainage activities of the bioabsorbent from *A. latus* B-16 and other polymers**

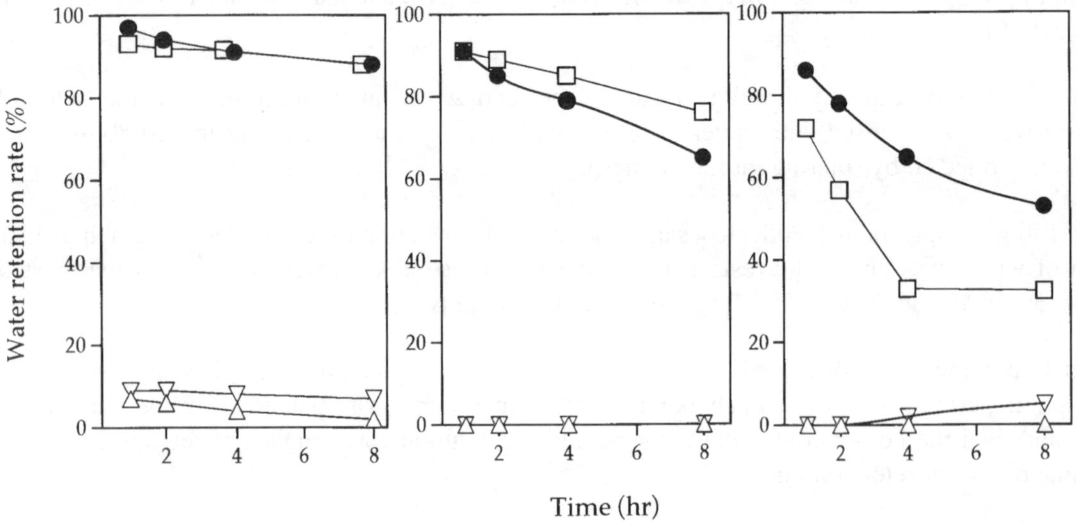

Notes: ● bioabsorbent 213 ppm; □ sodium alginate 3 077 ppm; Δ xanyhan gum 611 ppm; ∇ anionic polymer 868 ppm. Apparent viscosity of each sample was 50 cps. The particle diameter of sand was 50-80 mesh.

Source: Nohata and Kurane, 1996.

Table 3 clearly shows that sand with the bioabsorbent has high water-retention capacity, meaning that sand with the bioabsorbent retains much of the supplied water even at a high temperature (70 °C) like that of a hot desert surface in daytime.

Culture broth for production of bioabsorbent (Nohata and Kurane, 1994)

Glucose and sucrose were favourable carbon sources for the production of bioabsorbent from *A. latus* B-16. Yeast extract at a concentration of 0.5 per cent was also effective for growth. Biopolymers from *A. latus* were purified and separated into two types of biopolymers; a high-molecular weight component (acid-biopolymer), and a low-molecular weight component (neutral-biopolymer). These two types of biopolymers were shown to be homogeneous on electrophoresis and GPC (Gel Permeation Chromotography on HPLC). The high-molecular weight component was a bioabsorbent and was able to absorb water at more than 1 000 times its own weight. The total yield of acid and neutral-biopolymers from *A. latus* B-16 was 20-25 g (dry weight) per litre of culture broth, and they were produced at a ratio of 20:1 (acid:neutral)

Constituent sugars of bioabsorbent (Kurane and Nohata, 1995)

Constituent sugars of the polysaccharide bioabsorbent from *Alcaligenes latus* B-16, which can hold water at more than 1 000-fold (maximum 2 000-fold) its own weight, were identified by three methods: gas chromatography, high-pressure liquid chromatography, and gas mass spectroscopy. This polysaccharide bioabsorbent is composed of four different sugars: glucose, rhamnose, fucose and glucuronic acid. The molar ratio of the constituent sugars was glucose:rhamnose:glucuronic acid:fucose = 1.8:1.1:1:1.

Physiochemical properties of the bioabsorbent from Alcaligenes latus

The purified bioabsorbent obtained had the following physiochemical properties:

- Colour: white;
- Carbonisation temperature: 225-280 °C;
- Elementary analysis: C.40; H.6; 0.54;
- Solubility: very slightly soluble in water (neutral), soluble in alkalis; insoluble in methanol, ethanol, or acetone;
- UV absorption spectrum: No absorption was detected at 280 nm, which is characteristic of proteins, or at 260 nm, which is characteristic of nucleic acids.
- IR absorption spectrum: An absorption pattern characteristic of polysaccharides was observed near 800 - 1 200 cm^{-1}; an absorption pattern characteristic of uronic acid was observed at 1 620 ± 20 cm^{-1}; CH and CH_2 absorption patterns due to carbohydrates were observed near 2 950 cm^{-1}; and an OH absorption pattern due to carbohydrates was also observed near 3 400 ± 20 cm^{-1}. These results suggest that the purified bioabsorbent is an acid polysaccharide chiefly composed of sugars.

Constituent sugars of the bioabsorbent

Constituent sugars of the purified bioabsorbent were identified by TLC, LC, GC, and GC-MS.

As measured under the TLC experimental conditions, glucuronic acid alone had an Rf value of 0.17, but in the presence of galacturonic acid, even the standard sample had a higher Rf value of 0.21. Experiments under other conditions showed that the uronic acid was neither muramic acid nor mannuronic acid lactone. It therefore became apparent that the bioabsorbent contained glucuronic acid as the uronic acid. The Rf values of the hydrolyzate of the bioabsorbent were in agreement with those of glucose, rhamnose, fucose and glucuronic acid. With these results taken into consideration, four standard samples (glucose, rhamnose, fucose and glucuronic acid) in the admixture and the hydrolyzate of the bioabsorbent were tested by TLC. The four standard sugar samples had Rf values in very good agreement with those of the hydrolyzate of the bioabsorbent.

The results are shown in Figure 4A (standard example) and Figure 4B (the hydrolyzate of the sample before separation), Figure 4C (the hydrolyzate of the high-molecular weight bioabsorbent) and Figure 4D (the hydrolyzate of the low-molecular weight neutral biopolymer). It is apparent that the hydrolyzate of the bioabsorbent had peaks that corresponded to glucose, rhamnose, fucose, and glucuronic acid. As shown in Figure 4C, the HPLC chart had another peak (retention time = 35 minutes). This peak was fractionated and obtained, re-hydrolysed, and then analysed by HPLC. This peak was an oligosaccharide composed of glucose and glucuronic acid (data not shown). The above results provided further confirmation that the bioabsorbent, as analysed by HPLC, was composed of glucose, rhamnose, fucose, and glucuronic acid.

Figure 4. **HPLC charts for standard samples and the hydrolyzate of the bioabsorbent**

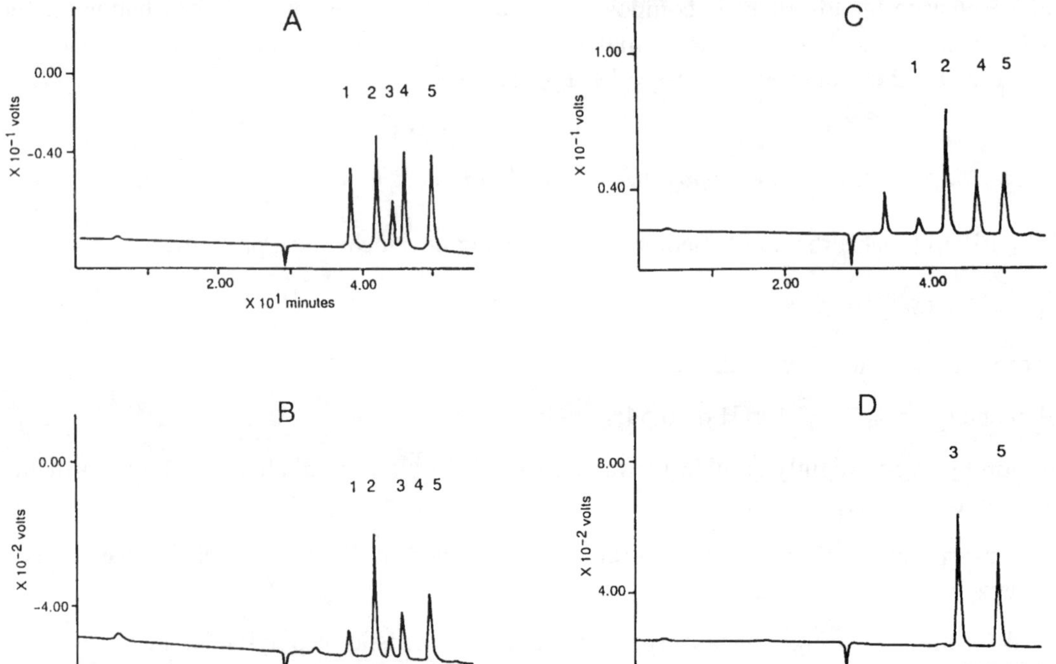

Notes: A: Standard samples. 1) glucuronic acid; 2) glucose; 3) mannose; 4) rhamnose; 5) fucose.
B: Hydrolyzate of polysaccharide component including high and low-molecular weight biopolymer before separation.
C: Hydrolyzate of the high-molecular weight acid biopolymer (the bioabsorbent).
D: Hydrolyzate of the low-molecular weight neutral biopolymer.

Source: Kurane and Nohata, 1995.

Figure 4D shows that the constituent sugars of the low-molecular-weight neutral biopolymer were mannose and fucose. As shown in Figure 4D, no other retention peaks appeared in the HPLC chart. The same results were obtained using NH2P-50 HPLC column chromatography (data not shown). Based on the previously mentioned results, the molar ratio of the constituent sugars of the low-molecular-weight neutral biopolymer was as follows: mannose:fucose = 3:2 (Wt. ratio), i.e. 4:3 (μ mole ratio).

The high-molecular-weight bioabsorbent has rhamnose as the constituent sugar, and the low-molecular-weight neutral biopolymer has mannose. Therefore, the yield ratio of two kinds of high and low-molecular-weight biopolymers was calculated from Figure 4B. The yield ratio of the high-molecular-weight bioabsorbent and low-molecular-weight neutral biopolymer was 7:1 (Wt. ratio).

To do a triple check of the constituent sugars identified by TLC and HPLC, the hydrolyzate of the bioabsorbent was analysed by GC. A gas chromotographic pattern of the reference samples of the trimethylsilylated derivatives of glucose, rhamnose, fucose and uronic acid (glucuronic acid) is shown in Figure 5A. A gas chromotographic pattern of the trimethylsilylated hydrolyzate of the purified bioabsorbent obtained is shown in Figure 5B. As these figures show, the silylated derivative of the sample hydrolyzate of interest was in complete agreement with the silylated derivatives of glucose, rhamnose, fucose, and glucuronic acid.

Figure 5. **Gas chromatography patterns of trimethylsilylated derivatives of standard samples and trimethylsilylated derivatives of the HCl hydrolyzate of the bioabsorbent**

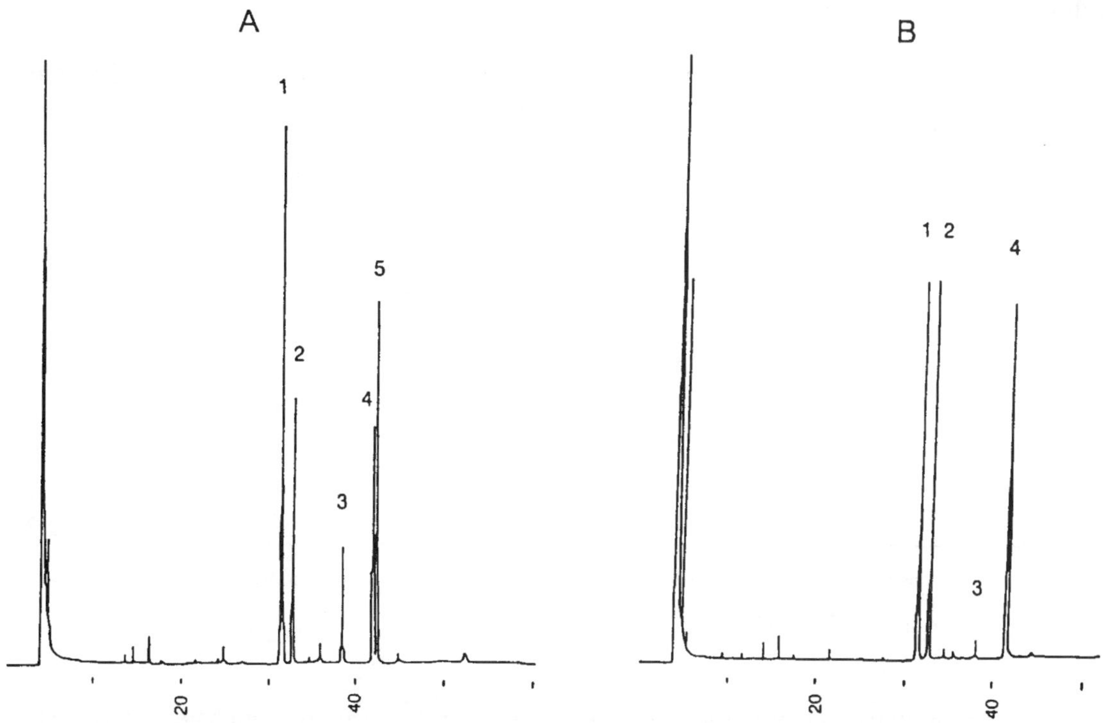

Notes: A: Standard samples. 1) rhamnose; 2) fucose; 3) glucuronic acid; 4) glucose; 5) mannose.
B: Hydrolyzate of the bioabsorbent.

Source: Kurane and Nohata, 1995.

Three peaks that were comparatively large in the gas chromotographic analysis (peak 1: rhamnose; peak 2: fucose; peak 5: glucose) were introduced into a mass spectograph and analysed by GC-MS. These mass spectra show that the fragments of peaks 1, 2, and 5 from the GC analysis are in complete agreement with those of the standard samples of rhamnose, fucose, and glucose, respectively (data not shown).

In summary, the results of TLC, HPLC, GC, and GC-MS showed that the bioabsorbent (high-molecular-weight biopolymer, MW about 5 x 10^9) contained glucose, rhamnose, fucose, and glucuronic acid as the constituent sugars. The low-molecular-weight neutral biopolymer (MW 50 000-100 000) had mannose and fucose (molar ratio 4:3) as the constituent sugars.

Molar ratio of constituent sugars of the high-molecular-weight bioabsorbent

The number of moles of the hydrolyzate after various different hydrolyses are shown in Table 4. The values were estimated by using the HPLC peak areas. Table 4 shows that the molar ratio of the four constituent sugars has many different unities after various weak hydrolyses. For example, the oligosaccharide was still detected after adding 2 M H_2SO_4 at 100 °C for two hours of hydrolysis, and then only two kinds of constituent sugars, i.e. glucose and glucuronic acid, were detected from the rehydrolyzate of that oligosaccharide. As previously mentioned, when the hydrolysis conditions were more severe, the molar ratio value of glucose was larger. However, the molar ratio of fucose and rhamnose remained at 1:1.1. Fucose and rhamnose were not detected from the rehydrolyzate of the oligosaccharide after adding 2 M H_2SO_4 at 100 °C for two hours hydrolysing. Also, we unfortunately detected the remaining oligosaccharide even after using 4 M H_2SO_4 at 100 °C for four hours of hydrolysis.

Table 4. **Molar ratio of constituent sugars of bioabsorbent treated with various different hydrolyses**

Concentration of H_2SO_4	Condition of hydrolysis		Molar ratio of		
(N)	temp (°C)	time (hr)	Fucose	Rhamnose	Glucose
4	90	5 (min)	1	0.1	0.1
4	90	1	1	0.7	0.5
4	90	2	1	0.9	0.9
4	90	4	1	1	1.2
4	100	0.5	1	0.7	0.6
4	100	1	1	1.1	1
4	100	2	1	1.1	1.7
8	100	2	1	1.1	2.1
8	100	4	1	1.1	2.2

Note: The peak areas from HPLC (column; KS801 + SC1011) were calculated and estimated.

Source: Kurane and Nohata, 1995.

From the quantitative analysis for glucuronic acid using the carbazole sulphate method, 21 ppm of glucuronic acid was contained in 100 ppm of the purified bioabsorbent sample. From the L-cystein sulphate procedure (Yamakawa, 1982), 39 ppm Dox-saccharide was contained in 100 ppm of the purified bioabsorbent sample. From the quantitative analysis for glucose using the 5-hydroxytetralon-sulfate method (Fukui, 1969), 36 ppm of glucose was contained in 100 ppm of the purified bioabsorbent sample. Based on these results, the bioabsorbent might consist of glucose (36 per cent), rhamnose (20 per cent), glucuronic acid (21 per cent) and fucose (18 per cent).

Glucose had a maximum absorption at 490 nm when the phenol sulphate method was used during the spectrophotometer scanning, while fucose and glucuronic acid were detected at 480 nm and 483 nm, respectively. By using these differences in spectrophotometric absorption, various mixtures of the four kinds of sugars in the molar ratio range from 1.6 - 2.2:1:1:0.9 (Glc:Rha:GluUA:Fuc) were measured.

The absorption spectrum of the molar ratio (Glc:Rha:GlcUA:Fuc = 1.8:1:1:0.9) was in complete agreement with the absorption spectrum of the purified bioabsorbent sample. This quantitative analysis and the absorption results show that the molar ratio for the bioabsorbent may be as follows: glucose:rhamnose:glucuronic acid:fucose = 36:22:21:20.

From the quantitative analysis for total sugars using the phenol sulphate method, 100 ppm of the purified bioabsorbent contained 97.2 ppm of total sugars. This value (97.2 ppm) is in good agreement with the 96.1 ppm for the total values from each sugar's quantitative analysis. This purified bioabsorbent contains 2.8 per cent water and also a very small amount of trace ash.

These results indicate that the molar ratio of the constituent sugars may be concluded to be as follows, with fucose taken as unity: glucose:rhamnose:glucuronic acid:fucose = 1.8:1.1:1:1.

To recognise the super absorbency (maximum 2 000 fold of water per its own weight), the polysaccharide bioabsorbent was treated with 88 per cent formic acid at 90 °C for one minute in order to release the fucosyl residues. After a mild weak hydrolysis of the bioabsorbent, the shape of the gel type of the bioabsorbent which absorbed water was immediately destroyed, and the viscosity was also rapidly decreased. The sample was dried after this mild weak hydrolysis, but its water absorption capacity was reduced. Moreover, only fucose was detected as a released monosaccharide after this mild weak hydrolysis, and the released fucose molar ratio was almost the same as for the total constituent sugars. This bioabsorbent contained a high ratio of fucose (about 20 per cent ratio of total sugar). Fucose has not been found in polysaccharides such as wheat starch from plants on land, but can easily be found in seaweed (marine algae) in which fucose is one of the main constituent sugars of the very high-molecular-weight polysaccharide. A very small amount of dried seaweed is able to absorb a lot of water very quickly.

The molecular weight of this bioabsorbent is huge ($MW = 5 \times 10^9$ calculated from the polysaccharide), as reported in a previous paper (Nohata and Kurane, 1994). The constituent sugars of the low-molecular-weight biopolymer (MW about $4.8\text{-}85 \times 10^4$) were also identified by TLC, HPLC, and GC methods as fucose and mannose (molar ratio 3:4) (data not shown). Therefore, we think that both fucose as a constituent component sugar and the very high molecular weight including its structure might contribute to the water absorption capacity of this bioabsorbent.

Application of the new super-bioabsorbent to greening of arid soil (Nohata and Kurane, 1996)

Highly absorbent biopolymers, on which great hopes are being placed to prevent desertification, is also applied for greening, by using sandy arid soil. Because of the greater absorbency, plants can grow more easily when seedings have their roots coated with biopolymer before they are planted. Biopolymer is made from a bacterium called *Alcaligenes latus*, and since it is more absorbent than synthetic polymer and more easily decomposed, it causes no damage to the soil.

A small amount of bioabsorbent was mixed with sandy arid soil (50 mg bioabsorbent / 1 litre soil). A very small amount of water was supplied. We have excellent results.

The bioabsorbent had excellent characteristics of water retention capacity through sand. In fact, when sand was sprinkled with water, more of the added water was lost to the underground than to the atmosphere. The osmotic water became a capillary phenomenon underground, and the salt was drawn up to the surface of the soil, and thus the concentration of salt on the surface of the soil increased. When the bioabsorbent is mixed with sand, the supplied water is retained near the surface of the soil, and thus we expect this bioabsorbent to be used to help promote greening of desert areas.

Also, the bioabsorbent showed the most advantageous physical characteristics which maintained excellent water retention capacity even at 70 °C, which other synthetic and natural polymers did not have, when we imagine the use of this bioabsorbent as an environmentally friendly water retention agent for irrigation of desert soil. Another area of global ecological concern is the increasing expansion of the world's deserts, a serious environmental protection problem, since this area is reportedly expanding by about 60 000 km^2 each year. In some countries, soil aridity is being prevented and desert greening experiments are being performed using synthetic high-molecular-weight polymer absorbent materials (Touyama, 1990*b*). Two major disadvantages of this technique are apprehension that these virtually non-biodegradable synthetic high-molecular-weight polymer materials will move from the soil to rivers, and the increase in the concentration of salt on the surface of the soil due to a capillary phenomenon.

Thus contributing to this bioabsorbent was believed to be an attractive alternative to the prevention of desertification, since the roots of plant seeding can be coated with biodegradable material like the bioabsorbent prior to planting. To solve these ecological problems, the development of biodegradable alternatives, which are more compatible with the earth's environment, is urgently needed. The bioabsorbent is not associated with the problems of conventional synthetic high-molecular weight polymer water absorbents because it is easily biodegradable, and therefore does not pose a secondary environmental pollution hazard.

Conclusions and prospects

Environmental impacts occur at all stages of a product's life cycle. Design can be employed to reduce these impacts by changing the amount and type of materials used in the product, by creating more efficiently designed products, by reducing material consumed during use, and by improving recovery materials.

Biopolymer(s) produced by micro-organism(s) is composed of polysaccharide (or protein) and is completely biodegradable. Therefore, it is one of the safer, more ideal and promising "green products" to preserve environmental quality for the next generation. For biotechnology R&D concentrated on "bioprevention", the "green design concept" will play an important role.

Questions, comments and answers

Q: Will the plant take the water from the bioabsorbent saturated with water or from added water?

A: Plant roots can take water from the absorbent. This is not the case with synthetic polymers, which can take water from roots.

Q: Can *Alcaligenes* grow in wastewater? What kind of substrates do you need?

A: So far, we have only used a complete defined medium.

Q: What are the comparable costs of these absorbents?

A: They are not yet commercialised but some companies are interested. They are now at pilot plant scale. The cost may be ¥ 2 000-3 000/kg dry weight ($ 20-30).

Q: Since the absorbent is biodegradable, will it remain in the desert long enough to be useful?

A: We believe that will not be a problem.

REFERENCES

FAO/UNEP (1984), "Provisional Methodology for Assessment and Mapping of Desertification", *FAO/UNEP* 80.

FUKUI, S. (1969), "Quantitative analysis method for reduced sugar", *Seikagaku Jiken-Hou* 1, pp. 59-66 (in Japanese).

KURANE, R. and Y. NOHATA (1991), "Microbial flocculation of waste fluids and oil emulsion by bioflocculant from *Alcaligenes latus*", *Agric. Biol. Chem.* 55, pp. 1 127-1 129.

KURANE, R. and Y. NOHATA (1994), "A new water-absorbing polysaccharide from *Alcaligenes latus*", *Biosci. Biotech. Biochem.* 58, pp. 235-238.

KURANE, R. and Y. NOHATA (1995), "Identification of constituent sugars of polysaccharide bioabsorbent from *Alcaligenes latus*", *Biosci. Biotech. Biochem.* 59, pp. 908-911.

KURANE, R. and H. MITA (1996), "Simple and cost-effective for purification of biopolymer from highly viscous *Alcaligenes latus* culture broth", *J. Ferment. Bioeng.* 81, pp. 90-92.

NOHATA, Y. and R. KURANE (1994), "Culture conditions for production and purification of bioabsorbent from *Alcaligenes latus* B-16", *J. Ferment. Bioeng.* 77, pp. 390-393.

NOHATA, Y. and R. KURANE (1996), "Physical characteristics of the acid biopolymer (bioabsorbent) from *Alcaligenes latus* B-16", *J. Ferment. Bioeng.* 82, pp 22-27.

TOUYAMA, M. (1990*a*), "Greening by synthetic polymer-absorbent", *Zousuigizyutu* 16:2, pp. 34-39 (in Japanese).

TOUYAMA, M. (1990*b*), "Synthetic high polymer absorbent to make greening of arid soil", *Chemistry* 45, pp. 108-110 (in Japanese).

UNEP (1990), "Global Assessment of Desertification: World Atlas of Thematic Indicators of Desertification", Proposal Document 11 February, *DC/PAC, UNEP* 47.

VANHORICK, M. and W. MOENS (1983), "Carcinogen of acrylamide", *Carcinogenesis* 4, pp. 1 459-1 463.

YAMAKAWA, T. (1982), "Chemistry for Carbohydrate", *Seikagaku Jiken Koza* 4, p. 371 (in Japanese).

PRESENTATION OF MEXICAN CASE STUDIES ON BIOTECHNOLOGY AND WATER

INTRODUCTION OF MEXICAN CASE STUDIES

by

Dr. Alberto Jaime Paredes
Subdirector General of the National Water Commission (CNA), Mexico

On behalf of Mr. Guerrero Villalobos, the Director General of the CNA, I should like to extend a warm welcome to you all, which should almost by now be a farewell, but I wanted to do it today in particular for the presentation of these four cases which we shall be looking at concerning case studies in the Republic of Mexico.

I should also like to express my thanks for this opportunity to take part in this workshop with you, since we feel it is of the highest importance on account of the subject it is dealing with, namely biotechnology for the use and conservation of water.

We should like to begin by making a few comments about the water situation in Mexico. Mexico has significant and varied sources of natural resources, which make it viable as a country. Nevertheless, we Mexicans have the greatest difficulty arriving at a correct evaluation, use and exploitation of these resources. One of the most important of the country's natural resources is water, which is unevenly distributed over time and space on our territory. It is scarce in the centre, north and north-eastern parts of the country and extremely abundant in the south-east. In these circumstances, the federal government and a number of institutions have made efforts to produce a reliable supply of water and have taken steps to save water, to re-use it and to control its quality. This last aspect is particularly important, considering that water is a natural resource which for a long time was considered to be renewable. Nowadays, however, owing to the degradation of its quality, it has to be considered practically as a non-renewable resource, which is no longer the basis for the development and conservation of life. Instead, it can become one of the most common ways of spreading epidemic diseases of all kinds.

In the light of the above, we might add some comments on the general situation as regards water in Mexico, which could provide a background reference for the four case studies which are to be presented at this session, namely contamination of the river Lerma, treatment of wastewater in the valley of Mexico, treatment of water in rural areas and contamination of coastal and maritime waters.

Availability of water. Mexico has around 90 million inhabitants and receives an average annual rainfall of 777 mm, which is equivalent to a volume of 1 522 km^3 of water. Twenty-seven per cent of this volume, i.e. around 412 km^3, runs off in surface waterways along the 314 basins which make up the national territory. Another part of the rainfall infiltrates the ground and refills the aquifers, the annual renewal requirements of which are estimated at 48 km^3. The occurrence and distribution of water in the country are also determined by its great variety of climates and natural ecosystems.

Nation-wide, available water resources amount to a volume of approximately 5 125 m^3 annually for every Mexican. This figure may appear excessive if compared with per capita availability in other countries, such as Egypt (30), Saudi Arabia (140), Libya (150), Jordan (190) or Israel (330), even though it represents only half the availability of water in the United States and one twentieth of what it is in Canada. For the country as a whole, 188 km^3 of surface and groundwater are spread over the different uses. Sixty-one per cent of the volume is used to generate hydroelectric power, 33 per cent for agricultural irrigation, 4.5 per cent for supplies of drinking water to the population and 1.3 per cent for industry. The use of water for agricultural purposes is concentrated in the northern and north-eastern states and in the Bajío region. Industrial uses are concentrated in the valley of Mexico, in the basin of the river Lerma, which will be referred to later, and in the north-east of the country, while uses for hydroelectric power generation arise mainly in the south-east. Forty-nine per cent of the volume for urban use is concentrated in the main metropolitan areas, that is, the cities of Mexico, Guadalajara and Monterrey, in other medium-sized towns and in the northern frontier area. Water consumption which does not return to the waterways after use amounts to 53 km^3. Of this almost 90 per cent is attributable to the agricultural sector, six per cent to the industrial sector and five per cent to population use. As everyone knows, hydroelectric power generation consumes practically no water at all. The wastewater generated in the country with a greater or lesser degree of contaminants amounts to less than 20 km^3 per year. The agricultural sector generates 62 per cent of this volume, with residual agro-chemical content which is discharged in the rivers or infiltrates in the aquifers. Industry generates 10.3 per cent, mainly in the form of material of organic origin, with some heavy metals and other toxic substances. 28.5 per cent is dumped in urban sewers and mainly contains substances of organic and bacteriological origin, as well as some toxic substances coming from industrial waste connected to the urban sewerage system.

With regard to the contamination of surface waters, in the studies we have carried out in the 218 largest basins, which cover 80 per cent of the territory, where about 93 per cent of the population lives and 72 per cent of industrial production is located, and which contain practically the whole of the area under irrigation, we have been able to establish a preliminary classification of the country's basins according to the degree of deterioration of the natural quality of the water. According to these studies, 89 per cent of the total organic load measured in terms of biochemical oxygen demand (BOD) is generated in 15 basins. Four basins only, Pánuco, Lerma, San Juan and Balsas, receive 50 per cent of the wastewater discharges, including those of the main townships. To these may be added the basins of the rivers Blanco, Cullacán and Coatzacoalcos, owing to the size and characteristics of the industrial contamination they receive, as well as basins running into the Gulf of Mexico in view of the agrochemicals they receive from agricultural waste.

Deforestation, unsuitable agricultural practices and urbanisation contribute to the degradation and deterioration of soils, which in turn affect the quality of water. One indicator which reflects the deterioration in the quality of water is the infestation of water bodies with aquatic weeds. According to research, almost 68 000 hectares are infested with weeds in 268 of the country's streams and waterways, in addition to 10 000 km of canals and 14 km of ditches in irrigation districts.

The contamination of groundwater has also been studied, and it has been found that the areas with the greatest degradation in the quality of groundwater are the Comarca Lagunera, the Valle de México and other aquifers in urban areas, chiefly in the Bajío region and in the Mezquital river valley. The main sources of contamination are leachates of solid waste, discharges of wastewater outside urban sewers, and insoluble minerals and rock formations. There is also a general problem of diffuse contamination in aquifers underlying agricultural areas, for which a systematic programme of monitoring will need to be developed. The overexploitation of aquifers in coastal areas induces contamination by saline infusion in coastal aquifers, mainly in the peninsulas of Baja California,

Yucatán and the Sonora coast. These problems have been dealt with by prohibiting the extraction of groundwater.

The problem of wastewater discharge may be summarised by saying that most of the contamination affecting the national hydrological network and the environment originates in urban, industrial and agricultural uses. Nevertheless, there is an increasing impact of natural water contamination, mainly affecting groundwater near the coasts by saline infusion and other aquifers through the dissolution of minerals from the rocks where they are situated.

With regard to arsenic, there are two main types of contamination, spot contamination and diffuse or dispersed contamination. The first of these can be pinpointed accurately and can be controlled through specific actions. The second infiltrates generally across extensive areas into aquifers through river banks and the sides of water reservoirs. As there is no concentration point, it is very difficult either to identify the contamination or to control it. It is estimated that there are 33 000 waste discharges in the country which empty into public recipient bodies. Waste disposal is governed by regulations in the form of disposal permits, which indicate specific quality conditions which have to be met, according to the quality of the water where the disposal takes place and the existing and potential uses of the water.

Outlook for control of water contamination

National development is linked simultaneously to environmental objectives and to the desire to eradicate exclusion and poverty. This situation leads on naturally to the concept of sustainable development, that is to say, we have to manage natural resources in such a way that we can ensure the constant satisfaction of human and environmental requirements for present and future generations. For the administration of wastewater, a strategy is being established in principle for the gradual regulation of discharges, with emphasis on three basic criteria:

- firstly, controlling the greatest proportion possible of the total volume of wastewater by regulating major users;
- secondly, controlling discharges which independently of volume produce the greatest impact on recipient bodies; and
- thirdly, controlling the greatest percentage possible of discharges in priority basins and frontier areas.

Technologies applicable in Mexico's case. From the above, it may be seen that there are opportunities in Mexico to use various technologies to preserve water quality, as well as to improve the quality of wastewater discharges. It is worth pointing out that such technologies should be suitable for regions with climates ranging from almost desert to tropical. Out of the near 300 m^3 of wastewater discharged in the country, only some 15 per cent is treated. There are some 700 treatment plants in the country, many out of service. The plants include around 37 anaerobic reactors, which have provided some firsthand experience in biotechnology. It should be borne in mind that it is important to think of or imagine solutions which are not merely an end-of-pipe process, as is the case with treatment plants. We must do more to change industrial processes and agricultural methods and habits, in order to generate and emit fewer contaminants in water, as well as to pre-treat industrial waste dumped in urban sewerage systems. There is no doubt that biotechnological solutions are gaining increasing acceptance, insofar as they fulfil some key requirements: they are competitive with alternative solutions, they are secure, and they are highly reliable to operate.

In conclusion, I should like to say that the improvement of water basins in Mexico should ideally be tackled through joint actions by the public and the private sectors, with tax and economic incentives, based on the principle that the polluter pays and that whoever fails to respect the rules will be penalised. Water strategies and policies try to satisfy environmental requirements, while making the best use of water for agriculture, households, industry, electricity generation, recreation, tourism, aquaculture, fishing and navigation, with a view eventually to facilitating the country's transition to a state of sustainable development.

Mexican Case Studies on Biotechnology and Water

CASE 1: CONTAMINATION OF THE LERMA RIVER

[ABSTRACTS]

LERMA RIVER POLLUTION (TURBIO RIVER/SILVA RESERVOIR) AND ITS SANITATION PROGRAM

by

José Eduardo Mestre Rodríguez
Lerma-Balsas Regional Administration, National Water Commission (CNA), Querétaro, México

The Lerma River basin is located in the central part of the Mexican Republic. It has an extension of 54 421 square kilometres and a population of 9.5 million inhabitants. It is one of the most developed urban-industrial basins in Mexico. However, this growth has resulted in a strong deterioration of the ecosystems, which is significantly reflected in water availability in the basin, in terms of quantity and quality. To deal with this situation, the Federal government, in co-ordination with the State governments and municipalities, began in 1989 an actions program, through a Basin Council which assembles the society in the basin sanitation and recovery.

The present document describes the main features of the basin, and its problems related to water pollution, as well as the overtaking, achievements and difficulties in the implantation of the Sanitation Program.

In the winter of 1994, there was a massive mortality of water birds in the Silva Reservoir, which is located in the northern portion of the state of Guanajuato, south of León city, in the Turbio River sub-basin, in itself a part of the Lerma River basin. This sub-basin is characterised by a strong industrial-urban presence, and a significant pollution degree. This presentation gives special attention to this sub-basin situation, taking into account its characteristics and the significance of the Silva Reservoir event.

TECHNOLOGY FOR THE ANALYSIS AND CONTROL OF POLLUTION IN THE LERMA RIVER AND CHAPALA LAKE, MEXICO

by

Alvaro Muñoz Mendoza
Mexican Institute of Water Technology, Jiutepec, Morelos, Mexico

The Lerma River and Chapala Lake water-basins are introduced by means of their water balance, uses, pollution indicators, and administrative organisation of government institutions in a water-basin council to manage them. The Mexican Institute of Water Technology (IMTA) is presented as the technological arm of the Mexican National Commission of Water (CNA). The technologies used for the analysis and eventual control and remediation of water acquifers are described, as well as their application to the Lerma-Chapala, DeSilva dam-reservoir and Nichupte coastal lagoon in the Mexican Caribbean. Such technologies include numerical modelling of the lake hydrodynamics, identification of stagnation areas, concentration of pollutants, forced circulation, aeration, pollutant transport, the use of the RAISON system for information analysis and management, hydrogeochemistry technology, such as reactions and experimental studies for pollutant migration, including heavy metals, geochemical modelling, the use of biotechnology in various problems, such as weed control, presence of toxicity, acquifer remediation, control of fossil fuel contamination and potential residual effects. Finally, conclusions and recommendations regarding the use of adequate technologies for Mexico are presented, for the rational use and conservation of the water resources.

Mexican Case Studies on Biotechnology and Water

CASE 2: MARINE AND COASTAL WATER CONTAMINATION

[ABSTRACTS]

WATER QUALITY IN MEXICAN COASTAL ZONES

by

Ignacio Castillo Escalante
Comisión Nacional del Agua (National Water Commission), México D.F., Mexico

Water quality results for coastal zones are presented, along with their evaluation, considering recreational use with primary contact (particularly on touristic beaches) and fishing (oysters), with the diagnosis of the actual situation.

Mexico's strategy for improving water quality in coastal zones is presented. The objective is to ensure water quality for recreational uses, either with or without direct contact, and for fishing. Goals that have been reached since its implementation are presented. Regulatory modifications to protect the resources are also included.

The main water quality problems in coastal areas due to wastewater discharges (point and non-point sources) from industry, agriculture, and municipal sources are highlighted, as well as possible solutions.

It is concluded that the main water quality problem in coastal zones is the sanitary quality, and in some specific cases, the presence of agrochemical and organic compounds. In this case, the main cause is the petroleum and petrochemical industry activities.

EFFECTS OF PETROLEUM INDUSTRY ACTIVITIES IN THE SEA SHORE ENVIRONMENT OF SALINA CRUZ, OAXACA, MEXICO

by

M.C. González Macías*, M. Tapia García, M.C. González Lozano*, V.M. García Vázquez*, and J.L. Mondragón***

*SPA, Instituto Mexicano del Petróleo, **Universidad Autónoma Metropolitana - Iztapalapa

This research has as one objective to obtain a global view of the environmental conditions of La Ventosa Bay, identifying the effects in the marine environment due to activities from the petroleum industry, mainly because of the presence of two chemical wastewaters submarine pipe discharge of the Antonio Dovalí Jaime Refinery.

From 1992 to 1988, seven intensive oceanographic campaigns were carried out by the ecology research group of the Mexican Petroleum Institute. The following aspects were studied: environmental conditions, pollutants levels in water column and sediments, as well as the biotic components, analysing plankton and benthos communities, and performing general inference regarding fish communities abundance and distribution patterns.

This environmental base line was updated with yearly monitoring during 1988-1990. Recently, in December 1995 and May 1996, a new field evaluation took place, which allowed the gathering of environmental information in La Ventosa Bay and Estuary, and also in Superior and Inferior Lagoons.

The integration and systematic handling of the information will supply an environmental diagnostic at present natural conditions in a regional approach, including also the levels of petroleum industry related compounds in spatial and temporal patterns, which allows in a regional framework to detect information gaps, propose additional studies and help as a decision tool for the sustainable development of the Tehuantepec Isthmus.

REMEDIATION OF THE BOJORQUEZ LAGOON, MEXICO

by

Ariosto Aguilar
Mexican Institute of Water Technology, Jiutepec, Morelos, Mexico

The Nichupté system, located in Cancun, Quintana Roo, Mexico, is made up of several interconnected lagoons. The Bojorquez lagoon, in the northern part of the system, has been stressed by intensive water use. Morphological changes, such as dredging, filling and wastewater discharges, have modified the natural conditions. The lagoon system can be described as shallow, with little communication between the lagoons due to low water levels and restricted access to the ocean. The water quality has deteriorated and affected the ecosystem. To reverse this situation, the discharge of wastewater must be stopped and the circulation within the system increased. This study will begin with the Bojorquez lagoon, as it is the most severely affected and currently shows eutrophication secondary to the high nutrient levels in the water. To improve circulation and increase the dissolved oxygen levels, IMTA has carried out hydrodynamic studies and field measurements of water quality. Based on this data, forced circulation, aeration and algal harvesting have been proposed. The results of the hydrodynamic study and the simulated short term behaviour of the forced circulation and aeration are presented in this paper for the Bojorquez lagoon during the first stage, and for future application in the entire lagoon system in a later stage.

Mexican Case Studies on Biotechnology and Water

CASE 3: MEXICO VALLEY'S WASTEWATER TREATMENT

[ABSTRACTS]

MEXICO VALLEY'S WASTEWATER TREATMENT

by

A. Capella Viscaino
Comisión Nacional del Agua (National Water Commission), México D.F., Mexico

With the objective of reducing the risk of floods, protecting the health of the population in the irrigated zones, and diminishing the restrictions for the use of residual waters, a sanitation project is being proposed for Mexico Valley's wastewater. This sanitation project basically consists of seven components: 1) rehabilitation of the drainage system; 2) construction of treatment plants; 3) sanitation of irrigated zones; 4) control of discharges; 5) institutional development; 6) sanitary training; and 7) studies, projects and designs.

The expanded project involves, besides Mexico City D.F., other authorities like the State of Hidalgo, the State of Mexico, the DDF, and the CNA, with whom it will establish agreements for the financing and execution of the works.

Among the benefits resulting from the project are the reduction of infectious disease, the diversification of crops and its economic impact, the improvement of water quality of the Tula river, and the amplification in the capacity and functionality of Mexico City's drainage system.

Mexican Case Studies on Biotechnology and Water

CASE 4: TREATMENT OF WATER IN RURAL AREAS

[ABSTRACTS]

TECHNOLOGICAL PACKAGES FOR THE TREATMENT OF NIGHT SOIL AND DOMESTIC WASTEWATER IN RURAL COMMUNITIES

by

Roberto Contreras-Martínez*, Antonio Fernández-Esparza* and José Colli-Misset**
*National Water Commission (CNA), **Mexican Institute of Water Technology, Mexico

In Mexico, 70 per cent of the communities have fewer than 100 inhabitants. These populations are widely separated and often accessed by poor roads, making integral sanitation from centralised drinking water and sewerage networks, as found in urban centres, difficult and costly. The National Water Commission has focused on rural sanitation, and in particular treatment of night soil and wastewater, using on-site treatment systems. Five technological packages were developed for use in nine population ranges under 100 inhabitants. Each package includes information concerning system operation, design criteria, and then functional, hydraulic and structural plans. An itemised list of materials needed for the project, including volumes and unit prices is provided in a computerised worksheet that may be modified to contemplate regional variations. The initial projects available under this program are double chamber ventilated latrine, multrum latrine, septic tank/facultative pond/maturation pond, septic tank/facultative pond/intermittent sand filter, and septic tank/facultative pond/wetlands. The selection of the treatment system to adopt is dictated by physiographic and socio-cultural conditions, and the expected effluent quality is based on re-use considerations.

RURAL WATER SUPPLY AND SANITATION: A CASE STUDY AT "EL TEPEGUAJE", MORELOS, MEXICO

by

José Colli-Misset, Arturo González-Herrera, Lina Cardoso-Vigueros and Lydia Marquez-Bravo
Mexican Institute of Water Technology, Jiutepec, Morelos, México

According to the ninth population census carried out in 1990, 20 per cent of the 81 million inhabitants in the nation do not receive drinking water from networks, and 39 per cent do not have drainage. Much of this population lives in 11 800 rural and underprivileged communities, dispersed throughout the country, and with fewer than 1 000 inhabitants. These towns and villages lack the basic sanitation facilities resulting in feces disposal on soil, and the consequent high incidence of gastrointestinal infections the fourth most common cause of death according to the Health Ministry.

Of the 1 200 000 persons living in the state of Morelos, 45 per cent are not connected to a drainage system and 22 per cent do not receive drinking water from a network. The small communities, where most of these people live, have fewer than 1 000 residents and make up 87 per cent of all towns and villages in the state.

The Mexican Institute of Water Technology proposed a pilot sanitation project for settlements of this size. The site chosen was El Tepeguaje, Tepalcingo, Morelos.

The project included tests of sanitation technology designed to protect the fresh water supply sources, and treat the water for household consumption on-site and the nightsoil produced. The technological alternatives contemplated in the project were presented to the community leaders and heads of household using graphic and audio-visual material and prototypes. Sixteen families agreed to participate in the project and jointly selected the technology they thought was most appropriate. The methods accepted included hewn sandstone filters for water purification, two-chamber dry latrines for excrement and cement-covered rims around the well shafts used to supply drinking water. The community obtained a greater per head supply of water as a result of the protection and rehabilitation given to the springs, improved bacteriological water quality with removal of up to 90 per cent of the organisms, community acceptance and participation in the construction of the latrines and the wooden frames for the sandstone filters, and government financing to continue the project. These results illustrate the benefits that may be obtained through integral sanitation projects where the communities to be served participate in the planning, adaptation and development of technologies.

CLOSING SESSION

MAIN CONCLUSIONS AND RECOMMENDATIONS

by

Dr. Adalberto Noyola
Instituto de Ingeniería, Universidad Nacional Autónoma de México, Mexico

Biotechnology provides a great variety of tools, which in combination with other physical and chemical methods, are better suited to meeting water quality and human health requirements. The improvement in public health standards is a priority for the greater part of the world's population. Advances in biotechnology are bringing us more and better techniques for the detection, identification, monitoring, modelling, prediction and evaluation of the risks of both existing and emerging diseases transmitted by water.

Since biotechnology is applied within a comprehensive, interdisciplinary framework, it offers better chances of successfully resolving problems related to contamination and public health. As a matter of high priority, therefore, suitable individuals need to be given the necessary training in the public, private and academic sectors. The public sector should be informed about areas of biotechnology which receive limited attention in the media but which are currently making good progress, as is the case with biotechnological applications in bioremediation and public health.

Governments should pay particular attention to the role played by environmental biotechnology for the protection of human health. The public must be informed and must participate if biotechnology is to be accepted and used, especially when genetically modified organisms are involved.

Environmental biotechnology will make more rapid progress if a uniform framework of rules is adopted world-wide. This recommendation should be given the highest priority. The immediate benefits of biotechnology stem mainly from its contribution to sustainable, environment-friendly technologies, which not only imply low cost, but which can also be applied in the field.

New urban developments must introduce different ways of handling water, taking account of health protection, the prevention of contamination and the recovery of resources, water and nutrients within a framework of sustainability.

Some of the more specific conclusions and recommendations would be as follows:

In marginal urban areas, human excreta constitute a decisive vector for the propagation of gastrointestinal diseases. One option which could solve this problem and which deserves particular attention owing to its simplicity is the dry latrine. More research and development is required, however, into methods of disposing of excreta which are efficient, cheap and well accepted by users.

The traditional aerobic treatment of wastewater produces great quantities of residual sludge. There is therefore a need to increase the use of anaerobic treatments, one of the advantages of which is to produce less sludge. This is particularly applicable to industrial effluents containing concentrated organic matter, but also for residual urban waters in countries with a warm climate and a residual water temperature above 20 °C. Biosolids, which are stabilised residues of wastewater treatment, can be used in agriculture, although this may present a number of difficulties. The biological treatment of wastewater produces less sludge than traditional physical and chemical processes and therefore presents certain advantages in this respect. In order to make better use of these biosolids in agriculture, there are two precautions which should be taken. Methods should be used to prevent the entry of metals and toxic substances into wastewater, thereby avoiding their accumulation in the biosolids produced, and suitable methods must be introduced to improve the sanitary quality of the biosolids.

Finally, and in more general terms, we would like to emphasize, or our group would like to emphasize, that in order to advance towards sustainability, governments must without delay support the application of biotechnology for the prevention of contamination and the recovery of resources, and in the longer term they must promote the development of cleaner biotechnologies.

These are the conclusions and recommendations that were reached at the end of our two days of intensive work and which we consider may in some manner be applied to the cases we shall be hearing about, and which will undoubtedly give rise to discussion when compared with reality.

CONCLUDING REMARKS

by

Bill L. Long
Director, Environment Directorate, OECD

I have the honour of presenting these closing remarks on behalf of the OECD.

First, I want to thank Environment Minister Carabias for being with us this afternoon. I know she made a special effort to be here -- and we look forward to her remarks.

I recall, Madam Minister, your participation in the OECD Environmental Ministerial last February in Paris. You emphasized to other Ministers the importance of strong science and technology to the pursuit of sustainable development -- and also challenged the OECD to think beyond exotic environmental problems of the upper atmosphere, and of pollution problems of the OECD cities, to include the day-to-day environmental conditions of the poor people of developing countries and of rural areas. You reminded Ministers that millions of people are involved in "survival agriculture".

I believe, Madam Minister, you will thus be pleased with the scope and conclusion of this workshop -- with the attention given to biotechnology as a contributor to the solution of water resources management problems faced by rich and poor in both developed and developing countries.

As I listened to the papers and discussions, from my perspective as the OECD's Environment Director, I was struck by several recurrent points:

1) First, the extent of public health threats from both traditional and newer sources of waterborne disease -- pathogens and others -- is striking. This is certainly not fully appreciated by most policy-makers and citizens.

2) Biotechnology is already contributing to improved water use and conservation, and its potential is great. And, the technological breakthroughs that the experts at this workshop envision as having potential for solving environmental problems can apply well beyond the water management field. There are multiple benefits from biotechnology for the environment that should be better delineated and articulated to policy-makers.

3) While water conservation is a high priority, the rush to recycle and re-use this resource raises some important water quality issues, for example in food processing. This requires further study and attention.

4) The workshop reinforced concern that has been growing about the link between climate change -- either natural or man-induced -- and the transmission of waterborne diseases.

Madam Minister, Distinguished Director of CONACYT, I hope you are pleased by the quality of this workshop. We from the OECD certainly are:

- The papers and discussions were very rich in information, and of very high quality;
- We were able to attract outstanding expertise -- from within Mexico and abroad;
- These wonderful surroundings in Cocoyoc contributed to building personal relationships that will no doubt continue beyond this event; and
- We learned about the special problems, interests and approaches of our host country regarding biotechnology and water problems.

These were major objectives for this Workshop, and I believe they have been achieved.

The conclusions and recommendations of this workshop -- as presented here this afternoon -- are very important. Our plans at the OECD for disseminating and addressing them are the following:

- We will publish a proceedings, as we did with the Tokyo and Amsterdam workshops. Collectively these documents will provide an excellent overview of the state of knowledge, and opportunities and needs for the future.
- The OECD Secretariat will also report on this meeting in Sydney, Australia, in February -- at a conference on "Sustainable Water Consumption", which will bring together officials and experts largely from Asia and the Pacific (developed and developing countries). We will send each of you copies of the conclusions of that event.
- The conclusions and recommendations will be examined by the OECD Working Party on Biotechnology to help define the OECD's future work. It is very useful and important in that regard that Dr. David Harper, Chair of the Working Party, was here throughout.
- We will also assess back in Paris -- my Directorate and DSTI -- what we can do to address some of the issues and goals identified in this workshop on water resources management in its broadest dimensions.
- Finally, we hope that the Mexican authorities will evaluate the results, and determine what they might mean for the solution of water problems in this country -- over both the near and longer terms.

It is unfortunate that Risaburo Nezu, DSTI's Director, had to leave early, to go to Korea (which has just been invited to become the OECD's 29th Member). He should be conveying this message, because it was he and his Directorate that provided the leadership in the OECD. We have had with us others from that Directorate -- Michael Oborne, Salomon Wald, and other DSTI staff -- and I want to acknowledge their contribution.

To our Mexican hosts, I would say:

We owe a special debt of gratitude to the Minister of Education, Dr. Limon Rojas, and Environment Minister Carabias for their support. Similarly, we have excellent contributions and assistance from the officials and experts of the National Water Commission and other Mexican institutions.

Special appreciation goes to out co-sponsor, CONACYT and its Director, Dr. Carlos Bazdresch. Their care, effort, quality of work and attention to detail is symbolised by these outstanding logos (signs) that I have looked at with admiration for the past four days.

And, we extend the OECD's gratitude to CONACYT officials and staff for their intellectual and administrative support to the meeting secretariat; ushers; interpreters; people who handled travel; and all others who performed such exceptional work.

Finally, I know I can speak for all foreign participants in thanking our Mexican hosts for the excellent planning and support we enjoyed.

I am particularly pleased that so many from abroad were able to experience for the first time the warmth and outstanding hospitality of Mexican society that others of us have been typically exposed to in the past -- and also to sample the excellent scientific and technological expertise that resides in one of the OECD's newest Member countries.

So, Minister Carabias and Dr. Carlos Bazdresch, Director of CONACYT, I wish to convey to you, and through you, to the others who are responsible for this most successful event, the OECD's deepest appreciation and gratitude.

Thank you.

CLOSING ADDRESS

by

Carlos Bazdresch Parada
Director General of Conacyt, Mexico

It is an honour for me to be with you once again at this closing ceremony of the Workshop on Biotechnology for Water Use and Conservation, organised jointly by the Organisation for Economic Co-operation and Development and the National Council of Science and Technology.

Like all other countries in the world, Mexico is aware of the importance of water conservation methods at the present time. Our government was therefore particularly pleased that our country should be chosen as the venue for this important event. In co-ordination with the OECD, through its Directorate for Science, Technology and Industry, the Mexico Workshop provided a follow-up to the two previous meetings. As you will remember, technological applications were reviewed in Tokyo in 1994, and subsequently in Amsterdam, the focus shifted to soil and air analysis. We trust that the recommendations produced within the framework of this workshop will assist with the co-ordination of policies for improving the quality of water and the rational use of this indispensable resource.

Considering that discussions here were conducted at such a high scientific and technical level, I feel sure, moreover, that the next workshop, to be held in Sydney in 1997, will start from a solid basis to discuss another theme of global interest, namely sustainable water consumption.

We were very satisfied at the quality and dedication of participants, who included 150 Mexicans and 65 foreigners, as well as a group of representatives of different states of the Republic, all of whom are officials involved in decision-making related to water. I hope that their attendance here will lead to the application of the biotechnologies we have discussed in state drinking water and sewerage systems.

Through its research programmes, CONACYT has decided to encourage further thought on national problems affecting major sectors of the population. Through the agreements we have signed with the Ministries of Health, the Environment and Agriculture, we have tried to ensure that the most competent national and international researchers will be producing solutions to our problems, and in that way, will make an effective contribution to the development of our country. With these instruments, we believe that we shall be able to support our researchers and experts in the areas of water and health, water and sustainability, and the distribution, re-use and conservation of water for agricultural uses.

Among its other purposes, the workshop has served to exhibit new technologies, which were not necessarily unknown in Mexico, but for which more specific knowledge regarding their introduction, and the ways in which they are currently marketed, will have encouraged more reflection regarding their applicability in Mexico. The challenge now is for the scientific community, the civil service and the private sector to select and adapt the most appropriate technologies for their environments.

Confirming CONACYT's interest in this matter, I am pleased to announce that in the coming months, three or four regional workshops are to be held in Mexico for the purpose of disseminating and implementing the proposals we have considered here, and we are hopeful that as many of you as possible will accept the invitation which you will be receiving in due course.

To sum up, this workshop has facilitated an international exchange between specialists of 14 countries regarding their research and development for bioremediation and the biotreatment of marine, coastal and surface waters. In addition, we have discussed techniques for combating the contamination of aquifers caused by agricultural and industrial activities, and by human settlements. As a result, the proceedings of this workshop which will be published shortly will represent a fairly complete compendium of the state-of-the-art in these areas.

We hope that this compendium, which will be very useful, will give a complete review of the activity developed at this workshop.

Lastly, I should like to emphasize that as links are strengthened between the academic world, industry and governmental agencies belonging to the three levels of government, it will become possible to improve and increase the treatment of wastewater, with all the benefits that represents for the country and for its future generations.

Ladies and gentlemen, we depend on water for our personal, public and environmental lives. We depend on our scientific knowledge to keep it clean. More than that, we now have the scientific and technological capacity to avoid further contamination of water, while it is our duty to ensure that the seas, coastlines, rivers, lakes and estuaries remain clean for future generations.

The lack of optimum co-ordination has proved an obstacle to the application of technologies. Research institutes have not met with sufficient co-operation among their counterparts in the public and private sectors. Industry requires a technology made to its own measure. That is to say that industry must invest in order to maximise its resources and to contribute to the protection of the environment. Biotechnology offers us the possibility of applying a sustainable approach to the treatment of water, and its use is therefore particularly recommendable.

This Workshop has taken some significant steps forward towards the solution of water problems. From now on, the recommendations issued at Cocoyoc will have to be taken into consideration for the purpose of improving water quality in all our countries.

Thank you very much.

CLOSING ADDRESS

by

Julia Carabias Lillo
Minister for the Environment, Natural Resources and Fisheries, Mexico

It is a great honour for me to preside at this Workshop on Biotechnology for Water Use and Conservation. The combined efforts of academic specialists, water users, industrialists and the expert authorities of various countries have provided us with an opportunity to localise this problem and find out what recommendations and decisions are required in order to ensure a more appropriate and efficient use of water. We have also had the opportunity to bring the scientific community closer to decision-making, to obtain a more useful contribution from current global scientific research, and to take the necessary steps to ensure co-operation, on both a national and an international level. This has been the essence of this workshop, in terms of our discussions regarding the biological quality of water and public health, the prevention of contamination, bioremediation and the biotreatment of aquifers, surface waters, marine and coastal waters, as well as desertification and environmental change.

It is a great honour for Mexico to have hosted this workshop, which is the third of a series of three workshops. The first was in Tokyo in 1994. That was followed by another in Amsterdam in 1995, and then by this one in Mexico. We may on this occasion confirm our satisfaction as Mexicans with our membership in the OECD, since it has given us the opportunity to organise a series of technical and scientific meetings, which are helping us to analyse current problems and to establish institutional strategies and government mechanisms of public management, at both national and international level.

This is one of the central purposes of the organisation, which brings together the experience of developed countries. Our own country, as a new member, is very rapidly beginning to benefit from its accumulated experience.

On several occasions, we have already worked in the OECD in areas related to our Ministry, namely forestry, fisheries, the environment and now water. We have made significant advances, and we are convinced that without the support and participation of the OECD we would have taken much more time to achieve the objectives we had set for ourselves. The possibility of developing a national strategy for sustainability in the use of natural resources has benefited greatly from our membership in this organisation, a fact which I would like to recognise, while expressing my heartfelt thanks to Bill Long for all the support he has been providing now for months, and for helping us to make the most of our membership in the organisation.

Whereas we might have expected, as a new member of the OECD, to have been free of obligations and rather able to benefit from the organisation, I believe it has been quite the opposite, and in this sense, I should like to reaffirm our commitment to work to the best of our ability.

I should like to take a little of your time to express some of our concerns regarding the problem of water. Of course I realise that the workshop is coming to an end, but I should like to let you know how and why a workshop of this kind contributes to our national priorities in the handling of water, and what the main problems are that we are currently facing.

As everywhere in the world, the availability of water is a serious problem in our country. From a planetary standpoint, although three quarters of the earth is covered by water, only one per cent of that water is available for use. This is really a tiny part, and when we see how unevenly distributed water is across the world, we realise what enormous limitations it presents for the development of humankind.

Mexico's position is an intermediate one. Fortunately, the country receives a quantity of rainfall each year, averaging some 770 millimetres, which is sufficient to avoid being classified as an extremely dry country. If we compare ourselves to countries in Africa and Asia, we find that we enjoy nearly 5 000 cubic metres per inhabitant, whereas some of the African and Asian countries make do with some 30 cubic metres. Our situation may be said to be fairly satisfactory, even though when we compare it to some other countries such as Canada, which uses 20 times more than we do, it does not look so good. This is even more apparent when we find that the distribution of the resource within the country is very uneven, with 80 per cent of the territory receiving only 20 per cent of the water, and the remaining 20 per cent receiving 80 per cent of the water, while in that remaining 20 per cent of the territory there is only 20 per cent of the population and of productive activity. Most of our production in fact is located in the drier areas of our country, which gives rise to some very real problems of availability, one of the worst being the distribution of water.

In the greater part of our country during the year, except in the most humid areas of the south-east, water is more plentiful in periods of rain, leading in fact to severe flooding and surface drainage.

For the purposes of the country's development, then, how can we ensure that we have enough water so that it does not become a limiting factor, either for health, or for production, or for the economy? One of our requirements is obviously the construction of infrastructures, for irrigation, for drinking water, for sewerage and for dams. Mexico occupies the seventh place in the world in terms of infrastructure. Although it is well equipped in that respect, we still have a number of infrastructure problems.

This takes me to my second consideration, namely the question of efficiency. We have 2 270 dams, and we have a huge agricultural irrigation network in the country, but we still lose 50 per cent of available agricultural water through leaks, while the same figure for towns is 40 per cent. This is unacceptable. We have few resources; they are badly distributed; we have difficulty getting hold of them, and when we do, we lose them through poor handling of this inadequate infrastructure. We have an enormous amount of ground to catch up. There are 12 million Mexicans today who have no drinking water, despite the fact that this is an OECD Member country, and we have 27 million Mexicans who have no sewerage. Only 10 per cent of our waters are correctly treated, while the remainder are discharged and contaminate our country's main basins. The 31 largest of these have very serious contamination problems.

The third problem we have is that in regions where there is little water, this has to be obtained from somewhere, and it tends to be taken from underground aquifers and water tables, which means that we are having serious problems of over-exploitation in areas of the high plateau and in the desert-like parts of our country. The consequences are the infiltration of salts and the over-exploitation of aquifers.

The fourth problem is the quality of water, only 10 per cent of which, as I said, is treated. We are using the waters of some basins such as the river Lerma, which receives the bulk of industrial waste. We have to make enormous efforts to try to catch up on our infrastructure in order to provide drinking water and sewerage for the community and catch up on sanitation.

You have heard about some major programmes at this workshop, such as the clean-up of the Mexico Valley basin, which will cost millions of dollars and will be expensive for a country which is already very indebted. We must do something about it. We cannot simply be one of the largest cities in the world and be dumping everything up towards the irrigation areas in the north-eastern part of the country. It is equally unacceptable that we should be the largest city in our country in economic terms, and that we are not investing resources to clean its water. We should be ashamed as a federal district, and we have to tackle the problem.

Another very important project is the clean-up of the Lerma-Chapala basin, where the greater part of industry is located. As you have been told, a great deal of planning has been done, as that seems to us to be the only way of making progress through all the basin councils involved. It is a common saying that you cannot sweep steps from the bottom step upwards. The same applies to water, in the sense that it is no use cleaning up the water downstream and not upstream. So we have to organise a proper planning structure to regulate the use and re-use and cleaning of water all along the basin, which requires a combined effort by many states of the Republic. These states have economic and social dynamics which differ considerably; some of them are poor and rural, while others have a very high gross domestic product, and are basically industrialised, thereby having a completely different relationship with water. There has to be agreement among federal, state and local authorities, and among agricultural, industrial and domestic users. Reaching agreement throughout the whole of the basin is a very complex task, but we are convinced that it is the only way we can resolve the matter and avoid what Dr. Noyola was telling us about, namely the danger that water can give rise to conflict within countries and between countries. Like few other natural resources, water is an indispensable resource in scarce supply and can generate powerful social and political conflict in our countries. We have to deal with the problem, and find a solution.

Another very important project which I would like to mention is the effort which is being made on the northern frontier of our country, which we share with the United States. There, a programme known as frontier 21 is looking into the infrastructure requirements for environmental improvement in terms of water, air and solid waste. A significant planning effort is being made in the case of water and a large economic investment is being made to meet urban demands. Much joint work has been undertaken, with academic institutions, such as the COLEF in particular, with non-governmental organisations, and with the local, state and federal governments of both countries. We are hoping that this programme 21 will provide us with a model of how to deal with the problem on our northern frontier.

How can biotechnology help us in these major challenges? With regard to availability of water, I am not sure. Perhaps not much, but you are the ones who should know, and I can only feel sorry that I was unable to listen in during the three days of the workshop, as I would have been delighted to be with you and to hear what you were saying. These are luxuries which I cannot afford right now, however, but I still hope that one day I shall be able to return to workshops as a listener. Where there seems to be no doubt is that in terms of quality, in terms of the possibility of bioremediation, and in terms of providing much better access for the use of some bodies of water by communities which have difficulty with their drinking water supplies, biotechnology can provide a valuable solution. It should also contribute to solving other problems of a technical nature, which should be dealt with technically, before they become political, social, health or ecological problems. Knowing which technology is best is not easy, but you are the people in charge. We shall also have to try to find the answer in biotechnology to all the questions which are now emerging. It is still very complicated and I think we are only at the beginning of this process.

To conclude, I should like to refer very briefly to three problems which I consider important and which we should study and not lose sight of in the development of biotechnology. One is the regulatory framework, which you have discussed. To the extent that these rules and regulatory frameworks are clearer, more homogeneous and more universal, we can achieve better ways of making common use of water. This is a very complex issue, which comes up in all discussions and all the forums. As soon as access to technology is involved, problems arise with regard to patents and rights and hence the use of technology. The questions are who pays for the technology, who develops it, and to whom do the micro-organisms used belong. The micro-organisms and organisms are universal, and technology is only a part; the question is also who buys the micro-organism and how we can have access to all that technology.

Even though clear rules are available, it is not always possible to reflect the social inequalities which we experience. I feel that the economic and social inequalities in our countries can be an important factor in the use of technology. We have to improve this situation and especially the technical aspect. We have to make every effort to ensure that it does not become a political problem.

Also with regard to biotechnology, another important risk to avoid, in the same connection, is that it should become a weapon, a new power of biotechnology, which may become an instrument for controlling the handling of water, since there are countries with access to it and countries which are dependent on it. For this reason, we are really pleased that this workshop has been held in Mexico. CONACYT organised the workshop in order to provide an opportunity for the universities and for our academicians to have access to this type of discussion, and so that we may continue to support research in biotechnology. We must ensure that Mexico does not fall behind or does not fall further behind, because I think we are already lagging, and for this reason I am pleased that we are tackling this problem in our country. We must find a way to develop our own technologies and participate in the biotechnology which is being generated in the world.

The third problem is how to ensure that this activity in biotechnology is turned into concrete, specific projects, which will benefit those who most need these technical solutions, those who have the most water problems, those who are paying most for it today, and are living in worse conditions of health. The major diseases which we have in our countries and in the third world are due basically to the quality of water, and this is really a problem which can be solved, provided that we have the necessary infrastructure and technology. This technology is expensive; it has been developed for years and will continue excluding large layers of the population and many countries. Biotechnology can provide an alternative solution for rural communities which can really have access to it, but this must be translated into decisions. We represent the authority at the highest level in this country, but the matter does not depend only on us. The problem of the handling of water is very much involved with many sectors of society, such as the 31 and soon to be 32 governors, and over 2 500 local authorities, sectors working in agriculture, energy, industry, households, and academic circles; these are all players who are involved in the decisions which have to be taken before practical action is possible, and before any specific programmes can be established. Here, I feel that the academic contribution can be significant in terms of spreading and generalising these solutions. I think we must not only focus on this matter from the point of view of sustainability, and not only from the environmental point of view, but we must also realise that this is a problem of development; it is a problem of catering to the needs of the whole world's population. In our own country in particular, as I said, we have fallen behind considerably in various ways, and if we are to improve our quality of living then we must have a better quality of water. You are yourselves aware of much of the solution, and we may only hope that we will have the opportunity to put it into practice.

I should like to congratulate all the organisers and all those who have been directly involved in this workshop, and I hope that its results will be really positive for all our countries.

Thank you very much for your attention.

At 5:30 p.m. on 23 October, I declare the Workshop on Biotechnology for Water Use and Conservation formally closed, and I extend my congratulations to everyone and my thanks to the OECD and to CONACYT.

LIST OF PARTICIPANTS

LIST OF PARTICIPANTS

CONFERENCE CHAIR

Mr. Miguel Limón ROJAS
Secretario de Educación Pública
Secretaría de Educación Pública, SEP
Argentina 28, 2do. Piso. Oficina 3011
Col. Centro, Del. Cuauhtémoc
06029 México D.F.
MEXICO

Tel: (52 5) 10 89 17; 21 66 90
Fax: (52 3) 29 68 73; 29 68 76

SCIENTIFIC CHAIR

Mr. Francisco BOLÍVAR Zapata
Director General
Instituto de Biotecnología de la UNAM
Av. Universidad 2001
Col. Chamilpa
Cuernavaca, Morelos
MEXICO

Tel: (52) 91 73 17 2399
Fax: (52) 91 73 17 2388

DEPUTY SCIENTIFIC CHAIRS

Mr. Adalberto NOYOLA
Coordinador de Bioprocesos Ambientales
Instituto de Ingeniería, UNAM
Circuito Interior
Edificio 5 de Hidráulica 2o. Piso
Ciudad Universitaria
04510 México D.F.
MEXICO

Tel: (52 5) 622 3324
Fax: (52 5) 616 2164

Mr. David HARPER
Chief Scientific Officer
Department of Health
Skipton House, Room 539B
80 London Road
London SE1 6LW
UNITED KINGDOM

Tel: (44 171) 972 5353
Fax: (44 171) 972 5155

CO-CHAIRS OF PARALLEL SESSIONS

Ms. Rita COLWELL
President
University of Maryland Biotechnology Institute
4321 Hartwick Road, Suite 550
College Park, MD 20740
UNITED STATES

Tel: (1 301) 403 0501
Fax: (1 301) 454 8123
Email: colwell@umbimail.umd.edu

Mr. Shigetoh MIYACHI
Executive Managing Director
Marine Biotechnology Institute Co., Ltd.
1-28-10 Hongo, Bunkyo-ku
Tokyo 113
JAPAN

Tel: (81 3) 5684 6211
Fax: (81 3) 5684 6200
Email: miyachit@super.win.or.jp

Mr. Juan L. RAMOS
Investigador Científico
Department of Plant Biochemistry, C.S.I.C.
Profesor Albareda 1, APDO 419
18008 Granada
SPAIN

Tel: (34) 581 21011
Fax: (34) 581 29600
Email: j.l.ramos@samba.cnb.uam.es

Mr. Willy VERSTRAETE
Laboratory of Microbial Ecology
Faculty of Agricultural & Applied Biological Sciences
Gent University
Coupure Links 653
9000 Gent
BELGIUM

Tel: (32 92) 64 59 76
Fax: (32 92) 64 62 48
Email: willy.verstraete@rug.ac.be

RAPPORTEURS

Mr. Mike H. GRIFFITHS
Mike Griffiths Associates
The Pantiles
Ivy Lane, Woking
Surrey GU22 7BY
UNITED KINGDOM

Tel: (44) 1483 767 818
Fax: (44) 1483 756 136
Email: 100044.2525@compuserve.com

Ms. Debbie POOLE
Biotechnology and Biological Sciences Research Council (BBSRC)
Chemicals and Pharmaceuticals Directorate
Polaris House, North Star Avenue
Swindon SN2 1UH
UNITED KINGDOM

Tel: (44) 17 93 414 656
Fax: (44) 17 93 414 674

OPENING AND CLOSING SESSION SPEAKERS

Mr. Carlos BAZDRESCH Parada
Director General of CONACYT
National Council of Science and Technology
Constituyentes No. 1046, 3er Piso
Col. Lomas Altas
Delegación Miguel Hidalgo
11950 México D.F.
MEXICO
Tel: (52 5) 327 75 75
Fax: (52 5) 327 76 09

Mr. Francisco BOLÍVAR Zapata
see under Scientific Chair

Ms. Julia CARABIAS Lillo
Secretaría del Medio Ambiente
Recursos Naturales y Pesca
Periférico Sur 4209 6o. Piso
Co. Fracc. Jardines en la Montaña
14210 México D.F.
MEXICO
Tel: (52 5) 628 0602
Fax: (52 5) 628 0643

Mr. Robert J. HUGGETT
Assistant Administrator for Research and Development
U.S. Environmental Protection Agency
401 M Street SW
Washington, DC 20460
UNITED STATES
Tel: (1 202) 260 7676
Fax: (1 202) 260 9761

Mr. Bill LONG
Director, Environment Directorate
OECD
15, Boulevard Amiral Bruix
75016 Paris Cedex 16
FRANCE
Tel: (33 1) 45 24 93 00
Fax: (33 1) 45 24 78 76
Email: bill.long@oecd.org

Mr. Risaburo NEZU
Director, Directorate for Science, Technology
and Industry
OECD
2, rue André-Pascal
75775 Paris Cedex 16
FRANCE
Tel: (33 1) 45 24 94 20
Fax: (33 1) 45 24 93 99
Email: risaburo.nezu@oecd.org

Mr. Adalberto NOYOLA
see under Deputy Scientific Chairs

Mr. Kikuji TATEISHI
Deputy Director-General
Basic Industries Bureau
Ministry of International Trade and Industry
1-3-1, Kasumigaseki, Chiyoda-ku
Tokyo 100
JAPAN
Tel: (81 3) 3501 8625
Fax: (81 3) 3501 0197

SPEAKERS IN SESSIONS I, II, AND CASE STUDIES

BELGIUM

Mr. Willy VERSTRAETE — see under Co-chairs of Parallel Sessions

CANADA

Mr. Michael DUBOW
Dept. Microbiology and Immunology
McGill University
3775 University Street
Lyman Duff Building, Room 511
Montreal, H3A 2B4 Quebec
Tel: (1 514) 398 3913
Fax: (1 514) 398 7052

Mr. Bruce E. JANK
Water Technology International Corporation
867 Lakeshore Road
P.O. Box 5068
Burlington, Ontario L7R 4L7
Tel: (1 905) 336 4740
Fax: (1 905) 336 8912

Mr. Sam JEYANAYAGAM
Project Manager, Environmental Engineering
Stanley Consulting Group Ltd.
Suite 1007
7445-132 Street Surrey
V3W 1J8 Surrey, B.C.
Tel: (1 604) 597 0422
Fax: (1 604) 591 1856
Email: saelensy@stantech.com

JAPAN

Mr. Shigeaki HARAYAMA
Research Director of Kamaishi Laboratories
Marine Biotechnology Institute
75-1 Dai-San chiwari Heita
Kamaishi City
Iwate 026
Tel: (81) 1 9326 6581
Fax: (81) 1 9326 6592

JAPAN (cont'd.)

Mr. Masayoshi KITAGAWA
Manager, Environmental Remediation Division
Biotechnology Department
EBARA Research Co., Ltd.
4-2-1 Honfujisawa
Fujisawa-shi 251
Tel: (81) 466 83 7740
Fax: (81) 466 81 7220

Mr. Ryuichiro KURANE
Director
Applied Microbiology Department
National Institute of BioScience and Human Technology
Agency of Industrial Science and Technology
Ministry of International Trade and Industry
1-1 Higashi, Tsukuba
Ibaraki 305
Tel: (81) 298 54 6030
Fax: (81) 298 54 6005
Email: rkurane@ccmail.nibh.go.jp

Mr. Yusaku MIYAKE
General Manager of Global Environmental
Protection Department
ORGANO Corporation
5-16, Hongo, 5-Chome
Bunkyo-ku
Tokyo 113
Tel: (81 3) 5689 5114
Fax: (81 3) 3812 5174

Mr. Sosuke NISHIMURA
Kurita Water Industries Ltd.
Kurita Centra Laboratories
7-1 Wakamiya, Morinosato
Atsugi City 24801
Tel: (81) 462 70 2134
Fax: (81) 462 70 2159

Mr. Osami YAGI
Chief
Water Quality Science Section
National Institute for Environmental Studies
Environment Agency
16-2 Onogawa
Tsukuba, Ibaraki 305
Tel: (81) 298 51 6111
Fax: (81) 298 51 4732

MEXICO

Mr. I. Ariosto AGUILAR Chávez
Subcoordinador de Hidráulica Ambiental
Instituto Mexicano de Tecnología del Agua
Paseo Cuauhnahuac No. 8532
Col. Progreso
62550 Jiutepec, MOR
Tel: (52) 73 20 8904
Fax: (52) 73 20 8725
Email: aaguilar@tlaloc.imta.mx

Mr. Antonio CAPELLA Viscaino
Asesor del Director General de la CNA
Comisión Nacional del Agua
Av. Insurgentes Sur. No. 1443, Piso 13
Col. Insurgentes Mixcoac
03920 México D.F.
Tel: (52 5) 237 4294; 237 4295
Fax: (52 5) 611 3504

Mr. Ignacio CASTILLO Escalante
Gerente de Saneamiento y Calidad del Agua
Comisión Nacional del Agua
Av. San Bernabé 549
Col. San Jerónimo Lidice
10200 México D.F.
Tel: (52 5) 595 2455
Fax: (52 5) 595 3950

Mr. Rubén CHÁVEZ Guillén
Gerente de Aguas Subterráneas
Comisión Nacional del Agua
Insurgentes Sur No. 1960, 5o. Piso
Col. Florida
01030 México D.F.
Tel: (52 5) 663 2217; 663 2255
Fax: (52 5) 663 3131

Mr. José COLLI-Misset
Especialista en Hidráulica
Subcoordinación de Tratamiento de Aguas Residuales
Instituto Mexicano de Tecnología del Agua
Paseo Cuauhnahuac No. 8532
Col. Progreso
62550 Jiutepec, MOR
Tel: (52) 73 19 43 66
Fax: (52) 73 19 43 81
Email: jcolli@chac.imta.mx

Mr. Roberto CONTRERAS Martínez
Subgerente de Proyecto y Saneamiento
Comisión Nacional del Agua
Cerrada Juan Sánchez Azcona 1723
Col. del Valle
03100 México D.F.
Tel: (52 5) 524 5117
Fax: (52 5) 524 5297

Ms. Alejandra COVARRUBIAS Robles
Jefe del Depto. de Biología Molecular de Plantas
Instituto de Biotecnologiá, UNAM
Apdo. Postal 510-3
62250 Cuernavaca, MOR
Tel: (52) 73 114 900
Fax: (52) 73 172 388
Email: crobles@ibt.unam.mx

Ms. del Carmen GONZALEZ Macías
Jefe de la Línea de Especialidad de Estudios Ecológicos
Instituto Mexicano del Petróleo
Eje Central Lázaro Cárdenas 152
Col. San Bartolo Atepehuacan
07730 México D.F.
Tel: (52 5) 368 3788
Fax: (52 5) 567 6047
Email: cgm@tsekuv.imp.mx

MEXICO (cont'd.)

Mr. José Eduardo MESTRE Rodríguez
Jefe de la Unidad de Consejos de Cuenca
Comisión Nacional del Agua
Insurgentes Sur 1960, 9o. Piso
Col. Florida
01050 México D.F.
Tel: (52 5) 662 5448; 662 5466
Fax: (52 5) 662 5446

Ms. Gabriela MOELLER Chávez
Subcoordinadora de Tratamiento de Aguas Residuales
Instituto Mexicano de Tecnología del Agua
Paseo Cuauhnahuac 8532
Col. Progreso
62550 Jiutepec, MOR
Tel: (52) 73 194 366
Fax: (52) 73 194 366; 194 381
Email: gmoeller@tlaloc.imta.mx

Mr. Alvaro MUÑOZ Mendoza
Especialista en Hidráulica Ambiental
Instituto Mexicano de Tecnología del Agua
Paseo Cuauhnahuac No. 8532
Col. Progreso
62550 Jiutepec, MOR
Tel: (52) 73 20 8904
Fax: (52) 73 20 8725
Email: alvaromm@tlaloc.imta.mx

Mr. Alberto JAIME Paredes
Subdirector General Técnico
Comisión Naciónal del Agua
Av. Insurgentes Sur No. 2140, 1er. Piso
Col. Ermita
01070 México D.F.
Tel: (52 5) 237 4023; 237 4024
Fax: (52 5) 661 5430

Mr. Oscar ROMO Ruíz
Director de Ecoparque
Colegio de la Frontera Norte, A.C.
Boulevard Abelardo L. Rodríguez No. 2925
Zona del Río
22320 Tijuana, Baja California
Tel: (52) 66 24 0531; 82 0088
Fax: (52) 66 24 0531
Email: oromo@colef.mx

Mr. Carlos SANTOS Burgoa
Director
Instituto de Salud, Ambiente y Trabajo (ISAT)
Coapa 160
Col. Toriello Guerra
14050 México D.F.
Tel: (52 5) 606 4066
Fax: (52 5) 665 0959
Email: rtn0523@rtn.net.mx

Ms. Susana SAVAL Bohorquez
Investigadora de Bioprocesos Ambientales
Instituto de Ingeniería, UNAM
Edif. 5, 2o. piso
Ciudad Universitaria
04510 México D.F.
Tel: (52 5) 622 3324
Fax: (52 5) 616 2164
Email: ssb@pumas.iingen.unam.mx

SWITZERLAND

Mr. Thomas EGLI
Senior Researcher
Swiss Federal Institute for Environmental Science and Technology (EAWAG)
Überlandstraße 133
8600 Dübendorf

Tel: (41 1) 823 5158
Fax: (41 1) 823 5547
Email: egli@eawag.ch

Mr. F. SCHERTENLEIB
Swiss Federal Institute for Environmental Science and Technology (EAWAG)
Überlandstraße 133
8600 Dübendorf

Tel: (41 1) 823 50 18
Fax: (41 1) 812 53 99
Email: schertenleib@eawag.ch

Mr. Martin WEGELIN
Swiss Federal Institute for Environmental Science and Technology (EAWAG)
Überlandstraße 133
8600 Dübendorf

Tel: (41 1) 823 50 19
Fax: (41 1) 823 53 99
Email: wegelin@eawag.ch

UNITED KINGDOM

Mr. Issy CAFFOOR
Yorkshire Water Services Ltd.
Research & Process Development
Western House
P.O. Box 500
Halifax Road
Bradford, West Yorkshire BD6 2LZ

Tel: (44) 1274 691111
Fax: (44) 1274 804550

Mr. Harry ECCLES
British Nuclear Fuels Plc.
Remediation Technologies Group
6 Riversway Business Village
Navigation Way, Ashton-on-Ribble,
Preston, Lancashire PR2 2YF

Tel: (44) 1 772 732 713
Fax: (44) 1 772 733 184

Mr. Colin FRICKER
Thames Water Utilities
Spencer House Laboratory
Manor Farm Road
Reading, Berkshire RG2 OJN

Tel: (44) 1734 236211
Fax: (44) 1734 236311

Mr. C. William KEEVIL
Microbial Technology Department
Centre for Applied Microbiology & Research (CAMR)
Porton Down
Salisbury, Wiltshire SP4 0JG

Tel: (44) 1980 612100
Fax: (44) 1980 611310

UNITED KINGDOM (cont'd.)

Mr. John UPTON
Process Technology Manager
Avon House
Finham Sewage Treatment Works
St. Martins Road
Coventry CV3 6PR
Tel: (44) 1203 693333
Fax: (44) 1203 642929

UNITED STATES

Mr. Ron ATLAS
Professor
University of Louisville
Department of Biology
139 Life Science Building
Belknap Campus
Louisville, KY 40292
Tel: (1 502) 852 8962
Fax: (1 502) 852 0725
Email: rmatla01@ulkyvm.louisville.edu

Mr. William S. BUSCH
Director of Emerging Technologies
Office of Global Programs
National Oceanic and Atmospheric Administration
1100 Wayne Av, Suite 1225
Silver Spring, MD 20910-5603
Tel: (1 301) 427 2089
Fax: (1 301) 427 2222

Mr. Russel R. CHIANELLI
Chairman, Department of Chemistry
University of Texas at El Paso
El Paso, TX 79968-0513
Tel: (1 915) 747 7555
Fax: (1 915) 747 5748

Ms. Rita COLWELL
see under Co-chairs of Parallel Sessions

Mr. John A. GLASER
U.S. Environmental Protection Agency
Office of Research and Development
National Risk Management Research Laboratory
26 W. Martin Luther King Dr.
Cincinnati, OH 45268
Tel: (1 513) 569 7568
Fax: (1 513) 569 7105
Email: john@epamail.epa.gov

Mr. Barry MARRS
Arres Enterprises
7 Possum Tree Lane
Kennett Square, PA 19348
Tel: (1 610) 388 7103
Fax: (1 610) 388 5884
Email: barrym4212@aol.com

Mr. Richard MEAGHER
Chair, Department of Genetics
Life Sciences Building, Room B402A
University of Georgia
Athens, GA 30602-7223
Tel: (1 706) 542 1444
Fax: (1 706) 542 1387
Email: meagher@bscr.uga.edu

UNITED STATES (cont'd.)

Mr. Greg PAYNE
Department of Chemical and Biochemical Engineering and Technology Research Center Building
University of Maryland Baltimore County
5401 Wilkens Av
Baltimore, MD 21228-5398
Tel: (1 410) 455 3413
Fax: (1 410) 455 6500
Email: payne@research.umbc.edu

Mr. Frederic PFAENDER
Department of Environmental Sciences and Engineering
University of North Carolina CB7400
Chapel Hill, NC 27599-7400
Tel: (1 919) 966 3842
Fax: (1 919) 966 7911

Mr. Ilya RASKIN
Center for Agricultural Molecular Biology
Foran Hall, Dudley Road
Rutgers, The State University of New Jersey
Cook College, P.O. Box 231
New Brunswick, NJ 08903-0231
Tel: (1 908) 932 8734
Fax: (1 908) 932 6535

Ms. Joan ROSE
Associate Professor
Department of Marine Science
University of South Florida
140 Seventh Av South
St. Petersburg, FL 33701-5016
Tel: (1 813) 553 3928
Fax: (1 813) 893 9189
Email: jrose@seas.marine.usf.edu

Mr. Bradley M. TEBO
Marine Biology Research Division and Center for Marine Biotechnology and Biomedecine, A-0202
Scripps Institution of Oceanography
University of California San Diego
9500 Gilman Drive
La Jolla, CA 92093-0202
Tel: (1 619) 534 5470
Fax: (1 619) 534 7313
Email: btebo@ucsd.edu

EC

Mr. Claude DANGLOT
Mairie de Paris - Direction de la Protection de l'Environnement
Centre de Recherche et de Contrôle des Eaux de Paris
Laboratoire de Biotechnologie
144 Avenue Paul Vaillant Couturier
75014 Paris
FRANCE
Tel: (33 1) 40 84 77 06
Fax: (33 1) 40 84 77 09
Email: danglot@infobiogen.fr

EC (cont'd.)

Mr. H. DIESTEL
Technical University of Berlin
Institute for Land Development
Albrecht-Thaer-Weg 2
14195 Berlin
GERMANY
Tel: (49) 30 3147 1225
Fax: (49) 30 3147 1228

Ms. M. FERDINANDY - VAN VLERKEN
Head, Biotechnology Program
Institute for Inland Water Management and
Waste Water Treatment (RIZA)
P.O. Box 17
8200AA Lelystad
THE NETHERLANDS
Tel: (31) 320 298 533
Fax: (31) 320 249 218

Ms. Freda HAWKES
Department of Science and Chemical Engineering
University of Glamorgan
Pontypridd
Mid-Glamorgan CF37 1DL
UNITED KINGDOM
Tel: (44) 1443 482116
Fax: (44) 1443 482285

Mr. Torben Moth IVERSEN
Director of Research Department
Ministry of the Environment
National Environmental Research Institute
PO Box 314
Vejlsøvey 25
8600 Silkeborg
DENMARK
Tel: (45) 89 20 14 00
Fax: (45) 89 20 14 14

Mr. Steven MYINT
University of Leicester
Department of Microbiology & Immunology
Maurice Shock Building
University Road
Leicester LE19HN
UNITED KINGDOM
Tel: (44) 1162 522951
Fax: (44) 1162 525030
Email: dsm@le.ac.uk

Ms. Montserrat PAGÈS
Consejo Superior de Investigaciones Cientificas
Centro de Investigación y Desarrollo
Jordi Girona 18-26
08034 Barcelona
SPAIN
Tel: (34) 3 4006131
Fax: (34) 3 2045904
Email: mptgmm@cid.csic.es

EC (cont'd.)

Ms. Brigitte SARRETTE
Chef du laboratoire de Virologie des Eaux
Mairie de Paris - Direction de la Protection
de l'Environnement
Centre de Recherche et de Contrôle des Eaux de Paris
144, Avenue Paul Vaillant Couturier
75014 Paris
FRANCE

Tel: (33 1) 40 84 77 25
Fax: (33 1) 40 84 77 26

Mr. Jacques J.M. VAN DE WORP
Department of Environmental Technology Studies
TNO Institute of Environmental Sciences,
Energy Research and Process Innovation
Laan van Westenenk 501
PO Box 342
7300 AH Apeldoorn
THE NETHERLANDS

Tel: (31) 55 549 3921
Fax: (31) 55 549 3410
Email: j.j.m.vandeworp@mep.tno.nl

OTHER PARTICIPANTS

AUSTRIA

Mr. Rudolf BRAUN

BELGIUM

Mr. Jan DE BRABANDERE

CZECH REPUBLIC

Mr. Jaroslav DROBNIK

DENMARK

Mr. H. PEDERSEN

IRELAND

Mr. Barry KIELY

JAPAN

Mr. K. KAWAMOTO
Mr. E. TAKEUCHI

MEXICO

Mr. Miguel A. AGUILAR Gómez
Mr. Rocío ALATORRE
Mr. Alvaro ALDAMA Rodríguez
Mr. José ALTAMIRANO Islas
Mr. Carlos ARIAS Castro
Ms. Aurora ARMIENTA
Mr. José Luis ARVIZU
Ms. Verónica BARRIOS
Ms. del Pilar BREMAUNTZ
Mr. Roberto CABRAL Bowling
Mr. Guillermo CABRERA López
Mr. Lenon CAJUSTA
Mr. Luis Enrique CALDERÓN Sánchez
Mr. Luis Enrique CÁRDENAS Sánchez
Ms. Lina CARDOSO Vigueros
Mr. José Alberto CASTAÑEDA Estrada
Mr. Raymundo CERVANTES
Mr. Enrique CIFUENTES
Mr. Gonzalo CHAPELA Mendoza
Mr. Jaime CRUZ Díaz
Mr. Luis DE LA SIERRA
Ms. Guadalupe DEL POZO
Mr. José Juan DOMINGUEZ Tapia
Ms. Ana Cecilia ESPINOZA
Ms. Violeta ESTRADA Escalante
Mr. Enrique ESTRADA Loera

Ms. Catalina FERAT
Mr. Luis FERNÁNDEZ Linareş
Mr. Luis M. FLORES Ordeña
Mr. Mauro GARCÍA Mendez
Mr. David GARDNER
Mr. Luis E. GARRIDO Sánchez
Mr. Ramón GILES López
Mr. Miguel Angel GODINEZ Antillon
Ms. Mariana GÓMEZ Camponovo
Mr. Edgar GONZÁLEZ Gaudiano
Mr. Arturo GONZÁLEZ Herrera
Mr. Carlos René GREEN
Ms. Anne GSCHAEDLER
Mr. Francisco GUZMÁR López
Ms. Patricia HERRERA Ascencio
Ms. Bianca JIMENEZ
Mr. Ilangovan KUPPUSAMY
Mr. Miguel G. LOMBERA González
Ms. Lourdes LÓPEZ Castro
Mr. Nicolás MANNING
Ms. Alejandra MARTÍN Dominguez
Mr. Gregorio MARTINEZ Ramirez
Mr. Alejandro MEDINA Chena
Ms. Guadalupe MEDINA Mejía
Ms. Elsa MENDOZA Amezquita
Ms. Petia MIJAVLOVA Nacheva
Mr. Yukio MIYAISHI
Ms. Patricia MOLINA Arcos
Mr. Enrique MONICA
Mr. Oscar MONROY
Ms. Leticia MONTELLANO Palacio
Mr. Alejandro MONTES DE OCA García
Ms. Leticia MONTOYA Herrera
Mr. Jesus Antonio MORENO Lugo
Mr. Arturo MURILLO
Mr. Juan Antonio MURO Ruiz
Mr. Luis ONTAÑÓN
Mr. Aniceto ORTEGA Caballero
Mr. Manuel ORTEGA Escobar
Mr. Victor OSEGUERA Green
Ms. Lourdes PARGA Mateos
Mr. Gustavo PAZ Soldan Cordova
Mr. Pedro RAMÍREZ
Mr. Victor RAMÍREZ Angulo
Ms. Esperanza RAMÍREZ Camperos
Mr. Sergio REVAH
Mr. René REYES Mazzoco
Mr. Jaime REYES Pérez
Mr. Gilberto REYES Sánchez
Mr. Michael ROBERTS
Mr. Armando ROMERO Rosales
Ms. Georgina SÁNCHEZ
Ms. Ana María SANDOVAL
Ms. Teresa SEVILLA Olquín
Mr. Pedro de Jesús TOLEDO Echegaray
Mr. David TORRES Mejía
Mr. Florencio TREVIÑO Rodríguez
Mr. Gildardo VAZQUEZ Tirado
Ms. Claudia VIRIDIANA Saldaña
Mr. Jésus María ZAVALA Ruiz
Mr. Rafael ZENDEJAS

PORTUGAL

Ms. Teresa AMARAL-COLLACO

SPAIN

Mr. Eloy GARCIA CALVO

UNITED KINGDOM

Mr. Gary SHANAHAN

ORGANISERS:

CONACYT SECRETARIAT

Mr. Efraín ACEVES Piña
Mr. Carlos BAZDRESCH Parada
Mr. José Luis CADENAS Palma
Ms. Clara MORÁN
Ms. Sylvía ORTEGA Salazar
Ms. Laura SIL Acosta
Mr. Arturo VELASCO
Mr. Rubén VENTURA Ramírez

OECD SECRETARIAT

Mr. Shinji FUJINO
Ms. Sonia GUIRAUD
Ms. Julie HARRIS
Mr. Tadashi HIRAKAWA
Mr. Bill LONG
Mr. Risaburo NEZU
Mr. Michael OBORNE
Mr. Salomon WALD

RESPONSIBLE CONTACT POINTS

Mr. Efraín ACEVES Piña
Director for International Co-operation
CONACYT
Av. Constituyentes 1046, 3rd floor
Col. Lomas Altas
11950 México D.F.
MEXICO

Tel: (52 5) 327 7615
Fax: (52 5) 327 7416
Email: aceves@buzon.main.conacyt.mx

Mr. Salomon WALD
Principal Administrator, Biotechnology Unit
Directorate for Science, Technology
and Industry
OECD
2, rue André-Pascal
75775 Paris Cedex 16
FRANCE

Tel: (33 1) 45 24 92 32
Fax: (33 1) 45 24 18 25
Email: salomon.wald@oecd.org

Mr. Tadashi HIRAKAWA
Principal Administrator, Biotechnology Unit
Directorate for Science, Technology
and Industry
OECD
2, rue André-Pascal
75775 Paris Cedex 16
FRANCE

Tel: (33 1) 45 24 93 42
Fax: (33 1) 45 24 18 25
Email: tadashi.hirakawa@oecd.org

PREPARATION OF PROCEEDINGS FOR PUBLICATION

Ms. Miriam KOREEN
OECD
2, rue André-Pascal
75775 Paris Cedex 16
FRANCE

Tel: (33 1) 45 24 93 70
Fax: (33 1) 45 24 18 25
Email: miriam.koreen@oecd.org

MAIN SALES OUTLETS OF OECD PUBLICATIONS
PRINCIPAUX POINTS DE VENTE DES PUBLICATIONS DE L'OCDE

AUSTRALIA – AUSTRALIE
D.A. Information Services
648 Whitehorse Road, P.O.B 163
Mitcham, Victoria 3132 Tel. (03) 9210.7777
Fax: (03) 9210.7788

AUSTRIA – AUTRICHE
Gerold & Co.
Graben 31
Wien I Tel. (0222) 533.50.14
Fax: (0222) 512.47.31.29

BELGIUM – BELGIQUE
Jean De Lannoy
Avenue du Roi, Koningslaan 202
B-1060 Bruxelles Tel. (02) 538.51.69/538.08.41
Fax: (02) 538.08.41

CANADA
Renouf Publishing Company Ltd.
5369 Canotek Road
Unit 1
Ottawa, Ont. K1J 9J3 Tel. (613) 745.2665
Fax: (613) 745.7660

Stores:
71 1/2 Sparks Street
Ottawa, Ont. K1P 5R1 Tel. (613) 238.8985
Fax: (613) 238.6041

12 Adelaide Street West
Toronto, QN M5H 1L6 Tel. (416) 363.3171
Fax: (416) 363.5963

Les Éditions La Liberté Inc.
3020 Chemin Sainte-Foy
Sainte-Foy, PQ G1X 3V6 Tel. (418) 658.3763
Fax: (418) 658.3763

Federal Publications Inc.
165 University Avenue, Suite 701
Toronto, ON M5H 3B8 Tel. (416) 860.1611
Fax: (416) 860.1608

Les Publications Fédérales
1185 Université
Montréal, QC H3B 3A7 Tel. (514) 954.1633
Fax: (514) 954.1635

CHINA – CHINE
Book Dept., China National Publications
Import and Export Corporation (CNPIEC)
16 Gongti E. Road, Chaoyang District
Beijing 100020 Tel. (10) 6506-6688 Ext. 8402
(10) 6506-3101

CHINESE TAIPEI – TAIPEI CHINOIS
Good Faith Worldwide Int'l. Co. Ltd.
9th Floor, No. 118, Sec. 2
Chung Hsiao E. Road
Taipei Tel. (02) 391.7396/391.7397
Fax: (02) 394.9176

CZECH REPUBLIC – RÉPUBLIQUE TCHÈQUE
National Information Centre
NIS – prodejna
Konviktská 5
Praha 1 – 113 57 Tel. (02) 24.23.09.07
Fax: (02) 24.22.94.33
E-mail: nkposp@dec.niz.cz
Internet: http://www.nis.cz

DENMARK – DANEMARK
Munksgaard Book and Subscription Service
35, Nørre Søgade, P.O. Box 2148
DK-1016 København K Tel. (33) 12.85.70
Fax: (33) 12.93.87

J. H. Schultz Information A/S,
Herstedvang 12,
DK – 2620 Albertslung Tel. 43 63 23 00
Fax: 43 63 19 69
Internet: s-info@inet.uni-c.dk

EGYPT – ÉGYPTE
The Middle East Observer
41 Sherif Street
Cairo Tel. (2) 392.6919
Fax: (2) 360.6804

FINLAND – FINLANDE
Akateeminen Kirjakauppa
Keskuskatu 1, P.O. Box 128
00100 Helsinki

Subscription Services/Agence d'abonnements :
P.O. Box 23
00100 Helsinki Tel. (358) 9.121.4403
Fax: (358) 9.121.4450

***FRANCE**
OECD/OCDE
Mail Orders/Commandes par correspondance :
2, rue André-Pascal
75775 Paris Cedex 16 Tel. 33 (0)1.45.24.82.00
Fax: 33 (0)1.49.10.42.76
Telex: 640048 OCDE
Internet: Compte.PUBSINQ@oecd.org

Orders via Minitel, France only/
Commandes par Minitel, France exclusivement :
36 15 OCDE

OECD Bookshop/Librairie de l'OCDE :
33, rue Octave-Feuillet
75016 Paris Tel. 33 (0)1.45.24.81.81
33 (0)1.45.24.81.67

Dawson
B.P. 40
91121 Palaiseau Cedex Tel. 01.89.10.47.00
Fax: 01.64.54.83.26

Documentation Française
29, quai Voltaire
75007 Paris Tel. 01.40.15.70.00

Economica
49, rue Héricart
75015 Paris Tel. 01.45.78.12.92
Fax: 01.45.75.05.67

Gibert Jeune (Droit-Économie)
6, place Saint-Michel
75006 Paris Tel. 01.43.25.91.19

Librairie du Commerce International
10, avenue d'Iéna
75016 Paris Tel. 01.40.73.34.60

Librairie Dunod
Université Paris-Dauphine
Place du Maréchal-de-Lattre-de-Tassigny
75016 Paris Tel. 01.44.05.40.13

Librairie Lavoisier
11, rue Lavoisier
75008 Paris Tel. 01.42.65.39.95

Librairie des Sciences Politiques
30, rue Saint-Guillaume
75007 Paris Tel. 01.45.48.36.02

P.U.F.
49, boulevard Saint-Michel
75005 Paris Tel. 01.43.25.83.40

Librairie de l'Université
12*a*, rue Nazareth
13100 Aix-en-Provence Tel. 04.42.26.18.08

Documentation Française
165, rue Garibaldi
69003 Lyon Tel. 04.78.63.32.23

Librairie Decitre
29, place Bellecour
69002 Lyon Tel. 04.72.40.54.54

Librairie Sauramps
Le Triangle
34967 Montpellier Cedex 2 Tel. 04.67.58.85.15
Fax: 04.67.58.27.36

A la Sorbonne Actual
23, rue de l'Hôtel-des-Postes
06000 Nice Tel. 04.93.13.77.75
Fax: 04.93.80.75.69

GERMANY – ALLEMAGNE
OECD Bonn Centre
August-Bebel-Allee 6
D-53175 Bonn Tel. (0228) 959.120
Fax: (0228) 959.12.17

GREECE – GRÈCE
Librairie Kauffmann
Stadiou 28
10564 Athens Tel. (01) 32.55.321
Fax: (01) 32.30.320

HONG-KONG
Swindon Book Co. Ltd.
Astoria Bldg. 3F
34 Ashley Road, Tsimshatsui
Kowloon, Hong Kong Tel. 2376.2062
Fax: 2376.0685

HUNGARY – HONGRIE
Euro Info Service
Margitsziget, Európa Ház
1138 Budapest Tel. (1) 111.60.61
Fax: (1) 302.50.35
E-mail: euroinfo@mail.matav.hu
Internet: http://www.euroinfo.hu//index.html

ICELAND – ISLANDE
Mál og Menning
Laugavegi 18, Pósthólf 392
121 Reykjavik Tel. (1) 552.4240
Fax: (1) 562.3523

INDIA – INDE
Oxford Book and Stationery Co.
Scindia House
New Delhi 110001 Tel. (11) 331.5896/5308
Fax: (11) 332.2639
E-mail: oxford.publ@axcess.net.in

17 Park Street
Calcutta 700016 Tel. 240832

INDONESIA – INDONÉSIE
Pdii-Lipi
P.O. Box 4298
Jakarta 12042 Tel. (21) 573.34.67
Fax: (21) 573.34.67

IRELAND – IRLANDE
Government Supplies Agency
Publications Section
4/5 Harcourt Road
Dublin 2 Tel. 661.31.11
Fax: 475.27.60

ISRAEL – ISRAËL
Praedicta
5 Shatner Street
P.O. Box 34030
Jerusalem 91430 Tel. (2) 652.84.90/1/2
Fax: (2) 652.84.93

R.O.Y. International
P.O. Box 13056
Tel Aviv 61130 Tel. (3) 546 1423
Fax: (3) 546 1442
E-mail: royil@netvision.net.il

Palestinian Authority/Middle East:
INDEX Information Services
P.O.B. 19502
Jerusalem Tel. (2) 627.16.34
Fax: (2) 627.12.19

ITALY – ITALIE
Libreria Commissionaria Sansoni
Via Duca di Calabria, 1/1
50125 Firenze Tel. (055) 64.54.15
Fax: (055) 64.12.57
E-mail: licosa@ftbcc.it

Via Bartolini 29
20155 Milano Tel. (02) 36.50.83

Editrice e Libreria Herder
Piazza Montecitorio 120
00186 Roma Tel. 679.46.28
Fax: 678.47.51

Libreria Hoepli
Via Hoepli 5
20121 Milano Tel. (02) 86.54.46
Fax: (02) 805.28.86

Libreria Scientifica
Dott. Lucio de Biasio 'Aeiou'
Via Coronelli, 6
20146 Milano Tel. (02) 48.95.45.52
Fax: (02) 48.95.45.48

JAPAN – JAPON
OECD Tokyo Centre
Landic Akasaka Building
2-3-4 Akasaka, Minato-ku
Tokyo 107 Tel. (81.3) 3586.2016
Fax: (81.3) 3584.7929

KOREA – CORÉE
Kyobo Book Centre Co. Ltd.
P.O. Box 1658, Kwang Hwa Moon
Seoul Tel. 730.78.91
Fax: 735.00.30

MALAYSIA – MALAISIE
University of Malaya Bookshop
University of Malaya
P.O. Box 1127, Jalan Pantai Baru
59700 Kuala Lumpur
Malaysia Tel. 756.5000/756.5425
Fax: 756.3246

MEXICO – MEXIQUE
OECD Mexico Centre
Edificio INFOTEC
Av. San Fernando no. 37
Col. Toriello Guerra
Tlalpan C.P. 14050
Mexico D.F. Tel. (525) 528.10.38
Fax: (525) 606.13.07
E-mail: ocde@rtn.net.mx

NETHERLANDS – PAYS-BAS
SDU Uitgeverij Plantijnstraat
Externe Fondsen
Postbus 20014
2500 EA's-Gravenhage Tel. (070) 37.89.880
Voor bestellingen: Fax: (070) 34.75.778

Subscription Agency/Agence d'abonnements :
SWETS & ZEITLINGER BV
Heereweg 347B
P.O. Box 830
2160 SZ Lisse Tel. 252.435.111
Fax: 252.415.888

NEW ZEALAND – NOUVELLE-ZÉLANDE
GPLegislation Services
P.O. Box 12418
Thorndon, Wellington Tel. (04) 496.5655
Fax: (04) 496.5698

NORWAY – NORVÈGE
NIC INFO A/S
Ostensjoveien 18
P.O. Box 6512 Etterstad
0606 Oslo Tel. (22) 97.45.00
Fax: (22) 97.45.45

PAKISTAN
Mirza Book Agency
65 Shahrah Quaid-E-Azam
Lahore 54000 Tel. (42) 735.36.01
Fax: (42) 576.37.14

PHILIPPINE – PHILIPPINES
International Booksource Center Inc.
Rm 179/920 Cityland 10 Condo Tower 2
HV dela Costa Ext cor Valero St.
Makati Metro Manila Tel. (632) 817 9676
Fax: (632) 817 1741

POLAND – POLOGNE
Ars Polona
00-950 Warszawa
Krakowskie Prezdmiescie 7 Tel. (22) 264760
Fax: (22) 265334

PORTUGAL
Livraria Portugal
Rua do Carmo 70-74
Apart. 2681
1200 Lisboa Tel. (01) 347.49.82/5
Fax: (01) 347.02.64

SINGAPORE – SINGAPOUR
Ashgate Publishing
Asia Pacific Pte. Ltd
Golden Wheel Building, 04-03
41, Kallang Pudding Road
Singapore 349316 Tel. 741.5166
Fax: 742.9356

SPAIN – ESPAGNE
Mundi-Prensa Libros S.A.
Castelló 37, Apartado 1223
Madrid 28001 Tel. (91) 431.33.99
Fax: (91) 575.39.98
E-mail: mundiprensa@tsai.es
Internet: http://www.mundiprensa.es

Mundi-Prensa Barcelona
Consell de Cent No. 391
08009 – Barcelona Tel. (93) 488.34.92
Fax: (93) 487.76.59

Libreria de la Generalitat
Palau Moja
Rambla dels Estudis, 118
08002 – Barcelona
(Suscripciones) Tel. (93) 318.80.12
(Publicaciones) Tel. (93) 302.67.23
Fax: (93) 412.18.54

SRI LANKA
Centre for Policy Research
c/o Colombo Agencies Ltd.
No. 300-304, Galle Road
Colombo 3 Tel. (1) 574240, 573551-2
Fax: (1) 575394, 510711

SWEDEN – SUÈDE
CE Fritzes AB
S–106 47 Stockholm Tel. (08) 690.90.90
Fax: (08) 20.50.21

For electronic publications only/
Publications électroniques seulement
STATISTICS SWEDEN
Informationsservice
S-115 81 Stockholm Tel. 8 783 5066
Fax: 8 783 4045

Subscription Agency/Agence d'abonnements :
Wennergren-Williams Info AB
P.O. Box 1305
171 25 Solna Tel. (08) 705.97.50
Fax: (08) 27.00.71

Liber distribution
Internatinal organizations
Fagerstagatan 21
S-163 52 Spanga

SWITZERLAND – SUISSE
Maditec S.A. (Books and Periodicals/Livres et périodiques)
Chemin des Palettes 4
Case postale 266
1020 Renens VD 1 Tel. (021) 635.08.65
Fax: (021) 635.07.80

Librairie Payot S.A.
4, place Pépinet
CP 3212
1002 Lausanne Tel. (021) 320.25.11
Fax: (021) 320.25.14

Librairie Unilivres
6, rue de Candolle
1205 Genève Tel. (022) 320.26.23
Fax: (022) 329.73.18

Subscription Agency/Agence d'abonnements :
Dynapresse Marketing S.A.
38, avenue Vibert
1227 Carouge Tel. (022) 308.08.70
Fax: (022) 308.07.99

See also – Voir aussi :
OECD Bonn Centre
August-Bebel-Allee 6
D-53175 Bonn (Germany) Tel. (0228) 959.120
Fax: (0228) 959.12.17

THAILAND – THAÏLANDE
Suksit Siam Co. Ltd.
113, 115 Fuang Nakhon Rd.
Opp. Wat Rajbopith
Bangkok 10200 Tel. (662) 225.9531/2
Fax: (662) 222.5188

TRINIDAD & TOBAGO, CARIBBEAN TRINITÉ-ET-TOBAGO, CARAÏBES
Systematics Studies Limited
9 Watts Street
Curepe
Trinidad & Tobago, W.I. Tel. (1809) 645.3475
Fax: (1809) 662.5654
E-mail: tobe@trinidad.net

TUNISIA – TUNISIE
Grande Librairie Spécialisée
Fendri Ali
Avenue Haffouz Imm El-Intilaka
Bloc B 1 Sfax 3000 Tel. (216-4) 296 855
Fax: (216-4) 298.270

TURKEY – TURQUIE
Kültür Yayinlari Is-Türk Ltd.
Atatürk Bulvari No. 191/Kat 13
06684 Kavaklidere/Ankara
Tel. (312) 428.11.40 Ext. 2458
Fax : (312) 417.24.90

Dolmabahce Cad. No. 29
Besiktas/Istanbul Tel. (212) 260 7188

UNITED KINGDOM – ROYAUME-UNI
The Stationery Office Ltd.
Postal orders only:
P.O. Box 276, London SW8 5DT
Gen. enquiries Tel. (171) 873 0011
Fax: (171) 873 8463

The Stationery Office Ltd.
Postal orders only:
49 High Holborn, London WC1V 6HB

Branches at: Belfast, Birmingham, Bristol, Edinburgh, Manchester

UNITED STATES – ÉTATS-UNIS
OECD Washington Center
2001 L Street N.W., Suite 650
Washington, D.C. 20036-4922 Tel. (202) 785.6323
Fax: (202) 785.0350
Internet: washcont@oecd.org

Subscriptions to OECD periodicals may also be placed through main subscription agencies.

Les abonnements aux publications périodiques de l'OCDE peuvent être souscrits auprès des principales agences d'abonnement.

Orders and inquiries from countries where Distributors have not yet been appointed should be sent to: OECD Publications, 2, rue André-Pascal, 75775 Paris Cedex 16, France.

Les commandes provenant de pays où l'OCDE n'a pas encore désigné de distributeur peuvent être adressées aux Éditions de l'OCDE, 2, rue André-Pascal, 75775 Paris Cedex 16, France.

12-1996

OECD PUBLICATIONS, 2, rue André-Pascal, 75775 PARIS CEDEX 16
PRINTED IN FRANCE
(93 97 05 1 P) ISBN 92-64-15594-5 – No. 49679 1997